Numerical Methods for Engineers

With Programming and Software Applications

Steven C. Chapra

Professor of Civil Engineering
The University of Colorado

Raymond P. Canale

Professor Emeritus of Civil Engineering
University of Michigan

Numerical Methods for Engineers

With Programming and Software Applications

Third Edition

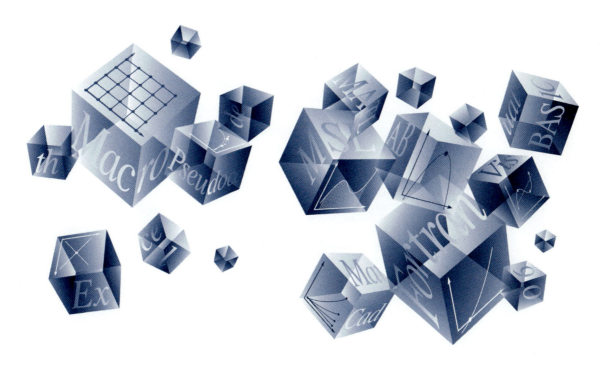

WCB
McGraw-Hill

Boston   Burr Ridge, IL   Dubuque, IA   Madison, WI   New York   San Francisco   St. Louis
Bangkok   Bogotá   Caracas   Lisbon   London   Madrid
Mexico City   Milan   New Delhi   Seoul   Singapore   Sydney   Taipei   Toronto

WCB/McGraw-Hill

A Division of The **McGraw·Hill** *Companies*

NUMERICAL METHODS FOR ENGINEERS: WITH PROGRAMMING AND SOFTWARE APPLICATIONS

This book is printed on acid-free paper.

1 2 3 4 5 6 7 8 9 0 VNH/VNH 9 0 9 8 7

ISBN 0-07-010938-9   ISBN 0-07-561254-2 (set number)

Vice president and editorial director: *Kevin T. Kane*
Publisher: *Tom Casson*
Executive editor: *Eric Munson*
Marketing manager: *John T. Wannemacher*
Senior project manager: *Susan Trentacosti*
Production supervisor: *Heather D. Burbridge*
Senior designer: *Crispin Prebys*
Designer: *Ellen Pettengell*
Typeface: *10/12 Times*
Printer: *Von Hoffman Press, Inc.*

Library of Congress Cataloging-in-Publication Data

Chapra, Steven C.
　　Numerical methods for engineers: with programming and software applications/Steven C. Chapra, Raymond P. Canale. —3rd ed.
　　　　p.　cm.
　　Includes index.
　　ISBN 0-07-010938-9
　　1. Engineering mathematics—Data processing.　2. Numerical calculations—Data processing.　3. Microcomputers—Programming.
I. Canale, Raymond P.　II. Title.
TA345.C47　　1998
519.4'024'62—dc21　　　　　　　　　　　　　　　　97-37136

http://www.mhhe.com

To

Margaret and Gabriel Chapra

Helen and Chester Canale

ABOUT THE AUTHORS

Steve Chapra teaches in the Civil, Environmental, and Architectural Engineering Department at the University of Colorado. His other books include *Surface Water-Quality Modeling* and *Introduction to Computing for Engineers*.

Dr. Chapra received engineering degrees from Manhattan College and the University of Michigan. Before joining the faculty at the University of Colorado, he worked for the Environmental Protection Agency and the National Oceanic and Atmospheric Administration, and was an Associate Professor at Texas A&M University. His general research interests focus on surface water-quality modeling and advanced computer applications in environmental engineering.

He has received a number of awards for his scholarly contributions, including the 1993 Rudolph Hering Medal (ASCE) and the 1987 Meriam-Wiley Distinguished Author Award (American Society for Engineering Education). He has also been recognized as the outstanding teacher among the engineering faculties at both Texas A&M University (1986 Tenneco Award) and the University of Colorado (1992 Hutchinson Award).

Raymond P. Canale is an emeritus professor at the University of Michigan. During his over 20-year career at the university, he taught numerous courses in the area of computers, numerical methods, and environmental engineering. He also directed extensive research programs in the area of mathematical and computer modeling of aquatic ecosystems. He has authored or co-authored several books and has published over 100 scientific papers and reports. He has also designed and developed personal computer software to facilitate engineering education and the solution of engineering problems. He has been given the Meriam-Wiley Distinguished Author Award by the American Society for Engineering Education for his books and software and several awards for his technical publications.

Professor Canale is now devoting his energies to applied problems, where he works with engineering firms and industry and governmental agencies as a consultant and expert witness.

PREFACE

This third edition of *Numerical Methods for Engineers* differs from the second edition in five key ways:

1. *Inclusion of sections on major software packages and libraries.* At appropriate points throughout the text, we present overviews of how numerical methods can be implemented on a variety of popular software packages and libraries: Mathcad, Excel, MATLAB, and the IMSL software library. The intent here is to acknowledge and support the increased use of these tools for numerical analysis. We provide students with the necessary guidance to implement the methods with these packages and relate them to the theory in the text. Several homework problems in each part are also designed to illustrate how the packages can be employed. Prerequisite information (e.g., "Getting Started with Mathcad") is included as appendices.

2. *Computer languages.* Our continued emphasis on algorithms and program structure may seem somewhat archaic to some instructors. Although we understand that canned software packages are important, we do not hold with those who contend that their use will make programming obsolete. In fact, even if packages were to become the primary tools for numerical calculations, advanced applications involving macros and scripts all require fundamental knowledge of programming and algorithm structure. As in the previous edition, we predominantly use pseudocode to describe our algorithms. In contrast to the previous edition, where the emphasis was on complete programs, this edition stresses a more modular approach using subroutines and functions.

3. *New material.* Although there are many minor changes and refinements throughout the text, there are only two major additions to the general coverage. First, we have included a major new part on optimization. Second, we have added a new chapter on determining the roots of polynomials. The optimization section was developed because (1) engineering students are increasingly using optimization and (2) optimization is used in a variety of numerical methods contexts such as root location and regression.

4. *New homework problems.* The homework problems have been significantly revamped. In particular, we have modified most of the problems from the previous edition as well as including a number of new problems in every chapter.

5. *Windows version of Numerical Methods TOOLKIT.* We have developed a Windows version of the Numerical Methods TOOLKIT that is included with the text on a 3 1/2-inch diskette. Aside from enhancing some of the methods (e.g., calculation of matrix inverse, integration of several simultaneous ODEs), the new version is much more user friendly because of its integration with Windows. Thus, built-in Windows utilities, such as printer output, can be exploited. Aside from student use, the software has been designed to expedite classroom demonstrations by the instructor.

Aside from these additions, the third edition is very similar to the second edition in most other respects. In particular, we have endeavored to maintain most of the features contributing to its pedagogical effectiveness. These include the overall organization, the use of introductions and epilogues to consolidate major topics, and the extensive use of worked examples and engineering applications.

It should be noted that our book has a web site. Its URL is www.mhhe.com/engcs/general/chapra/. Feel free to consult it for additional information on this book. In particular, you can use it to provide us with your feedback.

Finally, as with the previous editions, we have exerted a conscious effort to make this book as student friendly as possible. Thus, we have endeavored to keep our explanations straightforward and oriented practically. Although our primary intent is to provide students with a sound introduction to numerical methods, we have the ancillary objective of making this introduction exciting and pleasurable. We believe that motivated students who enjoy numerical methods, computers, and mathematics will, in the end, make better engineers. If our book fosters an enthusiasm for these subjects, we will consider our efforts a success.

Acknowledgments. Special thanks to Jery Stedinger of Cornell University, who generously shared many insights and suggestions. His comments on optimization were particularly useful to us. David Clough of the University of Colorado shared his wisdom and catholic understanding of numerical methods and software packages. In addition, useful suggestions and reviews were made by David V. Chase (The University of Dayton), Raymundo Cordero (ITESM), Theresa Good (Texas A&M University), Wallace Grant (Virginia Tech/Virgina Polytechnic Institute & State), James W. Hiestand (University of Tennessee at Chattanooga), Steve Klegka (U.S. Military Academy), James L. Kuester (Arizona State University), Karim Müci (ITESM), Robert L. Rankin (Arizona State University), Elisa D. Sotelino (Purdue University), and Hewlon Zimmer (U.S. Merchant Marine Academy).

Finally, it should be stressed that, although we received useful advice from the aforementioned individuals, we are responsible for any inaccuracies or mistakes you may detect in this edition.

Steven C. Chapra
Raymond P. Canale

CONTENTS

PART THREE

LINEAR ALGEBRAIC EQUATIONS 219

Numerical Methods for Engineers

With Programming and Software Applications

PART ONE

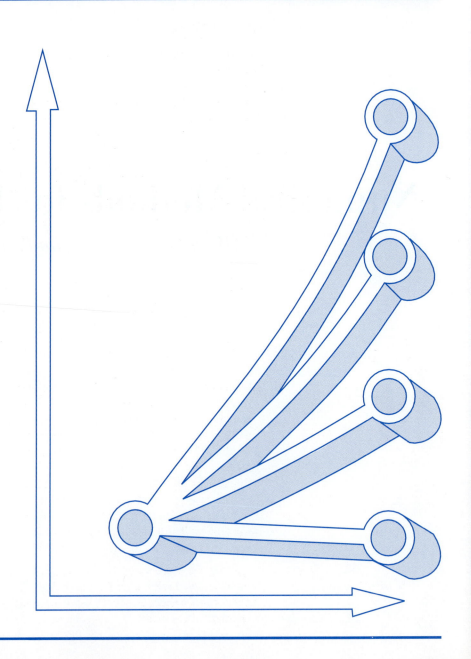

MODELING, COMPUTERS, AND ERROR ANALYSIS

PT1.1 MOTIVATION

Numerical methods are techniques by which mathematical problems are formulated so that they can be solved with arithmetic operations. Although there are many kinds of numerical methods, they have one common characteristic: they invariably involve large numbers of tedious arithmetic calculations. It is little wonder that with the development of fast, efficient digital computers, the role of numerical methods in engineering problem solving has increased dramatically in recent years.

PT1.1.1 Noncomputer Methods

Beyond providing increased computational firepower, the widespread availability of computers (especially personal computers) and their partnership with numerical methods has had a significant influence on the actual engineering problem-solving process. In the precomputer era there were generally three different ways in which engineers approached problem solving:

1. Solutions were derived for some problems using analytical, or exact, methods. These solutions were often useful and provided excellent insight into the behavior of some systems. However, analytical solutions can be derived for only a limited class of problems. These include those that can be approximated with linear models and those that have simple geometry and low dimensionality. Consequently, analytical solutions are of limited practical value because most real problems are nonlinear and involve complex shapes and processes.
2. Graphical solutions were used to characterize the behavior of systems. These graphical solutions usually took the form of plots or nomographs. Although graphical techniques can often be used to solve complex problems, the results are not very precise. Furthermore, graphical solutions (without the aid of computers) are extremely tedious and awkward to implement. Finally, graphical techniques are often limited to problems that can be described using three or fewer dimensions.
3. Calculators and slide rules were used to implement numerical methods manually. Although in theory such approaches should be perfectly adequate for solving complex problems, in actuality several difficulties are encountered. Manual calculations are slow and tedious. Furthermore, consistent results are elusive because of simple blunders that arise when numerous manual tasks are performed.

During the precomputer era, significant amounts of energy were expended on the solution technique itself, rather than on problem definition and interpretation (Fig. PT1.1a). This unfortunate situation existed because so much time and drudgery were required to obtain numerical answers using precomputer techniques.

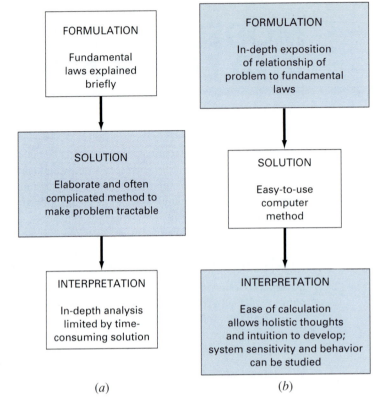

FIGURE PT1.1
The three phases of engineering problem solving in (a) the pre-computer and (b) the computer era. The sizes of the boxes indicate the level of emphasis directed toward each phase. Computers facilitate the implementation of solution techniques and thus allow more emphasis to be placed on the creative aspects of problem formulation and interpretation of results.

Today, computers and numerical methods provide an alternative for such complicated calculations. Using computer power to obtain solutions directly, you can approach these calculations without recourse to simplifying assumptions or time-intensive techniques. Although analytical solutions are still extremely valuable both for problem solving and for providing insight, numerical methods represent alternatives that greatly enlarge your capabilities to confront and solve problems. As a result, more time is available for the use of your creative skills. Thus, more emphasis can be placed on problem formulation and solution interpretation and the incorporation of total system, or "holistic," awareness (Fig. PT1.1b.)

PT1.1.2 Numerical Methods and Engineering Practice

Since the late 1940s the widespread availability of digital computers has led to a veritable explosion in the use and development of numerical methods. At first, this growth was somewhat limited by the cost of access to large mainframe computers, and, consequently, many engineers continued to use simple analytical approaches in a significant portion of their work. Needless to say, the recent evolution of inexpensive personal computers has given us ready access to powerful computational capabilities. There are several additional reasons why you should study numerical methods:

1. Numerical methods are extremely powerful problem-solving tools. They are capable of handling large systems of equations, nonlinearities, and complicated geometries that are not uncommon in engineering practice and that are often impossible to solve analytically. As such, they greatly enhance your problem-solving skills.

2. During your careers, you may often have occasion to use commercially available prepackaged, or "canned," computer programs that involve numerical methods. The intelligent use of these programs is often predicated on knowledge of the basic theory underlying the methods.

3. Many problems cannot be approached using canned programs. If you are conversant with numerical methods and are adept at computer programming, you can design your own programs to solve problems without having to buy or commission expensive software.

4. Numerical methods are an efficient vehicle for learning to use computers. It is well known that an effective way to learn programming is to actually write computer programs. Because numerical methods are for the most part designed for implementation on computers, they are ideal for this purpose. Further, they are especially well-suited to illustrate the power and the limitations of computers. When you successfully implement numerical methods on a computer and then apply them to solve otherwise intractable problems, you will be provided with a dramatic demonstration of how computers can serve your professional development. At the same time, you will also learn to acknowledge and control the errors of approximation that are part and parcel of large-scale numerical calculations.

5. Numerical methods provide a vehicle for you to reinforce your understanding of mathematics. Because one function of numerical methods is to reduce higher mathematics to basic arithmetic operations, they get at the "nuts and bolts" of some otherwise obscure topics. Enhanced understanding and insight can result from this alternative perspective.

PT1.2   MATHEMATICAL BACKGROUND

Every part in this book requires some mathematical background. Consequently, the introductory material for each part includes a section, such as the one you are reading, on mathematical background. Because Part One itself is devoted to background material on mathematics and computers, this section does not involve a review of a specific mathematical topic. Rather, we take this opportunity to introduce you to the types of mathematical subject areas covered in this book. As summarized in Fig. PT1.2, these are

1. *Roots of Equations* (Fig. PT1.2*a*). These problems are concerned with the value of a variable or a parameter that satisfies a single nonlinear equation. These problems are especially valuable in engineering design contexts where it is often impossible to explicitly solve design equations for parameters.

2. *Systems of Linear Algebraic Equations* (Fig. PT1.2*b*). These problems are similar in spirit to roots of equations in the sense that they are concerned with values that satisfy equations. However, in contrast to satisfying a single equation, a set of values is sought that simultaneously satisfies a set of linear algebraic equations. Such equations arise in a variety of problem contexts and in all disciplines of engineering. In particular, they

FIGURE PT1.2
Summary of the numerical methods covered in this book.

(a) *Part 2:* Roots of equations
Solve $f(x) = 0$ for x.

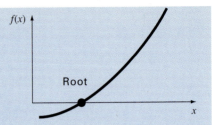

(b) *Part 3:* Linear algebraic equations
Given the a's and the c's, solve

$a_{11}x_1 + a_{12}x_2 = c_1$

$a_{21}x_1 + a_{22}x_2 = c_2$

for the x's.

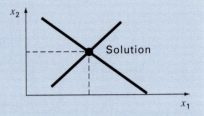

(c) *Part 4:* Optimization
Determine x that gives optimum $f(x)$.

(d) *Part 5:* Curve fitting

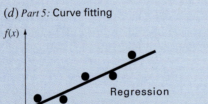

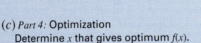

(e) *Part 6:* Integration
$I = \int_a^b f(x)\,dx$
Find the area under the curve.

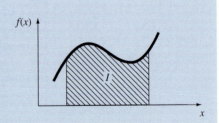

**FIGURE PT1.2
(concluded)**

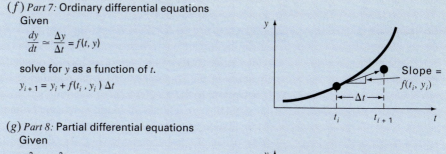

(f) *Part 7:* Ordinary differential equations
Given

$$\frac{dy}{dt} \simeq \frac{\Delta y}{\Delta t} = f(t, y)$$

solve for y as a function of t.

$$y_{i+1} = y_i + f(t_i, y_i) \, \Delta t$$

(g) *Part 8:* Partial differential equations
Given

$$\frac{\partial^2 u}{\partial x^2} + \frac{\partial^2 u}{\partial y^2} = f(x, y)$$

solve for u as a function of
x and y

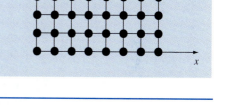

originate in the mathematical modeling of large systems of interconnected elements such as structures, electric circuits, and fluid networks. However, they are also encountered in other areas of numerical methods such as curve fitting and differential equations.

3. *Optimization* (Fig. PT1.2c). These problems involve determining a value or values of an independent variable that correspond to a "best" or optimal value of a function. Thus, as in Fig. PT1.2c, it involves identifying maxima and minima. Such problems occur routinely in engineering design contexts. They also arise in a number of other numerical methods. We address both single- and multi-variable unconstrained optimization. We also describe constrained optimization with particular emphasis on linear programming.

4. *Curve Fitting* (Fig. PT1.2d). You will often have occasion to fit curves to data points. The techniques developed for this purpose can be divided into two general categories: regression and interpolation. Regression is employed where there is a significant degree of error associated with the data. Experimental results are often of this kind. For these situations, the strategy is to derive a single curve that represents the general trend of the data without necessarily matching any individual points. In contrast, interpolation is used where the objective is to determine intermediate values between relatively error-free data points. Such is usually the case for tabulated information. For these situations, the strategy is to fit a curve directly through the data points and use the curve to predict the intermediate values.

5. *Integration* (Fig. PT1.2e). As depicted, a physical interpretation of numerical integration is the determination of the area under a curve. Integration has many applications in engineering practice, ranging from the determination of the centroids of oddly shaped objects to the calculation of total quantities based on sets of discrete measurements. In

addition, numerical integration formulas play an important role in the solution of differential equations.

6. *Ordinary Differential Equations* (Fig. PT1.2*f*). Ordinary differential equations are of great significance in engineering practice. This is because many physical laws are couched in terms of the rate of change of a quantity rather than the magnitude of the quantity itself. Examples range from population forecasting models (rate of change of population) to the acceleration of a falling body (rate of change of velocity). Two types of problems are addressed: initial-value and boundary-value problems. In addition, the computation of eigenvalues is covered.

7. *Partial Differential Equations* (Fig. PT1.2*g*). Partial differential equations are used to characterize engineering systems where the behavior of a physical quantity is couched in terms of its rate of change with respect to two or more independent variables. Examples include the steady-state distribution of temperature on a heated plate (two spatial dimensions) or the time-variable temperature of a heated rod (time and one spatial dimension). Two fundamentally different approaches are employed to solve partial differential equations numerically. In the present text, we will emphasize finite-difference methods that approximate the solution in a pointwise fashion (Fig. PT1.2*g*). However, we will also present an introduction to finite-element methods, which use a piecewise approach.

PT1.3 ORIENTATION

Some orientation might be helpful before proceeding with our introduction to numerical methods. The following is intended as an overview of the material in Part One. In addition, some objectives have been included to focus your efforts when studying the material.

PT1.3.1 Scope and Preview

Figure PT1.3 is a schematic representation of the material in Part One. We have designed this diagram to provide you with a global overview of this part of the book. We believe that a sense of the "big picture" is critical to developing insight into numerical methods. When reading a text, it is often possible to become lost in technical details. Whenever you feel that you are losing the big picture, refer back to Fig. PT1.3 to reorient yourself. Every part of this book includes a similar figure.

Figure PT1.3 also serves as a brief preview of the material covered in Part One. *Chapter 1* is designed to orient you to numerical methods and to provide motivation by demonstrating how these techniques can be used in the engineering modeling process. *Chapter 2* is an introduction and review of computer-related aspects of numerical methods and suggests the level of computer skills you should acquire to efficiently apply succeeding information. *Chapters 3* and *4* deal with the important topic of error analysis, which must be understood for the effective use of numerical methods. In addition, an *epilogue* is included that introduces the trade-offs that have such great significance for the effective implementation of numerical methods.

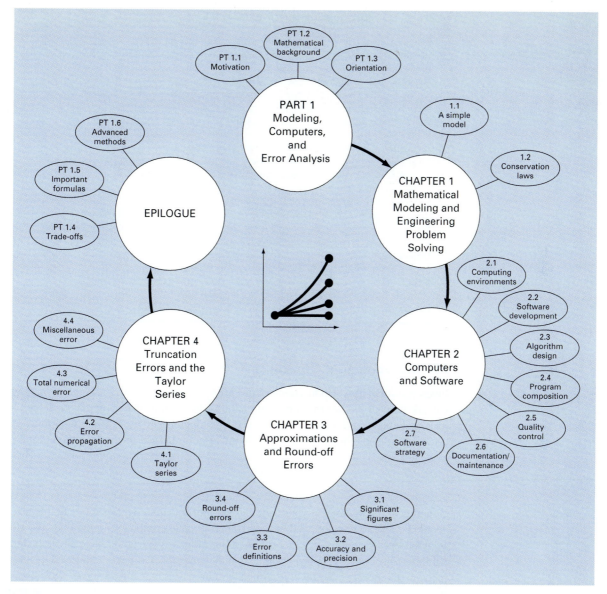

FIGURE PT1.3
Schematic of the organization of the material in Part One: Modeling, Computers, and Error Analysis.

TABLE PT1.1 Specific study objectives for Part One.

1. Recognize the difference between analytical and numerical solutions.
2. Understand how conservation laws are employed to develop mathematical models of physical systems.
3. Define top-down and modular design.
4. Delineate the rules that underlie structured programming.
5. Be capable of composing structured and modular programs in a high-level computer language.
6. Know how to translate structured flowcharts and pseudocode into code in a high-level language.
7. Start to familiarize yourself with any software packages that you will be using in conjunction with this text.
8. Recognize the distinction between truncation and round-off errors.
9. Understand the concepts of significant figures, accuracy, and precision.
10. Recognize the difference between true relative error ε_t, approximate relative error ε_a, and acceptable error ε_s, and understand how ε_a and ε_s are used to terminate an iterative computation.
11. Understand how numbers are represented in digital computers and how this representation induces round-off error. In particular, know the difference between single and extended precision.
12. Recognize how computer arithmetic can introduce and amplify round-off errors in calculations. In particular, appreciate the problem of subtractive cancellation.
13. Understand how the Taylor series and its remainder are employed to represent continuous functions.
14. Know the relationship between finite divided differences and derivatives.
15. Be able to analyze how errors are propagated through functional relationships.
16. Be familiar with the concepts of stability and condition.
17. Familiarize yourself with the trade-offs outlined in the Epilogue of Part One.

PT1.3.2 Goals and Objectives

Study Objectives.   Upon completing Part One, you should be adequately prepared to embark on your studies of numerical methods. In general, you should have gained a fundamental understanding of the importance of computers and the role of approximations and errors in the implementation and development of numerical methods. In addition to these general goals, you should have mastered each of the specific study objectives listed in Table PT1.1.

Computer Objectives.   Upon completing Part One, you should have mastered sufficient computer skills to develop your own software for the numerical methods in this text. You should be able to develop well-structured and reliable computer programs on the basis of pseudocode, flowcharts, or other forms of algorithms. You should have developed the capability to document your programs so that they may be effectively employed by users. Finally, in addition to your own programs, you may be using software packages along with this book. The Numerical Methods TOOLKIT, which has been provided along with the text, and other packages like Mathcad, MATLAB, or Excel are examples of such software. You should become familiar with these packages, so that you will be comfortable using them to solve numerical problems later in the text.

CHAPTER 1

Mathematical Modeling and Engineering Problem Solving

Knowledge and understanding are prerequisites for the effective implementation of any tool. No matter how impressive your tool chest, you will be hard-pressed to repair a car if you do not understand how it works.

This is particularly true when using computers to solve engineering problems. Although they have great potential utility, computers are practically useless without a fundamental understanding of how engineering systems work.

This understanding is initially gained by empirical means—that is, by observation and experiment. However, while such empirically derived information is essential, it is only half the story. Over years and years of observation and experiment, engineers and scientists have noticed that certain aspects of their empirical studies occur repeatedly. Such general behavior can then be expressed as fundamental laws that essentially embody the cumulative wisdom of past experience. Thus, most engineering problem solving employs the two-pronged approach of empiricism and theoretical analysis (Fig. 1.1).

It must be stressed that the two prongs are closely coupled. As new measurements are taken, the generalizations may be modified or new ones developed. Similarly, the generalizations can have a strong influence on the experiments and observations. In particular, generalizations can serve as organizing principles that can be employed to synthesize observations and experimental results into a coherent and comprehensive framework from which conclusions can be drawn. From an engineering problem-solving perspective, such a framework is most useful when it is expressed in the form of a mathematical model.

The primary objective of this chapter is to introduce you to mathematical modeling and its role in engineering problem solving. We will also illustrate how numerical methods figure in the process.

1.1 A SIMPLE MATHEMATICAL MODEL

A *mathematical model* can be broadly defined as a formulation or equation that expresses the essential features of a physical system or process in mathematical terms. In a very general sense, it can be represented as a functional relationship of the form

$$\begin{array}{c}\text{Dependent} \\ \text{variable}\end{array} = f\left(\begin{array}{c}\text{independent} \\ \text{variables}\end{array}, \text{parameters}, \begin{array}{c}\text{forcing} \\ \text{functions}\end{array}\right) \qquad (1.1)$$

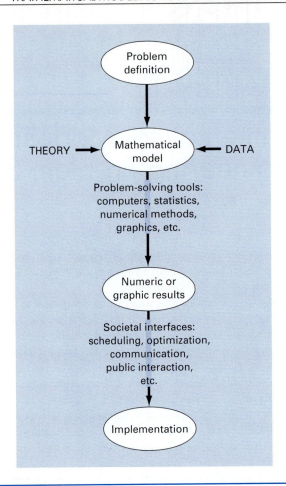

FIGURE 1.1
The engineering problem-solving process.

where the *dependent variable* is a characteristic that usually reflects the behavior or state of the system; the *independent variables* are usually dimensions, such as time and space, along which the system's behavior is being determined; the *parameters* are reflective of the system's properties or composition; and the *forcing functions* are external influences acting upon it.

The actual mathematical expression of Eq. (1.1) can range from a simple algebraic relationship to large complicated sets of differential equations. For example, on the basis of his observations, Newton formulated his second law of motion, which states that the time rate of change of momentum of a body is equal to the resultant force acting on it. The mathematical expression, or model, of the second law is the well-known equation

$$F = ma \tag{1.2}$$

where F = net force acting on the body (N, or kg m/s^2), m = mass of the object (kg), and a = its acceleration (m/s^2).

F_U

F_D

FIGURE 1.2
Schematic diagram of the forces acting on a falling parachutist. F_D is the downward force due to gravity. F_U is the upward force due to air resistance.

The second law can be recast in the format of Eq. (1.1) by merely dividing both sides by m to give

$$a = \frac{F}{m} \tag{1.3}$$

where a = the dependent variable reflecting the system's behavior, F = the forcing function, and m = a parameter representing a property of the system. Note that for this simple case there is no independent variable because we are not yet predicting how acceleration varies in time or space.

Equation (1.3) has several characteristics that are typical of mathematical models of the physical world:

1. It describes a natural process or system in mathematical terms.
2. It represents an idealization and simplification of reality. That is, the model ignores negligible details of the natural process and focuses on its essential manifestations. Thus, the second law does not include the effects of relativity that are of minimal importance when applied to objects and forces that interact on or about the earth's surface at velocities and on scales visible to humans.
3. Finally, it yields reproducible results and, consequently, can be used for predictive purposes. For example, if the force on an object and the mass of an object are known, Eq. (1.3) can be used to compute acceleration.

Because of its simple algebraic form, the solution of Eq. (1.2) can be obtained easily. However, other mathematical models of physical phenomena may be much more complex, and either cannot be solved exactly or require more sophisticated mathematical techniques than simple algebra for their solution. To illustrate a more complex model of this kind, Newton's second law can be used to determine the terminal velocity of a free-falling body near the earth's surface. Our falling body will be a parachutist (Fig. 1.2). A model for this case can be derived by expressing the acceleration as the time rate of change of the velocity (dv/dt) and substituting it into Eq. (1.3) to yield

$$\frac{dv}{dt} = \frac{F}{m} \tag{1.4}$$

where v is velocity (m/s). Thus, the mass multiplied by the rate of change of the velocity is equal to the net force acting on the body. If the net force is positive, the object will accelerate. If it is negative, the object will decelerate. If the net force is zero, the object's velocity will remain at a constant level.

Next, we will express the net force in terms of measurable variables and parameters. For a body falling within the vicinity of the earth (Fig. 1.2), the net force is composed of two opposing forces: the downward pull of gravity F_D and the upward force of air resistance F_U:

$$F = F_D + F_U \tag{1.5}$$

If the downward force is assigned a positive sign, the second law can be used to formulate the force due to gravity, as

$$F_D = mg \tag{1.6}$$

where g = the gravitational constant, or the acceleration due to gravity, which is approximately equal to 9.8 m/s^2.

Air resistance can be formulated in a variety of ways. A simple approach is to assume that it is linearly proportional to velocity[1] and acts in an upward direction, as in

$$F_U = -cv \tag{1.7}$$

where c = a proportionality constant called the *drag coefficient* (kg/s). Thus, the greater the fall velocity, the greater the upward force due to air resistance. The parameter c accounts for properties of the falling object, such as shape or surface roughness, that affect air resistance. For the present case, c might be a function of the type of jumpsuit or the orientation used by the parachutist during free-fall.

The net force is the difference between the downward and upward force. Therefore, Eqs. (1.4) through (1.7) can be combined to yield

$$\frac{dv}{dt} = \frac{mg - cv}{m} \tag{1.8}$$

or simplifying the right side,

$$\frac{dv}{dt} = g - \frac{c}{m}v \tag{1.9}$$

Equation (1.9) is a model that relates the acceleration of a falling object to the forces acting on it. It is a *differential equation* because it is written in terms of the differential rate of change (dv/dt) of the variable that we are interested in predicting. However, in contrast to the solution of Newton's second law in Eq. (1.3), the exact solution of Eq. (1.9) for the velocity of the falling parachutist cannot be obtained using simple algebraic manipulation. Rather, more advanced techniques such as those of calculus must be applied to obtain an exact or analytical solution. For example, if the parachutist is initially at rest ($v = 0$ at $t = 0$), calculus can be used to solve Eq. (1.9) for

$$v(t) = \frac{gm}{c}\left(1 - e^{-(c/m)t}\right) \tag{1.10}$$

Note that Eq. (1.10) is cast in the general form of Eq. (1.1), where $v(t)$ = the dependent variable, t = the independent variable, c and m = parameters, and g = the forcing function.

EXAMPLE 1.1   Analytical Solution to the Falling Parachutist Problem

Problem Statement.   A parachutist of mass 68.1 kg jumps out of a stationary hot air balloon. Use Eq. (1.10) to compute velocity prior to opening the chute. The drag coefficient is equal to 12.5 kg/s.

Solution.   Inserting the parameters into Eq. (1.10) yields

$$v(t) = \frac{9.8(68.1)}{12.5}\left(1 - e^{-(12.5/68.1)t}\right) = 53.39\left(1 - e^{-0.18355t}\right)$$

which can be used to compute

---

[1] In fact, the relationship is actually nonlinear and might better be represented by a power relationship such as $F_U = -cv^2$. We will explore how such nonlinearities affect the model in a problem at the end of this chapter.

t, s	v, m/s
0	0.00
2	16.40
4	27.77
6	35.64
8	41.10
10	44.87
12	47.49
∞	53.39

According to the model, the parachutist accelerates rapidly (Fig. 1.3). A velocity of 44.87 m/s (100.4 mi/h) is attained after 10 s. Note also that after a sufficiently long time, a constant velocity, called the *terminal velocity,* of 53.39 m/s (119.4 mi/h) is reached. This velocity is constant because, eventually, the force of gravity will be in balance with the air resistance. Thus, the net force is zero and acceleration has ceased.

Equation (1.10) is called an *analytical,* or *exact, solution* because it exactly satisfies the original differential equation. Unfortunately, there are many mathematical models that cannot be solved exactly. In many of these cases, the only alternative is to develop a numerical solution that approximates the exact solution.

As mentioned previously, *numerical methods* are those in which the mathematical problem is reformulated so it can be solved by arithmetic operations. This can be illustrated

FIGURE 1.3
The analytical solution to the falling parachutist problem as computed in Example 1.1. Velocity increases with time and asymptotically approaches a terminal velocity.

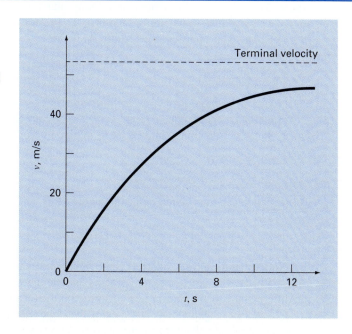

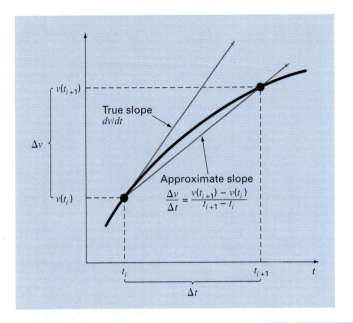

FIGURE 1.4
The use of a finite difference to approximate the first derivative of v with respect to t.

for Newton's second law by realizing that the time rate of change of velocity can be approximated by (Fig. 1.4):

$$\frac{dv}{dt} \cong \frac{\Delta v}{\Delta t} = \frac{v(t_{i+1}) - v(t_i)}{t_{i+1} - t_i} \tag{1.11}$$

where Δv and Δt = differences in velocity and time computed over finite intervals, $v(t_i)$ = velocity at an initial time t_i, and $v(t_{i+1})$ = velocity at some later time t_{i+1}. Note that $dv/dt \cong \Delta v/\Delta t$ is approximate because Δt is finite. Remember from calculus that

$$\frac{dv}{dt} = \lim_{\Delta t \to 0} \frac{\Delta v}{\Delta t}$$

Equation (1.11) represents the reverse process.

Equation (1.11) is called a *finite divided difference* approximation of the derivative at time t_i. It can be substituted into Eq. (1.9) to give

$$\frac{v(t_{i+1}) - v(t_i)}{t_{i+1} - t_i} = g - \frac{c}{m}v(t_i)$$

This equation can then be rearranged to yield

$$v(t_{i+1}) = v(t_i) + \left[g - \frac{c}{m}v(t_i) \right](t_{i+1} - t_i) \tag{1.12}$$

Notice that the term in brackets is the right-hand side of the differential equation itself [Eq. (1.9)]. That is, it provides a means to compute the rate of change or slope of v. Thus, the differential equation has been transformed into an equation that can be used to determine the velocity algebraically at t_{i+1} using the slope and previous values of v and t. If you are given an initial value for velocity at some time t_i, you can easily compute velocity at a later time t_{i+1}. This new value of velocity at t_{i+1} can in turn be employed to extend the computation to velocity at t_{i+2} and so on. Thus, at any time along the way,

New value = old value + slope × step size

EXAMPLE 1.2   Numerical Solution to the Falling Parachutist Problem

Problem Statement.   Perform the same computation as in Example 1.1 but use Eq. (1.12) to compute velocity. Employ a step size of 2 s for the calculation.

Solution.   At the start of the computation ($t_i = 0$), the velocity of the parachutist is zero. Using this information and the parameter values from Example 1.1, Eq. (1.12) can be used to compute velocity at $t_{i+1} = 2$ s:

$$v = 0 + \left[9.8 - \frac{12.5}{68.1}(0) \right] 2 = 19.60 \text{ m/s}$$

For the next interval (from $t = 2$ to 4 s), the computation is repeated, with the result

$$v = 19.60 + \left[9.8 - \frac{12.5}{68.1}(19.60) \right] 2 = 32.00 \text{ m/s}$$

The calculation is continued in a similar fashion to obtain additional values:

t, s	v, m/s
0	0.00
2	19.60
4	32.00
6	39.85
8	44.82
10	47.97
12	49.96
∞	53.39

The results are plotted in Fig. 1.5 along with the exact solution. It can be seen that the numerical method captures the essential features of the exact solution. However, because we have employed straight-line segments to approximate a continuously curving function, there is some discrepancy between the two results. One way to minimize such discrepancies is to use a smaller step size. For example, applying Eq. (1.12) at l-s intervals results in a smaller error, as the straight-line segments track closer to the true solution. Using hand calculations, the effort associated with using smaller and smaller step sizes would make such numerical solutions impractical. However, with the aid of the computer, large numbers of calculations can be performed easily. Thus, you can accurately model the velocity of the falling parachutist without having to solve the differential equation exactly.

As in the previous example, a computational price must be paid for a more accurate numerical result. Each halving of the step size to attain more accuracy leads to a doubling

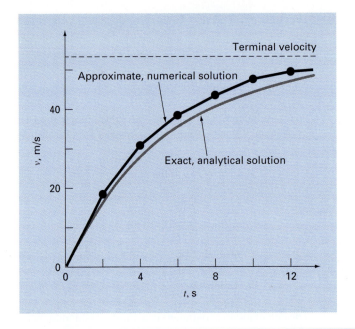

FIGURE 1.5
Comparison of the numerical
and analytical solutions for the
falling parachutist problem.

of the number of computations. Thus, we see that there is a trade-off between accuracy and computational effort. Such trade-offs figure prominently in numerical methods and constitute an important theme of this book. Consequently, we have devoted the Epilogue of Part One to an introduction to more of these trade-offs.

1.2 CONSERVATION LAWS AND ENGINEERING

Aside from Newton's second law, there are other major organizing principles in engineering. Among the most important of these are the conservation laws. Although they form the basis for a variety of complicated and powerful mathematical models, the great conservation laws of science and engineering are conceptually easy to understand. They all boil down to

$$\text{Change} = \text{increases} - \text{decreases} \tag{1.13}$$

This is precisely the format that we employed when using Newton's law to develop a force balance for the falling parachutist [Eq. (1.8)].

Although simple, Eq. (1.13) embodies one of the most fundamental ways in which conservation laws are used in engineering—that is, to predict changes with respect to time. We give Eq. (1.13) the special name *time-variable* (or *transient*) computation.

Aside from predicting changes, another way in which conservation laws are applied is for cases where change is nonexistent. If change is zero, Eq. (1.13) becomes

$$\text{Change} = 0 = \text{increases} - \text{decreases}$$

or

$$\text{Increases} = \text{decreases} \tag{1.14}$$

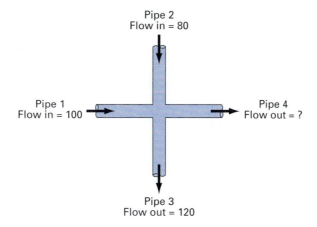

FIGURE 1.6
A flow balance for steady incompressible fluid flow at the junction of pipes.

Thus, if no change occurs, the increases and decreases must be in balance. This case, which is also given a special name—the *steady-state* computation—has many applications in engineering. For example, for steady-state incompressible fluid flow in pipes, the flow into a junction must be balanced by flow going out, as in

Flow in = flow out

For the junction in Fig. 1.6, the balance can be used to compute that the flow out of the fourth pipe must be 60.

For the falling parachutist, steady-state conditions would correspond to the case where the net force was zero, or [Eq. (1.8) with $dv/dt = 0$]

$$mg = cv \tag{1.15}$$

Thus, at steady state, the downward and upward forces are in balance and Eq. (1.15) can be solved for the terminal velocity

$$v = \frac{mg}{c}$$

Although Eqs. (1.13) and (1.14) might appear trivially simple, they embody the two fundamental ways that conservation laws are employed in engineering. As such, they will form an important part of our efforts in subsequent chapters to illustrate the connection between numerical methods and engineering. Our primary vehicles for making this connection are the engineering applications that appear at the end of each part of this book.

Table 1.1 summarizes some of the simple engineering models and associated conservation laws that will form the basis for many of these engineering applications. Most of the chemical engineering applications will focus on mass balances for reactors. The mass balance is derived from the conservation of mass. It specifies that the change of mass of a chemical in the reactor depends on the amount of mass flowing in minus the mass flowing out.

Both the civil and mechanical engineering applications will focus on models developed from the conservation of momentum. For civil engineering, force balances are utilized to analyze structures such as the simple truss in Table 1.1. The same principles are employed for the mechanical engineering applications to analyze the transient up-and-down motion or vibrations of an automobile.

TABLE 1.1 Devices and types of balances that are commonly used in the four major areas of engineering. For each case, the conservation law upon which the balance is based is specified.

Field	Device	Organizing Principle	Mathematical Expression
Chemical engineering	Reactors	Conservation of mass	Mass balance: Input → ☐ → Output Over a unit of time period Δmass = inputs − outputs
Civil engineering	Structure	Conservation of momentum	Force balance: $+F_v$ $-F_H \longleftarrow \bullet \longrightarrow +F_H$ $-F_v$ At each node Σ horizontal forces $(F_H) = 0$ Σ vertical forces $(F_V) = 0$
Mechanical engineering	Machine	Conservation of momentum	Force balance: Upward force $x = 0$ Downward force $m \dfrac{d^2x}{dt^2}$ = downward force − upward force
Electrical engineering	Circuit	Conservation of charge	Current balance: For each node Σ current $(i) = 0$ $+i_1 \longrightarrow \bullet \longrightarrow -i_3$ $+i_2$
		Conservation of energy	Voltage balance: i_1R_1 i_2R_2 ξ i_3R_3 Around each loop Σ emf's − Σ voltage drops for resistors = 0 $\Sigma\,\xi - \Sigma\,iR = 0$

TABLE 1.2 Some practical issues that will be explored in the engineering applications at the end of each part of this book.

1. *Nonlinear versus linear.* Much of classical engineering depends on linearization to permit analytical solutions. Although this is often appropriate, expanded insight can often be gained if nonlinear problems are examined.
2. *Large versus small systems.* Without a computer, it is often not feasible to examine systems with over three interacting components. With computers and numerical methods, more realistic multicomponent systems can be examined.
3. *Nonideal versus ideal.* Idealized laws abound in engineering. Often there are nonidealized alternatives that are more realistic but more computationally demanding. Approximate numerical approaches can facilitate the application of these nonideal relationships.
4. *Sensitivity analysis.* Because they are so involved, many manual calculations require a great deal of time and effort for successful implementation. This sometimes discourages the analyst from implementing the multiple computations that are necessary to examine how a system responds under different conditions. Such sensitivity analyses are facilitated when numerical methods allow the computer to assume the computational burden.
5. *Design.* It is often a straightforward proposition to determine the performance of a system as a function of its parameters. It is usually more difficult to solve the inverse problem—that is, determining the parameters when the required performance is specified. Numerical methods and computers often permit this task to be implemented in an efficient manner.

Finally, the electrical engineering applications employ both current and energy balances to model electric circuits. The current balance, which results from the conservation of charge, is similar in spirit to the flow balance depicted in Fig. 1.6. Just as flow must balance at the junction of pipes, electric current must balance at the junction of electric wires. The energy balance specifies that the changes of voltage around any loop of the circuit must add up to zero. The engineering applications are designed to illustrate how numerical methods are actually employed in the engineering problem-solving process. As such, they will permit us to explore practical issues (Table 1.2) that arise in real-world applications. Making these connections between mathematical techniques such as numerical methods and engineering practice is a critical step in tapping their true potential. Careful examination of the engineering applications will help you to take this step.

PROBLEMS

1.1 Answer true or false:

(a) The value of a variable that satisfies a single equation is called the root of the equation.

(b) Finite divided differences are used to represent derivatives in approximate terms.

(c) In the precomputer era, numerical methods were widely employed because they required little computational effort.

(d) Interpolation is employed for curve-fitting problems when there is significant error associated with the data points.

(e) Mathematical models should never be used for predictive purposes.

(f) The large systems of equations, nonlinearities, and complicated geometries that are common in engineering practice are easy to solve analytically.

(g) Newton's second law is a good example of the fact that most physical laws are based on the rate of change of quantities rather than on their magnitudes.

(h) A physical interpretation of integration is the area under a curve.

(i) Numerical methods are those in which a mathematical problem is reformulated so that it can be solved by arithmetic operations.

(j) Today more attention can be paid to problem formulation and interpretation because the computer and numerical methods facilitate the solution of engineering problems.

1.2 Read the following problem descriptions and identify which area of numerical methods (as outlined in Fig. PT1.2) relates to their solution.

(a) You are responsible for determining the flows in a large interconnected network of pipes to distribute natural gas to a series of communities spread out over a 20-mi^2 area.

(b) You are performing experiments to determine the voltage drop across a resistor as a function of current. You make measurements of voltage drop for a number of different values of current. Although there is some error associated with your data points, when you plot them they suggest a smooth curvilinear relationship. You are to derive an equation to characterize this relationship.

(c) You must develop a shock-absorber system for a racing car. Newton's second law can be used to derive an equation to predict the rate of change in position of the front wheel in response to external forces. You must compute the motion of the wheel as a function of time after it hits a 6-in bump at 150 mi/h.

(d) You have to determine the annual revenues required over a 20-year period for an entertainment center to be built for a client. Money can presently be borrowed at an interest rate of 10.37 percent. Although the information to perform this estimate is contained in economics tables, values are listed only for interest rates of 10 and 11 percent.

(e) You must determine the temperature distribution on the two-dimensional surface of a flat gasket as a function of the temperatures of its edges.

(f) For the falling parachutist problem, you must determine the value of the drag coefficient in order that a 200-lb parachutist not exceed 100 mi/h within 10 s of jumping. You must make this evaluation on the basis of the analytical solution [Eq. (1.10)]. The information will be used to design a jumpsuit.

(g) You are on a survey crew and must determine the area of a field that is bounded by two roads and a meandering stream.

1.3 Give one example of an engineering problem where each of the following classes of numerical methods can come in handy. If possible, draw from your experience in class and in readings or from any professional experience you have gathered to date.

(a) Roots of equations
(b) Linear algebraic equations
(c) Curve fitting: regression and interpolation
(d) Optimization
(e) Integration
(f) Ordinary differential equations
(g) Partial differential equations

1.4 What is the two-pronged approach to engineering problem solving? Into what category should the conservation laws be placed?

1.5 What is the form of the transient conservation law? What is it for steady state?

1.6 The following information is available for a bank account:

Date	Deposits	Withdrawals	Balance
5/1			512.33
	220.13	327.26	
6/1			
	216.80	378.61	
7/1			
	350.25	106.80	
8/1			
	127.31	450.61	
9/1			

Use the conservation of cash to compute the balance on 6/1, 7/1, 8/1, and 9/1. Show each step in the computation. Is this a steady-state or a transient computation?

1.7 Give examples of conservation laws in engineering and in everyday life.

1.8 Examine your engineering textbooks and find four examples where mathematical models are used to describe the behavior of physical systems. List the independent and dependent variables as well as the parameters and forcing functions.

1.9 Verify that Eq. (1.10) is a solution of Eq. (1.9).

1.10 Repeat Example 1.2. Compute the velocity to $t = 12$ s, with a step size of **(a)** 1 and **(b)** 0.5 s. Can you make any statement regarding the errors of the calculation based on the results?

1.11 Rather than the linear relationship of Eq. (1.7), the upward force on the falling parachutist is actually nonlinear and might better be represented by a power relationship such as

$$F_U = -c'v^2$$

where $c' =$ a second-order drag coefficient (kg/m). Using this relationship, repeat the calculation in Example 1.2 with the same initial condition and parameter values. Use a value of 0.23 kg/m for c'.

1.12 The amount of a uniformly distributed radioactive contaminant contained in a closed reactor is measured by its concentration c (becquerel/liter, or Bq/L). The contaminant decreases at a decay rate proportional to its concentration; that is,

$$\text{Decay rate} = -kc$$

where k = a constant with units of day^{-1}. Therefore, according to Eq. (1.13), a mass balance for the reactor can be written as

$$\frac{dc}{dt} = -kc$$

$$\left(\begin{array}{c}\text{change} \\ \text{in mass}\end{array}\right) = \left(\begin{array}{c}\text{decrease} \\ \text{by decay}\end{array}\right)$$

Use numerical methods to solve this equation from $t = 0$ to 1 d, with $k = 0.1$ d^{-1}. Employ a step size of $\Delta t = 0.1$. The concentration at $t = 0$ is 10 Bq/L.

1.13 A storage tank contains a liquid at depth y, where $y = 0$ when the tank is half full. Liquid is withdrawn at a constant flow rate Q to meet demands. The contents are resupplied at a sinusoidal rate $3Q \sin^2 t$ (see Fig. P1.13).

Equation (1.13) can be written for this system as

$$\frac{d(Ay)}{dt} = 3Q\sin^2 t - Q$$

$$\left(\begin{array}{c}\text{change in} \\ \text{volume}\end{array}\right) = (\text{inflow}) - (\text{outflow})$$

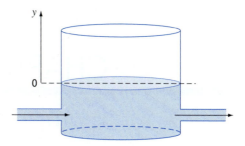

FIGURE P1.13

or, since the surface area, A, is constant

$$\frac{dy}{dt} = 3\frac{Q}{A}\sin^2 t - \frac{Q}{A}$$

Use a numerical method to solve for the height y from $t = 0$ to 5 d with a step size of 0.5 d. The parameter values are $A = 1200$ m^2 and $Q = 400$ m^3/d.

CHAPTER 2

Computers and Software

Numerical methods combine two of the most important tools in the engineering repertoire: mathematics and computers. In fact, numerical methods can be loosely defined as computer math. Good computer programming skills will enhance your studies of numerical methods. In particular, the powers and limitations of numerical techniques are best appreciated when you use the methods in tandem with a computer to solve engineering problems.

You will have the opportunity in using this book to develop your own software. Because of the widespread availability of personal computers and magnetic storage devices, this software can be easily retained and applied throughout your career. Therefore, one of the primary objectives of this text is that you come away from it possessing useful, high-quality software.

In addition, engineers are increasingly using software packages to implement numerical methods. Therefore, another objective of this text will be to acquaint yourself with the capabilities, operation, and limitations of some of these packages. In addition, advanced applications with such packages often involve some programming in the guise of *macros* or *scripts*. Thus, even though it might superficially appear that using the software is a simple "point and shoot" exercise, sophisticated engineers must also know how to develop short, well-structured programs to truly exploit these useful tools.

This text has a number of features that are intended to maximize this possibility. First, all the numerical techniques are accompanied by material related to effective computer implementation. Algorithms are provided for many of the methods. Second, several popular software packages and libraries are described and used to solve engineering-oriented problems. Finally, supplementary software is available for the application or illustration of several of the most elementary methods discussed in this book. This software, which is compatible with widely used personal computers, can serve as a starting point for your own program library.

This chapter provides background information that has utility if you plan to use this text as the basis for developing your own programs. It is written under the premise that you have had some prior exposure to computer programming. Because this book is not intended to support a computer programming course per se, the discussion pertains only to those aspects related to developing high-quality software for numerical methods. It is also intended to provide you with specific criteria for evaluating the quality of your efforts.

2.1   COMPUTING ENVIRONMENTS

We believe that a historical perspective is very useful in appreciating recent advances in software design. As shown in Fig. 2.1, the modern computer age can be conceptualized as a progression of generations. Prior to the early 1970s, computing was monolithic in the sense that large mainframe systems were the only available option. Because they were so expensive to own, operate, and maintain, computers could be acquired only by large organizations such as corporations, universities, government agencies, and the military. As might be expected, the centralized way in which computing was configured had a profound impact on the manner in which engineers and scientists interacted with the machines. In particular, a distinct separation existed between the user and the computer.

In the late 1960s, the introduction of integrated circuits resulted in machines that were significantly faster, cheaper, and smaller. This technological breakthrough in turn spawned a number of new developments that have led to a divergence in the previously one-dimensional evolution of the computing industry (Fig. 2.1).

FIGURE 2.1

The evolution of the modern computer era can be conceptualized as a succession of generations closely linked to technological advances. In the third generation, a divergence emerged with the advent of machines expressly designed for the needs of individuals. Today, a broad range of computers is available.

Source: Redrawn with permission from Steven Chapra and Raymond Canale, Introduction to Computing for Engineers (New York: McGraw-Hill, 1994).

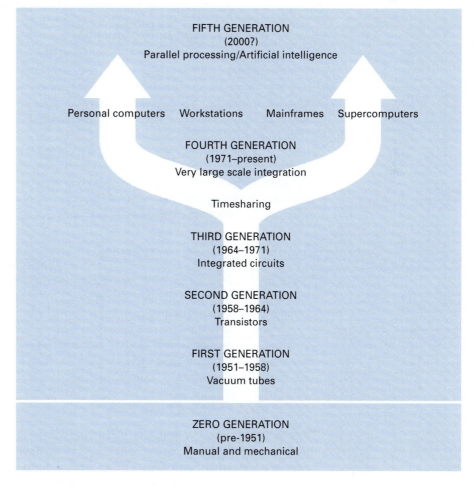

Whereas earlier computers had been developed expressly for large institutions and organizations, the fourth generation resulted in machines that were designed to meet the needs of individuals. Consequently, today's engineer has a broad spectrum of tools with which to implement computations. In terms of size, cost, and number of users, these tools can be broadly divided into four major categories: personal computers, workstations, mainframes, and supercomputers.

As the name implies, a *personal computer* (or *PC*) is a smaller machine principally intended for an individual user. The *workstation* is generally more powerful and, although often dedicated to an individual user, is typically part of a network of computers. The *mainframe* is a larger, more expensive machine usually owned by an institution and serves multiple users. Finally, the *supercomputer* is a large, very expensive, powerful machine that is principally employed for sophisticated applications by a select group of users.

It should be noted that these computer classifications are not hard and fast. For example, as more personal computers are being used as terminals, the boundaries between today's PC's and workstations are beginning to blur.

Whatever definitions are used, the beauty of today's situation is that we have potential access to all types of computers. For example, preliminary work on a large-scale computation might be performed on a personal computer. Then, a telecommunications link can be used to upload the program to a mainframe for fast implementation. In this way, the strengths of each type of computer are exploited. Engineers with access to a range of machines will have enormous capability at their command.

For this reason, and because we anticipate that you will be using different types of machines to learn numerical methods, we have designed this book so that it can be used with any of these systems. Thus, most of the material can be implemented on systems ranging from personal computers to mainframes.

At the same time, we have tried wherever possible to acknowledge and explore some of the exciting new applications that are directly associated with personal computers. In particular, we have devoted a significant portion of this edition to PC-oriented software packages. We believe that the "personal" aspect of the microcomputer revolution is a major source of the present excitement over computing among students and professionals alike.

2.2   THE SOFTWARE DEVELOPMENT PROCESS

No matter what type of computer you use, it has utility only if you provide it with careful instructions. These instructions are the *software*. The next sections deal with the process of composing your own high-quality software to implement numerical methods.

Figure 2.2 delineates the five steps that constitute the process. Each of these steps will be discussed in the subsequent sections. However, before embarking on this discussion, we must first mention some qualitative changes that have occurred in software development and that have great relevance to implementing numerical methods on computers.

2.2.1 Programming Style

Just as hardware has evolved dramatically, the software development process has also undergone a radical transformation in recent years. In the early years of computing, programming was somewhat more of an art than a science. Although all languages had very precise vocabularies and syntax, there were no standard definitions of good programming style.

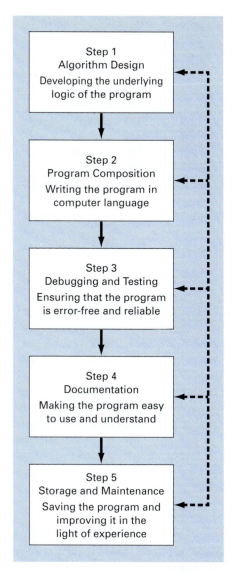

FIGURE 2.2
The five steps required to produce high-quality software. The feedback arrows indicate that the first four steps can be improved in the light of experience.

Consequently, individuals developed their own criteria for what constituted an excellent piece of software.

In the early years, these criteria were strongly influenced by the available hardware. Early computers were expensive and slow and had small memories. As a consequence, one mark of a good program was that it utilized as little memory as possible. Another might be that it executed rapidly. Still another was how quickly it could be written the first time.

The ultimate result was that programs were highly nonuniform. In particular, because they were strongly influenced by hardware limitations, programmers put little premium on the ease with which programs might be used and maintained. Thus, if you wanted to effectively utilize and maintain a large program over long periods, you usually had to keep

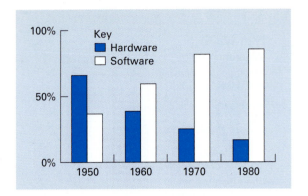

FIGURE 2.3

Trends in hardware and software costs in the computing industry.

Source: L. M. Branscomb, "Electronics and Computers: An Overview," *Science*, 215:755 (1982).

the original programmer close by. In other words, the program and the programmer formed a costly package.

Today many of the hardware limitations do not exist or are rapidly disappearing. In addition, since software costs constitute a proportionally larger fraction of the total computing budget (Fig. 2.3), there is a high premium on developing software that is easy to use and modify. In particular, there is a strong emphasis on clarity and readability rather than on terse, noncommunicative code. This emphasis has great significance in engineering, where work is often performed by teams and programs must be shared and modified by several individuals.

Computer scientists have systematically studied the factors and procedures needed to develop high-quality software of this kind. Although the resulting methods are somewhat oriented toward large-scale programming efforts, most of the general techniques are also extremely useful for the types of programs that engineers develop routinely in the course of their work. Collectively, we will call these techniques *structured design and programming*. In this section, we discuss three of these approaches: modular design, top-down design, and structured programming.

2.2.2 Modular Design

Imagine how difficult it would be to study a textbook that had no chapters, sections, or paragraphs. Breaking complicated tasks or subjects into more manageable parts is one way to make them easier to handle. In the same spirit, computer programs can be divided into small subprograms, or *modules,* that can be developed and tested separately. This approach is called *modular design.*

The most important attribute of modules is that they be as independent and self-contained as possible. In addition, they are typically designed to perform a specific, well-defined function and have one entry and one exit point. As such, they are usually short (generally 50 to 100 instructions in length) and highly focused.

One of the primary programming elements used to represent each module is the sub-program or procedure. A *subprogram* is a series of computer instructions that together perform a given task. Two types of procedures are commonly employed: *functions* and *sub-routines.* The former usually returns a single result, whereas the latter returns several. A *calling,* or *main, program* invokes these modules as they are needed. Thus, the main program orchestrates each of the parts in a logical fashion.

In addition, it should be mentioned that much of the programming related to software packages like Excel and MATLAB involves the development of functions. That is, Excel macros and MATLAB functions are designed to receive some information, perform a calculation, and return results. Thus, modular thinking is consistent with how programming is implemented in package environments.

Modular design has a number of advantages. The use of small, self-contained units makes the underlying logic easier to devise and to understand for both the developer and the user. Development is facilitated because each module can be perfected in isolation. In fact, for large projects, different programmers can work on individual parts. In the same spirit, you can maintain your own library of useful modules for later use in other programs. Modular design also increases the ease with which a program can be debugged and tested because errors can be more easily isolated. Finally, program maintenance and modification are facilitated. This is primarily because new modules can be developed to perform additional tasks and then easily incorporated into the already coherent and organized scheme.

2.2.3 Top-Down Design

Although the previous section introduced the notion of modular design, it did not specify how the actual modules are identified. The top-down design process is a particularly effective approach for accomplishing this objective.

Top-down design is a systematic development process that begins with the most general statement of a program's objectives and then successively divides it into more detailed segments. Thus, the design proceeds from the general to the specific. Most top-down designs start with a literal description of the program in its most general terms. This description is then broken down into a few elements that define the program's major functions. Then each major element is divided into subelements. This process is continued until well-defined modules have been identified.

Besides providing a method for dividing the algorithm into well-defined units, top-down design has other benefits. In particular, it ensures that the finished product is comprehensive. By starting with a broad definition and progressively adding detail, the programmer is less likely to overlook important operations.

2.2.4 Structured Programming

Modular and top-down approaches deal with effective ways to organize a program. In both cases, emphasis was placed on what the program is supposed to do. Although this usually guarantees that the program will be coherent and neatly organized, it does not ensure that the actual program instructions or code for each module will manifest an associated clarity and orderliness. Structured programming, on the other hand, deals with *how* the actual program code is developed so that it is easy to understand, correct, and modify.

In essence, *structured programming* is a set of rules that prescribe good style habits for the programmer. Although structured programming is flexible enough to allow considerable creativity and personal expression, its rules impose enough constraints to render resulting codes far superior to unstructured versions. In particular, the finished product is more elegant and easier to understand.

The rules that constitute structured programming will be described in a subsequent section (2.4.2). Before doing so, we will first turn to the initial step of the program development process—designing the underlying logic or algorithm of the program.

2.3   ALGORITHM DESIGN

The most common problems encountered by inexperienced programmers usually can be traced to the premature preparation of a program that does not encompass an overall strategy or plan. In the early days of computing, this was a particularly thorny problem because the design and programming components were not clearly separated. Programmers would often launch into a computing project without a clearly formulated design. This led to all sorts of problems and inefficiencies. For example, the programmer could get well into a job only to discover that some key preliminary factor had been overlooked. In addition, the logic underlying improvised programs was invariably obscure. This meant that it was usually difficult for someone to modify someone else's program. Today, because of the high costs of software development, much greater emphasis is placed on such aspects as preliminary planning that lead to a more coherent and efficient product. The focus of this planning is algorithm design.

An *algorithm* is the sequence of logical steps required to perform a specific task such as solving a problem. Aside from accomplishing its objectives, a good algorithm must feature several attributes:

1. Each step must be deterministic—that is, nothing can be left to chance. The final results cannot depend on who is following the algorithm. In this sense, an algorithm is analogous to a recipe. Two chefs working independently from a good recipe should end up with dishes that are essentially identical.
2. The process must always end after a finite number of steps. An algorithm cannot be open-ended.
3. The algorithm must be general enough to deal with any contingency.

Figure 2.4*a* shows an algorithm for the solution of the simple problem of adding a pair of numbers. Two independent programmers working from this algorithm might develop programs exhibiting somewhat different styles. However, given the same data, their programs should yield identical results.

Step-by-step literal descriptions of the sort depicted in Fig. 2.4*a* are one way to express an algorithm. These descriptions are particularly useful for small problems or for specifying the broad tasks of a large programming effort. However, for detailed representations of complicated programs, they rapidly become inadequate. For this reason, more versatile alternatives, called flowcharts and pseudocodes, have been developed.

2.3.1 Flowcharting

A *flowchart* is a visual or graphical representation of an algorithm. The flowchart employs a series of blocks and arrows, each of which represents a particular operation or step in the algorithm. The arrows represent the sequence in which the operations are implemented. Figure 2.4*b* shows a flowchart for the problem of adding two numbers.

Not everyone involved with computer programming agrees that flowcharting is a productive endeavor. In fact, some experienced programmers do not advocate flowcharts. However, we believe that there are three good reasons for studying them. First, they are still used for expressing and communicating algorithms. Second, even if they are not employed routinely, there will be times when they will prove useful in planning, unraveling, or communicating the logic of your own or someone else's program. Finally, and most

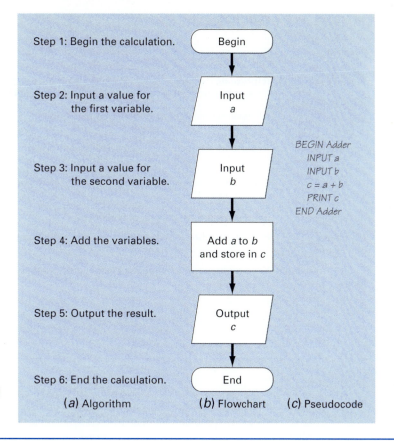

Step 1: Begin the calculation.   Begin

Step 2: Input a value for   Input
        the first variable.   *a*

Step 3: Input a value for   Input
        the second variable.   *b*

Step 4: Add the variables.   Add *a* to *b*
                                  and store in *c*

Step 5: Output the result.   Output
                                     c

Step 6: End the calculation.   End

BEGIN Adder
INPUT a
INPUT b
c = a + b
PRINT c
END Adder

(*a*) Algorithm   (*b*) Flowchart   (*c*) Pseudocode

FIGURE 2.4
(*a*) Algorithm, (*b*) flowchart, and (*c*) pseudocode for the solution of a simple addition program.

important for our purposes, they are excellent pedagogical tools. From a teaching perspective, they are ideal vehicles for visualizing some of the fundamental control structures employed in computer programming.

Flowchart Symbols.   Notice that in Fig. 2.4*b* the blocks have different shapes to distinguish different types of operations. In the early years of computing, flowchart symbols were nonstandardized. This led to confusion, because programmers used different symbols to represent the same operation. Today a number of standardized systems have been developed to facilitate communication. In this book, we use the set of symbols shown in Fig. 2.5.

A flowchart is read by starting at the beginning terminal and following the flowlines to trace the algorithm's logic. Thus, as shown in Fig. 2.4*b*, we would start at the "begin" block and then follow the arrow sequentially until we arrive at the "end" block. If the flowchart is written properly, there should never be any doubt regarding the correct path.

Top-Down Algorithm Design.   There are no set standards for how detailed a flowchart should be. During the development of a major algorithm, you will often draft flowcharts of various levels of complexity. As an example, Fig. 2.6 depicts a hierarchy of three charts that might be used in the development of an algorithm to determine the grade-point average (GPA) or cumulative index of a student. The system flowchart (Fig. 2.6*a*) delineates

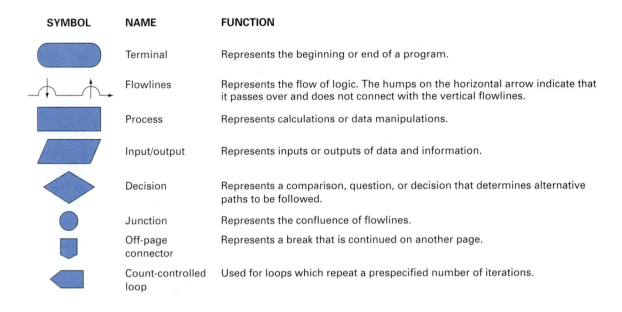

SYMBOL	NAME	FUNCTION
	Terminal	Represents the beginning or end of a program.
	Flowlines	Represents the flow of logic. The humps on the horizontal arrow indicate that it passes over and does not connect with the vertical flowlines.
	Process	Represents calculations or data manipulations.
	Input/output	Represents inputs or outputs of data and information.
	Decision	Represents a comparison, question, or decision that determines alternative paths to be followed.
	Junction	Represents the confluence of flowlines.
	Off-page connector	Represents a break that is continued on another page.
	Count-controlled loop	Used for loops which repeat a prespecified number of iterations.

FIGURE 2.5
Symbols used in flowcharts.

the big picture. It identifies the major tasks, or *modules,* and the sequence that is required to solve the entire problem. Such an overview flowchart is invaluable for ensuring that the total scheme is comprehensive enough to be successful. Next, the major modules can be broken off and charted in greater detail (Fig. 2.6*b*). Finally, it is sometimes advantageous to break the major modules down into even more manageable units, as in Fig. 2.6*c*. This is an example of the *top-down design process* that was introduced in Sec. 2.2.3.

In the following sections we will study and improve the flowchart in Fig. 2.6*c*. This will be done to illustrate how the symbols in Fig. 2.5 can be orchestrated to develop an efficient and effective algorithm. In the process we will demonstrate how the power of computers can be reduced to three fundamental types of operations. The first, which should already be apparent, is that computer logic can proceed in a definite *sequence.* As depicted in Fig. 2.6*a* and *c,* this is represented by the flow of logic from box to box. The other two fundamental operations are *selection* and *repetition.*

Structured Algorithms.   Notice that the operations in Fig. 2.6*a* and *c* specify a simple sequential flow. That is, the operations follow one after the other. However, notice how the first diamond-shaped decision symbol in Fig. 2.6*b* permits the flow to branch and follow one of two possible paths, depending on the answer to a question. For this case, if final grades are being determined, the flow follows the right branch, whereas if intermediate or midterm grades are being determined, the flow follows the left branch. This type of flowchart operation is called *selection.*

Aside from depicting a conditional branch, Fig. 2.6*b* demonstrates another flowchart operation that greatly increases the power of computer programs—*repetition.* Notice that

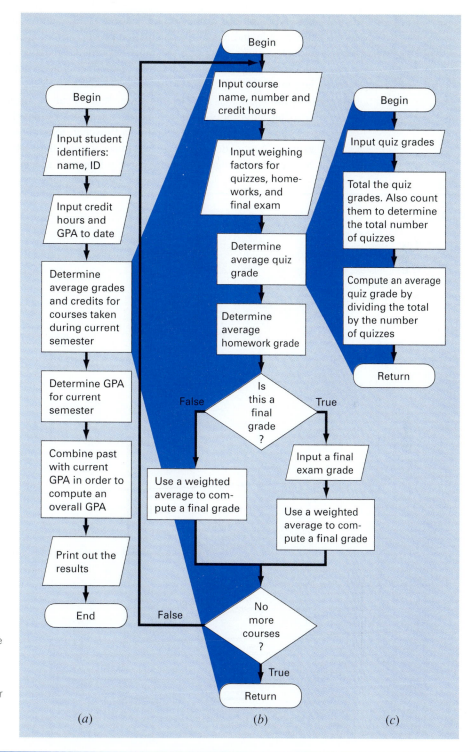

FIGURE 2.6

A hierarchy of flowcharts dealing with the problem of determining a student's grade-point average. The system flowchart in (a) provides a comprehensive plan. A more detailed chart for a major module is shown in (b). An even more detailed chart for a segment of the major module is seen in (c). This way of approaching a problem is called top-down design.

the "false" branch of the second decision symbol allows the flow to return to the head of the module and repeat the calculation process for another course. This ability to perform repetitive tasks is among the most important strengths of computers. The operation of "looping" back to repeat a set of operations is accordingly named a *loop* in computer jargon.

The operations of selection and repetition can be employed to great advantage when deriving algorithms. This can be demonstrated by developing an improved version of Fig. 2.6c. In Fig. 2.6c, we specified a flowchart to total the quiz grades and determine the average. A more detailed flowchart to accomplish these objectives is presented in Fig. 2.7. Notice that the algorithm consists of a loop and a decision construct. The loop inputs,

FIGURE 2.7

A detailed flowchart to compute the average quiz grade. Through the use of an accumulator and a counter, this example improves on the simple version shown in Fig. 2.6.

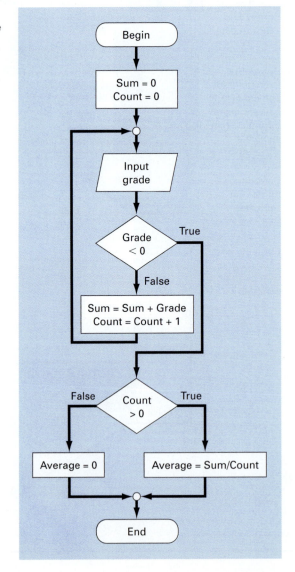

counts, and sums the grades. These operations are repeated until the input of a negative grade triggers the exit from the loop. At this point, a selection construct is used to compute the average grade. A selection construct is employed for this purpose to avoid division by zero in the event that no positive grade has been entered.

Structured Flowcharts.   Flowcharting symbols can be orchestrated to construct logic patterns of extreme complexity. However, programming experts have recognized that, without rules, flowcharts can become so complicated that they impede rather than facilitate clear thinking.

One of the main culprits that contributes to algorithm complexity is the unconditional transfer, also known as the GO TO. As the name implies, this control structure allows you to "go to" any other location in the algorithm. For flowcharts, it is simply represented by an arrow. Indiscriminate use of unconditional transfers can lead to algorithms that, at a distance, resemble a plate of spaghetti; hence, the derogatory label *spaghetti algorithm*.

In an effort to avoid this dilemma, advocates of structured programming have demonstrated that any program can be constructed with the three fundamental control structures depicted in Fig. 2.8. As can be seen, they correspond to the three operations—sequence, selection, and repetition—that we have just been discussing.

All flowcharts should be composed of the three fundamental control structures shown in Fig. 2.8. If we limit ourselves to these structures and avoid GO TOs, the resulting algorithm will be much easier to understand.

Before proceeding, we must make one practical revision to our "no GO TO" dictum. Unfortunately, some languages (notably Fortran 77) and some packages do not presently include a complete set of structured programming constructs. This is particularly true of the packages where simplified, primitive programming languages have often been adopted. As more and more users write macros and scripts, this situation is changing. However, at the time of this book's printing, there are still occasions where you must use a GO TO to implement a standard control structure. This said, if you avoid wild jumps using GO TOs, you can still write decent codes in these environments.

2.3.2 Pseudocode

The object of flowcharting is obviously to develop a quality computer program consisting of a set of step-by-step instructions called *code*. Inspection of Figs. 2.6 and 2.7 illustrates how top-down flowcharting moves in the direction of a computer program. Recall that in top-down design, more detailed flowcharts are developed as we progress from the general overview (Fig. 2.6a) to specific modules (Fig. 2.6c). At the highest level of detail, the module flowcharts are almost in the form of a computer program. That is, as seen in Fig. 2.7, each flowchart compartment represents a single well-defined instruction to the computer.

An alternative way to express an algorithm that bridges the gap between flowcharts and computer code is called *pseudocode*. This technique uses codelike statements in place of the graphical symbols of the flowchart. Figure 2.9 shows pseudocode representations for the fundamental control structures.

We have adopted some style conventions for the pseudocode in this book. Keywords such as IF, DO, etc., are capitalized, whereas the conditions, processing steps, and tasks are in lowercase. Additionally, notice that the processing steps are indented. Thus, the keywords form a "sandwich" around the steps to visually define the extent of each control structure.

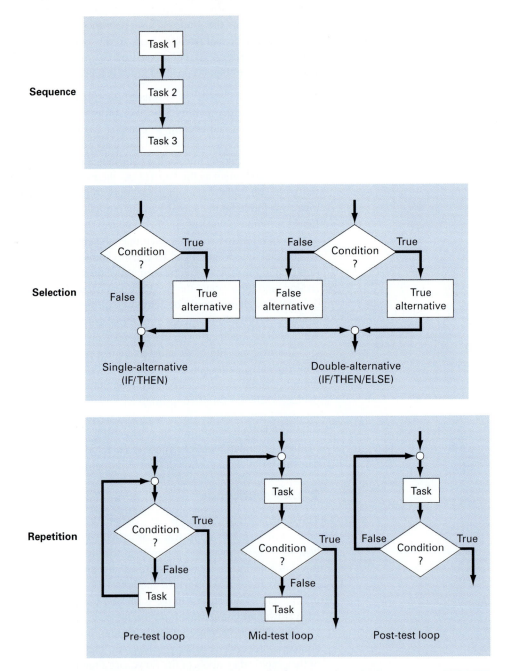

FIGURE 2.8
The three fundamental
control structures.

Figure 2.10 shows a pseudocode representation of the flowchart from Fig. 2.7. This version looks much more like a computer program than the flowchart. Thus, one advantage of pseudocode is that it is easier to develop a program with it than with a flowchart. The pseudocode is also easier to modify and share with others. However, because of their

Sequence	Task 1 Task 2 Task 3	

Selection	IF condition THEN
	   True alternative
	END IF
		IF condition THEN
		   True alternative
		ELSE
		   False alternative
		END IF

	Single-alternative (IF/THEN)	**Double-alternative (IF/THEN/ELSE)**

Repetition	DO
	  IF condition EXIT
	  Task
	END DO
		DO
		  Task
		  IF condition EXIT
		  Task
		END DO
			DO
			  Task
			  IF condition EXIT
			END DO

	Pre-test loop	**Mid-test loop**	**Post-test loop**

FIGURE 2.9
Pseudocode representations of the three fundamental control structures.

FIGURE 2.10
Pseudocode to compute the average quiz grade. This algorithm was represented previously by the flowchart in Fig. 2.7.

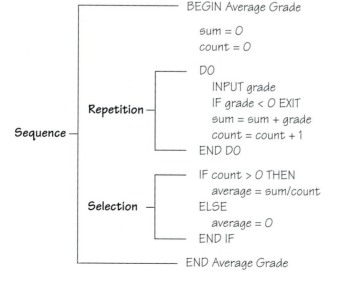

```
BEGIN Average Grade

    sum = 0
    count = 0

    DO
        INPUT grade
        IF grade < 0 EXIT
        sum = sum + grade
        count = count + 1
    END DO

    IF count > 0 THEN
        average = sum/count
    ELSE
        average = 0
    END IF

END Average Grade
```

Sequence
Repetition
Selection

graphic form, flowcharts sometimes are better suited for visualizing complex algorithms. In this text, pseudocode will be used to communicate algorithms related to numerical methods.

2.3.3 Other Useful Structures

Although any numerical algorithm can be developed from the three fundamental control structures, computer scientists have devised several additional structures that have great utility in numerical methods. The most important of these are the multialternative selection structure and the count-controlled loop.

The Multialternative Selection Structure. Suppose that the ELSE clause of an IF/THEN/ELSE contains another IF/THEN, as in the following pseudocode:

```
IF condition
    Task₁
ELSE
    IF condition
        Task₂
    ELSE
        Task₃
    END IF
END IF
```

For such cases, the ELSE and the IF can be combined, as in

```
IF condition
    Task₁
ELSE IF condition
    Task₂
ELSE
    Task₃
END IF
```

Notice how the ELSE IF makes the structure more concise. Not only are two lines consolidated, but one of the END IF statements is no longer needed. A general flowchart for the IF/THEN/ELSE IF structure is shown in Fig. 2.11.

Notice how there is a chain, or "cascade," of decisions. The first one is the IF statement, and each successive decision is an ELSE IF statement. Going down the chain, the first condition encountered that tests true will cause a branch to its corresponding code block followed by an exit of the structure. At the end of the chain of conditions, if all have tested false, there is an optional ELSE block.

The Count-Controlled Loop. The loops in Fig. 2.9 are called *logical loops* because they terminate on a logical condition. In contrast, *count-controlled loops* perform a specified number of repetitions, or iterations. Of course, such an operation could be programmed with a logical loop. For example, the following pseudocode is designed to repeat 10 times:

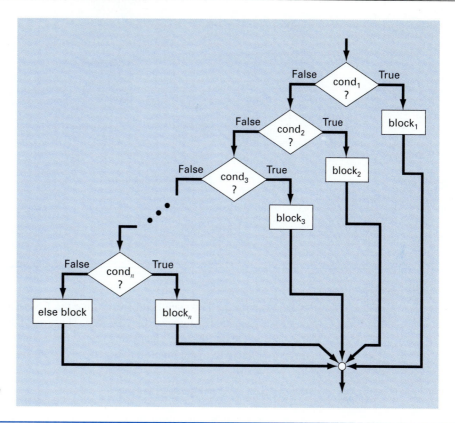

FIGURE 2.11
General multialternative structure
(IF/THEN/ELSE IF).

```
i = 1
DO
    IF i > 10 EXIT
    PRINT i
    i = i + 1
END DO
```

The variable i is a counter that keeps track of the number of iterations. If i is less than or equal to 10, an iteration is performed. On each pass, the value of i is displayed and i is incremented by 1. After the 10th iteration, i becomes 11; therefore, the logical condition will test true and the loop will terminate.

Although the pretest loop is certainly a feasible option for performing a specified number of iterations, the operation is so common that special statements are available in all common high-level languages for accomplishing the same objective in a more efficient manner. A flowchart and pseudocode for the count-controlled structure are depicted in Fig. 2.12.

The count-controlled loop works as follows. The *index* (represented as i in Fig. 2.12) is a variable that is set at an initial value of *start*. The program then tests whether the *index* is less than or equal to the final value, *finish*. If so, it executes the body of the loop and then

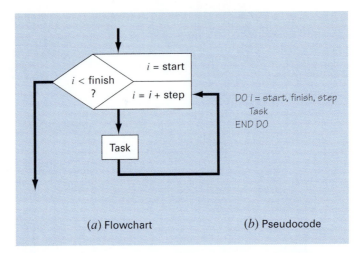

FIGURE 2.12
Flowchart and pseudocode representations of the count-controlled loop that are useful in implementing numerical methods.

(*a*) Flowchart   (*b*) Pseudocode

cycles back to the DO statement. Every time the END DO statement is encountered, the *index* is automatically increased by the *step*. Thus, the *index* acts as a counter. Then, when the *index* is greater than the final value (*finish*), the computer automatically transfers control out of the loop to the line following the END DO statement.[1]

2.4   PROGRAM COMPOSITION

After concocting the algorithm, the next step is to express it as code. Before discussing how this should be done, we will first discuss some computer languages.

2.4.1 High-Level and Macro Languages

Hundreds of computer languages have been developed since the computer age began. The question of which language is "best" has been the subject of heated debate. Unfortunately, many programmers have an almost emotional attachment to "their" language, and some go as far as to contend that all others are inferior. We believe that such parochial attitudes can be self-defeating. Although every language has limitations, each can be used to advantage in particular problem contexts.

In this text, we have chosen to emphasize pseudocode for expressing our numerical algorithms. These can be readily translated into the principal problem-solving languages used by engineers and scientists: Fortran and C. In addition, they can also be implemented in the macro/script languages of packages such as the Visual BASIC dialect used for Excel.

Fortran.   Fortran, standing for *for*mula *tran*slation, was introduced commercially by IBM in 1957. As the name implies, one of its distinctive features is that it uses a notation that fa-

---

[1]Although we believe the foregoing description has pedagogical value, it should be noted that computer languages such as Fortran 90 do not actually perform logical tests when implementing count-controlled loops. Rather, the number of iterations are precomputed during compilation. This is the reason why it is illegal to change the value of a counter variable within the body of a loop.

cilitates the writing of mathematical formulas. Because of such features, Fortran is the native computer language of many engineers and scientists.

Aside from mathematical formulas, Fortran has other features that are relevant to numerical methods. For example, it has standard capabilities to handle extended precision and special mathematical functions, including complex variables. Additionally, it is highly conducive to modular programming because its subroutines allow local variable scoping and the transfer of values between subprograms and the main program as arguments. Thus, libraries of numerically oriented subprograms can be developed or obtained commercially for use in a Fortran code.

C.   As noted in Fig. 2.3, software development constitutes the dominant share of today's computing costs. Developed at Bell Laboratories in 1972, C is the most popular language today for professional software development.

C possesses many of the features of a high-level programming language and hence can be employed by engineers for numerical calculations. It should be noted that it is generally inferior to Fortran 90 in this regard. However, it also offers the low-level access to hardware and software that is more characteristic of assembly language. At the same time, it is more generic than assembly language, which is usually hardware-specific. This combination of low-level capability and portability is what makes C appealing to software developers.

Although engineers and scientists use computers primarily to solve problems, a small but significant fraction develop software. Further, some universities choose to teach their engineering and science students C as their first computer language. Thus, it must be acknowledged that C will be used by a significant number of engineers and scientists over the next decade.

Visual BASIC.   BASIC was originally developed by John Kemeny and Thomas Kurtz as an instructional language for students at Dartmouth College in the mid-1960s. Over the ensuing years it has evolved and grown more sophisticated (e.g., Microsoft's Quick BASIC). Today, its primary implementation is Visual BASIC. This dialect is integrated closely with the Microsoft Windows operating system, the de facto computing environment for IBM PCs and compatibles. Today, it is one of the primary development tools used for software development on PCs. In addition, its utility is also reinforced because of its role as the macro language for the popular Excel spreadsheet program.

Dialects.   As time passes, new versions, or "dialects," of each language are being developed. Not surprisingly, the newer versions of each language often incorporate some of the advantages of the others. The specific dialects used in this book—Fortran 90 and Excel Visual BASIC—were chosen because of their availability, ease of use, and compatibility with a structured programming approach. Thus, all codes in the book are written in a highly structured style derived from pseudocode. Consequently, all can be easily translated into other languages such as Fortran 77 or C. Additionally, they can be readily implemented within software packages such as MATLAB functions. In the following section, we will elaborate on this structured approach.

2.4.2 Structured Programming

Not everyone agrees on a standard definition of structured programming. In general, the major idea is embodied in what can be called the structure principle (Dijkstra 1968):

The structure principle: The static structure (that is, spread out on the page) of the program should correspond in a simple way to the dynamic structure (that is, spread out in time) of the corresponding computation.

Thus, the "jumping around" due to indiscriminate branching and other poor programming habits will be avoided as a natural consequence of the orderly nature of well-conceived computations.

In addition to the structure principle, there are several other rules that can be considered as part of the structured programming philosophy. The following are among the most generally agreed upon rules:

1. Programs should consist solely of the three fundamental control structures of sequence, selection, and repetition (recall Fig. 2.8). Computer scientists have proved that any program can be constructed from these basic structures.
2. Each of the structures must have only one entrance and one exit.
3. Unconditional transfers (GO TOs) should be avoided.
4. The structures should be clearly identified with comments and visual devices such as indentation, blank lines, and blank spaces.

Although the rules may appear deceptively simple, adherence to them avoids the spaghettilike code that is the hallmark of indiscriminate branching. Because of their benefits, these rules and other structured programming practices will be stressed continually in subsequent parts of this book.

Figures 2.13 and 2.14 show how the selection and repetition constructs are programmed in Fortran 90, C, and Visual BASIC. Note how the codes strongly resemble pseudocode.

Figure 2.15 is a dramatic example of the contrast between unstructured and structured approaches. All three programs are designed to determine an average grade, the algorithm for which was delineated in Fig. 2.10. Although the programs compute identical values, their designs are in marked contrast. The purpose, logic, and organization of the unstructured program in Fig. 2.15a is not obvious. The user must delve into the code on a line-by-line basis to decipher the program.

In contrast, the major parts of Fig. 2.15b and c are immediately apparent. Program execution progresses in an orderly fashion from the top to the bottom with no major jumps. Every segment has one entrance and one exit, and each is clearly delineated by spaces and indentation. In essence, the programs are distinguished by a visual clarity that is reminiscent of a well-designed flowchart. Additionally, the structured programs are a natural evolution of the pseudocode shown in Fig. 2.10.

On the debit side, the programs in Fig. 2.15b and c may take longer to develop than the program in Fig. 2.15a. Because many pragmatic engineers are accustomed to "getting the job done" in an efficient and timely manner, the extra effort might not always be justified. Such would be the case for a short program you might develop for a "one-time-only" computation. However, for more important programs that are intended for application by yourself or others, there is no question that a structured approach yields long-term benefits.

Pseudocode	Fortran 90	C	Visual BASIC
(a) Single-alternative decision			
IF condition	IF (a < 0.) THEN	if (a<0) {	IF a < 0. THEN
True alternative	b = -a	b = -a;	b = -a
END IF	END IF	}	END IF
(b) Double-alternative decision			
IF condition	IF (a < 0.) THEN	if (a<0) {	IF a < 0. THEN
True alternative	b = -a	b = -a;	b = -a
ELSE	ELSE	} else {	ELSE
False alternative	b = a	b = a;	b = a
END IF	END IF	}	END IF
(c) Multialternative decision			
IF condition	IF (a < 0.) THEN	if (a<0) {	IF a < 0. THEN
Alternative$_1$	b = -a	b = -a;	b = -a
ELSE IF	ELSE IF (a > 0.) THEN	} else if (a>0) {	ELSE IF a > 0. THEN
Alternative$_2$	b = a	b = a;	b = a
ELSE	ELSE	} else {	ELSE
Else Alternative	b = 0.	b = 0;	b = 0.
END IF	END IF	}	END IF

FIGURE 2.13
Selection constructs as expressed in Fortran 90, ANSI C, and Visual BASIC.

2.5 QUALITY CONTROL

With the tools described up to this point, you would be capable of developing a program to implement a calculation on the computer. However, the fact that a program prints out information is no guarantee that these answers are correct. Consequently, we will now turn to some issues related to the reliability of your program. We are emphasizing this subject at this point to stress its importance in the program-development process. Since we are problem-solving engineers, most of our work will directly or indirectly be used by clients and sponsors. For this reason, it is essential that our programs be reliable—that is, do exactly what they are supposed to do. At best, lack of reliability can be embarrassing. At worst, for engineering problems involving public safety, it can be tragic.

2.5.1 Errors or "Bugs"

During the course of preparing and executing a program of any size, it is likely that errors will occur. These errors are sometimes called *bugs* in computer jargon. This label was

Pseudocode	Fortran 90	C	Visual BASIC
(a) Logical loop (midtest)			

DO
  Task
  IF condition EXIT
  Task
END DO
	`DO`
	  `a=a/2.`
	  `IF (a < 1.) EXIT`
	  `a=a/4.`
	`END DO`
		`while (1) {`
		  `a=a/2;`
		  `if (a<1) break;`
		  `a=a/4;`
		`}`
			`DO`
			  `a = a/2.`
			  `IF (a < 1.) EXIT`
			  `a = a/4.`
			`LOOP`

(b) Count-controlled loop			

DO i = start, finish, step
  Task
END DO
	`DO i= 1,10,2`
	  `x=x+i`
	`END DO`
		`for(i=1;i<=10;i=i+2) {`
		  `x=x+i;`
		`}`
			`FOR i= 1 TO 10 STEP`
			  `x=x+i`
			`NEXT i`

FIGURE 2.14
Repetition constructs as expressed in Fortran 90, ANSI C, and Visual BASIC.

coined by Admiral Grace Hopper, one of the pioneers of computer languages. In 1945, she was working with one of the earliest electromechanical computers when it went dead. She and her colleagues opened the machine and discovered that a moth had become lodged in one of the relays. Thus, the label "bug" came to be synonymous with problems and errors associated with computer operations, and the term *debugging* came to be associated with their solution.

When you are developing and running a program, four types of bugs can occur:

1. *Syntax errors* violate the rules of the language such as spelling, number formation, and other conventions. These errors are often the result of mistakes such as typing REED rather than READ. They usually result when the program is compiled and result in the computer printing out an error message. Such messages are called *diagnostics* because the computer is helping you to "diagnose" the problem.
2. *Link or build errors* occur during the link phase. A common example might be that you misspelled the name of an intrinsic function. In such cases, the computer would print out a message like, "unresolved external reference."
3. *Run-time errors* are those that occur during program execution. An example is when there are an insufficient number of data entries for the number of variables in an input statement. For such situations a diagnostic will usually be printed and you must then re-enter the data properly. For other run-time errors the computer may simply terminate the run or print a message providing you with information regarding the error. In any event, you will be cognizant of the fact that an error has occurred.
4. *Logic errors,* as the name implies, are due to faulty program logic. These are the worst type of error because they can occur without diagnostics. Thus, your program will appear to be working properly in the sense that it executes and generates output. However, the output will be incorrect.

(a) Fortran 77	**(b) Fortran 90**	**(c) ANSI C**

``
  S=0.
  I=0
1 READ *,G
  IF(G.LT.0) GO TO 2
  S=S+G
  I=I+1
  GO TO 1
2 IF (I.EQ.0) GO TO 3
  A=S/I
  GO TO 4
3 A=0.
4 PRINT *,A
  END
``
	``
	PROGRAM grader
	IMPLICIT none
	REAL :: avgrd
	PRINT *, avgrd()
	END
	
	FUNCTION avgrd()
	IMPLICIT none
	REAL :: sum,grade,avgrd
	INTEGER :: count
	
	count = 0
	sum = 0.
	READ *, grade
	
	DO
	  IF(grade < 0.) EXIT
	  sum = sum + grade
	  count = count + 1
	  READ *, grade
	END DO
	
	IF (count >0) THEN
	  avgrd = sum/count
	ELSE
	  avgrd = 0.
	END IF
	
	END
	``
		``
		main ()
		{
		    float avegrd ();
		    printf ("%f\n", avegrd());
		}
		
		float avegrd ()
		{
		    int count;
		    float sum, ave, grade;
		
		    count = 0;
		    sum = 0.0;
		    scanf ("%f", &grade);
		
		    while (1) {
		        if (grade < 0.0) break;
		        sum = sum + grade;
		        count++;
		        scanf ("%f", &grade);
		    }
		
		    if (count > 0) {
		        ave = sum/count;
		    } else {
		        ave = 0.0;
		    }
		    return (ave);
		}
		``

```
PROGRAM grader
```

FIGURE 2.15

(a) An unstructured Fortran 77 and structured (b) Fortran 90 and (c) C versions of the same program. Both versions perform identical computations but differ markedly in terms of clarity.

All the above types of errors must be removed before a program can be employed for engineering applications. We divide the process of quality control into two categories: debugging and testing. As mentioned above, *debugging* involves correcting known errors. In contrast, *testing* is a broader process intended to detect errors that are not known. Additionally, testing should also determine whether the program fulfills the needs for which it was designed.

As will be clear from the following discussion, debugging and testing are interrelated and are often carried out in tandem. For example, we might debug a module to remove the obvious errors. Then, after a test, we might uncover some additional bugs that require correction. This two-pronged process would be repeated until we were convinced that the program was absolutely reliable.

2.5.2 Debugging

As stated above, debugging deals with correcting known errors. There are three ways in which you will become aware of errors. First, an explicit diagnostic can provide you with the exact location and the nature of the error. Second, a diagnostic may occur but the exact location of the error is unclear. For example, the computer may indicate that an error has occurred in a line representing a long equation including some user-defined functions. This error could be due to syntax errors in the equation or it could also stem from errors in the statements defining the functions. Third, no diagnostics occur but the program does not operate properly. These are usually due to logic errors such as infinite loops or mathematical mistakes that do not yield syntax errors. Corrections are easy for the first type of error. However, for the second and third types, some detective work is in order to identify the errors and their exact locations.

Unfortunately, the analogy with detective work is all too apt. That is, ferreting out errors is often difficult and involves a lot of "leg work," collecting clues and running up some blind alleys. In particular, as with detective work, there is no clear-cut methodology involved. However, there are some ways of going about it that are more efficient than others.

One key to identifying bugs is to print out intermediate results. With this approach, it is often possible to determine at what point the computation went awry. Similarly the use of a modular approach will obviously help to localize errors in this way. You could put a different output statement at the start of each module and in this way focus in on the module in which the error occurred. In addition, it is always good practice to debug (and test) each module separately before integrating it with the total package. It should be noted that many programming environments include built-in tools that facilitate debugging.

Although the above procedures can help, there are often times when your only option will be to sit down and read the code line by line following the flow of logic and performing all the computations with a pencil, paper, and pocket calculator. Just as a detective sometimes has to think like a criminal to solve a case, there will be times you will have to think like a computer to debug a program.

Finally, we cannot stress enough that "an ounce of prevention is worth a pound of cure." That is, sound preliminary programming design and structured programming techniques are great ways to avoid errors in the first place. As with so many other aspects of your life, the sooner you adopt a "pay me now rather than pay me later" philosophy, the less grief you will experience as a computer programmer.

2.5.3 Testing

One of the greatest misconceptions of the novice programmer is the belief that, if a program runs and prints out results, it is correct. Of particular danger are those cases where the results appear to be "reasonable" but are in fact wrong. To ensure that such cases do not occur, the program must be subjected to a battery of tests for which the correct answer is known beforehand.

As mentioned in the previous section, it is good practice to debug and test modules prior to integrating them into the total program. Thus, testing usually proceeds in phases. These are module tests, development tests, whole system tests, and operational tests.

Module tests, as the name implies, deal with the reliability of individual modules. Because each is designed to perform a specific, well-defined function, the modules can be run in isolation to determine that they are executing properly. Sample input data can be developed for which the proper output is known. This data can be used to run the module and the outcome compared with the known result to verify successful performance.

Development tests are implemented as the modules are integrated into the total program package. That is, a test is performed after each module is integrated. An effective way to do this is with a top-down approach that starts with the first module and progresses down through the program in execution sequence. By performing a test after each module is incorporated, problems can be isolated more easily because new errors can usually be attributed to the latest module that has been added.

After the total program has been assembled, it can be subjected to a series of *whole-system tests*. These should be designed to expose the program to (1) typical data, (2) unusual but valid data, and (3) incorrect data to check the program's error-handling capabilities.

Note that all the foregoing tests are often collectively referred to as *alpha-testing*. Once they are successful, a *beta-* or *operational testing* phase is usually initiated. These tests are designed to check how the program performs in a realistic setting. A standard way to do this is to have a number of independent subjects (including the ultimate user) implement the program. Operational tests sometimes turn up bugs. In addition, they provide feedback to the developer regarding ways to improve the program so that it more closely meets the user's needs.

Although the ultimate objective of the testing process is to ensure that programs are error-free, there will be complicated cases where it is impossible to subject a program to every possible contingency. Also, it should be recognized that the level of testing depends on the importance and magnitude of the problem context for which the program is being developed. There will obviously be different levels of rigor applied to testing a program for determining the averages of your intramural softball team as compared to a program to regulate the operation of a nuclear reactor or a space station. In each case, however, adequate testing must be performed that is consistent with the liabilities connected with a particular problem context.

2.6 DOCUMENTATION AND MAINTENANCE

2.6.1 Documentation

After the program has been debugged and tested, it must be documented. Documentation is the addition of English-language descriptions to allow the user to implement the program more easily. Remember, along with other people who might employ your software, you also are a "user." Although a program may seem clear and simple when it is first composed, with time the same code may seem inscrutable. Therefore, sufficient documentation must be included to allow you and other users to immediately understand your program and how it can be implemented. This task has both internal and external aspects.

Internal Documentation.   Picture for a moment a text with just words and no paragraphs, subheadings, or chapters. Such a book would be more than a little intimidating to study. The way the pages of a text are structured—that is, all the devices that act to break up and organize the material—makes the text more effective and enjoyable. In the same sense, internal documentation of computer code can greatly enhance the user's understanding of a program and how it works.

The following are some general suggestions for documenting programs internally:

1. Include a module at the head of the program giving the program's title, your name, and the date. This is your signature and marks the program as your work.
2. Include a second module to define each of the key variables.
3. Select variable names that are reflective of the type of information the variables are to store.
4. Insert spaces within statements to make them easier to read.
5. Use comment statements liberally throughout the program to add explanation and to skip lines for the purposes of labeling, clarity, and separation of modules.
6. In particular, use comment statements to clearly label and set off all the modules.
7. Use indentation to clarify the program structures. In particular, indentation can be used to set off loops and decisions.

External Documentation.   This refers to instructions in the form of output messages and supplementary printed matter. The messages occur when the program runs and are intended to make the output attractive and "user-friendly." This involves the effective use of spaces, blank lines, or special characters to illuminate the logical sequence and structure of the program output. Attractive output simplifies the detection of errors and enhances the communication of program results. Finally, the program can be designed to output descriptive error messages to alert and inform the user regarding problems.

The supplementary printed matter can range from a single sheet to a comprehensive user's manual. The user's manual for your computer is an example of comprehensive documentation. This manual tells you how to run your computer system and disk-operating programs. When developing your own software, you may find it useful to develop a short "user's manual" that provides description and instruction.

2.6.2 Storage and Maintenance

Remember that when you sign off the computer, your active file will be destroyed. To retain a program for later use, you must transfer it to a secondary storage device such as a disk or tape before turning the machine off.

Aside from this physical act of retaining the program, storage and maintenance consists of two major tasks: (1) upgrading the program in light of experience and (2) ensuring that the program is safely stored. The former is a matter of communicating with users to gather feedback regarding suggested improvements to your design. The question of safe storage of programs is especially critical during development. A backup copy should always be created before making major code changes. Also, maintaining backups on external media such as diskettes is imperative.

2.7   SOFTWARE STRATEGY

One of your primary objectives in using this book should be to develop your own numerical methods software library. Whether you use them to build short macros or large sophisticated computer programs, such a library will be of immense value in your future acade-

TABLE 2.1 The programs in the Windows version of the Numerical Methods TOOLKIT.

Program	Capability
Roots of equations	Determines a real root of a single algebraic or transcendental equation using the bisection method. Can also be used to plot functions.
Systems of linear algebraic equations*	Solves or finds the inverse of up to 20 linear algebraic equations using an *LU* decomposition implementation of Gauss elimination.
Curve fitting*	Fits data with up to a 10th-order regression polynomial and statistically characterizes the adequacy of the fit.
Integration*	Integrates functions or unevenly spaced tabular data with the trapezoidal rule.
Systems of ordinary differential equations*	Solves up to five simultaneous ordinary differential equations using the 4th-order Runge-Kutta method.

*Denotes significant improvements compared to the previous DOS version.

mic and professional efforts. We have provided a diverse array of computer media to help you accomplish this objective. These include pseudocode and software.

Pseudocode and other descriptions of algorithms are provided for many methods in the book. If used as the basis for your own programs, these can constitute a fairly comprehensive numerical methods capability.

Over and above your own collection of programs, we have also developed a generic software product called the Numerical Methods TOOLKIT that is used in conjunction with this book. This package includes modules devoted to five of the more fundamental numerical methods (Table 2.1). Each of the numerical methods programs is fully illustrated by an example problem in the relevant chapter of the book. The examples can provide you with models for the design of input and output screens, including the graphical display of results. Homework problems are included in the relevant chapters to reinforce your capabilities to use the disk on your own computer. Additionally, the software can be employed to check the accuracy of your own programming efforts.

EXAMPLE 2.1   Install and use the Numerical Methods TOOLKIT on your computer

Problem Statement.   The purpose of this example is to use the software provided with the text (the Numerical Methods TOOLKIT).

Solution.   Run the install.exe program on the disk included with the text and start up the TOOLKIT. The screen should contain the TOOLKIT main menu as shown in Fig. 2.16a. First, notice the row of five control buttons on the bottom of the menu. The About button describes the authorship of the software. Click this button now and then click OK. The Help button has five messages in sequence that remind you of certain conventions and design features of the TOOLKIT. The first help message describes the user-defined functions that are recognized by the software. These include familiar trigonometric and exponential functions. Note that the arguments of the trigonometric functions are expressed in radians rather than degrees. Thus the sine of 90 degrees must be written as SIN(2*pi*90/360) or SIN(pi/2) or SIN(1.57) when using the TOOLKIT. The second Help message describes how plots are generated by the TOOLKIT. Four plot parameters

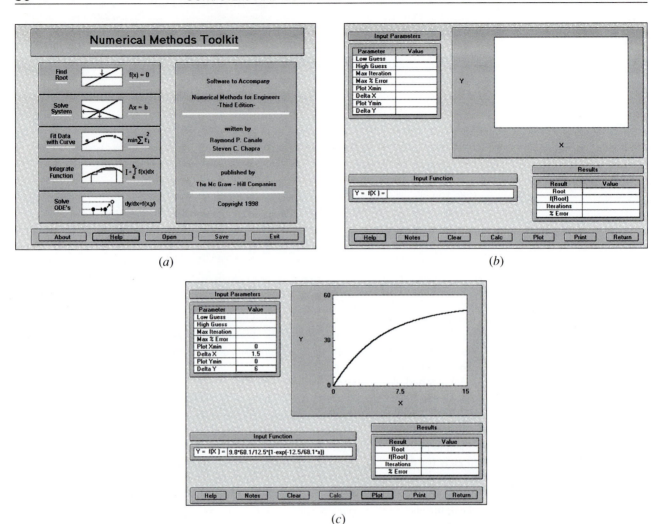

FIGURE 2.16
(a) Title screen from the Numerical Methods TOOLKIT software. (b) The root location screen showing a graph of the falling parachutist's velocity [Eq. (1.10)].

are required to construct a plot. These are the minimum values of X and Y and the interval between tic marks for both X and Y. There are always 10 Delta X and 10 Delta Y intervals on a plot such that the maximum values of X and Y are Xmin + 10 Delta X and Ymin + 10 Delta Y. This convention allows you to create plots of your liking. The third Help message explains the red warning button convention employed by the TOOLKIT. You will note that the TOOLKIT is designed to show all the inputs as well as numerical and graphical results for each method on a single screen. This allows you to view all aspects of a particular problem with a single glance. However, suppose you plot a function Y(X) with X and Y ranging from "0 to 10." Upon inspection of the plot you decide that it would be useful to change the range of Y from "0 to 10" to "0 to 20." When you change the value 10 to 20 in the input table, you will note that the Plot button changes color from black to red. This simply alerts you that the plot must be updated to reflect the new range of 20. Click the red Plot button and it changes to black, signaling that the plot and the input parameters are consistent. Thus, when using the software on an interactive basis, you should make sure that there are no red buttons when final results are recorded. The fourth Help message informs you that TOOLKIT computations can be saved and recovered as files with ".nmt" extensions. Thus, if you use the TOOLKIT to help with a homework problem in a mechanical engineering course, you might name your file, MEPROB4.nmt. The fifth Help message alerts you to the fact that the TOOLKIT Print function may not work properly on some network configurations.

The Open and Save buttons invoke typical Windows boxes that allow you to name, store, and recall files in locations of your choice. Any analyses performed with the TOOLKIT will be lost upon exit unless you take specific action using the Save button.

Next, notice the five large stacked buttons on the left side of the main menu. These buttons have names, figures, and mathematical symbols to remind you of the function of the numerical methods. We will carefully explore the program behind each one of these buttons in appropriate sections of the text.

For now we will use the Find Root (top button) program to plot the velocity of the falling parachutist as a function of time. To do this click the Find Root button, whereupon the screen should produce a pattern similar to that of Fig. 2.16b. The screen has five main parts: a table for the input parameters, a box for the input function, a plot window, a table for the results, and a bar containing control buttons. Let us ignore the root-finding capabilities of the program until Part Two and simply use the program to plot Eq. (1.10) with $m = 68.1$ kg, $c = 12.5$ kg/s, and $g = 9.8$ m/s^2. Figure 2.16c shows the results for Xmin and Ymin $= 0$ with Delta X $= 1.5$ and Delta Y $= 6$. Try various values for the plot parameters including negative Xmin and Ymin to gain familiarity with the plot operation and design. Note that the Plot button is red when the plot needs to be updated (i.e., the actual plot showing on the screen and the plot parameters are inconsistent). Experiment with the Clear button as an alternative way to remove the input parameters and function from the boxes. Also click the Notes button. A window appears that you can use to write notes concerning your work. These notes will be saved and retrieved when you use the Save and Open buttons on the main menu. Finally, use the Print button to generate a hard copy of the plot and input parameters and function. Remember, this may not work for some network printers. This hard copy might be submitted in a report to a client or as part of a homework assignment.

The ability of the TOOLKIT to create plots will have many other uses as you pursue your goal of understanding and applying numerical methods and computers to solve engineering problems. We will explore these uses as well as the other capabilities of the TOOLKIT in appropriate sections of the text.

2.7.1 Packages and Libraries

Throughout this book, we will endeavor to describe how numerical methods are implemented using program software packages and libraries. The former will stress Mathcad,

Excel, and MATLAB. The latter will focus on the IMSL library. A short description of each follows.

Mathcad.   Mathcad is the flagship product of MathSoft, Inc. This software package attempts to bridge the gap between spreadsheets like Excel and scratch pads. The result is an interactive worksheet that allows the user to perform a number of mathematical, data-handling, and graphical tasks commonly encountered in engineering and science. Information and equations are input to the worksheet as you would with paper and pencil rather than with a programming language or with spreadsheet notation. The first version of Mathcad was released in 1986, and version 7 was released in 1997.

Mathcad can perform tasks in either numeric or symbolic mode. In numeric mode, Mathcad functions and operators give numerical responses, whereas in symbolic mode results are given as general expressions or equations. Maple V, a comprehensive symbolic math package, is the basis of the symbolic mode and was incorporated into Mathcad in 1993.

Mathcad has a variety of functions and operators that allow convenient implementation of many of the numerical methods developed in this book. These will be described in detail in succeeding chapters. Because of its ease of use, we believe Mathcad is particularly useful as a pedagogical tool.

Excel.   Excel is the spreadsheet produced by Microsoft, Inc. Spreadsheets are a special type of mathematical software that allow the user to enter and perform calculations on rows and columns of data. As such, they are a computerized version of a large accounting worksheet on which large interconnected calculations can be implemented and displayed. Because the entire calculation is updated when any number on the sheet is changed, spreadsheets are ideal for "what if?" sorts of analysis.

Excel has some built-in numerical capabilities including equation solving, curve fitting, and optimization. It also includes Visual BASIC as a macro language that can be used to implement numerical calculations. Finally, it has several visualization tools (such as three-dimensional surface plots) that serve as valuable adjuncts for numerical analysis.

MATLAB.   MATLAB is the flagship software product of Mathworks, Inc. co-founded by the numerical analyst Cleve Moler and John N. Little. As the name implies, MATLAB was originally developed as a *mat*rix *lab*oratory. To this day, the major element of MATLAB is still the matrix. Mathematical manipulations of matrices are very conveniently implemented in an easy-to-use, interactive environment. To these matrix manipulations, MATLAB has added a variety of numerical functions, symbolic computations, and visualization tools. As a consequence, the present version represents a fairly comprehensive technical computing environment.

MATLAB has a variety of functions and operators that allow convenient implementation of many of the numerical methods developed in this book. These will be described in detail in succeeding chapters. In addition, programs can be written as so-called scripts, functions, and M-files that can be used to implement numerical calculations.

IMSL.   The IMSL (*I*nternational *M*athematical and *S*tatistical *L*ibrary, Inc.) consists of three separate but coordinated libraries of Fortran subroutines and functions: the MATH (general applied mathematics), STAT (statistics), and SFUN (special function) libraries. This text focuses on the MATH library, which contains over 700 subprograms spanning all the numerical areas covered in this text. Most are available in both single and double precision versions. As such, it constitutes an immense capability for numerical analysis. Although the routines themselves are written in Fortran, they can be accessed from other languages such as C.

It should be noted that most of our Mathcad, Excel, and MATLAB applications will stress their built-in numerical functions (e.g., matrix manipulations, regression, root solving). However, it should be noted that our pseudocodes can be readily translated into macro programs for direct implementation in these environments.

PROBLEMS

2.1 List and define five steps in the production of high-quality software in engineering.

2.2 Define top-down design, modular design, and structured programming.

2.3 Write pseudocode to implement the flowchart depicted in Fig. P2.3. Make sure that proper indentation is included to make the structure clear.

2.4 A value for the concentration of a pollutant in a lake is recorded on each in a set of index cards. A card marked "end of data" is placed at the end of the set. Write an algorithm to determine the sum and the average of these values.

2.5 Write a structured flowchart for Prob. 2.4.

2.6 Write an algorithm to print out the roots (either real or complex) of a quadratic equation $ax^2 + bx + c = 0$, where a, b, and c are real coefficients. Design your algorithm so that it deals with all possible contingencies that might occur during the computation.

2.7 Develop a structured flowchart for Prob. 2.6.

2.8 Write pseudocode for Prob. 2.6.

2.9 Develop, debug, and document a subprogram for Prob. 2.6 in either a high-level language or a macro language of your choice. Use a subroutine to compute the roots. Perform test runs for the cases **(a)** $a = 1, b = 5, c = 2$; **(b)** $a = 0, b = -3, c = 1.5$; **(c)** $a = 1$, $b = 2.5, c = 7$.

2.10 The sine function can be evaluated by the following infinite series:

$$\sin x = x - \frac{x^3}{3!} + \frac{x^5}{5!} - \cdots$$

Write an algorithm to implement this formula so that it computes and prints out the values of $\sin x$ as each term in the series is added. In other words, compute and print in sequence the values for

$$\sin x = x$$

$$\sin x = x - \frac{x^3}{3!}$$

$$\sin x = x - \frac{x^3}{3!} + \frac{x^5}{5!}$$

$$\vdots$$

up to the order term of your choosing. For each of the above, compute and print out the percent relative error as

$$\% \text{ error} = \frac{\text{true} - \text{series approximation}}{\text{true}} \times 100\%$$

2.11 Develop a structured flowchart for Prob. 2.10.

2.12 Write pseudocode for Prob. 2.10.

FIGURE P2.3

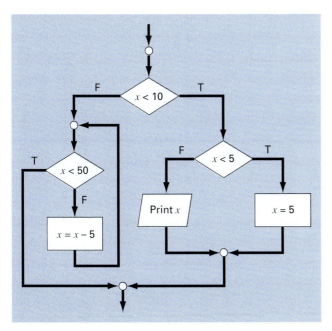

2.13 Develop, debug, and document a subprogram for Prob. 2.10 in either a high-level language or a macro language of your choice. Employ the library function for the sine in your computer to determine the true value. Have the program print out the series approximation and the error at each step. As a test case, employ the program to compute sin 1.5 for up to and including the term $x^{15}/15!$. Interpret your results.

2.14 The following algorithm is designed to determine a grade for a course that consists of quizzes, homework, and a final exam:

Step 1: Input course number and name.
Step 2: Input weighting factors for quizzes (WQ), homework (WH), and the final exam (WF).
Step 3: Input quiz grades and determine an average quiz grade (AQ).
Step 4: Input homework grades and determine an average homework grade (AH).
Step 5: If this course has a final grade, continue to step 6. If not, go to step 9.
Step 6: Input final exam grade (FE).
Step 7: Determine average grade AG according to

$$AG = \frac{WQ \times AQ + WH \times AH + WF * FE}{WQ + WH + WF}$$

Step 8: Go to step 10.
Step 9: Determine average grade AG according to

$$AG = \frac{WQ \times AQ + WH \times AH}{WQ + WH}$$

Step 10: Print out course number, name, and average grade.
Step 11: Terminate computation.

Write, debug, and document a structured computer program based on the algorithm. Test it using the following data to calculate a grade without the final exam and a grade with the final exam: WQ = 30; WH = 30; WF = 40; quizzes = 98, 95, 90, 60, 99; homework = 95, 90, 86, 100, 100, 77; and final exam = 91.

2.15 An amount of money P is invested in an account where interest is compounded at the end of the period. The future worth F yielded at an interest rate i after n periods may be determined from the following formulation:

$$F = P(1 + i)^n$$

Write a subprogram that will calculate the future worth of an investment in either a high-level language or a macro language of your choice. The input to the program should include the initial investment P, the interest rate i (as a decimal), and the number of years n for which the future worth is to be calculated. The output should also include these values. The output should also include, in a labeled table format, the future worth for each year up to and including the nth year. Run the program for $P = \$100,000$, $i = 0.08$, and $n = 25$ years.

2.16 Figure P2.16 shows the reverse of a checking account statement. The bank has developed this to help you balance your checkbook. If you look at it closely, you will realize that it is an algorithm. Develop a step-by-step algorithm (in a format similar to Fig. 2.6a) to accomplish the same task.

2.17 Write a structured flowchart for Prob. 2.16.

2.18 Develop, debug, and document a subprogram for Prob. 2.16 in either a high-level language or a macro language of your choice. Test it by verifying that it duplicates the results shown in Fig. P2.16.

2.19 Write, debug, and document a subprogram to determine statistics for your favorite sport in either a high-level language or a macro language of your choice. Pick anything from football to jogging to bowling. If you play intramural sports, make up a subprogram for your team. Design the program so that it is user-friendly and provides valuable and interesting information to anyone (for example, a coach or a player) who might use it to evaluate athletic performance.

2.20 The average daily temperature for an area can be approximated by the following function,

$$T = T_{mean} + (T_{peak} - T_{mean}) \cos\left(\omega(t - t_{peak})\right)$$

where T_{mean} = the average annual temperature, T_{peak} = the peak temperature, ω = the frequency of the annual variation ($= 2\pi/365$), and t_{peak} = day of the peak temperature ($\cong 205$ d). Parameters for some U.S. towns are listed in Table P2.20.

TABLE P2.20 Mean daily air temperature parameters for some selected U.S. locations.

City	T_{mean} (°C)	T_{peak} (°C)
Miami, FL	22.1	28.3
Yuma, AZ	23.1	33.6
Bismarck, ND	5.2	22.1
Seattle, WA	10.6	17.6
Boston, MA	10.7	22.9

Develop a subprogram in either a high-level language or a macro language of your choice that computes the average temperature between two days of the year entered by the user. Test it for (a) January–February in Bismarck, North Dakota ($t = 0$ to 59), and the July–August temperature in Yuma, Arizona ($t = 180$ to 242).

2.21 Economic formulas are available to compute annual payments for loans. Suppose that you borrow an amount of money P and agree to repay it in n annual payments at an interest rate of i. The formula to compute the annual payment A is

$$A = P \frac{i(1 + i)^n}{(1 + i)^n - 1}$$

CHECKS OUTSTANDING — NOT
CHARGED TO ACCOUNT

NO.	$	
453	25	00
458	5	68
460	14	33
461	150	00
463	16	74
464	9	32
465	44	15
466	50	00
TOTAL $		

MONTH _____April_____ 19 _97_

NEW BALANCE $ _____643.59_____
AS SHOWN ON THIS STATEMENT

ADD   _____250.00_____
DEPOSITS NOT CREDITED IN
THIS STATEMENT   _____22.15_____

_____

_____

TOTAL $ _____

SUBTRACT
TOTAL CHECKS OUTSTANDING   _____

YOUR CHECKBOOK BALANCE $ _____600.52_____
AFTER SUBTRACTING CURRENT
MONTH'S SERVICE CHARGE AND ADDING INTEREST EARNED
(EARNER ACCOUNTS ONLY) TO YOUR BALANCE

FIGURE P2.16

Write a subprogram in either a high-level language or a macro language of your choice to compute A. Test it with $P = \$35,000$ and an interest rate of 15 percent ($i = 0.15$). Set up the program so that you can evaluate as many values of n as you like. Compute results for $n = 1, 2, 3, 4$, and 5.

2.22 Develop, debug, and test a subprogram in either a high-level language or a macro language of your choice to compute the velocity of the falling parachutist as outlined in Example 1.2. Design the program so that it allows the user to input values for the drag coefficient and mass. Test the program by duplicating the results from Example 1.2. Repeat the computation but employ step sizes of 1 and 0.5 s. Compare your results with the analytical solution obtained previously in Example 1.1. Does a smaller step size make the results better or worse? Explain your results.

2.23 Use the bisection program on the TOOLKIT to plot several functions of your choice. Try polynomials and transcendental functions (e.g., sines, logs) whose behavior may be difficult to visualize

in advance of plotting. Try functions that may not be continuous or otherwise unusual. Utilize several choices for the scales of both the x and y axes to facilitate your exploration. Make permanent copies of the plots if you have a printer. Help us beta-test the program by sending us the output of any examples that failed to perform properly.

2.24 Read the appendices concerning the capabilities and operation of packages useful for numerical methods. Select one or more packages and repeat Prob. 2.23.

2.25 Attain the capability to plot functions in a manner similar to the TOOLKIT. Use your own computer and its utility software or library subroutines if appropriate. If your computer does not have programs to help you, then write your own code using the capabilities of your computer. Develop this program in the form of a subroutine and save it on a magnetic storage disk. Be prepared to update this program as you apply it to various problems covered in other parts of this book.

CHAPTER 3

Approximations and Round-Off Errors

Because so many of the methods in this book are straightforward in description and application, it would be very tempting at this point for us to proceed directly to the main body of the text and teach you how to use these techniques. However, understanding the concept of error is so important to the effective use of numerical methods that we have chosen to devote the next two chapters to this topic.

The importance of error was introduced in our discussion of the falling parachutist in Chap. 1. Recall that we determined the velocity of a falling parachutist by both analytical and numerical methods. Although the numerical technique yielded estimates that were close to the exact analytical solution, there was a discrepancy, or *error,* because the numerical method involved an approximation. Actually, we were fortunate in that case because the availability of an analytical solution allowed us to compute the error exactly. For many applied engineering problems we cannot obtain analytical solutions. Therefore, we cannot compute exactly the errors associated with our numerical methods. In these cases we must settle for approximations or estimates of the errors.

Such errors are characteristic of most of the techniques described in this book. This statement might at first seem contrary to what one normally conceives of as sound engineering. Students and practicing engineers constantly strive to limit errors in their work. When taking examinations or doing homework problems, you are penalized, not rewarded, for your errors. In professional practice, errors can be costly and sometimes catastrophic. If a structure or device fails, lives can be lost.

Although perfection is a laudable goal, it is rarely, if ever, attained. For example, despite the fact that the model developed from Newton's second law is an excellent approximation, it would never in practice exactly predict the parachutist's fall. A variety of factors such as winds and slight variations in air resistance would result in deviations from the prediction. If these deviations are systematically high or low, then we might need to develop a new model. However, if they are randomly distributed and tightly grouped around the prediction, then the deviations might be considered negligible and the model deemed adequate. Numerical approximations also introduce similar discrepancies into the analysis. Again, the question is: How much error is present in our calculations and is it tolerable?

This chapter and the next cover basic topics related to the identification, quantification, and minimization of these errors. In this chapter, general information concerned with the quantification of error is reviewed in the first sections. This is followed by a section on one of the two major forms of numerical error: round-off error. *Round-off error* is due to the fact that computers can represent only quantities with a finite number of digits. Then the next chapter deals with the other major form: truncation error. *Truncation error* is the discrepancy introduced by the fact that numerical methods may employ approximations to represent exact mathematical operations and quantities. Finally, we briefly discuss errors not directly connected with the numerical methods themselves. These include blunders, formulation or model errors, and data uncertainty.

3.1 SIGNIFICANT FIGURES

This book deals extensively with approximations connected with the manipulation of numbers. Consequently, before discussing the errors associated with numerical methods, it is useful to review basic concepts related to approximate representation of the numbers themselves.

Whenever we employ a number in a computation, we must have assurance that it can be used with confidence. For example, Fig. 3.1 depicts a speedometer and odometer from an automobile. Visual inspection of the speedometer indicates that the car is traveling between 48 and 49 km/h. Because the indicator is higher than the midpoint between the markers on the gauge, we can say with assurance that the car is traveling at approximately 49 km/h. We have confidence in this result because two or more reasonable individuals reading this gauge would arrive at the same conclusion. However, let us say that we insist that the speed be estimated to one decimal place. For this case, one person might say 48.8,

FIGURE 3.1
An automobile speedometer and odometer illustrating the concept of a significant figure.

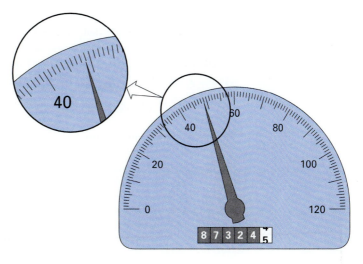

whereas another might say 48.9 km/h. Therefore, because of the limits of this instrument, only the first two digits can be used with confidence. Estimates of the third digit (or higher) must be viewed as approximations. It would be ludicrous to claim, on the basis of this speedometer, that the automobile is traveling at 48.8642138 km/h. In contrast, the odometer provides up to six certain digits. From Fig. 3.1, we can conclude that the car has traveled slightly less than 87,324.5 km during its lifetime. In this case, the seventh digit (and higher) is uncertain.

The concept of a significant figure, or digit, has been developed to formally designate the reliability of a numerical value. The *significant digits* of a number are those that can be used with confidence. They correspond to the number of certain digits plus one estimated digit. For example, the speedometer and the odometer in Fig. 3.1 yield readings of three and seven significant figures, respectively. For the speedometer, the two certain digits are 48. It is conventional to set the estimated digit at one-half of the smallest scale division on the measurement device. Thus the speedometer reading would consist of the three significant figures: 48.5. In a similar fashion, the odometer would yield a seven-significant-figure reading of 87,324.45.

Although it is usually a straightforward procedure to ascertain the significant figures of a number, some cases can lead to confusion. For example, zeros are not always significant figures because they may be necessary just to locate a decimal point. The numbers 0.00001845, 0.0001845, and 0.001845 all have four significant figures. Similarly, when trailing zeros are used in large numbers, it is not clear how many, if any, of the zeros are significant. For example, at face value the number 45,300 may have three, four, or five significant digits, depending on whether the zeros are known with confidence. Such uncertainty can be resolved by using scientific notation, where 4.53×10^4, 4.530×10^4, 4.5300×10^4 designate that the number is known to three, four, and five significant figures, respectively.

The concept of significant figures has two important implications for our study of numerical methods:

1. As introduced in the falling parachutist problem, numerical methods yield approximate results. We must therefore develop criteria to specify how confident we are in our approximate result. One way to do this is in terms of significant figures. For example, we might decide that our approximation is acceptable if it is correct to four significant figures.
2. Although quantities such as π, e, or $\sqrt{7}$ represent specific quantities, they cannot be expressed exactly by a limited number of digits. For example,

$$\pi = 3.14159265358979323846264 3\ldots$$

ad infinitum. Because computers retain only a finite number of significant figures, such numbers can never be represented exactly. The omission of the remaining significant figures is called round-off error.

Both round-off error and the use of significant figures to express our confidence in a numerical result will be explored in detail in subsequent sections. In addition, the concept of significant figures will have relevance to our definition of accuracy and precision in the next section.

3.2 ACCURACY AND PRECISION

The errors associated with both calculations and measurements can be characterized with regard to their accuracy and precision. *Accuracy* refers to how closely a computed or measured value agrees with the true value. *Precision* refers to how closely individual computed or measured values agree with each other.

These concepts can be illustrated graphically using an analogy from target practice. The bullet holes on each target in Fig. 3.2 can be thought of as the predictions of a numerical technique, whereas the bull's-eye represents the truth. *Inaccuracy* (also called *bias*) is defined as systematic deviation from the truth. Thus, although the shots in Fig. 3.2c are more tightly grouped than those in Fig. 3.2a, the two cases are equally biased because they are both centered on the upper left quadrant of the target. *Imprecision* (also called *uncertainty*), on the other hand, refers to the magnitude of the scatter. Therefore, although Fig. 3.2b and d are equally accurate (that is, centered on the bull's-eye), the latter is more precise because the shots are tightly grouped.

Numerical methods should be sufficiently accurate or unbiased to meet the requirements of a particular engineering problem. They also should be precise enough for adequate

FIGURE 3.2
An example from marksmanship illustrating the concepts of accuracy and precision. (a) Inaccurate and imprecise; (b) accurate and imprecise; (c) inaccurate and precise; (d) accurate and precise.

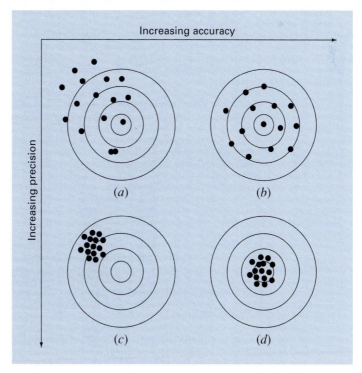

engineering design. In this book, we will use the collective term *error* to represent both the inaccuracy and the imprecision of our predictions. With these concepts as background, we can now discuss the factors that contribute to the error of numerical computations.

3.3 ERROR DEFINITIONS

Numerical errors arise from the use of approximations to represent exact mathematical operations and quantities. These include *truncation errors,* which result when approximations are used to represent exact mathematical procedures, and *round-off errors,* which result when numbers having limited significant figures are used to represent exact numbers. For both types, the relationship between the exact, or true, result and the approximation can be formulated as

$$\text{True value} = \text{approximation} + \text{error} \tag{3.1}$$

By rearranging Eq. (3.1), we find that the numerical error is equal to the discrepancy between the truth and the approximation, as in

$$E_t = \text{true value} - \text{approximation} \tag{3.2}$$

where E_t is used to designate the exact value of the error. The subscript t is included to designate that this is the "true" error. This is in contrast to other cases, as described shortly, where an "approximate" estimate of the error must be employed.

A shortcoming of this definition is that it takes no account of the order of magnitude of the value under examination. For example, an error of a centimeter is much more significant if we are measuring a rivet rather than a bridge. One way to account for the magnitudes of the quantities being evaluated is to normalize the error to the true value, as in

$$\text{True fractional relative error} = \frac{\text{true error}}{\text{true value}}$$

where, as specified by Eq. (3.2), error = true value − approximation. The relative error can also be multiplied by 100 percent to express it as

$$\varepsilon_t = \frac{\text{true error}}{\text{true value}} 100\% \tag{3.3}$$

where ε_t designates the true percent relative error.

EXAMPLE 3.1   Calculation of Errors

Problem Statement.   Suppose that you have the task of measuring the lengths of a bridge and a rivet and come up with 9999 and 9 cm, respectively. If the true values are 10,000 and 10 cm, respectively, compute (*a*) the true error and (*b*) the true percent relative error for each case.

Solution.

(a) The error for measuring the bridge is [Eq. (3.2)]

$$E_t = 10,000 - 9999 = 1 \text{ cm}$$

and for the rivet it is

$$E_t = 10 - 9 = 1 \text{ cm}$$

(b) The percent relative error for the bridge is [Eq. (3.3)]

$$\varepsilon_t = \frac{1}{10,000} 100\% = 0.01\%$$

and for the rivet it is

$$\varepsilon_t = \frac{1}{10} 100\% = 10\%$$

Thus, although both measurements have an error of 1 cm, the relative error for the rivet is much greater. We would conclude that we have done an adequate job of measuring the bridge, whereas our estimate for the rivet leaves something to be desired.

Notice that for Eqs. (3.2) and (3.3), E and ε are subscripted with a t to signify that the error is normalized to the true value. In Example 3.1, we were provided with this value. However, in actual situations such information is rarely available. For numerical methods, the true value will be known only when we deal with functions that can be solved analytically. Such will typically be the case when we investigate the theoretical behavior of a particular technique for simple systems. However, in real-world applications, we will obviously not know the true answer a priori. For these situations, an alternative is to normalize the error using the best available estimate of the true value, that is, to the approximation itself, as in

$$\varepsilon_a = \frac{\text{approximate error}}{\text{approximation}} 100\% \tag{3.4}$$

where the subscript a signifies that the error is normalized to an approximate value. Note also that for real-world applications, Eq. (3.2) cannot be used to calculate the error term for Eq. (3.4). One of the challenges of numerical methods is to determine error estimates in the absence of knowledge regarding the true value. For example, certain numerical methods use an *iterative approach* to compute answers. In such an approach, a present approximation is made on the basis of a previous approximation. This process is performed repeatedly, or iteratively, to successively compute (we hope) better and better approximations. For such cases, the error is often estimated as the difference between previous and current approximations. Thus, percent relative error is determined according to

$$\varepsilon_a = \frac{\text{current approximation} - \text{previous approximation}}{\text{current approximation}} 100\% \tag{3.5}$$

This and other approaches for expressing errors will be elaborated on in subsequent chapters.

The signs of Eqs. (3.2) through (3.5) may be either positive or negative. If the approximation is greater than the true value (or the previous approximation is greater than the current approximation), the error is negative; if the approximation is less than the true value, the error is positive. Also, for Eqs. (3.3) to (3.5), the denominator may be less than zero, which can also lead to a negative error. Often, when performing computations, we may not be concerned with the sign of the error, but we are interested in whether the percent absolute value is lower than a prespecified percent tolerance ε_s. Therefore, it is often useful to employ the absolute value of Eqs. (3.2) through (3.5). For such cases, the computation is repeated until

$$|\varepsilon_a| < \varepsilon_s \tag{3.6}$$

If this relationship holds, our result is assumed to be within the prespecified acceptable level ε_s. Note that for the remainder of this text, we will almost exclusively employ absolute values when we use relative errors.

It is also convenient to relate these errors to the number of significant figures in the approximation. It can be shown (Scarborough, 1966) that if the following criterion is met, we can be assured that the result is correct to *at least n* significant figures.

$$\varepsilon_s = (0.5 \times 10^{2-n})\% \tag{3.7}$$

EXAMPLE 3.2   Error Estimates for Iterative Methods

Problem Statement.   In mathematics, functions can often be represented by infinite series. For example, the exponential function can be computed using

$$e^x = 1 + x + \frac{x^2}{2} + \frac{x^3}{3!} + \cdots + \frac{x^n}{n!} \tag{E3.2.1}$$

Thus, as more terms are added in sequence, the approximation becomes a better and better estimate of the true value of e^x. Equation (E3.2.1) is called a *Maclaurin series expansion.*

Starting with the simplest version, $e^x = 1$, add terms one at a time to estimate $e^{0.5}$. After each new term is added, compute the true and approximate percent relative errors with Eqs. (3.3) and (3.5), respectively. Note that the true value is $e^{0.5} = 1.648721\ldots$. Add terms until the absolute value of the approximate error estimate ε_a falls below a prespecified error criterion ε_s conforming to three significant figures.

Solution.   First, Eq. (3.7) can be employed to determine the error criterion that ensures a result is correct to at least three significant figures:

$$\varepsilon_s = (0.5 \times 10^{2-3})\% = 0.05\%$$

Thus, we will add terms to the series until ε_a falls below this level.

The first estimate is simply equal to Eq. (E3.2.1) with a single term. Thus, the first estimate is equal to 1. The second estimate is then generated by adding the second term, as in

$$e^x = 1 + x$$

or for $x = 0.5$,

$$e^{0.5} = 1 + 0.5 = 1.5$$

This represents a true percent relative error of [Eq. (3.3)]

$$\varepsilon_t = \frac{1.648721 - 1.5}{1.648721} 100\% = 9.02\%$$

Equation (3.5) can be used to determine an approximate estimate of the error, as in

$$\varepsilon_a = \frac{1.5 - 1}{1.5} 100\% = 33.3\%$$

Because ε_a is not less than the required value of ε_s, we would continue the computation by adding another term, $x^2/2!$ and repeating the error calculations. The process is continued until $\varepsilon_a < \varepsilon_s$. The entire computation can be summarized as

Terms	Result	ε_t (%)	ε_a (%)
1	1	39.3	
2	1.5	9.02	33.3
3	1.625	1.44	7.69
4	1.645833333	0.175	1.27
5	1.648437500	0.0172	0.158
6	1.648697917	0.00142	0.0158

Thus, after six terms are included, the approximate error falls below $\varepsilon_s = 0.05\%$ and the computation is terminated. However, notice that, rather than three significant figures, the result is accurate to five! This is because, for this case, both Eqs. (3.5) and (3.7) are conservative. That is, they ensure that the result is at least as good as they specify. Although, as discussed in Chap. 6, this is not always the case for Eq. (3.5), it is true most of the time.

With the preceding definitions as background, we can now proceed to the two types of error connected directly with numerical methods: round-off errors and truncation errors.

3.4 ROUND-OFF ERRORS

As mentioned previously, round-off errors originate from the fact that computers retain only a fixed number of significant figures during a calculation. Numbers such as π, e, or $\sqrt{7}$ cannot be expressed by a fixed number of significant figures. Therefore, they cannot be represented exactly by the computer. In addition, because computers use a base-2 representation, they cannot precisely represent certain exact base-10 numbers. The discrepancy introduced by this omission of significant figures is called *round-off error*.

3.4.1 Computer Representation of Numbers

Numerical round-off errors are directly related to the manner in which numbers are stored in a computer. The fundamental unit whereby information is represented is called a *word*. This is an entity that consists of a string of *binary digits*, or *bits*. Numbers are typically stored in one or more words. To understand how this is accomplished, we must first review some material related to number systems.

Number Systems.   A *number system* is merely a convention for representing quantities. Because we have 10 fingers and 10 toes, the number system that we are most familiar with is the *decimal*, or *base-10*, number system. A base is the number used as the reference for constructing the system. The base-10 system uses the 10 digits—0, 1, 2, 3, 4, 5, 6, 7, 8, 9—to represent numbers. By themselves, these digits are satisfactory for counting from 0 to 9.

For larger quantities, combinations of these basic digits are used, with the position or *place value* specifying the magnitude. The rightmost digit in a whole number represents a number from 0 to 9. The second digit from the right represents a multiple of 10. The third digit from the right represents a multiple of 100 and so on. For example, if we have the number 86,409 then we have eight groups of 10,000, six groups of 1000, four groups of 100, zero groups of 10, and nine more units, or

$$(8 \times 10^4) + (6 \times 10^3) + (4 \times 10^2) + (0 \times 10^1) + (9 \times 10^0) = 86,409$$

Figure 3.3*a* provides a visual representation of how a number is formulated in the base-10 system. This type of representation is called *positional notation.*

Because the decimal system is so familiar, it is not commonly realized that there are alternatives. For example, if human beings happened to have had eight fingers and eight toes, we would undoubtedly have developed an *octal,* or *base-8,* representation. In the same sense, our friend the computer is like a two-fingered animal who is limited to two states—either 0 or 1. This relates to the fact that the primary logic units of digital comput-

FIGURE 3.3
How the (*a*) decimal (base 10) and the (*b*) binary (base 2) systems work. In (*b*), the binary number 10101101 is equivalent to the decimal number 173.

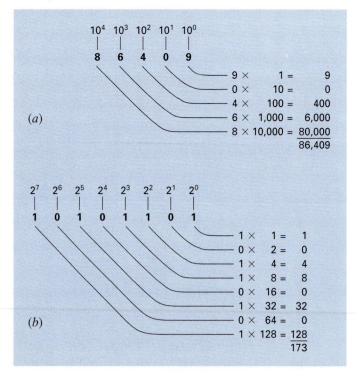

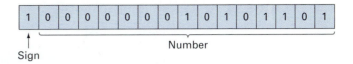

FIGURE 3.4
The representation of the decimal integer −173 on a 16-bit computer using the signed magnitude method.

ers are on/off electronic components. Hence, numbers on the computer are represented with a *binary,* or *base-2,* system. Just as with the decimal system, quantities can be represented using positional notation. For example, the binary number 11 is equivalent to $(1 \times 2^1) + (1 \times 2^0) = 2 + 1 = 3$ in the decimal system. Figure 3.3*b* illustrates a more complicated example.

Integer Representation. Now that we have reviewed how base-10 numbers can be represented in binary form, it is simple to conceive of how integers are represented on a computer. The most straightforward approach, called the *signed magnitude method,* employs the first bit of a word to indicate the sign, with a 0 for positive and a 1 for negative. The remaining bits are used to store the number. For example, the integer value of −173 would be stored on a 16-bit computer, as in Fig. 3.4.

EXAMPLE 3.3   Range of Integers

Problem Statement.   Determine the range of integers in base-10 that can be represented on a 16-bit computer.

Solution.   Of the 16 bits, the first bit holds the sign. The remaining 15 bits can hold binary numbers from 0 to 111111111111111. The upper limit can be converted to a decimal integer, as in

$$(1 \times 2^{14}) + (1 \times 2^{13}) + \cdots + (1 \times 2^1) + (1 \times 2^0)$$

which equals 32,767 (note that this expression can be simply evaluated as $2^{15} - 1$). Thus, a 16-bit computer word can store decimal integers ranging from −32,767 to 32,767. In addition, because zero is already defined as 0000000000000000, it is redundant to use the number 1000000000000000 to define a "minus zero." Therefore, it is usually employed to represent an additional negative number: −32,768, and the range is from −32,768 to 32,767.

Note that the signed magnitude method described above is not used to represent integers on conventional computers. A preferred approach called the *2's complement* technique directly incorporates the sign into the number's magnitude rather than providing a separate bit to represent plus or minus (see Chapra and Canale 1994). However, Example 3.3 still serves to illustrate how all digital computers are limited in their capability to represent integers. That is, numbers above or below the range cannot be represented. A more serious

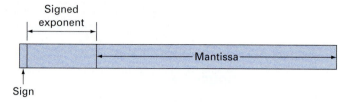

FIGURE 3.5
The manner in which a floating-point number is stored in a word.

limitation is encountered in the storage and manipulation of fractional quantities as described next.

Floating-Point Representation. Fractional quantities are typically represented in computers using floating-point form. In this approach, the number is expressed as a fractional part, called a *mantissa* or *significand*, and an integer part, called an *exponent* or *characteristic*, as in

$$m \cdot b^e$$

where $m =$ the mantissa, $b =$ the base of the number system being used, and $e =$ the exponent. For instance, the number 156.78 could be represented as 0.15678×10^3 in a floating-point base-10 system.

Figure 3.5 shows one way that a floating-point number could be stored in a word. The first bit is reserved for the sign, the next series of bits for the signed exponent, and the last bits for the mantissa.

Note that the mantissa is usually *normalized* if it has leading zero digits. For example, suppose the quantity $1/34 = 0.029411765...$ was stored in a floating-point base-10 system that allowed only four decimal places to be stored. Thus, $1/34$ would be stored as

$$0.0294 \times 10^0$$

However, in the process of doing this, the inclusion of the useless zero to the right of the decimal forces us to drop the digit 1 in the fifth decimal place. The number can be normalized to remove the leading zero by multiplying the mantissa by 10 and lowering the exponent by 1 to give

$$0.2941 \times 10^{-1}$$

Thus, we retain an additional significant figure when the number is stored.

The consequence of normalization is that the absolute value of m is limited. That is,

$$\frac{1}{b} \leq m < 1 \tag{3.8}$$

where $b =$ the base. For example, for a base-10 system, m would range between 0.1 and 1, and for a base-2 system, between 0.5 and 1.

Floating-point representation allows both fractions and very large numbers to be expressed on the computer. However, it has some disadvantages. For example, floating-point numbers take up more room and take longer to process than integer numbers. More significantly, however, their use introduces a source of error because the mantissa holds only a finite number of significant figures. Thus, a round-off error is introduced.

EXAMPLE 3.4   Hypothetical Set of Floating-Point Numbers

Problem Statement.   Create a hypothetical floating-point number set for a machine that stores information using 7-bit words. Employ the first bit for the sign of the number, the next three for the sign and the magnitude of the exponent, and the last three for the magnitude of the mantissa (Fig. 3.6).

Solution.   The smallest possible positive number is depicted in Fig. 3.6. The initial 0 indicates that the quantity is positive. The 1 in the second place designates that the exponent has a negative sign. The 1's in the third and fourth places give a maximum value to the exponent of

$$1 \times 2^1 + 1 \times 2^0 = 3$$

Therefore, the exponent will be -3. Finally, the mantissa is specified by the 100 in the last three places, which conforms to

$$1 \times 2^{-1} + 0 \times 2^{-2} + 0 \times 2^{-3} = 0.5$$

Although a smaller mantissa is possible (e.g., 000, 001, 010, 011), the value of 100 is used because of the limit imposed by normalization [Eq. (3.8)]. Thus, the smallest possible positive number for this system is $+0.5 \times 2^{-3}$, which is equal to 0.0625 in the base-10 system. The next highest numbers are developed by increasing the mantissa, as in

$$0111101 = (1 \times 2^{-1} + 0 \times 2^{-2} + 1 \times 2^{-3}) \times 2^{-3} = (0.078125)_{10}$$
$$0111110 = (1 \times 2^{-1} + 1 \times 2^{-2} + 0 \times 2^{-3}) \times 2^{-3} = (0.093750)_{10}$$
$$0111111 = (1 \times 2^{-1} + 1 \times 2^{-2} + 1 \times 2^{-3}) \times 2^{-3} = (0.109375)_{10}$$

Notice that the base-10 equivalents are spaced evenly with an interval of 0.015625.

At this point, to continue increasing, we must decrease the exponent to 10, which gives a value of

$$1 \times 2^1 + 0 \times 2^0 = 2$$

FIGURE 3.6
The smallest possible positive floating-point number from Example 3.4.

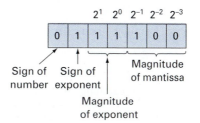

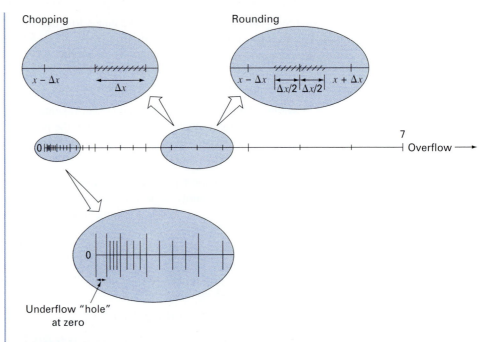

FIGURE 3.7
The hypothetical number system developed in Example 3.4. Each value is indicated by a tick mark. Only the positive numbers are shown. An identical set would also extend in the negative direction.

The mantissa is decreased back to its smallest value of 100. Therefore, the next number is

$$0110100 = (1 \times 2^{-1} + 0 \times 2^{-2} + 0 \times 2^{-3}) \times 2^{-2} = (0.125000)_{10}$$

This still represents a gap of $0.125000 - 0.109375 = 0.015625$. However, now when higher numbers are generated by increasing the mantissa, the gap is lengthened to 0.03125,

$$0110101 = (1 \times 2^{-1} + 0 \times 2^{-2} + 1 \times 2^{-3}) \times 2^{-2} = (0.156250)_{10}$$
$$0110110 = (1 \times 2^{-1} + 1 \times 2^{-2} + 0 \times 2^{-3}) \times 2^{-2} = (0.187500)_{10}$$
$$0110111 = (1 \times 2^{-1} + 1 \times 2^{-2} + 1 \times 2^{-3}) \times 2^{-2} = (0.218750)_{10}$$

This pattern is repeated as each larger quantity is formulated until a maximum number is reached,

$$0011111 = (1 \times 2^{-1} + 1 \times 2^{-2} + 1 \times 2^{-3}) \times 2^{3} = (7)_{10}$$

The final number set is depicted graphically in Fig. 3.7.

Figure 3.7 manifests several aspects of floating-point representation that have significance regarding computer round-off errors:

1. *There Is a Limited Range of Quantities That May Be Represented.* Just as for the integer case, there are large positive and negative numbers that cannot be represented. At-

tempts to employ numbers outside the acceptable range will result in what is called an *overflow* error. However, in addition to large quantities, the floating-point representation has the added limitation that very small numbers cannot be represented. This is illustrated by the *underflow* "hole" between zero and the first positive number in Fig. 3.7. It should be noted that this hole is enlarged because of the normalization constraint of Eq. (3.8).

2. *There Are Only a Finite Number of Quantities That Can Be Represented Within the Range.* Thus, the degree of precision is limited. Obviously, irrational numbers cannot be represented exactly. Furthermore, rational numbers that do not exactly match one of the values in the set also cannot be represented precisely. The errors introduced by approximating both these cases are referred to as *quantizing* errors. The actual approximation is accomplished in either of two ways: chopping or rounding. For example, suppose that the value of $\pi = 3.14159265358. . .$ is to be stored on a base-10 number system carrying seven significant figures. One method of approximation would be to merely omit, or "chop off," the eighth and higher terms, as in $\pi = 3.141592$, with the introduction of an associated error of [Eq. (3.2)]

$$E_t = 0.00000065. . .$$

This technique of retaining only the significant terms was originally dubbed "truncation" in computer jargon. We prefer to call it *chopping* to distinguish it from the truncation errors discussed in Chap. 4. Note that for the base-2 number system in Fig. 3.7, chopping means that any quantity falling within an interval of length Δx will be stored as the quantity at the lower end of the interval. Thus, the upper error bound for chopping is Δx. Additionally, a bias is introduced because all errors are positive. The shortcomings of chopping are attributable to the fact that the higher terms in the complete decimal representation have no impact on the shortened version. For instance, in our example of π, the first discarded digit is 6. Thus, the last retained digit should be rounded up to yield 3.141593. Such *rounding* reduces the error to

$$E_t = -0.00000035. . .$$

Consequently, rounding yields a lower absolute error than chopping. Note that for the base-2 number system in Fig. 3.7, rounding means that any quantity falling within an interval of length Δx will be represented as the nearest allowable number. Thus, the upper error bound for rounding is $\Delta x/2$. Additionally, no bias is introduced because some errors are positive and some are negative. Some computers employ rounding. However, this adds to the computational overhead, and, consequently, many machines use simple chopping. This approach is justified under the supposition that the number of significant figures is large enough that resulting round-off error is usually negligible.

3. *The Interval between Numbers, Δx, Increases as the Numbers Grow in Magnitude.* It is this characteristic, of course, that allows floating-point representation to preserve significant digits. However, it also means that quantizing errors will be proportional to the magnitude of the number being represented. For normalized floating-point numbers, this proportionality can be expressed, for cases where chopping is employed, as

$$\frac{|\Delta x|}{|x|} \leq \mathcal{E} \tag{3.9}$$

and, for cases where rounding is employed, as

$$\frac{|\Delta x|}{|x|} \leq \frac{\mathscr{E}}{2} \qquad (3.10)$$

where $\mathscr{E}$ is referred to as the *machine epsilon*, which can be computed as

$$\mathscr{E} = b^{1-t} \qquad (3.11)$$

where b is the number base and t is the number of significant digits in the mantissa. Notice that the inequalities in Eqs. (3.9) and (3.10) signify that these are error bounds. That is, they specify the worst cases.

EXAMPLE 3.5   Machine Epsilon

Problem Statement.   Determine the machine epsilon and verify its effectiveness in characterizing the errors of the number system from Example 3.4. Assume that chopping is used.

Solution.   The hypothetical floating-point system from Example 3.4 employed values of the base $b = 2$, and the number of mantissa bits $t = 3$. Therefore, the machine epsilon would be [Eq. (3.11)]

$$\mathscr{E} = 2^{1-3} = 0.25$$

Consequently, the relative quantizing error should be bounded by 0.25 for chopping. The largest relative errors should occur for those quantities that fall just below the upper bound of the first interval between successive equispaced numbers (Fig. 3.8). Those numbers falling in the succeeding higher intervals would have the same value of Δx but a greater value of x and, hence, would have a lower relative error. An example of a maximum error would be a value falling just below the upper bound of the interval between $(0.125000)_{10}$ and $(0.156250)_{10}$. For this case, the error would be less than

$$\frac{0.03125}{0.125000} = 0.25$$

Thus, the error is as predicted by Eq. (3.9).

FIGURE 3.8
The largest quantizing error will occur for those values falling just below the upper bound of the first of a series of equispaced intervals.

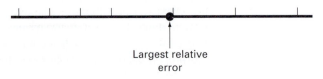

Largest relative
error

```
epsilon = 1
DO
  IF (epsilon+1 ≤ 1) EXIT
  epsilon = epsilon/2
END DO
epsilon = 2 × epsilon
```

FIGURE 3.9
Pseudocode to determine machine epsilon for a binary computer.

The magnitude dependence of quantizing errors has a number of practical applications in numerical methods. Most of these relate to the commonly employed operation of testing whether two numbers are equal. This occurs when testing convergence of quantities as well as in the stopping mechanism for iterative processes (recall Example 3.2). For these cases, it should be clear that, rather than test whether the two quantities are equal, it is advisable to test whether their difference is less than an acceptably small tolerance. Further, it should also be evident that normalized rather than absolute difference should be compared, particularly when dealing with numbers of large magnitude. In addition, the machine epsilon can be employed in formulating stopping or convergence criteria. This ensures that programs are portable—that is, they are not dependent on the computer on which they are implemented. Figure 3.9 lists pseudocode to automatically determine the machine epsilon of a binary computer.

Extended Precision. It should be noted at this point that, although round-off errors can be important in contexts such as testing convergence, the number of significant digits carried on most computers allows most engineering computations to be performed with more than acceptable precision. For example, the hypothetical number system in Fig. 3.7 is a gross exaggeration that was employed for illustrative purposes. Commercial computers use much larger words and, consequently, allow numbers to be expressed with more than adequate precision. For example, computers that use IEEE format allow 24 bits to be used for the mantissa, which translates into about seven significant base-10 digits of precision[1] with a range of about 10^{-38} to 10^{39}.

With this acknowledged, there are still cases where round-off error becomes critical. For this reason most computers allow the specification of extended precision. The most common of these is double precision, in which the number of words used to store floating-point numbers is doubled. It provides about 15 to 16 decimal digits of precision and a range of approximately 10^{-308} to 10^{308}.

In many cases, the use of double-precision quantities can greatly mitigate the effect of round-off errors. However, a price is paid for such remedies in that they also require more memory and execution time. The difference in execution time for a small calculation might seem insignificant. However, as your programs become larger and more complicated, the added execution time could become considerable and have a negative impact on your effectiveness as a problem solver. Therefore, extended precision should not be used frivolously. Rather, it should be selectively employed where it will yield the maximum benefit at the least cost in terms of execution time. In the following sections, we will look closer at how round-off errors affect computations, and in so doing provide a foundation of understanding to guide your use of the double-precision capability.

Before proceeding, it should be noted that some of the commonly used software packages (e.g., Excel, Mathcad) routinely use double precision to represent numerical quantities. Thus, the developers of these packages decided that mitigating roundoff errors would take precedence over any loss of speed incurred by using extended precision. Others, like MATLAB, allow you to use extended precision, if you desire.

---

[1] Note that only 23 bits are actually used to store the mantissa. However, because of normalization, the first bit of the mantissa is always 1 and is, therefore, not stored. Thus, this first bit together with the 23 stored bits gives the 24 total bits of precision for the mantissa.

3.4.2 Arithmetic Manipulations of Computer Numbers

Aside from the limitations of a computer's number system, the actual arithmetic manipulations involving these numbers can also result in round-off error. In the following section, we will first illustrate how common arithmetic operations affect round-off errors. Then we will investigate a number of particular manipulations that are especially prone to round-off errors.

Common Arithmetic Operations. Because of their familiarity, normalized base-10 numbers will be employed to illustrate the effect of round-off errors on simple addition, subtraction, multiplication, and division. Other number bases would behave in a similar fashion. To simplify the discussion, we will employ a hypothetical decimal computer with a 4-digit mantissa and a 1-digit exponent. In addition, chopping is used. Rounding would lead to similar though less dramatic errors.

When two floating-point numbers are added, the mantissa of the number with the smaller exponent is modified so that the exponents are the same. This has the effect of aligning the decimal points. For example, suppose we want to add $0.1557 \cdot 10^1 + 0.4381 \cdot 10^{-1}$. The decimal of the mantissa of the second number is shifted to the left a number of places equal to the difference of the exponents $[1 - (-1) = 2]$, as in

$$0.4381 \cdot 10^{-1} \rightarrow 0.004381 \cdot 10^1$$

Now the numbers can be added,

$$
\begin{array}{r}
0.1557 \quad \cdot 10^1 \\
0.004381 \cdot 10^1 \\
\hline
0.160081 \cdot 10^1
\end{array}
$$

and the result chopped to $0.1600 \cdot 10^1$. Notice how the last two digits of the second number that were shifted to the right have essentially been lost from the computation.

Subtraction is performed identically to addition except that the sign of the subtrahend is reversed. For example, suppose that we are subtracting 26.86 from 36.41. That is,

$$
\begin{array}{r}
0.3641 \cdot 10^2 \\
- \; 0.2686 \cdot 10^2 \\
\hline
0.0955 \cdot 10^2
\end{array}
$$

For this case the result is not normalized, and so we must shift the decimal one place to the right to give $0.9550 \cdot 10^1 = 9.550$. Notice that the zero added to the end of the mantissa is not significant but is merely appended to fill the empty space created by the shift. Even more dramatic results would be obtained when the numbers are very close, as in

$$
\begin{array}{r}
0.7642 \cdot 10^3 \\
- \; 0.7641 \cdot 10^3 \\
\hline
0.0001 \cdot 10^3
\end{array}
$$

which would be converted to $0.1000 \cdot 10^0 = 0.1000$. Thus, for this case, three nonsignificant zeros are appended. This introduces a substantial computational error because subsequent manipulations would act as if these zeros were significant. As we will see in a later section, the loss of significance during the subtraction of nearly equal numbers is among the greatest source of round-off error in numerical methods.

Multiplication and division are somewhat more straightforward than addition or subtraction. The exponents are added and the mantissas multiplied. Because multiplication of two n-digit mantissas will yield a $2n$-digit result, most computers hold intermediate results in a double-length register. For example,

$$0.1363 \cdot 10^3 \times 0.6423 \cdot 10^{-1} = 0.08754549 \cdot 10^2$$

If, as in this case, a leading zero is introduced, the result is normalized,

$$0.08754549 \cdot 10^2 \rightarrow 0.8754549 \cdot 10^1$$

and chopped to give

$$0.8754 \cdot 10^1$$

Division is performed in a similar manner, but the mantissas are divided and the exponents are subtracted. Then the results are normalized and chopped.

Large Computations. Certain methods require extremely large numbers of arithmetic manipulations to arrive at their final results. In addition, these computations are often interdependent. That is, the later calculations are dependent on the results of earlier ones. Consequently, even though an individual round-off error could be small, the cumulative effect over the course of a large computation can be significant.

EXAMPLE 3.6   Large Numbers of Interdependent Computations

Problem Statement. Investigate the effect of round-off error on large numbers of interdependent computations. Develop a program to sum a number 100,000 times. Sum the number 1 in single precision, and 0.00001 in single and double precision.

Solution. Figure 3.10 shows a Fortran 90 program that performs the summation. Whereas the single-precision summation of 1 yields the expected result, the single-precision

FIGURE 3.10

Fortran 90 program to sum a number 10^5 times. The case sums the number 1 in single precision and the number 10^{-5} in single and double precision.

```
PROGRAM fig0310
IMPLICIT none
INTEGER::i
REAL::sum1, sum2, x1, x2
DOUBLE PRECISION::sum3, x3
sum1=0.
sum2=0.
sum3=0.
x1=1.
x2=1.e-5
x3=1.d-5
DO i=1,100000
    sum1=sum1+x1
    sum2=sum2+x2
    sum3=sum3+x3
END DO
PRINT *, sum1
PRINT *, sum2
PRINT *, sum3
END
output:
100000.000000
        1.000990
  9.999999999980838E-001
```

summation of 0.00001 yields a large discrepancy. This error is reduced significantly when 0.00001 is summed in double precision.

Quantizing errors are the source of the discrepancies. Because the integer 1 can be represented exactly within the computer, it can be summed exactly. In contrast, 0.00001 cannot be represented exactly and is quantized by a value that is slightly different from its true value. Whereas this very slight discrepancy would be negligible for a small computation, it accumulates after repeated summations. The problem still occurs in double precision but is greatly mitigated because the quantizing error is much smaller.

Note that the type of error illustrated by the previous example is somewhat atypical in that all the errors in the repeated operation are of the same sign. In most cases the errors of a long computation alternate sign in a random fashion and, thus, often cancel out. However, there are also instances where such errors do not cancel but, in fact, lead to a spurious final result. The following sections are intended to provide insight into ways in which this may occur.

Adding a Large and a Small Number. Suppose we add a small number, 0.0010, to a large number, 4000, using a hypothetical computer with the 4-digit mantissa and the 1-digit exponent. We modify the smaller number so that its exponent matches the larger,

$$
\begin{array}{r}
0.4000 \quad\ \cdot 10^4 \\
0.0000001 \cdot 10^4 \\
\hline
0.4000001 \cdot 10^4
\end{array}
$$

which is chopped to $0.4000 \cdot 10^4$. Thus, we might as well have not performed the addition!

This type of error can occur in the computation of an infinite series. The initial terms in such series are often relatively large in comparison with the later terms. Thus, after a few terms have been added, we are in the situation of adding a small quantity to a large quantity.

One way to mitigate this type of error is to sum the series in reverse order—that is, in ascending rather than descending order. In this way, each new term will be of comparable magnitude to the accumulated sum (see Prob. 3.4).

Subtractive Cancellation. This term refers to the round-off induced when subtracting two nearly equal floating-point numbers.

One common instance where this can occur involves finding the roots of a quadratic equation or parabola with the quadratic formula,

$$
\begin{array}{r}
x_1 \\
x_2
\end{array}
= \frac{-b \pm \sqrt{b^2 - 4ac}}{2a}
\tag{3.12}
$$

For cases where $b^2 \gg 4ac$, the difference in the numerator can be very small. In such cases, double precision can mitigate the problem. In addition, an alternative formulation can be used to minimize subtractive cancellation,

$$
\begin{array}{r}
x_1 \\
x_2
\end{array}
= \frac{-2c}{b \pm \sqrt{b^2 - 4ac}}
\tag{3.13}
$$

An illustration of the problem and the use of this alternative formula are provided in the following example.

EXAMPLE 3.7   Subtractive Cancellation

Problem Statement.   Compute the values of the roots of a quadratic equation with $a = 1, b = 3000.001$, and $c = 3$. Check the computed values versus the true roots of $x_1 = -0.001$ and $x_2 = -3000$.

Solution.   Figure 3.11 shows a Fortran 90 program that computes the roots x_1 and x_2 on the basis of the quadratic formula [(Eq. (3.12)]. Note that both single- and double-precision versions are given. Whereas the results for x_2 are adequate, the percent relative errors for x_1 are poor for the single-precision version, $\varepsilon_t = 2.4\%$. This level could be inadequate for many applied engineering problems. This result is particularly surprising because we are employing an analytical formula to obtain our solution!

The loss of significance occurs in the line of both programs where two relatively large numbers are subtracted. Similar problems do not occur when the same numbers are added.

On the basis of the above, we can draw the general conclusion that the quadratic formula will be susceptible to subtractive cancellation whenever $b^2 \gg 4ac$. One way to circumvent this problem is to use double precision. Another is to recast the quadratic formula in the format of Eq. (3.13). As in the program output, both options give a much smaller error because the subtractive cancellation is minimized or avoided.

FIGURE 3.11
Fortran 90 programs to determine the roots of a quadratic. (a) Single and (b) double precision.

```
PROGRAM fig0311
IMPLICIT none
REAL::a,b,c,d,x1,x2,x1r
DOUBLE PRECISION::aa,bb,cc,dd,x11,x22
a = 1.
b = 3000.001
c = 3.
d = SQRT(b * b - 4. * a * c)
x1 = (-b + d) / (2. * a)
x2 = (-b - d) / (2. * a)
PRINT *, 'Single-precision results:'
PRINT '(1x,a10,f20.14)', 'x1 = ', x1
PRINT '(1x,a10,f10.4)', 'x2 = ', x2
PRINT *
aa = 1.
bb = 3000.001
cc = 3.
dd = SQRT(bb * bb - 4. * aa * cc)
x11 = (-bb + dd) / (2. * aa)
x22 = (-bb - dd) / (2. * aa)
```

```
PRINT *, 'Double-precision results:'
PRINT '(1x,a10,f20.14)', 'x1 = ', x11
PRINT '(1x,a10,f10.4)', 'x2 = ', x22
PRINT *
PRINT *, 'Modified formula for first root:'
x1r = -2. * c / (b + d)
PRINT '(1x,a10,f20.14)', 'x1 = ', x1r
END


OUTPUT:
Single-precision results:
    x1 =      -.00097656250000
    x2 = -3000.0000
Double-precision results:
    x1 =      -.00100000000771
    x2 = -3000.0000
Modified formula for first root:
    x1 =      -.00100000000000
```

Note that, as in the foregoing example, there are times where subtractive cancellation can be circumvented by using a transformation. However, the only general remedy is to employ extended precision.

Smearing.   Smearing occurs whenever the individual terms in a summation are larger than the summation itself. As in the following example, one case where this occurs is in series of mixed signs.

EXAMPLE 3.8   Evaluation of e^x using Infinite Series

Problem Statement.   The exponential function $y = e^x$ is given by the infinite series

$$y = 1 + x + \frac{x^2}{2} + \frac{x^3}{3!} + \cdots$$

Evaluate this function for $x = 10$ and $x = -10$, and be attentive to the problems of round-off error.

Solution.   Figure 3.12a gives a Fortran 90 program that uses the infinite series to evaluate e^x. The variable i is the number of terms in the series, *term* is the value of the current term added to the series, and *sum* is the accumulative value of the series. The variable *test* is the preceding accumulative value of the series prior to adding *term*. The series is terminated when the computer cannot detect the difference between *test* and *sum*.

Figure 3.12b shows the results of running the program for $x = 10$. Note that this case is completely satisfactory. The final result is achieved in 31 terms with the series identical to the library function value within seven significant figures.

Figure 3.12c shows similar results for $x = -10$. However, for this case, the results of the series calculation are not even the same sign as the true result. As a matter of fact, the negative results are open to serious question because e^x can never be less than zero. The problem here is caused by round-off error. Note that many of the terms that make up the sum are much larger than the final result of the sum. Furthermore, unlike the previous case, the individual terms vary in sign. Thus, in effect we are adding and subtracting large numbers (each with some small error) and placing great significance on the differences—that is, subtractive cancellation. Thus, we can see that the culprit behind this example of smearing is, in fact, subtractive cancellation. For such cases it is appropriate to seek some other computational strategy. For example, one might try to compute $y = e^{10}$ as $y = (e^{-1})^{10}$. Other than such a reformulation, the only general recourse is extended precision.

Inner Products.   As should be clear from the last sections, some infinite series are particularly prone to round-off error. Fortunately, the calculation of series is not one of the more common operations in numerical methods. A far more ubiquitous manipulation is the calculation of inner products, as in

$$\sum_{i=1}^{n} x_i y_i = x_1 y_1 + x_2 y_2 + \cdots + x_n y_n$$

This operation is very common, particularly in the solution of simultaneous linear algebraic equations. Such summations are prone to round-off error. Consequently, it is often desirable to compute such summations in extended precision.

(a) Program

```
PROGRAM fig0312
IMPLICIT none
REAL::term, test, sum,x
INTEGER::i
i = 0
term = 1.
sum = 1.
test = 0.
PRINT *, 'x = '
READ *, x
PRINT *, 'i', 'term', 'sum'
DO
  IF (sum.EQ.test) EXIT
  PRINT *, i, term, sum
  i = i + 1
  term = term*x/i
  test = sum
  sum = sum+term
END DO
PRINT *, 'exact value =',exp(x)
END
```

(b) Evaluation of e^{10}

```
x =
10
i          term              sum
0          1.000000          1.000000
1         10.000000         11.000000
2         50.000000         61.000000
3        166.666700        227.666700
4        416.666700        644.333400
5        833.333400       1477.667000
              .
              .
              .
27    9.183693E-02       22026.420000
28    3.279890E-02       22026.450000
29    1.130997E-02       22026.460000
30    3.769989E-03       22026.470000
31    1.216126E-03       22026.470000
exact value =    22026.460000
```

(c) Evaluation of e^{-10}

```
x =
-10
i          term              sum
0          1.000000          1.000000
1        -10.000000         -9.000000
2         50.000000         41.000000
3       -166.666700       -125.666700
4        416.666700        291.000000
5       -833.333400       -542.333400
              .
              .
              .
41   -2.989312E-09        8.137590E-05
42    7.117410E-10        8.137661E-05
43   -1.655212E-10        8.137644E-05
44    3.761845E-11        8.137648E-05
45   -8.359655E-12        8.137647E-05
exact value =    4.539993E-05
```

FIGURE 3.12

(a) A Fortran 90 program to evaluate e^x using an infinite series. (b) Evaluation of e^x.
(c) Evaluation of e^{-x}.

Although the foregoing sections should provide rules of thumb to mitigate round-off error, they do not provide a direct means beyond trial and error to actually determine the effect of such errors on a computation. In the next chapter, we will introduce the Taylor series, which will provide a mathematical approach for estimating these effects.

PROBLEMS

3.1 Compose your own program based on Fig. 3.9 and use it to determine your computer's machine epsilon.

3.2 In a fashion similar to that in Fig. 3.9, write a short program to determine the smallest number, x_{min}, used on the computer you will be employing along with this book. Note that your computer will be unable to reliably distinguish between zero and a quantity that is smaller than this number.

3.3 Determine a theoretical relationship to predict the smallest floating-point number for a digital computer on the basis of parameters such as its word size, number of mantissa bits, number of exponent bits, etc.

3.4 The infinite series

$$f(N) = \sum_{n=1}^{N} \frac{1}{n^2}$$

converges on a value of $f(N) = \pi^2/6$ as N approaches infinity. Write a program to compute $f(N)$ for $N = 10,000$ by computing the sum from $n = 1$ to 10,000. Then repeat the computation but in reverse order—that is, from $n = 10,000$ to 1 using increments of -1. Explain the results.

3.5 How can the machine epsilon be employed to formulate a stopping criterion, ε_s, for your programs? Provide an example.

3.6 Evaluate $e^{-8.3}$ using two approaches

$$e^{-x} = 1 - x + \frac{x^2}{2} - \frac{x^3}{3!} + \cdots$$

and

$$e^{-x} = \frac{1}{e^x} = \frac{1}{1 + x + \frac{x^2}{2} + \frac{x^3}{3!} + \cdots}$$

and compare with the true value of 2.485168×10^{-4} and discuss your results. Use 25 terms to evaluate each series.

3.7 The derivative of $f(x) = 1/(3 - 2x^2)$ is given by

$$\frac{4x}{(3 - 2x^2)^2}$$

Do you expect to have difficulties evaluating this function at $x = 1.22$? Try it using 3- and 4-digit arithmetic with chopping.

3.8 (a) Evaluate the polynomial

$$y = x^3 - 5x^2 + 6x + 0.55$$

at $x = 2.73$. Use 3-digit arithmetic with chopping. Evaluate the error.

(b) Repeat **(a)** but express y as

$$y = [(x - 5) + 6]x + 0.55$$

Evaluate the percent relative error and compare with part **(a)**.

CHAPTER 4

Truncation Errors and the Taylor Series

Truncation errors are those that result from using an approximation in place of an exact mathematical procedure. For example, in Chap. 1 we approximated the derivative of velocity of a falling parachutist by a finite-divided-difference equation of the form [Eq. (1.11)]

$$\frac{dv}{dt} \cong \frac{\Delta v}{\Delta t} = \frac{v(t_{i+1}) - v(t_i)}{t_{i+1} - t_i} \tag{4.1}$$

A truncation error was introduced into the numerical solution because the difference equation only approximates the true value of the derivative (recall Fig. 1.4). In order to gain insight into the properties of such errors, we now turn to a mathematical formulation that is used widely in numerical methods to express functions in an approximate fashion—the Taylor series.

4.1 THE TAYLOR SERIES

Taylor's theorem (Box 4.1) and its associated formula, the Taylor series, is of great value in the study of numerical methods. In essence, the *Taylor series* provides a means to predict a function value at one point in terms of the function value and its derivatives at another point. In particular, the theorem states that any smooth function can be approximated as a polynomial.

A useful way to gain insight into the Taylor series is to build it term by term. For example, the first term in the series is

$$f(x_{i+1}) \cong f(x_i) \tag{4.2}$$

This relationship, called the *zero-order approximation*, indicates that the value of f at the new point is the same as its value at the old point. This result makes intuitive sense because if x_i and x_{i+1} are close to each other, it is likely that the new value is probably similar to the old value.

Equation (4.2) provides a perfect estimate if the function being approximated is, in fact, a constant. However, if the function changes at all over the interval, additional terms

79

Box 4.1 Taylor's Theorem

Taylor's theorem: If the function f and its first $n + 1$ derivatives are continuous on an interval containing a and x, then the value of the function at x is given by

$$f(x) = f(a) + f'(a)(x - a) + \frac{f''(a)}{2!}(x - a)^2$$

$$+ \frac{f^{(3)}(a)}{3!}(x - a)^3 + \cdots$$

$$+ \frac{f^{(n)}(a)}{n!}(x - a)^n + R_n \qquad \text{(B4.1.1)}$$

where the remainder R_n is defined as

$$R_n = \int_a^x \frac{(x - t)^n}{n!} f^{(n+1)}(t)\, dt \qquad \text{(B4.1.2)}$$

where $t = $ a dummy variable. Equation (B4.1.1) is called the *Taylor series* or *Taylor's formula*. If the remainder is omitted, the right side of Eq. (B4.1.1) is the Taylor polynomial approximation to $f(x)$. In essence, the theorem states that any smooth function can be approximated as a polynomial.

Equation (B4.1.2) is but one way, called the *integral form,* by which the remainder can be expressed. An alternative formulation can be derived on the basis of the integral mean-value theorem.

First theorem of mean for integrals: If the function g is continuous and integrable on an interval containing a and x, then there exists a point ξ between a and x such that

$$\int_a^x g(t)\, dt = g(\xi)(x - a) \qquad \text{(B4.1.3)}$$

In other words, this theorem states that the integral can be represented by an average value for the function $g(\xi)$ times the interval length $x - a$. Because the average must occur between the minimum and maximum values for the interval, there is a point $x = \xi$ at which the function takes on the average value.

The first theorem is in fact a special case of a second mean-value theorem for integrals.

Second theorem of mean for integrals: If the functions g and h are continuous and integrable on an interval containing a and x, and h does not change sign in the interval, then there exists a point ξ between a and x such that

$$\int_a^x g(t)h(t)\, dt = g(\xi) \int_a^x h(t)\, dt \qquad \text{(B4.1.4)}$$

Thus, Eq. (B4.1.3) is equivalent to Eq. (B4.1.4) with $h(t) = 1$.

The second theorem can be applied to Eq. (B4.1.2) with

$$g(t) = f^{(n+1)}(t) \qquad h(t) = \frac{(x - t)^n}{n!}$$

As t varies from a to x, $h(t)$ is continuous and does not change sign. Therefore, if $f^{(n+1)}(t)$ is continuous, then the integral mean-value theorem holds and

$$R_n = \frac{f^{(n+1)}(\xi)}{(n + 1)!}(x - a)^{n+1}$$

This equation is referred to as the *derivative* or *Lagrange form* of the remainder.

---

of the Taylor series are required to provide a better estimate. For example, the *first-order approximation* is developed by adding another term to yield

$$f(x_{i+1}) \cong f(x_i) + f'(x_i)(x_{i+1} - x_i) \qquad (4.3)$$

The additional first-order term consists of a slope $f'(x_i)$ multiplied by the distance between x_i and x_{i+1}. Thus, the expression is now in the form of a straight line and is capable of predicting an increase or decrease of the function between x_i and x_{i+1}.

Although Eq. (4.3) can predict a change, it is exact only for a straight-line, or *linear,* trend. Therefore, a *second-order* term is added to the series to capture some of the curvature that the function might exhibit:

$$f(x_{i+1}) \cong f(x_i) + f'(x_i)(x_{i+1} - x_i) + \frac{f''(x_i)}{2!}(x_{i+1} - x_i)^2 \qquad (4.4)$$

In a similar manner, additional terms can be included to develop the complete Taylor series expansion:

$$f(x_{i+1}) = f(x_i) + f'(x_i)(x_{i+1} - x_i) + \frac{f''(x_i)}{2!}(x_{i+1} - x_i)^2$$

$$+ \frac{f^{(3)}(x_i)}{3!}(x_{i+1} - x_i)^3 + \cdots + \frac{f^{(n)}(x_i)}{n!}(x_{i+1} - x_i)^n + R_n \qquad (4.5)$$

Note that because Eq. (4.5) is an infinite series, an equal sign replaces the approximate sign that was used in Eqs. (4.2) through (4.4). A remainder term is included to account for all terms from $n + 1$ to infinity:

$$R_n = \frac{f^{(n+1)}(\xi)}{(n+1)!}(x_{i+1} - x_i)^{n+1} \qquad (4.6)$$

where the subscript n connotes that this is the remainder for the nth-order approximation and ξ is a value of x that lies somewhere between x_i and x_{i+1}. The introduction of the ξ is so important that we will devote an entire section (Sec. 4.1.1) to its derivation. For the time being, it is sufficient to recognize that there is such a value that provides an exact determination of the error.

It is often convenient to simplify the Taylor series by defining a step size $h = x_{i+1} - x_i$ and expressing Eq. (4.5) as

$$f(x_{i+1}) = f(x_i) + f'(x_i)h + \frac{f''(x_i)}{2!}h^2 + \frac{f^{(3)}(x_i)}{3!}h^3 + \cdots + \frac{f^{(n)}(x_i)}{n!}h^n + R_n$$

$$(4.7)$$

where the remainder term is now

$$R_n = \frac{f^{(n+1)}(\xi)}{(n+1)!}h^{n+1} \qquad (4.8)$$

EXAMPLE 4.1   Taylor Series Approximation of a Polynomial

Problem Statement.   Use zero- through fourth-order Taylor series expansions to approximate the function

$$f(x) = -0.1x^4 - 0.15x^3 - 0.5x^2 - 0.25x + 1.2$$

from $x_i = 0$ with $h = 1$. That is, predict the function's value at $x_{i+1} = 1$.

Solution.   Because we are dealing with a known function, we can compute values for $f(x)$ between 0 and 1. The results (Fig. 4.1) indicate that the function starts at $f(0) = 1.2$ and then curves downward to $f(1) = 0.2$. Thus, the true value that we are trying to predict is 0.2.

The Taylor series approximation with $n = 0$ is [Eq. (4.2)]

$$f(x_{i+1}) \simeq 1.2$$

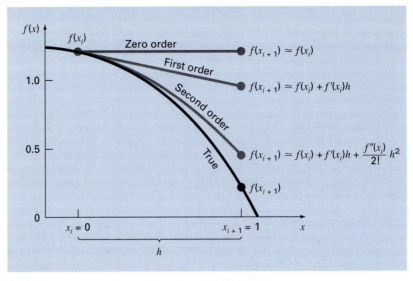

FIGURE 4.1
The approximation of $f(x) = -0.1x^4 - 0.15x^3 - 0.5x^2 - 0.25x + 1.2$ at $x = 1$ by zero-order, first-order, and second-order Taylor series expansions.

Thus, as in Fig. 4.1, the zero-order approximation is a constant. Using this formulation results in a truncation error [recall Eq. (3.2)] of

$$E_t = 0.2 - 1.2 = -1.0$$

at $x = 1$.

For $n = 1$, the first derivative must be determined and evaluated at $x = 0$:

$$f'(0) = -0.4(0.0)^3 - 0.45(0.0)^2 - 1.0(0.0) - 0.25 = -0.25$$

Therefore, the first-order approximation is [Eq. (4.3)]

$$f(x_{i+1}) \simeq 1.2 - 0.25h$$

which can be used to compute $f(1) = 0.95$. Consequently, the approximation begins to capture the downward trajectory of the function in the form of a sloping straight line (Fig. 4.1). This results in a reduction of the truncation error to

$$E_t = 0.2 - 0.95 = -0.75$$

For $n = 2$, the second derivative is evaluated at $x = 0$:

$$f'(0) = -1.2(0.0)^2 - 0.9(0.0) - 1.0 = -1.0$$

Therefore, according to Eq. (4.4),

$$f(x_{i+1}) \simeq 1.2 - 0.25h - 0.5h^2$$

and substituting $h = 1, f(1) = 0.45$. The inclusion of the second derivative now adds some downward curvature resulting in an improved estimate, as in Fig. 4.1. The truncation error is reduced further to $0.2 - 0.45 = -0.25$.

Additional terms would improve the approximation even more. In fact, the inclusion of the third and the fourth derivatives results in exactly the same equation we started with:

$$f(x) = 1.2 - 0.25h - 0.5h^2 - 0.15h^3 - 0.1h^4$$

where the remainder term is

$$R_4 = \frac{f^{(5)}(\xi)}{5!} h^5 = 0$$

because the fifth derivative of a fourth-order polynomial is zero. Consequently, the Taylor series expansion to the fourth derivative yields an exact estimate at $x_{i+1} = 1$:

$$f(1) = 1.2 - 0.25(1) - 0.5(1)^2 - 0.15(1)^3 - 0.1(1)^4 = 0.2$$

In general, the nth-order Taylor series expansion will be exact for an nth-order polynomial. For other differentiable and continuous functions, such as exponentials and sinusoids, a finite number of terms will not yield an exact estimate. Each additional term will contribute some improvement, however slight, to the approximation. This behavior will be demonstrated in Example 4.2. Only if an infinite number of terms are added will the series yield an exact result.

Although the above is true, the practical value of Taylor series expansions is that, in most cases, the inclusion of only a few terms will result in an approximation that is close enough to the true value for practical purposes. The assessment of how many terms are required to get "close enough" is based on the remainder term of the expansion. Recall that the remainder term is of the general form of Eq. (4.8). This relationship has two major drawbacks. First, ξ is not known exactly but merely lies somewhere between x_i and x_{i+1}. Second, to evaluate Eq. (4.8), we need to determine the $(n + 1)$th derivative of $f(x)$. To do this, we need to know $f(x)$. However, if we knew $f(x)$, there would be no need to perform the Taylor series expansion in the present context!

Despite this dilemma, Eq. (4.8) is still useful for gaining insight into truncation errors. This is because we *do* have control over the term h in the equation. In other words, we can choose how far away from x we want to evaluate $f(x)$, and we can control the number of terms we include in the expansion. Consequently, Eq. (4.8) is usually expressed as

$$R_n = O(h^{n+1})$$

where the nomenclature $O(h^{n+1})$ means that the truncation error is of the order of h^{n+1}. That is, the error is proportional to the step size h raised to the $(n + 1)$th power. Although this approximation implies nothing regarding the magnitude of the derivatives that multiply h^{n+1}, it is extremely useful in judging the comparative error of numerical methods based on Taylor series expansions. For example, if the error is $O(h)$, halving the step size will halve the error. On the other hand, if the error is $O(h^2)$, halving the step size will quarter the error.

In general, we can usually assume that the truncation error is decreased by the addition of terms to the Taylor series. In many cases, if h is sufficiently small, the first and other lower-order terms usually account for a disproportionately high percent of the error. Thus, only a few terms are required to obtain an adequate estimate. This property is illustrated by the following example.

EXAMPLE 4.2　Use of Taylor Series Expansion to Approximate a Function with an Infinite Number of Derivatives

Problem Statement.　Use Taylor series expansions with $n = 0$ to 6 to approximate $f(x) = \cos x$ at $x_{i+1} = \pi/3$ on the basis of the value of $f(x)$ and its derivatives at $x_i = \pi/4$. Note that this means that $h = \pi/3 - \pi/4 = \pi/12$.

Solution.　As with Example 4.1, our knowledge of the true function means that we can determine the correct value $f(\pi/3) = 0.5$.

The zero-order approximation is [Eq. (4.3)]

$$f\left(\frac{\pi}{3}\right) \cong \cos\left(\frac{\pi}{4}\right) = 0.707106781$$

which represents a percent relative error of

$$\varepsilon_t = \frac{0.5 - 0.707106781}{0.5} 100\% = -41.4\%$$

For the first-order approximation, we add the first derivative term where $f'(x) = -\sin x$:

$$f\left(\frac{\pi}{3}\right) \cong \cos\left(\frac{\pi}{4}\right) - \sin\left(\frac{\pi}{4}\right)\left(\frac{\pi}{12}\right) = 0.521986659$$

which has $\varepsilon_t = -4.40$ percent.

For the second-order approximation, we add the second derivative term where $f''(x) = -\cos x$:

$$f\left(\frac{\pi}{3}\right) \cong \cos\left(\frac{\pi}{4}\right) - \sin\left(\frac{\pi}{4}\right)\left(\frac{\pi}{12}\right) - \frac{\cos(\pi/4)}{2}\left(\frac{\pi}{12}\right)^2 = 0.497754491$$

with $\varepsilon_t = 0.449$ percent. Thus, the inclusion of additional terms results in an improved estimate.

The process can be continued and the results listed, as in Table 4.1. Notice that the derivatives never go to zero as was the case with the polynomial in Example 4.1. Therefore, each additional term results in some improvement in the estimate. However, also notice how most of the improvement comes with the initial terms. For this case, by the time we

TABLE 4.1　Taylor series approximation of $f(x) = \cos x$ at $x_{i+1} = \pi/3$ using a base point of $\pi/4$. Values are shown for various orders (n) of approximation.

Order n	$f^{(n)}(x)$	$f(\pi/3)$	ε_t
0	$\cos x$	0.707106781	-41.4
1	$-\sin x$	0.521986659	-4.4
2	$-\cos x$	0.497754491	0.449
3	$\sin x$	0.499869147	2.62×10^{-2}
4	$\cos x$	0.500007551	-1.51×10^{-3}
5	$-\sin x$	0.500000304	-6.08×10^{-5}
6	$-\cos x$	0.499999988	2.40×10^{-6}

have added the third-order term, the error is reduced to 2.62×10^{-2} percent, which means that we have attained 99.9738 percent of the true value. Consequently, although the addition of more terms will reduce the error further, the improvement becomes negligible.

4.1.1 The Remainder for the Taylor Series Expansion

Before demonstrating how the Taylor series is actually used to estimate numerical errors, we must explain why we included the argument ξ in Eq. (4.8). A mathematical derivation is presented in Box 4.1. We will now develop an alternative exposition based on a somewhat more visual interpretation. Then we can extend this specific case to the more general formulation.

Suppose that we truncated the Taylor series expansion [Eq. (4.7)] after the zero-order term to yield

$$f(x_{i+1}) \cong f(x_i)$$

A visual depiction of this zero-order prediction is shown in Fig. 4.2. The remainder, or error, of this prediction, which is also shown in the illustration, consists of the infinite series of terms that were truncated:

$$R_0 = f'(x_i)h + \frac{f''(x_i)}{2!}h^2 + \frac{f^{(3)}(x_i)}{3!}h^3 + \cdots$$

It is obviously inconvenient to deal with the remainder in this infinite series format. One simplification might be to truncate the remainder itself, as in

$$R_0 \cong f'(x_i)h \tag{4.9}$$

FIGURE 4.2
Graphical depiction of a zero-order Taylor series prediction and remainder.

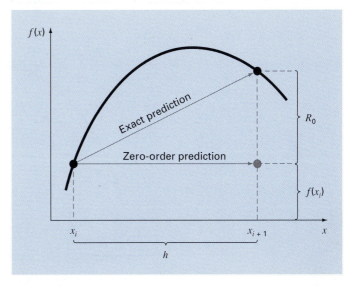

Although, as stated in the previous section, lower-order derivatives usually account for a greater share of the remainder than the higher-order terms, this result is still inexact because of the neglected second- and higher-order terms. This "inexactness" is implied by the approximate equality symbol ($\cong$) employed in Eq. (4.9).

An alternative simplification that transforms the approximation into an equivalence is based on a graphical insight. As in Fig. 4.3, *the derivative mean-value theorem* states that if a function $f(x)$ and its first derivative are continuous over an interval from x_i to x_{i+1}, then there exists at least one point on the function that has a slope, designated by $f'(\xi)$, that is parallel to the line joining $f(x_i)$ and $f(x_{i+1})$. The parameter ξ marks the x value where this slope occurs (Fig. 4.3). A physical illustration of this theorem is that, if you travel between two points with an average velocity, there will be at least one moment during the course of the trip when you will be moving at that average velocity.

By invoking this theorem it is simple to realize that, as illustrated in Fig. 4.3, the slope $f'(\xi)$ is equal to the rise R_0 divided by the run h, or

$$f'(\xi) = \frac{R_0}{h}$$

which can be rearranged to give

$$R_0 = f'(\xi)h \tag{4.10}$$

Thus, we have derived the zero-order version of Eq. (4.8). The higher-order versions are merely a logical extension of the reasoning used to derive Eq. (4.10). The first-order version is

$$R_1 = \frac{f''(\xi)}{2!}h^2 \tag{4.11}$$

FIGURE 4.3
Graphical depiction of the derivative mean-value theorem.

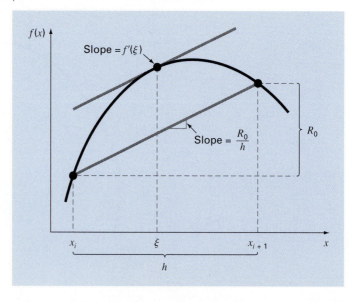

For this case, the value of ξ conforms to the x value corresponding to the second derivative that makes Eq. (4.11) exact. Similar higher-order versions can be developed from Eq. (4.8).

4.1.2 Using the Taylor Series to Estimate Truncation Errors

Although the Taylor series will be extremely useful in estimating truncation errors throughout this book, it may not be clear to you how the expansion can actually be applied to numerical methods. In fact, we have already done so in our example of the falling parachutist. Recall that the objective of both Examples 1.1 and 1.2 was to predict velocity as a function of time. That is, we were interested in determining $v(t)$. As specified by Eq. (4.5), $v(t)$ can be expanded in a Taylor series:

$$v(t_{i+1}) = v(t_i) + v'(t_i)(t_{i+1} - t_i) + \frac{v''(t_i)}{2!}(t_{i+1} - t_i)^2 + \cdots + R_n \tag{4.12}$$

Now let us truncate the series after the first derivative term:

$$v(t_{i+1}) = v(t_i) + v'(t_i)(t_{i+1} - t_i) + R_1 \tag{4.13}$$

Equation (4.13) can be solved for

$$v'(t_i) = \underbrace{\frac{v(t_{i+1}) - v(t_i)}{t_{i+1} - t_i}}_{\substack{\text{First-order} \\ \text{approximation}}} - \underbrace{\frac{R_1}{t_{i+1} - t_1}}_{\substack{\text{Truncation} \\ \text{error}}} \tag{4.14}$$

The first part of Eq. (4.14) is exactly the same relationship that was used to approximate the derivative in Example 1.2 [Eq. (1.11)]. However, because of the Taylor series approach, we have now obtained an estimate of the truncation error associated with this approximation of the derivative. Using Eqs. (4.6) and (4.14) yields

$$\frac{R_1}{t_{i+1} - t_i} = \frac{v''(\xi)}{2!}(t_{i+1} - t_i) \tag{4.15}$$

or

$$\frac{R_1}{t_{i+1} - t_i} = O(t_{i+1} - t_i) \tag{4.16}$$

Thus, the estimate of the derivative [Eq. (1.11) or the first part of Eq. (4.14)] has a truncation error of order $t_{i+1} - t_i$. In other words, the error of our derivative approximation should be proportional to the step size. Consequently, if we halve the step size, we would expect to halve the error of the derivative.

EXAMPLE 4.3   The Effect of Nonlinearity and Step Size on the Taylor Series Approximation

Problem Statement.   Figure 4.4 is a plot of the function

$$f(x) = x^m \tag{E4.3.1}$$

for $m = 1, 2, 3,$ and 4 over the range from $x = 1$ to 2. Notice that for $m = 1$ the function is linear, and as m increases, more curvature or nonlinearity is introduced into the function.

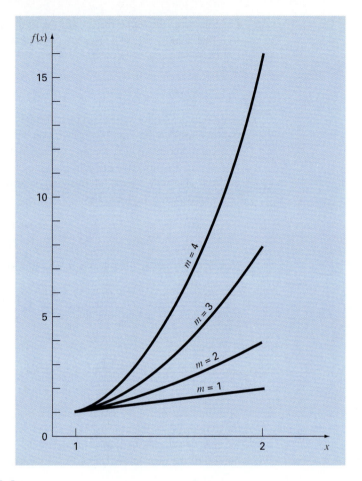

FIGURE 4.4
Plot of the function $f(x) = x^m$ for $m = 1, 2, 3,$ and 4. Notice that the function becomes more nonlinear as m increases.

Employ the first-order Taylor series to approximate this function for various values of the exponent m and the step size h.

Solution.   Equation (E4.3.1) can be approximated by a first-order Taylor series expansion, as in

$$f(x_{i+1}) = f(x_i) + mx_i^{m-1}h \tag{E4.3.2}$$

which has a remainder

$$R_1 = \frac{f''(x_i)}{2!}h^2 + \frac{f^{(3)}(x_i)}{3!}h^3 + \frac{f^{(4)}(x_i)}{4!}h^4 + \cdots$$

First, we can examine how the approximation performs as m increases—that is, as the function becomes more nonlinear. For $m = 1$, the actual value of the function at $x = 2$ is 2.

The Taylor series yields

$$f(2) = 1 + 1(1) = 2$$

and

$$R_1 = 0$$

The remainder is zero because the second and higher derivatives of a linear function are zero. Thus, as expected, the first-order Taylor series expansion is perfect when the underlying function is linear.

For $m = 2$, the actual value is $f(2) = 2^2 = 4$. The first-order Taylor series approximation is

$$f(2) = 1 + 2(1) = 3$$

and

$$R_1 = \tfrac{2}{2}(1)^2 + 0 + 0 + \cdots = 1$$

Thus, because the function is a parabola, the straight-line approximation results in a discrepancy. Note that the remainder is determined exactly.

For $m = 3$, the actual value is $f(2) = 2^3 = 8$. The Taylor series approximation is

$$f(2) = 1 + 3(1)^2(1) = 4$$

and

$$R_1 = \tfrac{6}{2}(1)^2 + \tfrac{6}{6}(1)^3 + 0 + 0 + \cdots = 4$$

Again, there is a discrepancy that can be determined exactly from the Taylor series.

For $m = 4$, the actual value is $f(2) = 2^4 = 16$. The Taylor series approximation is

$$f(2) = 1 + 4(1)^3(1) = 5$$

and

$$R_1 = \tfrac{12}{2}(1)^2 + \tfrac{24}{6}(1)^3 + \tfrac{24}{24}(1)^4 + 0 + 0 + \cdots = 11$$

On the basis of these four cases, we observe that R_1 increases as the function becomes more nonlinear. Furthermore, R_1 accounts exactly for the discrepancy. This is because Eq. (E4.3.1) is a simple monomial with a finite number of derivatives. This permits a complete determination of the Taylor series remainder.

Next, we will examine Eq. (E4.3.2) for the case $m = 4$ and observe how R_1 changes as the step size h is varied. For $m = 4$, Eq. (E4.3.2) is

$$f(x + h) = f(x) + 4x_i^3 h$$

If $x = 1, f(1) = 1$ and this equation can be expressed as

$$f(1 + h) = 1 + 4h$$

with a remainder of

$$R_1 = 6h^2 + 4h^3 + h^4$$

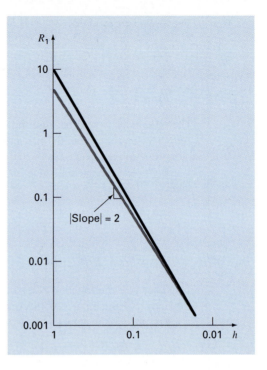

FIGURE 4.5
Log-log plot of the remainder R_1 of the first-order Taylor series approximation of the function $f(x) = x^4$ versus step size h. A line with a slope of 2 is also shown to indicate that as h decreases, the error becomes proportional to h^2.

TABLE 4.2 Comparison of the exact value of the function $f(x) = x^4$ with the first-order Taylor series approximation. Both the function and the approximation are evaluated at $x + h$, where $x = 1$.

h	True	First-Order Approximation	R_1
1	16	5	11
0.5	5.0625	3	2.0625
0.25	2.441406	2	0.441406
0.125	1.601807	1.5	0.101807
0.0625	1.274429	1.25	0.024429
0.03125	1.130982	1.125	0.005982
0.015625	1.063980	1.0625	0.001480

This leads to the conclusion that the discrepancy will decrease as h is reduced. Also, at sufficiently small values of h, the error should become proportional to h^2. That is, as h is halved, the error will be quartered. This behavior is confirmed by Table 4.2 and Fig. 4.5.

Thus, we conclude that the error of the first-order Taylor series approximation decreases as m approaches 1 and as h decreases. Intuitively, this means that the Taylor series becomes more accurate when the function we are approximating becomes more like a straight line over the interval of interest. This can be accomplished either by reducing the size of the interval or by "straightening" the function by reducing m. Obviously, the latter option is usually not available in the real world because the functions we analyze are typically dictated by the physical problem context. Consequently, we do not have control of their lack of linearity, and our only recourse is reducing the step size or including additional terms in the Taylor series expansion.

4.1.3 Numerical Differentiation

Equation (4.14) is given a formal label in numerical methods—it is called a *finite divided difference*. It can be represented generally as

$$f'(x_i) = \frac{f(x_{i+1}) - f(x_i)}{x_{i+1} - x_i} + O(x_{i+1} - x_i) \tag{4.17}$$

or

$$f'(x_i) = \frac{\Delta f_i}{h} + O(h) \tag{4.18}$$

where Δf_i is referred to as the *first forward difference* and h is called the step size, that is, the length of the interval over which the approximation is made. It is termed a "forward" difference because it utilizes data at i and $i+1$ to estimate the derivative (Fig. 4.6a). The entire term $\Delta f / h$ is referred to as a *first finite divided difference*.

This forward divided difference is but one of many that can be developed from the Taylor series to approximate derivatives numerically. For example, backward and centered difference approximations of the first derivative can be developed in a fashion similar to the derivation of Eq. (4.14). The former utilizes values at x_{i-1} and x_i (Fig. 4.6b), whereas the latter uses values that are equally spaced around the point at which the derivative is estimated (Fig. 4.6c). More accurate approximations of the first derivative can be developed by including higher-order terms of the Taylor series. Finally, all the above versions can also be developed for second, third, and higher derivatives. The following sections provide brief summaries illustrating how some of these cases are derived.

Backward Difference Approximation of the First Derivative. The Taylor series can be expanded backward to calculate a previous value on the basis of a present value, as in

$$f(x_{i-1}) = f(x_i) - f'(x_i)h + \frac{f''(x_i)}{2!}h^2 - \cdots \tag{4.19}$$

Truncating this equation after the first derivative and rearranging yields

$$f'(x_i) \cong \frac{f(x_i) - f(x_{i-1})}{h} = \frac{\nabla f_1}{h} \tag{4.20}$$

where the error is $O(h)$ and ∇f_i is referred to as the *first backward difference*. See Fig. 4.6b for a graphical representation.

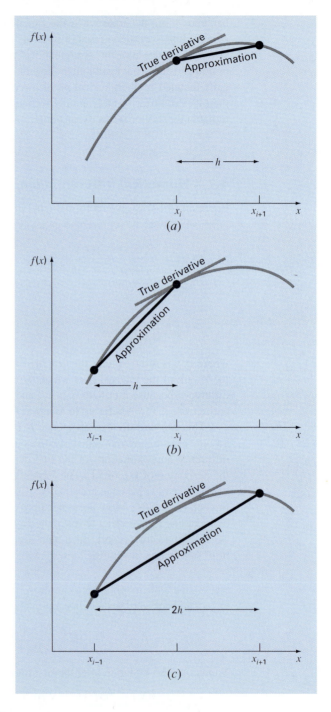

FIGURE 4.6
Graphical depiction of (a) forward, (b) backward, and (c) centered finite-divided-difference approximations of the first derivative.

Centered Difference Approximation of the First Derivative.   A third way to approximate the first derivative is to subtract Eq. (4.19) from the forward Taylor series expansion:

$$f(x_{i+1}) = f(x_i) + f'(x_i)h + \frac{f''(x_i)}{2!}h^2 + \cdots \tag{4.21}$$

to yield

$$f(x_{i+1}) = f(x_{i-1}) + 2f'(x_i)h + \frac{f^{(3)}(x_i)}{3!}h^3 + \cdots$$

which can be solved for

$$f'(x_i) = \frac{f(x_{i+1}) - f(x_{i-1})}{2h} - \frac{f^{(3)}(x_i)}{6}h^2 + \cdots$$

or

$$f'(x_i) = \frac{f(x_{i+1}) - f(x_{i-1})}{2h} - O(h^2) \tag{4.22}$$

Equation (4.22) is a *centered difference* representation of the first derivative. Notice that the truncation error is of the order of h^2 in contrast to the forward and backward approximations that were of the order of h. Consequently, the Taylor series analysis yields the practical information that the centered difference is a more accurate representation of the derivative (Fig. 4.6c). For example, if we halve the step size using a forward or backward difference, we would approximately halve the truncation error, whereas for the central difference, the error would be quartered.

EXAMPLE 4.4   Finite-Divided-Difference Approximations of Derivatives

Problem Statement.   Use forward and backward difference approximations of $O(h)$ and a centered difference approximation of $O(h^2)$ to estimate the first derivative of

$$f(x) = -0.1x^4 - 0.15x^3 - 0.5x^2 - 0.25x + 1.2$$

at $x = 0.5$ using a step size $h = 0.5$. Repeat the computation using $h = 0.25$. Note that the derivative can be calculated directly as

$$f'(x) = -0.4x^3 - 0.45x^2 - 1.0x - 0.25$$

and can be used to compute the true value as $f'(0.5) = -0.9125$.

Solution.   For $h = 0.5$, the function can be employed to determine

$$
\begin{aligned}
x_{i-1} &= 0 & f(x_{i-1}) &= 1.2 \\
x_i &= 0.5 & f(x_i) &= 0.925 \\
x_{i+1} &= 1.0 & f(x_{i+1}) &= 0.2
\end{aligned}
$$

These values can be used to compute the forward divided difference [Eq. (4.17)],

$$f'(0.5) \cong \frac{0.2 - 0.925}{0.5} = -1.45 \qquad |\varepsilon_t| = 58.9\%$$

the backward divided difference [Eq. (4.20)],

$$f'(0.5) \cong \frac{0.925 - 1.2}{0.5} = -0.55 \qquad |\varepsilon_t| = 39.7\%$$

and the centered divided difference [Eq. (4.22)],

$$f'(0.5) \cong \frac{0.2 - 1.2}{1.0} = -1.0 \qquad |\varepsilon_t| = 9.6\%$$

For $h = 0.25$,

$$
\begin{aligned}
x_{i-1} &= 0.25 & f(x_{i-1}) &= 1.10351563 \\
x_i &= 0.5 & f(x_i) &= 0.925 \\
x_{i+1} &= 0.75 & f(x_{i+1}) &= 0.63632813
\end{aligned}
$$

which can be used to compute the forward divided difference,

$$f'(0.5) \cong \frac{0.63632813 - 0.925}{0.25} = -1.155 \qquad |\varepsilon_t| = 26.5\%$$

the backward divided difference,

$$f'(0.5) \cong \frac{0.925 - 1.10351563}{0.25} = -0.714 \qquad |\varepsilon_t| = 21.7\%$$

and the centered divided difference,

$$f'(0.5) \cong \frac{0.63632813 - 1.10351563}{0.5} = -0.934 \qquad |\varepsilon_t| = 2.4\%$$

For both step sizes, the centered difference approximation is more accurate than forward or backward differences. Also, as predicted by the Taylor series analysis, halving the step size approximately halves the error of the backward and forward differences and quarters the error of the centered difference.

Finite Difference Approximations of Higher Derivatives. Besides first derivatives, the Taylor series expansion can be used to derive numerical estimates of higher derivatives. To do this, we write a forward Taylor series expansion for $f(x_{i+2})$ in terms of $f(x_i)$:

$$f(x_{i+2}) = f(x_i) + f'(x_i)(2h) + \frac{f''(x_i)}{2!}(2h)^2 + \cdots \tag{4.23}$$

Equation (4.21) can be multiplied by 2 and subtracted from Eq. (4.23) to give

$$f(x_{i+2}) - 2f(x_{i+1}) = -f(x_i) + f''(x_i)h^2 + \cdots$$

which can be solved for

$$f''(x_i) = \frac{f(x_{i+2}) - 2f(x_{i+1}) + f(x_i)}{h^2} + O(h) \tag{4.24}$$

This relationship is called the *second forward finite divided difference*. Similar manipulations can be employed to derive a backward version

$$f''(x_i) = \frac{f(x_i) - 2f(x_{i-1}) + f(x_{i-2})}{h^2} + O(h)$$

and a centered version

$$f''(x_i) = \frac{f(x_{i+1}) - 2f(x_i) + f(x_{i-1})}{h^2} + O(h^2)$$

As was the case with the first-derivative approximations, the centered case is more accurate. Notice also that the centered version can be alternatively expressed as

$$f''(x_i) \cong \frac{\dfrac{f(x_{i+1}) - f(x_i)}{h} - \dfrac{f(x_i) - f(x_{i-1})}{h}}{h}$$

Thus, just as the second derivative is a derivative of a derivative, the second divided difference approximation is a difference of two first divided differences.

We will return to the topic of numerical differentiation in Chap. 23. We have introduced you to the topic at this point because it is a very good example of why the Taylor series is important in numerical methods. In addition, several of the formulas introduced in this section will be employed prior to Chap. 23.

4.2 ERROR PROPAGATION

The purpose of this section is to study how errors in numbers can propagate through mathematical functions. For example, if we multiply two numbers that have errors, we would like to estimate the error in the product.

4.2.1 Functions of a Single Variable

Suppose that we have a function $f(x)$ that is dependent on a single independent variable x. Assume that $\tilde{x}$ is an approximation of x. We, therefore, would like to assess the effect of the discrepancy between x and $\tilde{x}$ on the value of the function. That is, we would like to estimate

$$\Delta f(\tilde{x}) = |f(x) - f(\tilde{x})|$$

The problem with evaluating $\Delta f(\tilde{x})$ is that $f(x)$ is unknown because x is unknown. We can overcome this difficulty if $\tilde{x}$ is close to x and $f(\tilde{x})$ is continuous and differentiable. If these conditions hold, a Taylor series can be employed to compute $f(x)$ near $f(\tilde{x})$, as in

$$f(x) = f(\tilde{x}) + f'(\tilde{x})(x - \tilde{x}) + \frac{f''(\tilde{x})}{2}(x - \tilde{x})^2 + \cdots$$

Dropping the second- and higher-order terms and rearranging yields

$$f(x) - f(\tilde{x}) \cong f'(\tilde{x})(x - \tilde{x})$$

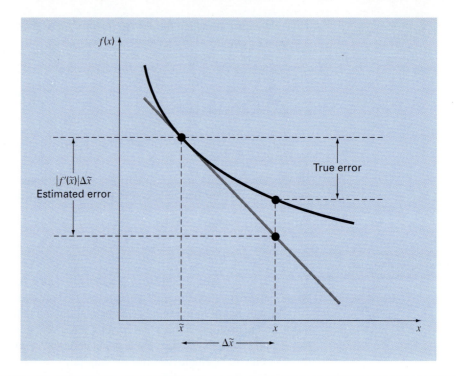

FIGURE 4.7
Graphical depiction of first-order error propagation.

or

$$\Delta f(\tilde{x}) = \left| f'(\tilde{x}) \right| (x - \tilde{x}) \tag{4.25}$$

where $\Delta f(\tilde{x}) = |f(x) - f(\tilde{x})|$ represents an estimate of the error of the function and $\Delta \tilde{x} = |x - \tilde{x}|$ represents an estimate of the error of x. Equation (4.25) provides the capability to approximate the error in $f(x)$ given the derivative of a function and an estimate of the error in the independent variable. Figure 4.7 is a graphical illustration of the operation.

EXAMPLE 4.5   Error Propagation in a Function of a Single Variable

Problem Statement.   Given a value of $\tilde{x} = 2.5$ with an error of $\Delta \tilde{x} = 0.01$, estimate the resulting error in the function, $f(x) = x^3$.

Solution.   Using Eq. (4.25),

$$\Delta f(\tilde{x}) \cong 3(2.5)^2 (0.01) = 0.1875$$

Because $f(2.5) = 15.625$, we predict that

$$f(2.5) = 15.625 \pm 0.1875$$

or that the true value lies between 15.4375 and 15.8125. In fact, if x were actually 2.49, the function could be evaluated as 15.4382 and if x were 2.51, it would be 15.8132. For this case, the first-order error analysis provides a fairly close estimate of the true error.

4.2.2 Functions of More than One Variable

The foregoing approach can be generalized to functions that are dependent on more than one independent variable. This is accomplished with a multivariable version of the Taylor series. For example, if we have a function of two independent variables u and v, the Taylor series can be written as

$$f(u_{i+1}, v_{i+1}) = f(u_i, v_i) + \frac{\partial f}{\partial u}(u_{i+1} - u_i) + \frac{\partial f}{\partial v}(v_{i+1} - v_i)$$

$$+ \frac{1}{2!}\left[\frac{\partial^2 f}{\partial u^2}(u_{i+1} - u_i)^2 + 2\frac{\partial^2 f}{\partial u \partial v}(u_{i+1} - u_i)(v_{i+1} - v_i)\right.$$

$$\left. + \frac{\partial^2 f}{\partial v^2}(v_{i+1} - v_i)^2\right] + \cdots \tag{4.26}$$

where all partial derivatives are evaluated at the base point i. If all second-order and higher terms are dropped, Eq. (4.26) can be solved for

$$\Delta f(\tilde{u}, \tilde{v}) = \left|\frac{\partial f}{\partial u}\right| \Delta\tilde{u} + \left|\frac{\partial f}{\partial v}\right| \Delta\tilde{v}$$

where $\Delta\tilde{u}$ and $\Delta\tilde{v}$ = estimates of the errors in u and v, respectively.

For n independent variables $\tilde{x}_1, \tilde{x}_2, \ldots, \tilde{x}_n$ having errors $\Delta\tilde{x}_1, \Delta\tilde{x}_2, \ldots, \Delta x_n$ the following general relationship holds:

$$\Delta f(\tilde{x}_1, \tilde{x}_2, \ldots, \tilde{x}_n) \cong \left|\frac{\partial f}{\partial x_1}\right| \Delta\tilde{x}_1 + \left|\frac{\partial f}{\partial x_2}\right| \Delta\tilde{x}_2 + \cdots + \left|\frac{\partial f}{\partial x_n}\right| \Delta\tilde{x}_n \tag{4.27}$$

EXAMPLE 4.6   Error Propagation in a Multivariable Function

Problem Statement.   The deflection y of the top of a sailboat mast is

$$y = \frac{FL^4}{8EI}$$

where F = a uniform side loading (lb/ft), L = height (ft), E = the modulus of elasticity (lb/ft^2), and I = the moment of inertia (ft^4). Estimate the error in y given the following data:

$\tilde{F} = 50$ lb/ft $\qquad\qquad \Delta\tilde{F} = 2$ lb/ft

$\tilde{L} = 30$ ft $\qquad\qquad \Delta\tilde{L} = 0.1$ ft

$\tilde{E} = 1.5 \times 10^8$ lb/ft^2 $\qquad \Delta\tilde{E} = 0.01 \times 10^8$ lb/ft^2

$\tilde{I} = 0.06$ ft^4 $\qquad\qquad \Delta\tilde{I} = 0.0006$ ft^4

Solution.   Employing Eq. (4.27) gives

$$\Delta y(\tilde{F}, \tilde{L}, \tilde{E}, \tilde{I}) = \left|\frac{\partial y}{\partial F}\right| \Delta\tilde{F} + \left|\frac{\partial y}{\partial L}\right| \Delta\tilde{L} + \left|\frac{\partial y}{\partial E}\right| \Delta\tilde{E} + \left|\frac{\partial y}{\partial I}\right| \Delta\tilde{I}$$

or

$$\Delta y(\tilde{F}, \tilde{L}, \tilde{E}, \tilde{I}) \cong \frac{\tilde{L}^4}{8\tilde{E}\tilde{I}}\Delta\tilde{F} + \frac{\tilde{F}\tilde{L}^3}{2\tilde{E}\tilde{I}}\Delta\tilde{L} + \frac{\tilde{F}\tilde{L}^4}{8\tilde{E}^2\tilde{I}}\Delta\tilde{E} + \frac{\tilde{F}\tilde{L}^4}{8\tilde{E}\tilde{I}^2}\Delta\tilde{I}$$

Substituting the appropriate values gives

$$\Delta y = 0.0225 + 0.0075 + 0.00375 + 0.005625 = 0.039375$$

Therefore, $y = 0.5625 \pm 0.039375$. In other words, y is between 0.523125 and 0.601875 ft. The validity of these estimates can be verified by substituting the extreme values for the variables into the equation to generate an exact minimum of

$$y_{\min} = \frac{48(29.9)^4}{8(1.51 \times 10^8)0.0606} = 0.52407$$

and

$$y_{\max} = \frac{52(30.1)^4}{8(1.49 \times 10^8)0.0594} = 0.60285$$

Thus, the first-order estimates are reasonably close to the exact values.

Equation (4.27) can be employed to define error propagation relationships for common mathematical operations. The results are summarized in Table 4.3. We will leave the derivation of these formulas as a homework exercise.

4.2.3 Stability and Condition

The *condition* of a mathematical problem relates to its sensitivity to changes in its input values. We say that a computation is *numerically unstable* if the uncertainty of the input values is grossly magnified by the numerical method.

These ideas can be studied using a first-order Taylor series

$$f(x) = f(\tilde{x}) + f'(\tilde{x})(x - \tilde{x})$$

This relationship can be employed to estimate the *relative error of* $f(x)$ as in

$$\frac{f(x) - f(\tilde{x})}{f(x)} \cong \frac{f'(\tilde{x})(x - \tilde{x})}{f(\tilde{x})}$$

The *relative error of* x is given by

$$\frac{x - \tilde{x}}{\tilde{x}}$$

TABLE 4.3 Estimated error bounds associated with common mathematical operations using inexact numbers $\tilde{u}$ and $\tilde{v}$.

Operation		Estimated Error
Addition	$\Delta(\tilde{u} + \tilde{v})$	$\Delta\tilde{u} + \Delta\tilde{v}$
Subtraction	$\Delta(\tilde{u} - \tilde{v})$	$\Delta\tilde{u} + \Delta\tilde{v}$
Multiplication	$\Delta(\tilde{u} \times \tilde{v})$	$\lvert\tilde{u}\rvert\Delta\tilde{v} + \lvert\tilde{v}\rvert\Delta\tilde{u}$
Division	$\Delta\left(\dfrac{\tilde{u}}{\tilde{v}}\right)$	$\dfrac{\lvert\tilde{u}\rvert\Delta\tilde{v} + \lvert\tilde{v}\rvert\Delta\tilde{u}}{\lvert\tilde{v}\rvert^2}$

A *condition number* can be defined as the ratio of these relative errors

$$\text{Condition number} = \frac{\tilde{x} f'(\tilde{x})}{f(\tilde{x})} \tag{4.28}$$

The condition number provides a measure of the extent to which an uncertainty in x is magnified by $f(x)$. A value of 1 tells us that the function's relative error is identical to the relative error in x. A value greater than 1 tells us that the relative error is amplified, whereas a value less than 1 tells us that it is attenuated. Functions with very large values are said to be *ill-conditioned*. Any combination of factors in Eq. (4.28) that increases the numerical value of the condition number will tend to magnify uncertainties in the computation of $f(x)$.

EXAMPLE 4.7   Condition Number

Problem Statement.   Compute and interpret the condition number for

$$f(x) = \tan x \qquad \text{for } \tilde{x} = \frac{\pi}{2} + 0.1\left(\frac{\pi}{2}\right)$$

$$f(x) = \tan x \qquad \text{for } \tilde{x} = \frac{\pi}{2} + 0.01\left(\frac{\pi}{2}\right)$$

Solution.   The condition number is computed as

$$\text{Condition number} = \frac{\tilde{x}(1/\cos^2 x)}{\tan \tilde{x}}$$

For $\tilde{x} = \pi/2 + 0.1(\pi/2)$,

$$\text{Condition number} = \frac{1.7279(40.86)}{-6.314} = -11.2$$

Thus, the function is ill-conditioned. For $\tilde{x} = \pi/2 + 0.01(\pi/2)$, the situation is even worse:

$$\text{Condition number} = \frac{1.5865(4053)}{-63.66} = -101$$

For this case, the major cause of ill conditioning appears to be the derivative. This makes sense because in the vicinity of $\pi/2$, the tangent approaches both positive and negative infinity.

4.3   TOTAL NUMERICAL ERROR

The *total numerical error* is the summation of the truncation and round-off errors. In general, the only way to minimize round-off errors is to increase the number of significant figures of the computer. Further, we have noted that round-off error will *increase* due to subtractive cancellation or due to an increase in the number of computations in an analysis. In contrast, Example 4.4 demonstrated that the truncation error can be reduced by decreasing the step size. Because a decrease in step size can lead to subtractive cancellation or to an increase in computations, the truncation errors are *decreased* as the round-off errors are *increased*. Therefore, we are faced by the following dilemma: The strategy for decreasing

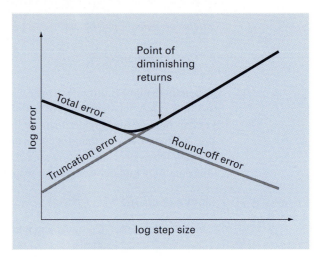

FIGURE 4.8
A graphical depiction of the trade-off between round-off and truncation error that sometimes comes into play in the course of a numerical method. The point of diminishing returns is shown, where round-off error begins to negate the benefits of step-size reduction.

one component of the total error leads to an increase of the other component. In a computation, we could conceivably decrease the step size to minimize truncation errors only to discover that in doing so, the round-off error begins to dominate the solution and the total error grows! Thus, our remedy becomes our problem (Fig. 4.8). One challenge that we face is to determine an appropriate step size for a particular computation. We would like to choose a large step size in order to decrease the amount of calculations and round-off errors without incurring the penalty of a large truncation error. If the total error is as shown in Fig. 4.8, the challenge is to identify the point of diminishing returns where round-off error begins to negate the benefits of step-size reduction.

In actual cases, however, such situations are relatively uncommon because most computers carry enough significant figures that round-off errors do not predominate. Nevertheless, they sometimes do occur and suggest a sort of "numerical uncertainty principle" that places an absolute limit on the accuracy that may be obtained using certain computerized numerical methods.

4.3.1 Control of Numerical Errors

For most practical cases, we do not know the exact error associated with numerical methods. The exception, of course, is when we have obtained the exact solution that makes our numerical approximations unnecessary. Therefore, for most engineering applications we must settle for some estimate of the error in our calculations.

There are no systematic and general approaches to evaluating numerical errors for all problems. In many cases error estimates are based on the experience and judgment of the engineer.

Although error analysis is to a certain extent an art, there are several practical programming guidelines we can suggest. First and foremost, avoid subtracting two nearly equal numbers. Loss of significance almost always occurs when this is done. Sometimes you can rearrange or reformulate the problem to avoid subtractive cancellation. If this is not possible, you may want to use extended-precision arithmetic. Furthermore, when adding and subtracting numbers, it is best to sort the numbers and work with the smallest numbers first. This avoids loss of significance.

Beyond these computational hints, one can attempt to predict total numerical errors using theoretical formulations. The Taylor series is our primary tool for analysis of both truncation and round-off errors. Several examples have been presented in this chapter. Prediction of total numerical error is very complicated for even moderately sized problems and tends to be pessimistic. Therefore, it is usually attempted for only small-scale tasks.

The tendency is to push forward with the numerical computations and try to estimate the accuracy of your results. This can sometimes be done by seeing if the results satisfy some condition or equation as a check. Or it may be possible to substitute the results back into the original equation to check that it is actually satisfied.

Finally you should be prepared to perform numerical experiments to increase your awareness of computational errors and possible ill-conditioned problems. Such experiments may involve repeating the computations with a different step size or method and comparing the results. We may employ sensitivity analysis to see how our solution changes when we change model parameters or input values. We may want to try different numerical algorithms that have different theoretical foundations, are based on different computational strategies, or have different convergence properties and stability characteristics.

When the results of numerical computations are extremely critical and may involve loss of human life or have severe economic ramifications, it is appropriate to take special precautions. This may involve the use of two or more independent groups to solve the same problem so that their results can be compared.

The roles of errors will be a topic of concern and analysis in all sections of this book. We will leave these investigations to specific sections.

4.4 BLUNDERS, FORMULATION ERRORS, AND DATA UNCERTAINTY

Although the following sources of error are not directly connected with most of the numerical methods in this book, they can sometimes have great impact on the success of a modeling effort. Thus, they must always be kept in mind when applying numerical techniques in the context of real-world problems.

4.4.1 Blunders

We are all familiar with gross errors, or blunders. In the early years of computers, erroneous numerical results could sometimes be attributed to malfunctions of the computer itself. Today, this source of error is highly unlikely, and most blunders must be attributed to human imperfection.

Blunders can occur at any stage of the mathematical modeling process and can contribute to all the other components of error. They can be avoided only by sound knowledge of fundamental principles and by the care with which you approach and design your solution to a problem.

Blunders are usually disregarded in discussions of numerical methods. This is no doubt due to the fact that, try as we may, mistakes are to a certain extent unavoidable. However, we believe that there are a number of ways in which their occurrence can be minimized. In particular, the good programming habits that were outlined in Chap. 2 are extremely useful for mitigating programming blunders. In addition, there are usually simple ways to check whether a particular numerical method is working properly. Throughout this book, we discuss ways to check the results of numerical calculations.

4.4.2 Formulation Errors

Formulation, or *model, errors* relate to bias that can be ascribed to incomplete mathematical models. An example of a negligible formulation error is the fact that Newton's second law does not account for relativistic effects. This does not detract from the adequacy of the solution in Example 1.1 because these errors are minimal on the time and space scales associated with the falling parachutist problem.

However, suppose that air resistance is not linearly proportional to fall velocity, as in Eq. (1.7), but is a function of the square of velocity. If this were the case, both the analytical and numerical solutions obtained in the first chapter would be erroneous because of formulation error. Further consideration of formulation error is included in some of the engineering applications in the remainder of the book. You should be cognizant of these problems and realize that, if you are working with a poorly conceived model, no numerical method will provide adequate results.

4.4.3 Data Uncertainty

Errors sometimes enter into an analysis because of uncertainty in the physical data upon which a model is based. For instance, suppose we wanted to test the falling parachutist model by having an individual make repeated jumps and then measuring his or her velocity after a specified time interval. Uncertainty would undoubtedly be associated with these measurements, since the parachutist would fall faster during some jumps than during others. These errors can exhibit both inaccuracy and imprecision. If our instruments consistently underestimate or overestimate the velocity, we are dealing with an inaccurate, or biased, device. On the other hand, if the measurements are randomly high and low, we are dealing with a question of precision.

Measurement errors can be quantified by summarizing the data with one or more well-chosen statistics that convey as much information as possible regarding specific characteristics of the data. These descriptive statistics are most often selected to represent (1) the location of the center of the distribution of the data and (2) the degree of spread of the data. As such, they provide a measure of the bias and imprecision, respectively. We will return to the topic of characterizing data uncertainty in Part Five.

Although you must be cognizant of blunders, formulation errors, and uncertain data, the numerical methods used for building models can be studied, for the most part, independently of these errors. Therefore, for most of this book, we will assume that we have not made gross errors, we have a sound model, and we are dealing with error-free measurements. Under these conditions, we can study numerical errors without complicating factors.

PROBLEMS

4.1 The infinite series

$$e^x = 1 + x + \frac{x^2}{2} + \frac{x^3}{3!} + \cdots + \frac{x^n}{n!}$$

can be used to approximate e^x.

(a) Prove that this Maclaurin series expansion is a special case of the Taylor series expansion [Eq. (4.7)] with $x_i = 0$ and $h = x$.

(b) Use the Taylor series to estimate $f(x) = e^{-x}$ at $x_{i+1} = 1$ for $x_i = 0.25$. Employ the zero-, first-, second-, and third-order versions and compute the $|\varepsilon_t|$ for each case.

4.2 The Maclaurin series expansion for $\cos x$ is

$$\cos x = 1 - \frac{x^2}{2} + \frac{x^4}{4!} - \frac{x^6}{6!} + \frac{x^8}{8!} - \cdots$$

Starting with the simplest version, $\cos x = 1$, add terms one at a time to estimate $\cos(\pi/4)$. After each new term is added, compute the true and approximate percent relative errors. Use your pocket calculator to determine the true value. Add terms until the absolute value of the approximate error estimate falls below an error criterion conforming to two significant figures.

4.3 Perform the same computation as in Prob. 4.2, but use the Maclaurin series expansion for $\sin x$,

$$\sin x = x - \frac{x^3}{3!} + \frac{x^5}{5!} - \frac{x^7}{7!} + \cdots$$

to estimate $\sin(\pi/4)$.

4.4 Use zero- through third-order Taylor series expansions to predict $f(2)$ for

$$f(x) = 25x^3 - 6x^2 + 7x - 88$$

using a base point at $x = 1$. Compute the true percent relative error ε_t for each approximation.

4.5 Use zero- through fourth-order Taylor series expansions to predict $f(3)$ for $f(x) = \ln x$ using a base point at $x = 1$. Compute the true percent relative error ε_t for each approximation. Discuss the meaning of the results.

4.6 Use forward and backward difference approximations of $O(h)$ and a centered difference approximation of $O(h^2)$ to estimate the first derivative of the function examined in Prob. 4.4. Evaluate the derivative at $x = 2$ using a step size of $h = 0.25$. Compare your results with the true value of the derivative. Interpret your results on the basis of the remainder term of the Taylor series expansion.

4.7 Use a centered difference approximation of $O(h^2)$ to estimate the second derivative of the function examined in Prob. 4.4. Perform the evaluation at $x = 2$ using step sizes of $h = 0.2$ and 0.1.

Compare your estimates with the true value of the second derivative. Interpret your results on the basis of the remainder term of the Taylor series expansion.

4.8 Recall that the velocity of the falling parachutist can be computed by [Eq. (1.10)],

$$v(t) = \frac{gm}{c}\left(1 - e^{-(c/m)t}\right)$$

Use a first-order error analysis to estimate the error of v at $t = 6$, if $g = 9.8$ and $m = 50$ but $c = 12.5 \pm 2$.

4.9 Repeat Prob. 4.8 with $g = 9.8$, $t = 6$, $c = 12.5 \pm 2$, and $m = 50 \pm 0.5$.

4.10 The Stefan-Boltzmann law can be employed to estimate the rate of radiation of energy H from a surface, as in

$$H = Ae\sigma T^4$$

where H is in watts, $A =$ the surface area (m²), $e =$ the emissivity that characterizes the emitting properties of the surface (dimensionless), $\sigma =$ a universal constant called the Stefan-Boltzmann constant ($= 5.67 \times 10^{-8}$ W m⁻² K⁻⁴), and $T =$ absolute temperature (K). Determine the error of H for a steel plate with $A = 0.15$ m², $e = 0.90$, and $T = 650 \pm 25$. Compare your results with the exact error. Repeat the computation but with $T = 650 \pm 50$. Interpret your results.

4.11 Repeat Prob. 4.10 but for a copper sphere with radius $= 0.15 \pm 0.02$ m, $e = 0.90 \pm 0.05$, and $T = 550 \pm 25$.

4.12 Evaluate and interpret the condition numbers for

(a) $f(x) = \sqrt{|x - 1|} + 1$ for $x = 1.0001$

(b) $f(x) = e^{-x}$ for $x = 9$

(c) $f(x) = \sqrt{x^2 + 1} - x$ for $x = 200$

(d) $f(x) = \dfrac{e^x - 1}{x}$ for $x = 0.01$

(e) $f(x) = \dfrac{\sin x}{1 + \cos x}$ for $x = 1.001\pi$

4.13 Employing ideas from Sec. 4.2, derive the relationships from Table 4.3.

4.14 Prove that Eq. (4.4) is exact for all values of x if $f(x) = ax^2 + bx + c$.

4.15 Manning's formula for a rectangular channel can be written as

$$Q = \frac{1}{n} \frac{(BH)^{5/3}}{(B + 2H)^{2/3}} S^{1/2}$$

where $Q =$ flow (m³/s), $n =$ a roughness coefficient, $B =$ width (m), $H =$ depth (m), and $S =$ slope. You are applying this formula to a stream where you know that the width $= 20$ m and the depth $= 0.3$ m. Unfortunately, you know the roughness and the slope to only a $\pm 10\%$ precision. That is, you know that the roughness is about 0.03 with a range from 0.027 to 0.033 and the slope is 0.0003 with a range from 0.00027 to 0.00033. Use a first-order error analysis to determine the sensitivity of the flow prediction to each of these two factors. Which one should you attempt to measure with more precision?

4.16 If $|x| < 1$, it is known that

$$\frac{1}{1-x} = 1 + x + x^2 + x^3 + \cdots$$

Repeat Prob. 4.2 for this series for $x = 0.1$.

4.17 A missile leaves the ground with an initial velocity v_0 forming an angle ϕ_0 with the vertical as shown in Fig. P4.17. The maximum desired altitude is αR where R is the radius of the earth. The laws of mechanics can be used to show that

$$\sin \phi_0 = (1+\alpha)\sqrt{1 - \frac{\alpha}{1+\alpha}\left(\frac{v_e}{v_0}\right)^2}$$

where v_e = the escape velocity of the missile. It is desired to fire the missile and reach the design maximum velocity within an accuracy of $\pm 1\%$. Determine the range of values for ϕ_0 if $v_e/v_0 = 2$ and $\alpha = 0.2$.

FIGURE P4.17

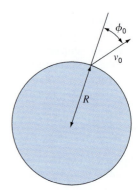

EPILOGUE: PART ONE

PT1.4 TRADE-OFFS

Numerical methods are scientific in the sense that they represent systematic techniques for solving mathematical problems. However, there is a certain degree of art, subjective judgment, and compromise associated with their effective use in engineering practice. For each problem, you may be confronted with several alternative numerical methods and many different types of computers. Thus, the elegance and efficiency of different approaches to problems is highly individualistic and correlated with your ability to choose wisely among options. Unfortunately, as with any intuitive process, the factors influencing this choice are difficult to communicate. Only by experience can these skills be fully comprehended and honed. However, because these skills play such a prominent role in the effective implementation of the methods, we have included this section as an introduction to some of the trade-offs that you must consider when selecting a numerical method and the tools for implementing the method. It is hoped that the discussion that follows will influence your orientation when approaching subsequent material. Also, it is hoped that you will refer back to this material when you are confronted with choices and trade-offs in the remainder of the book.

1. *Type of Mathematical Problem.* As delineated previously in Fig. PT1.2, several types of mathematical problems are discussed in this book:
   (a) Roots of equations.
   (b) Systems of simultaneous linear algebraic equations.
   (c) Optimization.
   (d) Curve fitting.
   (e) Numerical integration.
   (f) Ordinary differential equations.
   (g) Partial differential equations.

You will probably be introduced to the applied aspects of numerical methods by confronting a problem in one of the above areas. Numerical methods will be required because the problem cannot be solved efficiently using analytical techniques. You should be cognizant of the fact that your professional activities will eventually involve problems in all the above areas. Thus, the study of numerical methods and the selection of automatic computation equipment should, at the minimum, consider these basic types of problems. More advanced problems may require capabilities of handling areas such as functional approximation, integral equations, etc. These areas typically demand greater computation power or advanced methods not covered in this text. Other references such as Carnahan, Luther, and Wilkes (1969); Hamming (1973); Ralston and Rabinowitz (1978); and Burden and Faires (1993) should be consulted for problems beyond the scope of this book. In addition, at the end of each part of this text, we include a brief summary and references for advanced methods to provide you with avenues for pursuing further studies of numerical methods.

2. *Type, Availability, Precision, Cost, and Speed of Computer.* You may have the option of working with a variety of computation tools. These range from pocket calculators to large mainframe computers. Of course, any of the tools can be used to implement any numerical method (including simple paper and pencil). It is usually not a question of ultimate capability but rather of cost, convenience, speed, dependability, repeatability, and precision. Although each of the tools will continue to have utility, the recent rapid advances in the performance of personal computers have already had a major impact on the engineering profession. We expect this revolution will spread as technological improvements continue, because personal computers offer an excellent compromise in convenience, cost, precision, speed, and storage capacity. Furthermore, they can be readily applied to most practical engineering problems.

3. *Program Development Cost versus Software Cost versus Run-Time Cost.* Once the types of mathematical problems to be solved have been identified and the computer system has been selected, it is appropriate to consider software and run-time costs. Software development may represent a substantial effort in many engineering projects and may therefore be a significant cost. In this regard, it is particularly important that you be very well acquainted with the theoretical and practical aspects of the relevant numerical methods. In addition, you should be familiar with professionally developed software. Low-cost software is widely available to implement numerical methods that may be readily adapted to a broad variety of problems.

4. *Characteristics of the Numerical Method.* When computer hardware and software costs are high, or if computer availability is limited (for example, on some timeshare systems), it pays to choose carefully the numerical method to suit the situation. On the other hand, if the problem is still at the exploratory stage and computer access and cost are not problems, it may be appropriate for you to select a numerical method that always works but may not be the most computationally efficient. The numerical methods available to solve any particular type of problem involve the types of trade-offs just discussed and others:

   (a) *Number of Initial Guesses or Starting Points.* Some of the numerical methods for finding roots of equations or solving differential equations require the user to specify initial guesses or starting points. Simple methods usually require one value, whereas complicated methods may require more than one value. The advantages of complicated methods that are computationally efficient may be offset by the requirement for multiple starting points. You must use your experience and judgment to assess the trade-offs for each particular problem.

   (b) *Rate of Convergence.* Certain numerical methods converge more rapidly than others. However, this rapid convergence may require more refined initial guesses and more complex programming than a method with slower convergence. Again, you must use your judgment in selecting a method. Faster is not always better.

   (c) *Stability.* Some numerical methods for finding roots of equations or solutions for systems of linear equations may diverge rather than converge on the correct answer for certain problems. Why would you tolerate this possibility when confronted with design or planning problems? The answer is that these methods may be highly efficient when they work. Thus, trade-offs again emerge. You must decide if your problem requirements justify the effort needed to apply a method that may not always converge.

(d) *Accuracy and Precision.* Some numerical methods are simply more accurate or precise than others. Good examples are the various equations available for numerical integration. Usually, the performance of low-accuracy methods can be improved by decreasing the step size or increasing the number of applications over a given interval. Is it better to use a low-accuracy method with small step sizes or a high-accuracy method with large step sizes? This question must be addressed on a case-by-case basis taking into consideration the additional factors such as cost and ease of programming. In addition, you must also be concerned with round-off errors when you are using multiple applications of low-accuracy methods and when the number of computations becomes large. Here the number of significant figures handled by the computer may be the deciding factor.

(e) *Breadth of Application.* Some numerical methods can be applied to only a limited class of problems or to problems that satisfy certain mathematical restrictions. Other methods are not affected by such limitations. You must evaluate whether it is worth your effort to develop programs that employ techniques that are appropriate for only a limited number of problems. The fact that such techniques may be widely used suggests that they have advantages that will often outweigh their disadvantages. Obviously, trade-offs are occurring.

(f) *Special Requirements.* Some numerical techniques attempt to increase accuracy and rate of convergence using additional or special information. An example would be to use estimated or theoretical values of errors to improve accuracy. However, these improvements are generally not achieved without some inconvenience in terms of added computing costs or increased program complexity.

(g) *Programming Effort Required.* Efforts to improve rates of convergence, stability, and accuracy can be creative and ingenious. When improvements can be made without increasing the programming complexity, they may be considered elegant and will probably find immediate use in the engineering profession. However, if they require more complicated programs, you are again faced with a trade-off situation that may or may not favor the new method.

It is clear that the above discussion concerning a choice of numerical methods reduces to one of cost and accuracy. The costs are those involved with computer time and program development. Appropriate accuracy is a question of professional judgment and ethics.

5. *Mathematical Behavior of the Function, Equation, or Data.* In selecting a particular numerical method, type of computer, and type of software, you must consider the complexity of your functions, equations, or data. Simple equations and smooth data may be appropriately handled by simple numerical algorithms and inexpensive computers. The opposite is true for complicated equations and data exhibiting discontinuities.

6. *Ease of Application (User-Friendly?).* Some numerical methods are easy to apply; others are difficult. This may be a consideration when choosing one method over another. This same idea applies to decisions regarding program development costs versus professionally developed software. It may take considerable effort to convert a difficult program to one that is user-friendly. Ways to do this were introduced in Chap. 2 and are elaborated throughout the book. In addition, the Numerical Methods TOOLKIT software supplement is an example of user-friendly programming.

7. *Maintenance.* Programs for solving engineering problems require maintenance because during application, difficulties invariably occur. Maintenance may require changing the program code or expanding the documentation. Simple programs and numerical algorithms are simpler to maintain.

The chapters that follow involve the development of various types of numerical methods for various types of mathematical problems. Several alternative methods will be given in each chapter. These various methods (rather than a single method chosen by the authors) are presented because there is no single "best" method. There is no best method because there are many trade-offs that must be considered when applying the methods to practical problems. A table that highlights the trade-offs involved in each method will be found at the end of each part of the book. This table should assist you in selecting the appropriate numerical procedure for your particular problem context.

PT1.5   IMPORTANT RELATIONSHIPS AND FORMULAS

Table PT1.2 summarizes important information that was presented in Part One. The table can be consulted to quickly access important relationships and formulas. The epilogue of each part of the book will contain such a summary.

PT1.6   ADVANCED METHODS AND ADDITIONAL REFERENCES

The epilogue of each part of the book will also include a section designed to facilitate and encourage further studies of numerical methods. This section will reference other books on the subject as well as material related to more advanced methods.[1]

To extend the background provided in Part One, numerous manuals on computer programming are available. It would be difficult to reference all the excellent books and manuals pertaining to specific languages and computers. In addition, you probably already have material from your previous exposure to programming. However, if this is your first experience with computers, Chapra and Canale (1994) provide a general introduction to BASIC and Fortran. Your instructor and fellow students should also be able to advise you regarding good reference books for the machines and languages available at your school.

As for error analysis, any good introductory calculus book will include supplementary material related to subjects such as the Taylor series expansion. Texts by Swokowski (1979), Thomas and Finney (1979), and Simmons (1985) provide very readable discussions of these subjects. In addition, Taylor (1982) presents a nice introduction to error analysis.

---

[1]Books are referenced only by author here; a complete bibliography will be found at the back of this text.

TABLE PT1.2 Summary of important information presented in Part One.

The Structure Principle

The static structure of the program should correspond in a simple way to the dynamic structure of the corresponding computation.

Error Definitions

True error

$$E_t = \text{true value} - \text{approximation}$$

True percent relative error

$$\epsilon_t = \frac{\text{true value} - \text{approximation}}{\text{true value}} 100\%$$

Approximate percent relative error

$$\epsilon_a = \frac{\text{present approximation} - \text{previous approximation}}{\text{present approximation}} 100\%$$

Stopping criterion

Terminate computation when

$$\epsilon_a < \epsilon_s$$

where ϵ_s is the desired percent relative error

Taylor Series

Taylor series expansion

$$f(x_{i+1}) = f(x_i) + f'(x_i)h + \frac{f''(x_i)}{2!} h^2$$

$$+ \frac{f'''(x_i)}{3!}h^3 + \cdots + \frac{f^{(n)}(x_i)}{n!}h^n + R_n$$

where

Remainder

$$R_n = \frac{f^{(n+1)}(\xi)}{(n+1)!} h^{n+1}$$

or

$$R_n = O(h^{n+1})$$

Numerical Differentiation

First forward finite divided difference

$$f'(x_i) = \frac{f(x_{i+1}) - f(x_i)}{h} + O(h)$$

(Other divided differences are summarized in Chaps. 3 and 17.)

Error Propagation

For n independent variables $x_1, x_2, \ldots, x_n$ having errors $\Delta\tilde{x}_1, \Delta\tilde{x}_2, \ldots, \Delta\tilde{x}_n$, the error in the function f can be estimated via

$$\Delta f = \left| \frac{\partial f}{\partial x_1} \right| \Delta\tilde{x}_1 + \left| \frac{\partial f}{\partial x_2} \right| \Delta\tilde{x}_2 + \cdots + \left| \frac{\partial f}{\partial x_n} \right| \Delta\tilde{x}_n$$

Finally, although we hope that our book serves you well, it is always good to consult other sources when trying to master a new subject. Burden and Faires (1993); Ralston and Rabinowitz (1978); Hoffman (1992); and Carnahan, Luther, and Wilkes (1969) provide comprehensive discussions of most numerical methods, including some advanced methods that are beyond our scope. Other enjoyable books on the subject are Gerald and Wheatley (1989); Rice (1983); and Cheney and Kincaid (1985). In addition, Press et al. (1992) include computer codes to implement a variety of methods.

PART TWO

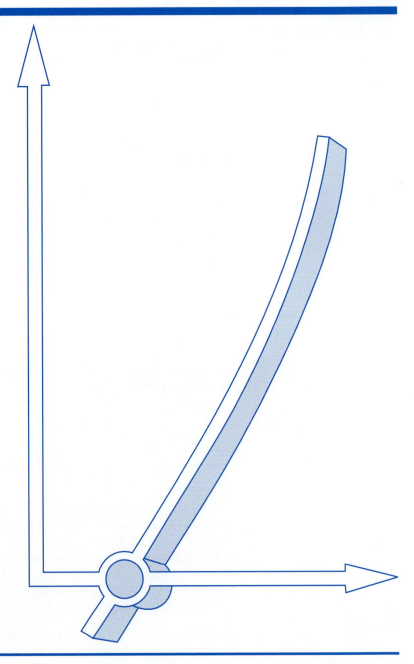

ROOTS OF EQUATIONS

PT2.1 MOTIVATION

Years ago, you learned to use the quadratic formula

$$x = \frac{-b \pm \sqrt{b^2 - 4ac}}{2a} \tag{PT2.1}$$

to solve

$$f(x) = ax^2 + bx + c = 0 \tag{PT2.2}$$

The values calculated with Eq. (PT2.1) are called the "roots" of Eq. (PT2.2). They represent the values of x that make Eq. (PT2.2) equal to zero. Thus, we can define the root of an equation as the value of x that makes $f(x) = 0$. For this reason, roots are sometimes called the *zeros* of the equation.

Although the quadratic formula is handy for solving Eq. (PT2.2), there are many other functions for which the root cannot be determined so easily. For these cases, the numerical methods described in Chaps. 5, 6, and 7 provide efficient means to obtain the answer.

PT2.1.1 Noncomputer Methods for Determining Roots

Before the advent of digital computers, there were several ways to solve for roots of algebraic and transcendental equations. For some cases, the roots could be obtained by direct methods, as was done with Eq. (PT2.1). Although there were equations like this that could be solved directly, there were many more that could not. For example, even an apparently simple function such as $f(x) = e^{-x} - x$ cannot be solved analytically. In such instances, the only alternative is an approximate solution technique.

One method to obtain an approximate solution is to plot the function and determine where it crosses the x axis. This point, which represents the x value for which $f(x) = 0$, is the root. Graphical techniques are discussed at the beginning of Chaps. 5 and 6.

Although graphical methods are useful for obtaining rough estimates of roots, they are limited because of their lack of precision. An alternative approach is to use trial and error. This "technique" consists of guessing a value of x and evaluating whether $f(x)$ is zero. If not (as is almost always the case), another guess is made, and $f(x)$ is again evaluated to determine whether the new value provides a better estimate of the root. The process is repeated until a guess is obtained that results in an $f(x)$ that is close to zero.

Such haphazard methods are obviously inefficient and inadequate for the requirements of engineering practice. The techniques described in Part Two represent alternatives that are also approximate but employ systematic strategies to home in on the true root. As elaborated on in the following pages, the combination of these systematic methods and computers makes the solution of most applied roots-of-equations problems a simple and efficient task.

PT2.1.2 Roots of Equations and Engineering Practice

Although they arise in other problem contexts, roots of equations frequently occur in the area of engineering design. Table PT2.1 lists several fundamental principles that are routinely used in design work. As introduced in Chap. 1, mathematical equations or models derived from these principles are employed to predict dependent variables as a function of independent variables, forcing functions, and parameters. Note that in each case, the dependent variables reflect the state or performance of the system, whereas the parameters represent its properties or composition.

An example of such a model is the equation, derived from Newton's second law, used in Chap. 1 for the parachutist's velocity:

$$v = \frac{gm}{c} \left(1 - e^{-(c/m)t} \right) \qquad\qquad (PT2.3)$$

TABLE PT2.1   Fundamental principles used in engineering design problems.

Fundamental Principle	Dependent Variable	Independent Variable	Parameters
Heat balance	Temperature	Time and position	Thermal properties of material and geometry of system
Mass balance	Concentration or quantity of mass	Time and position	Chemical behavior of material, mass transfer coefficients, and geometry of system
Force balance	Magnitude and direction of forces	Time and position	Strength of material, structural properties, and geometry of system
Energy balance	Changes in the kinetic- and potential-energy states of the system	Time and position	Thermal properties, mass of material, and system geometry
Newton's laws of motion	Acceleration, velocity, or location	Time and position	Mass of material, system geometry, and dissipative parameters such as friction or drag
Kirchhoff's laws	Currents and voltages in electric circuits	Time	Electrical properties of systems such as resistance, capacitance, and inductance

where velocity v = the dependent variable, time t = the independent variable, the gravitational constant g = the forcing function, and the drag coefficient c and mass m = parameters. If the parameters are known, Eq. (PT2.3) can be used to predict the parachutist's velocity as a function of time. Such computations can be performed directly because v is expressed *explicitly* as a function of time. That is, it is isolated on one side of the equal sign.

However, suppose we had to determine the drag coefficient for a parachutist of a given mass to attain a prescribed velocity in a set time period. Although Eq. (PT2.3) provides a mathematical representation of the interrelationship among the model variables and parameters, it cannot be solved explicitly for the drag coefficient. Try it. There is no way to rearrange the equation so that c is isolated on one side of the equal sign. In such cases, c is said to be *implicit*.

This represents a real dilemma, because many engineering design problems involve specifying the properties or composition of a system (as represented by its parameters) to ensure that it performs in a desired manner (as represented by its variables). Thus, these problems often require the determination of implicit parameters.

The solution to the dilemma is provided by numerical methods for roots of equations. To solve the problem using numerical methods, it is conventional to re-express Eq. (PT2.3). This is done by subtracting the dependent variable v from both sides of the equation to give

$$f(c) = \frac{gm}{c} \left(1 - e^{-(c/m)t}\right) - v \tag{PT2.4}$$

The value of c that makes $f(c) = 0$ is, therefore, the root of the equation. This value also represents the drag coefficient that solves the design problem.

Part Two of this book deals with a variety of numerical and graphical methods for determining roots of relationships such as Eq. (PT2.4). These techniques can be applied to engineering design problems that are based on the fundamental principles outlined in Table PT2.1 as well as to many other problems confronted routinely in engineering practice.

PT2.2   MATHEMATICAL BACKGROUND

For most of the subject areas in this book, there is usually some prerequisite mathematical background needed to successfully master the topic. For example, the concepts of error estimation and the Taylor series expansion discussed in Chaps. 3 and 4 have direct relevance to our discussion of roots of equations. Additionally, prior to this point we have mentioned the terms "algebraic" and "transcendental" equations. It might be helpful to formally define these terms and discuss how they relate to the scope of this part of the book.

By definition, a function given by $y = f(x)$ is algebraic if it can be expressed in the form

$$f_n y^n + f_{n-1} y^{n-1} + \cdots + f_1 y + f_0 = 0 \tag{PT2.5}$$

where f_i = an ith-order polynomial in x. *Polynomials* are a simple class of algebraic functions that are represented generally by

$$f_n(x) = a_0 + a_1 x + a_2 x^2 + \cdots + a_n x^n \tag{PT2.6}$$

where n = the *order* of the polynomial and the a's = constants. Some specific examples are

$$f_2(x) = 1 - 2.37x + 7.5x^2 \tag{PT2.7}$$

and

$$f_6(x) = 5x^2 - x^3 + 7x^6 \tag{PT2.8}$$

A *transcendental* function is one that is nonalgebraic. These include trigonometric, exponential, logarithmic, and other, less familiar, functions. Examples are

$$f(x) = \ln x^2 - 1 \tag{PT2.9}$$

and

$$f(x) = e^{-0.2x} \sin (3x - 0.5) \tag{PT2.10}$$

Roots of equations may be either real or complex. Although there are cases where complex roots of nonpolynomials are of interest, such situations are less common than for polynomials. As a consequence, the standard methods for locating roots typically fall into two somewhat related but primarily distinct problem areas:

1. *The determination of the real roots of algebraic and transcendental equations.* These techniques are usually designed to determine the value of a single real root on the basis of foreknowledge of its approximate location.
2. *The determination of all real and complex roots of polynomials.* These methods are specifically designed for polynomials. They systematically determine all the roots of the polynomial rather than determining a single real root given an approximate location.

In this book we discuss both. Chapters 5 and 6 are devoted to the first category. Chapter 7 deals with polynomials.

PT2.3 ORIENTATION

Some orientation is helpful before proceeding to the numerical methods for determining roots of equations. The following is intended to give you an overview of the material in Part Two. In addition, some objectives have been included to help you focus your efforts when studying the material.

PT2.3.1 Scope and Preview

Figure PT2.1 is a schematic representation of the organization of Part Two. Examine this figure carefully, starting at the top and working clockwise.

After the present introduction, *Chap. 5* is devoted to *bracketing methods* for finding roots. These methods start with guesses that bracket, or contain, the root and then systematically reduce the width of the bracket. Two specific methods are covered: *bisection* and *false position*. Graphical methods are used to provide visual insight into the techniques. Error formulations are developed to help you determine how much computational effort is required to estimate the root to a prespecified level of precision.

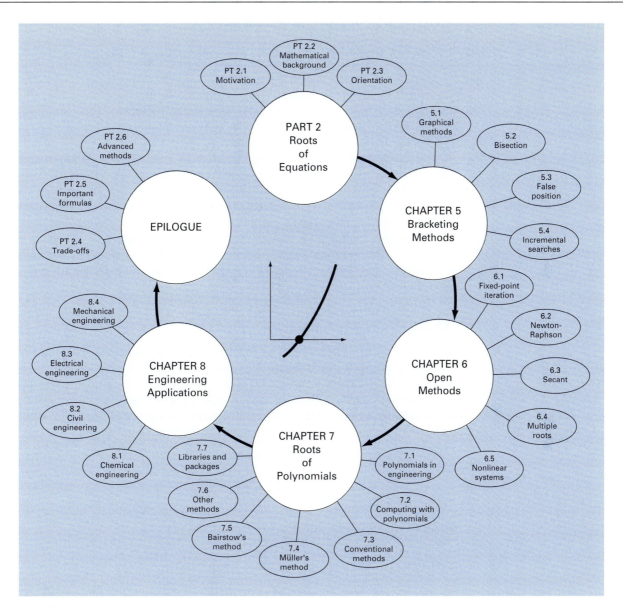

FIGURE PT2.1
Schematic of the organization of the material in Part Two: Roots of Equations.

Chapter 6 covers *open methods*. These methods also involve systematic trial-and-error iterations but do not require that the initial guesses bracket the root. We will discover that these methods are usually more computationally efficient than bracketing methods, but that they do not always work. *One-point iteration, Newton-Raphson,* and *secant* methods are described. Graphical methods are used to provide geometric insight into cases where the open methods do not work. Formulas are developed that provide an idea of how fast open methods home in on the root. In addition, an approach to extend the Newton-Raphson method to *systems of nonlinear equations* is explained.

Chapter 7 is devoted to finding the *roots of polynomials*. After background sections on polynomials, the use of conventional methods (in particular the open methods from Chap. 6) are discussed. Then two special methods for locating polynomial roots are described: Müller's and Bairstow's methods. The chapter ends with information related to finding roots with program libraries and software packages.

Chapter 8 extends the above concepts to actual engineering problems. Engineering applications are used to illustrate the strengths and weaknesses of each method and to provide insight into the application of the techniques in professional practice. The applications also highlight the trade-offs (as discussed in Part One) associated with the various methods.

An epilogue is included at the end of Part Two. It contains a detailed comparison of the methods discussed in Chaps. 5, 6, and 7. This comparison includes a description of trade-offs related to the proper use of each technique. This section also provides a summary of important formulas, along with references for some numerical methods that are beyond the scope of this text.

PT2.3.2 Goals and Objectives

Study Objectives.   After completing Part Two, you should have sufficient information to successfully approach a wide variety of engineering problems dealing with roots of equations. In general, you should have mastered the techniques, have learned to assess their reliability, and be capable of choosing the best method (or methods) for any particular problem. In addition to these general goals, the specific concepts in Table PT2.2 should be assimilated for a comprehensive understanding of the material in Part Two.

Computer Objectives.   The book provides you with software and simple computer algorithms to implement the techniques discussed in Part Two. All have utility as learning tools.

TABLE PT2.2   Specific study objectives for Part Two.

1. Understand the graphical interpretation of a root
2. Know the graphical interpretation of the false-position method and why it is usually superior to the bisection method
3. Understand the difference between bracketing and open methods for root location
4. Understand the concepts of convergence and divergence. Use the two-curve graphical method to provide a visual manifestation of the concepts
5. Know why bracketing methods always converge, whereas open methods may sometimes diverge
6. Realize that convergence of open methods is more likely if the initial guess is close to the true root
7. Understand the concepts of linear and quadratic convergence and their implications for the efficiencies of the fixed-point-iteration and Newton-Raphson methods
8. Know the fundamental difference between the false-position and secant methods and how it relates to convergence
9. Understand the problems posed by multiple roots and the modifications available to mitigate them
10. Know how to extend the single-equation Newton-Raphson approach to solve systems of nonlinear equations

The Numerical Methods TOOLKIT contains the bisection method to determine the real roots of algebraic and transcendental equations. The graphics associated with this software will enable you to easily visualize the behavior of the function being analyzed. The software can be used to conveniently determine roots of equations to a desired degree of precision. The TOOLKIT is easy to apply in solving many practical problems and can be used to check the results of any computer programs you may develop yourself.

Pseudocodes for several methods are also supplied directly in the text. This information will allow you to expand your software library to include programs that are more efficient than the bisection method. For example, you may also want to have your own software for the false-position, Newton-Raphson, and secant techniques, which are often more efficient than the bisection method.

Finally, software packages such as Mathcad, Excel, and MATLAB and program libraries like IMSL have powerful capabilities for locating roots. You can use this part of the book to become familiar with these capabilities.

CHAPTER 5

Bracketing Methods

This chapter on roots of equations deals with methods that exploit the fact that a function typically changes sign in the vicinity of a root. These techniques are called *bracketing methods* because two initial guesses for the root are required. As the name implies, these guesses must "bracket," or be on either side of, the root. The particular methods described herein employ different strategies to systematically reduce the width of the bracket and, hence, home in on the correct answer.

As a prelude to these techniques, we will briefly discuss graphical methods for depicting functions and their roots. Beyond their utility for providing rough guesses, graphical techniques are also useful for visualizing the properties of the functions and the behavior of the various numerical methods.

5.1 GRAPHICAL METHODS

A simple method for obtaining an estimate of the root of the equation $f(x) = 0$ is to make a plot of the function and observe where it crosses the x axis. This point, which represents the x value for which $f(x) = 0$, provides a rough approximation of the root.

EXAMPLE 5.1   The Graphical Approach

Problem Statement.   Use the graphical approach to determine the drag coefficient c needed for a parachutist of mass $m = 68.1$ kg to have a velocity of 40 m/s after free-falling for time $t = 10$ s. *Note:* The acceleration due to gravity is 9.8 m/s^2.

Solution.   This problem can be solved by determining the root of Eq. (PT2.4) using the parameters $t = 10$, $g = 9.8$, $v = 40$, and $m = 68.1$:

$$f(c) = \frac{9.8(68.1)}{c}\left(1 - e^{-(c/68.1)10}\right) - 40$$

or

$$f(c) = \frac{667.38}{c}\left(1 - e^{-0.146843c}\right) - 40 \tag{E5.1.1}$$

Various values of c can be substituted into the right-hand side of this equation to compute

c	f(c)
4	34.115
8	17.653
12	6.067
16	−2.269
20	−8.401

These points are plotted in Fig. 5.1. The resulting curve crosses the c axis between 12 and 16. Visual inspection of the plot provides a rough estimate of the root of 14.75. The validity of the graphical estimate can be checked by substituting it into Eq. (E5.1.1) to yield

$$f(14.75) = \frac{667.38}{14.75}\left(1 - e^{-0.146843(14.75)}\right) - 40 = 0.059$$

which is close to zero. It can also be checked by substituting it into Eq. (PT2.4) along with the parameter values from this example to give

$$v = \frac{9.8(68.1)}{14.75}\left(1 - e^{-(14.75/68.1)10}\right) = 40.059$$

which is very close to the desired fall velocity of 40 m/s.

FIGURE 5.1
The graphical approach for determining the roots of an equation.

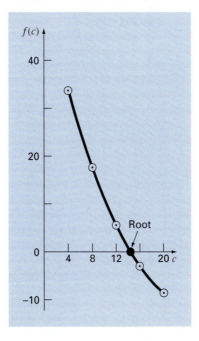

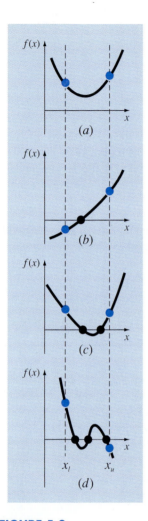

Graphical techniques are of limited practical value because they are not precise. However, graphical methods can be utilized to obtain rough estimates of roots. These estimates can be employed as starting guesses for numerical methods discussed in this and the next chapter. For example, the Numerical Methods TOOLKIT software that supplements this text allows you to plot the function over a specified range. This plot can be used to select guesses that bracket the root prior to implementing the numerical method. The plotting option greatly enhances the utility of the software.

Aside from providing rough estimates of the root, graphical interpretations are important tools for understanding the properties of the functions and anticipating the pitfalls of the numerical methods. For example, Fig. 5.2 shows a number of ways in which roots can occur (or be absent) in an interval prescribed by a lower bound x_l and an upper bound x_u. Figure 5.2b depicts the case where a single root is bracketed by negative and positive values of $f(x)$. However, Fig. 5.2d, where $f(x_l)$ and $f(x_u)$ are also on opposite sides of the x axis, shows three roots occurring within the interval. In general, if $f(x_l)$ and $f(x_u)$ have opposite signs, there are an odd number of roots in the interval. As indicated by Fig. 5.2a and c, if $f(x_l)$ and $f(x_u)$ have the same sign, there are either no roots or an even number of roots between the values.

Although these generalizations are usually true, there are cases where they do not hold. For example, functions that are tangential to the x axis (Fig. 5.3a) and discontinuous functions (Fig. 5.3b) can violate these principles. An example of a function that is tangential to the axis is the cubic equation $f(x) = (x - 2)(x - 2)(x - 4)$. Notice that $x = 2$ makes two terms in this polynomial equal to zero. Mathematically, $x = 2$ is called a *multiple root*. At the end of Chap. 6, we will present techniques that are expressly designed to locate multiple roots.

FIGURE 5.3
Illustration of some exceptions to the general cases depicted in Fig. 5.2. (a) Multiple root that occurs when the function is tangential to the x axis. For this case, although the end points are of opposite signs, there are an even number of axis intersections for the interval. (b) Discontinuous function where end points of opposite sign bracket an even number of roots. Special strategies are required for determining the roots for these cases.

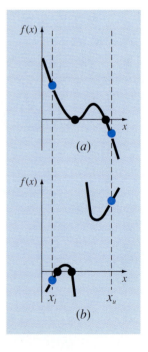

The existence of cases of the type depicted in Fig. 5.3 makes it difficult to develop general computer algorithms guaranteed to locate all the roots in an interval. However, when used in conjunction with graphical approaches, the methods described in the following sections are extremely useful for solving many roots of equations problems confronted routinely by engineers and applied mathematicians.

EXAMPLE 5.2   Use of Computer Graphics to Locate Roots

Problem Statement.   Computer graphics can expedite and improve your efforts to locate roots of equations. The present example was developed using the Numerical Methods TOOLKIT software package. We have chosen the TOOLKIT because of the ease with which the axis scaling can be modified with this tool. Thus, it is ideal for the type of visual, exploratory analysis we are illustrating here. It should be noted, however, that while it might be a tad more inconvenient, other software with graphical capabilities could be used

FIGURE 5.4

The progressive enlargement of $f(x) = \sin 10x + \cos 3x$ by the computer. Such interactive graphics permits the analyst to determine that two distinct roots exist between $x = 4.2$ and $x = 4.3$.

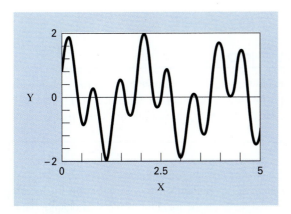

(a)

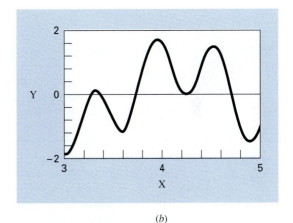

(b)

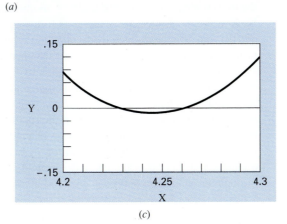

(c)

to perform a similar analysis. Further, the insights and issues raised are not software-specific, but are relevant to computer graphics in general.

The function

$$f(x) = \sin 10x + \cos 3x$$

has several roots over the range $x = 0$ to $x = 5$. Use computer graphics to gain insight into the behavior of this function.

Solution.   As previously illustrated in Example 2.1, the TOOLKIT can be used to generate plots. Figure 5.4*a* is a plot of $f(x)$ from $x = 0$ to $x = 5$. This plot suggests the presence of several roots, including a possible double root at about $x = 4.2$, where $f(x)$ appears to be tangent to the x axis. A more detailed picture of the behavior of $f(x)$ is obtained by changing the plotting range from $x = 3$ to $x = 5$, as shown in Fig. 5.4*b*. Finally, in Fig. 5.4*c*, the vertical scale is narrowed further to $f(x) = -0.15$ to $f(x) = 0.15$ and the horizontal scale is narrowed to $x = 4.2$ to $x = 4.3$. This plot shows clearly that a double root does not exist in this region and that in fact there are two distinct roots at about $x = 4.23$ and $x = 4.26$.

Computer graphics will have great utility in your studies of numerical methods. This capability will also find many other applications in your other classes and professional activities as well.

5.2   THE BISECTION METHOD

When applying the graphical technique in Example 5.1, you have observed (Fig. 5.1) that $f(x)$ changed sign on opposite sides of the root. In general, if $f(x)$ is real and continuous in the interval from x_l to x_u and $f(x_l)$ and $f(x_u)$ have opposite signs, that is,

$$f(x_l)f(x_u) < 0 \tag{5.1}$$

then there is at least one real root between x_l and x_u.

Incremental search methods capitalize on this observation by locating an interval where the function changes sign. Then the location of the sign change (and consequently, the root) is identified more precisely by dividing the interval into a number of subintervals. Each of these subintervals is searched to locate the sign change. The process is repeated and the root estimate refined by dividing the subintervals into finer increments. We will return to the general topic of incremental searches in Sec. 5.4.

The *bisection method,* which is alternatively called binary chopping, interval halving, or Bolzano's method, is one type of incremental search method in which the interval is always divided in half. If a function changes sign over an interval, the function value at the midpoint is evaluated. The location of the root is then determined as lying at the midpoint of the subinterval within which the sign change occurs. The process is repeated to obtain refined estimates. A simple algorithm for the bisection calculation is listed in Fig. 5.5, and a graphical depiction of the method is provided in Fig. 5.6. The following example goes through the actual computations involved in the method.

Step 1: Choose lower x_l and upper x_u guesses for the root such that the function changes sign over the interval. This can be checked by ensuring that $f(x_l)f(x_u) < 0$.

Step 2: An estimate of the root x_r is determined by

$$x_r = \frac{x_l + x_u}{2}$$

Step 3: Make the following evaluations to determine in which subinterval the root lies:

(a) If $f(x_l)f(x_r) < 0$, the root lies in the lower subinterval. Therefore, set $x_u = x_r$ and return to step 2.

(b) If $f(x_l)f(x_r) > 0$, the root lies in the upper subinterval. Therefore, set $x_l = x_r$ and return to step 2.

(c) If $f(x_l)f(x_r) = 0$, the root equals x_r; terminate the computation.

FIGURE 5.5

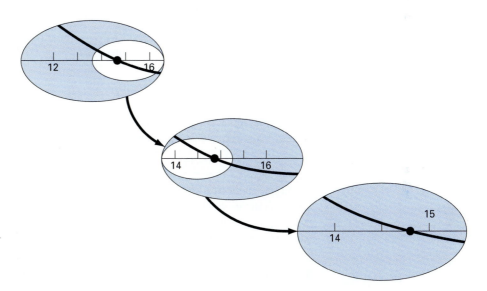

FIGURE 5.6
A graphical depiction of the bisection method. This plot conforms to the first three iterations from Example 5.3.

EXAMPLE 5.3   Bisection

Problem Statement.   Use bisection to solve the same problem approached graphically in Example 5.1.

Solution.   The first step in bisection is to guess two values of the unknown (in the present problem, c) that give values for $f(c)$ with different signs. From Fig. 5.1, we can see that the function changes sign between values of 12 and 16. Therefore, the initial estimate of the root x_r lies at the midpoint of the interval

$$x_r = \frac{12 + 16}{2} = 14$$

This estimate represents a true percent relative error of $E_t = 5.3\%$ (note that the true value of the root is 14.7802). Next we compute the product of the function value at the lower bound and at the midpoint:

$$f(12)\,f(14) = 6.067(1.569) = 9.517$$

which is greater than zero, and hence no sign change occurs between the lower bound and the midpoint. Consequently, the root must be located between 14 and 16. Therefore, we create a new interval by redefining the lower bound as 14 and determining a revised root estimate as

$$x_r = \frac{14 + 16}{2} = 15$$

which represents a true percent error of $\varepsilon_t = 1.5\%$. The process can be repeated to obtain refined estimates. For example,

$$f(14)\,f(15) = 1.569(-0.425) = -0.666$$

Therefore, the root is between 14 and 15. The upper bound is redefined as 15, and the root estimate for the third iteration is calculated as

$$x_r = \frac{14 + 15}{2} = 14.5$$

which represents a percent relative error of $\varepsilon_t = 1.9\%$. The method can be repeated until the result is accurate enough to satisfy your needs.

In the previous example, you may have noticed that the true error does not decrease with each iteration. However, the interval within which the root is located is halved with each step in the process. As discussed in the next section, the interval width provides an exact estimate of the upper bound of the error for the bisection method.

5.2.1 Termination Criteria and Error Estimates

We ended Example 5.3 with the statement that the method could be continued to obtain a refined estimate of the root. We must now develop an objective criterion for deciding when to terminate the method.

An initial suggestion might be to end the calculation when the true error falls below some prespecified level. For instance, in Example 5.3, the relative error dropped from 5.3 to 1.9 percent during the course of the computation. We might decide that we should terminate when the error drops below, say, 0.1 percent. This strategy is flawed because the error estimates in the example were based on knowledge of the true root of the function. This would not be the case in an actual situation because there would be no point in using the method if we already knew the root.

Therefore, we require an error estimate that is not contingent on foreknowledge of the root. As developed previously in Sec. 3.3, an approximate percent relative error ε_a can be calculated, as in [recall Eq. (3.5)]

$$\varepsilon_a = \left| \frac{x_r^{new} - x_r^{old}}{x_r^{new}} \right| 100\% \tag{5.2}$$

where x_r^{new} is the root for the present iteration and x_r^{old} is the root from the previous iteration. The absolute value is used because we are usually concerned with the magnitude of ε_a rather than with its sign. When ε_a becomes less than a prespecified stopping criterion ε_s, the computation is terminated.

EXAMPLE 5.4   Error Estimates for Bisection

Problem Statement.   Continue Example 5.3 until the approximate error falls below a stopping criterion of $\varepsilon_s = 0.5\%$. Use Eq. (5.2) to compute the errors.

Solution.   The results of the first two iterations for Example 5.3 were 14 and 15. Substituting these values into Eq. (5.2) yields

$$|\varepsilon_a| = \left| \frac{15 - 14}{14} \right| 100\% = 6.667\%$$

Recall that the true percent relative error for the root estimate of 15 was 1.5%. Therefore, ε_a is greater than ε_s. This behavior is manifested for the other iterations:

Iteration	x_l	x_u	x_r	ε_a (%)	ε_t (%)
1	12	16	14		5.279
2	14	16	15	6.667	1.487
3	14	15	14.5	3.448	1.896
4	14.5	15	14.75	1.695	0.204
5	14.75	15	14.875	0.840	0.641
6	14.75	14.875	14.8125	0.422	0.219

Thus, after six iterations ε_a finally falls below $\varepsilon_s = 0.5\%$, and the computation can be terminated.

These results are summarized in Fig. 5.7. The "ragged" nature of the true error is due to the fact that, for bisection, the true root can lie anywhere within the bracketing interval. The true and approximate errors are far apart when the interval happens to be centered on the true root. They are close when the true root falls at either end of the interval.

Although the approximate error does not provide an exact estimate of the true error, Fig. 5.7 suggests that ε_a captures the general downward trend of ε_t. In addition, the plot exhibits the extremely attractive characteristic that ε_a is always greater than ε_t. Thus, when ε_a falls below ε_s, the computation could be terminated with confidence that the root is known to be at least as accurate as the prespecified acceptable level.

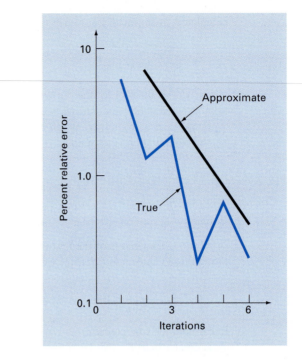

FIGURE 5.7
Errors for the bisection method. True and estimated errors are plotted versus the number of iterations.

Although it is always dangerous to draw general conclusions from a single example, it can be demonstrated that ε_a will always be greater than ε_t for the bisection method. This is because each time an approximate root is located using bisection as $x_r = (x_l + x_u)/2$, we know that the true root lies somewhere within an interval of $(x_u - x_l)/2 = \Delta x/2$. Therefore, the root must lie within $\pm\Delta x/2$ of our estimate (Fig. 5.8). For instance, when Example 5.3 was terminated, we could make the definitive statement that

$$x_r = 14.5 \pm 0.5$$

Because $\Delta x/2 = x_r^{\text{new}} - x_r^{\text{old}}$ (Fig. 5.9), Eq. (5.2) provides an exact upper bound on the true error. For this bound to be exceeded, the true root would have to fall outside the bracketing interval, which, by definition, could never occur for the bisection method. As illustrated in a subsequent example (Example 5.7), other root-locating techniques do not always behave as nicely. Although bisection is generally slower than other methods, the neatness of its error analysis is certainly a positive aspect that could make it attractive for certain engineering applications.

Before proceeding to the computer program for bisection, we should note that the relationships (Fig. 5.9)

$$x_r^{\text{new}} - x_r^{\text{old}} = \frac{x_u - x_l}{2}$$

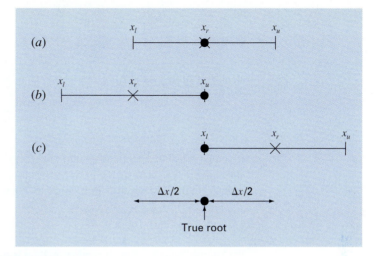

FIGURE 5.8
Three ways in which the interval may bracket the root. In (a) the true value lies at the center of the interval, whereas in (b) and (c) the true value lies near the extreme. Notice that the discrepancy between the true value and the midpoint of the interval never exceeds half the interval length, or $\Delta x/2$.

FIGURE 5.9
Graphical depiction of why the error estimate for bisection ($\Delta x/2$) is equivalent to the root estimate for the present iteration (x_r^{new}) minus the root estimate for the previous iteration (x_r^{old}).

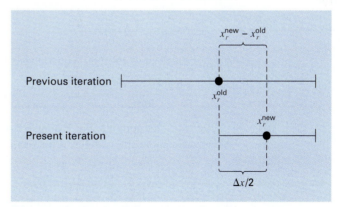

and

$$x_r^{new} = \frac{x_l + x_u}{2}$$

can be substituted into Eq. (5.2) to develop an alternative formulation for the approximate percent relative error

$$\varepsilon_a = \left| \frac{x_u - x_l}{x_u + x_l} \right| 100\% \tag{5.3}$$

This equation yields identical results to Eq. (5.2) for bisection. In addition, it allows us to calculate an error estimate on the basis of our initial guesses—that is, on our first iteration. For instance, on the first iteration of Example 5.2, an approximate error can be computed as

$$\varepsilon_a = \left| \frac{16 - 12}{16 + 12} \right| 100\% = 14.29\%$$

Another benefit of the bisection method is that the number of iterations required to attain an absolute error can be computed *a priori*—that is, before starting the iterations. This can be seen by recognizing that before starting the technique, the absolute error is

$$E_a^0 = x_u^0 - x_l^0 = \Delta x^0$$

where the superscript designates the iteration. Hence, before starting the method, we are at the "zero iteration." After the first iteration, the error becomes

$$E_a^1 = \frac{\Delta x^0}{2}$$

Because each succeeding iteration halves the error, a general formula relating the error and the number of iterations, n, is

$$E_a^n = \frac{\Delta x^0}{2^n} \tag{5.4}$$

If $E_{a,d}$ is the desired error, this equation can be solved for

$$n = \frac{\log\left(\Delta x^0 / E_{a,d}\right)}{\log 2} = \log_2\left(\frac{\Delta x^0}{E_{a,d}}\right) \tag{5.5}$$

Let's test the formula. For Example 5.4, the initial interval was $\Delta x_0 = 16 - 12 = 4$. After six iterations, the absolute error was

$$E_a = \frac{|14.875 - 14.75|}{2} = 0.0625$$

We can substitute these values into Eq. (5.5) to give

$$n = \frac{\log(4/0.0625)}{\log 2} = 6$$

Thus, if we knew beforehand that an error of less than 0.0625 was acceptable, the formula tells us that six iterations would yield the desired result.

Although we have emphasized the use of relative errors for obvious reasons, there will be cases where (usually through knowledge of the problem context) you will be able to specify an absolute error. For these cases, bisection along with Eq. (5.5) can provide a useful root-location algorithm. We will explore such applications in the end-of-chapter problems.

5.2.2 Bisection Algorithm

The algorithm in Fig. 5.5 can now be expanded to include the error check (Fig. 5.10). The algorithm employs user-defined functions to make root location and function evaluation more efficient. In addition, an upper limit is placed on the number of iterations. Finally, an error check is included to avoid division by zero during the error evaluation. Such would be the case when the bracketing interval is centered on zero. For this situation Eq. (5.2) becomes infinite. If this occurs, the program skips over the error evaluation for that iteration.

The algorithm in Fig. 5.10 is not user-friendly; it is designed strictly to come up with the answer. In Prob. 5.14 at the end of this chapter, you will have the task of making it easier to use and understand. An example of a user-friendly program for implementing bisection is included in the Numerical Methods TOOLKIT software associated with this text. The following example demonstrates the use of this software for root location. It also provides a good reference for assessing and testing your own software.

FIGURE 5.10

Pseudocode for function to implement bisection.

```
FUNCTION Bisect (xl, xu, es, imax, xr, iter, ea)
    iter = 0
    DO
        xrold = xr
        xr = (xl + xu) / 2
        iter = iter + 1
        IF xr ≠ 0 THEN
            ea = ABS((xr − xrold) / xr) * 100
        END IF
        test = f(xl) * f(xr)
        IF test < 0 THEN
            xu = xr
        ELSE IF test > 0 THEN
            xl = xr
        ELSE
            ea = 0
        END IF
        IF ea < es OR iter ≥ imax EXIT
    END DO
    Bisect = xr
END Bisect
```

EXAMPLE 5.5   Root Location Using the Computer

Problem Statement.   A user-friendly computer program to implement the bisection method is contained in the Numerical Methods TOOLKIT software associated with the text. We can use this software to find the smallest positive root of

$$f(x) = x - \cos x \qquad\qquad (E5.5.1)$$

Solution.   Press the "Find Root" button from the main menu of the Numerical Methods TOOLKIT to obtain a blank screen similar to Fig. 5.11. The first step is to input the equation that you want to solve in the form of $f(x) = 0$ in the input box provided. The Help button on the main menu lists the functions and constants recognized by the TOOLKIT.

To enter Eq. (E5.5.1), simply click on the input function box and enter the function. Let's assume we have no idea what the root might be, and therefore decide to construct a plot to provide some guidance regarding selection of appropriate upper and lower guesses. A plot is constructed by supplying values for the minimum X and Y and a Delta X and Y to the Parameter Input table. This generates a plot with maximum values of (Xmin + 10 Delta X) and (Ymin + 10 Delta Y) when the Plot button is clicked. This is the convention that is used for all the plots in the TOOLKIT and is further explained from the Help button on the main menu. A plot with Xmin = −2, Delta X = 0.4, Ymin = −5, and Delta Y = 1 is shown in Fig. 5.11. Inspection of this plot shows that a root exists at about $x = 0.7$. This is the insight needed to provide low and high guesses for the root. The final inputs required are the maximum number of iterations and the maximum allowable percent error (ε_s).

Iterations stop either when the iterations equal the input Max Iteration or when the estimated error calculated with Eq. (5.3) is less than the input Max % Error (ε_s). The actual calculation is initiated by clicking the Calc button. Note that this button is red when any input to the screen renders the numerical values in the Results Table obsolete.

FIGURE 5.11

Finding the root of an equation using bisection and the TOOLKIT.

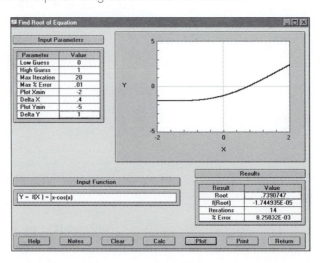

In this case a root equal to 0.7390747 was found in 14 iterations that gave an error of about 0.008%. The function is also evaluated at the root as a check. The function evaluated at the root is close to zero (-1.745×10^{-5}). The screen design facilitates experimentation by allowing you to vary the Max % Error and Max Iteration. The overall procedure for finding roots is summarized by the Help button. In addition, you may write yourself some notes regarding the analysis by clicking the Notes button. Finally, you can save the results for further reference by using the Save and Open buttons on the main menu.

These results are based on a simple algorithm for the bisection method with user-friendly input and output routines. The algorithm employed is similar to the one shown in Fig. 5.10. You should be able to write your own program for the bisection method and use our software as a model and to check the adequacy of your own program.

5.2.3 Minimizing Function Evaluations

The bisection algorithm in Fig. 5.10 is just fine if you are performing a single root evaluation for a function that is easy to evaluate. However, there are many instances in

FIGURE 5.12
Pseudocode for bisection subprogram which minimizes function evaluations.

```
FUNCTION Bisect (xl, xu, es, imax, xr, iter, ea)
    iter = 0
    fl = f(xl)
    DO
        xrold = xr
        xr = (xl + xu) / 2
        fr = f(xr)
        iter = iter + 1
        F xr ≠ 0 THEN
            ea = ABS((xr − xrold) / xr) * 100
        END IF
        test = fl * fr
        IF test < 0 THEN
            xu = xr
        ELSE IF test > 0 THEN
            xl = xr
            fl = fr
        ELSE
            ea = 0
        END IF
        IF ea < es OR iter ≥ imax EXIT
    END DO
    Bisect = xr
END Bisect
```

engineering when this is not the case. For example, suppose that you develop a computer program that must locate a root numerous times. In such cases you could call the algorithm from Fig. 5.10 thousands and even millions of times in the course of a single run.

Further, in its most general sense, a univariate function is merely an entity that returns a single value in return for a single value you send to it. Perceived in this sense, functions are not always simple formulas like the one-line equations solved in the preceding examples in this chapter. For example, a function might consist of many lines of code that could take a significant amount of execution time to evaluate. In some cases, the function might even represent an independent computer program.

Because of both these factors, it is imperative that numerical algorithms minimize function evaluations. In this light, the algorithm from Fig. 5.10 is deficient. In particular, notice that in making two function evaluations per iteration, it recalculates one of the functions that was determined on the previous iteration.

Figure 5.12 provides a modified algorithm that does not have this deficiency. We have highlighted the lines that differ from Fig. 5.10. In this case, only the new function value at the root estimate is calculated. Previously calculated values are saved and merely reassigned as the bracket shrinks. Thus, rather than $2n$, the function evaluations are reduced to $n + 1$.

5.3 THE FALSE-POSITION METHOD

Although bisection is a perfectly valid technique for determining roots, its "brute-force" approach is relatively inefficient. False position is an alternative based on a graphical insight.

A shortcoming of the bisection method is that, in dividing the interval from x_l to x_u into equal halves, no account is taken of the magnitudes of $f(x_l)$ and $f(x_u)$. For example, if $f(x_l)$ is much closer to zero than $f(x_u)$, it is likely that the root is closer to x_l, than to x_u (Fig. 5.13). An alternative method that exploits this graphical insight is to join $f(x_l)$ and $f(x_u)$ by a straight line. The intersection of this line with the x axis represents an improved estimate of the root. The fact that the replacement of the curve by a straight line gives a "false position" of the root is the origin of the name, *method of false position*, or in Latin, *regula falsi*. It is also called the *linear interpolation method*.

Using similar triangles (Fig. 5.13), the intersection of the straight line with the x axis can be estimated as

$$\frac{f(x_l)}{x_r - x_l} = \frac{f(x_u)}{x_r - x_u} \tag{5.6}$$

which can be solved for (see Box 5.1 for details).

$$x_r = x_u - \frac{f(x_u)(x_l - x_u)}{f(x_l) - f(x_u)} \tag{5.7}$$

This is the *false-position formula*. The value of x_r computed with Eq. (5.7) then replaces whichever of the two initial guesses, x_l or x_u, yields a function value with the same sign as $f(x_r)$. In this way, the values of x_l and x_u always bracket the true root. The process is repeated until the root is estimated adequately. The algorithm is identical to the one for bisection (Fig. 5.5) with the exception that Eq. (5.7) is used for step 2. In addition, the same stopping criterion [Eq. (5.2)] is used to terminate the computation.

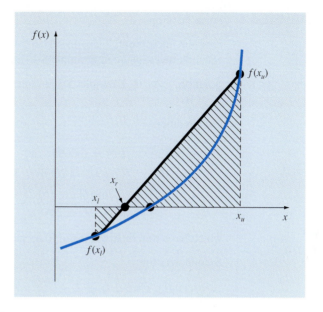

FIGURE 5.13
A graphical depiction of the method of false position. Similar triangles used to derive the formula for the method are shaded.

Box 5.1 Derivation of the Method of False Position

Cross-multiply Eq. (5.6) to yield

$$f(x_l)(x_r - x_u) = f(x_u)(x_r - x_l)$$

Collect terms and rearrange:

$$x_r [f(x_l) - f(x_u)] = x_u f(x_l) - x_l f(x_u)$$

Divide by $f(x_l) - f(x_u)$:

$$x_r = \frac{x_u f(x_l) - x_l f(x_u)}{f(x_l) - f(x_u)} \qquad (B5.1.1)$$

This is one form of the method of false position. Note that it allows the computation of the root x_r as a function of the lower and upper guesses x_l and x_u. It can be put in an alternative form by expanding it:

$$x_r = \frac{x_u f(x_l)}{f(x_l) - f(x_u)} - \frac{x_l f(x_u)}{f(x_l) - f(x_u)}$$

then adding and subtracting x_u on the right-hand side:

$$x_r = x_u + \frac{x_u f(x_l)}{f(x_l) - f(x_u)} - x_u - \frac{x_l f(x_u)}{f(x_l) - f(x_u)}$$

Collecting terms yields

$$x_r = x_u + \frac{x_u f(x_u)}{f(x_l) - f(x_u)} - \frac{x_l f(x_u)}{f(x_l) - f(x_u)}$$

or

$$x_r = x_u - \frac{f(x_u)(x_l - x_u)}{f(x_l) - f(x_u)}$$

which is the same as Eq. (5.7). We use this form because it involves one less function evaluation and one less multiplication than Eq. (B5.1.1). In addition, it is directly comparable with the secant method which will be discussed in Chap. 6.

EXAMPLE 5.6   False Position

Problem Statement.   Use the false-position method to determine the root of the same equation investigated in Example 5.1 [Eq. (E5.1.1)].

Solution.   As in Example 5.3, initiate the computation with guesses of $x_l = 12$ and $x_u = 16$.

First iteration:

$$x_l = 12 \qquad f(x_l) = 6.0699$$
$$x_l = 16 \qquad f(x_u) = -2.2688$$
$$x_r = 16 - \frac{-2.2688(12 - 16)}{6.0669 - (-2.2688)} = 14.9113$$

which has a true relative error of 0.89 percent.

Second iteration:

$$f(x_l)f(x_r) = -1.5426$$

Therefore, the root lies in the first subinterval, and x_r becomes the upper limit for the next iteration, $x_u = 14.9113$:

$$x_l = 12 \qquad\qquad f(x_l) = 6.0699$$
$$x_u = 14.9113 \qquad f(x_u) = -0.2543$$
$$x_r = 14.9113 - \frac{-0.2543(12 - 14.9113)}{6.0669 - (-0.2543)} = 14.7942$$

which has true and approximate relative errors of 0.09 and 0.79 percent. Additional iterations can be performed to refine the estimate of the roots.

A feeling for the relative efficiency of the bisection and false-position methods can be appreciated by referring to Fig. 5.14, where we have plotted the true percent relative errors for Examples 5.4 and 5.6. Note how the error for false position decreases much faster than for bisection because of the more efficient scheme for root location in the false-position method.

Recall in the bisection method that the interval between x_l and x_u grew smaller during the course of a computation. The interval, as defined by $\Delta x/2 = |x_u - x_l|/2$ for the first iteration, therefore provided a measure of the error for this approach. This is not the case for the method of false position because one of the initial guesses may stay fixed throughout the computation as the other guess converges on the root. For instance, in Example 5.6 the lower guess x_l remained at 12 while x_u, converged on the root. For such cases, the interval does not shrink but rather approaches a constant value.

Example 5.6 suggests that Eq. (5.2) represents a very conservative error criterion. In fact, Eq. (5.2) actually constitutes an approximation of the discrepancy of the previous

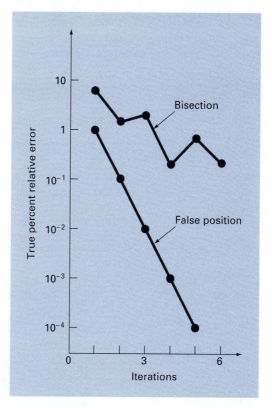

FIGURE 5.14
Comparison of the relative errors of the bisection and the false-position methods.

iteration. This is because for a case such as Example 5.6, where the method is converging quickly (for example, the error is being reduced nearly an order of magnitude per iteration), the root for the present iteration x_r^{new} is a much better estimate of the true value than the result of the previous iteration x_r^{old}. Thus, the quantity in the numerator of Eq. (5.2) actually represents the discrepancy of the previous iteration. Consequently, we are assured that satisfaction of Eq. (5.2) ensures that the root will be known with greater accuracy than the prescribed tolerance. However, as described in the next section, there are cases where false position converges slowly. For these cases, Eq. (5.2) becomes unreliable, and an alternative stopping criterion must be developed.

5.3.1 Pitfalls of the False-Position Method

Although the false-position method would seem to always be the bracketing method of preference, there are cases where it performs poorly. In fact, as in the following example, there are certain cases where bisection yields superior results.

EXAMPLE 5.7   A Case Where Bisection Is Preferable to False Position

Problem Statement.   Use bisection and false position to locate the root of

$$f(x) = x^{10} - 1$$

between $x = 0$ and 1.3.

Solution.   Using bisection, the results can be summarized as

Iteration	x_l	x_u	x_r	ε_a (%)	ε_t (%)
1	0	1.3	0.65	100.0	35
2	0.65	1.3	0.975	33.3	2.5
3	0.975	1.3	1.1375	14.3	13.8
4	0.975	1.1375	1.05625	7.7	5.6
5	0.975	1.05625	1.015625	4.0	1.6

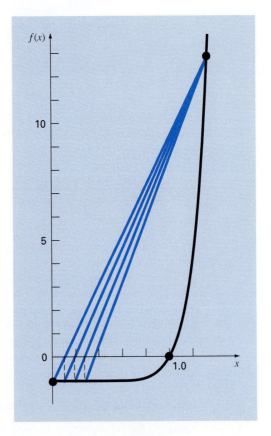

FIGURE 5.15
Plot of $f(x) = x^{10} - 1$, illustrating slow convergence of the false-position method.

Thus, after five iterations, the true error is reduced to less than 2 percent. For false position, a very different outcome is obtained:

Iteration	x_l	x_u	x_r	ε_a (%)	ε_t (%)
1	0	1.3	0.09430		90.6
2	0.09430	1.3	0.18176	48.1	81.8
3	0.18176	1.3	0.26287	30.9	73.7
4	0.26287	1.3	0.33811	22.3	66.2
5	0.33811	1.3	0.40788	17.1	59.2

After five iterations, the true error has only been reduced to about 59 percent. In addition, note that $\varepsilon_a < \varepsilon_t$. Thus, the approximate error is misleading. Insight into these results can be gained by examining a plot of the function. As in Fig. 5.15, the curve violates the premise upon which false position was based—that is, if $f(x_l)$ is much closer to zero than $f(x_u)$ is, then the root is closer to x_l than to x_u (recall Fig. 5.13). Because of the shape of the present function, the opposite is true.

The foregoing example illustrates that blanket generalizations regarding root-location methods are usually not possible. Although a method such as false position is often superior to bisection, there are invariably cases that violate this general conclusion. Therefore, in addition to using Eq. (5.2), the results should always be checked by substituting the root estimate into the original equation and determining whether the result is close to zero. Such a check should be incorporated into all computer programs for root location.

5.3.2 False-Position Algorithm

An algorithm for the false-position method can be developed directly from the bisection algorithm in Fig. 5.10. The only modification is to substitute Eq. (5.7). In addition, the zero check suggested in the last section should also be incorporated into the code.

An alternative version that minimizes function evaluation can also be patterned after Fig. 5.12. For this case, additional modifications are needed to evaluate and save the necessary function evaluation per iteration. This will be left as a homework exercise.

5.4   INCREMENTAL SEARCHES AND DETERMINING INITIAL GUESSES

Besides checking an individual answer, you must determine whether all possible roots have been located. As mentioned previously, a plot of the function is usually very useful in guiding you in this task. Another option is to incorporate an incremental search at the beginning of the computer program. This consists of starting at one end of the region of interest and then making function evaluations at small increments across the region. When the function changes sign, it is assumed that a root falls within the increment. The x values at the beginning and the end of the increment can then serve as the initial guesses for one of the bracketing techniques described in this chapter.

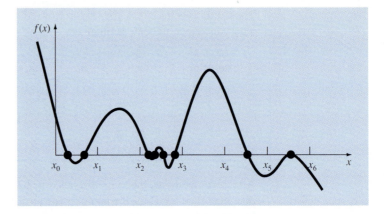

FIGURE 5.16
Cases where roots could be missed because the increment length of the search procedure is too large. Note that the last root on the right is multiple and would be missed regardless of increment length.

A potential problem with an incremental search is the choice of the increment length. If the length is too small, the search can be very time-consuming. On the other hand, if the length is too great, there is a possibility that closely spaced roots might be missed (Fig. 5.16). The problem is compounded by the possible existence of multiple roots. A partial remedy for such cases is to compute the first derivative of the function $f'(x)$ at the beginning and the end of each interval. If the derivative changes sign, it suggests that a minimum or maximum may have occurred and that the interval should be examined more closely for the existence of a possible root.

Although such modifications or the employment of a very fine increment can alleviate the problem, it should be clear that brute-force methods such as incremental search are not foolproof. You would be wise to supplement such automatic techniques with any other information that provides insight into the location of the roots. Such information can be found in plotting and in understanding the physical problem from which the equation originated.

PROBLEMS

5.1 Determine the real roots of $f(x) = -0.4x^2 + 2.2x + 4.7$:
(a) Graphically.
(b) Using the quadratic formula.
(c) Using three iterations of the bisection method to determine the highest root. Employ initial guesses of $x_l = 5$ and $x_u = 10$. Compute the estimated error ε_a and the true error ε_t after each iteration.

5.2 Determine the real roots of $f(x) = -2 + 7x - 5x^2 + 6x^3$:
(a) Graphically.

(b) Using bisection to locate the lowest root. Employ initial guesses of $x_l = 0$ and $x_u = 1$ and iterate until the estimated error ε_a falls below a level of $\varepsilon_s = 10\%$.

5.3 Determine the real roots of $f(x) = -26 + 82.3x - 88x^2 + 45.4x^3 - 9x^4 + 0.65x^5$:
(a) Graphically.
(b) Using bisection to determine the highest root to $\varepsilon_s = 10\%$. Employ initial guesses of $x_l = 0.5$ and $x_u = 1.0$.

(c) Perform the same computation as in (b) but use the false-position method and $\varepsilon_s = 0.1\%$.

5.4 Determine the real roots of $f(x) = -11 - 22x + 17x^2 - 2.5x^3$:

(a) Graphically.

(b) Using the false-position method with a value of ε_s corresponding to three significant figures to determine the lowest root.

5.5 Locate the first nontrivial root of $\sin x = x^2$, where x is in radians. Use a graphical technique and bisection with the initial interval from 0.5 to 1. Perform the computation until ε_a is less than $\varepsilon_s = 2\%$. Also perform an error check by substituting your final answer into the original equation.

5.6 Determine the real root of $\ln x^2 = 0.7$:

(a) Graphically.

(b) Using three iterations of the bisection method, with initial guesses of $x_l = 0.5$ and $x_u = 2$.

(c) Using three iterations of the false-position method, with the same initial guesses as in (b).

5.7 Determine the real root of $f(x) = (0.9 - 0.4x)/x$:

(a) Analytically.

(b) Graphically.

(c) Using three iterations of the false-position method and initial guesses of 1 and 3. Compute the approximate error ε_a and the true error ε_t after each iteration.

5.8 Find the positive square root of 15 using the false-position method to within $\varepsilon_s = 0.5\%$. Employ initial guesses of $x_l = 3$ and $x_u = 4$.

5.9 Find the smallest positive root of the function (x is in radians) $x^2 \left| \sin \sqrt{x} \right| = 5$ using the false-position method. To locate the region in which the root lies, first plot this function for values of x between 0 and 5. Perform the computation until ε_a falls below $\varepsilon_s = 1\%$. Check your final answer by substituting it into the original function.

5.10 Find the positive real root of $f(x) = x^4 - 8x^3 - 36x^2 + 462x - 1010$ using the false-position method. Use a plot to make your initial guess, and perform the computation to within $\varepsilon_s = 1.0\%$.

5.11 Determine the real root of $x^{3.3} = 79$:

(a) Analytically.

(b) With the false-position method to within $\varepsilon_s = 0.1\%$. Use initial guesses of 3.0 and 4.0.

5.12 The velocity v of a falling parachutist is given by

$$v = \frac{gm}{c}\left(1 - e^{-(c/m)t}\right)$$

where $g = 9.8$. For a parachutist with a drag coefficient $c = 14$ kg/s, compute the mass m so that the velocity is $v = 35$ m/s at $t = 7$ s. Use the false-position method to determine m to a level of $\varepsilon_s = 0.1\%$.

5.13 The saturation concentration of dissolved oxygen in freshwater can be calculated with the equation (APHA, 1992)

$$\ln o_{sf} = -139.34411 + \frac{1.575701 \times 10^5}{T_a}$$
$$- \frac{6.642308 \times 10^7}{T_a^2} + \frac{1.243800 \times 10^{10}}{T_a^3}$$
$$- \frac{8.621949 \times 10^{11}}{T_a^4}$$

where o_{sf} = the saturation concentration of dissolved oxygen in freshwater at 1 atm (mg/L) and T_a = absolute temperature (K). Remember that $T_a = T + 273.15$, where T = temperature (°C). According to this equation, saturation decreases with increasing temperature. For typical natural waters in temperate climates, the equation can be used to determine that oxygen concentration ranges from 14.621 mg/L at 0°C to 6.949 mg/L at 35°C. Given a value of oxygen concentration, this formula and the bisection method can be used to solve for temperature in °C.

(a) If the initial guesses are set as 0 and 35°C, how many bisection iterations would be required to determine temperature to an absolute error of 0.05°C?

(b) Based on (a), develop and test a bisection program to determine T as a function of a given oxygen concentration. Test your program for $o_{sf} = 8$, 10, and 14 mg/L. Check your results.

5.14 Integrate the algorithm outlined in Fig. 5.10 into a complete, user-friendly bisection subprogram. Among other things:

(a) Place documentation statements throughout the subprogram to identify what each section is intended to accomplish.

(b) Label the input and output.

(c) Add an answer check that substitutes the root estimate into the original function to verify whether the final result is close to zero.

(d) Test the subprogram by duplicating the computations from Example 5.3.

5.15 Use the subprogram you developed in Prob. 5.14 to repeat Probs. 5.1 through 5.6.

5.16 Repeat Prob. 5.15 except use the TOOLKIT software available with the text. Use the plotting capabilities of this program to verify your results.

5.17 Use the TOOLKIT software to find the real roots of two third-order-polynomial functions of your choice. Plot each function over a range you specify to obtain upper and lower bounds on the roots.

5.18 Repeat Prob. 5.17 except use two transcendental functions of your choice.

5.19 This problem uses only the graphics capability of the TOOLKIT software available with the text. The software plots the function over smaller and smaller intervals to increase the number of significant figures to which a root can be estimated. Start with $f(x) = e^{-x} \cos 8x$. Plot the function with a full-scale range of $x = 0$ to 2.5. Estimate the lowest root in this interval. Plot the function again with $x = 0$ to 0.5. Estimate the root. Finally, plot the function over a range of 0.1 to 0.2. This permits you to find the root to two significant figures and estimate a third.

5.20 Develop a user-friendly subprogram for the false-position method based on Sec. 5.3.2 and Fig. 5.10. Test the program by duplicating Example 5.6.

5.21 Use the subprogram you developed in Prob. 5.20 to duplicate the computation from Example 5.7. Perform a number of runs until the true percent relative error falls below 0.01%. Plot the true and approximate percent relative errors versus number of iterations on semilog paper. Interpret your results.

5.22 Develop a subprogram for the false-position method that minimizes function evaluations in a fashion similar to Fig. 5.11. Determine the number n of function evaluations per total iterations. Test the program by duplicating Example 5.6.

CHAPTER 6
Open Methods

For the bracketing methods in the previous chapter, the root is located within an interval prescribed by a lower and an upper bound. Repeated application of these methods always results in closer estimates of the true value of the root. Such methods are said to be *convergent* because they move closer to the truth as the computation progresses (Fig. 6.1*a*).

In contrast, the *open methods* described in this chapter are based on formulas that require only a single starting value of *x* or two starting values that do not necessarily bracket

FIGURE 6.1
Graphical depiction of the fundamental difference between the (*a*) bracketing and (*b*) and (*c*) open methods for root location. In (*a*), which is the bisection method, the root is constrained within the interval prescribed by x_l and x_u. In contrast, for the open method depicted in (*b*) and (*c*), a formula is used to project from x_i to x_{i+1} in an iterative fashion. Thus, the method can either (*b*) diverge or (*c*) converge rapidly, depending on the value of the initial guess.

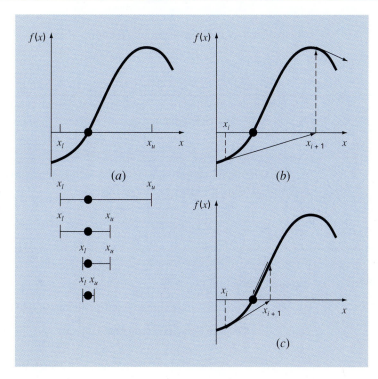

the root. As such, they sometimes *diverge* or move away from the true root as the computation progresses (Fig. 6.1*b*). However, when the open methods converge (Fig. 6.1*c*) they usually do so much more quickly than the bracketing methods. We will begin our discussion of open techniques with a simple version that is useful for illustrating their general form and also for demonstrating the concept of convergence.

6.1   SIMPLE FIXED-POINT ITERATION

As mentioned above, open methods employ a formula to predict the root. Such a formula can be developed for simple *fixed-point iteration* (or, as it is also called, one-point iteration, or successive substitution) by rearranging the function $f(x) = 0$ so that x is on the left-hand side of the equation:

$$x = g(x) \tag{6.1}$$

This transformation can be accomplished either by algebraic manipulation or by simply adding x to both sides of the original equation. For example,

$$x^2 - 2x + 3 = 0$$

can be simply manipulated to yield

$$x = \frac{x^2 + 3}{2}$$

whereas $\sin x = 0$ could be put into the form of Eq. (6.1) by adding x to both sides to yield

$$x = \sin x + x$$

The utility of Eq. (6.1) is that it provides a formula to predict a new value of x as a function of an old value of x. Thus, given an initial guess at the root x_i, Eq. (6.1) can be used to compute a new estimate x_{i+1} as expressed by the iterative formula

$$x_{i+1} = g(x_i) \tag{6.2}$$

As with other iterative formulas in this book, the approximate error for this equation can be determined using the error estimator [Eq. (3.5)]:

$$\varepsilon_a = \left| \frac{x_{i+1} - x_i}{x_{i+1}} \right| 100\%$$

EXAMPLE 6.1   Simple Fixed-Point Iteration

Problem Statement.   Use simple fixed-point iteration to locate the root of $f(x) = e^{-x} - x$.

Solution.   The function can be separated directly and expressed in the form of Eq. (6.2) as

$$x_{i+1} = e^{-x_i}$$

Starting with an initial guess of $x_0 = 0$, this iterative equation can be applied to compute

i	x_i	ε_a (%)	ε_t (%)
0	0		100.0
1	1.000000	100.0	76.3
2	0.367879	171.8	35.1
3	0.692201	46.9	22.1
4	0.500473	38.3	11.8
5	0.606244	17.4	6.89
6	0.545396	11.2	3.83
7	0.579612	5.90	2.20
8	0.560115	3.48	1.24
9	0.571143	1.93	0.705
10	0.564879	1.11	0.399

Thus, each iteration brings the estimate closer to the true value of the root: 0.56714329.

6.1.1 Convergence

Notice that the true percent relative error for each iteration of Example 6.1 is roughly proportional (by a factor of about 0.5 to 0.6) to the error from the previous iteration. This property, called *linear convergence,* is characteristic of fixed-point iteration.

Aside from the "rate" of convergence, we must comment at this point about the "possibility" of convergence. The concepts of convergence and divergence can be depicted graphically. Recall that in Sec. 5.1, we graphed a function to visualize its structure and behavior (Example 5.1). Such an approach is employed in Fig. 6.2a for the function $f(x) = e^{-x} - x.$ An alternative graphical approach is to separate the equation into two component parts, as in

$$f_1(x) = f_2(x)$$

Then the two equations

$$y_1 = f_1(x) \tag{6.3}$$

and

$$y_2 = f_2(x) \tag{6.4}$$

can be plotted separately (Fig. 6.2b). The x values corresponding to the intersections of these functions represent the roots of $f(x) = 0$.

EXAMPLE 6.2   The Two-Curve Graphical Method

Problem Statement.   Separate the equation $e^{-x} - x = 0$ into two parts and determine its root graphically.

Solution.   Reformulate the equation as $y_1 = x$ and $y_2 = e^{-x}$. The following values can be computed:

x	y₁	y₂
0.0	0.0	1.000
0.2	0.2	0.819
0.4	0.4	0.670
0.6	0.6	0.549
0.8	0.8	0.449
1.0	1.0	0.368

These points are plotted in Fig. 6.2b. The intersection of the two curves indicates a root estimate of approximately $x = 0.57$, which corresponds to the point where the single curve in Fig. 6.2a crosses the x axis.

FIGURE 6.2
Two alternative graphical methods for determining the root of $f(x) = e^{-x} - x$. (a) Root at the point where it crosses the x axis; (b) root at the intersection of the component functions.

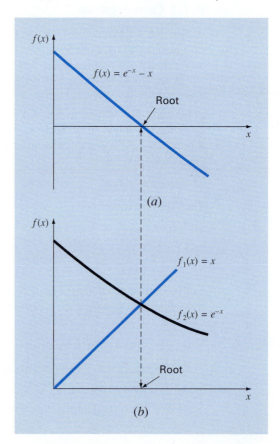

The two-curve method can now be used to illustrate the convergence and divergence of fixed-point iteration. First, Eq. (6.1) can be re-expressed as a pair of equations $y_1 = x$ and $y_2 = g(x)$. These two equations can then be plotted separately. As was the case with Eqs. (6.3) and (6.4), the roots of $f(x) = 0$ correspond to the abscissa value at the intersection of the two curves. The function $y_1 = x$ and four different shapes for $y_2 = g(x)$ are plotted in Fig. 6.3.

For the first case (Fig. 6.3a), the initial guess of x_0 is used to determine the corresponding point on the y_2 curve $[x_0, g(x_0)]$. The point (x_1, x_1) is located by moving left horizontally to the y_1 curve. These movements are equivalent to the first iteration in the fixed-point method:

$$x_1 = g(x_0)$$

Thus, in both the equation and in the plot, a starting value of x_0 is used to obtain an estimate of x_1. The next iteration consists of moving to $[x_1, g(x_1)]$ and then to (x_2, x_2). This iteration

FIGURE 6.3

Graphical depiction of (a) and (b) convergence and (c) and (d) divergence of simple fixed-point iteration. Graphs (a) and (c) are called monotone patterns, whereas (b) and (d) are called oscillating or spiral patterns. Note that convergence occurs when $|g'(x)| < 1$.

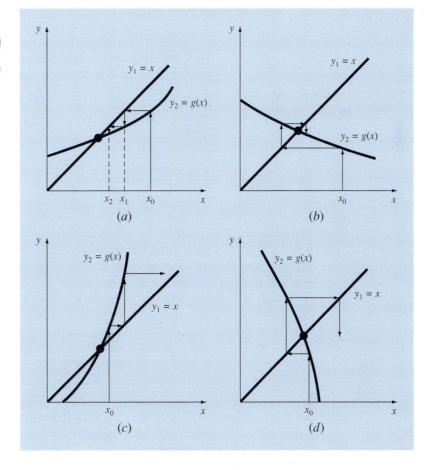

Box 6.1 Convergence of Fixed-Point Iteration

From studying Fig. 6.3 it should be clear that fixed-point iteration converges if, in the region of interest, $|g'(x)| < 1$. In other words, convergence occurs if the magnitude of the slope of $g(x)$ is less than the slope of the line $f(x) = x$. This observation can be demonstrated theoretically. Recall that the iterative equation is

$$x_{i+1} = g(x_i)$$

Suppose that the true solution is

$$x_r = g(x_r)$$

Subtracting these equations yields

$$x_r - x_{i+1} = g(x_r) - g(x_i) \qquad \text{(B6.1.1)}$$

The *derivative mean-value theorem* (recall Sec. 4.1.1) states that if a function $g(x)$ and its first derivative are continuous over an interval $a \leq x \leq b$, then there exists at least one value of $x = \xi$ within the interval such that

$$g'(\xi) = \frac{g(b) - g(a)}{b - a} \qquad \text{(B6.1.2)}$$

The right-hand side of this equation is the slope of the line joining $g(a)$ and $g(b)$. Thus, the mean-value theorem states that there is at least one point between a and b that has a slope, designated by $g'(\xi)$, which is parallel to the line joining $g(a)$ and $g(b)$ (recall Fig. 4.3).

Now, if we let $a = x_i$ and $b = x_r$, the right-hand side of Eq. (B6.1.1) can be expressed as

$$g(x_r) - g(x_i) = (x_r - x_i)g'(\xi)$$

where ξ is somewhere between x_i and x_r. This result can then be substituted into Eq. (B6.1.1) to yield

$$x_r - x_{i+1} = (x_r - x_i)g'(\xi) \qquad \text{(B6.1.3)}$$

If the true error for iteration i is defined as

$$E_{t,i} = x_r - x_i$$

then Eq. (B6.1.3) becomes

$$E_{t,i+1} = g'(\xi)E_{t,i}$$

Consequently, if $|g'(x)| < 1$, the errors decrease with each iteration. For $|g'(x)| > 1$, the errors grow. Notice also that if the derivative is positive, the errors will be positive, and hence, the iterative solution will be monotonic (Fig. 6.3a and c). If the derivative is negative, the errors will oscillate (Fig. 6.3b and d).

An offshoot of the analysis is that it also demonstrates that when the method converges, the error is roughly proportional to and less than the error of the previous step. For this reason, simple fixed-point iteration is said to be *linearly convergent*.

is equivalent to the equation

$$x_2 = g(x_1)$$

The solution in Fig. 6.3a is *convergent* because the estimates of x move closer to the root with each iteration. The same is true for Fig. 6.3b. However, this is not the case for Fig. 6.3c and d, where the iterations diverge from the root. Notice that convergence seems to occur only when the absolute value of the slope of $y_2 = g(x)$ is less than the slope of $y_1 = x$, that is, when $|g'(x)| < 1$. Box 6.1 provides a theoretical derivation of this result.

6.1.2 Algorithm for Fixed-Point Iteration

The computer algorithm for fixed-point iteration is extremely simple. It consists of a loop to iteratively compute new estimates until the termination criterion has been met. Figure 6.4 presents pseudocode for the algorithm. Other open methods can be programmed in a similar way, the major modification being to change the iterative formula that is used to compute the new root estimate.

```
FUNCTION Onept (x0, es, imax, iter, ea)
   xr = x0
   iter = 0
   DO
      xrold = xr
      xr = g(xrold)
      iter = iter + 1
      IF xr ≠ 0 THEN
```

$$ea = \left| \frac{xr - xrold}{xr} \right| \cdot 100$$

```
      END IF
      IF ea < es OR iter ≥ imax EXIT
   END DO
   Onept = xr
END OnePt
```

FIGURE 6.4
Pseudocode for fixed-point iteration. Note that other open methods can be cast in this general format.

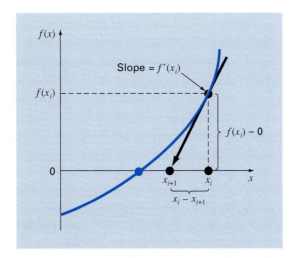

FIGURE 6.5
Graphical depiction of the Newton-Raphson method. A tangent to the function of x_i [that is, $f'(x_i)$] is extrapolated down to the x axis to provide an estimate of the root at x_{i+1}.

6.2   THE NEWTON-RAPHSON METHOD

Perhaps the most widely used of all root-locating formulas is the Newton-Raphson equation (Fig. 6.5). If the initial guess at the root is x_i, a tangent can be extended from the point $[x_i, f(x_i)]$. The point where this tangent crosses the x axis usually represents an improved estimate of the root.

The Newton-Raphson method can be derived on the basis of this geometrical interpretation (an alternative method based on the Taylor series is described in Box 6.2). As in Fig. 6.5, the first derivative at x is equivalent to the slope:

$$f'(x_i) = \frac{f(x_i) - 0}{x_i - x_{i+1}} \tag{6.5}$$

which can be rearranged to yield

$$x_{i+1} = x_i - \frac{f(x_i)}{f'(x_i)} \tag{6.6}$$

which is called the *Newton-Raphson formula*.

EXAMPLE 6.3   Newton-Raphson Method

Problem Statement.   Use the Newton-Raphson method to estimate the root of $f(x) = e^{-x} - x$ employing an initial guess of $x_0 = 0$.

Solution.   The first derivative of the function can be evaluated as

$$f'(x) = -e^{-x} - 1$$

which can be substituted along with the original function into Eq. (6.6) to give

$$x_{i+1} = x_i - \frac{e^{-x_i} - x_i}{-e^{-x_i} - 1}$$

Starting with an initial guess of $x_0 = 0$, this iterative equation can be applied to compute

i	x_i	ε_t (%)
0	0	100
1	0.500000000	11.8
2	0.566311003	0.147
3	0.567143165	0.0000220
4	0.567143290	$< 10^{-8}$

Thus, the approach rapidly converges on the true root. Notice that the true percent relative error at each iteration decreases much faster than it does in simple fixed-point iteration (compare with Example 6.1).

6.2.1 Termination Criteria and Error Estimates

As with other root-location methods, Eq. (3.5) can be used as a termination criterion. In addition, however, the Taylor series derivation of the method (Box 6.2) provides theoretical insight regarding the rate of convergence as expressed by $E_{i+1} = O(E_i^2)$. Thus the error should be roughly proportional to the square of the previous error. In other words, the

Box 6.2 Derivation and Error Analysis of the Newton-Raphson Method

Aside from the geometric derivation [Eqs. (6.5) and (6.6)], the Newton-Raphson method may also be developed from the Taylor series expansion. This alternative derivation is useful in that it also provides insight into the rate of convergence of the method.

Recall from Chap. 4 that the Taylor series expansion can be represented as

$$f(x_{i+1}) = f(x_i) + f'(x_i)(x_{i+1} - x_i) \\ + \frac{f''(\xi)}{2!}(x_{i+1} - x_i)^2 \tag{B6.2.1}$$

where ξ lies somewhere in the interval from x_i to x_{i+1}. An approximate version is obtainable by truncating the series after the first derivative term:

$$f(x_{i+1}) \cong f(x_i) + f'(x_i)(x_{i+1} - x_i)$$

At the intersection with the x axis, $f(x_{i+1})$ would be equal to zero, or

$$0 = f(x_i) + f'(x_i)(x_{i+1} - x_i) \tag{B6.2.2}$$

which can be solved for

$$x_{i+1} = x_i - \frac{f(x_i)}{f'(x_i)}$$

which is identical to Eq. (6.6). Thus, we have derived the Newton-Raphson formula using a Taylor series.

Aside from the derivation, the Taylor series can also be used to estimate the error of the formula. This can be done by realizing that if the complete Taylor series were employed, an exact result would be obtained. For this situation $x_{i+1} = x_r$, where x is the true value of the root. Substituting this value along with $f(x_r) = 0$ into Eq. (B6.2.1) yields

$$0 = f(x_i) + f'(x_i)(x_r - x_i) + \frac{f''(\xi)}{2!}(x_r - x_i)^2 \tag{B6.2.3}$$

Equation (B6.2.2) can be subtracted from Eq. (B6.2.3) to give

$$0 = f'(x_i)(x_r - x_{i+1}) + \frac{f''(\xi)}{2!}(x_r - x_i)^2 \tag{B6.2.4}$$

Now, realize that the error is equal to the discrepancy between x_{i+1} and the true value x_r, as in

$$E_{t,i+1} = x_r - x_{i+1}$$

and Eq. (B6.2.4) can be expressed as

$$0 = f'(x_i)E_{t,i+1} + \frac{f''(\xi)}{2!}E_{t,i}^2 \tag{B6.2.5}$$

If we assume convergence, both x_i and ξ should eventually be approximated by the root x_r, and Eq. (B6.2.5) can be rearranged to yield

$$E_{t,i+1} = \frac{-f''(x_r)}{2f'(x_r)}E_{t,i}^2 \tag{B6.2.6}$$

According to Eq. (B6.2.6), the error is roughly proportional to the square of the previous error. This means that the number of correct decimal places approximately doubles with each iteration. Such behavior is referred to as *quadratic convergence*. Example 6.4 manifests this property.

number of significant figures of accuracy approximately doubles with each iteration. This behavior is examined in the following example.

EXAMPLE 6.4   Error Analysis of Newton-Raphson Method

Problem Statement.   As derived in Box 6.2, the Newton-Raphson method is quadratically convergent. That is, the error is roughly proportional to the square of the previous error, as in

$$E_{t,i+1} \cong \frac{-f''(x_r)}{2f'(x_r)}E_{t,i}^2 \tag{E6.4.1}$$

Examine this formula and see if it applies to the results of Example 6.3.

Solution.   The first derivative of $f(x) = e^{-x} - x$ is

$$f'(x) = -e^{-x} - 1$$

which can be evaluated at $x_r = 0.56714329$ as $f'(0.56714329) = -1.56714329$. The second derivative is

$$f''(x) = e^{-x}$$

which can be evaluated as $f''(0.56714329) = 0.56714329$. These results can be substituted into Eq. (E6.4.1) to yield

$$E_{t,i+1} \cong -\frac{0.56714329}{2(-1.56714329)} E_{t,i}^2 = 0.18095 E_{t,i}^2$$

From Example 6.3, the initial error was $E_{t,0} = 0.56714329$, which can be substituted into the error equation to predict

$$E_{t,1} \cong 0.18095(0.56714329)^2 = 0.0582$$

which is close to the true error of 0.06714329. For the next iteration,

$$E_{t,2} \cong 0.18095(0.06714329)^2 = 0.0008158$$

which also compares favorably with the true error of 0.0008323. For the third iteration,

$$E_{t,3} \cong 0.18095(0.0008323)^2 = 0.000000125$$

which is the error obtained in Example 6.3. The error estimate improves in this manner because, as we come closer to the root, x and ξ are better approximated by x_r [recall our assumption in going from Eq. (B6.2.5) to Eq. (B6.2.6) in Box 6.2]. Finally,

$$E_{t,4} \cong 0.18095(0.000000125)^2 = 2.83 \times 10^{-15}$$

Thus, this example illustrates that the error of the Newton-Raphson method for this case is, in fact, roughly proportional (by a factor of 0.18095) to the square of the error of the previous iteration.

6.2.2 Pitfalls of the Newton-Raphson Method

Although the Newton-Raphson method is often very efficient, there are situations where it performs poorly. A special case—multiple roots—will be addressed later in this chapter. However, even when dealing with simple roots, difficulties can also arise, as in the following example.

EXAMPLE 6.5   Example of a Slowly Converging Function with Newton-Raphson

Problem Statement.   Determine the positive root of $f(x) = x^{10} - 1$ using the Newton-Raphson method and an initial guess of $x = 0.5$.

Solution.   The Newton-Raphson formula for this case is

$$x_{i+1} = x_i - \frac{x_i^{10} - 1}{10 x_i^9}$$

which can be used to compute

Iteration	x
0	0.5
1	51.65
2	46.485
3	41.8365
4	37.65285
5	33.887565
.	
.	
.	
∞	1.0000000

Thus, after the first poor prediction, the technique is converging on the true root of 1, but at a very slow rate.

Aside from slow convergence due to the nature of the function, other difficulties can arise, as illustrated in Fig. 6.6. For example, Fig. 6.6a depicts the case where an inflection point [that is, $f''(x) = 0$] occurs in the vicinity of a root. Notice that iterations beginning at x_0 progressively diverge from the root. Figure 6.6b illustrates the tendency of the Newton-Raphson technique to oscillate around a local maximum or minimum. Such oscillations may persist, or as in Fig. 6.6b, a near-zero slope is reached, whereupon the solution is sent far from the area of interest. Figure 6.6c shows how an initial guess that is close to one root can jump to a location several roots away. This tendency to move away from the area of interest is because near-zero slopes are encountered. Obviously, a zero slope [$f'(x) = 0$] is truly a disaster because it causes division by zero in the Newton-Raphson formula [Eq. (6.6)]. Graphically (see Fig 6.6d), it means that the solution shoots off horizontally and never hits the x axis.

Thus, there is no general convergence criterion for Newton-Raphson. Its convergence depends on the nature of the function and on the accuracy of the initial guess. The only remedy is to have an initial guess that is "sufficiently" close to the root. And for some functions, no guess will work! Good guesses are usually predicated on knowledge of the physical problem setting or on devices such as graphs that provide insight into the behavior of the solution. The lack of a general convergence criterion also suggests that good computer software should be designed to recognize slow convergence or divergence. The next section addresses some of these issues.

6.2.3 Algorithm for Newton-Raphson

An algorithm for the Newton-Raphson method is readily obtained by substituting Eq. (6.6) for the predictive formula [Eq. (6.2)] in Fig. 6.4. Note, however, that the program must also be modified to compute the first derivative. This can be simply accomplished by the inclusion of a user-defined function.

Additionally, in light of the foregoing discussion of potential problems of the Newton-Raphson method, the program would be improved by incorporating several additional features:

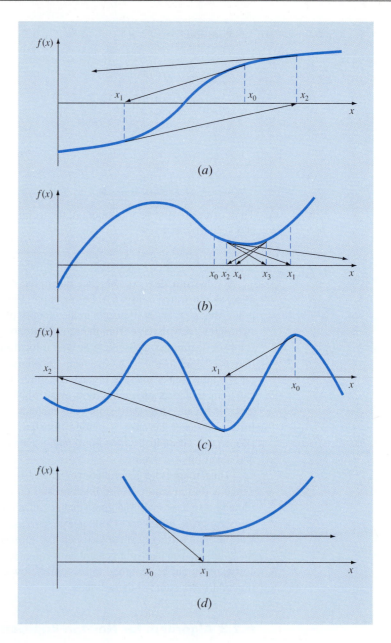

FIGURE 6.6
Four cases where the Newton-Raphson method exhibits poor convergence.

1. A plotting routine should be included in the program.
2. At the end of the computation, the final root estimate should always be substituted into the original function to compute whether the result is close to zero. This check partially guards against those cases where slow or oscillating convergence may lead to a small value of ε_a while the solution is still far from a root.
3. The program should always include an upper limit on the number of iterations to guard against oscillating, slowly convergent, or divergent solutions that could persist interminably.
4. The program should alert the user and take account of the possibility that $f'(x)$ might equal zero at any time during the computation.

6.3   THE SECANT METHOD

A potential problem in implementing the Newton-Raphson method is the evaluation of the derivative. Although this is not inconvenient for polynomials and many other functions, there are certain functions whose derivatives may be extremely difficult or inconvenient to evaluate. For these cases, the derivative can be approximated by a backward finite divided difference, as in (Fig. 6.7)

$$f'(x_i) \cong \frac{f(x_{i-1}) - f(x_i)}{x_{i-1} - x_i}$$

FIGURE 6.7

Graphical depiction of the secant method. This technique is similar to the Newton-Raphson technique (Fig. 6.5) in the sense that an estimate of the root is predicted by extrapolating a tangent of the function to the x axis. However, the secant method uses a difference rather than a derivative to estimate the slope.

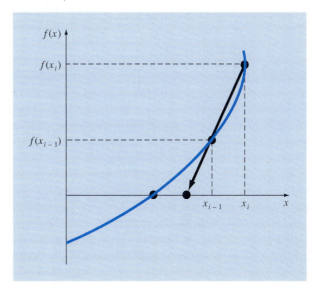

This approximation can be substituted into Eq. (6.6) to yield the following iterative equation:

$$x_{i+1} = x_i - \frac{f(x_i)(x_{i-1} - x_i)}{f(x_{i-1}) - f(x_i)} \tag{6.7}$$

Equation (6.7) is the formula for the *secant method*. Notice that the approach requires two initial estimates of x. However, because $f(x)$ is not required to change signs between the estimates, it is not classified as a bracketing method.

EXAMPLE 6.6 The Secant Method

Problem Statement. Use the secant method to estimate the root of $f(x) = e^{-x} - x$. Start with initial estimates of $x_{-1} = 0$ and $x_0 = 1.0$.

Solution. Recall that the true root is 0.56714329. . . .

First iteration:

$$x_{-1} = 0 \qquad f(x_{-1}) = 1.00000$$
$$x_0 = 1 \qquad f(x_0) \quad = -0.63212$$
$$x_1 = 1 - \frac{-0.63212(0 - 1)}{1 - (-0.63212)} = 0.61270 \qquad \varepsilon_t = 8.0\%$$

Second iteration:

$$x_0 = 1 \qquad\qquad f(x_0) = -0.63212$$
$$x_1 = 0.61270 \qquad f(x_1) = -0.07081$$

(Note that both estimates are now on the same side of the root.)

$$x_2 = 0.61270 - \frac{-0.07081(1 - 0.61270)}{-0.63212 - (-0.07081)} = 0.56384 \qquad \varepsilon_t = 0.58\%$$

Third iteration:

$$x_1 = 0.61270 \qquad f(x_1) = -0.07081$$
$$x_2 = 0.56384 \qquad f(x_2) = 0.00518$$
$$x_3 = 0.56384 - \frac{0.00518(0.61270 - 0.56384)}{-0.07081 - (-0.00518)} = 0.56717 \qquad \varepsilon_t = 0.0048\%$$

6.3.1 The Difference Between the Secant and False-Position Methods

Note the similarity between the secant method and the false-position method. For example, Eqs. (6.7) and (5.7) are identical on a term-by-term basis. Both use two initial estimates to compute an approximation of the slope of the function that is used to project to the x axis for a new estimate of the root. However, a critical difference between the methods is how one of the initial values is replaced by the new estimate. Recall that in the false-position method the latest estimate of the root replaces whichever of the original values yielded a function value with the same sign as $f(x_r)$. Consequently, the two estimates always bracket the root. Therefore, for all practical purposes, the method always converges because the root is kept within the bracket. In contrast, the secant method replaces the values in strict sequence, with the new value x_{i+1} replacing x_i and x_i replacing x_{i-1}. As a result, the two values can sometimes lie on the same side of the root. For certain cases, this can lead to divergence.

EXAMPLE 6.7   Comparison of Convergence of the Secant and False-Position Techniques

Problem Statement.   Use the false-position and secant methods to estimate the root of $f(x) = \ln x$. Start the computation with values of $x_l = x_{i-1} = 0.5$ and $x_u = x_i = 5.0$.

FIGURE 6.8
Comparison of the false-position and the secant methods. The first iterations (a) and (b) for both techniques are identical. However, for the second iterations (c) and (d), the points used differ. As a consequence, the secant method can diverge, as indicated in (d).

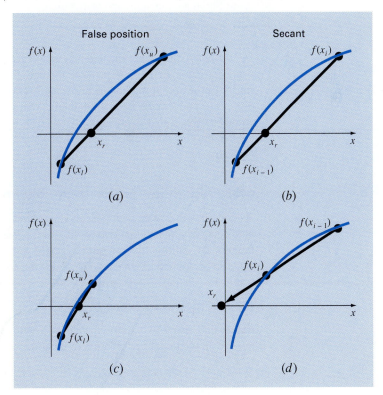

Solution.   For the false-position method the use of Eq. (5.7) and the bracketing criterion for replacing estimates results in the following iterations:

Iteration	x_l	x_u	x_r
1	0.5	5.0	1.8546
2	0.5	1.8546	1.2163
3	0.5	1.2163	1.0585

As can be seen (Fig. 6.8a and c), the estimates are converging on the true root which is equal to 1.

For the secant method, using Eq. (6.7) and the sequential criterion for replacing estimates results in

Iteration	x_{i-1}	x_i	x_{i+1}
1	0.5	5.0	1.8546
2	5.0	1.8546	−0.10438

As in Fig. 6.8d, the approach is divergent.

Although the secant method may be divergent, when it converges it usually does so at a quicker rate than the false-position method. For instance, Fig. 6.9 demonstrates the superiority of the secant method in this regard. The inferiority of the false-position method is due to one end staying fixed to maintain the bracketing of the root. This property, which is an advantage in that it prevents divergence, is a shortcoming with regard to the rate of convergence; it makes the finite-difference estimate a less-accurate approximation of the derivative.

6.3.2 Algorithm for the Secant Method

As with the other open methods, an algorithm for the secant method is obtained simply by modifying Fig. 6.4 so that two initial guesses are input and by using Eq. (6.7) to calculate the root. In addition, the options suggested in Sec. 6.2.3 for the Newton-Raphson method can also be applied to good advantage for the secant program.

6.3.3 Modified Secant Method

Rather than using two arbitrary values to estimate the derivative, an alternative approach involves a fractional perturbation of the independent variable to estimate $f'(x)$,

$$f'(x_i) \cong \frac{f(x_i + \delta x_i) - f(x_i)}{\delta x_i}$$

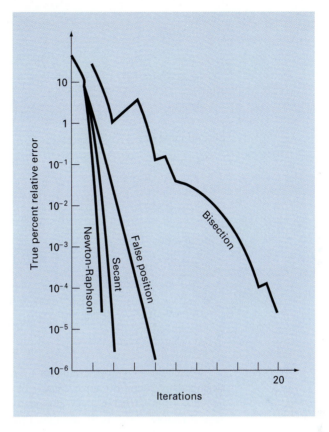

FIGURE 6.9
Comparison of the true percent relative errors ε_t for the methods to determine the roots of $f(x) = e^{-x} - x$.

where $\delta = $ a small perturbation fraction. This approximation can be substituted into Eq. (6.6) to yield the following iterative equation:

$$x_{i+1} = x_i - \frac{\delta x_i\, f(x_i)}{f(x_i + \delta x_i) - f(x_i)} \qquad (6.8)$$

EXAMPLE 6.8   Modified Secant Method

Problem Statement.   Use the modified secant method to estimate the root of $f(x) = e^{-x} - x$. Use a value of 0.01 for δ and start with $x_0 = 1.0$. Recall that the true root is 0.56714329. . . .

Solution.
First iteration:

$$x_0 = 1 \qquad\qquad\qquad f(x_0) = -0.63212$$

$$x_0 + \delta x_0 = 1.01 \qquad f(x_0 + \delta x_0) = -0.64578$$

$$x_1 = 1 - \frac{0.01(-0.63212)}{-0.64578 - (-0.63212)} = 0.537263 \qquad |\varepsilon_t| = 5.3\%$$

Second iteration:

$$x_0 = 0.537263 \qquad\qquad\qquad f(x_0) = 0.047083$$

$$x_0 + \delta x_0 = 0.542635 \qquad f(x_0 + \delta x_0) = 0.038579$$

$$x_1 = 0.537263 - \frac{0.005373(0.047083)}{0.038579 - 0.047083} = 0.56701 \qquad |\varepsilon_t| = 0.0236\%$$

Third iteration:

$$x_0 = 0.56701 \qquad\qquad\qquad f(x_0) = 0.000209$$

$$x_0 + \delta x_0 = 0.567143 \qquad f(x_0 + \delta x_0) = -0.00867$$

$$x_1 = 0.56701 - \frac{0.00567(0.000209)}{-0.00867 - 0.000209} = 0.567143 \qquad |\varepsilon_t| = 2.365 \times 10^{-5}\%$$

The choice of a proper value for δ is not automatic. If δ is too small, the method can be swamped by round-off error caused by subtractive cancellation in the denominator of Eq. (6.8). If it is too big, the technique can become inefficient and even divergent. However, if chosen correctly, it provides a nice alternative for cases where evaluating the derivative is difficult and developing two initial guesses is inconvenient.

6.4   MULTIPLE ROOTS

A *multiple root* corresponds to a point where a function is tangent to the x axis. For example, a double root results from

$$f(x) = (x - 3)(x - 1)(x - 1) \tag{6.9}$$

or, multiplying terms, $f(x) = x^3 - 5x^2 + 7x - 3$. The equation has a *double root* because one value of x makes two terms in Eq. (6.9) equal to zero. Graphically, this corresponds to the curve touching the x axis tangentially at the double root. Examine Fig. 6.10a at $x = 1$. Notice that the function touches the axis but does not cross it at the root.

A *triple root* corresponds to the case where one x value makes three terms in an equation equal to zero, as in

$$f(x) = (x - 3)(x - 1)(x - 1)(x - 1)$$

or, multiplying terms, $f(x) = x^4 - 6x^3 + 12x^2 - 10x + 3$. Notice that the graphical depiction (Fig. 6.10b) again indicates that the function is tangent to the axis at the root, but that for this case the axis is crossed. In general, odd multiple roots cross the axis, whereas even ones do not. For example, the quadruple root in Fig. 6.10c does not cross the axis.

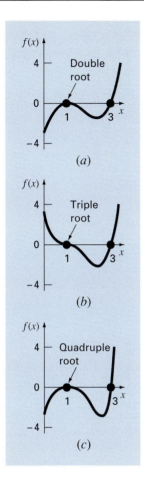

FIGURE 6.10
Examples of multiple roots that are tangential to the x axis. Notice that the function does not cross the axis on either side of even multiple roots (a) and (c), whereas it crosses the axis for odd cases (b).

Multiple roots pose some difficulties for many of the numerical methods described in Part Two:

1. The fact that the function does not change sign at even multiple roots precludes the use of the reliable bracketing methods that were discussed in Chap. 5. Thus, of the methods covered in this book, you are limited to the open methods that may diverge.
2. Another possible problem is related to the fact that not only $f(x)$ but also $f'(x)$ goes to zero at the root. This poses problems for both the Newton-Raphson and secant methods, which both contain the derivative (or its estimate) in the denominator of their respective formulas. This could result in division by zero when the solution converges very close to the root. A simple way to circumvent these problems is based on the fact that it can be demonstrated theoretically (Ralston and Rabinowitz, 1978) that $f(x)$ will always reach zero before $f'(x)$. Therefore, if a zero check for $f(x)$ is incorporated into the computer program, the computation can be terminated before $f'(x)$ reaches zero.
3. It can be demonstrated that the Newton-Raphson and secant methods are linearly, rather than quadratically, convergent for multiple roots (Ralston and Rabinowitz, 1978). Modifications have been proposed to alleviate this problem. Ralston and Rabinowitz (1978) have indicated that a slight change in the formulation returns it to quadratic convergence, as in

$$x_{i+1} = x_i - m \frac{f(x_i)}{f'(x_i)} \tag{6.9a}$$

where m is the multiplicity of the root (that is, $m = 2$ for a double root, $m = 3$ for a triple root, etc.). Of course, this may be an unsatisfactory alternative because it hinges on foreknowledge of the multiplicity of the root.

Another alternative, also suggested by Ralston and Rabinowitz (1978), is to define a new function $u(x)$, that is, the ratio of the function to its derivative, as in

$$u(x) = \frac{f(x)}{f'(x)} \tag{6.10}$$

It can be shown that this function has roots at all the same locations as the original function. Therefore, Eq. (6.10) can be substituted into Eq. (6.6) to develop an alternative form of the Newton-Raphson method:

$$x_{i+1} = x_i - \frac{u(x_i)}{u'(x_i)} \tag{6.11}$$

Equation (6.10) can be differentiated to give

$$u'(x) = \frac{f'(x)f'(x) - f(x)f''(x)}{[f'(x)]^2} \tag{6.12}$$

Equations (6.10) and (6.12) can be substituted into Eq. (6.11) and the result simplified to yield

$$x_{i+1} = x_i - \frac{f(x_i)f'(x_i)}{[f'(x_i)]^2 - f(x_i)f''(x_i)} \tag{6.13}$$

EXAMPLE 6.9   Modified Newton-Raphson Method for Multiple Roots

Problem Statement.   Use both the standard and modified Newton-Raphson methods to evaluate the multiple root of Eq. (6.9), with an initial guess of $x_0 = 0$.

Solution.   The first derivative of Eq. (6.9) is $f'(x) = 3x^2 - 10x + 7$, and therefore, the standard Newton-Raphson method for this problem is [Eq. (6.6)]

$$x_{i+1} = x_i - \frac{x_i^3 - 5x_i^2 + 7x_i - 3}{3x_i^2 - 10x_i + 7}$$

which can be solved iteratively for

i	x_i	ε_t (%)
0	0	100
1	0.4285714	57
2	0.6857143	31
3	0.8328654	17
4	0.9133290	8.7
5	0.9557833	4.4
6	0.9776551	2.2

As anticipated, the method is linearly convergent toward the true value of 1.0.

For the modified method, the second derivative is $f''(x) = 6x - 10$, and the iterative relationship is [Eq. (6.13)]

$$x_{i+1} = x_i - \frac{(x_i^3 - 5x_i^2 + 7x_i - 3)(3x_i^2 - 10x_i + 7)}{(3x_i^2 - 10x_i + 7)^2 - (x_i^3 - 5x_i^2 + 7x_i - 3)(6x_i - 10)}$$

which can be solved for

i	x_i	ε_t (%)
0	0	100
1	1.105263	11
2	1.003082	0.31
3	1.000002	0.00024

Thus, the modified formula is quadratically convergent. We can also use both methods to search for the single root at $x = 3$. Using an initial guess of $x_0 = 4$ gives the following results:

i	Standard	ε_t (%)	Modified	ε_t (%)
0	4	33	4	33
1	3.4	13	2.636364	12
2	3.1	3.3	2.820225	6.0
3	3.008696	0.29	2.961728	1.3
4	3.000075	0.0025	2.998479	0.051
5	3.000000	2×10^{-7}	2.999998	7.7×10^{-5}

Thus, both methods converge quickly, with the standard method being somewhat more efficient.

The above example illustrates the trade-offs involved in opting for the modified Newton-Raphson method. Although it is preferable for multiple roots, it is somewhat less efficient and requires more computational effort than the standard method for simple roots.

It should be noted that a modified version of the secant method suited for multiple roots can also be developed by substituting Eq. (6.10) into Eq. (6.7). The resulting formula is (Ralston and Rabinowitz, 1978)

$$x_{i+1} = x_i - \frac{u(x_i)(x_{i-1} - x_i)}{u(x_{i-1}) - u(x_i)}$$

6.5 SYSTEMS OF NONLINEAR EQUATIONS

To this point, we have focused on the determination of the roots of a single equation. A related problem is to locate the roots of a set of simultaneous equations,

$$f_1(x_1, x_2, \ldots, x_n) = 0$$
$$f_2(x_1, x_2, \ldots, x_n) = 0$$

$$\cdot \qquad \cdot$$
$$\cdot \qquad \cdot \qquad (6.14)$$
$$\cdot \qquad \cdot$$

$$f_n(x_1, x_2, \ldots, x_n) = 0$$

The solution of this system consists of a set of x values that simultaneously result in all the equations equaling zero.

In Part Three, we will present methods for the case where the simultaneous equations are linear—that is, they can be expressed in the general form

$$f(x) = a_1 x_1 + a_2 x_2 + \cdots + a_n x_n - b = 0 \qquad (6.15)$$

where the b and the a's are constants. Algebraic and transcendental equations that do not fit this format are called *nonlinear equations*. For example,

$$x^2 + xy = 10$$

and

$$y + 3xy^2 = 57$$

are two simultaneous nonlinear equations with two unknowns, x and y. They can be expressed in the form of Eq. (6.14) as

$$u(x, y) = x^2 + xy - 10 = 0 \qquad (6.16a)$$
$$v(x, y) = y + 3xy^2 - 57 = 0 \qquad (6.16b)$$

Thus, the solution would be the values of x and y that make the functions $u(x, y)$ and $v(x, y)$ equal to zero. Most approaches for determining such solutions are extensions of the open methods for solving single equations. In this section, we will investigate two of these: fixed-point iteration and Newton-Raphson.

6.5.1 Fixed-Point Iteration

The fixed-point-iteration approach (Sec. 6.1) can be modified to solve two simultaneous, nonlinear equations. This approach will be illustrated in the following example.

EXAMPLE 6.10   Fixed-Point Iteration for a Nonlinear System

Problem Statement.   Use fixed-point iteration to determine the roots of Eq. (6.16). Note that a correct pair of roots is $x = 2$ and $y = 3$. Initiate the computation with guesses of $x = 1.5$ and $y = 3.5$.

Solution.   Equation (6.16a) can be solved for

$$x_{i+1} = \frac{10 - x_i^2}{y_i} \tag{E6.10.1}$$

and Eq. (6.16b) can be solved for

$$y_{i+1} = 57 - 3x_i y_i^2 \tag{E6.10.2}$$

Note that we will drop the subscripts for the remainder of the example.

On the basis of the initial guesses, Eq. (E6.10.1) can be used to determine a new value of x:

$$x = \frac{10 - (1.5)^2}{3.5} = 2.21429$$

This result and the initial value of $y = 3.5$ can be substituted into Eq. (E6.10.2) to determine a new value of y:

$$y = 57 - 3(2.21429)(3.5)^2 = -24.37516$$

Thus, the approach seems to be diverging. This behavior is even more pronounced on the second iteration:

$$x = \frac{10 - (2.21429)^2}{-24.37516} = -0.20910$$

$$y = 57 - 3(-0.20910)(-24.37516)^2 = 429.709$$

Obviously, the approach is deteriorating.

Now we will repeat the computation but with the original equations set up in a different format. For example, an alternative formulation of Eq. (6.16a) is

$$x = \sqrt{10 - xy}$$

and of Eq. (6.16b) is

$$y = \sqrt{\frac{57 - y}{3x}}$$

Now the results are more satisfactory:

$$x = \sqrt{10 - 1.5(3.5)} = 2.17945$$

$$y = \sqrt{\frac{57 - 3.5}{3(2.17945)}} = 2.86051$$

$$x = \sqrt{10 - 2.17945(2.86051)} = 1.94053$$

$$y = \sqrt{\frac{57 - 2.86051}{3(1.94053)}} = 3.04955$$

Thus, the approach is converging on the true values of $x = 2$ and $y = 3$.

The previous example illustrates the most serious shortcoming of simple fixed-point iteration—that is, convergence often depends on the manner in which the equations are formulated. Additionally, even in those instances where convergence is possible, divergence can occur if the initial guesses are insufficiently close to the true solution. Using reasoning similar to that in Box 6.1, it can be demonstrated that sufficient conditions for convergence for the two-equation case are

$$\left| \frac{\partial u}{\partial x} \right| + \left| \frac{\partial v}{\partial x} \right| < 1$$

and

$$\left| \frac{\partial u}{\partial y} \right| + \left| \frac{\partial v}{\partial y} \right| < 1$$

These criteria are so restrictive that fixed-point iteration has limited utility for solving nonlinear systems. However, as we will describe later in the book, it can be very useful for solving linear systems.

6.5.2 Newton-Raphson

Recall that the Newton-Raphson method was predicated on employing the derivative (that is, the slope) of a function to estimate its intercept with the axis of the independent variable—that is, the root (Fig. 6.5). This estimate was based on a first-order Taylor series expansion (recall Box 6.2),

$$f(x_{i+1}) = f(x_i) + (x_{i+1} - x_i) f'(x_i) \tag{6.17}$$

where x_i is the initial guess at the root and x_{i+1} is the point at which the slope intercepts the x axis. At this intercept, $f(x_{i+1})$ by definition equals zero and Eq. (6.17) can be rearranged to yield

$$x_{i+1} = x_i - \frac{f(x_i)}{f'(x_i)} \tag{6.18}$$

which is the single-equation form of the Newton-Raphson method.

The multiequation form is derived in an identical fashion. However, a multivariable Taylor series must be used to account for the fact that more than one independent variable contributes to the determination of the root. For the two-variable case, a first-order Taylor series can be written [recall Eq. (4.26)] for each nonlinear equation as

$$u_{i+1} = u_i + (x_{i+1} - x_i)\frac{\partial u_i}{\partial x} + (y_{i+1} - y_i)\frac{\partial u_i}{\partial y} \tag{6.19a}$$

and

$$v_{i+1} = v_i + (x_{i+1} - x_i)\frac{\partial v_i}{\partial x} + (y_{i+1} - y_i)\frac{\partial v_i}{\partial y} \tag{6.19b}$$

Just as for the single-equation version, the root estimate corresponds to the values of x and y, where u_{i+1} and v_{i+1} equal zero. For this situation, Eq. (6.19) can be rearranged to give

$$\frac{\partial u_i}{\partial x}x_{i+1} + \frac{\partial u_i}{\partial y}y_{i+1} = -u_i + x_i\frac{\partial u_i}{\partial x} + y_i\frac{\partial u_i}{\partial y} \tag{6.20a}$$

$$\frac{\partial v_i}{\partial x}x_{i+1} + \frac{\partial v_i}{\partial y}y_{i+1} = -v_i + x_i\frac{\partial v_i}{\partial x} + y_i\frac{\partial v_i}{\partial y} \tag{6.20b}$$

Because all values subscripted with i's are known (they correspond to the latest guess or approximation), the only unknowns are x_{i+1} and y_{i+1}. Thus, Eq. (6.20) is a set of two linear equations with two unknowns [compare with Eq. (6.15)]. Consequently, algebraic manipulations (for example, Cramer's rule) can be employed to solve for

$$x_{i+1} = x_i - \frac{u_i\dfrac{\partial v_i}{\partial y} - v_i\dfrac{\partial u_i}{\partial y}}{\dfrac{\partial u_i}{\partial x}\dfrac{\partial v_i}{\partial y} - \dfrac{\partial u_i}{\partial y}\dfrac{\partial v_i}{\partial x}} \tag{6.21a}$$

$$y_{i+1} = y_i - \frac{v_i\dfrac{\partial u_i}{\partial x} - u_i\dfrac{\partial v_i}{\partial x}}{\dfrac{\partial u_i}{\partial x}\dfrac{\partial v_i}{\partial y} - \dfrac{\partial u_i}{\partial y}\dfrac{\partial v_i}{\partial x}} \tag{6.21b}$$

The denominator of each of these equations is formally referred to as the determinant of the *Jacobian* of the system.

Equation (6.21) is the two-equation version of the Newton-Raphson method. As in the following example, it can be employed iteratively to home in on the roots of two simultaneous equations.

EXAMPLE 6.11　Newton-Raphson for a Nonlinear System

Problem Statement.　Use the multiple-equation Newton-Raphson method to determine roots of Eq. (6.16). Note that a correct pair of roots is $x = 2$ and $y = 3$. Initiate the computation with guesses of $x = 1.5$ and $y = 3.5$.

Solution.   First compute the partial derivatives and evaluate them at the initial guesses of x and y:

$$\frac{\partial u_0}{\partial x} = 2x + y = 2(1.5) + 3.5 = 6.5 \qquad \frac{\partial u_0}{\partial y} = x = 1.5$$

$$\frac{\partial v_0}{\partial x} = 3y^2 = 3(3.5)^2 = 36.75 \qquad \frac{\partial v_0}{\partial y} = 1 + 6xy = 1 + 6(1.5)(3.5) = 32.5$$

Thus, the determinant of the Jacobian for the first iteration is

$$6.5(32.5) - 1.5(36.75) = 156.125$$

The values of the functions can be evaluated at the initial guesses as

$$u_0 = (1.5)^2 + 1.5(3.5) - 10 = -2.5$$

$$v_0 = 3.5 + 3(1.5)(3.5)^2 - 57 = 1.625$$

These values can be substituted into Eq. (6.21) to give

$$x = 1.5 - \frac{-2.5(32.5) - 1.625(1.5)}{156.125} = 2.03603$$

$$y = 3.5 - \frac{1.625(6.5) - (-2.5)(36.75)}{156.125} = 2.84388$$

Thus, the results are converging to the true values of $x = 2$ and $y = 3$. The computation can be repeated until an acceptable accuracy is obtained.

Just as with fixed-point iteration, the Newton-Raphson approach will often diverge if the initial guesses are not sufficiently close to the true roots. Whereas graphical methods could be employed to derive good guesses for the single-equation case, no such simple procedure is available for the multiequation version. Although there are some advanced approaches for obtaining acceptable first estimates, often the initial guesses must be obtained on the basis of trial and error and knowledge of the physical system being modeled.

The two-equation Newton-Raphson approach can be generalized to solve n simultaneous equations. Because the most efficient way to do this involves matrix algebra and the solution of simultaneous linear equations, we will defer discussion of the general approach to Part Three.

PROBLEMS

6.1 Use simple fixed-point iteration to locate the root of

$$f(x) = \sin(\sqrt{x}) - x$$

Use an initial guess of $x_0 = 0.5$ and iterate until $\varepsilon_a \leq 0.01\%$.

6.2 Use (a) fixed-point iteration and (b) the Newton-Raphson method to determine a root of $f(x) = -0.9x^2 + 1.7x + 2.5$ using $x_0 = 5$. Perform the computation until ε_a is less than $\varepsilon_s = 0.01\%$. Also perform an error check of your final answer.

6.3 Determine the real roots of $f(x) = -2.0 + 6x - 4x^2 + 0.5x^3$: (a) graphically and (b) using the Newton-Raphson method to within $\varepsilon_s = 0.01\%$.

6.4 Employ the Newton-Raphson method to determine a real root for $f(x) = -2.0 + 6x - 4x^2 + 0.5x^3$ using initial guesses of (a) 4.2 and (b) 4.43. Discuss and use graphical and analytical methods to explain any peculiarities in your results.

6.5 Determine the lowest real root of $f(x) = -11 - 22x + 17x^2 - 2.5x^3$: (a) graphically and (b) using the secant method to a value of ε_s corresponding to three significant figures.

6.6 Locate the first positive root of

$$f(x) = \sin x + \cos (1 + x^2) - 1$$

where x is in radians. Use four iterations of the Newton-Raphson method with initial guesses of (a) $x_{i-1} = 1.0$ and $x_i = 3.0$, and (b) $x_{i-1} = 1.5$ and $x_i = 2.5$, to locate the root. (c) Use the graphical method to explain your results.

6.7 Determine the real root of $x^{3.3} = 79$, with the modified secant method to within $\varepsilon_s = 0.1\%$. Try different values of δ and discuss your results.

6.8 Determine the highest real root of $f(x) = x^3 - 6x^2 + 11x - 6.1$:
(a) Graphically.
(b) Using the Newton-Raphson method (three iterations, $x_i = 3.5$).
(c) Using the secant method (three iterations, $x_{i-1} = 2.5$ and $x_i = 3.5$).
(d) Using the modified secant method (three iterations, $x_i = 3.5$, $\delta = 0.02$).

6.9 Determine the lowest positive root of $f(x) = 7 \sin (x)e^{-x} - 1$:
(a) Graphically.
(b) Using the Newton-Raphson method (three iterations, $x_i = 0.3$).
(c) Using the secant method (three iterations, $x_{i-1} = 0.5$ and $x_i = 0.4$).
(d) Using the modified secant method (five iterations, $x_i = 0.5$, $\delta = 0.03$).

6.10 The function $x^3 + 2x^2 - 5x + 3$ has a double root at $x = 1$. Use (a) the standard Newton-Raphson [Eq. (6.6)], (b) the modified Newton-Raphson [Eq. (6.9a)], and (c) the modified Newton-Raphson [Eq. (6.13)] to solve for the root at $x = 1$. Compare and discuss the rate of convergence using $x_0 = 0.2$.

6.11 Determine the roots of the following simultaneous nonlinear equations using (a) fixed-point iteration and (b) the Newton-Raphson method:

$$x = y + x^2 - 0.5$$
$$y = x^2 - 5xy$$

Employ initial guesses of $x = y = 1.0$ and discuss the results.

6.12 Determine the roots of the simultaneous nonlinear equations

$$(x - 4)^2 + (y - 4)^2 = 4$$
$$x^2 + y^2 = 16$$

Use a graphical approach to obtain your initial guesses. Determine refined estimates with the two-equation Newton-Raphson method described in Sec. 6.5.2.

6.13 Repeat Prob. 6.12 except for

$$y = x^2 + 1$$
$$y = 3 \cos x$$

6.14 A mass balance for a pollutant in a well-mixed lake can be written as

$$V\frac{dc}{dt} = W - Qc - kV \sqrt{c}$$

Given the parameter values $V = 1 \times 10^6$ m³, $Q = 1 \times 10^5$ m³/yr, $W = 1 \times 10^6$ g/yr, and $k = 0.2$ m$^{0.5}$/g$^{0.5}$/yr, use the modified secant method to solve for the steady-state concentration. Employ an initial guess of $c = 4$ g/m³ and an $\delta = 0.5$. Perform two iterations and determine the percent relative error after the second iteration.

6.15 For Prob. 6.14, the root can be located with fixed-point iteration as

$$c = \left(\frac{W - Qc}{kV}\right)^2$$

or as

$$c = \frac{W - kV \sqrt{c}}{Q}$$

Only one will work all the time for initial guesses of $c > 1$. Select the correct one and demonstrate why it will always work.

6.16 Develop a user-friendly subprogram for the Newton-Raphson method based on Fig. 6.4 and Sec. 6.2.3. Test it by duplicating the computation from Example 6.3.

6.17 Develop a user-friendly subprogram for the secant method based on Fig. 6.4 and Sec. 6.3.2. Test it by duplicating the computation from Example 6.6.

6.18 Develop a user-friendly subprogram for the modified secant method based on Fig. 6.4 and Sec. 6.3.2. Test it by duplicating the computation from Example 6.8.

6.19 Develop a user-friendly subprogram for the two-equation Newton-Raphson method based on Sec. 6.5. Test it by solving Example 6.10.

6.20 Use the subprogram you developed in Prob. 6.19 to solve Probs. 6.11 and 6.12 to within a tolerance of $\varepsilon_s = 0.01\%$.

6.21 The "divide and average" method, an old-time method for approximating the square root of any positive number a, can be formulated as

$$x = \frac{x + a/x}{2}$$

Prove that this formula is based on the Newton-Raphson algorithm.

CHAPTER 7

Roots of Polynomials

In this chapter, we will discuss methods to find the roots of polynomial equations of the general form

$$f_n(x) = a_0 + a_1 x + a_2 x^2 + \cdots + a_n x^n \tag{7.1}$$

where n = the order of the polynomial and the a's = constant coefficients. Although the coefficients can be complex numbers, we will limit our discussion to cases where they are real. For such cases, the roots can be real and/or complex.

The roots of such polynomials follow these rules:

1. For an nth-order equation, there are n real or complex roots. It should be noted that these roots will not necessarily be distinct.
2. If n is odd, there is at least one real root.
3. If complex roots exist, they exist in conjugate pairs (that is, $\lambda + \mu i$ and $\lambda - \mu i$), where $i = \sqrt{-1}$.

Before describing the techniques for locating the roots of polynomials, we will provide some background. The first section offers some motivation for studying the techniques; the second deals with some fundamental computer manipulations involving polynomials.

7.1 POLYNOMIALS IN ENGINEERING AND SCIENCE

Polynomials have many applications in engineering and science. For example, they are used extensively in curve-fitting. However, we believe that one of their most interesting and powerful applications is in characterizing dynamic systems and, in particular, linear systems. Examples include mechanical devices, structures, and electrical circuits. We will be exploring specific examples throughout the remainder of this text. In particular, they will be the focus of several of the engineering applications throughout the remainder of this text.

For the time being, we will keep the discussion simple and general by focusing on a simple second-order system defined by the following linear ordinary differential equation (or ODE):

$$a_2 \frac{d^2 y}{dt^2} + a_1 \frac{dy}{dt} + a_0 y = F(t) \tag{7.2}$$

167

where y and t are the dependent and independent variables, respectively, the a's are constant coefficients, and $F(t)$ is the forcing function. If seeing how this equation is derived from a physical system helps motivate your study of the mathematics, you might peruse the beginning of Sec. 8.4 before proceeding.

In addition, it should be noted that Eq. (7.2) can be alternatively expressed as a pair of first-order ODEs by defining a new variable z,

$$z = \frac{dy}{dt} \tag{7.3}$$

Equation (7.3) can be substituted along with its derivative into Eq. (7.2) to remove the second-derivative term. This reduces the problem to solving

$$\frac{dz}{dt} = \frac{F(t) - a_1 z - a_0 y}{a_2} \tag{7.4}$$

$$\frac{dy}{dt} = z \tag{7.5}$$

In a similar fashion, an nth-order linear ODE can always be expressed as a system of n first-order ODEs.

Now let's look at the solution. The forcing function represents the effect of the external world on the system. The homogeneous or *general solution* of the equation deals with the case when the forcing function is set to zero,

$$a_2 \frac{d^2 y}{dt^2} + a_1 \frac{dy}{dt} + a_0 y = 0 \tag{7.6}$$

Thus, as the name implies, the *general solution* should tell us something very fundamental about the system being simulated—that is, how the system responds in the absence of external stimuli.

Now, the general solution to all unforced linear systems is of the form $y = e^{rt}$. If this function is differentiated and substituted into Eq. (7.6), the result is

$$a_2 r^2 e^{rt} + a_1 r e^{rt} + a_0 e^{rt} = 0$$

or canceling the exponential terms,

$$a_2 r^2 + a_1 r + a_0 = 0 \tag{7.7}$$

Notice that the result is a polynomial called the *characteristic equation*. The roots of this polynomial are the values of r that satisfy Eq. (7.7). These r's are referred to as the system's characteristic values, or *eigenvalues*.

So, here is the connection between roots of polynomials and engineering and science. The eigenvalue tells us something fundamental about the system we're modeling, and finding the eigenvalues involves finding the roots of polynomials. And, whereas finding the root of a second-order equation is easy with the quadratic formula, finding roots of higher-order systems (and hence, higher-order polynomials) is arduous analytically. Thus, the best general approach requires numerical methods of the type described in this chapter.

Before proceeding to these methods, let's take our analysis a bit farther by investigating what specific values of the eigenvalues might imply about the behavior of physical systems. First, let's evaluate the roots of Eq. (7.7) with the quadratic formula,

$$\frac{r_1}{r_2} = \frac{-a_1 \pm \sqrt{a_1^2 - 4a_2a_0}}{a_0}$$

Thus, we get two roots. If the *discriminant* ($a_1^2 - 4a_2a_0$) is positive, the roots are real and the general solution can be represented as

$$y = c_1 e^{r_1 t} + c_2 e^{r_2 t} \tag{7.8}$$

where the c's = constants that can be determined from the initial conditions. This is called the *overdamped case.*

If the discriminant is zero, a single real root results, and the general solution can be formulated as

$$y = (c_1 + c_2 t)e^{\lambda t} \tag{7.9}$$

This is called the *critically damped case.*

If the discriminant is negative, the roots will be complex conjugate numbers,

$$\frac{r_1}{r_2} = \lambda \pm \mu i$$

and the general solution can be formulated as

$$y = c_1 e^{(\lambda + \mu i)t} + c_2 e^{(\lambda - \mu i)t}$$

FIGURE 7.1
The general solution for linear ODEs can be composed of (a) exponential and (b) sinusoidal components. The combination of the two shapes results in the damped sinusoid shown in (c).

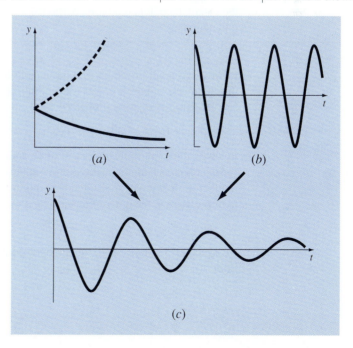

The physical behavior of this solution can be elucidated by using Euler's formula

$$e^{\mu i t} = \cos \mu t + i \sin \mu t$$

to reformulate the general solution as (see Boyce and DiPrima, 1992, for details of the derivation)

$$y = c_1 e^{\lambda t} \cos \mu t + c_2 e^{\lambda t} \sin \mu t \tag{7.10}$$

This is called the *underdamped case.*

Equations (7.8), (7.9), and (7.10) express the possible ways that linear systems respond dynamically. The exponential terms mean that the solutions are capable of decaying (negative real part) or growing (positive real part) exponentially with time (Fig. 7.1*a*). The sinusoidal terms (imaginary part) mean that the solutions can oscillate (Fig. 7.1*b*). If the eigenvalue has both real and imaginary parts, the exponential and sinusoidal shapes are combined (Fig. 7.1*c*) Because such knowledge is a key element in understanding, designing, and controlling the behavior of a physical system, characteristic polynomials are very important in engineering and many branches of science. We will explore the dynamics of several engineering systems in the applications covered in Chap. 8.

7.2 COMPUTING WITH POLYNOMIALS

Before describing root-location methods we will discuss some fundamental computer operations involving polynomials. These have utility in their own right, as well as providing support for root finding.

7.2.1 Polynomial Evaluation and Differentiation

Although it's the most common format, Eq. (7.1) provides a poor means for determining the value of a polynomial for a particular value of x. For example, evaluating a third-order polynomial as

$$f_3(x) = a_3 x^3 + a_2 x^2 + a_1 x + a_0 \tag{7.11}$$

involves six multiplications and three additions. In general, for an nth-order polynomial, this approach requires $n(n + 1)/2$ multiplications and n additions.

In contrast, a nested format,

$$f_3(x) = ((a_3 x + a_2)x + a_1)x + a_0 \tag{7.12}$$

involves three multiplications and three additions. For an nth-order polynomial, this approach requires n multiplications and n additions. Because the nested format minimizes the number of operations, it also tends to minimize round-off errors. Note that, depending on your preference, the order of nesting can be reversed:

$$f_3(x) = a_0 + x(a_1 + x(a_2 + x a_3)) \tag{7.13}$$

Succinct pseudocode to implement the nested form can be written simply as

```
DO j = n, 0, −1
  p = p * x + a(j)
END DO
```

where p holds the value of the polynomial (defined by its coefficients, the a's) evaluated at x.

There are cases (such as in the Newton-Raphson method) where you might want to evaluate both the function and its derivative. This evaluation can also be neatly included by adding a single line to the preceding pseudocode,

```
DO j = n, 0, −1

    df = df * x+p

    p = p * x+a(j)
END DO
```

where df holds the first derivative of the polynomial.

7.2.2 Polynomial Deflation

Suppose that you determine a single root of an nth-order polynomial. If you repeat your root location procedure, you might find the same root. Therefore, it would be nice to remove the found root before proceeding. This removal process is referred to as *polynomial deflation*.

Before we show how this is done, some orientation might be useful. Polynomials are typically represented in the format of Eq. (7.1). For example, a fifth-order polynomial could be written as

$$f_5(x) = -120 - 46x + 79x^2 - 3x^3 - 7x^4 + x^5 \qquad (7.14)$$

Although this is a familiar format, it is not necessarily the best expression to understand the polynomial's mathematical behavior. For example, this fifth-order polynomial might be expressed alternatively as

$$f_5(x) = (x + 1)(x - 4)(x - 5)(x + 3)(x - 2) \qquad (7.15)$$

This is called the *factored* form of the polynomial. If multiplication is completed and like terms collected, Eq. (7.14) would be obtained. However, the format of Eq. (7.15) has the advantage that it clearly indicates the function's roots. Thus, it is apparent that $x = -1$, 4, 5, −3, and 2 are all roots because each causes an individual term in Eq. (7.15) to become zero.

Now, suppose that we divide this fifth-order polynomial by any of its factors, for example, $x + 3$. For this case the result would be a fourth-order polynomial

$$f_4(x) = (x + 1)(x - 4)(x - 5)(x - 2) = -40 - 2x + 27x^2 - 10x^3 + x^4 \qquad (7.16)$$

with a remainder of zero.

In the distant past, you probably learned to divide polynomials using the approach called *synthetic division*. Several computer algorithms (based on both synthetic division, as well as other methods) are available for performing the operation. One simple scheme is provided by the following pseudocode, which divides an nth-order polynomial by a monomial factor $x - t$:

```
r = a(n)
a(n) = 0
DO i = n-1, 0, -1
  s = a(i)
  a(i) = r
  r = s+r*t
END DO
```

If the monomial is a root of the polynomial, the remainder r will be zero, and the coefficients of the quotient stored in a, at the end of the loop.

EXAMPLE 7.1   Polynomial Deflation

Problem Statement.   Divide the second-order polynomial,

$$f(x) = (x - 4)(x + 6) = x^2 + 2x - 24$$

by the factor $x - 4$.

Solution.   Using the approach outlined in the above pseudocode, the parameters are $n = 2$, $a_0 = -24$, $a_1 = 2$, $a_2 = 1$, and $t = 4$. These can be used to compute

$$r = a_2 = 1$$
$$a_2 = 0$$

The loop is then iterated from $i = 2 - 1 = 1$ to 0. For $i = 1$,

$$s = a_1 = 2$$
$$a_1 = r = 1$$
$$r = s + rt = 2 + 1(4) = 6$$

For $i = 0$,

$$s = a_0 = 24$$
$$a_0 = r = 6$$
$$r = -24 + 6(4) = 0$$

Thus, the result is as expected—the quotient is $a_0 + a_1 x = 6 + x$, with a remainder of zero.

It is also possible to divide by polynomials of higher order. As we will see later in this chapter, the most common task involves dividing by a second-order polynomial or parabola. The subroutine in Fig. 7.2 addresses the more general problem of dividing an nth-order polynomial a by an mth-order polynomial d. The result is an $(n - m)$th-order polynomial q, with an $(m - 1)$th-order polynomial as the remainder.

Because each calculated root is known only approximately, it should be noted that deflation is sensitive to round-off error. In some cases, they can grow to the point that the results can become meaningless.

Some general strategies can be applied to minimize this problem. For example, round-off error is affected by the order in which the terms are evaluated. *Forward deflation* refers

```
SUB poldiv (a, n, d, m, q, r)
DO j = 0, n
    r(j) = a(j)
    q(j) = 0
END DO
DO k = n−m, 0, −1
    q(k+1) = r(m+k) / d(m)
    DO j = m+k−1, k, −1
        r(j) = r(j)−q(k+1) ∗ b(j−k)
    END DO
END DO
DO j = m, n
    r(j) = 0
END DO
n = n−m
DO i = 0, n
    a(i) = q(i+1)
END DO
END SUB
```

FIGURE 7.2
Algorithm to divide a polynomial (defined by its coefficients a) by a lower-order polynomial d.

to the case where new polynomial coefficients are in order of descending powers of x (that is, from the highest-order to the zero-order term). For this case, it is preferable to divide by the roots of smallest absolute value first. Conversely, for *backward deflation* (that is, from the zero-order to the highest-order term), it is preferable to divide by the roots of smallest absolute value first.

Another way to reduce round-off errors is to consider each successive root estimate obtained during deflation as a good first guess. These can then be used as a starting guess, and the root determined again with the original nondeflated polynomial. This is referred to as *root polishing*.

Finally, a problem arises when two deflated roots are inaccurate enough that they both converge on the same undeflated root. In that case, you might be erroneously led to believe that the polynomial has a multiple root (recall Sec. 6.4). One way to detect this problem is to compare each polished root with those that were located previously. Press et al. (1992) discuss this problem in more detail.

7.3 CONVENTIONAL METHODS

Now that we've covered some background material on polynomials, we can begin to describe methods to locate their roots. The obvious first step would be to investigate the viability of the bracketing and open approaches described in Chaps. 5 and 6.

The efficacy of these approaches depends on whether the problem being solved involves complex roots. If only real roots exist, any of the previously described methods could have utility. However, the problem of finding good initial guesses complicates both the bracketing and the open methods, whereas the open methods could be susceptible to divergence.

When complex roots are possible, the bracketing methods cannot be used because of the obvious problem that the criterion for defining a bracket (that is, sign change) does not translate to complex guesses.

Of the open methods, the conventional Newton-Raphson method would provide a viable approach. In particular, concise code including deflation can be developed. If a language that accommodates complex variables (like Fortran) is used, such an algorithm will locate both real and complex roots. However, as might be expected, it would be susceptible to convergence problems. For this reason, special methods have been developed to find the real and complex roots of polynomials. We describe two—the Müller and Bairstow methods—in the following sections. As you will see, both are related to the more conventional open approaches described in Chap. 6.

7.4 MÜLLER'S METHOD

Recall that the secant method obtains a root estimate by projecting a straight line to the x axis through two function values (Fig. 7.3a). Müller's method takes a similar approach, but projects a parabola through three points (Fig. 7.3b).

The method consists of deriving the coefficients of the parabola that goes through the three points. These coefficients can then be substituted into the quadratic formula to obtain the point where the parabola intercepts the x axis—that is, the root estimate. The approach is facilitated by writing the parabolic equation in a convenient form,

$$f_2(x) = a(x - x_2)^2 + b(x - x_2) + c \tag{7.17}$$

We want this parabola to intersect the three points $[x_0, f(x_0)]$, $[x_1, f(x_1)]$, and $[x_2, f(x_2)]$. The coefficients of Eq. (7.17) can be evaluated by substituting each of the three points to give

$$f(x_0) = a(x_0 - x_2)^2 + b(x_0 - x_2) + c \tag{7.18}$$

$$f(x_1) = a(x_1 - x_2)^2 + b(x_1 - x_2) + c \tag{7.19}$$

$$f(x_2) = a(x_2 - x_2)^2 + b(x_2 - x_2) + c \tag{7.20}$$

FIGURE 7.3

A comparison of two related approaches for locating roots: (a) the secant method and (b) Müller's method.

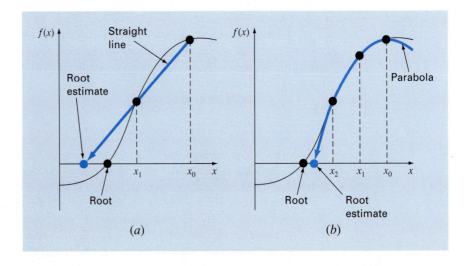

(a) (b)

Note that we have dropped the subscript "2" from the function for conciseness. Because we have three equations, we can solve for the three unknown coefficients, a, b, and c. Because two of the terms in Eq. (7.20) are zero, it can be immediately solved for $c = f(x_2)$. Thus, the coefficient c is merely equal to the function value evaluated at the third guess, x_2. This result can then be substituted into Eqs. (7.18) and (7.19) to yield two equations with two unknowns:

$$f(x_0) - f(x_2) = a(x_0 - x_2)^2 + b(x_0 - x_2) \tag{7.21}$$

$$f(x_1) - f(x_2) = a(x_1 - x_2)^2 + b(x_1 - x_2) \tag{7.22}$$

Algebraic manipulation can then be used to solve for the remaining coefficients, a and b. One way to do this involves defining a number of differences,

$$h_0 = x_1 - x_0 \qquad\qquad h_1 = x_2 - x_1$$
$$\delta_0 = \frac{f(x_1) - f(x_0)}{x_1 - x_0} \qquad \delta_1 = \frac{f(x_2) - f(x_1)}{x_2 - x_1} \tag{7.23}$$

These can be substituted into Eqs. (7.21) and (7.22) to give

$$(h_0 + h_1)b - (h_0 + h_1)^2 a = h_0\delta_0 + h_1\delta_1$$
$$h_1 \quad b - \quad h_1^2 \quad a = \quad h_1\delta_1$$

which can be solved for a and b. The results can be summarized as

$$a = \frac{\delta_1 - \delta_0}{h_1 + h_0} \tag{7.24}$$

$$b = ah_1 + \delta_1 \tag{7.25}$$

$$c = f(x_2) \tag{7.26}$$

To find the root, we apply the quadratic formula to Eq. (7.17). However, because of potential round-off error, rather than using the conventional form, we use the alternative formulation [Eq. (3.13)] to yield

$$x_3 - x_2 = \frac{-2c}{b \pm \sqrt{b^2 - 4ac}} \tag{7.27a}$$

or isolating the unknown x_3 on the left side of the equal sign,

$$x_3 = x_2 + \frac{-2c}{b \pm \sqrt{b^2 - 4ac}} \tag{7.27b}$$

Note that the use of the quadratic formula means that both real and complex roots can be located. This is a major benefit of the method.

In addition, Eq. (7.27a) provides a neat means to determine the approximate error. Because the left side represents the difference between the present (x_3) and the previous (x_2) root estimate, the error can be calculated as

$$\varepsilon_a = \left| \frac{x_3 - x_2}{x_3} \right| 100\%$$

Now, a problem with Eq. (7.27a) is that it yields two roots, corresponding to the $\pm$ term in the denominator. In Müller's method, the sign is chosen to agree with the sign of b. This choice will result in the largest denominator, and hence, will give the root estimate that is closest to x_2.

Once x_3 is determined, the process is repeated. This brings up the issue of which point is discarded. Two general strategies are typically used:

1. If only real roots are being located, we choose the two original points that are nearest the new root estimate, x_3.
2. If both real and complex roots are being evaluated, a sequential approach is employed. That is, just like the secant method, x_1, x_2, and x_3 take the place of x_0, x_1, and x_2.

EXAMPLE 7.2   Müller's Method

Problem Statement.   Use Müller's method with guesses of x_0, x_1, and $x_2 = 4.5$, 5.5, and 5 to determine a root of the equation

$$f(x) = x^3 - 13x - 12$$

Note that the roots of this equation are -3, -1, and 4.

Solution.   First, we evaluate the function at the guesses

$$f(4.5) = 20.625 \qquad f(5.5) = 82.875 \qquad f(5) = 48$$

which can be used to calculate

$$h_0 = 5.5 - 4.5 = 1 \qquad\qquad h_1 = 5 - 5.5 = -0.5$$

$$\delta_0 = \frac{82.875 - 20.625}{5.5 - 4.5} = 62.25 \qquad \delta_1 = \frac{48 - 82.875}{5 - 5.5} = 69.75$$

These values in turn can be substituted into Eqs. (7.24) through (7.26) to compute

$$a = \frac{69.75 - 62.25}{-0.5 + 1} = 15 \qquad b = 15(-0.5) + 69.75 = 62.25 \qquad c = 48$$

The square root of the discriminant can be evaluated as

$$\sqrt{62.25^2 - 4(15)48} = 31.54461$$

Then, because $|62.25 + 31.54451| > |62.25 - 31.54451|$, a positive sign is employed in the denominator of Eq. (7.27b), and the new root estimate can be determined as

$$x_3 = 5 + \frac{-2(48)}{62.25 + 31.54451} = 3.976487$$

and develop the error estimate

$$\varepsilon_a = \left| \frac{-1.023513}{3.976487} \right| 100\% = 25.74\%$$

Because the error is large, new guesses are assigned; x_0 is replaced by x_1, x_1 is replaced by x_2, and x_2 is replaced by x_3. Therefore, for the new iteration,

$$x_0 = 5.5 \qquad x_1 = 5 \qquad x_2 = 3.976487$$

and the calculation is repeated. The results, tabulated below, show that the method converges rapidly on the root, $x_r = 4$:

i	x_r	$\varepsilon_a(\%)$
0	5	
1	3.976487	25.74
2	4.00105	0.6139
3	4	0.0262
4	4	0.0000119

Pseudocode to implement Müller's method for real roots is presented in Fig. 7.4. Notice that this routine is set up to take a single initial nonzero guess that is then perturbed to

FIGURE 7.4
Pseudocode for Müller's method.

```
SUB Muller(xr, h, eps, maxit)
x₂ = xᵣ
x₁ = xᵣ + h*xᵣ
x₀ = xᵣ − h*xᵣ
DO
    iter = iter + 1
    h₀ = x₁ − x₀
    h₁ = x₂ − x₁
    d₀ = (f(x₁) − f(x₀)) / h₀
    d₁ = (f(x₂) − f(x₁)) / h₁
    a = (d₁ − d₀) / (h₁ + h₀)
    b = a*h₁ + d₁
    c = f(x₂)
    rad = SQRT(b*b − 4*a*c)
    If |b+rad| > |b−rad| THEN
        den = b + rad
    ELSE
        den = b − rad
    END IF
    dxᵣ = −2*c / den
    xᵣ = x₂ + dxᵣ
    PRINT iter, xᵣ
    IF ( |dxᵣ| < eps*xᵣ   OR   iter > maxit) EXIT
    x₀ = x₁
    x₁ = x₂
    x₂ = xᵣ
END DO
END Müller
```

develop the other two guesses. Of course, the algorithm can also be programmed to accommodate three guesses. For languages like Fortran, the code will find complex roots if the proper variables are declared as complex.

7.5 BAIRSTOW'S METHOD

Bairstow's method is an iterative approach related loosely to both the Müller and Newton-Raphson methods. Before launching into a mathematical description of the technique, recall the factored form of the polynomial,

$$f_5(x) = (x + 1)(x - 4)(x - 5)(x + 3)(x - 2) \tag{7.28}$$

If we divided by a factor that is not a root (for example, $x + 6$), the quotient would be a fourth-order polynomial. However, for this case, a remainder would result.

On the basis of the above, we can elaborate on an algorithm for determining a root of a polynomial: (1) guess a value for the root $x = t$, (2) divide the polynomial by the factor $x - t$, and (3) determine whether there is a remainder. If not, the guess was perfect and the root is equal to t. If there is a remainder, the guess can be systematically adjusted and the procedure repeated until the remainder disappears and a root is located. After this is accomplished, the entire procedure can be repeated for the quotient to locate another root.

Bairstow's method is generally based on this approach. Consequently, it hinges on the mathematical process of dividing a polynomial by a factor. Recall from our discussion of polynomial deflation (Sec. 7.2.2) that synthetic division involves dividing a polynomial by a factor $x - t$. For example, the general polynomial [Eq. (7.1)]

$$f_n(x) = a_0 + a_1x + a_2x^2 + \cdots + a_nx^n \tag{7.29}$$

can be divided by the factor $x - t$ to yield a second polynomial that is one order lower,

$$f_{n-1}(x) = b_1 + b_2x + b_3x^2 + \cdots + b_nx^{n-1} \tag{7.30}$$

with a remainder $R = b_0$, where the coefficients can be calculated by the recurrence relationship

$$b_n = a_n$$
$$b_i = a_i + b_{i+1}t \qquad \text{for } i = n - 1 \text{ to } 0$$

Note that if t were a root of the original polynomial, the remainder b_0 would equal zero.

To permit the evaluation of complex roots, Bairstow's method divides the polynomial by a quadratic factor $x^2 - rx - s$. If this is done to Eq. (7.29), the result is a new polynomial

$$f_{n-2}(x) = b_2 + b_3x + \cdots + b_{n-1}x^{n-3} + b_nx^{n-2}$$

with a remainder

$$R = b_1(x - r) + b_0 \tag{7.31}$$

As with normal synthetic division, a simple recurrence relationship can be used to perform the division by the quadratic factor:

$$b_n = a_n \tag{7.32a}$$

$$b_{n-1} = a_{n-1} + rb_n \tag{7.32b}$$

$$b_i = a_i + rb_{i+1} + sb_{i+2} \qquad \text{for } i = n-2 \text{ to } 0 \tag{7.32c}$$

The quadratic factor is introduced to allow the determination of complex roots. This relates to the fact that, if the coefficients of the original polynomial are real, the complex roots occur in conjugate pairs. If $x^2 - rx - s$ is an exact divisor of the polynomial, complex roots can be determined by the quadratic formula. Thus, the method reduces to determining the values of r and s that make the quadratic factor an exact divisor. In other words, we seek the values that make the remainder term equal to zero.

Inspection of Eq. (7.31) leads us to conclude that for the remainder to be zero, b_0 and b_1 must be zero. Because it is unlikely that our initial guesses at the values of r and s will lead to this result, we must determine a systematic way to modify our guesses so that b_0 and b_1 approach zero. To do this, Bairstow's method uses a strategy similar to the Newton-Raphson approach. Because both b_0 and b_1 are functions of both r and s, they can be expanded using a Taylor series, as in [recall Eq. (4.26)]

$$b_1(r + \Delta r, s + \Delta s) = b_1 + \frac{\partial b_1}{\partial r}\Delta r + \frac{\partial b_1}{\partial s}\Delta s$$

$$b_0(r + \Delta r, s + \Delta s) = b_0 + \frac{\partial b_0}{\partial r}\Delta r + \frac{\partial b_0}{\partial s}\Delta s \tag{7.33}$$

where the values on the right-hand side are all evaluated at r and s. Notice that second- and higher-order terms have been neglected. This represents an implicit assumption that $-r$ and $-s$ are small enough that the higher-order terms are negligible. Another way of expressing this assumption is to say that the initial guesses are adequately close to the values of r and s at the roots.

The changes, Δr and Δs, needed to improve our guesses can be estimated by setting Eq. (7.33) equal to zero to give

$$\frac{\partial b_1}{\partial r}\Delta r + \frac{\partial b_1}{\partial s}\Delta s = -b_1 \tag{7.34}$$

$$\frac{\partial b_0}{\partial r}\Delta r + \frac{\partial b_0}{\partial s}\Delta s = -b_0 \tag{7.35}$$

If the partial derivatives of the b's can be determined, these are a system of two equations that can be solved simultaneously for the two unknowns, Δr and Δs. Bairstow showed that the partial derivatives can be obtained by a synthetic division of the b's in a fashion similar to the way in which the b's themselves were derived:

$$c_n = b_n \tag{7.36a}$$

$$c_{n-1} = b_{n-1} + rc_n \tag{7.36b}$$

$$c_i = b_i + rc_{i+1} + sc_{i+2} \qquad \text{for } i = n-2 \text{ to } 1 \tag{7.36c}$$

where $\partial b_0/\partial r = c_1$, $\partial b_0/\partial s = \partial b_1/\partial r = c_2$, and $\partial b_1/\partial s = c_3$. Thus, the partial derivatives are obtained by synthetic division of the b's. Then the partial derivatives can be substituted into Eqs. (7.34) and (7.35) along with the b's to give

$$c_2\Delta r + c_3\Delta s = -b_1$$

$$c_1\Delta r + c_2\Delta s = -b_0$$

These equations can be solved for Δr and Δs, which can in turn be employed to improve the initial guesses of r and s. At each step, an approximate error in r and s can be estimated, as in

$$|\varepsilon_{a,r}| = \left|\frac{\Delta r}{r}\right| 100\%$$

(7.37)

and

$$|\varepsilon_{a,s}| = \left|\frac{\Delta s}{s}\right| 100\%$$

(7.38)

When both of these error estimates fall below a prespecified stopping criterion ε_s, the values of the roots can be determined by

$$x = \frac{r \pm \sqrt{r^2 + 4s}}{2}$$

(7.39)

At this point, three possibilities exist:

1. *The quotient is a third-order polynomial or greater.* For this case, Bairstow's method would be applied to the quotient to evaluate new values for r and s. The previous values of r and s can serve as the starting guesses for this application.
2. *The quotient is a quadratic.* For this case, the remaining two roots could be evaluated directly with Eq. (7.39).
3. *The quotient is a first-order polynomial.* For this case, the remaining single root can be evaluated simply as

$$x = -\frac{s}{r}$$

(7.40)

EXAMPLE 7.3   Bairstow's Method

Problem Statement.   Employ Bairstow's method to determine the roots of the polynomial

$$f_5(x) = x^5 - 3.5x^4 + 2.75x^3 + 2.125x^2 - 3.875x + 1.25$$

Use initial guesses of $r = s = -1$ and iterate to a level of $\varepsilon_s = 1\%$.

Solution.   Equations (7.32) and (7.36) can be applied to compute

$$b_5 = 1 \qquad b_4 = -4.5 \qquad b_3 = 6.25 \qquad b_2 = 0.375 \qquad b_1 = -10.5$$
$$b_0 = 11.375$$
$$c_5 = 1 \qquad c_4 = -5.5 \qquad c_3 = 10.75 \qquad c_2 = -4.875 \qquad c_1 = -16.375$$

Thus, the simultaneous equations to solve for Δr and Δs are

$$-4.875\Delta r + 10.75\Delta s = 10.5$$
$$-16.375\Delta r - 4.875\Delta s = -11.375$$

which can be solved for $\Delta r = 0.3558$ and $\Delta s = 1.1381$. Therefore, our original guesses can be corrected to

$$r = -1 + 0.3558 = -0.6442$$
$$s = -1 + 1.1381 = 0.1381$$

and the approximate errors can be evaluated by Eqs. (7.37) and (7.38),

$$\left|\varepsilon_{a,r}\right| = \left|\frac{0.3558}{-0.6442}\right|100\% = 55.23\% \qquad \left|\varepsilon_{a,s}\right| = \left|\frac{1.1381}{0.1381}\right|100\% = 824.1\%$$

Next, the computation is repeated using the revised values for r and s. Applying Eqs. (7.32) and (7.36) yields

$$b_5 = 1 \qquad b_4 = -4.1442 \qquad b_3 = 5.5578 \qquad b_2 = -2.0276 \qquad b_1 = -1.8013$$
$$b_0 = 2.1304$$

$$c_5 = 1 \qquad c_4 = -4.7884 \qquad c_3 = 8.7806 \qquad c_2 = -8.3454 \qquad c_1 = 4.7874$$

Therefore, we must solve

$$-8.3454\Delta r + 8.7806\Delta s = 1.8013$$
$$4.7874\Delta r - 8.3454\Delta s = -2.1304$$

for $\Delta r = 0.1331$ and $\Delta s = 0.3316$, which can be used to correct the root estimates as

$$r = -0.6442 + 0.1331 = -0.5111 \qquad \left|\varepsilon_{a,r}\right| = 26.0\%$$
$$s = 0.1381 + 0.3316 = 0.4697 \qquad \left|\varepsilon_{a,s}\right| = 70.6\%$$

The computation can be continued, with the result that after four iterations the method converges on values of $r = -0.5$ ($\left|\varepsilon_{a,r}\right| = 0.063\%$) and $s = 0.5$ ($\left|\varepsilon_{a,s}\right| = 0.040\%$). Equation (7.39) can then be employed to evaluate the roots as

$$x = \frac{-0.5 \pm \sqrt{(-0.5)^2 + 4(0.5)}}{2} = 0.5, -1.0$$

At this point, the quotient is the cubic equation

$$f(x) = x^3 - 4x^2 + 5.25x - 2.5$$

Bairstow's method can be applied to this polynomial using the results of the previous step, $r = -0.5$ and $s = 0.5$, as starting guesses. Five iterations yield estimates of $r = 2$ and $s = -1.249$, which can be used to compute

$$x = \frac{2 \pm \sqrt{2^2 + 4(-1.249)}}{2} = 1 \pm 0.499i$$

At this point, the quotient is a first-order polynomial that can be directly evaluated by Eq. (7.40) to determine the fifth root: 2.

Note that the heart of Bairstow's method is the evaluation of the b's and c's via Eqs. (7.32) and (7.36). One of the primary strengths of the method is the concise way in which these recurrence relationships can be programmed.

Figure 7.5 lists pseudocode to implement Bairstow's method. The heart of the algorithm consists of the loop to evaluate the b's and c's. Also notice that the code to solve the simultaneous equations checks to prevent division by zero. If this is the case, the values of r and s are perturbed slightly and the procedure is begun again. In addition, the algorithm

(a) Bairstow Algorithm

```
SUB Bairstow (a,nn,es,rr,ss,maxit,re,im,ier)
DIMENSION b(nn), c(nn)
r = rr; s = ss; n = nn
ier = 0; ea1 = 1; ea2 = 1
DO
  IF n < 3 OR iter ≥ maxit EXIT
  iter = 0
  DO
    iter = iter + 1
    b(n) = a(n)
    b(n − 1) = a(n − 1) + r * b(n)
    c(n) = b(n)
    c(n − 1) = b(n − 1) + r * c(n)
    DO i = n − 2, 0, −1
      b(i) = a(i) + r * b(i + 1) + s * b(i + 2)
      c(i) = b(i) + r * c(i + 1) + s * c(i + 2)
    END DO
    det = c(2) * c(2) − c(3) * c(1)
    IF det ≠ 0 THEN
      dr = (−b(1) * c(2) + b(0) * c(3))/det
      ds = (−b(0) * c(2) + b(1) * c(1))/det
      r = r + dr
      s = s + ds
      IF r≠0 THEN ea1 = ABS(dr/r) * 100
      IF s≠0 THEN ea2 = ABS(ds/s) * 100
    ELSE
      r = r + 1
      s = s + 1
      iter = 0
    END IF
    IF ea1 ≤ es AND ea2 ≤ es OR iter ≥ maxit EXIT
  END DO
  CALL Quadroot(r,s,r1,i1,r2,i2)
  re(n) = r1
  im(n) = i1
  re(n − 1) = r2
  im(n − 1) = i2
  n = n − 2
  DO   i = 0, n
    a(i) = b(i + 2)
  END DO
END DO
```

```
IF iter < maxit   THEN
  IF n = 2   THEN
    r = −a(1)/a(2)
    s = −a(0)/a(2)
    CALL Quadroot(r,s,r1,i1,r2,i2)
    re(n) = r1
    im(n) = i1
    re(n − 1) = r2
    im(n − 1) = i2
  ELSE
    re(n) = −a(0)/a(1)
    im(n) = 0
  END IF
ELSE
  ier = 1
END IF
END Bairstow
```

(b) Roots of Quadratic Algorithm

```
SUB Quadroot (r,s,r1,i1,r2,i2)
disc = r ^ 2 + 4 * s
IF disc > 0   THEN
  r1 = (r + SQRT(disc))/2
  r2 = (r − SQRT(disc))/2
  i1 = 0
  i2 = 0
ELSE
  r1 = r/2
  r2 = r1
  i1 = SQRT(ABS(disc))/2
  i2 = −i1
END IF
END QuadRoot
```

FIGURE 7.5

(a) Algorithm for implementing Bairstow's method, along with (b) an algorithm to determine the roots of a quadratic.

places a user-defined upper limit on the number of iterations (MAXIT) and should be designed to avoid division by zero while calculating the error estimates. Finally, the algorithm requires initial guesses for r and s (rr and ss in the code). If no prior knowledge of the roots exist, they can be set to zero in the calling program.

7.6   OTHER METHODS

Other methods are available to locate the roots of polynomials. The *Jenkins-Traub method* (Jenkins and Traub, 1970) is commonly used in software libraries like IMSL. It is fairly complicated, and a good starting point to understanding it is found in Ralston and Rabinowitz (1978).

   Laguerre's method, which approximates both real and complex roots and has cubic convergence, is among the best approaches. A complete discussion can be found in Householder (1970). In addition, Press et al. (1992) present a nice algorithm to implement the method.

7.7   ROOT LOCATION WITH LIBRARIES AND PACKAGES

Software libraries and packages have great capabilities for locating roots. In this section, we will give you a taste of some of the more useful ones.

7.7.1 Mathcad

Mathcad has a numeric mode function called **root** that can be used to solve an equation of a single variable using the secant method. The method requires that you supply a function $f(x)$ and an initial guess. It iterates until the magnitude of $f(x)$ at the proposed root is less than the predefined value of **TOL.** The Mathcad implementation has similar advantages and disadvantages as the conventional secant method such as issues concerning the accuracy of the initial guess and the rate of convergence.

   Mathcad can find all the real or complex roots of polynomials with **polyroots.** This numeric or symbolic mode function is based on the Laguerre method. This function does not require initial guesses, and all the roots are returned at the same time.

   Mathcad contains a numeric mode function called **Find** that can be used to solve up to 50 simultaneous nonlinear algebraic equations. Acceptable values for the solution may be unconstrained or constrained to fall within specified limits. If **Find** fails to locate a solution that satisfies the equations and constraints, it returns the error message "did not find solution." However, Mathcad also contains a similar function called **Minerr.** This function gives solution results that minimize the errors in the constraints even when exact solutions cannot be found. Thus, the problem of solving for the roots of nonlinear equations is closely related to both nonlinear least-squares and optimization problems. These problems and **Minerr** are covered in detail in Parts Four and Five.

   Figure 7.6 shows a typical Mathcad worksheet. The menus at the top give you quick access to common arithmetic operators and functions, various two- and three-dimensional plot types, and the environment to create subprograms. Equations, text, data, or graphs can be placed anywhere on the screen. You can use a variety of fonts, colors, and styles to construct worksheets with almost any design and format that pleases you. Consult the

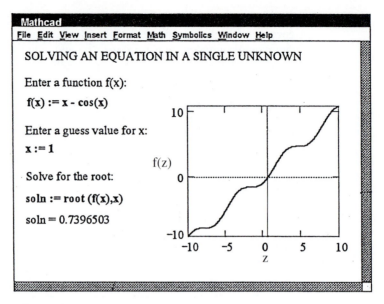

FIGURE 7.6   Mathcad screen to solve single equation.

summary of the Mathcad User's manual in App. B or the full manual available from Math-soft. Note that in all our Mathcad examples, we have tried to fit the entire Mathcad session onto a single screen. You should realize that the graph would have to be placed below the commands to work properly.

Let's start with an example that solves for the root of $f(x) = x - \cos x$. The first step is to enter the function. This is done by typing $f(x)$: which is automatically converted to $f(x) :=$ by Mathcad. The $:=$ is called the definition symbol. Next an initial guess is input in a similar manner using the definition symbol. Now, soln is defined as **root** $(f(x), x)$, which invokes the secant method with a starting value of 1.0. Iteration is continued until $f(x)$ evaluated at the proposed root is less than **TOL**. The value of **TOL** is set from the Math/Options pull down menu. Finally the value of soln is displayed using a normal equal sign ($=$). The number of significant figures is set from the Format/Number pull down menu. The text labels and equation definitions can be placed anywhere on the screen in a number of different fonts, styles, sizes, and colors. The graph can be placed anywhere on the worksheet by clicking to the desired location. This places a red cross hair at that location. Then use the Insert/Graph/X-Y Plot pull down menu to place an empty plot on the worksheet with placeholders for the expressions to be graphed and for the ranges of the x and y axes. Simply type $f(z)$ in the placeholder on the y axis and -10 and 10 for the z-axis range. Mathcad does all the rest to produce the graph shown in Fig. 7.6. Once the graph has been created you can use the Format/Graph/X-Y Plot pull down menu to vary the type of graph; change the color, type, and weight of the trace of the function; and add titles, labels and other features.

Figure 7.7 shows how Mathcad can be used to find the roots of a polynomial using the **polyroots** function. First, $p(x)$ and v are input using the $:=$ definition symbol. Note that v is a vector that contains the coefficients of the polynomial starting with zero-order term and ending in this case with the third-order term. Next, r is defined (using $:=$) as **polyroots**(v),

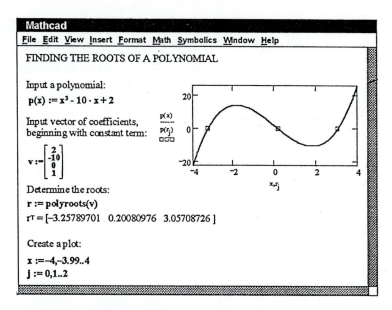

FIGURE 7.7 Mathcad screen to solve for roots of polynomial.

which invokes the Laguerre method. The roots contained in r are displayed as r^T using a normal equal sign ($=$). Next, a plot is constructed in a manner similar to the above, except that now two range variables, j and x, are used to define the range of the x axis and the location of the roots. The range variable for x is constructed by typing x and then ":" (which appears as :=) and then -4, and then "," and then -3.99, and then ";" (which is transformed into .. by Mathcad), and finally 4. This creates a vector of values of x ranging from -4 to 4 with an increment of 0.01 for the x axis with corresponding values for $p(x)$ on the y axis. The j range variable is used to create three values for r and $p(r)$ that are plotted as individual small squares. Note that again, in our effort to fit the entire Mathcad session onto a single screen, we have placed the graph above the commands. You should realize that the graph would have to be below the commands to work properly.

The last example shows the solution of a system of nonlinear equations using a Mathcad Solve Block (Fig. 7.8). The process begins with using the definition symbol to create initial guesses for x and y. The word **Given** then alerts Mathcad that what follows is a system of equations. Then comes the equations and inequalities (not used here). Note that for this application Mathcad requires the use of a symbolic equal sign typed as **[Ctrl]=** or $<$ and $>$ to separate left and right sides of an equation. Now, the variable vec is defined as **Find** (x,y) and the value of vec is shown using an equal sign.

7.7.2 Excel

A spreadsheet like Excel can be used to locate a root by *trial and error.* For example, if we want to find a root of

$$f(x) = x - \cos x$$

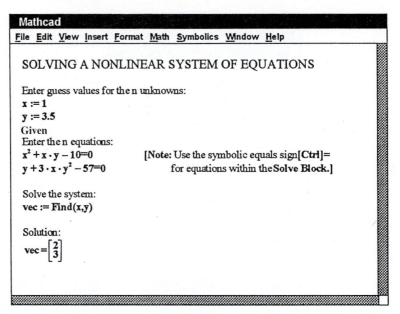

FIGURE 7.8 Mathcad screen to solve a system of nonlinear equations.

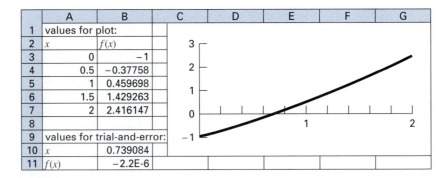

FIGURE 7.9

A spreadsheet set up to determine the root of $f(x) = x - \cos x$ by trial and error. The plot is used to obtain a good initial guess.

first, you can enter a value for x in a cell. Then set up another cell for $f(x)$ that would obtain its value for x from the first cell. You can then vary the x cell until the $f(x)$ cell approaches zero. This process can be further enhanced by using Excel's plotting capabilities to obtain a good initial guess (Fig. 7.9).

Although Excel does facilitate a trial-and-error approach, it also has two standard tools that can be employed for root location: *Goal Seek* and *Solver*. Both these tools can be employed to systematically adjust the initial guesses. Goal Seek is expressly used to drive an equation to a value (in our case, zero) by varying a single parameter.

EXAMPLE 7.4   Using Excel's Goal Seek Tool to Locate a Single Root

Problem Statement.   Employ Goal Seek to determine the root of the transcendental function

$$f(x) = x - \cos x$$

Solution.   As in Fig. 7.9, the key to solving a single equation with Excel is creating a cell to hold the value of the function in question, and then making the value dependent on another cell. Once this is done, the selection Goal Seek is chosen from the Tools menu. At this point a dialogue box will be displayed, asking you to set a cell to a value by changing another cell. For the example, suppose that as in Fig. 7.9 your guess is entered in cell B10 and your function result in cell B11. The Goal Seek dialogue box would be filled out as

Set cell:	B11
To value:	0
By changing cell:	B10

When the OK button is selected, a message box displays the results,

Goal seek with cell B11
found a solution.
Target value: 0
Current value: $-2.19892E{-}06$

The cells on the spreadsheet would also be modified to the new values (as shown in Fig. 7.9).

The *Solver* tool is more sophisticated than Goal Seek in that (1) it can vary several cells simultaneously and (2) along with driving a target cell to a value, it can minimize and maximize its value. The next example illustrates how it can be used to solve a system of nonlinear equations.

EXAMPLE 7.5   Using Excel's Solver for a Nonlinear System

Problem Statement.   Recall that in Sec. 6.5 we obtained the solution of the following set of simultaneous equations,

$$u(x, y) = x^2 + xy - 10 = 0$$

$$v(x, y) = y + 3xy^2 - 57 = 0$$

Note that a correct pair of roots is $x = 2$ and $y = 3$. Use Solver to determine the roots using initial guesses of $x = 1$ and $y = 3.5$.

Solution. As shown below, two cells (B1 and B2) can be created to hold the guesses for x and y. The function values themselves, $u(x, y)$ and $v(x, y)$ can then be entered into two other cells (B3 and B4). As can be seen, the initial guesses result in function values that are far from zero.

	A	B	C	D
1	x	1		
2	y	3.5		
3	u(x, y)	−5.5		
4	v(x, y)	−16.75		
5				
6	Sum square	310.8125		

Next, another cell can be created that contains a single value reflecting how close both functions are to zero. One way to do this is to sum the squares of the function values. This is done and the result entered in cell B6. If both functions are at zero, this function should also be at zero. Further, using the squared functions avoids the possibility that both functions could have the same nonzero value, but with opposite signs. For this case, the target cell (B6) would be zero, but the roots would be incorrect.

Once the spreadsheet is created, the selection **Solver** is chosen from the **Tools** menu. At this point a dialogue box will be displayed, querying you for pertinent information. The pertinent cells of the Solver dialogue box would be filled out as

Set target cell:	B6	
Equal to ○ max ○ min ● equal to	0
By changing cells:		
B1:B2		

When the OK button is selected, a dialogue box will open with a report on the success of the operation. For the present case, the Solver obtains the correct solution:

	A	B	C	D
1	x	1.999476		
2	y	3.0003865		
3	u(x, y)	−0.0029		
4	v(x, y)	0.000112		
5				
6	Sum square	8.42E−06		

It should be noted that the Solver can fail. Its success depends on (1) the condition of the system of equations and/or (2) the quality of the initial guesses. Thus, the successful outcome of the previous example is not guaranteed. Despite this, we have found Solver useful enough to make it a feasible option for quickly obtaining roots in a wide range of engineering applications.

7.7.3 Matlab

Although add-ons are available that can expand its fundamental capabilities, the basic MATLAB package did not have a general root-location capability at the time this book was published. However, it is superb at manipulating and locating the roots of polynomials. Some of the key MATLAB functions related to polynomial manipulation are listed in Table 7.1. The following example illustrates how some of these functions can be employed to both manipulate polynomials and determine their roots.

TABLE 7.1 Common functions in MATLAB related to polynomial manipulation and root location.

Function	Description
roots	Find polynomial roots.
poly	Construct polynomial with specified roots.
polyval	Evaluate polynomial.
polyvalm	Evaluate polynomial with matrix argument.
residue	Partial-fraction expansion (residues).
polyder	Differentiate polynomial.
conv	Multiply polynomials.
deconv	Divide polynomials.

EXAMPLE 7.6   Using MATLAB to Manipulate and Determine the Roots of Polynomials

Problem Statement.   Explore how MATLAB can be employed to manipulate and determine the roots of polynomials. Use the following equation from Example 7.3,

$$f_5(x) = x^5 - 3.5x^4 + 2.75x^3 + 2.125x^2 - 3.875x + 1.25 \qquad \text{(E7.6.1)}$$

which has three real roots: 0.5, -1.0, and 2, and one pair of complex roots: $-1 \pm 0.5i$.

Solution.   Polynomials are entered into MATLAB by storing the coefficients as a vector. For example, at the MATLAB prompt ($\gg$) typing and entering the follow line stores the coefficients in the vector **a**,

```
>> a=[1 -3.5 2.75 2.125 -3.875 1.25];
```

We can then proceed to manipulate the polynomial. For example we can evaluate it at $x = 1$ by typing

```
>> polyval(a,1)
```

with the result $1(1)^5 - 3.5(1)^4 + 2.75(1)^3 + 2.125(1)^2 - 3.875(1) + 1.25 = -0.25$,

```
ans =
   -0.2500
```

We can evaluate the derivative $f'(x) = 5x^4 - 14x^3 + 8.25x^2 + 4.25x - 3.875$ by

```
>> polyder (a)
ans =
     5.0000   -14.0000    8.2500    4.2500   -3.8750
```

Next, let's create a quadratic polynomial that has roots corresponding to two of the original roots of Eq. (E7.6.1): 0.5 and -1. This quadratic is $(x - 0.5)(x + 1) = x^2 + 0.5x - 0.5$ and can be entered into MATLAB as the vector **b**,

```
>> b=[1 0.5 -0.5];
```

We can divide this polynomial into the original polynomial by

```
>> [d,e]=deconv(a,b)
```

with the result being a quotient (a third-order polynomial **d**) and a remainder (**e**),

```
d =
     1.0000   -4.0000    5.2500   -2.5000
e =
     0     0     0     0     0     0
```

Because the polynomial is a perfect divisor, the remainder polynomial has zero coefficients. Now, the roots of the quotient polynomial can be determined as

```
>> roots(d)
```

with the expected result that the remaining roots of the original polynomial (E7.6.1) are found,

```
ans =
     2.0000
     1.0000 + 0.5000i
     1.0000 - 0.5000i
```

We can now multiply **d** by **b** to come up with the original polynomial,

```
>> conv(d,b)
ans =
     1.0000   -3.5000    2.7500    2.1250   -3.8750    1.2500
```

Finally, we can determine all the roots of the original polynomial by

```
>> r=roots(a)
r =
    -1.0000
     2.0000
     1.0000 + 0.5000i
     1.0000 - 0.5000i
     0.5000
```

TABLE 7.2   IMSL routines to locate roots.

Category	Routine	Capability
Roots of a function		
	ZREAL	Finds the real zeros of a real function using Müller's method.
	ZBREN	Finds a zero of a real function that changes sign in a given interval.
	ZANLY	Finds the zeros of a univariate complex function using Müller's method.
Root of a system of equations		
	NEQNF	Solves a system of nonlinear equations using a modified Powell hybrid algorithm (a variation of Newton's method) and a finite-difference approximation of the Jacobian.
	NEQNJ	Solves a system of nonlinear equations using a modified Powell hybrid algorithm (a variation of Newton's method) with a user-supplied Jacobian.
	NEQBF	Solves a system of nonlinear equations using factored secant update and a finite-difference approximation of the Jacobian.
	NEQBJ	Solves a system of nonlinear equations using factored secant update with a user-supplied Jacobian.
Roots of polynomials		
	ZPORC	Finds the zeros of polynomials with real coefficients with the Jenkins-Traub algorithm.
	ZPLRC	Finds the zeros of polynomials with real coefficients with the Laguerre method.
	ZPOCC	Finds the zeros of polynomials with complex coefficients with the Jenkins-Traub algorithm.

7.7.4 IMSL

IMSL has several subroutines for determining roots of equations (Table 7.2). In the present discussion, we will focus on the ZREAL routine. This routine locates the real zeros of a real function using Müller's method.

ZREAL is implemented by the following CALL statement,

```
CALL ZREAL(F, ERABS, ERREL, EPS, ETA, NR, IMAX, XO, X, INFO)
```

where F = A user-defined function for which the zeros are to be found.

ERABS = First stopping criterion, terminates if $|f(x_i)| <$ ERABS. (Input)

ERREL = Second stopping criterion; terminates if $|(x_i - x_{i-1})/x_i| <$ ERREL. (Input)

EPS = See ETA. (Input)

ETA = Spread criteria for multiple roots. (Input)

> If the zero, x_i, has been computed and $|x_i - x_j| <$ EPS, where x_j is a previously computed zero, the computation is restarted with a new guess of $x_i +$ ETA.

NR = Number of roots to be found. (Input)

IMAX = Maximum allowable iterations per zero. (Input)

X0 = A vector of length NROOT containing the initial guesses. (Input)

X = A vector of length NROOT containing the computed zeros. (Output)

INFO = An integer vector of length NROOT. (Output)
Contains the number of iterations to find each root.

Note that the iterations terminate when either of the stopping criteria are met, or the maximum iterations are exceeded. The function F has the general format

```
FUNCTION F(X)
REAL F,X
F = ...
END
```

where the line "F = ..." is where the function of the unknown variable X is written.

EXAMPLE 7.7   Using IMSL to Locate a Single Root

Problem Statement.   Use ZREAL to determine the root of the transcendental function

$$f(x) = x - \cos x$$

Solution.   An example of a main Fortran 90 program and function using ZREAL to solve this problem can be written as

```
PROGRAM Root
IMPLICIT NONE
INTEGER::nroot
PARAMETER(nroot=1)
INTEGER::itmax=50
REAL::errabs=0.,errrel=1.E-5,eps=0.,eta=0.
REAL::f,x0(nroot),x(nroot)
EXTERNAL f
INTEGER::info(nroot)
PRINT *, "Enter initial guess"
READ *, x0
CALL ZREAL(f,errabs,errrel,eps,eta,nroot,itmax,x0,x,info)
PRINT *, "root = ", x
PRINT *, "iterations = ", info
END PROGRAM

FUNCTION f(x)
IMPLICIT NONE
REAL::f,x
f = x - cos(x)
END FUNCTION
```

Output:

```
Enter initial guess
0.5
root =      7.390851E-01
iterations =            5
```

PROBLEMS

7.1 Divide a polynomial

$$f(x) = x^4 - 5x^3 + 5x^2 + 5x - 6$$

by the monomial factor $x - 2$. Is $x = 2$ a root?

7.2 Divide a polynomial

$$f(x) = x^5 - 6x^4 + x^3 - 7x^2 - 7x + 12$$

by the monomial factor $x - 2$.

7.3 Use Müller's method to determine the positive real root of
(a) $f(x) = x^3 + x^2 - 4x - 4$
(b) $f(x) = x^3 - 0.5x^2 + 4x - 2$

7.4 Use Müller's method to determine the real and complex roots of
(a) $f(x) = x^3 - x^2 + 2x - 2$
(b) $f(x) = 2x^4 + 6x^2 + 8$
(c) $f(x) = x^4 - 2x^3 + 6x^2 - 2x + 5$

7.5 Use Bairstow's method to determine the roots of
(a) $f(x) = -2 + 6.2x - 4x^2 + 0.7x^3$
(b) $f(x) = 9.34 - 21.97x + 16.3x^2 - 3.704x^3$
(c) $f(x) = x^4 - 2x^3 + 6x^2 - 2x + 5$

7.6 Develop a subprogram to implement Müller's method. Test it by duplicating Example 7.2.

7.7 Use the subprogram developed in Prob. 7.6 to determine the real roots of Prob. 7.4a. Construct a graph (by hand or with the TOOLKIT, Excel, or some other graphics package) to develop suitable starting guesses.

7.8 Develop a subprogram to implement Bairstow's method. Test it by duplicating Example 7.3.

7.9 Use the subprogram developed in Prob. 7.8 to determine the roots of the equations in Prob. 7.5.

7.10 Determine the real root of $x^{3.3} = 79$ with the Goal Seek capability of Excel or a library or package of your choice.

7.11 The velocity of a falling parachutist v is given by

$$v = \frac{gm}{c}(1 - e^{-(c/m)t})$$

where $g = 9.8$ m/s². For a parachutist with a drag coefficient $c = 14$ kg/s, compute the mass m so that the velocity is $v = 35$ m/s at $t = 7$ s. Use the Goal Seek capability of Excel or a library or package of your choice to determine m.

7.12 Determine the roots of the simultaneous nonlinear equations

$$y = -x^2 + x + 0.5$$
$$y + 5xy = x^2$$

Employ initial guesses of $x = y = 1.2$ and use the Solver tool from Excel or a library or package of your choice.

7.13 Determine the roots of the simultaneous nonlinear equations

$$y + 1 = x^2$$
$$x^2 = 5 - y^2$$

Use a graphical approach to obtain your initial guesses. Determine refined estimates with the Solver tool from Excel or a library or package of your choice.

7.14 Perform the identical MATLAB operations as those in Example 7.6, or use a library or package of your choice to find all the roots of the polynomial

$$f_5(x) = (x + 2)(x - 6)(x - 1)(x + 4)(x - 8)$$

7.15 Use MATLAB or a library or package of your choice to determine the roots for the equations in Prob. 7.5.

7.16 Develop a subprogram to solve for the roots of a polynomial using the IMSL routine, ZREAL, or a library or package of your choice. Test it by determining the roots of the equations from Probs. 7.4 and 7.5.

CHAPTER 8

Engineering Applications: Roots of Equations

The purpose of this chapter is to use the numerical procedures discussed in Chaps. 5, 6, and 7 to solve actual engineering problems. Numerical techniques are important for practical applications because engineers frequently encounter problems that cannot be approached using analytical techniques. For example, simple mathematical models that can be solved analytically may not be applicable when real problems are involved. Thus, more complicated models must be employed. For these cases, it is appropriate to implement a numerical solution on a computer. In other situations, engineering design problems may require solutions for implicit variables in complicated equations.

The following applications are typical of those that are routinely encountered during upper-class and graduate studies. Furthermore, they are representative of problems you will address professionally. The problems are drawn from the four major disciplines of engineering: chemical, civil, electrical, and mechanical. These applications also serve to illustrate the trade-offs among the various numerical techniques.

The first application, taken from chemical engineering, provides an excellent example of how root-location methods allow you to use realistic formulas in engineering practice. In addition, it also demonstrates how the efficiency of the Newton-Raphson technique is used to advantage when a large number of root-location computations is required.

The following engineering design problems are taken from civil, electrical, and mechanical engineering. Section 8.2 uses both bracketing and open methods to determine the depth and velocity of water flowing in an open channel. Section 8.3 shows how the roots of transcendental equations can be used in the design of an electrical circuit. Sections 8.2 and 8.3 also illustrate how graphical methods provide insight into the root-location process. Finally, Sec. 8.4 uses polynomial root location to analyze the vibrations of an automobile.

8.1 IDEAL AND NONIDEAL GAS LAWS (CHEMICAL/PETROLEUM ENGINEERING)

Background. The *ideal gas law* is given by

$$pV = nRT \tag{8.1}$$

where p is the absolute pressure, V is the volume, n is the number of moles, R is the universal gas constant, and T is the absolute temperature. Although this equation is widely used by engineers and scientists, it is accurate over only a limited range of pressure and temperature. Furthermore, Eq. (8.1) is more appropriate for some gases than for others.

An alternative equation of state for gases is given by

$$\left(p + \frac{a}{v^2}\right)(v - b) = RT \tag{8.2}$$

known as the *van der Waals equation*, where $v = V/n$ is the molal volume and a and b are empirical constants that depend on the particular gas.

A chemical engineering design project requires that you accurately estimate the molal volume (v) of both carbon dioxide and oxygen for a number of different temperature and pressure combinations so that appropriate containment vessels can be selected. It is also of interest to examine how well each gas conforms to the ideal gas law by comparing the molal volume as calculated by Eqs. (8.1) and (8.2). The following data are provided:

$$R = 0.082054 \text{ L atm/(mol K)}$$

$$\left.\begin{array}{l} a = 3.592 \\ b = 0.04267 \end{array}\right\} \text{ carbon dioxide}$$

$$\left.\begin{array}{l} a = 1.360 \\ b = 0.03183 \end{array}\right\} \text{ oxygen}$$

The design pressures of interest are 1, 10, and 100 atm for temperature combinations of 300, 500, and 700 K.

Solution.   Molal volumes for both gases are calculated using the ideal gas law, with $n = 1$. For example, if $p = 1$ atm and $T = 300$ K,

$$v = \frac{V}{n} = \frac{RT}{p} = 0.082054 \frac{\text{L atm}}{\text{mol K}} \frac{300 \text{ K}}{1 \text{ atm}} = 24.6162 \text{ L/mol}$$

These calculations are repeated for all temperature and pressure combinations and presented in Table 8.1.

TABLE 8.1   Computations of molal volume.

Temperature, K	Pressure, atm	Molal Volume (Ideal Gas Law), L/mol	Molal Volume (van der Waals) Carbon Dioxide, L/mol	Molal Volume (van der Waals) Oxygen, L/mol
300	1	24.6162	24.5126	24.5928
	10	2.4616	2.3545	2.4384
	100	0.2462	0.0795	0.2264
500	1	41.0270	40.9821	41.0259
	10	4.1027	4.0578	4.1016
	100	0.4103	0.3663	0.4116
700	1	57.4378	57.4179	57.4460
	10	5.7438	5.7242	5.7521
	100	0.5744	0.5575	0.5842

The computation of molal volume from the van der Waals equation can be accomplished using any of the numerical methods for finding roots of equations discussed in Chaps. 5, 6, and 7, with

$$f(v) = \left(p + \frac{a}{v^2}\right)(v - b) - RT \tag{8.3}$$

In this case, the derivative of $f(v)$ is easy to determine and the Newton-Raphson method is convenient and efficient to implement. The derivative of $f(v)$ with respect to v is given by

$$f'(v) = p - \frac{a}{v^2} + \frac{2ab}{v^3} \tag{8.4}$$

The Newton-Raphson method is described by Eq. (6.6):

$$v_{i+1} = v_i - \frac{f(v_i)}{f'(v_i)}$$

which can be used to estimate the root. For example, using the initial guess of 24.6162, the molal volume of carbon dioxide at 300 K and 1 atm is computed as 24.5126 L/mol. This result was obtained after just two iterations and has an ε_a of less than 0.001 percent.

Similar computations for all combinations of pressure and temperature for both gases are presented in Table 8.1. It is seen that the results for the ideal gas law differ from those for van der Waals equation for both gases, depending on specific values for p and T. Furthermore, because some of these results are significantly different, your design of the containment vessels would be quite different, depending on which equation of state was used.

In this case, a complicated equation of state was examined using the Newton-Raphson method. The results varied significantly from the ideal gas law for several cases. From a practical standpoint, the Newton-Raphson method was appropriate for this application because $f'(v)$ was easy to calculate. Thus, the rapid convergence properties of the Newton-Raphson method could be exploited.

In addition to demonstrating its power for a single computation, the present design problem also illustrates how the Newton-Raphson method is especially attractive when numerous computations are required. Because of the speed of digital computers, the efficiency of various numerical methods for most roots of equations are indistinguishable for a single computation. Even a 1-s difference between the crude bisection approach and the efficient Newton-Raphson does not amount to a significant time loss when only one computation is performed. However, suppose that millions of root evaluations are required to solve a problem. In this case, the efficiency of the method could be a deciding factor in the choice of a technique.

For example, suppose that you are called upon to design an automatic computerized control system for a chemical production process. This system requires accurate estimates of molal volumes on an essentially continuous basis to properly manufacture the final product. Gauges are installed that provide instantaneous readings of pressure and temperature. Evaluations of v must be obtained for a variety of gases that are used in the process.

For such an application, bracketing methods such as bisection or false position would probably be too time-consuming. In addition, the two initial guesses that are required for these approaches may also interject a critical delay in the procedure. This shortcoming is relevant to the secant method, which also needs two initial estimates.

In contrast, the Newton-Raphson method requires only one guess for the root. The ideal gas law could be employed to obtain this guess at the initiation of the process. Then, assuming that the time frame is short enough so that pressure and temperature do not vary wildly between computations, the previous root solution would provide a good guess for the next application. Thus, the close guess that is often a prerequisite for convergence of the Newton-Raphson method would automatically be available. All the above considerations would greatly favor the Newton-Raphson technique for such problems.

8.2   OPEN-CHANNEL FLOW (CIVIL/ENVIRONMENTAL ENGINEERING)

Background.   Civil engineering is a broad field that includes such diverse areas as structural, geotechnical, transportation, environmental, and water-resources engineering. The last two specialties deal with both water pollution and water supply, and hence, make extensive use of the science of fluid mechanics.

One general problem relates to the flow of water in open channels such as rivers and canals. The flow rate, which is routinely measured in most major rivers and streams, is defined as the volume of water passing a particular point in a channel per unit time, Q (m³/s).

Although the flow rate is a useful quantity, a further question relates to what happens when you put a specific flow rate into a sloping channel (Fig. 8.1). In fact, two things happen: the water will reach a specific depth H (m) and move at a specific velocity U (m/s). Environmental engineers might be interested in knowing these quantities to predict the transport and fate of pollutants in a river. So the general question is: if you are given the flow rate for a channel, how do you compute the depth and velocity?

Solution.   The most fundamental relationship between flow and depth is the *continuity equation:*

$$Q = UA_c \tag{8.5}$$

where A_c = the cross-sectional area of the channel (m²). Depending on the channel shape, the area can be related to the depth by some functional relationship. For the rectangular

FIGURE 8.1

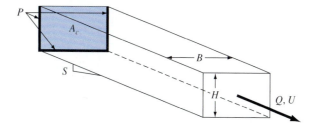

channel depicted in Fig. 8.1, $A_c = BH$. Substituting this relationship into Eq. (8.5) gives

$$Q = UBH \tag{8.6}$$

where B = the width (m). It should be noted that the continuity equation derives from the *conservation of mass* (recall Table 1.1).

Now, although Eq. (8.6) certainly relates the channel parameters, it is not sufficient to answer our question. Assuming that B is specified, we have one equation and two unknowns (U and H). We therefore require an additional equation. For uniform flow (meaning that the flow does not vary spatially and temporally) the Irish engineer Robert Manning proposed the following semiempirical formula (appropriately called the *Manning equation*)

$$U = \frac{1}{n} R^{2/3} S^{1/2} \tag{8.7}$$

where n = the Manning roughness coefficient (a dimensionless number used to parameterize the channel friction), S = the channel slope (dimensionless, meters drop per meter length), and R = the hydraulic radius (m), which is related to more fundamental parameters by

$$R = \frac{A_c}{P} \tag{8.8}$$

where P = the wetted perimeter (m). As the name implies, the wetted perimeter is the length of the channel sides and bottom that is under water. For example, for a rectangular channel, it is defined as

$$P = B + 2H \tag{8.9}$$

It should be noted that just as the continuity equation derives from the conservation of mass, the Manning equation is an expression of the *conservation of momentum*. In particular, it indicates how velocity is dependent on roughness, a manifestation of friction.

Although the system of nonlinear equations (8.6 and 8.7) can be solved simultaneously (for example, using the multidimensional Newton-Raphson approach described in Sec. 6.5.2), an easier approach would be to combine the equations. Equation (8.7) can be substituted into Eq. (8.6) to give

$$Q = \frac{BH}{n} R^{2/3} S^{1/2} \tag{8.10}$$

Then the hydraulic radius, Eq. (8.8), along with the various relationships for the rectangular channel can be substituted,

$$Q = \frac{S^{1/2}}{n} \frac{(BH)^{5/3}}{(B + 2H)^{2/3}} \tag{8.11}$$

Thus, the equation now contains a single unknown H along with the given value for Q and the channel parameters (n, S, and B).

Although we have one equation with an unknown, it is impossible to solve explicitly for H. However, the depth can be determined numerically by reformulating the equation as a roots problem,

$$f(H) = \frac{S^{1/2}}{n} \frac{(BH)^{5/3}}{(B + 2H)^{2/3}} - Q = 0 \tag{8.12}$$

Equation (8.12) can be solved readily with any of the root-location methods described in Chaps. 5 and 6. For example, if $Q = 5$ m³/s, $B = 20$ m, $n = 0.03$, and $S = 0.0002$, the equation is

$$f(H) = 0.471405 \frac{(20H)^{5/3}}{(20 + 2H)^{2/3}} - 5 = 0 \tag{8.13}$$

It can be solved for $H = 0.7023$ m. The result can be checked by substitution into Eq. (8.13) to give

$$f(H) = 0.471405 \frac{(20 \times 0.7023)^{5/3}}{(20 + 2 \times 0.7023)^{2/3}} - 5 = 7.8 \times 10^{-5} \tag{8.14}$$

which is quite close to zero.

Our other unknown, the velocity, can now be determined by substitution back into Eq. (8.6),

$$U = \frac{Q}{BH} = \frac{5}{20(0.7023)} = 0.356 \text{ m/s} \tag{8.15}$$

Thus, we have successfully solved for the depth and velocity.

Now let's delve a little deeper into the numerical aspects of this problem. One pertinent question might be: how do we come up with good initial guesses for our numerical method? The answer depends on the type of method.

For bracketing methods such as bisection and false position, one approach would be to determine whether we can estimate lower and upper guesses that always bracket a single root. A conservative approach might be to choose zero as our lower bound. Then, if we knew the maximum possible depth that could occur, this value could serve as the upper guess. For example, all but the world's biggest rivers are less than about 10 m deep. Therefore, we might choose 0 and 10 as our bracket for H.

If $Q > 0$ and $H = 0$, Eq. (8.12) will always be negative for the lower guess. As H is increased, Eq. (8.12) will increase monotonically and eventually become positive. Therefore, the guesses should bracket a single root for most cases routinely confronted in natural rivers and streams.

Now, a technique like bisection should very reliably home in on the root. But what price is paid? By using such a wide bracket and a technique like bisection, the number of iterations to attain a desirable precision could be computationally excessive. For example, if a tolerance of 0.001 m were chosen, Eq. (5.5) could be used to calculate

$$n = \frac{\log{(10/0.001)}}{\log 2} = 13.3$$

Thus, 14 iterations would be required. Although this would certainly not be costly for a single calculation, it could be exorbitant if many such evaluations were made. The alternatives would be to narrow the initial bracket (based on system-specific knowledge), change to a more efficient bracketing technique (like false position), or accept a coarser precision.

Another way to get better efficiency would be to use an open method like the Newton-Raphson or secant methods. Of course, for these cases, the problem of initial guesses is complicated by the issue of convergence.

Insight into these issues can be attained by examining the least efficient of the open approaches—fixed-point iteration. Examining Eq. (8.11), there are two straightforward ways to solve for H; that is, we can solve for either the H in the numerator,

$$H = \frac{(Qn)^{3/5}(B + 2H)^{2/5}}{BS^{3/10}} \tag{8.16}$$

or the H in the denominator

$$H = \frac{1}{2}\left[\frac{S^3(BH)^{5/2}}{(Qn)^{3/2}} - B\right] \tag{8.17}$$

Now, here's where physical reasoning can be helpful. For most rivers and streams the width is much greater than the depth. Thus, the quantity $B + 2H$ should not vary much. In fact, it should be roughly equal to B. In comparison, BH is directly proportional to H. Consequently, Eq. (8.16) should home in more rapidly on the root. This can be verified by substituting the brackets of $H = 0$ and 10 into both equations. For Eq. (8.16), the results are 0.6834 and 0.9012, which are both close to the true value of 0.7023. In contrast, the results for Eq. (8.17) are -10 and 8,178, which clearly are distant from the root.

The superiority of Eq. (8.16) is further supported by component plots (recall Fig. 6.3). As in Fig. 8.2, the $g(H)$ component for Eq. (8.16) is almost flat. Thus, it will not only converge, but should do so rapidly. In contrast, the $g(H)$ component for Eq. (8.17) is almost vertical, connoting strong and rapid divergence.

There are two practical benefits to such an analysis:

1. In the event that a more refined open method were used, Eq. (8.16) provides a means to develop an excellent starting guess. For example, if H is chosen as zero, Eq. (8.12) becomes

FIGURE 8.2
Component plots for two cases of fixed-point iteration, one that will converge [(a), Eq. (8.16)] and one that will diverge [(b), Eq. (8.17)].

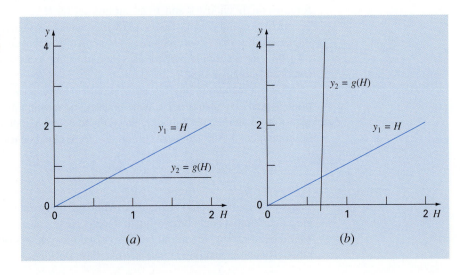

$$H_0 = \frac{(Qn/B)^{3/5}}{S^{3/10}}$$

where H_0 would be the initial value used in the Newton-Raphson or secant methods.

2. We have demonstrated that fixed-point iteration provides a viable option for this particular problem. For example, using an initial guess of $H = 0$, Eq. (8.16) attains six digits of precision in four iterations for the case we are examining. One setting where a fixed-point formula might come in handy would be a spreadsheet application. That is, spreadsheets are ideal for a convergent, iterative formula that depends on a single cell.

8.3 DESIGN OF AN ELECTRIC CIRCUIT (ELECTRICAL ENGINEERING)

Background. Electrical engineers often use Kirchhoff's laws to study the steady-state (not time-varying) behavior of electric circuits. Such steady-state behavior will be examined in Sec. 12.3. Another important problem involves circuits that are transient in nature where sudden temporal changes take place. Such a situation occurs following the closing of the switch in Fig. 8.3. In this case, there will be a period of adjustment following the closing of the switch as a new steady state is reached. The length of this adjustment period is closely related to the storage properties of the capacitor and the inductor. Energy storage may oscillate between these two elements during a transient period. However, resistance in the circuit will dissipate the magnitude of the oscillations.

The flow of current through the resistor causes a voltage drop (V_R) given by

$$V_R = iR$$

where i = the current and R = the resistance of the resistor. When R and i have units of ohms and amperes, V_R has units of volts.

Similarly, an inductor resists changes in current, such that the voltage drop V_L across it is

$$V_L = L\frac{di}{dt}$$

FIGURE 8.3
An electric circuit. When the switch is closed, the current will undergo a series of oscillations until a new steady state is reached.

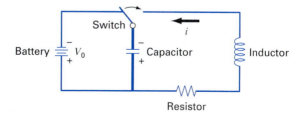

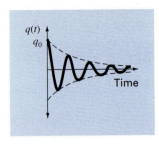

FIGURE 8.4
The charge on a capacitor as a
function of time following the
closing of the switch in
Fig. 8.3.

where L = the inductance. When L and i have units of henrys and amperes, V_L has units of volts and t has units of seconds.

The voltage drop across the capacitor (V_C) depends on the charge (q) on it:

$$V_C = \frac{q}{C}$$

where C = the capacitance. When the charge is expressed in units of coulombs, the unit of C is the farad.

Kirchhoff's second law states that the algebraic sum of voltage drops around a closed circuit is zero. After the switch is closed we have

$$L\frac{di}{dt} + Ri + \frac{q}{C} = 0$$

However, the current is related to the charge according to

$$i = \frac{dq}{dt}$$

Therefore,

$$L\frac{d^2q}{dt^2} + R\frac{dq}{dt} + \frac{1}{C}q = 0 \tag{8.18}$$

This is a second-order linear ordinary differential equation that can be solved using the methods of calculus (see Sec. 8.4). This solution is given by

$$q(t) = q_0 e^{-Rt/(2L)} \cos\left[\sqrt{\frac{1}{LC} - \left(\frac{R}{2L}\right)^2}\, t\right] \tag{8.19}$$

where at $t = 0$, $q = q_0 = V_0 C$, and V_0 = the voltage from the charging battery. Equation (8.19) describes the time variation of the charge on the capacitor. The solution $q(t)$ is plotted in Fig. 8.4.

A typical electrical engineering design problem might involve determining the proper resistor to dissipate energy at a specified rate, with known values for L and C. For this problem, assume the charge must be dissipated to 1 percent of its original value ($q/q_0 = 0.01$) in $t = 0.05$ s, with $L = 5$ H and $C = 10^{-4}$ F.

Solution. It is necessary to solve Eq. (8.19) for R, with known values of q, q_0, L, and C. However, a numerical approximation technique must be employed because R is an implicit variable in Eq. (8.19). The bisection method will be used for this purpose. The other methods discussed in Chaps. 5 and 6 are also appropriate, although the Newton-Raphson method might be deemed inconvenient because the derivative of Eq. (8.19) is a little cumbersome. Rearranging Eq. (8.19),

$$f(R) = e^{-Rt/(2L)} \cos\left[\sqrt{\frac{1}{LC} - \left(\frac{R}{2L}\right)^2}\, t\right] - \frac{q}{q_0}$$

or using the numerical values given,

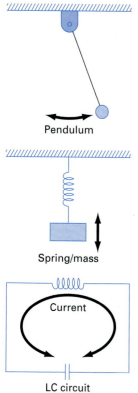

FIGURE 8.5
Plot of Eq. (8.20) used to obtain initial guesses for R that bracket the root.

FIGURE 8.6
Three examples of simple harmonic oscillators. The two-way arrows illustrate the oscillations for each system.

$$f(R) = e^{-0.005R} \cos{[\sqrt{2000 - 0.01R^2}\,(0.05)]} - 0.01 \qquad (8.20)$$

Examination of this equation suggests that a reasonable initial range for R is 0 to 400 Ω (because $2000 - 0.01R^2$ must be greater than zero). Figure 8.5, a plot of Eq. (8.20), confirms this. Twenty-one iterations of the bisection method give $R = 328.1515\ \Omega$, with an error of less than 0.0001 percent.

Thus, you can specify a resistor with this rating for the circuit shown in Fig. 8.6 and expect to achieve a dissipation performance that is consistent with the requirements of the problem. This design problem could not be solved efficiently without using the numerical methods in Chaps. 5 and 6.

8.4 VIBRATION ANALYSIS (MECHANICAL/AEROSPACE ENGINEERING)

Background.   Differential equations are often used to model the vibration of engineering systems. Some examples (Fig. 8.6) are a simple pendulum, a mass on a spring, and an inductance-capacitance electric circuit (recall Sec. 8.3). The vibration of these systems may be damped by some energy-absorbing mechanism. In addition, the vibration may be free or subject to some external periodic disturbance. In the latter case the motion is said to be *forced*. In this section, we will examine the free and forced vibration of the automo-

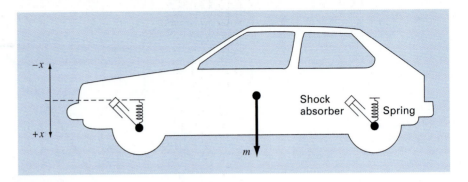

FIGURE 8.7
A car of mass m.

bile shown in Fig. 8.7. The general approach is applicable to various other engineering problems.

As shown in Fig. 8.7, a car of mass m is supported by springs and shock absorbers. Shock absorbers offer resistance to the motion that is proportional to the vertical speed (up-and-down motion). Free vibrations result when the car is disturbed from equilibrium, such as after encountering a pothole. At any instant after hitting the pothole the net forces acting on m are the resistance of the springs and the damping force of the shock absorbers. These forces tend to return the car to the original equilibrium state. According to *Hooke's law*, the resistance of the spring is proportional to the spring constant k and the distance from the equilibrium position, x. Therefore,

$$\text{Spring force} = -kx$$

where the negative sign indicates that the restoring force acts to return the car toward the position of equilibrium (that is, the negative x direction). The damping force of the shock absorbers is given by

$$\text{Damping force} = -c\frac{dx}{dt}$$

where c is a damping coefficient and dx/dt is the vertical velocity. The negative sign indicates that the damping force acts in the opposite direction against the velocity.

The equations of motion for the system are given by Newton's second law ($F = ma$), which for the present problem is expressed as

$$\underbrace{m}_{\text{Mass}} \times \underbrace{\frac{d^2x}{dt^2}}_{\text{acceleration}} = \underbrace{-c\frac{dx}{dt}}_{\text{damping force}} + \underbrace{(-kx)}_{\text{spring force}}$$

or

$$m\frac{d^2x}{dt^2} + c\frac{dx}{dt} + kx = 0$$

Notice the similarity with Eq. (8.18) that was developed in Sec. 8.3 for an electrical circuit.

If we assume that the solution takes the form $x(t) = e^{rt}$, we can write the *characteristic equation*

$$mr^2 + cr + k = 0 \qquad (8.21)$$

The unknown, r, is the solution of the quadratic characteristic equation that can be attained either analytically or numerically. In this design problem, we will first use the analytical solution to give us general insight into the way the system motion is affected by the model coefficients—m, k, and c. We will also use various numerical methods to attain solutions and check the accuracy of the results using the analytical solution. Finally, we will set the stage for more complex problems to be described later in the text, where analytical results are difficult or impossible to obtain.

The solution of Eq. (8.21) for r is given by the quadratic formula

$$\genfrac{}{}{0pt}{}{r_1}{r_2} = \frac{-c \pm \sqrt{c^2 - 4mk}}{2m} \qquad (8.22)$$

Note the significance of magnitude of c compared to $2\sqrt{km}$. If $c > 2\sqrt{km}$, r_1 and r_2 are negative real numbers, and the solution is of the form

$$x(t) = Ae^{r_1 t} + Be^{r_2 t} \qquad (8.23)$$

where A and B = constants to be determined from the initial conditions of x and dx/dt. Such systems are called *overdamped*.

If $c < 2\sqrt{km}$, the roots are complex,

$$\genfrac{}{}{0pt}{}{r_1}{r_2} = \lambda \pm \mu i$$

where

$$\mu = \frac{\sqrt{|c^2 - 4mk|}}{2m}$$

and solution is of the form

$$x(t) = e^{-\lambda t}(A \cos \mu t + B \sin \mu t) \qquad (8.24)$$

Such systems are called *underdamped*.

Finally, if $c = 2\sqrt{km}$, the characteristic equation has a double root and the solution is of the form

$$x(t) = (A + Bt)e^{-\lambda t} \qquad (8.25)$$

where

$$\lambda = \frac{c}{2m}$$

Such systems are called *critically damped*.

In all three cases, $x(t)$ approaches zero as t approaches infinity. This means that the car always returns to the equilibrium position after encountering the pothole (although this may seem unlikely in some cities we've visited!). These cases are illustrated in Fig. 8.8.

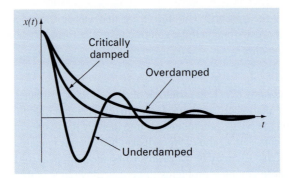

FIGURE 8.8
(a) Overdamped, (b) underdamped, and (c) critically
damped oscillations.

The *critical damping coefficient* c_c is the value of c that makes the radical in Eq. (8.22) equal to zero,

$$c_c = 2\sqrt{km} \qquad \text{or} \qquad c_c = 2mp \tag{8.26}$$

where

$$p = \sqrt{\frac{k}{m}} \tag{8.27}$$

The ratio c/c_c is called the *damping factor* and p is called the *natural frequency* of the undamped free vibration.

Now, let us consider the case where the car is subject to a periodic force given by

$$P = P_m \sin \omega t \qquad \text{or} \qquad d = d_m \sin \omega t$$

where $d_m = P_m/k =$ the static deflection of the car subject to a force P_m. The governing differential equation for this case is

$$m\frac{d^2x}{dt^2} + c\frac{dx}{dt} + kx = P_m \sin \omega t$$

The general solution of this equation is obtained by adding a particular solution to the free vibration solution given by Eqs. (8.23) through (8.25). Let us consider the steady-state motion of the forced system where the initial transient motion has been damped out. If we assume that this steady-state solution has the form

$$x_{ss}(t) = x_m \sin (\omega t - \phi)$$

it can be shown that

$$\frac{x_m}{P_m/k} = \frac{x_m}{d_m} = \frac{1}{\sqrt{[1 - (\omega/p)]^2 + 4(c/c_c)^2(\omega/p)^2}} \tag{8.28}$$

The quantity, x_m/d_m, called the *amplitude magnification factor*, depends only on the ratio of the actual damping to the critical damping and the ratio of the forcing frequency to the natural frequency. Note that when the forcing frequency ω approaches zero, the magnification

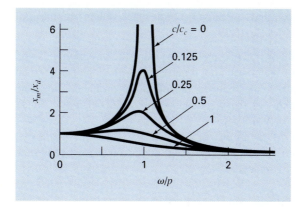

FIGURE 8.9
Plot of the amplitude magnification factor x_m/x_d [Eq. (8.28)] versus the frequency ω over the natural frequency p for various values of the damping coefficient c over the critical damping coefficient c_c.

factor approaches 1. Also if the system is lightly damped, that is, if c/c_c is small, then the magnification factor becomes large if ω is close to p. If the damping is zero, then the magnification factor becomes infinite when $\omega = p$, and the forcing function is said to be in *resonance* with the system. Finally, as ω/p becomes very large, the magnification factor approaches zero. Figure 8.9 shows a plot of magnification factor as a function of ω/p for various damping factors.

Observe that the magnification factor may be kept small by choosing a large damping factor or by keeping the natural and the forced frequencies far apart.

The design of the car suspension system involves a trade-off between comfort and stability for all driving conditions and speeds. You are asked to determine the stability of a proposed design that has good comfort on rough roads. The mass of the car is $m = 1.2 \times 10^6$ g, and it has a shock system with a dampening coefficient of $c = 1 \times 10^7$ g/s.

Assume that the public's expectation of comfort is satisfied if the free vibration of the car is underdamped and the first crossing of the equilibrium position takes place in 0.05 s. If, at $t = 0$, the car is suddenly displaced, x_0, from equilibrium and the velocity is zero ($dx/dt = 0$), the solution of the equation of motion is given by Eq. (8.24), with $A = x_0$ and $B = x_0\lambda/\mu$. Therefore,

$$x(t) = x_0 e^{-\lambda t}\left(\cos \mu t + \frac{\lambda}{\mu}\sin \mu t\right)$$

Our design conditions are met if

$$x(t) = 0 = \cos(0.05\mu) + \frac{\lambda}{\mu}\sin(0.05\mu)$$

or

$$0 = \cos\left(0.05\sqrt{\frac{k}{m} - \frac{c^2}{4m^2}}\right) + \frac{c}{\sqrt{4km - c^2}}\sin\left(0.05\sqrt{\frac{k}{m} - \frac{c^2}{4m^2}}\right) \tag{8.29}$$

Since c and m are given, our design problem becomes finding an appropriate value of k that satisfies Eq. (8.29).

Solution.   This can be done using the bisection, false-position, or secant methods because these methods do not require the evaluation of the derivative of Eq. (8.29), which might be considered a bit inconvenient for this problem. The solution is $k = 1.397 \times 10^9$, with 12 iterations of the bisection method with an initial bracket from $k = 1 \times 10^9$ to 2×10^9 ($\varepsilon_a = 0.07305\%$).

Although this design satisfies our free vibration requirements (after hitting the pothole), it must also be tested under rough road conditions. The surface of the road can be approximated as

$$d = d_m \sin\left(\frac{2\pi x}{D}\right)$$

where d is the deflection, d_m is the maximum deflection of 0.1 m, and D is the distance between peaks equal to 20 m. If the horizontal speed of the car (m/s) is v, then the overall equation of motion for the system can be written as

$$m\frac{d^2x}{dt^2} + c\frac{dx}{dt} + kx = kd_m \sin\left(\frac{2\pi v}{D}t\right)$$

where $\omega = 2\pi v/D$ is the forcing frequency.

The stability of the car is considered satisfactory if at steady state the maximum distance x_m is below 0.2 m for all driving speeds. The damping factor is calculated according to Eq. (8.26),

$$\frac{c}{c_c} = \frac{10}{2\sqrt{km}} = \frac{1 \times 10^7}{2\sqrt{1.397 \times 10^9(1.2 \times 10^6)}} = 0.1221$$

Now, we seek values ω/p that satisfy Eq. (8.28),

$$2 = \frac{1}{\sqrt{\left[1 - (\omega/p)^2\right]^2 + 4(0.1221)^2(\omega/p)^2}} \tag{8.30}$$

When Eq. (8.30) is expressed as a roots problem,

$$f(\omega/p) = 2\sqrt{\left[1 - (\omega/p)^2\right]^2 + 4(0.1221)^2(\omega/p)^2} - 1 = 0 \tag{8.31}$$

We see that values of ω/p can be determined by finding the roots of Eq. (8.31).

A plot of Eq. (8.31) is shown in Fig. 8.10. This plot shows that Eq. (8.31) has two positive roots that can be determined with the bisection method using the TOOLKIT. The smaller value for ω/p is found to equal 0.7300 in 18 iterations, with an estimated error of 0.000525% with upper and lower guesses of 0 and 1. The higher value of ω/p is found to be 1.1864 in 17 iterations, with an estimated error of 0.00064% with upper and lower guesses of 1 and 2.

It is also possible to express Eq. (8.30) as a polynomial,

$$\left(\frac{\omega}{p}\right)^4 - 1.9404\left(\frac{\omega}{p}\right)^2 + 0.75 \tag{8.32}$$

and use MATLAB to determine the roots, as in the following:

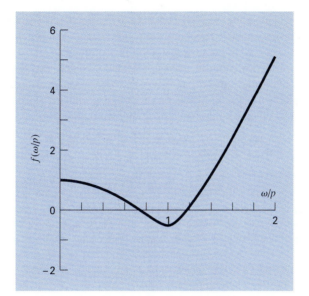

FIGURE 8.10
Plot of the amplitude magnification factor x_m/x_d versus the frequency ω over the natural frequency ρ for various values of the damping coefficient c over the critical damping coefficient c_c.

```
>> a=[1 0 -1.9404 0 .75];
>> roots (a)
ans =

    1.1864
   -1.1864
    0.7300
   -0.7300
```

This confirms the result obtained with bisection. It also suggests that, although it superficially appears to be a fourth-order equation in ω/p, Eq. (8.32) is actually a quadratic equation in $(\omega/p)^2$.

The value of the natural frequency p is given by Eq. (8.27),

$$p = \sqrt{\frac{1.397 \times 10^9}{1.2 \times 10^6}} = 34.12 \text{ s}^{-1}$$

The forcing frequencies, for which the maximum deflection is 0.2 m, is then calculated as

$$\omega = 0.7300(34.12) = 24.91 \text{ s}^{-1}$$
$$\omega = 1.1864(34.12) = 40.48 \text{ s}^{-1}$$

which yields

$$v = \frac{\omega D}{2\pi} = \frac{24.91(20)}{2(3.14159)} = 79.29 \frac{\text{m}}{\text{s}} \times \frac{3600 \text{ s}}{\text{hr}} \frac{\text{km}}{1000 \text{ m}} = 285 \text{ km/hr} \ (= 177 \text{ mi/hr})$$

$$v = \frac{\omega D}{2\pi} = \frac{40.48(20)}{2(3.14159)} = 128.85 \frac{\text{m}}{\text{s}} \times \frac{3600 \text{ s}}{\text{hr}} \frac{\text{km}}{1000 \text{ m}} = 464 \text{ km/hr} \ (= 288 \text{ mi/hr})$$

Thus, using the above results and Fig. 8.10, it is found that the proposed car design will behave acceptably for common driving speeds. At this point, the designer must be aware that the design would not meet suitability requirements at extremely high speeds (e.g., racing).

This design problem has presented an extremely simple example that has allowed us to obtain some analytical results that were used to evaluate the accuracy of our numerical methods for finding roots. Real cases can quickly become so complicated that solutions can be obtained only by using numerical methods.

PROBLEMS

Chemical Engineering

8.1 Perform the same computation as in Sec. 8.1, but for ethyl alcohol ($a = 12.02$ and $b = 0.08407$) at a temperature of 375 K and p of 2.0 atm. Compare your results with the ideal gas law. If possible, use your computer software to determine the molal volume. Otherwise, use any of the numerical methods discussed in Chaps. 5 and 6 to perform the computation. Justify your choice of technique.

8.2 In chemical engineering, plug flow reactors (that is, those in which fluid flows from one end to the other with minimal mixing along the longitudinal axis) are often used to convert reactants into products. It has been determined that the efficiency of the conversion can sometimes be improved by recycling a portion of the product stream so that it returns to the entrance for an additional pass through the reactor (Fig. P8.2). The recycle rate is defined as

$$R = \frac{\text{volume of fluid returned to entrance}}{\text{volume leaving the system}}$$

Suppose that we are processing a chemical A to generate a product B. For the case where A forms B according to an autocatalytic reaction (that is, in which one of the products acts as a catalyst or stimulus for the reaction), it can be shown that an optimal recycle rate must satisfy

$$\ln \frac{1 + R(1 - X_{Af})}{R(1 - X_{Af})} = \frac{R + 1}{R[1 + R(1 - X_{Af})]}$$

where X_{Af} = the fraction of reactant A that is converted to product B. The optimal recycle rate corresponds to the minimum-sized reactor needed to attain the desired level of conversion. Use a numer-

ical method to determine the recycle ratios needed to minimize reactor size for a fractional conversion of $X_{Af} = 0.9$.

8.3 In a chemical engineering process, water vapor (H_2O) is heated to sufficiently high temperatures that a significant portion of the water dissociates, or splits apart, to form oxygen (O_2) and hydrogen (H_2):

$$H_2O \rightleftharpoons H_2 + \tfrac{1}{2}O_2$$

If it is assumed that this is the only reaction involved, the mole fraction x of H_2O that dissociates can be represented by

$$K = \frac{x}{1 - x}\sqrt{\frac{2p_t}{2 + x}} \qquad \text{(P8.3)}$$

where K = the reaction equilibrium constant and p_t = the total pressure of the mixture. If $p_t = 3$ atm and $K = 0.05$, determine the value of x that satisfies Eq. (P8.3).

8.4 The following equation pertains to the concentration of a chemical in a completely mixed reactor:

$$c = c_{in}(1 - e^{-0.04t}) + c_0 e^{-0.04t}$$

If the initial concentration $c_0 = 4$ and the inflow concentration $c_{in} = 10$, compute the time required for c to be 93 percent of c_{in}.

8.5 A reversible chemical reaction

$$2A + B \rightleftharpoons C$$

can be characterized by the equilibrium relationship

$$K = \frac{c_c}{c_a^2 c_b}$$

where the nomenclature c_n represents the concentration of constituent N. Suppose that we define a variable x as representing the number of moles of C that are produced. Conservation of mass can be used to reformulate the equilibrium relationship as

$$K = \frac{(c_{c,0} + x)}{(c_{a,0} - 2x)^2(c_{b,0} - x)}$$

where the subscript 0 designates the initial concentration of each constituent. If $K = 0.015$, $c_{a,0} = 42$, $c_{b,0} = 30$, and $c_{c,0} = 4$, compute x.

FIGURE P8.2
Schematic representation of a plug flow reactor with recycle.

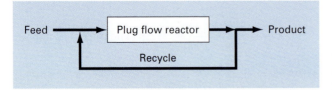

8.6 The following chemical reactions take place in a closed system

$$2A + B \rightleftharpoons C$$

$$A + D \rightleftharpoons C$$

At equilibrium, they can be characterized by

$$K_1 = \frac{c_c}{c_a^2 c_b}$$

$$K_2 = \frac{c_c}{c_a c_d}$$

where the nomenclature c_n represents the concentration of constituent N. If x_1 and x_2 are the number of moles of C that are produced due to the first and second reactions, respectively, use an approach similar to that of Prob. 8.5 to reformulate the equilibrium relationships in terms of the initial concentrations of the constituents. Then, use the Newton-Raphson method to solve the pair of simultaneous nonlinear equations for x_1 and x_2 if $K_1 = 4 \times 10^{-4}$, $K_2 = 3.7 \times 10^{-2}$, $c_{a,0} = 50$, $c_{b,0} = 20$, $c_{c,0} = 5$, and $c_{d,0} = 10$. Use a graphical approach to develop your initial guesses.

8.7 The Redlich-Kwong equation of state is given by

$$p = \frac{RT}{v - b} - \frac{a}{v(v + b)\sqrt{T}}$$

where R = the universal gas constant [= 0.518 kJ/(kg K)], T = absolute temperature (K), p = absolute pressure (kPa), and v = the volume of a kg of gas (m³/kg). The parameters a and b are calculated by

$$a = 0.427 \frac{R^2 T_c^{2.5}}{p_c} \qquad b = 0.0866R\frac{T_c}{p_c}$$

where $p_c = 4600$ kPa and $T_c = 191$ K. As a chemical engineer, you are asked to determine the amount of methane fuel that can be held in a 3-m³ tank at a temperature of $-40°$C with a pressure of 65,000 kPa. Use a root-locating method of your choice to calculate v and then determine the mass of methane contained in the tank.

8.8 The volume V of liquid, in a hollow horizontal cylinder of radius r and length L is related to the depth of the liquid h by

$$V = \left[r^2 \cos^{-1}\left(\frac{r - h}{r}\right) - (r - h)\sqrt{2rh - h^2} \right] L$$

Determine h given $r = 2$ m, $L = 5$ m³, and $V = 8$ m³. Note that if you are using a programming language or software tool that is not rich in trigonometric functions, the arc cosine can be computed with

$$\cos^{-1} x = \frac{\pi}{2} - \tan^{-1}\left(\frac{x}{\sqrt{1 - x^2}}\right)$$

8.9 The volume V of liquid in a spherical tank of radius r is related to the depth h of the liquid by

$$V = \frac{\pi h^2 (3r - h)}{3}$$

Determine h given $r = 1$ m and $V = 0.5$ m³.

8.10 For the spherical tank in Prob. 8.9, it is possible to develop the following two fixed-point formulas:

$$h = \sqrt{\frac{h^3 + (3V/\pi)}{3r}}$$

and

$$h = \sqrt[3]{3\left(rh^2 - \frac{V}{\pi}\right)}$$

If $r = 1$ m and $V = 0.5$ m³, determine whether either of these is stable, and the range of initial guesses for which they are stable.

Civil and Environmental Engineering

8.11 The displacement of a structure is defined by the following equation for a damped oscillation:

$$y = 8e^{-kt} \cos \omega t$$

where $k = 0.5$ and $\omega = 3$. (a) Use the graphical method to make an initial estimate of the time required for the displacement to decrease to 4. (b) Use the Newton-Raphson method to determine the root to $\varepsilon_s = 0.01\%$. (c) Use the secant method to determine the root to $\varepsilon_s = 0.01\%$.

8.12 The secant formula defines the force per unit area, P/A, that causes a maximum stress σ_m in a column of given slenderness ratio L_e/r:

$$\frac{P}{A} = \frac{\sigma_m}{1 + (ec/r^2)\sec[0.5\sqrt{P/(EA)}(L_e/r)]}$$

If $E = 200{,}000$ kPa, $ec/r^2 = 0.2$, and $\sigma_m = 250$ kPa, compute P/A for $L_e/r = 100$. Recall that $\sec x = 1/\cos x$.

8.13 A catenary cable is one that is hung between two points not in the same vertical line. As depicted in Fig. P8.13a, it is subject to no loads other than its own weight. Thus, its weight w (N/m) acts as a uniform load per unit length along the cable. A free-body diagram of a section AB is depicted in Fig. P8.13b, where T_A and T_B are the tension forces at the end. Based on horizontal and vertical force balances, the following differential equation model of the cable can be derived:

$$\frac{d^2 y}{dx^2} = \frac{w}{T_A}\sqrt{1 + \left(\frac{dy}{dx}\right)^2}$$

Calculus can be employed to solve this equation for the height y of the cable as a function of distance x,

$$y = \frac{T_A}{w}\cosh\left(\frac{w}{T_A}x\right) + y_0 - \frac{T_A}{w}$$

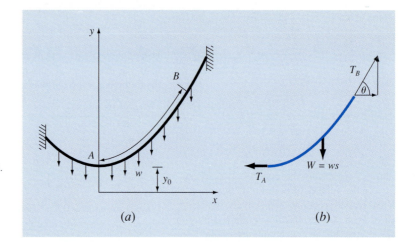

FIGURE P8.13
(a) Forces acting on a section AB of a flexible hanging cable. The load is uniform along the cable (but not uniform per the horizontal distance x). (b) A free-body diagram of section AB.

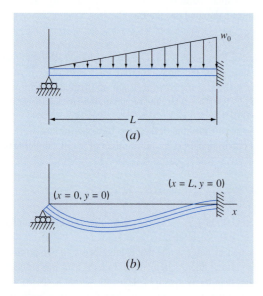

FIGURE P8.14

where the hyperbolic cosine can be computed by

$$\cosh x = \tfrac{1}{2}(e^x + e^{-x})$$

(a) Use a numerical method to calculate a value for the parameter T_A given values for the parameters $w = 10$ and $y_0 = 5$, such that the cable has a height of $y = 15$ at $x = 50$.
(b) Determine the location of the minimum height for the case described in (a).

8.14 Figure P8.14a shows a uniform beam subject to a linearly increasing distributed load. The equation for the resulting elastic curve is (see Fig. P8.14b)

$$y = \frac{w_0}{120EIL}(-x^5 + 2L^2x^3 - L^4x) \tag{P8.14}$$

Use bisection to determine the point of maximum deflection (that is, the value of x where $dy/dx = 0$). Then substitute this value into Eq. (P8.14) to determine the value of the maximum deflection. Use the following parameter values in your computation: $L = 450$ cm, $E = 50,000$ kN/cm^2, $I = 30,000$ cm^4, and $w_0 = 1.75$ kN/cm.

8.15 In environmental engineering (a specialty area in civil engineering), the following equation can be used to compute the oxygen level in a river downstream from a sewage discharge:

$$c = 10 - 20(e^{-0.2x} - e^{-0.75x})$$

where x is the distance downstream in kilometers. Determine the distance downstream where the oxygen level first falls to a reading of 5. (*Hint:* It is within 2 km of the discharge.) Determine you answer to a 1% error.

8.16 The concentration of pollutant bacteria c in a lake decreases according to

$$c = 70e^{-1.5t} + 25e^{-0.075t}$$

Determine the time required for the bacteria concentration to be reduced to 9 using (a) the graphical method and (b) the Newton-Raphson method.

8.17 In ocean engineering, the equation for a reflected standing wave in a harbor is given by $\lambda = 16$, $t = 12$, $v = 48$:

$$h = h_0 \left[\sin\left(\frac{2\pi x}{\lambda}\right) \cos\left(\frac{2\pi tv}{\lambda}\right) + e^{-x} \right]$$

Solve for the lowest positive value of x if $h = 0.4h_0$.

8.18 You buy a $20,000 piece of equipment for nothing down and $4000 per year for 6 years. What interest rate are you paying? The formula relating present worth P, annual payments A, number of years n, and interest rate i is

$$A = P \frac{i(1+i)^n}{(1+i)^n - 1}$$

8.19 Many fields of engineering require accurate population estimates. For example, transportation engineers might find it necessary to determine separately the population growth trends of a city and adjacent suburb. The population of the urban area is declining with time according to

$$P_u(t) = P_{u,\text{max}} e^{-k_u t} + P_{u,\text{min}}$$

while the suburban population is growing, as in

$$P_s(t) = \frac{P_{s,\text{max}}}{1 + \left[P_{s,\text{max}}/P_0 - 1\right] e^{k_s t}}$$

where $P_{u,\text{max}}$, k_u, $P_{s,\text{max}}$, P_0, and k_s = empirically derived parameters. Determine the time and corresponding values of $P_u(t)$ and $P_s(t)$ when the city is 20% larger than the suburban population. The parameter values are $P_{u,\text{max}} = 75{,}000$, $k_u = 005/\text{yr}$, $P_{u,\text{min}} = 100{,}000$ people, $P_{s,\text{max}} = 300{,}000$ people, $P_0 = 5000$ people, $k_s = 0.075/\text{yr}$. To obtain your solutions, use (a) graphical, (b) false-position, and (c) modified secant methods.

Electrical Engineering

8.20 Perform the same computation as in Sec. 8.3, but determine the value of L required for the circuit to dissipate to 1% of its original value in $t = 0.05$ s, given $R = 280\ \Omega$ and $C = 10^{-4}$ F.

8.21 An oscillating current in an electric circuit is described by $I = 9e^{-t} \sin(2\pi t)$, where t is in seconds. Determine all values of t such that $I = 3.5$.

8.22 The resistivity ρ of doped silicon is based on the charge q on an electron, the electron density n, and the electron mobility μ. The electron density is given in terms of the doping density N and the intrinsic carrier density n_i. The electron mobility is described by the temperature T, the reference temperature T_0, and the reference mobility μ_0. The equations required to compute the resistivity are

$$\rho = \frac{1}{qn\mu}$$

where

$$n = \tfrac{1}{2}\left(N + \sqrt{N^2 + 4n_i^2}\right) \qquad \text{and} \qquad \mu = \mu_0 \left(\frac{T}{T_0}\right)^{-2.42}$$

Determine N, given $T_0 = 300$ K, $T = 1000$ K, $\mu_0 = 1330$ cm²/(V s), $q = 1.6 \times 10^{-19}$ C, $n_i = 6.21 \times 10^9$ cm⁻³, and a desired $\rho = 6 \times 10^6$ V s cm/C. Use (a) bisection and (b) the modified secant method.

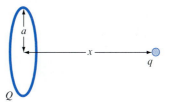

FIGURE P8.23

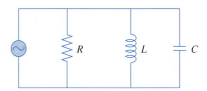

FIGURE P8.24

8.23 A total charge Q is uniformly distributed around a ring-shaped conductor with radius a. A charge q is located at a distance x from the center of the ring (Fig. P8.23). The force exerted on the charge by the ring is given by

$$F = \frac{1}{4\pi e_0} \frac{qQx}{(x^2 + a^2)^{3/2}}$$

where $e_0 = 8.85 \times 10^{-12}$ C²/(N m²). Find the distance x where the force is 1 N if q and Q are 2×10^{-5} C for a ring with a radius of 0.8 m.

8.24 Figure P8.24 shows a circuit with a resistor, an inductor, and a capacitor in parallel. Kirchhoff's rules can be used to express the impedance of the system as

$$\frac{1}{Z} = \sqrt{\frac{1}{R^2} + \left(\omega C - \frac{1}{\omega L}\right)^2}$$

where Z = impedance (Ω) and ω = the angular frequency. Find the ω that results in an impedance of 100 Ω using both bisection and false position with initial guesses of 1 and 1000 for the following parameters: $R = 225\ \Omega$, $C = 0.6 \times 10^{-6}$ F, and $L = 0.5$ H.

Mechanical and Aerospace Engineering

8.25 Repeat the calculations at the end of Sec. 8.4, but determine the speeds for which the maximum deflection is 0.15 m.

8.26 For fluid flow in pipes, friction is described by a dimensionless number, the *Fanning friction factor f*. The Fanning friction factor is dependent on a number of parameters related to the size of the pipe and the fluid, which can all be represented by another

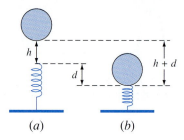

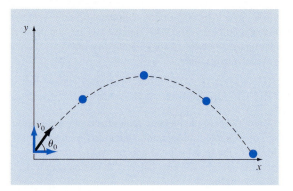

FIGURE P8.27

FIGURE P8.29

dimensionless quantity, the *Reynolds number* Re. A formula that predicts f given Re is the *von Karman* equation,

$$\frac{1}{\sqrt{f}} = 4 \log_{10} \left(\text{Re} \sqrt{f} \right) - 0.4$$

Typical values for the Reynolds number for turbulent flow are 10,000 to 500,000 and for the Fanning friction factor are 0.001 to 0.01. Develop a subprogram that solves for f given a user-supplied value between 2500 and 1,000,000.

8.27 Real mechanical systems may involve the deflection of non-linear springs. In Fig. P8.27, a block of mass m is released a distance h above a nonlinear spring. The resistance force F of the spring is given by

$$F = -(k_1 d + k_2 d^{3/2})$$

Conservation of energy can be used to show that

$$0 = \frac{2 k_2 d^{5/2}}{5} + \frac{1}{2} k_1 d^2 - mgd - mgh$$

Solve for d, given the following parameter values: $k_1 = 40,000 \text{ g/s}^2$, $k_2 = 40 \text{ g/(s}^2 \text{ m}^5)$, $m = 95 \text{ g}$, $g = 9.8 \text{ m/s}^2$, and $h = 0.43 \text{ m}$.

8.28 Mechanical engineers, as well as most other engineers, use thermodynamics extensively in their work. The following polynomial can be used to relate the zero-pressure specific heat of dry air, c_p kJ/(kg K), to temperature:

$$c_p = 0.99403 + 1.671 \times 10^{-4} T + 9.7215 \times 10^{-8} T^2 \\ -9.5838 \times 10^{-11} T^3 + 1.9520 \times 10^{-14} T^4$$

Determine the temperature that corresponds to a specific heat of 1.2 kJ/(kg K).

8.29 Aerospace engineers sometimes compute the trajectories of projectiles like rockets. A related problem deals with the trajectory

of a thrown ball. The trajectory of a ball thrown by a right fielder is defined by the (x, y) coordinates, as displayed in Fig. P8.29. The trajectory can be modeled as

$$y = (\tan \theta_0) x - \frac{g}{2 v_0^2 \cos^2 \theta_0} x^2 + 1.8$$

Find the appropriate initial angle θ_0 if $v_0 = 20$ m/s and the distance to second base is 40 m. Note that the throw leaves the right fielder's hand at an elevation of 1.8 m and the second baseman receives it at 1 m.

8.30 The upward velocity of a rocket can be computed by the following formula:

$$v = u \ln \frac{m_0}{m_0 - qt} - gt$$

where $v =$ upward velocity, $u =$ the velocity at which fuel is expelled relative to the rocket, $m_0 =$ the initial mass of the rocket at time $t = 0$, $q =$ the fuel consumption rate, and $g =$ the downward acceleration of gravity (assumed constant $= 9.8$ m/s^2). If $u = 2200$ m/s, $m_0 = 160,000$ kg, and $q = 2680$ kg/s, compute the time at which $v = 1000$ m/s. (*Hint: t* is somewhere between 10 and 50 s.) Determine your result so that it is within 1% of the true value. Check your answer.

8.31 In Sec. 8.4, the phase angle ϕ between the forced vibration caused by the rough road and the motion of the car is given by

$$\tan \phi = \frac{2(c/c_c)(\omega/p)}{1 - (\omega/p)^2}$$

As a mechanical engineer, you would like to know if there are cases where $\phi = \omega/2 + 1$. Use the other parameters from the section to set up the equation as a roots problem and solve for ω.

EPILOGUE: PART TWO

PT2.4 TRADE-OFFS

Table PT2.3 provides a summary of the trade-offs involved in solving for roots of algebraic and transcendental equations. Although graphical methods are time-consuming, they provide insight into the behavior of the function and are useful in identifying initial guesses and potential problems such as multiple roots. Therefore, if time permits, a quick sketch (or better yet, a computerized graph) can yield valuable information regarding the behavior of the function.

The numerical methods themselves are divided into two general categories: bracketing and open methods. The former requires two initial guesses that are on either side of a root.

TABLE PT2.3 Comparison of the characteristics of alternative methods for finding roots of algebraic and transcendental equations. The comparisons are based on general experience and do not account for the behavior of specific functions.

Method	Initial Guesses	Convergence Rate	Stability	Accuracy	Breadth of Application	Programming Effort	Comments
Direct	—	—	—	—	Limited		
Graphical	—	—	—	Poor	Real roots	—	May take more time than the numerical method
Bisection	2	Slow	Always	Good	Real roots	Easy	
False-position	2	Medium	Always	Good	Real roots	Easy	
Fixed-point iteration	1	Slow	Possibly divergent	Good	General	Easy	
Newton-Raphson	1	Fast	Possibly divergent	Good	General	Easy	Requires evaluation of $f'(x)$
Modified Newton-Raphson	1	Fast for multiple roots; medium for single	Possibly divergent	Good	General	Easy	Requires evaluation of $f''(x)$ and $f'(x)$
Secant	2	Medium to fast	Possibly divergent	Good	General	Easy	Initial guesses do not have to bracket the root
Modified secant	1	Medium to fast	Possibly divergent	Good	General	Easy	
Müller	2	Medium to fast	Possibly divergent	Good	Polynomials	Moderate	
Bairstow	2	Fast	Possibly divergent	Good	Polynomials	Moderate	

This "bracketing" is maintained as the solution proceeds, and thus, these techniques are always convergent. However, a price is paid for this property in that the rate of convergence is relatively slow.

Open techniques differ from bracketing methods in that they use information at a single point (or two values that need not bracket the root to extrapolate to a new root estimate). This property is a double-edged sword. Although it leads to quicker convergence, it also allows the possibility that the solution may diverge. In general, the convergence of open techniques is partially dependent on the quality of the initial guess and the nature of the function. The closer the guess is to the true root, the more likely the methods will converge.

Of the open techniques, the standard Newton-Raphson method is often used because of its property of quadratic convergence. However, its major shortcoming is that it requires the derivative of the function be obtained analytically. For some functions this is impractical. In these cases, the secant method, which employs a finite-difference representation of the derivative, provides a viable alternative. Because of the finite-difference approximation, the rate of convergence of the secant method is initially slower than for the Newton-Raphson method. However, as the root estimate is refined, the difference approximation becomes a better representation of the true derivative, and convergence accelerates rapidly. The modified Newton-Raphson technique can be used to attain rapid convergence for multiple roots. However, this technique requires an analytical expression for both the first and second derivative.

All the numerical methods are easy to program on computers and require minimal time to determine a single root. On this basis, you might conclude that simple methods such as bisection would be good enough for practical purposes. This would be true if you were exclusively interested in determining the root of an equation once. However, there are many cases in engineering where numerous root locations are required and where speed becomes important. For these cases, slow methods are very time-consuming and, hence, costly. On the other hand, the fast open methods may diverge, and the accompanying delays can also be costly. Some computer algorithms attempt to capitalize on the strong points of both classes of techniques by initially employing a bracketing method to approach the root, then switching to an open method to rapidly refine the estimate. Whether a single approach or a combination is used, the trade-offs between convergence and speed are at the heart of the choice of a root-location technique.

PT2.5   IMPORTANT RELATIONSHIPS AND FORMULAS

Table PT2.4 summarizes important information that was presented in Part Two. This table can be consulted to quickly access important relationships and formulas.

PT2.6   ADVANCED METHODS AND ADDITIONAL REFERENCES

The methods in this text have focused on determining a single real root of an algebraic or transcendental equation based on foreknowledge of its approximate location. In addition, we have also described methods expressly designed to determine both the real and complex roots of polynomials. Additional references on the subject are Ralston and Rabinowitz (1978) and Carnahan, Luther, and Wilkes (1969).

In addition to Müller's and Bairstow's methods, several techniques are available to determine all the roots of polynomials. In particular, the *quotient difference (QD) algorithm* (Henrici, 1964, and Gerald and Wheatley, 1984) determines all roots without initial

TABLE PT2.4 Summary of important information presented in Part Two.

Method	Formulation	Graphical Interpretation	Errors and Stopping Criteria
		Bracketing methods:	
Bisection	$x_r = \dfrac{x_l + x_u}{2}$ If $f(x_l)f(x_r) < 0, x_u = x_r$ $f(x_l)f(x_r) > 0, x_l = x_r$		Stopping criterion: $\left\|\dfrac{x_r^{new} - x_r^{old}}{x_r^{new}}\right\|100\% \leq \epsilon_s$
False Position	$x_r = x_u - \dfrac{f(x_u)(x_l - x_u)}{f(x_l) - f(x_u)}$ If $f(x_l)f(x_r) < 0, x_u = x_r$ $f(x_l)f(x_r) > 0, x_l = x_r$		Stopping criterion: $\left\|\dfrac{x_r^{new} - x_r^{old}}{x_r^{new}}\right\|100\% \leq \epsilon_s$
Newton-Raphson	$x_{i+1} = x_i - \dfrac{f(x_i)}{f'(x_i)}$		Stopping criterion: $\left\|\dfrac{x_{i+1} - x_i}{x_{i+1}}\right\|100\% \leq \epsilon_s$ Error: $E_{i+1} = O(E_i^2)$
Secant	$x_{i+1} = x_i - \dfrac{f(x_i)(x_{i-1} - x_i)}{f(x_{i-1}) - f(x_i)}$		Stopping criterion: $\left\|\dfrac{x_{i+1} - x_i}{x_{i+1}}\right\|100\% \leq \epsilon_s$

guesses. Ralston and Rabinowitz (1978) and Carnahan, Luther, and Wilkes (1969) contain discussions of this method as well as of other techniques for locating roots of polynomials. As discussed in the text, the *Jenkins-Traub* and *Laguerre's* methods are widely employed.

In summary, the foregoing is intended to provide you with avenues for deeper exploration of the subject. Additionally, all the above references provide descriptions of the basic techniques covered in Part Two. We urge you to consult these alternative sources to broaden your understanding of numerical methods for root location.[1]

[1]Books are referenced only by author here, a complete bibliography will be found at the back of this text.

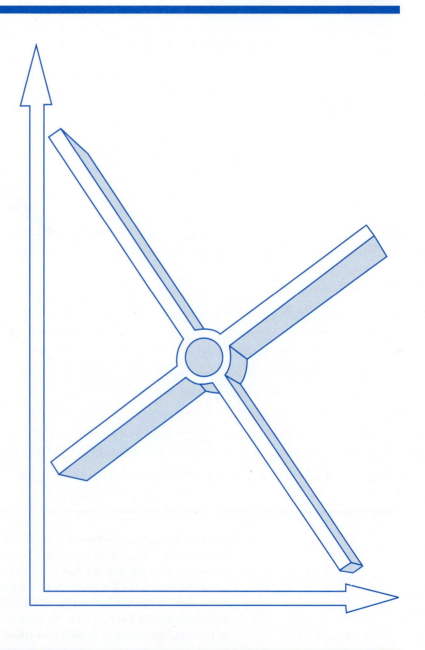

PART THREE

LINEAR ALGEBRAIC EQUATIONS

PT3.1 MOTIVATION

In Part Two, we determined the value x that satisfied a single equation, $f(x) = 0$. Now, we deal with the case of determining the values $x_1, x_2, \ldots, x_n$ that simultaneously satisfy a set of equations

$$f_1(x_1, x_2, \ldots, x_n) = 0$$
$$f_2(x_1, x_2, \ldots, x_n) = 0$$
$$\begin{matrix} \cdot & & \cdot \\ \cdot & & \cdot \\ \cdot & & \cdot \end{matrix}$$
$$f_n(x_1, x_2, \ldots, x_n) = 0$$

Such systems can be either linear or nonlinear. In Part Three, we deal with *linear algebraic equations* that are of the general form

$$a_{11}x_1 + a_{12}x_2 + \cdots + a_{1n}x_n = b_1$$
$$a_{21}x_1 + a_{22}x_2 + \cdots + a_{2n}x_n = b_2$$
$$\begin{matrix} \cdot & & \cdot \\ \cdot & & \cdot \\ \cdot & & \cdot \end{matrix} \qquad\qquad \text{(PT3.1)}$$
$$a_{n1}x_1 + a_{n2}x_2 + \cdots + a_{nn}x_n = b_n$$

where the a's are constant coefficients, the b's are constants, and n is the number of equations. All other equations are nonlinear. Nonlinear systems were discussed in Chap. 6 and will be covered briefly again in Chap. 9.

PT3.1.1 Noncomputer Methods for Solving Systems of Equations

For small numbers of equations ($n \le 3$), linear (and sometimes nonlinear) equations can be solved readily by simple techniques. Some of these methods will be reviewed at the beginning of Chap. 9. However, for four or more equations, solutions become arduous and computers must be utilized. Historically, the inability to solve all but the smallest sets of equations by hand has limited the scope of problems addressed in many engineering applications.

Before computers, techniques to solve linear algebraic equations were time-consuming and awkward. These approaches placed a constraint on creativity because the methods were often difficult to implement and understand. Consequently, the techniques were sometimes overemphasized at the expense of other aspects of the problem-solving process such as formulation and interpretation (recall Fig. PT1.1 and accompanying discussion).

The advent of easily accessible computers makes it possible and practical for you to solve large sets of simultaneous linear algebraic equations. Thus, you can approach more complex and realistic examples and problems. Furthermore, you will have more time to test your creative skills because you will be able to place more emphasis on problem formulation and solution interpretation.

PT3.1.2 Linear Algebraic Equations and Engineering Practice

Many of the fundamental equations of engineering are based on conservation laws (recall Table 1.1). Some familiar quantities that conform to such laws are mass, energy, and momentum. In mathematical terms, these principles lead to balance or continuity equations that relate system *behavior* as represented by the *levels* or *response* of the quantity being modeled to the *properties* or *characteristics* of the system and the external *stimuli* or *forcing functions* acting on the system.

As an example, the principle of mass conservation can be used to formulate a model for a series of chemical reactors (Fig. PT3.1a). For this case, the quantity being modeled is the mass of the chemical in each reactor. The system properties are the reaction characteristics of the chemical and the reactors' sizes and flow rates. The forcing functions are the feed rates of the chemical into the system.

In Part Two, you saw how single-component systems result in a single equation that can be solved using root-location techniques. Multicomponent systems result in a coupled set of mathematical equations that must be solved simultaneously. The equations are cou-

FIGURE PT3.1
Two types of systems that can be modeled using linear algebraic equations: (a) lumped variable system that involves coupled finite components and (b) distributed variable system that involves a continuum.

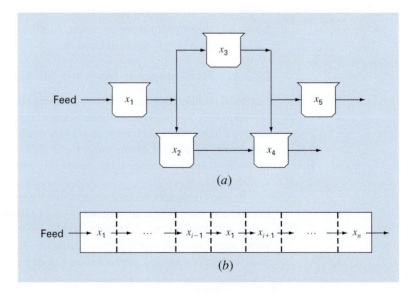

pled because the individual parts of the system are influenced by other parts. For example, in Fig. PT3.1a, reactor 4 receives chemical inputs from reactors 2 and 3. Consequently, its response is dependent on the quantity of chemical in these other reactors.

When these dependencies are expressed mathematically, the resulting equations are often of the linear algebraic form of Eq. (PT3.1). The x's are usually measures of the magnitudes of the responses of the individual components. Using Fig. PT3.1a as an example, x_1 might quantify the amount of mass in the first reactor, x_2 might quantify the amount in the second, and so forth. The a's typically represent the properties and characteristics that bear on the interactions between components. For instance, the a's for Fig. PT3.1a might be reflective of the flow rates of mass between the reactors. Finally, the b's usually represent the forcing functions acting on the system, such as the feed rate in Fig. PT3.1a. The applications in Chap. 12 provide other examples of such equations derived from engineering practice.

Multicomponent problems of the above types arise from both *lumped* (macro-) or *distributed* (micro-) variable mathematical models (Fig. PT3.1). Lumped variable problems involve coupled finite components. Examples include trusses (Sec. 12.2), reactors (Fig. PT3.1a and Sec. 12.1), and electric circuits (Sec. 12.3). These types of problems use models that provide little or no spatial detail.

Conversely, distributed variable problems attempt to describe spatial detail of systems on a continuous or semicontinuous basis. The distribution of chemicals along the length of an elongated, rectangular reactor (Fig. PT3.1b) is an example of a continuous variable model. Differential equations derived from the conservation laws specify the distribution of the dependent variable for such systems. These differential equations can be solved numerically by converting them to an equivalent system of simultaneous algebraic equations. The solution of such sets of equations represents a major engineering application area for the methods in the following chapters. These equations are coupled because the variables at one location are dependent on the variables in adjoining regions. For example, the concentration at the middle of the reactor is a function of the concentration in adjoining regions. Similar examples could be developed for the spatial distribution of temperature or momentum. We will address such problems when we discuss differential equations later in the book.

Aside from physical systems, simultaneous linear algebraic equations also arise in a variety of mathematical problem contexts. These result when mathematical functions are required to satisfy several conditions simultaneously. Each condition results in an equation that contains known coefficients and unknown variables. The techniques discussed in this part can be used to solve for the unknowns when the equations are linear and algebraic. Some widely used numerical techniques that employ simultaneous equations are regression analysis (Chap. 17) and spline interpolation (Chap. 18).

PT3.2 MATHEMATICAL BACKGROUND

All parts of this book require some mathematical background. For Part Three, matrix notation and algebra are useful because they provide a concise way to represent and manipulate linear algebraic equations. If you are already familiar with matrices, feel free to skip to Sec. PT3.3. For those who are unfamiliar or require a review, the following material provides a brief introduction to the subject.

Column 3

$$[A] = \begin{bmatrix} a_{11} & a_{12} & a_{13} & \cdots & a_{1m} \\ a_{21} & a_{22} & a_{23} & \cdots & a_{2m} \\ \cdot & \cdot & & & \cdot \\ \cdot & \cdot & & & \cdot \\ \cdot & \cdot & & & \cdot \\ a_{n} & a_{n2} & a_{n3} & \cdots & a_{nm} \end{bmatrix} \quad \text{Row 2}$$

FIGURE PT3.2
A matrix.

PT3.2.1 Matrix Notation

A *matrix* consists of a rectangular array of elements represented by a single symbol. As depicted in Fig. PT3.2, [A] is the shorthand notation for the matrix and a_{ij} designates an individual *element* of the matrix.

A horizontal set of elements is called a *row* and a vertical set is called a *column*. The first subscript i always designates the number of the row in which the element lies. The second subscript j designates the column. For example, element a_{23} is in row 2 and column 3.

The matrix in Fig. PT3.2 has n rows and m columns and is said to have a dimension of n by m (or $n \times m$). It is referred to as an n by m matrix.

Matrices with row dimension $n = 1$, such as

$$[B] = [b_1 \quad b_2 \quad \cdots \quad b_m]$$

are called *row vectors*. Note that for simplicity, the first subscript of each element is dropped. Also, it should be mentioned that there are times when it is desirable to employ a special shorthand notation to distinguish a row matrix from other types of matrices. One way to accomplish this is to employ special open-topped brackets, as in $\lfloor B \rfloor$.

Matrices with column dimension $m = 1$, such as

$$[C] = \begin{bmatrix} c_1 \\ c_2 \\ \cdot \\ \cdot \\ \cdot \\ c_n \end{bmatrix}$$

are referred to as *column vectors*. For simplicity, the second subscript is dropped. As with the row vector, there are occasions when it is desirable to employ a special shorthand notation to distinguish a column matrix from other types of matrices. One way to accomplish this is to employ special brackets, as in $\{C\}$.

Matrices where $n = m$ are called *square matrices*. For example, a 4 by 4 matrix is

$$[A] = \begin{bmatrix} a_{11} & a_{12} & a_{13} & a_{14} \\ a_{21} & a_{22} & a_{23} & a_{24} \\ a_{31} & a_{32} & a_{33} & a_{34} \\ a_{41} & a_{42} & a_{43} & a_{44} \end{bmatrix}$$

The diagonal consisting of the elements a_{11}, a_{22}, a_{33}, and a_{44} is termed the *principal*, or *main, diagonal* of the matrix.

Square matrices are particularly important when solving sets of simultaneous linear equations. For such systems, the number of equations (corresponding to rows) and the number of unknowns (corresponding to columns) must be equal for a unique solution to be possible. Consequently, square matrices of coefficients are encountered when dealing with such systems. Some special types of square matrices are described in Box PT3.1.

PT3.2.2 Matrix Operating Rules

Now that we have specified what we mean by a matrix, we can define some operating rules that govern its use. Two n by m matrices are equal if, and only if, every element in the first is equal to every element in the second; that is $[A] = [B]$ if $a_{ij} = b_{ij}$ for all i and j.

Box PT3.1   Special Types of Square Matrices

There are a number of special forms of square matrices that are important and should be noted:

A **symmetric matrix** is one where $a_{ij} = a_{ji}$ for all i's and j's. For example,

$$[A] = \begin{bmatrix} 5 & 1 & 2 \\ 1 & 3 & 7 \\ 2 & 7 & 8 \end{bmatrix}$$

is a 3 by 3 symmetric matrix.

A *diagonal matrix* is a square matrix where all elements off the main diagonal are equal to zero, as in

$$[A] = \begin{bmatrix} a_{11} & & & \\ & a_{22} & & \\ & & a_{33} & \\ & & & a_{44} \end{bmatrix}$$

Note that where large blocks of elements are zero, they are left blank.

An *identity matrix* is a diagonal matrix where all elements on the main diagonal are equal to 1, as in

$$[I] = \begin{bmatrix} 1 & & & \\ & 1 & & \\ & & 1 & \\ & & & 1 \end{bmatrix}$$

The symbol $[I]$ is used to denote the identity matrix. The identity matrix has properties similar to unity.

An *upper triangular matrix* is one where all the elements below the main diagonal are zero, as in

$$[A] = \begin{bmatrix} a_{11} & a_{12} & a_{13} & a_{14} \\ & a_{22} & a_{23} & a_{24} \\ & & a_{33} & a_{34} \\ & & & a_{44} \end{bmatrix}$$

A *lower triangular matrix* is one where all elements above the main diagonal are zero, as in

$$[A] = \begin{bmatrix} a_{11} & & & \\ a_{21} & a_{22} & & \\ a_{31} & a_{32} & a_{33} & \\ a_{41} & a_{42} & a_{43} & a_{44} \end{bmatrix}$$

A *banded matrix* has all elements equal to zero, with the exception of a band centered on the main diagonal:

$$[A] = \begin{bmatrix} a_{11} & a_{12} & & \\ a_{21} & a_{22} & a_{23} & \\ & a_{32} & a_{33} & a_{34} \\ & & a_{43} & a_{44} \end{bmatrix}$$

The above matrix has a bandwidth of 3 and is given a special name—the *tridiagonal matrix*.

Addition of two matrices, say, [A] and [B], is accomplished by adding corresponding terms in each matrix. The elements of the resulting matrix [C] are computed,

$$c_{ij} = a_{ij} + b_{ij}$$

for $i = 1, 2, \ldots, n$ and $j = 1, 2, \ldots, m$. Similarly, the *subtraction* of two matrices, say, [E] minus [F], is obtained by subtracting corresponding terms, as in

$$d_{ij} = e_{ij} - f_{ij}$$

for $i = 1, 2, \ldots, n$ and $j = 1, 2, \ldots, m$. It follows directly from the above definitions that addition and subtraction can be performed only between matrices having the same dimensions.

Both addition and subtraction are *commutative:*

$$[A] + [B] = [B] + [A]$$

Addition and subtraction are also *associative;* that is,

$$[E] - [F] = -[F] + [E]$$

The *multiplication* of a matrix [A] by a scalar g is obtained by multiplying every element of [A] by g, as in

$$[D] = g[A] = \begin{bmatrix} ga_{11} & ga_{12} & \cdots & ga_{1m} \\ ga_{21} & ga_{22} & \cdots & ga_{2m} \\ \cdot & \cdot & & \cdot \\ \cdot & \cdot & & \cdot \\ \cdot & \cdot & & \cdot \\ ga_{n1} & ga_{n2} & \cdots & ga_{nm} \end{bmatrix}$$

The *product* of two matrices is represented as $[C] = [A][B]$, where the elements of [C] are defined as (see Box PT3.2 for a simple way to conceptualize matrix multiplication)

$$c_{ij} = \sum_{k=1}^{n} a_{ik} b_{kj} \tag{PT3.2}$$

where n = the column dimension of [A] and the row dimension of [B]. That is, the c_{ij} element is obtained by adding the product of individual elements from the ith row of the first matrix, in this case [A], by the jth column of the second matrix [B].

According to this definition, multiplication of two matrices can be performed only if the first matrix has as many columns as the number of rows in the second matrix. Thus, if [A] is an n by m matrix, [B] could be an m by l matrix. For this case, the resulting [C] matrix would have the dimension of n by l. However, if [B] were an l by m matrix, the multiplication could not be performed. Figure PT3.3 provides an easy way to check whether two matrices can be multiplied.

Box PT3.2   A Simple Method for Multiplying Two Matrices

Although Eq. (PT3.2) is well-suited for implementation on a computer, it is not the simplest means for visualizing the mechanics of multiplying two matrices. What follows gives more tangible expression to the operation.

Suppose that we want to multiply $[X]$ by $[Y]$ to yield $[Z]$,

$$[Z] = [X][Y] = \begin{bmatrix} 3 & 1 \\ 8 & 6 \\ 0 & 4 \end{bmatrix} \begin{bmatrix} 5 & 9 \\ 7 & 2 \end{bmatrix}$$

A simple way to visualize the computation of $[Z]$ is to raise $[Y]$, as in

$$\begin{matrix} & \Uparrow & \\ & \begin{bmatrix} 5 & 9 \\ 7 & 2 \end{bmatrix} & \leftarrow [Y] \\ [X] \rightarrow \begin{bmatrix} 3 & 1 \\ 8 & 6 \\ 0 & 4 \end{bmatrix} & \begin{bmatrix} \; ? \; \end{bmatrix} & \leftarrow [Z] \end{matrix}$$

Now the answer $[Z]$ can be computed in the space vacated by $[Y]$. This format has utility because it aligns the appropriate rows and columns that are to be multiplied. For example, according to Eq. (PT3.2), the element z_{11} is obtained by multiplying the first row of $[X]$ by the first column of $[Y]$. This amounts to adding the product of x_{11} and y_{11} to the product of x_{12} and y_{21}, as in

$$\begin{matrix} & \begin{bmatrix} 5 & 9 \\ 7 & 2 \end{bmatrix} \\ & \downarrow \\ \begin{bmatrix} 3 & 1 \\ 8 & 6 \\ 0 & 4 \end{bmatrix} \rightarrow & \begin{bmatrix} 3 \times 5 + 1 \times 7 = 22 & \\ & \\ & \end{bmatrix} \end{matrix}$$

Thus, z_{11} is equal to 22. Element z_{21} can be computed in a similar fashion, as in

$$\begin{matrix} & \begin{bmatrix} 5 & 9 \\ 7 & 2 \end{bmatrix} \\ & \downarrow \\ \begin{bmatrix} 3 & 1 \\ 8 & 6 \\ 0 & 4 \end{bmatrix} \rightarrow & \begin{bmatrix} & 22 & \\ 8 \times 5 + 6 \times 7 = 82 & \\ & \end{bmatrix} \end{matrix}$$

The computation can be continued in this way, following the alignment of the rows and columns, to yield the result

$$[Z] = \begin{bmatrix} 22 & 29 \\ 82 & 84 \\ 28 & 8 \end{bmatrix}$$

Note how this simple method makes it clear why it is impossible to multiply two matrices if the number of columns of the first matrix does not equal the number of rows in the second matrix. Also, note how it demonstrates that the order of multiplication matters (that is, matrix multiplication is not commutative).

FIGURE PT3.3

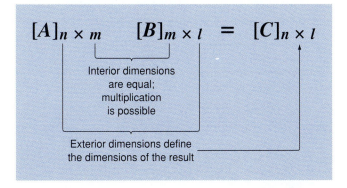

$$[A]_{n \times m} \quad [B]_{m \times l} \quad = \quad [C]_{n \times l}$$

Interior dimensions are equal; multiplication is possible

Exterior dimensions define the dimensions of the result

If the dimensions of the matrices are suitable, matrix multiplication is *associative*,

$$([A][B])[C] = [A]([B][C])$$

and *distributive*,

$$[A]([B] + [C]) = [A][B] + [A][C]$$

or

$$([A] + [B])[C] = [A][C] + [B][C]$$

However, multiplication is not generally *commutative*:

$$[A][B] \neq [B][A]$$

That is, the order of multiplication is important.

Figure PT3.4 shows pseudocode to multiply an n by m matrix $[A]$, by an m by l matrix $[B]$, and store the result in an n by l matrix $[C]$. Notice that, instead of the inner product being directly accumulated in $[C]$, it is collected in a temporary variable, sum. This is done for two reasons. First, it is a bit more efficient, because the computer need determine the location of $c_{i,j}$ only $n \times l$ times rather than $n \times l \times m$ times. Second, the precision of the multiplication can be greatly improved by declaring sum as a double precision variable (recall the discussion of inner products in Sec. 3.4.2).

Although multiplication is possible, matrix division is not a defined operation. However, if a matrix $[A]$ is square and nonsingular, there is another matrix $[A]^{-1}$, called the *inverse* of $[A]$, for which

$$[A][A]^{-1} = [A]^{-1}[A] = [I] \tag{PT3.3}$$

Thus, the multiplication of a matrix by the inverse is analogous to division, in the sense that a number divided by itself is equal to 1. That is, multiplication of a matrix by its inverse leads to the identity matrix (recall Box PT3.1).

The inverse of a two-dimensional square matrix can be represented simply by

$$[A]^{-1} = \frac{1}{a_{11}a_{22} - a_{12}a_{21}} \begin{bmatrix} a_{22} & -a_{12} \\ -a_{21} & a_{11} \end{bmatrix} \tag{PT3.4}$$

FIGURE PT3.4

```
SUBROUTINE Mmult (a, b, c, m, n, l)
DO i = 1, n
  DO j = 1, l
    sum = 0.
    DO k = 1, m
      sum = sum + a(i, k) · b(k, j)
    END DO
    c(i, j) = sum
  END DO
END DO
```

Similar formulas for higher-dimensional matrices are much more involved. Sections in Chaps. 10 and 11 will be devoted to techniques for using numerical methods and the computer to calculate the inverse for such systems.

Two other matrix manipulations that will have utility in our discussion are the transpose and the trace of a matrix. The transpose of a matrix involves transforming its rows into columns and its columns into rows. For example, for the 4×4 matrix,

$$[A] = \begin{bmatrix} a_{11} & a_{12} & a_{13} & a_{14} \\ a_{21} & a_{22} & a_{23} & a_{24} \\ a_{31} & a_{32} & a_{33} & a_{34} \\ a_{41} & a_{42} & a_{43} & a_{44} \end{bmatrix}$$

the transpose, designated $[A]^T$ is defined as

$$[A]^T = \begin{bmatrix} a_{11} & a_{21} & a_{31} & a_{41} \\ a_{12} & a_{22} & a_{32} & a_{42} \\ a_{13} & a_{23} & a_{33} & a_{43} \\ a_{14} & a_{24} & a_{34} & a_{44} \end{bmatrix}$$

In other words, the element a_{ij} of the transpose is equal to the a_{ji} element of the original matrix.

The *transpose* has a variety of functions in matrix algebra. One simple advantage is that it allows a column vector to be written as a row. For example, if

$$\{C\} = \begin{Bmatrix} c_1 \\ c_2 \\ c_3 \\ c_4 \end{Bmatrix}$$

then

$$\{C\}^T = \lfloor c_1 \quad c_2 \quad c_3 \quad c_4 \rfloor$$

where the superscript T designates the transpose. For example, this can save space when writing a column vector in a manuscript. In addition, the transpose has numerous mathematical applications.

The *trace* of a matrix is the sum of the elements on its principal diagonal. It is designated as tr $[A]$ and is computed as

$$\mathrm{tr}\,[A] = \sum_{i=1}^{n} a_{ii}$$

The trace will be used in our discussion of eigenvalues in Chap. 27.

The final matrix manipulation that will have utility in our discussion is augmentation. A matrix is augmented by the addition of a column (or columns) to the original matrix. For example, suppose we have a matrix of coefficients:

$$[A] = \begin{bmatrix} a_{11} & a_{12} & a_{13} \\ a_{21} & a_{22} & a_{23} \\ a_{31} & a_{32} & a_{33} \end{bmatrix}$$

We might wish to augment this matrix $[A]$ with an identity matrix (recall Box PT3.1) to yield a 3-by-6-dimensional matrix:

$$[A] = \begin{bmatrix} a_{11} & a_{12} & a_{13} & \vdots & 1 & 0 & 0 \\ a_{21} & a_{22} & a_{23} & \vdots & 0 & 1 & 0 \\ a_{31} & a_{32} & a_{33} & \vdots & 0 & 0 & 1 \end{bmatrix}$$

Such an expression has utility when we must perform a set of identical operations on two matrices. Thus, we can perform the operations on the single augmented matrix rather than on the two individual matrices.

PT3.2.3 Representing Linear Algebraic Equations in Matrix Form

It should be clear that matrices provide a concise notation for representing simultaneous linear equations. For example, Eq. (PT3.1) can be expressed as

$$[A]\{X\} = \{B\} \tag{PT3.5}$$

where $[A]$ is the n by n square matrix of coefficients,

$$[A] = \begin{bmatrix} a_{11} & a_{12} & \cdots & a_{1n} \\ a_{21} & a_{22} & \cdots & a_{2n} \\ . & . & & . \\ . & . & & . \\ . & . & & . \\ a_{n1} & a_{n2} & \cdots & a_{nn} \end{bmatrix}$$

$\{B\}$ is the n by 1 column vector of constants,

$$\{B\}^T = \lfloor b_1 \quad b_2 \quad \cdots \quad b_n \rfloor$$

and $\{X\}$ is the n by 1 column vector of unknowns:

$$\{X\}^T = \lfloor x_1 \quad x_2 \quad \cdots \quad x_n \rfloor$$

Recall the definition of matrix multiplication [Eq. (PT3.2) or Box PT3.2] to convince yourself that Eqs. (PT3.1) and (PT3.5) are equivalent. Also, realize that Eq. (PT3.5) is a valid matrix multiplication because the number of columns, n, of the first matrix $[A]$ is equal to the number of rows, n, of the second matrix $\{X\}$.

This part of the book is devoted to solving Eq. (PT3.5) for $\{X\}$. A formal way to obtain a solution using matrix algebra is to multiply each side of the equation by the inverse of $[A]$ to yield

$$[A]^{-1}[A]\{X\} = [A]^{-1}\{B\}$$

Because $[A]^{-1}[A]$ equals the identity matrix, the equation becomes

$$\{X\} = [A]^{-1}\{B\} \tag{PT3.6}$$

Therefore, the equation has been solved for $\{X\}$. This is another example of how the inverse plays a role in matrix algebra that is similar to division. It should be noted that this is not a very efficient way to solve a system of equations. Thus, other approaches are employed in numerical algorithms. However, as discussed in Chap. 10, the matrix inverse itself has great value in the engineering analyses of such systems.

Finally, we will sometimes find it useful to augment $[A]$ with $\{B\}$. For example, if $n = 3$, this results in a 3-by-4-dimensional matrix:

$$[A] = \begin{bmatrix} a_{11} & a_{12} & a_{13} & \vdots & b_1 \\ a_{21} & a_{22} & a_{23} & \vdots & b_2 \\ a_{31} & a_{32} & a_{33} & \vdots & b_3 \end{bmatrix} \qquad \text{(PT3.7)}$$

Expressing the equations in this form has utility, because several of the techniques for solving linear systems perform identical operations on a row of coefficients and the corresponding right-hand-side constant. As expressed in Eq. (PT3.7), we can perform the manipulation once on an individual row of the augmented matrix rather than separately on the coefficient matrix and the right-hand-side vector.

PT3.3 ORIENTATION

Before proceeding to the numerical methods, some further orientation might be helpful. The following is intended as an overview of the material discussed in Part Three. In addition, we have formulated some objectives to help focus your efforts when studying the material.

PT3.3.1 Scope and Preview

Figure PT3.5 provides an overview for Part Three. *Chapter 9* is devoted to the most fundamental technique for solving linear algebraic systems: *Gauss elimination.* Before launching into a detailed discussion of this technique, a preliminary section deals with simple methods for solving small systems. These approaches are presented to provide you with visual insight and because one of the methods—the elimination of unknowns—represents the basis for Gauss elimination.

After the preliminary material, "naive" Gauss elimination is discussed. We start with this "stripped-down" version because it allows the fundamental technique to be elaborated on without complicating details. Then, in subsequent sections, we discuss potential problems of the naive approach and present a number of modifications to minimize and circumvent these problems. The focus of this discussion will be the process of switching rows, or *partial pivoting.*

Chapter 10 begins by illustrating how Gauss elimination can be formulated as an *LU decomposition* solution. Such solution techniques are valuable for cases where many right-hand-side vectors need to be evaluated. It is shown how this attribute allows efficient calculation of the *matrix inverse,* which has tremendous utility in engineering practice. Finally, the chapter ends with a discussion of matrix condition. The *condition number* is introduced as a measure of the loss of significant digits of accuracy that can result when solving ill-conditioned matrices.

The beginning of Chap. 11 focuses on special types of systems of equations that have broad engineering application. In particular, efficient techniques for solving *tridiagonal systems* are presented. Then, the remainder of the chapter focuses on an alternative to

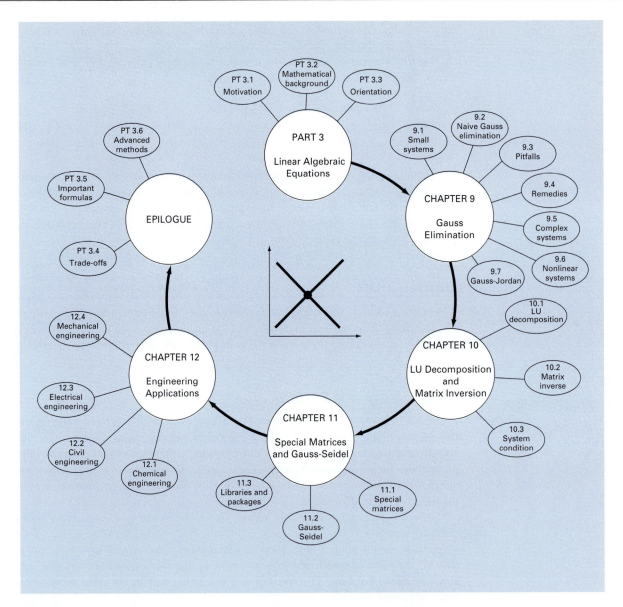

FIGURE PT3.5
Schematic of the organization of the material in Part Three: Systems of Linear Algebraic Equations.

elimination methods called the *Gauss-Seidel method*. This technique is similar in spirit to the approximate methods for roots of equations that were discussed in Chap. 6. That is, the technique involves guessing a solution and then iterating to obtain a refined estimate. The chapter ends with information related to solving linear algebraic equations with software packages and libraries.

Chapter 12 demonstrates how the methods can actually be applied for problem solving. As with other parts of the book, applications are drawn from all fields of engineering.

Finally, an epilogue is included at the end of Part Three. This review includes discussion of trade-offs that are relevant to implementation of the methods in engineering practice. This section also summarizes the important formulas and advanced methods related to linear algebraic equations. As such, it can be used before exams or as a refresher after you have graduated and must return to linear algebraic equations as a professional.

PT3.3.2 Goals and Objectives

Study Objectives.   After completing Part Three, you should be able to solve problems involving linear algebraic equations and appreciate the application of these equations in many fields of engineering. You should strive to master several techniques and assess their reliability. You should understand the trade-offs involved in selecting the "best" method (or methods) for any particular problem. In addition to these general objectives, the specific concepts listed in Table PT3.1 should be assimilated and mastered.

Computer Objectives.   Your most fundamental computer objectives are to be able to solve a system of linear algebraic equations and to evaluate the matrix inverse. You will want to have subprograms developed for LU decomposition of both full and tridiagonal matrices. You may also want to have your own software to implement the Gauss-Seidel method.

TABLE PT3.1   Specific study objectives for Part Three.

1. Understand the graphical interpretation of ill-conditioned systems and how it relates to the determinant.
2. Be familiar with terminology: forward elimination, back substitution, pivot equation, and pivot coefficient.
3. Understand the problems of division by zero, round-off error, and ill-conditioning.
4. Know how to compute the determinant using Gauss elimination.
5. Understand the advantages of pivoting; realize the difference between partial and complete pivoting.
6. Know the fundamental difference between Gauss elimination and the Gauss-Jordan method and which is more efficient.
7. Recognize how Gauss elimination can be formulated as an LU decomposition.
8. Know how to incorporate pivoting and matrix inversion into an LU decomposition algorithm.
9. Know how to interpret the elements of the matrix inverse in evaluating stimulus response computations in engineering.
10. Realize how to use the inverse and matrix norms to evaluate system condition.
11. Understand how banded and symmetric systems can be decomposed and solved efficiently.
12. Understand why the Gauss-Seidel method is particularly well suited for large, sparse systems of equations.
13. Know how to assess diagonal dominance of a system of equations and how it relates to whether the system can be solved with the Gauss-Seidel method.
14. Understand the rationale behind relaxation; know where underrelaxation and overrelaxation are appropriate.

You should know how to use packages to solve linear algebraic equations and find the matrix inverse. The Numerical Methods TOOLKIT that accompanies this book uses *LU* decomposition to perform these tasks. You should also become familiar with how the same evaluations can be implemented on popular packages such as Mathcad, Excel, and MATLAB as well as with software libraries such as IMSL.

CHAPTER 9

Gauss Elimination

This chapter deals with simultaneous linear algebraic equations that can be represented generally as

$$a_{11}x_1 + a_{12}x_2 + \cdots + a_{1n}x_n = b_1$$
$$a_{21}x_1 + a_{22}x_2 + \cdots + a_{2n}x_n = b_2$$
$$\vdots \qquad\qquad \vdots \qquad\qquad\qquad\qquad (9.1)$$
$$a_{n1}x_1 + a_{n2}x_2 + \cdots + a_{nn}x_n = b_n$$

where the a's are constant coefficients and the b's are constants.

The technique described in this chapter is called *Gauss elimination* because it involves combining equations to eliminate unknowns. Although it is one of the earliest methods for solving simultaneous equations, it remains among the most important algorithms in use today, and is the basis for linear equation solving on many popular software packages.

9.1 SOLVING SMALL NUMBERS OF EQUATIONS

Before proceeding to the computer methods, we will describe several methods that are appropriate for solving small ($n \leq 3$) sets of simultaneous equations and that do not require a computer. These are the graphical method, Cramer's rule, and the elimination of unknowns.

9.1.1 The Graphical Method

A graphical solution is obtainable for two equations by plotting them on Cartesian coordinates with one axis corresponding to x_1 and the other to x_2. Because we are dealing with linear systems, each equation is a straight line. This can be easily illustrated for the general equations

$$a_{11}x_1 + a_{12}x_2 = b_1$$
$$a_{21}x_1 + a_{22}x_2 = b_2$$

233

Both equations can be solved for x_2:

$$x_2 = -\left(\frac{a_{11}}{a_{12}}\right)x_1 + \frac{b_1}{a_{12}}$$

$$x_2 = -\left(\frac{a_{21}}{a_{22}}\right)x_1 + \frac{b_2}{a_{22}}$$

Thus, the equations are now in the form of straight lines; that is, $x_2 = $ (slope) $x_1 + $ intercept. These lines can be graphed on Cartesian coordinates with x_2 as the ordinate and x_1 as the abscissa. The values of x_1 and x_2 at the intersection of the lines represent the solution.

EXAMPLE 9.1   The Graphical Method for Two Equations

Problem Statement.   Use the graphical method to solve

$$3x_1 + 2x_2 = 18 \tag{E9.1.1}$$

$$-x_1 + 2x_2 = 2 \tag{E9.1.2}$$

Solution.   Let x_1 be the abscissa. Solve Eq (E9.1.1) for x_2:

$$x_2 = -\frac{3}{2}x_1 + 9$$

which, when plotted on Fig. 9.1, is a straight line with an intercept of 9 and a slope of $-3/2$.

FIGURE 9.1
Graphical solution of a set of two simultaneous linear algebraic equations. The intersection of the lines represents the solution.

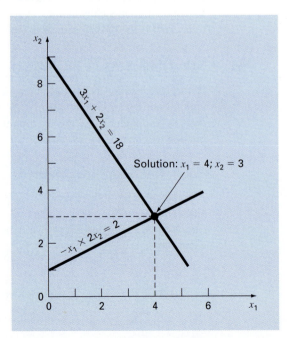

Equation (E9.1.2) can also be solved for x_2:

$$x_2 = \frac{1}{2}x_1 + 1$$

which is also plotted on Fig. 9.1. The solution is the intersection of the two lines at $x_1 = 4$ and $x_2 = 3$. This result can be checked by substituting these values into the original equations to yield

$$3(4) + 2(3) = 18$$
$$-(4) + 2(3) = 2$$

Thus, the results are equivalent to the right-hand sides of the original equations.

For three simultaneous equations, each equation would be represented by a plane in a three-dimensional coordinate system. The point where the three planes intersect would represent the solution. Beyond three equations, graphical methods break down and, consequently, have little practical value for solving simultaneous equations. However, they sometimes prove useful in visualizing properties of the solutions. For example, Fig. 9.2 depicts three cases that can pose problems when solving sets of linear equations. Figure 9.2*a* shows the case where the two equations represent parallel lines. For such situations, there is no solution because the lines never cross. Figure 9.2*b* depicts the case where the two lines are coincident. For such situations there is an infinite number of solutions. Both types of systems are said to be *singular*. In addition, systems that are very close to being singular (Fig. 9.2*c*) can also cause problems. These systems are said to be *ill-conditioned*. Graphically, this corresponds to the fact that it is difficult to identify the exact point at which the lines intersect. Ill-conditioned systems will also pose problems when they are encountered during the numerical solution of linear equations. This is because they will be extremely sensitive to round-off error (recall Sec. 4.2.3).

FIGURE 9.2
Graphical depiction of singular and ill-conditioned systems: (a) no solution, (b) infinite solutions, and (c) ill-conditioned system where the slopes are so close that the point of intersection is difficult to detect visually.

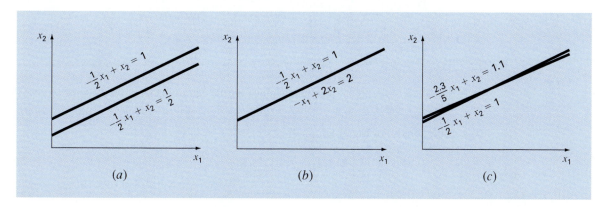

9.1.2 Determinants and Cramer's Rule

Cramer's rule is another solution technique that is best suited to small numbers of equations. Before describing this method, we will briefly introduce the concept of the determinant, which is used to implement Cramer's rule. In addition, the determinant has relevance to the evaluation of ill-conditioning of a matrix.

Determinants. The determinant can be illustrated for a set of three equations:

$$[A]\{X\} = \{B\}$$

where $[A]$ is the coefficient matrix:

$$[A] = \begin{bmatrix} a_{10.5} & a_{12} & a_{13} \\ a_{21} & a_{22} & a_{23} \\ a_{31} & a_{32} & a_{33} \end{bmatrix}$$

The determinant D of this system is formed from the coefficients of the equation, as in

$$D = \begin{vmatrix} a_{11} & a_{12} & a_{13} \\ a_{21} & a_{22} & a_{23} \\ a_{31} & a_{32} & a_{33} \end{vmatrix} \tag{9.2}$$

Although the determinant D and the coefficient matrix $[A]$ are composed of the same elements, they are completely different mathematical concepts. That is why they are distinguished visually by using brackets to enclose the matrix and straight lines to enclose the determinant. In contrast to a matrix, the determinant is a single number. For example, the value of the second-order determinant

$$D = \begin{vmatrix} a_{11} & a_{12} \\ a_{21} & a_{22} \end{vmatrix}$$

is calculated by

$$D = a_{11}a_{22} - a_{12}a_{21} \tag{9.3}$$

For the third-order case [Eq. (9.2)], a single numerical value for the determinant can be computed as

$$D = a_{11}\begin{vmatrix} a_{22} & a_{23} \\ a_{32} & a_{33} \end{vmatrix} - a_{12}\begin{vmatrix} a_{21} & a_{23} \\ a_{31} & a_{33} \end{vmatrix} + a_{13}\begin{vmatrix} a_{21} & a_{22} \\ a_{31} & a_{32} \end{vmatrix} \tag{9.4}$$

where the 2 by 2 determinants are called *minors.*

EXAMPLE 9.2   Determinants

Problem Statement.   Compute values for the determinants of the systems represented in Figs. 9.1 and 9.2.

Solution.   For Fig. 9.1:

$$D = \begin{vmatrix} 3 & 2 \\ -1 & 2 \end{vmatrix} = 3(2) - 2(-1) = 8$$

For Fig. 9.2*a*:

$$D = \begin{vmatrix} -1/2 & 1 \\ -1/2 & 1 \end{vmatrix} = \frac{-1}{2}(1) - 1\left(\frac{-1}{2}\right) = 0$$

For Fig. 9.2*b*:

$$D = \begin{vmatrix} -1/2 & 1 \\ -1 & 2 \end{vmatrix} = \frac{-1}{2}(2) - 1(-1) = 0$$

For Fig. 9.2*c*:

$$D = \begin{vmatrix} -1/2 & 1 \\ -2.3/5 & 1 \end{vmatrix} = \frac{-1}{2}(1) - 1\left(\frac{-2.3}{5}\right) = -0.04$$

In the foregoing example, the singular systems had zero determinants. Additionally, the results suggest that the system that is almost singular (Fig. 9.2*c*) has a determinant that is close to zero. These ideas will be pursued further in our subsequent discussion of ill-conditioning (Sec. 9.3.3).

Cramer's Rule.   This rule states that each unknown in a system of linear algebraic equations may be expressed as a fraction of two determinants with denominator D and with the numerator obtained from D by replacing the column of coefficients of the unknown in question by the constants $b_1, b_2, \ldots, b_n$. For example, x_1 would be computed as

$$x_1 = \frac{\begin{vmatrix} b_1 & a_{12} & a_{13} \\ b_2 & a_{22} & a_{23} \\ b_3 & a_{32} & a_{33} \end{vmatrix}}{D} \tag{9.5}$$

EXAMPLE 9.3   Cramer's Rule

Problem Statement.   Use Cramer's rule to solve

$$0.3x_1 + 0.52x_2 + x_3 = -0.01$$
$$0.5x_1 + x_2 + 1.9x_3 = 0.67$$
$$0.1x_1 + 0.3x_2 + 0.5x_3 = -0.44$$

Solution.   The determinant D can be written as [Eq. (9.2)]

$$D = \begin{vmatrix} 0.3 & 0.52 & 1 \\ 0.5 & 1 & 1.9 \\ 0.1 & 0.3 & 0.5 \end{vmatrix}$$

The minors are [Eq. (9.3)]

$$A_1 = \begin{vmatrix} 1 & 1.9 \\ 0.3 & 0.5 \end{vmatrix} = 1(0.5) - 1.9(0.3) = -0.07$$

$$A_2 = \begin{vmatrix} 0.5 & 1.9 \\ 0.1 & 0.5 \end{vmatrix} = 0.5(0.5) - 1.9(0.1) = 0.06$$

$$A_3 = \begin{vmatrix} 0.5 & 1 \\ 0.1 & 0.3 \end{vmatrix} = 0.5(0.3) - 1(0.1) = 0.05$$

These can be used to evaluate the determinant, as in [Eq. (9.4)]

$$D = 0.3(-0.07) - 0.52(0.06) + 1(0.05) = -0.0022$$

Applying Eq. (9.5), the solution is

$$x_1 = \frac{\begin{vmatrix} -0.01 & 0.52 & 1 \\ 0.67 & 1 & 1.9 \\ -0.44 & 0.3 & 0.5 \end{vmatrix}}{-0.0022} = \frac{0.03278}{-0.0022} = -14.9$$

$$x_2 = \frac{\begin{vmatrix} 0.3 & -0.01 & 1 \\ 0.5 & 0.67 & 1.9 \\ 0.1 & -0.44 & 0.5 \end{vmatrix}}{-0.0022} = \frac{0.0649}{-0.0022} = -29.5$$

$$x_3 = \frac{\begin{vmatrix} 0.3 & 0.52 & -0.01 \\ 0.5 & 1 & 0.67 \\ 0.1 & 0.3 & -0.44 \end{vmatrix}}{-0.0022} = \frac{-0.04356}{-0.0022} = 19.8$$

For more than three equations, Cramer's rule becomes impractical because, as the number of equations increases, the determinants are time-consuming to evaluate by hand (or by computer). Consequently, more efficient alternatives are used. Some of these alternatives are based on the last noncomputer solution technique covered in this book—the elimination of unknowns.

9.1.3 The Elimination of Unknowns

The elimination of unknowns by combining equations is an algebraic approach that can be illustrated for a set of two equations:

$$a_{11}x_1 + a_{12}x_2 = b_1 \tag{9.6}$$

$$a_{21}x_1 + a_{22}x_2 = b_2 \tag{9.7}$$

The basic strategy is to multiply the equations by constants so that one of the unknowns will be eliminated when the two equations are combined. The result is a single equation that can be solved for the remaining unknown. This value can then be substituted into either of the original equations to compute the other variable.

For example, Eq. (9.6) might be multiplied by a_{21} and Eq. (9.7) by a_{11} to give

$$a_{11}a_{21}x_1 + a_{12}a_{21}x_2 = b_1a_{21} \tag{9.8}$$

$$a_{21}a_{11}x_1 + a_{22}a_{11}x_2 = b_2a_{11} \tag{9.9}$$

Subtracting Eq. (9.8) from Eq. (9.9) will, therefore, eliminate the x_1 term from the equations to yield

$$a_{22}a_{11}x_2 + a_{12}a_{21}x_2 = b_2a_{11} - b_1a_{21}$$

which can be solved for

$$x_2 = \frac{a_{11}b_2 - a_{21}b_1}{a_{11}a_{22} - a_{12}a_{21}} \tag{9.10}$$

Equation (9.10) can then be substituted into Eq. (9.6), which can be solved for

$$x_1 = \frac{a_{22}b_1 - a_{12}b_2}{a_{11}a_{22} - a_{12}a_{21}} \tag{9.11}$$

Notice that Eqs. (9.10) and (9.11) follow directly from Cramer's rule, which states

$$x_1 = \frac{\begin{vmatrix} b_1 & a_{12} \\ b_2 & a_{22} \end{vmatrix}}{\begin{vmatrix} a_{11} & a_{12} \\ a_{21} & a_{22} \end{vmatrix}} = \frac{b_1a_{22} - a_{12}b_2}{a_{11}a_{22} - a_{12}a_{21}}$$

$$x_2 = \frac{\begin{vmatrix} a_{11} & b_1 \\ a_{21} & b_2 \end{vmatrix}}{\begin{vmatrix} a_{11} & a_{12} \\ a_{21} & a_{22} \end{vmatrix}} = \frac{a_{11}b_2 - b_1a_{21}}{a_{11}a_{22} - a_{12}a_{21}}$$

EXAMPLE 9.4 Elimination of Unknowns

Problem Statement.   Use the elimination of unknowns to solve (recall Example 9.1)

$$3x_1 + 2x_2 = 18$$
$$-x_1 + 2x_2 = 2$$

Solution.   Using Eqs. (9.11) and (9.10),

$$x_1 = \frac{2(18) - 2(2)}{3(2) - 2(-1)} = 4$$

$$x_2 = \frac{3(2) - (-1)18}{3(2) - 2(-1)} = 3$$

which is consistent with our graphical solution (Fig. 9.1).

The elimination of unknowns can be extended to systems with more than two or three equations. However, the numerous calculations that are required for larger systems make the method extremely tedious to implement by hand. However, as described in the next section, the technique can be formalized and readily programmed for the computer.

9.2 NAIVE GAUSS ELIMINATION

In the previous section, the elimination of unknowns was used to solve a pair of simultaneous equations. The procedure consisted of two steps:

1. The equations were manipulated to eliminate one of the unknowns from the equations. The result of this *elimination* step was that we had one equation with one unknown.
2. Consequently, this equation could be solved directly and the result *back-substituted* into one of the original equations to solve for the remaining unknown.

This basic approach can be extended to large sets of equations by developing a systematic scheme or algorithm to eliminate unknowns and to back-substitute. *Gauss elimination* is the most basic of these schemes.

This section includes the systematic techniques for forward elimination and back substitution that comprise Gauss elimination. Although these techniques are ideally suited for implementation on computers, some modifications will be required to obtain a reliable algorithm. In particular, the computer program must avoid division by zero. The following method is called *"naive" Gauss elimination* because it does not avoid this problem. Subsequent sections will deal with the additional features required for an effective computer program.

The approach is designed to solve a general set of n equations:

$$a_{11}x_1 + a_{12}x_2 + a_{13}x_3 + \cdots + a_{1n}x_n = b_1 \qquad (9.12a)$$

$$a_{21}x_1 + a_{22}x_2 + a_{23}x_3 + \cdots + a_{2n}x_n = b_2 \qquad (9.12b)$$

$$\cdot \qquad \qquad \cdot$$
$$\cdot \qquad \qquad \cdot$$
$$\cdot \qquad \qquad \cdot$$

$$a_{n1}x_1 + a_{n2}x_2 + a_{n3}x_3 + \cdots + a_{nn}x_n = b_n \qquad (9.12c)$$

As was the case with the solution of two equations, the technique for n equations consists of two phases: elimination of unknowns and solution through back substitution.

Forward Elimination of Unknowns. The first phase is designed to reduce the set of equations to an upper triangular system (Fig. 9.3). The initial step will be to eliminate the first unknown, x_1, from the second through the nth equations. To do this, multiply Eq. (9.12a) by a_{21}/a_{11} to give

$$a_{21}x_1 + \frac{a_{21}}{a_{11}}a_{12}x_2 + \cdots + \frac{a_{21}}{a_{11}}a_{1n}x_n = \frac{a_{21}}{a_{11}}b_1 \qquad (9.13)$$

Now, this equation can be subtracted from Eq. (9.12b) to give

$$\left(a_{22} - \frac{a_{21}}{a_{11}}a_{12}\right)x_2 + \cdots + \left(a_{2n} - \frac{a_{21}}{a_{11}}a_{1n}\right)x_n = b_2 - \frac{a_{21}}{a_{11}}b_1$$

or

$$a'_{22}x_2 + \cdots + a'_{2n}x_n = b'_2$$

where the prime indicates that the elements have been changed from their original values.

The procedure is then repeated for the remaining equations. For instance, Eq. (9.12a) can be multiplied by a_{31}/a_{11} and the result subtracted from the third equation. Repeating

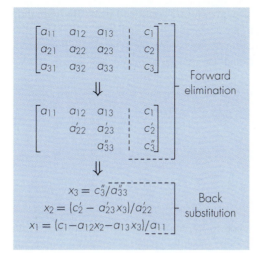

FIGURE 9.3
The two phases of Gauss elimination: forward elimination and back substitution. The primes indicate the number of times that the coefficients and constants have been modified.

the procedure for the remaining equations results in the following modified system:

$$a_{11}x_1 + a_{12}x_2 + a_{13}x_3 + \cdots + a_{1n}x_n = b_1 \qquad (9.14a)$$
$$a'_{22}x_2 + a'_{23}x_3 + \cdots + a'_{2n}x_n = b'_2 \qquad (9.14b)$$
$$a'_{32}x_2 + a'_{33}x_3 + \cdots + a'_{3n}x_n = b'_3 \qquad (9.14c)$$

$$\cdot \qquad \cdot$$
$$\cdot \qquad \cdot$$
$$\cdot \qquad \cdot$$

$$a'_{n2}x_2 + a'_{n3}x_3 + \cdots + a'_{nn}x_n = b'_n \qquad (9.14d)$$

For the foregoing steps, Eq. (9.12a) is called the *pivot equation*, and a_{11} is called the *pivot coefficient* or *element*. Note that the process of multiplying the first row by a_{21}/a_{11} is equivalent to dividing it by a_{11} and multiplying it by a_{21}. Sometimes the division operation is referred to as normalization. We make this distinction because a zero pivot element can interfere with normalization by causing a division by zero. We will return to this important issue after we complete our description of naive Gauss elimination.

Now repeat the above to eliminate the second unknown from Eq. (9.14c) through (9.14d). To do this multiply Eq. (9.14b) by a'_{32}/a'_{22} and subtract the result from Eq. (9.14c). Perform a similar elimination for the remaining equations to yield

$$a_{11}x_1 + a_{12}x_2 + a_{13}x_3 + \cdots + a_{1n}x_n = b_1$$
$$a'_{22}x_2 + a'_{23}x_3 + \cdots + a'_{2n}x_n = b'_2$$
$$a''_{33}x_3 + \cdots + a''_{3n}x_n = b''_3$$

$$\cdot \qquad \cdot$$
$$\cdot \qquad \cdot$$
$$\cdot \qquad \cdot$$

$$a''_{n3}x_3 + \cdots + a''_{nn}x_n = b''_n$$

where the double prime indicates that the elements have been modified twice.

The procedure can be continued using the remaining pivot equations. The final manipulation in the sequence is to use the $(n-1)$th equation to eliminate the x_{n-1} term from the nth equation. At this point, the system will have been transformed to an upper triangular system (recall Box PT3.1):

$$a_{11}x_1 + a_{12}x_2 + a_{13}x_3 + \cdots + a_{1n}x_n = b_1 \tag{9.15a}$$
$$a_{22}'x_2 + a_{23}'x_3 + \cdots + a_{2n}'x_n = b_2' \tag{9.15b}$$
$$a_{33}''x_3 + \cdots + a_{3n}''x_n = b_3'' \tag{9.15c}$$

$$\cdot \qquad\qquad \cdot$$
$$\cdot \qquad\qquad \cdot$$
$$\cdot \qquad\qquad \cdot$$

$$a_{nn}^{(n-1)}x_n = b_n^{(n-1)} \tag{9.15d}$$

Pseudocode to implement forward elimination is presented in Fig. 9.4a. Notice that three nested loops provide a concise representation of the process. The outer loop moves down the matrix from one pivot row to the next. The middle loop moves below the pivot row to each of the subsequent rows where elimination is to take place. Finally, the innermost loop progresses across the columns to eliminate or transform the elements of a particular row.

Back Substitution.   Equation (9.15d) can now be solved for x_n:

$$x_n = \frac{b_n^{(n-1)}}{a_{nn}^{(n-1)}} \tag{9.16}$$

FIGURE 9.4
Pseudocode to perform (a) forward elimination and (b) back substitution.

(a)
```
DO k = 1, n−1
    DO i = k+1, n
        factor = a_{i,k} / a_{k,k}
        DO j = k+1 to n
            a_{i,j} = a_{i,j} − factor · a_{k,j}
        END DO
        b_i = b_i − factor · b_k
    END DO
END DO
```

(b)
```
x_n = b_n / a_{n,n}
DO i = n − 1, 1, −1
    sum = 0
    DO j = i+1, n
        sum = sum + a_{i,j} · x_j
    END DO
    x_i = (b_i − sum) / a_{i,i}
END DO
```

This result can be back-substituted into the $(n-1)$th equation to solve for x_{n-1}. The procedure, which is repeated to evaluate the remaining x's, can be represented by the following formula:

$$x_i = \frac{b_i^{(i-1)} - \sum_{j=i+1}^{n} a_{ij}^{(i-1)} x_j}{a_{ii}^{(i-1)}} \qquad \text{for } i = n-1, n-2, \ldots, 1 \qquad (9.17)$$

Pseudocode to implement Eqs. (9.16) and (9.17) is presented in Fig. 9.4b. Notice the similarity between this pseudocode and that in Fig. PT3.4 for matrix multiplication. As with Fig. PT3.4, a temporary variable, *sum*, is used to accumulate the summation from Eq. (9.17). This results in a somewhat faster execution time than if the summation were accumulated in b_i. More importantly, it allows efficient improvement in precision if the variable, *sum*, is declared in double precision.

EXAMPLE 9.5 | Naive Gauss Elimination

Problem Statement.   Use Gauss elimination to solve

$$3x_1 - 0.1x_2 - 0.2x_3 = 7.85 \qquad (E9.5.1)$$

$$0.1x_1 + 7x_2 - 0.3x_3 = -19.3 \qquad (E9.5.2)$$

$$0.3x_1 - 0.2x_2 + 10x_3 = 71.4 \qquad (E9.5.3)$$

Carry six significant figures during the computation.

Solution.   The first part of the procedure is forward elimination. Multiply Eq. (E9.5.1) by $(0.1)/3$ and subtract the result from Eq. (E9.5.2) to give

$$7.00333x_2 - 0.293333x_3 = -19.5617$$

Then multiply Eq. (E9.5.1) by $(0.3)/3$ and subtract it from Eq. (E9.5.3) to eliminate x_1. After these operations, the set of equations is

$$3x_1 \qquad - 0.1x_2 \qquad - 0.2x_3 = 7.85 \qquad (E9.5.4)$$

$$7.00333x_2 - 0.293333x_3 = -19.5617 \qquad (E9.5.5)$$

$$-0.190000x_2 + 10.0200x_3 = 70.6150 \qquad (E9.5.6)$$

To complete the forward elimination, x_2 must be removed from Eq. (E9.5.6). To accomplish this, multiply Eq. (E9.5.5) by $-0.190000/7.00333$ and subtract the result from Eq. (E9.5.6). This eliminates x_2 from the third equation and reduces the system to an upper triangular form, as in

$$3x_1 \qquad - 0.1x_2 \qquad - 0.2x_3 = 7.85 \qquad (E9.5.7)$$

$$7.00333x_2 - 0.293333x_3 = -19.5617 \qquad (E9.5.8)$$

$$10.0200x_3 = 70.0843 \qquad (E9.5.9)$$

We can now solve these equations by back substitution. First, Eq. (E9.5.9) can be solved for

$$x_3 = \frac{70.0843}{10.0200} = 7.00003 \qquad (E9.5.10)$$

This result can be back-substituted into Eq. (E9.5.8):

$$7.00333x_2 - 0.293333(7.00003) = -19.5617$$

which can be solved for

$$x_2 = \frac{-19.5617 + 0.293333(7.00003)}{7.00333} = -2.50000 \tag{E9.5.11}$$

Finally, Eqs. (E9.5.10) and (E9.5.11) can be substituted into Eq. (E9.5.4):

$$3x_1 - 0.1(-2.50000) - 0.2(7.00003) = 7.85$$

which can be solved for

$$x_1 = \frac{7.85 + 0.1(-2.50000) + 0.2(7.00003)}{3} = 3.00000$$

Although there is a slight round-off error in Eq. (E9.5.10), the results are very close to the exact solution of $x_1 = 3$, $x_2 = -2.5$, and $x_3 = 7$. This can be verified by substituting the results into the original equation set

$$3(3) - 0.1(-2.5) - 0.2(7.00003) = 7.84999 \cong 7.85$$
$$0.1(3) + 7(-2.5) - 0.3(7.00003) = -19.3000 = -19.3$$
$$0.3(3) - 0.2(-2.5) + 10(7.00003) = 71.4003 \cong 71.4$$

9.2.1 Operation Counting

The execution time of Gauss elimination depends on the amount of *floating-point opera-tions* (or *FLOPs*) involved in the algorithm. In general, the time consumed to perform multiplications and divisions is about the same, and is larger than for additions and sub-tractions.

Before analyzing naive Gauss elimination, we will first define some quantities that fa-cilitate operation counting:

$$\sum_{i=1}^{m} cf(i) = c\sum_{i=1}^{m} f(i) \qquad \sum_{i=1}^{m} f(i) + g(i) = \sum_{i=1}^{m} f(i) + \sum_{i=1}^{m} g(i) \tag{9.18a,b}$$

$$\sum_{i=1}^{m} 1 = 1 + 1 + \cdots + 1 = m \qquad \sum_{i=k}^{m} 1 = m - k + 1 \tag{9.18c,d}$$

$$\sum_{i=1}^{m} i = 1 + 2 + 3 + \cdots + m = \frac{m(m+1)}{2} = \frac{m^2}{2} + O(m) \tag{9.18e}$$

$$\sum_{i=1}^{m} i^2 = 1^2 + 2^2 + 3^2 + \cdots + m^2 = \frac{m(m+1)(2m+1)}{6} = \frac{m^3}{3} + O(m^2) \tag{9.18f}$$

where $O(m^n)$ means "terms of order m^n and lower."

Now let's examine the naive Gauss elimination algorithm in detail. As in Fig. 9.4a, we will first count the multiplication/division FLOPs in the elimination stage. On the first pass through the outer loop, $k = 1$. Therefore, the limits on the middle loop are from $i = 2$ to n. According to Eq. (9.18d), this means that the number of iterations of the middle loop will be

$$\sum_{i=2}^{n} 1 = n - 2 + 1 = n - 1 \tag{9.19}$$

Now for every one of these iterations, there is 1 division to define $factor = a_{i,k}/a_{k,k}$. The interior loop then performs a single multiplication ($factor \cdot a_{k,j}$) for each iteration from $j = 2$ to n. Finally, there is one additional multiplication of the right-hand-side value ($factor \cdot b_k$). Thus, for every iteration of the middle loop, the number of multiplications is

$$1 + [n - 2 + 1] + 1 = 1 + n \tag{9.20}$$

The total for the first pass through the outer loop is therefore obtained by multiplying Eq. (9.19) by (9.20) to give $[n - 1](1 + n)$.

A similar procedure can be used to estimate the multiply/divide FLOPs for the subsequent iterations of the outer loop. These can be summarized as

Outer Loop k	Middle Loop i	FLOPs
1	2, n	$[n - 1](1 + n)$
2	3, n	$[n - 2](n)$
.	.	.
.	.	.
.	.	.
k	$k + 1, n$	$[n - k](n + 2 - k)$
.	.	.
.	.	.
.	.	.
$n - 1$	n, n	$[1] (3)$

Therefore, the total FLOPs for elimination can be computed as

$$\sum_{k=1}^{n-1} [n - k](n + 2 - k) = \sum_{k=1}^{n-1} \left\{ n(n + 2) - k(2n + 2) + k^2 \right\} \tag{9.21}$$

Applying some of the relationships from Eq. (9.18) yields

$$\left\{ n^3 + O(n^2) \right\} - \left\{ n^3 + O(n^2) \right\} + \left\{ \frac{2n^3}{6} + O(n^2) \right\} = \frac{n^3}{3} + O(n^2) \tag{9.22}$$

Thus, the total number of multiply/divide FLOPs is equal to $n^3/3$ plus an additional component proportional to terms of order n^2 and lower. The result is written in this way, because as n gets large, the $O(n^2)$ terms become negligible. We are therefore justified in concluding that for large n, the effort involved in forward elimination is $n^3/3$.

Because only a single loop is used, back substitution is much simpler to evaluate. The number of multiplication FLOPs can be directly taken from Eq. (9.18e),

$$\sum_{i=1}^{n} i = 1 + 2 + 3 + \cdots + n = \frac{n(n + 1)}{2} = \frac{n^2}{2} + O(n)$$

TABLE 9.1 Number of FLOPs for naive Gauss elimination.

n	Elimination	Back Substitution	Total FLOPs	$n^3/3$	Percent due to Elimination
10	375	55	430	333	87.21%
100	338250	5050	343300	333333	98.53%
1000	3.34E+08	500500	3.34×10^8	3.33×10^8	99.85%

Thus, the total effort in naive Gauss elimination can be represented as

$$\underbrace{\frac{n^3}{3} + O(n^2)}_{\substack{\text{Forward} \\ \text{elimination}}} + \underbrace{\frac{n^2}{2} + O(n)}_{\substack{\text{Back} \\ \text{substitution}}} \quad \xrightarrow{\text{as } n \text{ increases}} \quad \frac{n^3}{3} + O(n^2) \tag{9.23}$$

Two useful general conclusions can be drawn from this analysis:

1. As the system gets larger, the computation time increases greatly. As in Table 9.1, the amount of FLOPs increase nearly 3 orders of magnitude for every order of magnitude increase in the dimension.
2. Most of the effort is incurred in the elimination step. Thus, efforts to make the method more efficient should probably focus on this step.

Throughout the remainder of this part, we will make operation counts to compare alternative solution methods. Although we may not go into the detail of the above analysis, the same general approach will be employed.

9.3 PITFALLS OF ELIMINATION METHODS

Whereas there are many systems of equations that can be solved with naive Gauss elimination, there are some pitfalls that must be explored before writing a general computer program to implement the method. Although the following material relates directly to naive Gauss elimination, the information has relevance to other elimination techniques.

9.3.1 Division by Zero

The primary reason that the foregoing technique is called "naive" is that during both the elimination and the back-substitution phases, it is possible that a division by zero can occur. For example, if we use naive Gauss elimination to solve

$$2x_2 + 3x_3 = 8$$
$$4x_1 + 6x_2 + 7x_3 = -3$$
$$2x_1 + x_2 + 6x_3 = 5$$

the normalization of the first row would involve division by $a_{11} = 0$. Problems also can arise when a coefficient is very close to zero. The technique of *pivoting* has been developed to partially avoid these problems. It will be described in Sec. 9.4.2.

9.3.2 Round-Off Errors

Even though the solution in Example 9.5 was close to the true answer, there was a slight discrepancy in the result for x_3 [Eq. (E9.5.10)]. This discrepancy, which amounted to a relative error of -0.00043 percent, was due to our use of six significant figures during the computation. If we had used more significant figures, the error in the results would be reduced further. If we had used fractions instead of decimals (and consequently avoided round-off altogether), the answers would have been exact. However, because computers carry only a limited number of significant figures (recall Sec. 3.4.1), round-off errors can occur and must be considered when evaluating the results.

The problem of round-off error can become particularly important when large numbers of equations are to be solved. This is due to the fact that every result is dependent on previous results. Consequently, an error in the early steps will tend to propagate—that is, it will cause errors in subsequent steps.

Specifying the system size where round-off error becomes significant is complicated by the fact that the type of computer and the properties of the equations are determining factors. A rough rule of thumb is that round-off error may be important when dealing with 100 or more equations. In any event, you should always substitute your answers back into the original equations to check whether a substantial error has occurred. However, as discussed below, the magnitudes of the coefficients themselves can influence whether such an error check ensures a reliable result.

9.3.3 Ill-Conditioned Systems

The adequacy of the solution depends on the condition of the system. In Sec. 9.1.1, a graphical depiction of system condition was developed. As discussed in Sec. 4.2.3, *well-conditioned systems* are those where a small change in one or more of the coefficients results in a similar small change in the solution. *Ill-conditioned systems* are those where small changes in coefficients result in large changes in the solution. An alternative interpretation of ill-conditioning is that a wide range of answers can approximately satisfy the equations. Because round-off errors can induce small changes in the coefficients, these artificial changes can lead to large solution errors for ill-conditioned systems, as illustrated in the following example.

EXAMPLE 9.6   Ill-Conditioned Systems

Problem Statement.   Solve the following system:

$$x_1 + 2x_2 = 10 \qquad\qquad\qquad (E9.6.1)$$

$$1.1x_1 + 2x_2 = 10.4 \qquad\qquad\qquad (E9.6.2)$$

Then, solve it again, but with the coefficient of x_1 in the second equation modified slightly to 1.05.

Solution.   Using Eqs. (9.10) and (9.11), the solution is

$$x_1 = \frac{2(10) - 2(10.4)}{1(2) - 2(1.1)} = 4$$

$$x_2 = \frac{1(10.4) - 1.1(10)}{1(2) - 2(1.1)} = 3$$

However, with the slight change of the coefficient a_{21} from 1.1 to 1.05, the result is changed dramatically to

$$x_1 = \frac{2(10) - 2(10.4)}{1(2) - 2(1.05)} = 8$$

$$x_2 = \frac{1(10.4) - 1.1(10)}{1(2) - 2(1.05)} = 1$$

Notice that the primary reason for the discrepancy between the two results is that the denominator represents the difference of two almost-equal numbers. As illustrated previously in Sec. 3.4.2, such differences are highly sensitive to slight variations in the numbers being manipulated.

At this point, you might suggest that substitution of the results into the original equations would alert you to the problem. Unfortunately, for ill-conditioned systems this is often not the case. Substitution of the erroneous values of $x_1 = 8$ and $x_2 = 1$ into Eqs. (E9.6.1) and (E9.6.2) yields

$$8 + 2(1) = 10 = 10$$
$$1.1(8) + 2(1) = 10.8 \cong 10.4$$

Therefore, although $x_1 = 8$ and $x_2 = 1$ is not the true solution to the original problem, the error check is close enough to possibly mislead you into believing that your solutions are adequate.

As was done previously in the section on graphical methods, a visual representative of ill-conditioning can be developed by plotting Eqs. (E9.6.1) and (E9.6.2) (recall Fig. 9.2). Because the slopes of the lines are almost equal, it is visually difficult to see exactly where they intersect. This visual difficulty is reflected quantitatively in the nebulous results of Example 9.6. We can mathematically characterize this situation by writing the two equations in general form:

$$a_{11}x_1 + a_{12}x_2 = b_1 \tag{9.24}$$

$$a_{21}x_1 + a_{22}x_2 = b_2 \tag{9.25}$$

Dividing Eq. (9.24) by a_{12} and Eq. (9.25) by a_{22} and rearranging yields alternative versions that are in the format of straight lines [$x_2 = $ (slope) $x_1 + $ intercept]:

$$x_2 = -\frac{a_{11}}{a_{12}}x_1 + \frac{b_1}{a_{12}}$$

$$x_2 = -\frac{a_{21}}{a_{22}}x_1 + \frac{b_2}{a_{22}}$$

Consequently, if the slopes are nearly equal,

$$\frac{a_{11}}{a_{12}} \cong \frac{a_{21}}{a_{22}}$$

or, cross-multiplying,

$$a_{11}a_{22} \cong a_{12}a_{21}$$

which can be also expressed as

$$a_{11}a_{22} - a_{12}a_{21} \cong 0 \qquad (9.26)$$

Now, recalling that $a_{11}a_{22} - a_{12}a_{21}$ is the determinant of a two-dimensional system [Eq. (9.3)], we arrive at the general conclusion that an ill-conditioned system is one with a determinant close to zero. In fact, if the determinant is exactly zero, the two slopes are identical, which connotes either no solution or an infinite number of solutions, as is the case for the singular systems depicted in Fig. 9.2a and b.

It is difficult to specify how close to zero the determinant must be to indicate ill-conditioning. This is complicated by the fact that the determinant can be changed by multiplying one or more of the equations by a scale factor without changing the solution. Consequently, the determinant is a relative value that is influenced by the magnitude of the coefficients.

EXAMPLE 9.7   Effect of Scale on the Determinant

Problem Statement.   Evaluate the determinant of the following systems:

(a) From Example 9.1:

$$3x_1 + 2x_2 = 18 \qquad (E9.7.1)$$

$$-x_1 + 2x_2 = 2 \qquad (E9.7.2)$$

(b) From Example 9.6:

$$x_1 + 2x_2 = 10 \qquad (E9.7.3)$$

$$1.1x_1 + 2x_2 = 10.4 \qquad (E9.7.4)$$

(c) Repeat (b) but with the equations multiplied by 10.

Solution.

(a) The determinant of Eqs. (E9.7.1) and (E9.7.2), which are well-conditioned, is

$$D = 3(2) - 2(-1) = 8$$

(b) The determinant of Eqs. (E9.7.3) and (E9.7.4), which are ill-conditioned, is

$$D = 1(2) - 2(1.1) = -0.2$$

(c) The results of (a) and (b) seem to bear out the contention that ill-conditioned systems have near-zero determinants. However, suppose that the ill-conditioned system in (b) is multiplied by 10 to give

$$10x_1 + 20x_2 = 100$$
$$11x_1 + 20x_2 = 104$$

The multiplication of an equation by a constant has no effect on its solution. In addition, it is still ill-conditioned. This can be verified by the fact that multiplying by a constant has no effect on the graphical solution. However, the determinant is dramatically affected:

$$D = 10(20) - 20(11) = -20$$

Not only has it been raised two orders of magnitude, but it is now over twice as large as the determinant of the well-conditioned system in (*a*).

As illustrated by the previous example, the magnitude of the coefficients interjects a scale effect that complicates the relationship between system condition and determinant size. One way to partially circumvent this difficulty is to scale the equations so that the maximum element in any row is equal to 1.

EXAMPLE 9.8   Scaling

Problem Statement.   Scale the systems of equations in Example 9.7 to a maximum value of 1 and recompute their determinants.

Solution.

(a) For the well-conditioned system, scaling results in

$$x_1 + 0.667x_2 = 6$$
$$-0.5x_1 + \qquad x_2 = 1$$

for which the determinant is

$$D = 1(1) - 0.667(-0.5) = 1.333$$

(b) For the ill-conditioned system, scaling gives

$$0.5x_1 + x_2 = 5$$
$$0.55x_1 + x_2 = 5.2$$

for which the determinant is

$$D = 0.5(1) - 1(0.55) = -0.05$$

(c) For the last case, scaling changes the system to the same form as in (*b*) and the determinant is also -0.05. Thus, the scale effect is removed.

In a previous section (Sec. 9.1.2), we suggested that the determinant is difficult to compute for more than three simultaneous equations. Therefore, it might seem that it does not provide a practical means for evaluating system condition. However, as described in Box 9.1, there is a simple algorithm that results from Gauss elimination that can be used to evaluate the determinant.

Aside from the approach used in the previous example, there are a variety of other ways to evaluate system condition. For example, there are alternative methods for normalizing the elements (see Stark, 1970). In addition, as described in the next chapter (Sec. 10.3), the matrix inverse and matrix norms can be employed to evaluate system condition. Finally, a simple (but time-consuming) test is to modify the coefficients slightly and

Box 9.1 Determinant Evaluation Using Gauss Elimination

In Sec. 9.1.2, we stated that determinant evaluation by expansion of minors was impractical for large sets of equations. Thus, we concluded that Cramer's rule would be applicable only to small systems. However, as mentioned in Sec. 9.3.3, the determinant has value in assessing system condition. It would, therefore, be useful to have a practical method for computing this quantity.

Fortunately, Gauss elimination provides a simple way to do this. The method is based on the fact that the determinant of a triangular matrix can be simply computed as the product of its diagonal elements:

$$D = a_{11}a_{22}a_{33} \cdots a_{nn} \qquad (B9.1.1)$$

The validity of this formulation can be illustrated for a 3 by 3 system:

$$D = \begin{vmatrix} a_{11} & a_{12} & a_{13} \\ 0 & a_{22} & a_{23} \\ 0 & 0 & a_{33} \end{vmatrix}$$

where the determinant can be evaluated as [recall Eq. (9.4)]

$$D = a_{11}\begin{vmatrix} a_{22} & a_{23} \\ 0 & a_{33} \end{vmatrix} - a_{12}\begin{vmatrix} 0 & a_{23} \\ 0 & a_{33} \end{vmatrix} + a_{13}\begin{vmatrix} 0 & a_{22} \\ 0 & 0 \end{vmatrix}$$

or, by evaluating the minors (that is, the 2 by 2 determinants),

$$D = a_{11}a_{22}a_{33} - a_{12}(0) + a_{13}(0) = a_{11}a_{22}a_{33}$$

Recall that the forward-elimination step of Gauss elimination results in an upper triangular system. Because the value of the determinant is not changed by the forward-elimination process, the determinant can be simply evaluated at the end of this step via

$$D = a_{11}a'_{22}a''_{33} \cdots a^{(n-1)}_{nn} \qquad (B9.1.2)$$

where the superscripts signify the number of times that the elements have been modified by the elimination process. Thus, we can capitalize on the effort that has already been expended in reducing the system to triangular form and, in the bargain, come up with a simple estimate of the determinant.

There is a slight modification to the above approach when the program employs partial pivoting (Sec. 9.4.2). For this case, the determinant changes sign every time a row is pivoted. One way to represent this is to modify Eq. (B9.1.2):

$$D = a_{11}a'_{22}a''_{33} \cdots a^{(n-1)}_{nn}(-1)^p \qquad (B9.1.3)$$

where p represents the number of times that rows are pivoted. This modification can be incorporated simply into a program; merely keep track of the number of pivots that take place during the course of the computation and then use Eq. (B9.1.3) to evaluate the determinant.

---

repeat the solution. If such modifications lead to drastically different results, the system is likely to be ill-conditioned.

As you might gather from the foregoing discussion, ill-conditioned systems are problematic. Fortunately, most linear algebraic equations derived from engineering-problem settings are naturally well-conditioned. In addition, some of the techniques outlined in Sec. 9.4 help to alleviate the problem.

9.3.4 Singular Systems

In the previous section, we learned that one way in which a system of equations can be ill-conditioned is when two or more of the equations are nearly identical. Obviously, it's even worse when two are identical. In such cases, we would lose one degree of freedom, and would be dealing with the impossible case of $n - 1$ equations with n unknowns. Such cases might not be obvious to you, particularly when dealing with large equation sets. Consequently, it would be nice to have some way of automatically detecting singularity.

The answer to this problem is neatly offered by the fact that the determinant of a singular system is zero. This idea can, in turn, be connected to Gauss elimination by recognizing that after the elimination step, the determinant can be evaluated as the product of the diagonal elements (recall Box 9.1). Thus, a computer algorithm can test to discern whether a zero diagonal element is created during the elimination stage. If one is discovered, the calculation can be immediately terminated and a message displayed alerting the user. We will show the details of how this is done when we present a full algorithm for Gauss elimination later in this chapter.

9.4 TECHNIQUES FOR IMPROVING SOLUTIONS

The following techniques can be incorporated into the naive Gauss elimination algorithm to circumvent some of the pitfalls discussed in the previous section.

9.4.1 Use of More Significant Figures

The simplest remedy for ill-conditioning is to use more significant figures in the computation. If your computer has the capability of being extended to handle larger word size, such a feature will greatly reduce the problem. However, a price must be paid in the form of the computational and memory overhead connected with using extended precision (recall Sec. 3.4.1).

9.4.2 Pivoting

As mentioned at the beginning of Sec. 9.3, obvious problems occur when a pivot element is zero because the normalization step leads to division by zero. Problems may also arise when the pivot element is close to, rather than exactly equal to, zero because if the magnitude of the pivot element is small compared to the other elements, then round-off errors can be introduced.

Therefore, before each row is normalized, it is advantageous to determine the largest available coefficient in the column below the pivot element. The rows can then be switched so that the largest element is the pivot element. This is called *partial pivoting*. If columns as well as rows are searched for the largest element and then switched, the procedure is called *complete pivoting*. Complete pivoting is rarely used because switching columns changes the order of the x's and, consequently, adds significant and usually unjustified complexity to the computer program. The following example illustrates the advantages of partial pivoting. Aside from avoiding division by zero, pivoting also minimizes round-off error. As such, it also serves as a partial remedy for ill-conditioning.

EXAMPLE 9.9   Partial Pivoting

Problem Statement.   Use Gauss elimination to solve

$$0.0003x_1 + 3.0000x_2 = 2.0001$$
$$1.0000x_1 + 1.0000x_2 = 1.0000$$

Note that in this form the first pivot element, $a_{11} = 0.0003$, is very close to zero. Then repeat the computation, but partial pivot by reversing the order of the equations. The exact solution is $x_1 = 1/3$ and $x_2 = 2/3$.

Solution.   Multiplying the first equation by $1/(0.0003)$ yields

$$x_1 + 10,000x_2 = 6667$$

which can be used to eliminate x_1 from the second equation:

$$-9999x_2 = -6666$$

which can be solved for

$$x_2 = \frac{2}{3}$$

This result can be substituted back into the first equation to evaluate x_1:

$$x_1 = \frac{2.0001 - 3(2/3)}{0.0003} \tag{E9.9.1}$$

However, due to subtractive cancellation, the result is very sensitive to the number of significant figures carried in the computation:

Significant Figures	x_2	x_1	Absolute Value of Percent Relative Error for x_1
3	0.667	−3.33	1099
4	0.6667	0.0000	100
5	0.66667	0.30000	10
6	0.666667	0.330000	1
7	0.6666667	0.3330000	0.1

Note how the solution for x_1 is highly dependent on the number of significant figures. This is because in Eq. (E9.9.1), we are subtracting two almost-equal numbers. On the other hand, if the equations are solved in reverse order, the row with the larger pivot element is normalized. The equations are

$$1.0000x_1 + 1.0000x_2 = 1.0000$$
$$0.0003x_1 + 3.0000x_2 = 2.0001$$

Elimination and substitution yield $x_2 = 2/3$. For different numbers of significant figures, x_1 can be computed from the first equation, as in

$$x_1 = \frac{1 - (2/3)}{1} \tag{E9.9.2}$$

This case is much less sensitive to the number of significant figures in the computation:

Significant Figures	x_2	x_1	Absolute Value of Percent Relative Error for x_1
3	0.667	0.333	0.1
4	0.6667	0.3333	0.01
5	0.66667	0.33333	0.001
6	0.666667	0.333333	0.0001
7	0.6666667	0.3333333	0.00001

Thus, a pivot strategy is much more satisfactory.

$p = k$
$big = |a_{k,k}|$
DO $ii = k+1, n$
    $dummy = |a_{ii,k}|$
    IF $(dummy > big)$
        $big = dummy$
        $p = ii$
    END IF
END DO
IF $(p \neq k)$
    DO $jj = k, n$
        $dummy = a_{p,jj}$
        $a_{p,jj} = a_{k,jj}$
        $a_{k,jj} = dummy$
    END DO
    $dummy = b_p$
    $b_p = b_k$
    $b_k = dummy$
END IF

FIGURE 9.5
Pseudocode to implement partial pivoting.

General-purpose computer programs must include a pivot strategy. Figure 9.5 provides a simple algorithm to implement such a strategy. Notice that the algorithm consists of two major loops. After storing the current pivot element and its row number as the variables, *big* and *p*, the first loop compares the pivot element with the elements below it to check whether any of these is larger than the pivot element. If so, the new largest element and its row number are stored in *big* and *p*. Then, the second loop switches the original pivot row with the one with the largest element so that the latter becomes the new pivot row. This pseudocode can be integrated into a program based on the other elements of Gauss elimination outlined in Fig. 9.4. The best way to do this is to employ a modular approach and write Fig. 9.5 as a subroutine (or procedure) that would be called directly after the beginning of the first loop in Fig. 9.4*a*.

Note that the second IF/THEN construct in Fig. 9.5 physically interchanges the rows. For large matrices, this can become quite time-consuming. Consequently, most codes do not actually exchange rows but rather keep track of the pivot rows by storing the appropriate subscripts in a vector. This vector then provides a basis for specifying the proper row ordering during the forward-elimination and back-substitution operations. Thus, the operations are said to be implemented *in place*.

9.4.3 Scaling

In Sec. 9.3.3, we proposed that scaling had value in standardizing the size of the determinant. Beyond this application, it has utility in minimizing round-off errors for those cases where some of the equations in a system have much larger coefficients than others. Such situations are frequently encountered in engineering practice when widely different units are used in the development of simultaneous equations. For instance, in electric-circuit problems, the unknown voltages can be expressed in units ranging from microvolts to kilovolts. Similar examples can arise in all fields of engineering. As long as each equation is consistent, the system will be technically correct and solvable. However, the use of widely differing units can lead to coefficients of widely differing magnitudes. This, in turn, can have an impact on round-off error as it affects pivoting, as illustrated by the following example.

EXAMPLE 9.10   Effect of Scaling on Pivoting and Round-Off

Problem Statement.

(a) Solve the following set of equations using Gauss elimination and a pivoting strategy:

$$2x_1 + 100{,}000x_2 = 100{,}000$$
$$x_1 + \qquad x_2 = 2$$

(b) Repeat the solution after scaling the equations so that the maximum coefficient in each row is 1.

(c) Finally, use the scaled coefficients to determine whether pivoting is necessary. However, actually solve the equations with the original coefficient values. For all cases, retain only three significant figures. Note that the correct answers are $x_1 = 1.00002$ and $x_2 = 0.99998$ or, for three significant figures, $x_1 = x_2 = 1.00$.

Solution.

(a) Without scaling, forward elimination is applied to give

$$2x_1 + 100{,}000x_2 = 100{,}000$$
$$-50{,}000x_2 = -50{,}000$$

which can be solved by back substitution for

$$x_2 = 1.00$$
$$x_1 = 0.00$$

Although x_2 is correct, x_1 is 100 percent in error because of round-off.

(b) Scaling transforms the original equations to

$$0.00002x_1 + x_2 = 1$$
$$x_1 + x_2 = 2$$

Therefore, the rows should be pivoted to put the greatest value on the diagonal.

$$x_1 + x_2 = 2$$
$$0.00002x_1 + x_2 = 1$$

Forward elimination yields

$$x_1 + x_2 = 2$$
$$x_2 = 1.00$$

which can be solved for

$$x_1 = x_2 = 1$$

Thus, scaling leads to the correct answer.

(c) The scaled coefficients indicate that pivoting is necessary. We therefore pivot but retain the original coefficients to give

$$x_1 + \qquad x_2 = 2$$
$$2x_1 + 100{,}000x_2 = 100{,}000$$

Forward elimination yields

$$x_1 + \qquad x_2 = 2$$
$$100{,}000x_2 = 100{,}000$$

which can be solved for the correct answer: $x_1 = x_2 = 1$. Thus, scaling was useful in determining whether pivoting was necessary, but the equations themselves did not require scaling to arrive at a correct result.

```
SUB Gauss (a, b, n, x, tol, er)
    DIMENSION s (n)
    er = 0
    DO i = 1, n
        s_i = ABS(a_{i,1})
        DO j = 2, n
            IF ABS(a_{i,j})>s_i THEN s_i = ABS(a_{i,j})
        END DO
    END DO
    CALL Eliminate(a, s, n, b, tol, er)
    IF er ≠ −1 THEN
        CALL Substitute(a, n, b, x)
    END IF
END Gauss


SUB Eliminate (a, s, n, b, tol, er)
    DO k = 1, n − 1
        CALL Pivot (a, b, s, n, k)
        IF ABS (a_{k,k} / s_k) < tol THEN
            er = −1
            EXIT DO
        END IF
        DO i = k + 1, n
            factor = a_{i,k} / a_{k,k}
            DO j = k + 1, n
                a_{i,j} = a_{i,j} − factor*a_{k,j}
            END DO
            b_i = b_i − factor * b_k
        END DO
    END DO
    IF ABS(a_{k,k}/s_k) < to1 THEN er = −1
END Eliminate
```

```
SUB Pivot (a, b, s, n, k)
    p = k
    big = ABS(a_{k,k} / s_k)
    DO ii = k + 1, n
        dummy = ABS(a_{ii,k} / s_{ii})
        IF dummy > big THEN
            big = dummy
            p = ii
        END IF
    END DO
    IF p ≠ k THEN
        DO jj = k, n
            dummy = a_{p,jj}
            a_{p,jj} = a_{k,jj}
            a_{k,jj} = dummy
        END DO
        dummy = b_p
        b_p = b_k
        b_k = dummy
        dummy = s_p
        s_p = s_k
        s_k = dummy
    END IF
END pivot


SUB Substitute (a, n, b, x)
    x_n = b_n/a_{n,n}
    DO i = n − 1, 1, −1
        sum = 0
        DO j = i + 1, n
            sum = sum + a_{i,j} * x_j
        END DO
        x_i = (b_i − sum) / a_{i,i}
    END DO
END Substitute
```

FIGURE 9.6
Pseudocode to implement Gauss elimination with partial pivoting.

As in the previous example, scaling has utility in minimizing round-off. However, it should be noted that scaling itself also leads to round-off. For example, given the equation

$$2x_1 + 300{,}000x_2 = 1$$

and using three significant figures, scaling leads to

$$0.00000667x_1 + x_2 = 0.00000333$$

Thus, scaling introduces a round-off error to the first coefficient and the right-hand-side constant. For this reason, it is sometimes suggested that scaling should be employed only as in part (c) of the preceding example. That is, it is used to calculate scaled values for the coefficients solely as a criterion for pivoting, but the original coefficient values are retained for the actual elimination and substitution computations. This involves a trade-off if the determinant is being calculated as part of the program. That is, the resulting determinant will be unscaled. However, because many applications of Gauss elimination do not require determinant evaluation, it is the most common approach and will be used in the algorithm in the next section.

9.4.4 Computer Algorithm for Gauss Elimination

The algorithms from Figs. 9.4 and 9.5 can now be combined into a larger algorithm to implement the entire Gauss elimination algorithm. Figure 9.6 shows an algorithm for a general subroutine to implement Gauss elimination.

Note that the program includes modules for the three primary operations of the Gauss elimination algorithm: forward elimination, back substitution, and pivoting. In addition, there are several aspects of the code that differ and represent improvements over the pseudocodes from Figs. 9.4 and 9.5. These are:

- The equations are not scaled, but scaled values of the elements are used to determine whether pivoting is to be implemented.
- The diagonal term is monitored during the pivoting phase to detect near-zero occurrences in order to flag singular systems. If it passes back a value of $er = -1$, a singular matrix has been detected and the computation should be terminated. A parameter tol is set by the user to a small number in order to detect near-zero occurrences.

EXAMPLE 9.11   Solution of Linear Algebraic Equations Using the Computer

Problem Statement.   A computer program to solve linear algebraic equations is contained on the Numerical Methods TOOLKIT software that accompanies this book. We can use this software (or your own software based on Fig. 9.6) to solve a problem associated with the falling parachutist example discussed in Chap. 1. Suppose that a team of three parachutists is connected by a weightless cord while free-falling at a velocity of 5 m/s (Fig. 9.7). Calculate the tension in each section of cord and the acceleration of the team, given the following:

FIGURE 9.7
Three parachutists free-falling
while connected by weightless
cords.

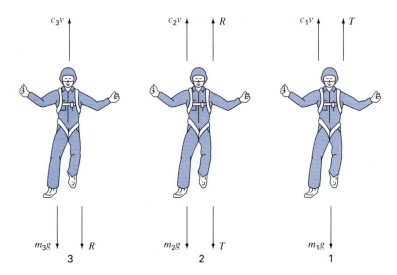

FIGURE 9.8
Free-body diagrams for each of the three falling parachutists.

Parachutist	Mass, kg	Drag Coefficient, kg/s
1	70	10
2	60	14
3	40	17

Solution. Free-body diagrams for each of the parachutists are depicted in Fig. 9.8. Summing the forces in the vertical direction and using Newton's second law gives a set of three simultaneous linear equations:

$$m_1 g - T - c_1 v \qquad = m_1 a$$
$$m_2 g + T - c_2 v - R = m_2 a$$
$$m_3 g \qquad - c_3 v + R = m_3 a$$

These equations have three unknowns: a, T, and R. After substituting the known values, the equations can be expressed in matrix form as ($g = 9.8 \text{ m/s}^2$),

$$\begin{bmatrix} 70 & 1 & 0 \\ 60 & -1 & 1 \\ 40 & 0 & -1 \end{bmatrix} \begin{Bmatrix} a \\ T \\ R \end{Bmatrix} = \begin{Bmatrix} 636 \\ 518 \\ 307 \end{Bmatrix}$$

This system can be solved using your own software or the Solve System option on the Numerical Methods TOOLKIT. The result is $a = 8.5941 \text{ m/s}^2$; $T = 34.4118 \text{ N}$; and $R = 36.7647 \text{ N}$.

9.5 COMPLEX SYSTEMS

In some problems, it is possible to obtain a complex system of equations

$$[C]\{Z\} = \{W\} \tag{9.27}$$

where

$$[C] = [A] + i[B]$$
$$\{Z\} = \{X\} + i\{Y\}$$
$$\{W\} = \{U\} + i\{V\} \tag{9.28}$$

where $i = \sqrt{-1}$.

The most straightforward way to solve such a system is to employ one of the algorithms described in this part of the book, but replace all real operations with complex ones. Of course, this is only possible for those languages, such as Fortran, that allow complex variables.

For languages that do not permit the declaration of complex variables, it is possible to write a code to convert real to complex operations. However, this is not a trivial task. An alternative is to convert the complex system into an equivalent one dealing with real variables. This can be done by substituting Eq. (9.28) into Eq. (9.27) and equating real and complex parts of the resulting equation to yield

$$[A]\{X\} - [B]\{Y\} = \{U\} \tag{9.29}$$

and

$$[B]\{X\} + [A]\{Y\} = \{V\} \tag{9.30}$$

Thus, the system of n complex equations is converted to a set of $2n$ real ones. This means that storage and execution time will be increased significantly. Consequently, a trade-off exists regarding this option. If you evaluate complex systems infrequently, it is preferable to use Eqs. (9.29) and (9.30) because of their convenience. However, if you use them often and desire to employ a language that does not allow complex data types, it may be worth the up-front programming effort to write a customized equation solver that converts real to complex operations.

9.6 NONLINEAR SYSTEMS OF EQUATIONS

Recall that at the end of Chap. 6 we presented an approach to solve two nonlinear equations with two unknowns. This approach can be extended to the general case of solving n simultaneous nonlinear equations.

$$
\begin{aligned}
f_1(x_1, x_2, \ldots, x_n) &= 0 \\
f_2(x_1, x_2, \ldots, x_n) &= 0 \\
&\ \ \vdots \\
f_n(x_1, x_2, \ldots, x_n) &= 0
\end{aligned}
\tag{9.31}
$$

The solution of this system consists of the set of x values that simultaneously result in all the equations equaling zero.

As described in Sec. 6.5.2, one approach to solving such systems is based on a multidimensional version of the Newton-Raphson method. Thus, a Taylor series expansion is written for each equation. For example, for the kth equation,

$$f_{k,i+1} = f_{k,i} + (x_{1,i+1} - x_{1,i})\frac{\partial f_{k,i}}{\partial x_1} + (x_{2,i+1} - x_{2,i})\frac{\partial f_{k,i}}{\partial x_2} + \cdots + (x_{n,i+1} - x_{n,i})\frac{\partial f_{k,i}}{\partial x_n}$$

$$(9.32)$$

where the first subscript, k, represents the equation or unknown and the second subscript denotes whether the value or function in question is at the present value (i) or at the next value ($i + 1$).

Equations of the form of (9.32) are written for each of the original nonlinear equations. Then, as was done in deriving Eq. (6.20) from (6.19), all $f_{k,i+1}$ terms are set to zero as would be the case at the root, and Eq. (9.32) can be written as

$$-f_{k,i} + x_{1,i}\frac{\partial f_{k,i}}{\partial x_1} + x_{2,i}\frac{\partial f_{k,i}}{\partial x_2} + \cdots + x_{n,i}\frac{\partial f_{k,i}}{\partial x_n}$$

$$= x_{1,i+1}\frac{\partial f_{k,i}}{\partial x_1} + x_{2,i+1}\frac{\partial f_{k,i}}{\partial x_2} + \cdots + x_{n,i+1}\frac{\partial f_{k,i}}{\partial x_n} \qquad (9.33)$$

Notice that the only unknowns in Eq. (9.33) are the $x_{k,i+1}$ terms on the right-hand side. All other quantities are located at the present value (i) and, thus, are known at any iteration. Consequently, the set of equations generally represented by Eq. (9.33) (that is, with $k = 1$, $2, \ldots, n$) constitutes a set of linear simultaneous equations that can be solved by methods elaborated in this part of the book.

Matrix notation can be employed to express Eq. (9.33) concisely. The partial derivatives can be expressed as

$$[Z] = \begin{bmatrix} \dfrac{\partial f_{1,i}}{\partial x_1} & \dfrac{\partial f_{1,i}}{\partial x_2} & \cdots & \dfrac{\partial f_{1,i}}{\partial x_n} \\[2mm] \dfrac{\partial f_{2,i}}{\partial x_1} & \dfrac{\partial f_{2,i}}{\partial x_2} & \cdots & \dfrac{\partial f_{2,i}}{\partial x_n} \\[2mm] . & . & & . \\ . & . & & . \\ . & . & & . \\ \dfrac{\partial f_{n,i}}{\partial x_1} & \dfrac{\partial f_{n,i}}{\partial x_2} & \cdots & \dfrac{\partial f_{n,i}}{\partial x_n} \end{bmatrix} \qquad (9.34)$$

The initial and final values can be expressed in vector form as

$$\{X_i\}^T = \lfloor x_{1,i} \quad x_{2,i} \quad \cdots \quad x_{n,i} \rfloor$$

and

$$\{X_{i+1}\}^T = \lfloor x_{1,i+1} \quad x_{2,i+1} \quad \cdots \quad x_{n,i+1} \rfloor$$

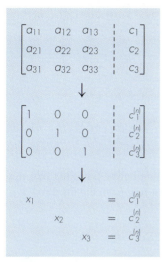

FIGURE 9.9

Graphical depiction of the Gauss-Jordan method. Compare with Fig. 9.3 to elucidate the differences between this technique and Gauss elimination. The superscript (n) means that the elements of the right-hand-side vector have been modified n times (for this case, n = 3).

Finally, the function values at i can be expressed as

$$\{F_i\}^T = \lfloor f_{1,i} \quad f_{2,i} \quad \cdots \quad f_{n,i} \rfloor$$

Using these relationships, Eq. (9.33) can be represented concisely as

$$[Z]\{X_{i+1}\} = -\{F_i\} + [Z]\{X_i\} \qquad (9.35)$$

Equation (9.35) can be solved using a technique such as Gauss elimination. This process can be repeated iteratively to obtain refined estimates in a fashion similar to the two-equation case in Sec. 6.5.2.

It should be noted that there are two major shortcomings to the foregoing approach. First, Eq. (9.34) is often inconvenient to evaluate. Therefore, variations of the Newton-Raphson approach have been developed to circumvent this dilemma. As might be expected, most are based on using finite-difference approximations for the partial derivatives that comprise $[Z]$.

The second shortcoming of the multiequation Newton-Raphson method is that excellent initial guesses are usually required to ensure convergence. Because these are often difficult to obtain, alternative approaches that are slower than Newton-Raphson but which have better convergence behavior have been developed. One common approach is to reformulate the nonlinear system as a single function

$$F(x) = \sum_{i=1}^{n} [f_i(x_1, x_2, \ldots, x_n)]^2 \qquad (9.36)$$

where $f_i(x_1, x_2, \ldots, x_n)$ is the ith member of the original system of Eq. (9.31). The values of x that minimize this function also represent the solution of the nonlinear system. As we will see in Chap. 17, this reformulation belongs to a class of problems called *nonlinear regression*. As such it can be approached with a number of optimization techniques such as the ones described later in this text (Part Four and specifically Chap. 14).

9.7 GAUSS-JORDAN

The Gauss-Jordan method is a variation of Gauss elimination. The major difference is that when an unknown is eliminated in the Gauss-Jordan method, it is eliminated from all other equations rather than just the subsequent ones. In addition, all rows are normalized by dividing them by their pivot elements. Thus, the elimination step results in an identity matrix rather than a triangular matrix (Fig. 9.9). Consequently, it is not necessary to employ back substitution to obtain the solution. The method is best illustrated by an example.

EXAMPLE 9.12 | Gauss-Jordan Method

Problem Statement.   Use the Gauss-Jordan technique to solve the same system as in Example 9.5:

$$3x_1 - 0.1x_2 - 0.2x_3 = 7.85$$
$$0.1x_1 + 7x_2 - 0.3x_3 = -19.3$$
$$0.3x_1 - 0.2x_2 + 10x_3 = 71.4$$

Solution.   First, express the coefficients and the right-hand side as an augmented matrix:

$$\begin{bmatrix} 3 & -0.1 & -0.2 & 7.85 \\ 0.1 & 7 & -0.3 & -19.3 \\ 0.3 & -0.2 & 10 & 71.4 \end{bmatrix}$$

Then normalize the first row by dividing it by the pivot element, 3, to yield

$$\begin{bmatrix} 1 & -0.0333333 & -0.066667 & 2.61667 \\ 0.1 & 7 & -0.3 & -19.3 \\ 0.3 & -0.2 & 10 & 71.4 \end{bmatrix}$$

The x_1 term can be eliminated from the second row by subtracting 0.1 times the first row from the second row. Similarly, subtracting 0.3 times the first row from the third row will eliminate the x_1 term from the third row:

$$\begin{bmatrix} 1 & -0.0333333 & -0.066667 & 2.61667 \\ 0 & 7.00333 & -0.293333 & -19.5617 \\ 0 & -0.190000 & 10.0200 & 70.6150 \end{bmatrix}$$

Next, normalize the second row by dividing it by 7.00333:

$$\begin{bmatrix} 1 & -0.0333333 & -0.066667 & 2.61667 \\ 0 & 1 & -0.0418848 & -2.79320 \\ 0 & -0.190000 & 10.0200 & 70.6150 \end{bmatrix}$$

Reduction of the x_2 terms from the first and third equations gives

$$\begin{bmatrix} 1 & 0 & -0.0680629 & 2.52356 \\ 0 & 1 & -0.0418848 & -2.79320 \\ 0 & 0 & 10.01200 & 70.0843 \end{bmatrix}$$

The third row is then normalized by dividing it by 10.0120:

$$\begin{bmatrix} 1 & 0 & -0.0680629 & 2.52356 \\ 0 & 1 & 0.0418848 & 2.79320 \\ 0 & 0 & 1 & 7.00003 \end{bmatrix}$$

Finally, the x_3 terms can be reduced from the first and the second equations to give

$$\begin{bmatrix} 1 & 0 & 0 & 3.00000 \\ 0 & 1 & 0 & -2.50001 \\ 0 & 0 & 1 & 7.00003 \end{bmatrix}$$

Thus, as depicted in Fig. 9.8; the coefficient matrix has been transformed to the identity matrix, and the solution is obtained in the right-hand-side vector. Notice that no back substitution was required to obtain the solution.

All the material in this chapter regarding the pitfalls and improvements in Gauss elimination also applies to the Gauss-Jordan method. For example, a similar pivoting strategy can be used to avoid division by zero and to reduce round-off error.

Although the Gauss-Jordan technique and Gauss elimination might appear almost identical, the former requires more work. Using a similar approach to Sec. 9.2.1, it can be determined that the number of multiply/divide FLOPs involved in naive Gauss-Jordan is

$$\frac{n^3}{2} + n^2 - \frac{n}{2} \xrightarrow{\text{as } n \text{ increases}} \frac{n^3}{2} + O(n^2) \tag{9.37}$$

Thus, Gauss-Jordan involves approximately 50 percent more operations than Gauss elimination [compare with Eq. (9.23)]. Therefore, Gauss elimination is the simple elimination method of preference for obtaining solutions of linear algebraic equations. One of the primary reasons that we have introduced the Gauss-Jordan, however, is that it is still used in engineering as well as in some numerical algorithms.

9.8 SUMMARY

In summary, we have devoted most of this chapter to Gauss elimination, the most fundamental method for solving simultaneous linear algebraic equations. Although it is one of the earliest techniques developed for this purpose, it is nevertheless an extremely effective algorithm for obtaining solutions for many engineering problems. Aside from this practical utility, this chapter also provided a context for our discussion of general issues such as round-off, scaling, and conditioning. In addition, we briefly presented material on the Gauss-Jordan method, as well as complex and nonlinear systems.

Answers obtained using Gauss elimination may be checked by substituting them into the original equations. However, this does not always represent a reliable check for ill-conditioned systems. Therefore, some measure of condition, such as the determinant of the scaled system, should be computed if round-off error is suspected. Using partial pivoting and more significant figures in the computation are two options for mitigating round-off error. In the next chapter, we will return to the topic of system condition when we discuss the matrix inverse.

PROBLEMS

9.1 Write the following set of equations in matrix form:

$$30 = 2x_2 + 6x_3$$
$$20 = 3x_2 + 8x_1$$
$$10 = x_1 + x_3$$

Write the transpose of the matrix of coefficients.

9.2 A number of matrices are defined as

$$[A] = \begin{bmatrix} 4 & 5 \\ 1 & 2 \\ 5 & 6 \end{bmatrix} \quad [B] = \begin{bmatrix} 4 & 3 & 7 \\ 1 & 2 & 6 \\ 1 & 0 & 4 \end{bmatrix}$$

$$\{C\} = \begin{Bmatrix} 2 \\ 6 \\ 1 \end{Bmatrix} \quad [D] = \begin{bmatrix} 5 & 4 & 3 & 6 \\ 2 & 1 & 7 & 5 \end{bmatrix}$$

$$[E] = \begin{bmatrix} 1 & 5 & 6 \\ 7 & 1 & 3 \\ 4 & 0 & 5 \end{bmatrix}$$

$$[F] = \begin{bmatrix} 2 & 0 & 1 \\ 1 & 6 & 3 \end{bmatrix} \quad \lfloor G \rfloor = \lfloor 8 \quad 6 \quad 4 \rfloor$$

Answer the following questions regarding these matrices:
(a) What are the dimensions of the matrices?
(b) Identify the square, column, and row matrices.
(c) What are the values of the elements: $a_{12}, b_{23}, d_{32}, e_{22}, f_{12}, g_{12}$?
(d) Perform the following operations:

(1) $[A] + [B]$ (3) $[A] + [F]$
(2) $[B] - [A]$ (4) $5 \times [B]$

(5) $[A] \times [B]$　　(8) $\{C\}^T$

(6) $[B] \times [A]$　　(9) $[D]^T$

(7) $[A] \times \{C\}$　　(10) $[I] \times [B]$

9.3 Three matrices are defined as

$$[X] = \begin{bmatrix} 1 & 6 \\ 3 & 10 \\ 7 & 4 \end{bmatrix} \quad [Y] = \begin{bmatrix} 6 & 0 \\ 1 & 4 \end{bmatrix} \quad [Z] = \begin{bmatrix} 1 & 1 \\ 6 & 8 \end{bmatrix}$$

(a) Perform all possible multiplications that can be computed between pairs of these matrices.

(b) Use the method in Box PT3.2 to justify why the remaining pairs cannot be multiplied.

(c) Use the results of (a) to illustrate why the order of multiplication is important.

9.4 Use the graphical method to solve

$$2x_1 - 6x_2 = -18$$
$$-x_1 + 8x_2 = 40$$

Check your results by substituting them back into the equations.

9.5 Given the system of equations

$$0.77x_1 + x_2 = 14.25$$
$$1.2x_1 + 1.7x_2 = 20$$

(a) Solve graphically.

(b) On the basis of the graphical solution, what do you expect regarding the condition of the system?

(c) Solve by the elimination of unknowns.

9.6 For the set of equations

$$2x_2 + 5x_3 = 1$$
$$2x_1 + x_2 + 2x_3 = 1$$
$$3x_1 + x_2 = 2$$

(a) Compute the determinant.

(b) Use Cramer's rule to solve for the x's.

(c) Substitute your results back into the original equation to check your results.

9.7 Given the equations

$$0.5x_1 - x_2 = -9.5$$
$$0.26x_1 - 0.5x_2 = -4.7$$

(a) Solve graphically.

(b) After scaling, compute the determinant.

(c) On the basis of (a) and (b) what would you expect regarding the system's condition?

(d) Solve by the elimination of unknowns.

(e) Solve again, but with a_{11} modified slightly to 0.55. Interpret your results in light of the discussion of ill conditioning in Sec. 9.3.3.

9.8 Given the system

$$-12x_1 + x_2 - x_3 = -20$$
$$-2x_1 - 4x_2 + 2x_3 = 10$$
$$x_1 + 2x_2 + 2x_3 = 25$$

(a) Solve by naive Gauss elimination. Show all steps of the computation.

(b) Substitute your results into the original equation to check your answers.

9.9 Use Gauss elimination to solve:

$$4x_1 + x_2 - x_3 = -2$$
$$5x_1 + x_2 + 2x_3 = 4$$
$$6x_1 + x_2 + x_3 = 6$$

Employ partial pivoting and check your answers by substituting them into the original equations.

9.10 Use Gauss-Jordan elimination to solve:

$$2x_1 + x_2 - x_3 = 1$$
$$5x_1 + 2x_2 + 2x_3 = -4$$
$$3x_1 + x_2 + x_3 = 5$$

Employ partial pivoting. Check your answers by substituting them into the original equations.

9.11 Solve:

$$x_1 + x_2 - x_3 = -3$$
$$6x_1 + 2x_2 + 2x_3 = 2$$
$$-3x_1 + 4x_2 + x_3 = 1$$

with (a) naive Gauss elimination, (b) Gauss elimination with partial pivoting, (c) naive Gauss-Jordan, (d) Gauss-Jordan with partial pivoting.

9.12 Perform the same computation as in Example 9.11, but use five parachutists with the following characteristics.

Parachutist	Mass, kg	Drag Coefficient, kg/s
1	50	12
2	80	12
3	60	14
4	70	16
5	90	10

The parachutists have a velocity of 9 m/s.

9.13 Solve

$$\begin{bmatrix} 2+2i & 4 \\ -i & 2 \end{bmatrix} \begin{Bmatrix} z_1 \\ z_2 \end{Bmatrix} = \begin{Bmatrix} 1+i \\ 3 \end{Bmatrix}$$

9.14 Develop, debug, and test a subprogram in either a high-level language or macro language of your choice to multiply two matrices, that is, $[X] = [Y][Z]$, where $[Y]$ is m by n and $[Z]$ is n by p. Test the program using the matrices from Prob. 9.3.

9.15 Develop, debug, and test a subprogram in either a high-level language or macro language of your choice to generate the transpose of a matrix. Test it on the matrices from Prob. 9.3.

9.16 Develop, debug, and test a subprogram in either a high-level language or macro language of your choice to solve a system of equations with Gauss elimination with partial pivoting. Base the subprogram on the pseudocodes from Fig. 9.6. Test the program using the following system (which has an answer of $x_1 = x_2 = x_3 = 1$),

$$x_1 + x_2 - x_3 = 1$$
$$6x_1 + 2x_2 + 2x_3 = 10$$
$$-3x_1 + 4x_2 + x_3 = 2$$

CHAPTER 10

LU Decomposition and Matrix Inversion

This chapter deals with a class of elimination methods called *LU* decomposition techniques. The primary appeal of *LU* decomposition is that the time-consuming elimination step can be formulated so that it involves only operations on the matrix of coefficients, [A]. Thus, it is well-suited for those situations where many right-hand-side vectors {B} must be evaluated for a single value of [A]. Although there are a variety of ways in which this is done, we will focus on showing how the Gauss elimination method can be implemented as an *LU* decomposition.

One motive for introducing *LU* decomposition is that it provides an efficient means to compute the matrix inverse. The inverse has a number of valuable applications in engineering practice. It also provides a means for evaluating system condition.

10.1 *LU* DECOMPOSITION

As described in the previous chapter, Gauss elimination is designed to solve systems of linear algebraic equations,

$$[A]\{X\} = \{B\} \tag{10.1}$$

Although it certainly represents a sound way to solve such systems, it becomes inefficient when solving equations with the same coefficients [A], but with different right-hand-side constants (the b's).

Recall that Gauss elimination involves two steps: forward elimination and back-substitution (Fig. 9.3). Of these, the forward-elimination step comprises the bulk of the computational effort (recall Table 9.1). This is particularly true for large systems of equations.

LU decomposition methods separate the time-consuming elimination of the matrix [A] from the manipulations of the right-hand side {B}. Thus, once [A] has been "decomposed," multiple right-hand-side vectors can be evaluated in an efficient manner.

Interestingly, Gauss elimination itself can be expressed as an *LU* decomposition. Before showing how this can be done, let's first provide a mathematical overview of the decomposition strategy.

10.1.1 Overview of *LU* Decomposition

Just as was the case with Gauss elimination, *LU* decomposition requires pivoting to avoid division by zero. However, to simplify the following description, we will defer the issue of pivoting until after the fundamental approach is elaborated. In addition, the following explanation is limited to a set of three simultaneous equations. The results can be directly extended to n-dimensional systems.

Equation (10.1) can be rearranged to give

$$[A]\{X\} - \{B\} = 0 \qquad (10.2)$$

Suppose that Eq. (10.2) could be expressed as an upper triangular system:

$$\begin{bmatrix} u_{11} & u_{12} & u_{13} \\ 0 & u_{22} & u_{23} \\ 0 & 0 & u_{33} \end{bmatrix} \begin{Bmatrix} x_1 \\ x_2 \\ x_3 \end{Bmatrix} = \begin{Bmatrix} d_1 \\ d_2 \\ d_3 \end{Bmatrix} \qquad (10.3)$$

Recognize that this is similar to the manipulation that occurs in the first step of Gauss elimination. That is, elimination is used to reduce the system to upper triangular form. Equation (10.3) can also be expressed in matrix notation and rearranged to give

$$[U]\{X\} - \{D\} = 0 \qquad (10.4)$$

Now, assume that there is a lower diagonal matrix with 1's on the diagonal,

$$[L] = \begin{bmatrix} 1 & 0 & 0 \\ l_{21} & 1 & 0 \\ l_{31} & l_{32} & 1 \end{bmatrix} \qquad (10.5)$$

that has the property that when Eq. (10.4) is premultiplied by it, Eq. (10.2) is the result. That is,

$$[L]\{[U]\{X\} - \{D\}\} = [A]\{X\} - \{B\} \qquad (10.6)$$

If this equation holds, it follows from the rules for matrix multiplication that

$$[L][U] = [A] \qquad (10.7)$$

and

$$[L]\{D\} = \{B\} \qquad (10.8)$$

A two-step strategy (see Fig. 10.1) for obtaining solutions can be based on Eqs. (10.4), (10.7), and (10.8):

1. *LU decomposition step.* [A] is factored or "decomposed" into lower [L] and upper [U] triangular matrices.
2. *Substitution step.* [L] and [U] are used to determine a solution {X} for a right-hand side {B}. This step itself consists of two steps. First, Eq. (10.8) is used to generate an intermediate vector {D} by forward substitution. Then, the result is substituted into Eq. (10.4) which can be solved by back substitution for {X}.

Now, let's show how Gauss elimination can be implemented in this way.

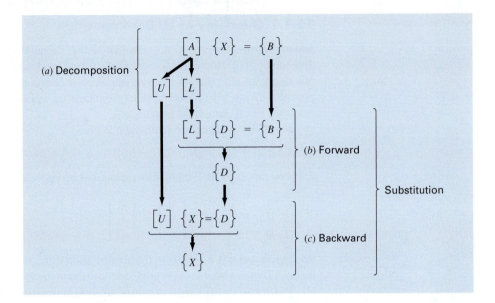

FIGURE 10.1
The steps in *LU* decomposition.

10.1.2 *LU* Decomposition Version of Gauss Elimination

Although it might appear at face value to be unrelated to *LU* decomposition, Gauss elimination can be used to decompose [*A*] into [*L*] and [*U*]. This can be easily seen for [*U*], which is a direct product of the forward elimination. Recall that the forward-elimination step is intended to reduce the original coefficient matrix [*A*] to the form

$$[U] = \begin{bmatrix} a_{11} & a_{12} & a_{13} \\ 0 & a'_{22} & a'_{23} \\ 0 & 0 & a''_{33} \end{bmatrix}$$

which is in the desired upper triangular format.

(10.9)

Though it might not be as apparent, the matrix [*L*] is also produced during the step. This can be readily illustrated for a three-equation system,

$$\begin{bmatrix} a_{11} & a_{12} & a_{13} \\ a_{21} & a_{22} & a_{23} \\ a_{31} & a_{32} & a_{33} \end{bmatrix} \begin{Bmatrix} x_1 \\ x_2 \\ x_3 \end{Bmatrix} = \begin{Bmatrix} b_1 \\ b_2 \\ b_3 \end{Bmatrix}$$

The first step in Gauss elimination is to multiply row 1 by the factor [recall Eq. (9.13)]

$$f_{21} = \frac{a_{21}}{a_{11}}$$

and subtract the result from the second row to eliminate a_{21}. Similarly, row 1 is multiplied by

$$f_{31} = \frac{a_{31}}{a_{11}}$$

and the result subtracted from the third row to eliminate a_{31}. The final step is to multiply the modified second row by

$$f_{32} = \frac{a'_{32}}{a'_{22}}$$

and subtract the result from the third row to eliminate a'_{32}.

Now suppose that we merely perform all these manipulations on the matrix $[A]$. Clearly, if we do not want to change the equation, we also have to do the same to the right-hand side $\{B\}$. But there's absolutely no reason that we have to perform the manipulations simultaneously. Thus, we could save the f's and manipulate $\{B\}$ later.

Where do we store the factors f_{21}, f_{31}, and f_{32}? Recall that the whole idea behind the elimination was to create zeros in a_{21}, a_{31}, and a_{32}. Thus, we can store f_{21} in a_{21}, f_{31} in a_{31}, and f_{32} in a_{32}. After elimination, the $[A]$ matrix can therefore be written as

$$\begin{bmatrix} a_{11} & a_{12} & a_{13} \\ f_{21} & a'_{22} & a'_{23} \\ f_{31} & f_{32} & a''_{33} \end{bmatrix} \tag{10.10}$$

This matrix, in fact, represents an efficient storage of the *LU* decomposition of $[A]$,

$$[A] \rightarrow [L][U] \tag{10.11}$$

where

$$[U] = \begin{bmatrix} a_{11} & a_{12} & a_{13} \\ 0 & a'_{22} & a'_{23} \\ 0 & 0 & a''_{33} \end{bmatrix}$$

and

$$[L] = \begin{bmatrix} 1 & 0 & 0 \\ f_{21} & 1 & 0 \\ f_{31} & f_{32} & 1 \end{bmatrix}$$

The following example confirms that $[A] = [L][U]$.

EXAMPLE 10.1   *LU* Decomposition with Gauss Elimination

Problem Statement.   Derive an *LU* decomposition based on the Gauss elimination performed in Example 9.5.

Solution.   In Example 9.5, we solved the matrix

$$[A] = \begin{bmatrix} 3 & -0.1 & -0.2 \\ 0.1 & 7 & -0.3 \\ 0.3 & -0.2 & 10 \end{bmatrix}$$

After forward elimination, the following upper triangular matrix was obtained:

$$[U] = \begin{bmatrix} 3 & -0.1 & -0.2 \\ 0 & 7.00333 & -0.293333 \\ 0 & 0 & 10.0120 \end{bmatrix}$$

The factors employed to obtain the upper triangular matrix can be assembled into a lower triangular matrix. The elements a_{21} and a_{31} were eliminated by using the factors

$$f_{21} = \frac{0.1}{3} = 0.03333333 \qquad f_{31} = \frac{0.3}{3} = 0.1000000$$

and the element a'_{32} was eliminated by using the factor

$$f_{32} = \frac{-0.19}{7.00333} = -0.0271300$$

Thus, the lower triangular matrix is

$$[L] = \begin{bmatrix} 1 & 0 & 0 \\ 0.0333333 & 1 & 0 \\ 0.100000 & -0.0271300 & 1 \end{bmatrix}$$

Consequently, the LU decomposition is

$$[A] = [L][U] = \begin{bmatrix} 1 & 0 & 0 \\ 0.0333333 & 1 & 0 \\ 0.100000 & -0.0271300 & 1 \end{bmatrix} \begin{bmatrix} 3 & -0.1 & -0.2 \\ 0 & 7.00333 & -0.293333 \\ 0 & 0 & 10.0120 \end{bmatrix}$$

This result can be verified by performing the multiplication of $[L][U]$ to give

$$[L][U] = \begin{bmatrix} 3 & -0.1 & -0.2 \\ 0.0999999 & 7 & -0.3 \\ 0.3 & -0.2 & 9.99996 \end{bmatrix}$$

where the minor discrepancies are due to round-off.

The following is pseudocode for a subroutine to implement the decomposition phase:

```
SUB Decompose (a, n)
  DO k = 1, n − 1
    DO i = k + 1, n
      factor = a_{i,k}/a_{k,k}
      a_{i,k} = factor
      DO j = k + 1, n
        a_{i,j} = a_{i,j} − factor * a_{k,j}
      END DO
    END DO
  END DO
END Decompose
```

Notice that this algorithm is "naive" in the sense that pivoting is not included. This feature will be added later when we develop the full algorithm for LU decomposition.

After the matrix is decomposed, a solution can be generated for a particular right-hand-side vector $\{B\}$. This is done in two steps. First, a forward-substitution step is executed by solving Eq. (10.8) for $\{D\}$. It is important to recognize that this merely amounts to performing the elimination manipulations on $\{B\}$. Thus, at the end of this step, the right-hand side will be in the same state that it would have been had we performed forward manipulation on $[A]$ and $\{B\}$ simultaneously.

The forward-substitution step can be represented concisely as

$$d_i = d_i - \sum_{j=1}^{i-1} a_{ij} b_j \qquad \text{for } i = 2, 3, \ldots, n \tag{10.12}$$

The second step then merely amounts to implementing back substitution, as in Eq. (10.4). Again, it is important to recognize that this is identical to the back-substitution phase of conventional Gauss elimination. Thus, in a fashion similar to Eqs. (9.16) and (9.17), the back-substitution step can be represented concisely as

$$x_n = d_n / a_{nn} \tag{10.13}$$

$$x_i = \frac{d_i - \sum_{j=i+1}^{n} a_{ij} d_j}{a_{ii}} \qquad \text{for } i = n - 1, n - 2, \ldots, 1 \tag{10.14}$$

EXAMPLE 10.2   The Substitution Steps

Problem Statement.   Complete the problem initiated in Example 10.1 by generating the final solution with forward and back substitution.

Solution.   As stated above, the intent of forward substitution is to impose the elimination manipulations, that we had formerly applied to $[A]$, on the right-hand-side vector $\{B\}$. Recall that the system being solved in Example 9.5 was

$$\begin{bmatrix} 3 & -0.1 & -0.2 \\ 0.1 & 7 & -0.3 \\ 0.1 & 0.3 & 10 \end{bmatrix} \begin{Bmatrix} x_1 \\ x_2 \\ x_3 \end{Bmatrix} = \begin{Bmatrix} 7.85 \\ -19.3 \\ 71.4 \end{Bmatrix}$$

and that the forward-elimination phase of conventional Gauss elimination resulted in

$$\begin{bmatrix} 3 & -0.1 & -0.2 \\ 0 & 7.00333 & -0.293333 \\ 0 & 0 & 10.0120 \end{bmatrix} \begin{Bmatrix} x_1 \\ x_2 \\ x_3 \end{Bmatrix} = \begin{Bmatrix} 7.85 \\ -19.5617 \\ 70.0843 \end{Bmatrix} \tag{E10.2.1}$$

The forward-substitution phase is implemented by applying Eq. (10.7) to our problem,

$$\begin{bmatrix} 1 & 0 & 0 \\ 0.0333333 & 1 & 0 \\ 0.100000 & -0.0271300 & 1 \end{bmatrix} \begin{Bmatrix} d_1 \\ d_2 \\ d_3 \end{Bmatrix} = \begin{Bmatrix} 7.85 \\ -19.3 \\ 71.4 \end{Bmatrix}$$

or multiplying out the left-hand side,

$$d_1 = 7.85$$
$$0.0333333 d_1 + d_2 = -19.3$$
$$0.1 d_1 - 0.02713 d_2 + d_3 = 71.4$$

We can solve the first equation for d_1,

$$d_1 = 7.85$$

which can be substituted into the second to solve for

$$d_2 = -19.3 - 0.0333333(7.85) = -19.5617$$

Both d_1 and d_2 can be substituted into the third equation to give

$$d_3 = 71.4 - 0.1(7.85) + 0.02713(-19.5617) = 70.0843$$

Thus,

$$\{D\} = \left\{ \begin{array}{c} 7.85 \\ -19.5617 \\ 70.0843 \end{array} \right\}$$

which is identical to the right-hand side of Eq. (E10.2.1).

This result can then be substituted into Eq. (10.4), $[U]\{X\} = \{D\}$, to give

$$\begin{bmatrix} 3 & -0.1 & -0.2 \\ 0 & 7.00333 & -0.293333 \\ 0 & 0 & 10.0120 \end{bmatrix} \left\{ \begin{array}{c} x_1 \\ x_2 \\ x_3 \end{array} \right\} = \left\{ \begin{array}{c} 7.85 \\ -19.5617 \\ 70.0843 \end{array} \right\}$$

which can be solved by back substitution (see Example 9.5 for details) for the final solution,

$$\{X\} = \left\{ \begin{array}{c} 3 \\ -2.5 \\ 7.00003 \end{array} \right\}$$

The following is pseudocode for a subroutine to implement both substitution phases:

```
SUB Substitute (a, n, b, x)
  'forward substitution
  DO i = 2, n
    sum = b_i
    FOR j = 1, i − 1
      sum = sum − a_{i,j} * b_j
    END DO
    b_i = sum
  END DO
  'back substitution
  x_n = b_n / a_{n,n}
  DO i = n − 1, 1, −1
    sum = 0
    DO j = i + 1, n
      sum = sum + a_{i,j} * x_j
    END DO
    x_i = (b_i − sum) / a_{i,i}
  END DO
END Substitute
```

The *LU* decomposition algorithm requires the same total multiply/divide FLOPS as for Gauss elimination. The only difference is that a little less effort is expended in the decomposition phase since the operations are not applied to the right-hand side. Thus, the number of multiply/divide FLOPs involved in the decomposition phase can be calculated as

$$\frac{n^3}{3} - \frac{n}{3} \xrightarrow{\text{as } n \text{ increases}} \frac{n^3}{3} + O(n) \tag{10.15}$$

Conversely, the substitution phase takes a little more effort. Thus, the number of FLOPs for forward and back substitution is n^2. The total effort is therefore identical to Gauss elimination

$$\frac{n^3}{3} - \frac{n}{3} + n^2 \xrightarrow{\text{as } n \text{ increases}} \frac{n^3}{3} + O(n^2) \tag{10.16}$$

10.1.3 *LU* Decomposition Algorithm

An algorithm to implement an *LU* decomposition expression of Gauss elimination is listed in Fig. 10.2. Four features of this algorithm bear mention:

- The factors generated during the elimination phase are stored in the lower part of the matrix. This can be done because these are converted to zeros anyway and are unnecessary for the final solution. This storage saves space.
- This algorithm keeps track of pivoting by using an order vector o. This greatly speeds up the algorithm because only the order vector (as opposed to the whole row) is pivoted.
- The equations are not scaled, but scaled values of the elements are used to determine whether pivoting is to be implemented.
- The diagonal term is monitored during the pivoting phase to detect near-zero occurrences in order to flag singular systems. If it passes back a value of $er = -1$, a singular matrix has been detected and the computation should be terminated. A parameter *tol* is set by the user to a small number in order to detect near-zero occurrences.

10.1.4 Crout Decomposition

Notice that for the *LU* decomposition implementation of Gauss elimination, the [*L*] matrix has 1's on the diagonal. This is formally referred to as a *Doolittle decomposition,* or factorization. An alternative approach involves a [*U*] matrix with 1's on the diagonal. This is called *Crout decomposition.* Although there are some differences between the approaches (Atkinson, 1978; Ralston and Rabinowitz, 1979), their performance is comparable.

The Crout decomposition approach generates [*U*] and [*L*] by sweeping through the matrix by columns and rows, as depicted in Fig. 10.3. It can be implemented by the following concise series of formulas:

$$l_{i,1} = a_{i,1} \qquad \text{for } i = 1, 2, \ldots, n \tag{10.17}$$

$$u_{ij} = \frac{a_{1j}}{l_{11}} \qquad \text{for } j = 2, 3, \ldots, n \tag{10.18}$$

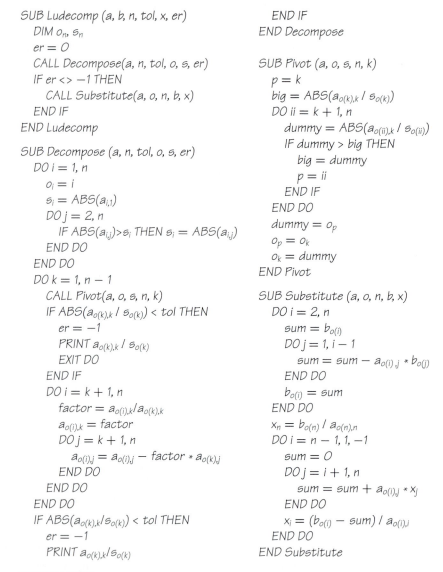

```
SUB Ludecomp (a, b, n, tol, x, er)
  DIM oₙ, sₙ
  er = O
  CALL Decompose(a, n, tol, o, s, er)
  IF er <> −1 THEN
    CALL Substitute(a, o, n, b, x)
  END IF
END Ludecomp

SUB Decompose (a, n, tol, o, s, er)
  DO i = 1, n
    oᵢ = i
    sᵢ = ABS(aᵢ,₁)
    DO j = 2, n
      IF ABS(aᵢ,ⱼ)>sᵢ THEN sᵢ = ABS(aᵢ,ⱼ)
    END DO
  END DO
  DO k = 1, n − 1
    CALL Pivot(a, o, s, n, k)
    IF ABS(a_o(k),k / s_o(k)) < tol THEN
      er = −1
      PRINT a_o(k),k / s_o(k)
      EXIT DO
    END IF
    DO i = k + 1, n
      factor = a_o(i),k/a_o(k),k
      a_o(i),k = factor
      DO j = k + 1, n
        a_o(i),j = a_o(i),j − factor * a_o(k),j
      END DO
    END DO
  END DO
  IF ABS(a_o(k),k/s_o(k)) < tol THEN
    er = −1
    PRINT a_o(k),k/s_o(k)
```

```
  END IF
END Decompose

SUB Pivot (a, o, s, n, k)
  p = k
  big = ABS(a_o(k),k / s_o(k))
  DO ii = k + 1, n
    dummy = ABS(a_o(ii),k / s_o(ii))
    IF dummy > big THEN
      big = dummy
      p = ii
    END IF
  END DO
  dummy = o_p
  o_p = o_k
  o_k = dummy
END Pivot

SUB Substitute (a, o, n, b, x)
  DO i = 2, n
    sum = b_o(i)
    DO j = 1, i − 1
      sum = sum − a_o(i),j * b_o(j)
    END DO
    b_o(i) = sum
  END DO
  xₙ = b_o(n) / a_o(n),n
  DO i = n − 1, 1, −1
    sum = O
    DO j = i + 1, n
      sum = sum + a_o(i),j * xⱼ
    END DO
    xᵢ = (b_o(i) − sum) / a_o(i),i
  END DO
END Substitute
```

FIGURE 10.2
Pseudocode for an *LU* decomposition algorithm.

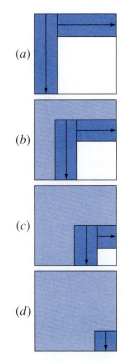

(a)

(b)

(c)

(d)

FIGURE 10.3
A schematic depicting the evaluations involved in Crout *LU* decomposition.

For $j = 2, 3, \ldots, n - 1$

$$l_{ij} = a_{ij} - \sum_{k=1}^{j-1} l_{ik}u_{kj} \qquad \text{for } i = j, j+1, \ldots, n \tag{10.19}$$

$$u_{jk} = \frac{a_{jk} - \sum_{i=1}^{j-1} l_{ji}u_{ik}}{l_{jj}} \qquad \text{for } k = j+1, j+2, \ldots, n \tag{10.20}$$

```
DO j = 2, n
   a_{1,j} = a_{1,j}/a_{1,1}
END DO
DO j = 2, n − 1
   DO i = j, n
      sum = 0
      DO k = 1, j − 1
         sum = sum + a_{i,k} · a_{k,j}
      END DO
      a_{i,j} = a_{i,j} − sum
   END DO
   DO k = j + 1, n
      sum = 0
      DO i = 1, j − 1
         sum = sum + a_{j,i} · a_{i,k}
      END DO
      a_{j,k} = (a_{j,k} − sum)/a_{j,j}
   END DO
END DO
sum = 0
DO k = 1, n − 1
   sum = sum + a_{n,k} · a_{k,n}
END DO
a_{n,n} = a_{n,n} − sum
```

FIGURE 10.4
Pseudocode for Crout's *LU* decomposition algorithm.

and

$$l_{nn} = a_{nn} - \sum_{k=1}^{n-1} l_{nk} u_{kn} \tag{10.21}$$

Aside from the fact that it consists of a few concise loops, the foregoing approach also has the benefit that storage space can be economized. There is no need to store the 1's on the diagonal of $[U]$ or the 0's for $[L]$ or $[U]$ because they are givens in the method. Consequently, the values of $[U]$ can be stored in the zero space of $[L]$. Further, close examination of the foregoing derivation makes it clear that after each element of $[A]$ is employed once, it is never used again. Therefore, as each element $[L]$ and $[U]$ is computed, it can be substituted for the corresponding element (as designated by its subscripts) of $[A]$.

Pseudocode to accomplish this is presented in Fig. 10.4. Notice that Eq. (10.17) is not included in the pseudocode because the first column of $[L]$ is already stored in $[A]$. Otherwise, the algorithm directly follows from Eqs. (10.18) through (10.21).

10.2   THE MATRIX INVERSE

In our discussion of matrix operations (Sec. PT3.2.2), we introduced the notion that if a matrix $[A]$ is square, there is another matrix, $[A]^{-1}$, called the inverse of $[A]$, for which [Eq. (PT3.3)]

$$[A][A]^{-1} = [A]^{-1}[A] = [I]$$

Now we will focus on how the inverse can be computed numerically. Then we will explore how it can be used for engineering analysis.

10.2.1 Calculating the Inverse

The inverse can be computed in a column-by-column fashion by generating solutions with unit vectors as the right-hand-side constants. For example, if the right-hand-side constant has a 1 in the first position and zeros elsewhere,

$$\{b\} = \left\{ \begin{array}{c} 1 \\ 0 \\ 0 \end{array} \right\}$$

the resulting solution will be the first column of the matrix inverse. Similarly, if a unit vector with a 1 at the second row is used

$$\{b\} = \left\{ \begin{array}{c} 0 \\ 1 \\ 0 \end{array} \right\}$$

the result will be the second column of the matrix inverse.

The best way to implement such a calculation is with the *LU* decomposition algorithm described at the beginning of this chapter. Recall that one of the great strengths of *LU* decomposition is that it provides a very efficient means to evaluate multiple right-hand-side vectors. Thus, it is ideal for evaluating the multiple unit vectors needed to compute the inverse.

EXAMPLE 10.3   Matrix Inversion

Problem Statement.   Employ *LU* decomposition to determine the matrix inverse for the system from Example 10.2.

$$[A] = \begin{bmatrix} 3 & -0.1 & -0.2 \\ 0.1 & 7 & -0.3 \\ 0.3 & -0.2 & 10 \end{bmatrix}$$

Recall that the decomposition resulted in the following lower and upper triangular matrices:

$$[U] = \begin{bmatrix} 3 & -0.1 & -0.2 \\ 0 & 7.00333 & -0.293333 \\ 0 & 0 & 10.0120 \end{bmatrix} \quad [L] = \begin{bmatrix} 1 & 0 & 0 \\ 0.0333333 & 1 & 0 \\ 0.100000 & -0.0271300 & 1 \end{bmatrix}$$

Solution.   The first column of the matrix inverse can be determined by performing the forward-substitution solution procedure with a unit vector (with 1 in the first row) as the right-hand-side vector. Thus, Eq. (10.8), the lower-diagonal system, can be set up as

$$\begin{bmatrix} 1 & 0 & 0 \\ 0.0333333 & 1 & 0 \\ 0.100000 & -0.0271300 & 1 \end{bmatrix} \left\{ \begin{array}{c} d_1 \\ d_2 \\ d_3 \end{array} \right\} = \left\{ \begin{array}{c} 1 \\ 0 \\ 0 \end{array} \right\}$$

and solved with forward substitution for $\{D\}^T = \lfloor 1 \quad -0.03333 \quad -0.1009 \rfloor$. This vector can then be used as the right-hand side of Eq. (10.3),

$$
\begin{bmatrix}
3 & -0.1 & -0.2 \\
0 & 7.00333 & -0.293333 \\
0 & 0 & 10.0120
\end{bmatrix}
\begin{Bmatrix}
x_1 \\
x_2 \\
x_3
\end{Bmatrix}
=
\begin{Bmatrix}
1 \\
-0.03333 \\
-0.1009
\end{Bmatrix}
$$

which can be solved by back substitution for $\{X\}^T = \lfloor 0.33249 \quad -0.00518 \quad -0.01008 \rfloor$, which is the first column of the matrix,

$$
[A]^{-1} =
\begin{bmatrix}
0.33249 & 0 & 0 \\
-0.00518 & 0 & 0 \\
-0.01008 & 0 & 0
\end{bmatrix}
$$

To determine the second column, Eq. (10.8) is formulated as

$$
\begin{bmatrix}
1 & 0 & 0 \\
0.0333333 & 1 & 0 \\
0.100000 & -0.0271300 & 1
\end{bmatrix}
\begin{Bmatrix}
d_1 \\
d_2 \\
d_3
\end{Bmatrix}
=
\begin{Bmatrix}
0 \\
1 \\
0
\end{Bmatrix}
$$

This can be solved for $\{D\}$ and the results used with Eqs. (10.3) to determine $\{X\}^T = \lfloor 0.004944 \quad 0.142903 \quad 0.00271 \rfloor$, which is the second column of the matrix,

$$
[A]^{-1} =
\begin{bmatrix}
0.33249 & 0.004944 & 0 \\
-0.00518 & 0.142903 & 0 \\
-0.01008 & 0.00271 & 0
\end{bmatrix}
$$

Finally, the forward- and back-substitution procedures can be implemented with $\{B\}^T = \lfloor 0 \quad 0 \quad 1 \rfloor$, to solve for $\{X\}^T = \lfloor 0.006798 \quad 0.004183 \quad 0.09988 \rfloor$, which is the final column of the matrix,

$$
[A]^{-1} =
\begin{bmatrix}
0.33249 & 0.004944 & 0.006798 \\
-0.00518 & 0.142903 & 0.004183 \\
-0.01008 & 0.00271 & 0.09988
\end{bmatrix}
$$

The validity of this result can be checked by verifying that $[A][A]^{-1} = [I]$.

Pseudocode to generate the matrix inverse is shown in Fig. 10.5. Notice how the decomposition subroutine from Fig. 10.2 is called to perform the decomposition and then generates the inverse by repeatedly calling the substitution algorithm with unit vectors.

The effort required for this algorithm is simply computed as

$$
\underbrace{\frac{n^3}{3} - \frac{n}{3}}_{\text{decomposition}} + \underbrace{n(n^2)}_{n \times \text{substitutions}} = \frac{4n^3}{3} - \frac{n}{3} \tag{10.22}
$$

where from Sec. 10.1.2 the decomposition is defined by Eq. (10.15) and the effort involved with every right-hand-side evaluation involves n^2 multiply/divide FLOPs.

```
CALL Decompose (a, n, tol, o, s, er)
IF er = O THEN
    DO i = 1, n
        DO j = 1, n
            IF i = j THEN
                b(j) = 1
            ELSE
                b(j) = O
            END IF
        END DO
        CALL Substitute (a, o, n, b, x)
        DO j = 1, n
            ai(j, i) = x(j)
        END DO
    END DO
    Output ai, if desired
ELSE
    PRINT "ill-conditioned system"
END IF
```

FIGURE 10.5
Driver program that uses some of the subprograms from Fig. 10.2 to generate a matrix inverse.

10.2.2 Stimulus-Response Computations

As discussed in Sec. PT3.1.2, many of the linear systems of equations confronted in engineering practice are derived from conservation laws. The mathematical expression of these laws is some form of balance equation to ensure that a particular property—mass, force, heat, momentum, or other—is conserved. For a force balance on a structure, the properties might be horizontal or vertical components of the forces acting on each node of the structure (see Sec. 12.2). For a mass balance, the properties might be the mass in each reactor of a chemical process (see Sec. 12.1). Other fields of engineering would yield similar examples.

A single balance equation can be written for each part of the system, resulting in a set of equations defining the behavior of the property for the entire system. These equations are interrelated, or coupled, in that each equation may include one or more of the variables from the other equations. For many cases, these systems are linear and, therefore, of the exact form dealt with in this chapter:

$$[A]\{X\} = \{B\} \tag{10.23}$$

Now, for balance equations, the terms of Eq. (10.23) have a definite physical interpretation. For example, the elements of $\{X\}$ are the levels of the property being balanced for each part of the system. In a force balance of a structure, they represent the horizontal and vertical forces in each member. For the mass balance, they are the mass of chemical in each reactor. In either case, they represent the system's state or response, which we are trying to determine.

The right-hand-side vector $\{B\}$ contains those elements of the balance that are independent of behavior of the system—that is, they are constants. As such, they often represent the external forces or stimuli that drive the system.

Finally, the matrix of coefficients $[A]$ usually contains the parameters that express how the parts of the system interact or are coupled. Consequently, Eq. (10.23) might be re-expressed as

$$[\text{Interactions}]\{\text{response}\} = \{\text{stimuli}\}$$

Thus, Eq. (10.23) can be seen as an expression of the fundamental mathematical model that we formulated previously as a single equation in Chap. 1 [recall Eq. (1.1)]. We can now see that Eq. (10.23) represents a version that is designed for coupled systems involving several dependent variables $\{X\}$.

As we know from this and the previous chapter, there are a variety of ways to solve Eq. (10.23). However, using the matrix inverse yields a particularly interesting result. The formal solution can be expressed as

$$\{X\} = [A]^{-1}\{B\}$$

or (recalling our definition of matrix multiplication from Box PT3.2)

$$x_1 = a_{11}^{-1}b_1 + a_{12}^{-1}b_2 + a_{13}^{-1}b_3$$
$$x_2 = a_{21}^{-1}b_1 + a_{22}^{-1}b_2 + a_{23}^{-1}b_3$$
$$x_3 = a_{31}^{-1}b_1 + a_{32}^{-1}b_2 + a_{33}^{-1}b_3$$

Thus, we find that the inverted matrix itself, aside from providing a solution, has extremely useful properties. That is, each of its elements represents the response of a single part of the system to a unit stimulus of any other part of the system.

Notice that these formulations are linear and, therefore, superposition and proportionality hold. *Superposition* means that if a system is subject to several different stimuli (the b's), the responses can be computed individually and the results summed to obtain a total response. *Proportionality* means that multiplying the stimuli by a quantity results in the response to those stimuli being multiplied by the same quantity. Thus, the coefficient a_{11}^{-1} is a proportionality constant that gives the value of x_1 due to a unit level of b_1. This result is independent of the effects of b_2 and b_3 on x_1, which are reflected in the coefficients a_{12}^{-1} and a_{13}^{-1}, respectively. Therefore, we can draw the general conclusion that the element a_{ij}^{-1} of the inverted matrix represents the value of x_i due to a unit quantity of b_j. Using the example of the structure, element a_{ij}^{-1} of the matrix inverse would represent the force in member i due to a unit external force at node j. Even for small systems, such behavior of individual stimulus-response interactions would not be intuitively obvious. As such, the matrix inverse provides a powerful technique for understanding the interrelationships of component parts of complicated systems. This power will be demonstrated in Secs. 12.1 and 12.2.

10.3 ERROR ANALYSIS AND SYSTEM CONDITION

Aside from its engineering applications, the inverse also provides a means to discern whether systems are ill-conditioned. Three methods are available for this purpose:

1. Scale the matrix of coefficients $[A]$ so that the largest element in each row is 1. Invert the scaled matrix and if there are elements of $[A]^{-1}$ that are several orders of magnitude greater than one, it is likely that the system is ill-conditioned (see Box 10.1).

Box 10.1   Interpreting the Elements of the Matrix Inverse as a Measure of Ill-Conditioning

One method for assessing a system's condition is to scale $[A]$ so that the largest element in each row is 1 and then compute $[A]^{-1}$. If elements of $[A]^{-1}$ are several orders of magnitude greater than the elements of the original scaled matrix, it is likely that the system is ill-conditioned.

Insight into this approach can be gained by recalling that a way to check whether an approximate solution $\{X\}$ is acceptable is to substitute it into the original equations and see whether the original right-hand-side constants result. This is equivalent to

$$\{R\} = \{B\} - [A]\{\tilde{X}\} \tag{B10.1.1}$$

where $\{R\}$ is the residual between the right-hand-side constants and the values computed with the solution $\{\tilde{X}\}$. If $\{R\}$ is small, we might conclude that the $\{\tilde{X}\}$ values are adequate. However, suppose that $\{X\}$ is the exact solution that yields a zero residual, as in

$$\{0\} = \{B\} - [A]\{X\} \tag{B10.1.2}$$

Subtracting Eq. (B10.1.2) from (B10.1.1) yields

$$\{R\} = [A]\left\{\{X\} - \{\tilde{X}\}\right\}$$

Multiplying both sides of this equation by $[A]^{-1}$ gives

$$\{X\} - \{\tilde{X}\} = [A]^{-1}\{R\}$$

This result indicates why checking a solution by substitution can be misleading. For cases where elements of $[A]^{-1}$ are large, a small discrepancy in the right-hand-side residual $\{R\}$ could correspond to a large error $\{X\} - \{\tilde{X}\}$ in the calculated value of the unknowns. In other words, a small residual does not guarantee an accurate solution. However, we can conclude that if the largest element of $[A]^{-1}$ is on the order of magnitude of unity, the system can be considered to be well-conditioned. Conversely, if $[A]^{-1}$ includes elements much larger than unity, we conclude that the system is ill-conditioned.

2. Multiply the inverse by the original coefficient matrix and assess whether the result is close to the identity matrix. If not, it indicates ill-conditioning.
3. Invert the inverted matrix and assess whether the result is sufficiently close to the original coefficient matrix. If not, it again indicates that the system is ill-conditioned.

Although these methods can indicate ill-conditioning, it would be preferable to obtain a single number (such as the condition number from Sec. 4.2.3) that could serve as an indicator of the problem. Attempts to formulate such a matrix condition number are based on the mathematical concept of the norm.

10.3.1 Vector and Matrix Norms

A *norm* is a real-valued function that provides a measure of the size or "length" of multi-component mathematical entities such as vectors and matrices (see Box 10.2).

A simple example is a vector in three-dimensional Euclidean space (Fig. 10.6) that can be represented as

$$\lfloor F \rfloor = \lfloor a \quad b \quad c \rfloor$$

where a, b, and c are the distances along the x, y, and z axes, respectively. The length of this vector—that is, the distance from the coordinate $(0, 0, 0)$ to (a, b, c)—can be simply computed as

$$\|F\|_e = \sqrt{a^2 + b^2 + c^2}$$

where the nomenclature $\|F\|_e$ indicates that this length is referred to as the Euclidean norm of $[F]$.

Box 10.2 Matrix Norms

As developed in this section, Euclidean norms can be employed to quantify the size of a vector,

$$\|X\|_e = \sqrt{\sum_{i=1}^{n} x_i^2}$$

or matrix,

$$\|A\|_e = \sqrt{\sum_{i=1}^{n}\sum_{j=1}^{n} a_{i,j}^2}$$

For vectors, there are alternatives called *p* norms that can be represented generally by

$$\|X\|_p = \left(\sum_{i=1}^{n} |x_i|^p\right)^{1/p}$$

We can also see that the Euclidean norm and the 2 norm, $\|X\|_2$ are identical for vectors.

Other important examples are

$$\|X\|_1 = \sum_{i=1}^{n} |x_i|$$

which represents the norm as the sum of the absolute values of the elements. Another is the *maximum-magnitude* or *uniform-vector norm*.

$$\|X\|_\infty = \max_{1 \le i \le n} |x_i|$$

which defines the norm as the element with the largest absolute value.

Using a similar approach, norms can be developed for matrices. For example,

$$\|A\|_\infty = \max_{1 \le j \le n} \sum_{i=1}^{n} |a_{ij}|$$

That is, a summation of the absolute values of the coefficients is performed for each column and the largest of these summations is taken as the norm. This is called the *column-sum norm*.

A similar determination can be made for the rows, resulting in a *uniform-matrix* or *row-sum norm*,

$$\|A\|_\infty = \max_{1 \le i \le n} \sum_{j=1}^{n} |a_{ij}|$$

It should be noted that, in contrast to vectors, the 2 norm and the Euclidean norm for a matrix are not the same. Whereas the Euclidean norm $\|A\|_e$ can be easily determined by Eq. (10.24), the matrix 2 norm $\|A\|_2$ is calculated as

$$\|A\|_2 = (\mu_{max})^{1/2}$$

where μ_{max} is the largest eigenvalue of $[A]^T[A]$. In Chap. 27, we will learn more about eigenvalues. For the time being, the important point is that the $\|A\|_2$, or *spectral*, norm is the minimum norm and, therefore, provides the tightest measure of size (Ortega 1972).

FIGURE 10.6
Graphical depiction of a vector $\lfloor F \rfloor = \lfloor a \quad b \quad c \rfloor$ in Euclidean space.

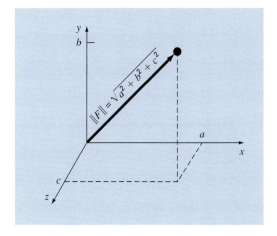

Similarly, for an *n*-dimensional vector $\lfloor X \rfloor = \lfloor x_1 \quad x_2 \quad \cdots \quad x_n \rfloor$, a Euclidean norm would be computed as

$$\|X\|_e = \sqrt{\sum_{i=1}^{n} x_i^2}$$

The concept can be extended further to a matrix [*A*], as in

$$\|A\|_e = \sqrt{\sum_{i=1}^{n} \sum_{j=1}^{n} a_{i,j}^2} \tag{10.24}$$

which is given a special name—the *Frobenius norm*. However, as with the other vector norms, it provides a single value to quantify the "size" of [*A*].

It should be noted that there are alternatives to the Euclidean and Frobenius norms (see Box 10.2). For example, a *uniform vector norm* is defined as

$$\|X\|_\infty = \max_{1 \le i \le n} |x_i|$$

That is, the element with the largest absolute value is taken as the measure of the vector's size. Similarly, a *uniform matrix norm* or *row-sum norm* is defined as

$$\|A\|_\infty = \max_{1 \le i \le n} \sum_{j=1}^{n} |a_{ij}| \tag{10.25}$$

In this case, the sum of the absolute value of the elements is computed for each row, and the largest of these is taken as the norm.

Although there are theoretical benefits for using certain of the norms, the choice is sometimes influenced by practical considerations. For example, the uniform-row norm is widely used because of the ease with which it can be calculated and the fact that it usually provides an adequate measure of matrix size.

10.3.2 Matrix Condition Number

Now that we have introduced the concept of the norm, we can use it to define

$$\text{Cond}\,[A] = \|A\| \cdot \|A^{-1}\| \tag{10.26}$$

where Cond [*A*] is called the *matrix condition number*. Note that for a matrix [*A*], this number will be greater than or equal to 1. It can be shown (Ralston and Rabinowitz, 1978; Gerald and Wheatley, 1984) that

$$\frac{\|\Delta X\|}{\|X\|} \le \text{Cond}\,[A] \frac{\|\Delta A\|}{\|A\|}$$

That is, the relative error of the norm of the computed solution can be as large as the relative error of the norm of the coefficients of [*A*] multiplied by the condition number. For example, if the coefficients of [*A*] are known to *t*-digit precision (that is, rounding errors are on the order of 10^{-t}) and Cond [*A*] $= 10^c$, the solution [*X*] may be valid to only $t - c$ digits (rounding errors $\sim 10^{c-t}$).

EXAMPLE 10.4   Matrix Condition Evaluation

Problem Statement.   The Hilbert matrix, which is notoriously ill-conditioned, can be represented generally as

$$
\begin{bmatrix}
1 & 1/2 & 1/3 & \cdots & 1/n \\
1/2 & 1/3 & 1/4 & \cdots & 1/(n+1) \\
\cdot & \cdot & \cdot & & \cdot \\
\cdot & \cdot & \cdot & & \cdot \\
\cdot & \cdot & \cdot & & \cdot \\
1/n & 1/(n+1) & 1/(n+2) & \cdots & 1/(2n)
\end{bmatrix}
$$

Use the row-sum norm to estimate the matrix condition number for the 3×3 Hilbert matrix,

$$
[A] = \begin{bmatrix}
1 & 1/2 & 1/3 \\
1/2 & 1/3 & 1/4 \\
1/3 & 1/4 & 1/5
\end{bmatrix}
$$

Solution.   First, the matrix can be normalized so that the maximum element in each row is 1,

$$
[A] = \begin{bmatrix}
1 & 1/2 & 1/3 \\
1 & 2/3 & 1/2 \\
1 & 3/4 & 3/5
\end{bmatrix}
$$

Summing each of the rows gives 1.833, 2.1667, and 2.35. Thus, the third row has the largest sum and the row-sum norm is

$$
\|A\|_\infty = 1 + \frac{3}{4} + \frac{3}{5} = 2.35
$$

The inverse of the scaled matrix can be computed as

$$
[A]^{-1} = \begin{bmatrix}
9 & -18 & 10 \\
-36 & 96 & -60 \\
30 & -90 & 60
\end{bmatrix}
$$

Note that the elements of this matrix are larger than the original matrix. This is also reflected in its row-sum norm, which is computed as

$$
\|A\|_\infty = |-36| + |96| + |-60| = 192
$$

Thus, the condition number can be calculated as

$$
\text{Cond } [A] = 2.35(192) = 451.2
$$

The fact that the condition number is considerably greater than unity suggests that the system is ill-conditioned. The extent of the ill-conditioning can be quantified by calculating $c = \log 451.2 = 2.65$. Computers using IEEE floating-point representation have approximately $t = \log 2^{-24} = 7.2$ significant base-10 digits (recall Sec. 3.4.1). Therefore, the solution could exhibit rounding errors of up to $10^{(2.65-7.2)} = 3 \times 10^{-5}$. Note that such estimates almost always overpredict the actual error. However, they are useful in alerting you to the possibility that round-off errors may be significant.

Practically speaking, the problem with implementing Eq. (10.26) is the computational price required to obtain $\|A^{-1}\|$. Rice (1983) outlines some possible strategies to mitigate this problem. Further, he suggests an alternative way to assess system condition: run the same solution on two different compilers. Because the resulting codes will likely implement the arithmetic differently, the effect of ill-conditioning should be evident from such an experiment. Finally, it should be mentioned that software packages and libraries such as MATLAB and Mathcad have the capability to conveniently compute matrix condition. We will review these capabilities when we review such packages at the end of Chap. 11.

EXAMPLE 10.5   Linear Equation Solving Using the Computer

Problem Statement.   A user-friendly computer program to implement *LU* decomposition with matrix inversion is contained in the Numerical Methods TOOLKIT software associated with the text. We can use this software to solve and determine the matrix inverse for the following Hilbert system:

$$
\begin{bmatrix} 1 & 0.5 & 0.3333333 \\ 1 & 0.6666667 & 0.5 \\ 1 & 0.75 & 0.6 \end{bmatrix} \begin{Bmatrix} x_1 \\ x_2 \\ x_3 \end{Bmatrix} = \begin{Bmatrix} 1.833333 \\ 2.166667 \\ 2.35 \end{Bmatrix} \tag{E10.5.1}
$$

The correct solution is $x_1 = x_2 = x_3 = 1$.

Solution.   Press the Solve System button on the TOOLKIT main menu to obtain a screen similar to Fig. 10.7*a*. This screen contains the input and output information needed to find the solution of a system of linear algebraic equations or to find the matrix inverse using *LU* decomposition.

The first step is to input the coefficients of the matrix and the right-hand-side vector. This is done by simply clicking on the appropriate table to activate it and then typing the numbers in appropriate cell locations. Up to 20 simultaneous equations may be solved by the TOOLKIT. In this case Eq. (E10.5.1) has been entered.

Next, it is necessary to specify the number of equations you want to solve. In this case we will solve three equations although it is also possible to solve a 2 by 2 or 1 by 1 subsystem. Finally, click on either Find Solution Vector or Find Matrix Inverse option button. Figure 10.7*a* shows the calculated solution vector. The input coefficients were input using seven significant figures and the solution is accurate to about five significant figures, which is consistent with the analysis of the matrix condition number (Example 10.4).

Now click on the Find Matrix Inverse option button and then the Calc button. The result is shown in Fig. 10.7*b*. Note that the magnitude of the matrix inverse coefficients are high, as would be expected for an ill-conditioned system. The Notes, Save, and Open buttons can be used to preserve comments and analyses you might wish to remember.

FIGURE 10.7
Screen from Numerical Methods TOOLKIT for solving linear algebraic equations: (*a*) solution and (*b*) matrix inverse of Hilbert system.

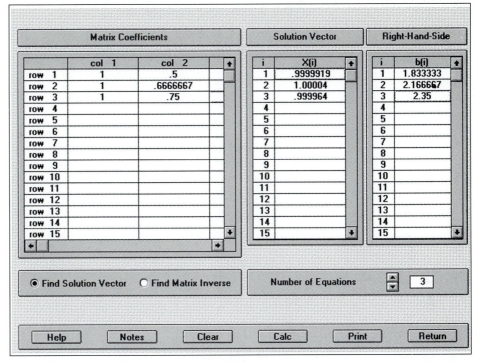

(*a*)

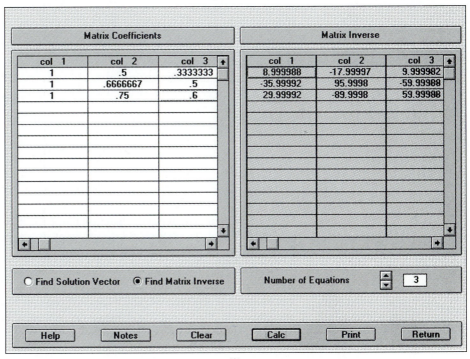

(*b*)

10.3.3 Iterative Refinement

In some cases, round-off errors can be reduced by the following procedure. Suppose that we are solving the following set of equations:

$$
\begin{aligned}
a_{11}x_1 + a_{12}x_2 + a_{13}x_3 &= b_1 \\
a_{21}x_1 + a_{22}x_2 + a_{23}x_3 &= b_2 \\
a_{31}x_1 + a_{32}x_2 + a_{33}x_3 &= b_3
\end{aligned}
\tag{10.27}
$$

For conciseness, we will limit the following discussion to this small (3×3) system. However, the approach is generally applicable to larger sets of linear equations.

Suppose an approximate solution vector is given by $\{\tilde{X}\}^T = \lfloor \tilde{x}_1 \quad \tilde{x}_2 \quad \tilde{x}_3 \rfloor$. This solution can be substituted into Eq. (10.27) to give

$$
\begin{aligned}
a_{11}\tilde{x}_1 + a_{12}\tilde{x}_2 + a_{13}\tilde{x}_3 &= \tilde{b}_1 \\
a_{21}\tilde{x}_1 + a_{22}\tilde{x}_2 + a_{23}\tilde{x}_3 &= \tilde{b}_2 \\
a_{31}\tilde{x}_1 + a_{32}\tilde{x}_2 + a_{33}\tilde{x}_3 &= \tilde{b}_3
\end{aligned}
\tag{10.28}
$$

Now, suppose that the exact solution $\{X\}$ is expressed as a function of the approximate solution and a vector of correction factors $\{\Delta X\}$, where

$$
\begin{aligned}
x_1 &= \tilde{x}_1 + \Delta x_1 \\
x_2 &= \tilde{x}_2 + \Delta x_2 \\
x_3 &= \tilde{x}_3 + \Delta x_3
\end{aligned}
\tag{10.29}
$$

If these results are substituted into Eq. (10.27), the following system results:

$$
\begin{aligned}
a_{11}(\tilde{x}_1 + \Delta x_1) + a_{12}(\tilde{x}_2 + \Delta x_2) + a_{13}(\tilde{x}_3 + \Delta x_3) &= b_1 \\
a_{21}(\tilde{x}_1 + \Delta x_1) + a_{22}(\tilde{x}_2 + \Delta x_2) + a_{23}(\tilde{x}_3 + \Delta x_3) &= b_2 \\
a_{31}(\tilde{x}_1 + \Delta x_1) + a_{32}(\tilde{x}_2 + \Delta x_2) + a_{33}(\tilde{x}_3 + \Delta x_3) &= b_3
\end{aligned}
\tag{10.30}
$$

Now, Eq. (10.28) can be subtracted from Eq. (10.30) to yield

$$
\begin{aligned}
a_{11}\Delta x_1 + a_{12}\Delta x_2 + a_{13}\Delta x_3 &= b_1 - \tilde{b}_1 = E_1 \\
a_{21}\Delta x_1 + a_{22}\Delta x_2 + a_{23}\Delta x_3 &= b_2 - \tilde{b}_2 = E_2 \\
a_{31}\Delta x_1 + a_{32}\Delta x_2 + a_{33}\Delta x_3 &= b_3 - \tilde{b}_3 = E_3
\end{aligned}
\tag{10.31}
$$

This system itself is a set of simultaneous linear equations that can be solved to obtain the correction factors. The factors can then be applied to improve the solution, as specified by Eq. (10.29).

It is relatively straightforward to integrate an iterative refinement procedure into computer programs for elimination methods. It is especially effective for the *LU* decomposition approaches described earlier, which are designed to evaluate different right-hand-side vectors efficiently. Note that to be effective for correcting ill-conditioned systems, the E's in Eq. (10.31) must be expressed in double precision.

PROBLEMS

10.1 Use the rules of matrix multiplication to prove that Eqs. (10.7) and (10.8) follow from Eq. (10.6).

10.2 Use naive Gauss elimination to decompose the following system according to the description in Sec. 10.2.

$$7x_1 + 2x_2 - 3x_3 = -12$$
$$2x_1 + 5x_2 - 3x_3 = -20$$
$$x_1 - x_2 - 6x_3 = -26$$

Then, multiply the resulting $[L]$ and $[U]$ matrices to determine that $[A]$ is produced.

10.3 Use LU decomposition to solve the system of equations in Prob. 10.2. Show all the steps in the computation. Also solve the system for an alternative right-hand-side vector

$$\{B\}^T = \lfloor 12 \quad 18 \quad -6 \rfloor$$

10.4 Determine the matrix inverse for Prob. 10.2. Check your results by verifying that $[A][A]^{-1} = [I]$. Do not use a pivoting strategy.

10.5 Solve the following system of equations using LU decomposition with partial pivoting:

$$x_1 + 7x_2 - 4x_3 = -51$$
$$4x_1 - 4x_2 + 9x_3 = 62$$
$$12x_1 - x_2 + 3x_3 = 8$$

10.6 Solve the following system of equations using LU decomposition with partial pivoting,

$$-5x_1 + 12x_3 = 60$$
$$4x_1 - x_2 - x_3 = -2$$
$$x_1 - 2x_2 + 12x_3 = -86$$

10.7 Determine the matrix inverse for Prob. 10.5. Use the inverse to solve the original problem as well as to solve the additional case where the right-hand-side vector is $\{B\}^T = [110 \quad 55 \quad -105]$.

10.8 Perform Crout decomposition on

$$2x_1 - 5x_2 + x_3 = 12$$
$$-x_1 + 3x_2 - x_3 = -8$$
$$3x_1 - 4x_2 + 2x_3 = 16$$

Then, multiply the resulting $[L]$ and $[U]$ matrices to determine that $[A]$ is produced.

10.9 The following system of equations is designed to determine concentrations (the c's in g/m^3) in a series of coupled reactors as a function of the amount of mass input to each reactor (the right-hand

sides in g/d),

$$17c_1 - 2c_2 - 3c_3 = 500$$
$$-5c_1 + 21c_2 - 2c_3 = 200$$
$$-5c_1 - 5c_2 + 22c_3 = 30$$

(a) Use any means at your disposal to determine the matrix inverse (e.g., the Numerical Methods TOOLKIT, Excel, MATLAB, your own program, etc.).
(b) Use the matrix inverse to determine the concentrations.
(c) Determine how much the rate of mass input to reactor 3 must be increased to induce a 5 g/m^3 rise in the concentration of reactor 1.
(d) How much will the concentration in reactor 3 be reduced if the rate of mass inputs to reactors 1 and 2 is reduced by 50 and 100 g/d, respectively?

10.10 Determine $\|A\|_e$, $\|A\|_1$, and $\|A\|_\infty$ for

$$[A] = \begin{bmatrix} 6 & -2 & 5 \\ 8.5 & 1.1 & -2.5 \\ -1.6 & -1 & 10.3 \end{bmatrix}$$

Scale the matrix prior to making your evaluations.

10.11 Determine the Euclidean and the row-sum norms for the systems in Probs. 10.5 and 10.6.

10.12 Determine the condition number for the following system using the row-sum norm. Do not normalize the system.

$$\begin{bmatrix} 1 & 4 & 9 & 16 \\ 4 & 9 & 16 & 25 \\ 9 & 16 & 25 & 36 \\ 16 & 25 & 36 & 49 \end{bmatrix}$$

How many significant digits of precision will be lost due to ill-conditioning?

10.13 Determine the condition number based on the row-sum norm for the normalized 4×4 Hilbert matrix. How many significant digits of precision will be lost due to ill-conditioning?

10.14 Develop a user-friendly program for LU decomposition based on the codes from Fig. 10.2.

10.15 Develop a user-friendly program for LU decomposition, including the capability to evaluate the matrix inverse. Base the program on Figs. 10.2 and 10.5.

10.16 Use iterative refinement techniques to improve $x_1 = 2$, $x_2 = -5$, and $x_3 = 12$ which are approximate solutions of

$$2x_1 + 4x_2 + x_3 = -5$$
$$5x_1 + 2x_2 + x_3 = 12$$
$$x_1 + 2x_2 + x_3 = 3$$

CHAPTER 11

Special Matrices and Gauss-Seidel

Certain matrices have a particular structure that can be exploited to develop efficient solution schemes. The first part of this chapter is devoted to two such systems: *banded* and *symmetric matrices.* Efficient elimination methods are described for both.

The second part of the chapter turns to an alternative to elimination methods; that is, approximate, iterative methods. The focus is on the *Gauss-Seidel method,* which employs initial guesses and then iterates to obtain refined estimates of the solution. The Gauss-Seidel method is particularly well-suited for large numbers of equations. In these cases, elimination methods can be subject to round-off errors. Because the error of the Gauss-Seidel method is controlled by the number of iterations, round-off error is not an issue of concern with this method. However, there are certain instances where the Gauss-Seidel technique will not converge on the correct answer. These and other trade-offs between elimination and iterative methods will be discussed in subsequent pages.

11.1 SPECIAL MATRICES

As mentioned in Box PT3.1, a *banded matrix* is a square matrix that has all elements equal to zero, with the exception of a band centered on the main diagonal. Banded systems are frequently encountered in engineering and scientific practice. For example, they typically occur in the solution of differential equations. In addition, other numerical methods such as cubic splines (Sec. 18.5) involve the solution of banded systems.

The dimensions of a banded system can be quantified by two parameters: the bandwidth BW and the half-bandwidth HBW (Fig. 11.1). These two values are related by $BW = 2HBW + 1$. In general, then, a banded system is one for which $a_{ij} = 0$ if $|i - j| >$ HBW.

Although Gauss elimination or conventional *LU* decomposition can be employed to solve banded equations, they are inefficient, because if pivoting is unnecessary none of the elements outside the band would change from their original values of zero. Thus, unnecessary space and time would be expended on the storage and manipulation of these useless zeros. If it is known beforehand that pivoting is unnecessary, very efficient algorithms can be developed that do not involve the zero elements outside the band. Because many problems involving banded systems do not require pivoting, these alternative algorithms, as described next, are the methods of choice.

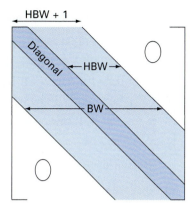

FIGURE 11.1
Parameters used to quantify the dimensions of a banded system. BW and HBW designate the bandwidth and the half-bandwidth, respectively.

(a) Decomposition

$$DO\ k = 2, n$$
$$e_k = e_k / f_{k-1}$$
$$f_k = f_k - e_k \cdot g_{k-1}$$
$$END\ DO$$

(b) Forward substitution

$$DO\ k = 2, n$$
$$r_k = r_k - e_k \cdot r_{k-1}$$
$$END\ DO$$

(c) Back substitution

$$x_n = r_n / f_n$$
$$DO\ k = n - 1, 1, -1$$
$$x_k = (r_k - g_k \cdot x_{k+1}) / f_k$$
$$END\ DO$$

FIGURE 11.2
Pseudocode to implement the Thomas algorithm, an *LU* decomposition method for tridiagonal systems.

11.1.1 Tridiagonal Systems

A tridiagonal system—that is, one with a bandwidth of 3—can be expressed generally as

$$
\begin{bmatrix}
f_1 & g_1 & & & & \\
e_2 & f_2 & g_2 & & & \\
 & e_3 & f_3 & g_3 & & \\
 & & \cdot & \cdot & \cdot & \\
 & & & \cdot & \cdot & \cdot \\
 & & & & e_{n-1} & f_{n-1} & g_{n-1} \\
 & & & & & e_n & f_n
\end{bmatrix}
\begin{Bmatrix}
x_1 \\ x_2 \\ x_3 \\ \cdot \\ \cdot \\ x_{n-1} \\ x_n
\end{Bmatrix}
=
\begin{Bmatrix}
r_1 \\ r_2 \\ r_3 \\ \cdot \\ \cdot \\ r_{n-1} \\ r_n
\end{Bmatrix}
\tag{11.1}
$$

Notice that we have changed our notation for the coefficients from a's and b's to e's, f's, g's, and r's. This was done to avoid storing large numbers of useless zeros in the square matrix of a's. This space-saving modification is advantageous because the resulting algorithm requires less computer memory.

Figure 11.2 shows pseudocode for an efficient method, called the *Thomas algorithm*, to solve Eq. (11.1). As with conventional *LU* decomposition, the algorithm consists of three steps: decomposition and forward and back substitution. Thus, all the advantages of *LU* decomposition, such as convenient evaluation of multiple right-hand-side vectors and the matrix inverse, can be accomplished by proper application of this algorithm.

EXAMPLE 11.1   Tridiagonal Solution with the Thomas Algorithm

Problem Statement.   Solve the following tridiagonal system with the Thomas algorithm.

$$
\begin{bmatrix}
2.04 & -1 & & \\
-1 & 2.04 & -1 & \\
 & -1 & 2.04 & -1 \\
 & & -1 & 2.04
\end{bmatrix}
\begin{Bmatrix}
T_1 \\ T_2 \\ T_3 \\ T_4
\end{Bmatrix}
=
\begin{Bmatrix}
40.8 \\ 0.8 \\ 0.8 \\ 200.8
\end{Bmatrix}
$$

Solution.   First, the decomposition is implemented as

$$e_2 = -1/2.04 = -0.49$$
$$f_2 = 2.04 - (-0.49)(-1) = 1.550$$
$$e_3 = -1/1.550 = -0.645$$
$$f_3 = 2.04 - (-0.645)(-1) = 1.395$$
$$e_4 = -1/1.395 = -0.717$$
$$f_4 = 2.04 - (-0.717)(-1) = 1.323$$

Thus, the matrix has been transformed to

$$\begin{bmatrix} 2.04 & -1 & & \\ -0.49 & 1.550 & -1 & \\ & -0.645 & 1.395 & -1 \\ & & -0.717 & 1.323 \end{bmatrix}$$

and the *LU* decomposition is

$$[A] = [L][U] = \begin{bmatrix} 1 & & & \\ -0.49 & 1 & & \\ & -0.645 & 1 & \\ & & -0.717 & 1 \end{bmatrix} \begin{bmatrix} 2.04 & -1 & & \\ & 1.550 & -1 & \\ & & 1.395 & -1 \\ & & & 1.323 \end{bmatrix}$$

You can verify that this is correct by multiplying $[L][U]$ to yield $[A]$.

The forward substitution is implemented as

$$r_2 = 0.8 - (-0.49)40.8 = 20.8$$
$$r_3 = 0.8 - (-0.645)20.8 = 14.221$$
$$r_4 = 200.8 - (-0.717)14.221 = 210.996$$

Thus the right-hand-side vector has been modified to

$$\left\{ \begin{array}{c} 40.8 \\ 20.8 \\ 14.221 \\ 210.996 \end{array} \right\}$$

which then can be used in conjunction with the $[U]$ matrix to perform back substitution and obtain the solution

$$T_4 = 210.996/1.323 = 159.480$$
$$T_3 = [14.221 - (-1)159.48]/1.395 = 124.538$$
$$T_2 = [20.800 - (-1)124.538]/1.550 = 93.778$$
$$T_1 = [40.800 - (-1)93.778]/2.040 = 65.970$$

11.1.2 Cholesky Decomposition

Recall from Box PT3.1 that a symmetric matrix is one where $a_{ij} = a_{ji}$ for all i and j. In other words, $[A] = [A]^T$. Such systems occur commonly in both mathematical and engineering problem contexts. They offer computational advantages because only half the storage is needed and, in most cases, only half the computation time is required for their solution.

One of the most popular approaches involves *Cholesky decomposition*. This algorithm is based on the fact that a symmetric matrix can be decomposed, as in

$$[A] = [L][L]^T \tag{11.2}$$

That is, the resulting triangular factors are the transpose of each other.

The terms of Eq. (11.2) can be multiplied out and set equal to each other (see Prob. 11.4 at the end of the chapter). The result can be expressed simply by recurrence relations. For the kth row,

$$l_{kj} = \frac{a_{ki} - \sum_{j=1}^{i-1} l_{ij} l_{kj}}{l_{ii}} \qquad \text{for } i = 1, 2, \ldots, k-1 \tag{11.3}$$

and

$$l_{kk} = \sqrt{a_{kk} - \sum_{j=1}^{k-1} l_{kj}^2} \tag{11.4}$$

EXAMPLE 11.2   Cholesky Decomposition

Problem Statement.   Apply Cholesky decomposition to the symmetric matrix

$$[A] = \begin{bmatrix} 6 & 15 & 55 \\ 15 & 55 & 225 \\ 55 & 225 & 979 \end{bmatrix}$$

Solution.   For the first row ($k = 1$), Eq. (11.3) is skipped and Eq. (11.4) is employed to compute

$$l_{11} = \sqrt{a_{11}} = \sqrt{6} = 2.4495$$

For the second row ($k = 2$), Eq. (11.3) gives

$$l_{21} = \frac{a_{21}}{l_{11}} = \frac{15}{2.4495} = 6.1237$$

and Eq. (11.4) yields

$$l_{22} = \sqrt{a_{11} - l_{21}^2} = \sqrt{55 - (6.1237)^2} = 4.1833$$

For the third row ($k = 3$), Eq. (11.3) gives ($i = 1$)

$$l_{31} = \frac{a_{31}}{l_{11}} = \frac{55}{2.4495} = 22.454$$

and ($i = 2$)

$$l_{32} = \frac{a_{32} - l_{21}l_{31}}{l_{22}} = \frac{225 - 6.1237(22.454)}{4.1833} = 20.916$$

```
DO k = 1, n
  DO i = 1, k − 1
    sum = 0.
    DO j = 1, i − 1
      sum = sum + aᵢⱼ · aₖⱼ
    END DO
    aₖᵢ = (aₖᵢ − sum)/aᵢᵢ
  END DO
  sum = 0.
  DO j = 1, k − 1
    sum = sum + a²ₖⱼ
  END DO
  aₖₖ = √(aₖₖ − sum)
END DO
```

FIGURE 11.3
Pseudocode for Cholesky's *LU* decomposition algorithm.

and Eq. (11.4) yields

$$l_{33} = \sqrt{a_{33} - l_{31}^2 - l_{32}^2} = \sqrt{979 - (22.454)^2 - (20.916)^2} = 6.1106$$

Thus, the Cholesky decomposition yields

$$[L] = \begin{bmatrix} 2.4495 & & \\ 6.1237 & 4.1833 & \\ 22.454 & 20.916 & 6.1106 \end{bmatrix}$$

The validity of this decomposition can be verified by substituting it and its transpose into Eq. (11.2) to see if their product yields the original matrix [A]. This is left for an exercise.

Figure 11.3 presents pseudocode for implementing the Cholesky decomposition algorithm. It should be noted that the algorithm in Fig. 11.3 could result in an execution error if the evaluation of a_{kk} involves taking the square root of a negative number. However, for cases where the matrix is *positive definite*,[1] this will never occur. Because many symmetric matrices dealt with in engineering are, in fact, positive definite, the Cholesky algorithm has wide application. Another benefit of dealing with positive definite symmetric matrices is that pivoting is not required to avoid division by zero. Thus, we can implement the algorithm in Fig. 11.3 without the complication of pivoting.

11.2 GAUSS-SEIDEL

Iterative or approximate methods provide an alternative to the elimination methods described to this point. Such approaches are similar to the techniques we developed to obtain the roots of a single equation in Chap. 6. Those approaches consisted of guessing a value and then using a systematic method to obtain a refined estimate of the root. Because the present part of the book deals with a similar problem—obtaining the values that simultaneously satisfy a set of equations—we might suspect that such approximate methods could be useful in this context.

The *Gauss-Seidel method* is the most commonly used iterative method. Assume that we are given a set of *n* equations:

$$[A]\{X\} = \{B\}$$

Suppose that for conciseness we limit ourselves to a 3×3 set of equations. If the diagonal elements are all nonzero, the first equation can be solved for x_1, the second for x_2, and the third for x_3 to yield

$$x_1 = \frac{b_1 - a_{12}x_2 - a_{13}x_3}{a_{11}} \tag{11.5a}$$

---

[1] A *positive definite matrix* is one for which the product $\{X\}^T[A]\{X\}$ is greater than zero for all nonzero vectors $\{X\}$.

$$x_2 = \frac{b_2 - a_{21}x_1 - a_{23}x_3}{a_{22}} \tag{11.5b}$$

$$x_3 = \frac{b_3 - a_{31}x_1 - a_{32}x_2}{a_{33}} \tag{11.5c}$$

Now, we can start the solution process by choosing guesses for the x's. A simple way to obtain initial guesses is to assume that they are all zero. These zeros can be substituted into Eq. (11.5a), which can be used to calculate a new value for $x_1 = b_1/a_{11}$. Then, we substitute this new value of x_1 along with the previous guess of zero for x_3 into Eq. (11.5b) to compute a new value for x_2. The process is repeated for Eq. (11.5c) to calculate a new estimate for x_3. Then we return to the first equation and repeat the entire procedure until our solution converges closely enough to the true values. Convergence can be checked using the criterion [recall Eq. (3.5)]

$$|\varepsilon_{a,i}| = \left| \frac{x_i^j - x_i^{j-1}}{x_i^j} \right| 100\% < \varepsilon_s \tag{11.6}$$

for all i, where j and $j - 1$ are the present and previous iterations.

EXAMPLE 11.3   Gauss-Seidel Method

Problem Statement.   Use the Gauss-Seidel method to obtain the solution of the same system used in Example 11.1:

$$3x_1 - 0.1x_2 - 0.2x_3 = 7.85$$
$$0.1x_1 + 7x_2 - 0.3x_3 = -19.3$$
$$0.3x_1 - 0.2x_2 + 10x_3 = 71.4$$

Recall that the true solution is $x_1 = 3$, $x_2 = -2.5$, and $x_3 = 7$.

Solution.   First, solve each of the equations for its unknown on the diagonal.

$$x_1 = \frac{7.85 + 0.1x_2 + 0.2x_3}{3} \tag{E11.3.1}$$

$$x_2 = \frac{-19.3 - 0.1x_1 + 0.3x_3}{7} \tag{E11.3.2}$$

$$x_3 = \frac{71.4 - 0.3x_1 + 0.2x_2}{10} \tag{E11.3.3}$$

By assuming that x_2 and x_3 are zero, Eq. (E11.3.1) can be used to compute

$$x_1 = \frac{7.85 + 0 + 0}{3} = 2.616667$$

This value, along with the assumed value of $x_3 = 0$, can be substituted into Eq. (E11.3.2) to calculate

$$x_2 = \frac{-19.3 - 0.1(2.616667) + 0}{7} = -2.794524$$

The first iteration is completed by substituting the calculated values for x_1 and x_2 into Eq. (E11.3.3) to yield

$$x_3 = \frac{71.4 - 0.3(2.616667) + 0.2(-2.794524)}{10} = 7.005610$$

For the second iteration, the same process is repeated to compute

$$x_1 = \frac{7.85 + 0.1(-2.794524) + 0.2(7.005610)}{3} = 2.990557 \qquad |\varepsilon_t| = 0.31\%$$

$$x_2 = \frac{-19.3 - 0.1(2.990557) + 0.3(7.005610)}{7} = -2.499625 \qquad |\varepsilon_t| = 0.015\%$$

$$x_3 = \frac{71.4 - 0.3(2.990557) + 0.2(-2.499625)}{10} = 7.000291 \qquad |\varepsilon_t| = 0.0042\%$$

The method is, therefore, converging on the true solution. Additional iterations could be applied to improve the answers. However, in an actual problem, we would not know the true answer a priori. Consequently, Eq. (11.6) provides a means to estimate the error. For example, for x_1,

$$|\varepsilon_{a,1}| = \left| \frac{2.990557 - 2.616667}{2.990557} \right| 100\% = 12.5\%$$

For x_2 and x_3, the error estimates are $|\varepsilon_{a,2}| = 11.8\%$ and $|\varepsilon_{a,3}| = 0.076\%$. Note that, as was the case when determining roots of a single equation, formulations such as Eq. (11.6) usually provide a conservative appraisal of convergence. Thus, when they are met, they ensure that the result is known to at least the tolerance specified by ε_s.

As each new x value is computed for the Gauss-Seidel method, it is immediately used in the next equation to determine another x value. Thus, if the solution is converging, the best available estimates will be employed. An alternative approach, called *Jacobi iteration,* utilizes a somewhat different tactic. Rather than using the latest available x's, this technique uses Eq. (11.5) to compute a set of new x's on the basis of a set of old x's. Thus, as new values are generated, they are not immediately used but rather are retained for the next iteration.

The difference between the Gauss-Seidel method and Jacobi iteration is depicted in Fig. 11.4. Although there are certain cases where the Jacobi method is useful, Gauss-Seidel's utilization of the best available estimates usually makes it the method of preference.

11.2.1 Convergence Criterion for the Gauss-Seidel Method

Note that the Gauss-Seidel method is similar in spirit to the technique of simple fixed-point iteration that was used in Sec. 6.1 to solve for the roots of a single equation. Recall that simple fixed-point iteration had two fundamental problems: (1) it was sometimes nonconvergent and (2) when it converged it often did so very slowly. The Gauss-Seidel method can also exhibit these shortcomings.

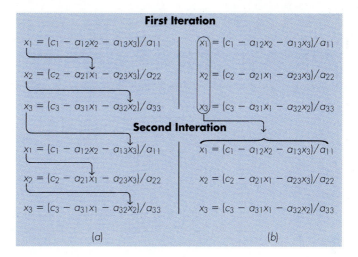

FIGURE 11.4
Graphical depiction of the difference between (a) the Gauss-Seidel and (b) the Jacobi iterative methods for solving simultaneous linear algebraic equations.

Convergence criteria can be developed by recalling from Sec. 6.5.1 that sufficient conditions for convergence of two nonlinear equations, $u(x, y)$ and $v(x, y)$, are

$$\left|\frac{\partial u}{\partial x}\right| + \left|\frac{\partial v}{\partial x}\right| < 1 \tag{11.7a}$$

and

$$\left|\frac{\partial u}{\partial y}\right| + \left|\frac{\partial v}{\partial y}\right| < 1 \tag{11.7b}$$

These criteria also apply to linear equations of the sort we are solving with the Gauss-Seidel method. For example, in the case of two simultaneous equations, the Gauss-Seidel algorithm [Eq. (11.5)] can be expressed as

$$u(x_1,\ x_2) = \frac{c_1}{a_{11}} - \frac{a_{12}}{a_{11}}x_2 \tag{11.8a}$$

and

$$v(x_1,\ x_2) = \frac{c_2}{a_{22}} - \frac{a_{21}}{a_{22}}x_1 \tag{11.8b}$$

The partial derivatives of these equations can be evaluated with respect to each of the unknowns as

$$\frac{\partial u}{\partial x_1} = 0 \qquad \frac{\partial v}{\partial x_1} = -\frac{a_{21}}{a_{22}}$$

and

$$\frac{\partial u}{\partial x_2} = -\frac{a_{12}}{a_{11}} \qquad \frac{\partial v}{\partial x_2} = 0$$

which can be substituted into Eq. (11.7) to give

$$\left|\frac{a_{21}}{a_{22}}\right| < 1 \qquad (11.9a)$$

and

$$\left|\frac{a_{12}}{a_{11}}\right| < 1 \qquad (11.9b)$$

In other words, the absolute values of the slopes of Eq. (11.8) must be less than unity to ensure convergence. This is displayed graphically in Fig. 11.5. Equation (11.9) can also be reformulated as

$$|a_{22}| > |a_{21}|$$

and

$$|a_{11}| > |a_{12}|$$

That is, the diagonal element must be greater than the off-diagonal element for each row.

The extension of the above to n equations is straightforward and can be expressed as

$$|a_{ii}| > \sum_{\substack{j=1 \\ j\neq i}}^{n} |a_{i,j}| \qquad (11.10)$$

FIGURE 11.5
Illustration of (a) convergence and (b) divergence of the Gauss-Seidel method. Notice that the same functions are plotted in both cases (u: $11x_1 + 13x_2 = 286$; v: $11x_1 - 9x_2 = 99$). Thus, the order in which the equations are implemented (as depicted by the direction of the first arrow from the origin) dictates whether the computation converges.

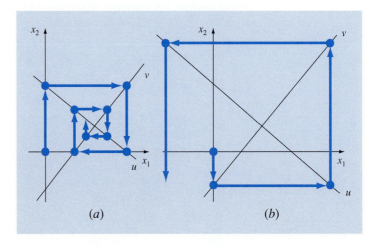

(a)   (b)

That is, the diagonal coefficient in each of the equations must be larger than the sum of the absolute values of the other coefficients in the equation. This criterion is sufficient but not necessary for convergence. That is, although the method may sometimes work if Eq. (11.10) is not met, convergence is guaranteed if the condition is satisfied. Systems where Eq. (11.10) holds are called *diagonally dominant*. Fortunately, many engineering problems of practical importance fulfill this requirement.

11.2.2 Improvement of Convergence Using Relaxation

Relaxation represents a slight modification of the Gauss-Seidel method and is designed to enhance convergence. After each new value of x is computed using Eq. (11.5), that value is modified by a weighted average of the results of the previous and the present iterations:

$$x_i^{new} = \lambda x_i^{new} + (1 - \lambda)x_i^{old} \tag{11.11}$$

where λ is a weighting factor that is assigned a value between 0 and 2.

If $\lambda = 1$, $(1 - \lambda)$ is equal to 0 and the result is unmodified. However, if λ is set at a value between 0 and 1, the result is a weighted average of the present and the previous results. This type of modification is called *underrelaxation*. It is typically employed to make a nonconvergent system converge or to hasten convergence by dampening out oscillations.

For values of λ from 1 to 2, extra weight is placed on the present value. In this instance, there is an implicit assumption that the new value is moving in the correct direction toward the true solution but at too slow a rate. Thus, the added weight of λ is intended to improve the estimate by pushing it closer to the truth. Hence, this type of modification, which is called *overrelaxation,* is designed to accelerate the convergence of an already convergent system. The approach is also called *successive* or *simultaneous overrelaxation,* or *SOR.*

The choice of a proper value for λ is highly problem-specific and is often determined empirically. For a single solution of a set of equations it is often unnecessary. However, if the system under study is to be solved repeatedly, the efficiency introduced by a wise choice of λ can be extremely important. Good examples are the very large systems of partial differential equations that often arise when modeling continuous variations of variables (recall the distributed system depicted in Fig. PT3.1*b*). We will return to this topic in Part Eight.

11.2.3 Algorithm for Gauss-Seidel

An algorithm for the Gauss-Seidel method, with relaxation, is depicted in Fig. 11.6. Note that this algorithm is not guaranteed to converge if the equations are not input in a diagonally dominant form.

The pseudocode has two features that bear mentioning. First, there is an initial set of nested loops to divide each equation by its diagonal element. This reduces the total number of operations in the algorithm. Second, notice that the error check is designated by a variable called *sentinel*. If any of the equations has an approximate error greater than the stopping criterion (e_s), then the iterations are allowed to continue. The use of the sentinel allows us to circumvent unnecessary calculations of error estimates once one of the equations exceeds the criterion.

```
SUBROUTINE Gseid (a,b,n,x,imax,es,lambda)
  DO i = 1,n
    dummy = a_{i,i}
    DO j = 1,n
      a_{i,j} = a_{i,j}/dummy
    END DO
    b_i = b_i /dummy
  END DO
  DO i = 1, n
    sum = b_i
    DO j = 1, n
      IF i≠j THEN sum = sum − a_{i,j}*x_j
    END DO
    x_i=sum
  END DO
  iter=1
  DO
    sentinel = 1
    DO i = 1, n
      old = x_i
      sum = b_i
      DO j = 1, n
        IF i≠j THEN sum = sum − a_{i,j}*x_j
      END DO
      x_i = lambda*sum+(1.−lambda)*old
      IF sentinel = 1 AND x_i ≠ 0. THEN
        ea = ABS((x_i − old) /x_i)*100.
        IF ea > es THEN sentinel = 0
      END IF
    END DO
    iter = iter + 1
    IF sentinel = 1 OR (iter ≥ imax) EXIT
  END DO
END Gseid
```

FIGURE 11.6
Pseudocode for Gauss-Seidel with relaxation.

11.2.4 Problem Contexts for the Gauss-Seidel Method

Aside from circumventing the round-off dilemma, the Gauss-Seidel technique has a number of other advantages that make it particularly attractive in the context of certain engineering problems. For example, when the matrix in question is very large and very sparse (that is, most of the elements are zero), elimination methods waste large amounts of computer memory by storing zeros.

At the beginning of this chapter, we saw how this shortcoming could be circumvented if the coefficient matrix is banded. For nonbanded systems, there is usually no simple way to avoid large memory requirements when using elimination methods. Because all computers have a finite amount of memory, this inefficiency can place a constraint on the size of systems for which elimination methods are practical.

Although a general algorithm such as the one in Fig. 11.6 is prone to the same constraint, the structure of the Gauss-Seidel equations [Eq. (11.5)] permits concise programs to be developed for specific systems. Because only nonzero coefficients need be included in Eq. (11.5), large savings of computer memory are possible. Although this entails more up-front investment in software development, the long-term advantages are substantial when dealing with large systems for which many simulations are to be performed. Both lumped- and distributed-variable systems can result in large, sparse matrices for which the Gauss-Seidel method has utility.

11.3   LINEAR ALGEBRAIC EQUATIONS WITH LIBRARIES AND PACKAGES

Software libraries and packages have great capabilities for solving systems of linear algebraic equations. Before describing these tools, we should mention that the approaches described in Chap. 7 for solving nonlinear systems can be applied to linear systems. However, in this section, we will focus on the approaches that are expressly designed for linear equations.

11.3.1 Mathcad

Mathcad contains many special functions that manipulate vectors and matrices. These include common operations such as dot product, matrix transpose, matrix addition, and matrix multiplication. In addition, it allows calculation of the matrix inverse, determinant, trace, various types of norms, and condition numbers based on different norms. It also has several functions that decompose matrices.

Systems of linear equations can be solved in two ways by Mathcad. First, it is possible to use matrix inversion and subsequent multiplication by the right-hand side as discussed in Chap. 10. In addition Mathcad has a special function called **Isolve(M,b)** that is specifically designed to solve linear equations. You can use other built-in functions to evaluate the condition of [**M**] to determine if **M** is nearly singular and thus possibly subject to round-off errors.

As an example, let's use the **Isolve** Mathcad function to solve a system of linear equations. As shown in Fig. 11.7, the first step is to enter the coefficients of the [**M**] matrix using the definition symbol and the Insert/Matrix pull down menu. This gives a box that allows you to specify the dimensions of the matrix. For our case, we will select a dimension of 4×4, and Mathcad places a blank 4-by-4-size matrix on screen. Now, simply click the appropriate cell location and enter values. Repeat similar operations to create the right-hand-side **v** vector. Now the variable **soln** is defined as **Isolve(M,v)** and the value of **soln** is shown using an equal sign.

Next, let's use some other functions from Mathcad to find the inverse and the condition number of the Hilbert matrix. The scaled matrix can be entered using the definition

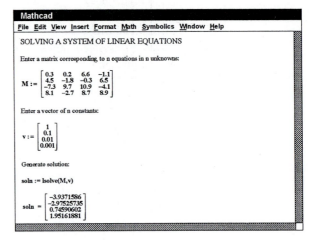

FIGURE 11.7
Mathcad screen to solve a system of linear algebraic equations.

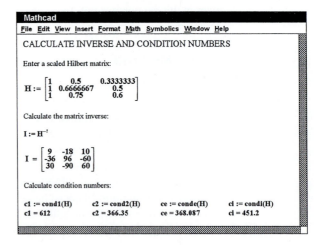

FIGURE 11.8
Mathcad screen to determine the matrix inverse and condition numbers of a scaled 3×3 Hilbert matrix.

symbol and the Insert/Matrix pull down menu. The inverse can then be computed by merely raising **H** to the exponent -1. The result is shown in Fig. 11.8.

Some other Mathcad functions can be employed to determine condition numbers by using the definition symbol to define variables **c1**, **c2**, **ce**, and **ci** as the condition number based on the column-sum (l_1), spectral (l_2), Euclidean, and the row-sum (l_∞) norms, respectively. The resulting values are shown at the bottom of Fig. 11.8. As expected, the spectral norm provides the smallest measure of magnitude.

11.3.2 Excel

There are two ways to solve linear algebraic equations with Excel: (1) using the Solver tool or (2) using matrix inversion and multiplication functions.

Recall that one way to determine the solution of linear algebraic equations is

$$\{X\} = [A]^{-1}\{B\} \tag{11.12}$$

Excel has built-in functions for both matrix inversion and multiplication that can be used to implement this formula.

EXAMPLE 11.4   Using Excel to Solve Linear Systems

Problem Statement.   Recall in Chap. 10 we introduced the Hilbert matrix. The following system is based on the Hilbert matrix. Note that it is scaled, as was done previously in Example 10.3, so that the maximum coefficient in each row is unity.

$$\begin{bmatrix} 1 & 1/2 & 1/3 \\ 1 & 2/3 & 1/2 \\ 1 & 3/4 & 3/5 \end{bmatrix} \begin{Bmatrix} x_1 \\ x_2 \\ x_3 \end{Bmatrix} = \begin{Bmatrix} 1.833333 \\ 2.166667 \\ 2.35 \end{Bmatrix}$$

The solution to this system is $\{X\}^T = \lfloor 1\ 1\ 1 \rfloor$. Use Excel to obtain this solution.

Solution.   The spreadsheet to solve this problem is displayed in Fig. 11.9. First, the matrix $[A]$ and the right-hand-side constants $\{B\}$ are entered into the spreadsheet cells. Then, a set of cells of the proper dimensions (in our example 3×3) is highlighted by either clicking and dragging the mouse or by using the arrow keys while depressing the shift key. As in Fig. 11.9, we highlight the range: B5. .D7.

Next, a formula invoking the matrix inverse function is entered,

```
=minverse(B1..D3)
```

Note that the argument is the range holding the elements of $[A]$. The *Ctrl* and *Shift* keys are held down while the *Enter* key is depressed. The resulting inverse of $[A]$ will be calculated by Excel and displayed in the range B5. .D7 as shown in Fig. 11.9.

A similar approach is used to multiply the inverse by the right-hand-side vector. For this case, the range from F5..F7 is highlighted and the following formula is entered

```
=mmult(B7..D7,F1..F3)
```

FIGURE 11.9
Excel spreadsheet solution of simultaneous linear algebraic equations using matrix inversion and multiplication.

	A	B	C	D	E	F
1		1	0.5	0.333333		1.833333333
2	[A]=	1	0.666667	0.5	{B}=	2.166666667
3		1	0.75	0.6		2.35
4						
5		9	−18	10		1.000000000000010
6	[A]−1=	−36	96	−60	{X}=	0.999999999999972
7		30	−90	60		1.000000000000030

=minverse(B1..D3)   =mmult(B7..D7,F1..F3)

where the first range is the first matrix to be multiplied, $[A]^{-1}$, and the second range is the second matrix to be multiplied, $\{B\}$. By again using the *Ctrl-Shift-Enter* combination, the solution $\{X\}$ will be calculated by Excel and displayed in the range B5. .D7, as shown in Fig. 11.9. As can be seen, the correct answer results.

Notice that we deliberately reformatted the results in Example 11.4 to show 15 digits. We did this because Excel uses double-precision to store numerical values. Thus, we see that round-off error occurs in the last two digits. This implies a condition number on the order of 100, which agrees with the result of 451.2 originally calculated in Example 10.3. Excel does not have the capability to calculate a condition number. In most cases, particularly because it employs double-precision numbers, this does not represent a problem. However, for cases where you suspect that the system is ill-conditioned, determination of the condition number is useful. Mathcad, MATLAB, and IMSL are capable of computing this quantity.

11.3.3 MATLAB

As the name implies, MATLAB (short for MATrix LABoratory) was designed to facilitate matrix manipulations. Thus, as might be expected, its capabilities in this area are excellent. Some of the key MATLAB functions related to matrix operations are listed in Table 11.1. The following example illustrates a few of these capabilities.

TABLE 11.1   MATLAB functions to implement matrix analysis and numerical linear algebra.

Matrix Analysis		Linear Equations	
Function	**Description**	**Function**	**Description**
cond	Matrix condition number	\ and /	Linear equation solution; use "help slash"
norm	Matrix or vector norm	chol	Cholesky factorization
rcond	LINPACK reciprocal condition estimator	lu	Factors from Gauss elimination
rank	Number of linearly independent rows or columns	inv	Matrix inverse
det	Determinant	qr	Orthogonal-triangular decomposition
trace	Sum of diagonal elements	qrdelete	Delete a column from the QR factorization
null	Null space	qrinsert	Insert a column in the QR factorization
orth	Orthogonalization	nnls	Nonnegative least squares
rref	Reduced row echelon form	pinv	Pseudoinverse
		lscov	Least squares in the presence of known covariance

EXAMPLE 11.5　Using MATLAB to Manipulate Linear Algebraic Equations

Problem Statement. Explore how MATLAB can be employed to solve and analyze linear algebraic equations. Use the same system as in Example 11.4.

Solution. First, we can enter the [A] matrix and the {B} vector. MATLAB actually has a special function to generate Hilbert matrices,

```
>> A=hilb(3)

A =
    1.0000    0.5000    0.3333
    1.0000    0.6667    0.5000
    1.0000    0.7500    0.6000

>> B=[1+1/2+1/3;1+2/3+1/2;1+3/4+3/5]

B =
    1.8333
    2.1667
    2.3500
```

Next, we can determine the condition number for [A], as in

```
>> Cond(A)

ans =
    366.3503
```

This result is based on the spectral, or $\|A\|_2$, norm discussed in Box 10.2. Note that it is of the same order of magnitude as the condition number = 451.2 based on the row-sum norm in Example 10.3. Both results imply that between 2 and 3 digits of precision could be lost.

Now we can solve the system of equations in two different ways. The most direct and efficient way is to employ backslash, or "left division":

```
>> X=A\B

X =
    1.0000
    1.0000
    1.0000
```

For cases such as ours, MATLAB uses Gauss elimination to solve such systems.

As an alternative, we can implement Eq. (PT3.6) directly, as in

```
>> X=inv(A)*B

X =
    1.0000
    1.0000
    1.0000
```

This approach actually determines the matrix inverse first, and then performs the matrix multiplication. Hence, it is more time-consuming than using the backslash approach.

11.3.4 IMSL

IMSL has numerous routines for linear systems, matrix inversion, and determinant evaluation. Table 11.2 lists the categories that are covered.

As listed in Table 11.3, eight routines are devoted to the specific case of real general matrices. In the present discussion, we will focus on two routines: LFCRG and LFIRG.

The LFCRG performs an *LU* decomposition of the matrix [*A*] and computes its condition number. LFCRG is implemented by the following CALL statement:

```
CALL LFCRG(N, A, LDA, FAC, LDFAC, IPVT, RCOND)
```

where N = Order of the matrix. (Input)

A = N × N matrix to be decomposed. (Input)

LDA = Leading dimension of A as specified in the dimension statement of the calling program. (Input)

TABLE 11.2 Categories of IMSL routines for solution of linear systems, matrix inversion, and determinant evaluation.

- Real general matrices
- Complex general matrices
- Real triangular matrices
- Complex triangular matrices
- Real positive definite matrices
- Real symmetric matrices
- Complex Hermitian positive definite matrices
- Complex Hermitian matrices
- Real band matrices in band storage
- Real band symmetric positive definite matrices in band storage
- Complex band matrices in band storage
- Complex band positive definite matrices in band storage
- Real sparse linear equation solvers
- Complex sparse linear equation solvers
- Real sparse symmetric positive definite linear equation solvers
- Complex sparse Hermitian positive definite linear equation solvers
- Real Toeplitz matrices in Toeplitz storage
- Complex Toeplitz matrices in Toeplitz storage
- Complex Circulant matrices in Circulant storage
- Iterative methods
- Linear least squares and matrix factorization
- Least squares, QR decomposition, and generalized inverse
- Cholesky factorization
- Singular value decomposition (SVD)
- Mathematical support for statistics

TABLE 11.3 IMSL routines for the solution of real general matrices.

Routine	Capability
LSARG	High-accuracy linear system solution
LSLRG	Solve a linear system
LFCRG	Factor and compute condition number
LFTRG	Factor
LFSRG	Solve after factoring
LFIRG	High-accuracy linear system solution after factoring
LFDRG	Compute determinant after factoring
LINRG	Invert

FAC = N × N matrix containing the LU decomposition of A. (Output)

LDFAC = Leading dimension of FAC as specified in the dimension statement of the calling program. (Input)

IPVT = Vector of length N containing the pivoting information for the LU decomposition. (Output)

RCOND = Scalar containing the reciprocal of the condition number of A. (Output)

The LFIRG uses the LU decomposition and a particular right-hand-side vector to generate a high-accuracy solution using iterative refinement. LFIRG is implemented by the following CALL statement:

```
CALL LFIRG(N, A, LDA, FAC, LDFAC, IPVT, B, IPATH, X, RES)
```

where N = Order of the matrix. (Input)

A = N × N matrix to be decomposed. (Input)

LDA = Leading dimension of A as specified in the dimension statement of the calling program. (Input)

FAC = N × N matrix containing the LU decomposition of A. (Input)

LDFAC = Leading dimension of FAC as specified in the dimension statement of the calling program. (Input)

IPVT = Vector of length N containing the pivoting information for the LU decomposition. (Input)

B = Vector of length N containing the right-hand side of the linear system

IPATH = Path indicator. (Input)

= 1 means the system AX = B is solved

= 2 means the system $A^TX = B$ is solved

X = Vector of length N containing the solution to the linear system. (Output)

RES = Vector of length N containing the residual vector at the improved solution. (Output)

These two routines are used in tandem in the following example. First, LFCRG is called to decompose the matrix and to return the condition number. Then LFIRG is called

N times with the B vector containing each column of the identity matrix to generate the columns of the matrix inverse. Finally, LFIRG can be called one additional time to obtain the solution for a single right-hand-side vector.

EXAMPLE 11.6   Using IMSL to Analyze and Solve a Hilbert Matrix

Problem Statement.   Use LFCRG and LFIRG to determine the condition number, the matrix inverse and the solution for the following Hilbert matrix system,

$$\begin{bmatrix} 1 & 1/2 & 1/3 \\ 1/2 & 2/3 & 1/2 \\ 1 & 3/4 & 3/5 \end{bmatrix} \begin{Bmatrix} x_1 \\ x_2 \\ x_3 \end{Bmatrix} = \begin{Bmatrix} 1.833333 \\ 2.166667 \\ 2.35 \end{Bmatrix}$$

Solution.   An example of a main Fortran 90 program using LFCRG and LFIRG to solve this problem can be written as

```
PROGRAM Lineqs
USE msimsl
IMPLICIT NONE
INTEGER::ipath,lda,n,ldfac
PARAMETER(ipath=1,lda=3,ldfac=3,n=3)
INTEGER::ipvt(n),i,j,itmax=50
REAL::A(lda,lda),Ainv(lda,lda),factor(ldfac,ldfac),Rcond,Res(n)
REAL::Rj(n),B(n),x(n)

DATA A/1.0,0.5,0.3333333,0.5,0.3333333,0.25,0.3333333,0.25,0.2/
DATA B/1.833333,1.083333,0.783333/

!Perform LU decomposition; determine and display condition number

CALL LFCRG(n,A,lda,factor,ldfac,ipvt,Rcond)
PRINT *, "Condition number = ", 1.0E0/Rcond
PRINT *

!Initialize vector Rj to zero
DO i = 1,n
  Rj(i) = 0.
END DO

!Feed columns of identity matrix through LFIRG to generate
!inverse and store result in Ainv. Display Ainv

DO j = 1, n
  Rj(j) = 1.0
  CALL LFIRG(n,A,lda,factor,ldfac,ipvt,Rj,ipath,ainv(1,j),Res)
  Rj(j) = 0.0
END DO
PRINT *, "Matrix inverse:"
DO i = 1,n
  PRINT *, (Ainv(i,j),j=1,n)
END DO
PRINT *
```

```
!Use LFIRG to obtain solution for B. Display results

PRINT *, "Solution:"
DO I = 1,n
  PRINT *, x(i)
END DO

END PROGRAM
```

Output:

```
Condition number =      680.811600

Matrix inverse:
        9.000033        -36.000180        30.000160
      -36.000180        192.000900      -180.000800
       30.000160       -180.000800       180.000800

Solution:
    9.999986E-01
    1.000010
    9.999884E-01
```

Again, the condition number is of the same order as the condition number based on the row-sum norm in Example 10.3 (451.2). Both results imply that between 2 and 3 digits of precision could be lost. This is confirmed by the solution where we see that round-off error occurs in the last two or three digits.

PROBLEMS

11.1 Perform the same calculations as in Example 11.1, but for the tridiagonal system,

$$\begin{bmatrix} 2 & -1 & \\ -1 & 2 & -1 \\ & -1 & 2 \end{bmatrix} \begin{Bmatrix} x_1 \\ x_2 \\ x_3 \end{Bmatrix} = \begin{Bmatrix} 124 \\ 4 \\ 14 \end{Bmatrix}$$

11.2 Determine the matrix inverse for Example 11.1 based on the LU decomposition and unit vectors.

11.3 The following tridiagonal system must be solved as part of a larger algorithm (Crank-Nicolson) for solving partial differential equations:

$$\begin{bmatrix} 2.01475 & -0.02875 & & \\ -0.02875 & 2.01475 & -0.02875 & \\ & -0.02875 & 2.01475 & -0.02875 \\ & & -0.02875 & 2.01475 \end{bmatrix}$$

$$\times \begin{Bmatrix} T_1 \\ T_2 \\ T_3 \\ T_4 \end{Bmatrix} = \begin{Bmatrix} 4.175 \\ 0 \\ 0 \\ 2.0875 \end{Bmatrix}$$

Use the Thomas algorithm to obtain a solution.

11.4 Confirm the validity of the Cholesky decomposition of Example 11.2 by substituting the results into Eq. (11.2) to see if the product of $[L]$ and $[L]^T$ yields $[A]$.

11.5 Perform the same calculations as in Example 11.2, but for the symmetric system,

$$\begin{bmatrix} 6 & 16.5 & 14 \\ 16.5 & 76.25 & 48 \\ 14 & 48 & 54 \end{bmatrix} \begin{Bmatrix} a_0 \\ a_1 \\ a_2 \end{Bmatrix} = \begin{Bmatrix} 54 \\ 243.5 \\ 100 \end{Bmatrix}$$

In addition to solving for the LU decomposition, employ it to solve for the a's.

11.6 Use the Gauss-Seidel method to solve the tridiagonal system from Prob. 11.1 ($\varepsilon_s = 5\%$).

11.7 Recall from Prob. 10.9, that the following system of equations is designed to determine concentrations (the c's in g/m^3) in a series of coupled reactors as a function of amount of mass input to each reactor (the right-hand sides in g/d),

$$17c_1 - 2c_2 - 3c_3 = 500$$
$$-5c_1 + 21c_2 - 2c_3 = 200$$
$$-5c_1 - 5c_2 + 22c_3 = 30$$

Solve this problem with the Gauss-Seidel method to $\varepsilon_s = 5\%$.

11.8 Repeat Prob. 11.7, but use Jacobi iteration.

11.9 Use the Gauss-Seidel method with relaxation to solve ($\lambda = 0.90$ and $\varepsilon_s = 5\%$):

$$-5x_1 + 12x_3 = 80$$
$$4x_1 - x_2 - x_3 = -2$$
$$6x_1 + 8x_2 = 45$$

If necessary, make sure to rearrange the equations to achieve convergence.

11.10 Redraw Fig. 11.5 for the case where the slopes of the equations are plus and minus unity. What is the result of applying Gauss-Seidel to such a system?

11.11 Use the software library or package of your choice to (a) obtain a solution, (b) calculate the inverse, and (c) determine the condition number (without scaling) for,

(i)

$$\begin{bmatrix} 1 & 4 & 9 \\ 4 & 9 & 16 \\ 9 & 16 & 25 \end{bmatrix} \begin{Bmatrix} x_1 \\ x_2 \\ x_3 \end{Bmatrix} = \begin{Bmatrix} 14 \\ 29 \\ 50 \end{Bmatrix}$$

(ii)

$$\begin{bmatrix} 1 & 4 & 9 & 16 \\ 4 & 9 & 16 & 25 \\ 9 & 16 & 25 & 36 \\ 16 & 25 & 36 & 49 \end{bmatrix} \begin{Bmatrix} x_1 \\ x_2 \\ x_3 \\ x_4 \end{Bmatrix} = \begin{Bmatrix} 30 \\ 54 \\ 86 \\ 126 \end{Bmatrix}$$

In both cases (i) and (ii), the answers for all the x's should be 1.

11.12 Modify the program from Example 11.6 so that, rather than calculating the inverse, it is designed as a subroutine to return an individual solution for a single right-hand side vector.

11.13 Develop a user-friendly subprogram in either a high-level or macro language of your choice for Cholesky decomposition based on Fig. 11.3. Test your program by duplicating the results of Example 11.2

11.14 Develop a user-friendly subprogram in either a high-level or macro language of your choice for the Gauss-Seidel method based on Fig. 11.6. Test your program by duplicating the results of Example 11.3.

CHAPTER 12

Engineering Applications: Linear Algebraic Equations

The purpose of this chapter is to use the numerical procedures discussed in Chaps. 9, 10, and 11 to solve systems of linear algebraic equations for some engineering applications. These systematic numerical techniques have practical significance because engineers frequently encounter problems involving systems of equations that are too large to solve by hand. The numerical algorithms in these applications are particularly convenient to implement on personal computers.

Section 12.1 shows how a mass balance can be employed to model a system of reactors. *Section 12.2* places special emphasis on the use of the matrix inverse to determine the complex cause-effect interactions between forces in the members of a truss. *Section 12.3* is an example of the use of Kirchhoff's laws to compute the currents and voltages in a resistor circuit. Finally, *Section 12.4* is an illustration of how linear equations are employed to determine the steady-state configuration of a mass-spring system.

12.1 STEADY-STATE ANALYSIS OF A SYSTEM OF REACTORS (CHEMICAL/PETROLEUM ENGINEERING)

Background. One of the most important organizing principles in chemical engineering is the *conservation of mass* (recall Table 1.1). In quantitative terms, the principle is expressed as a mass balance that accounts for all sources and sinks of a material that pass in and out of a volume (Fig. 12.1). Over a finite period of time, this can be expressed as

$$\text{Accumulation} = \text{inputs} - \text{outputs} \qquad (12.1)$$

The mass balance represents a bookkeeping exercise for the particular substance being modeled. For the period of the computation, if the inputs are greater than the outputs, the mass of the substance within the volume increases. If the outputs are greater than the inputs, the mass decreases. If inputs are equal to the outputs, accumulation is zero and mass remains constant. For this stable condition, or steady state, Eq. (12.1) can be expressed as

$$\text{Inputs} = \text{outputs} \qquad (12.2)$$

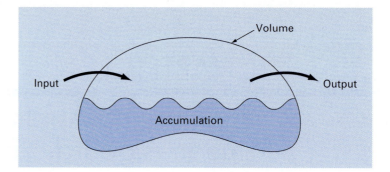

FIGURE 12.1
A schematic representation of mass balance.

Employ the conservation of mass to determine the steady-state concentrations of a system of coupled reactors.

Solution.   The mass balance can be used for engineering problem solving by expressing the inputs and outputs in terms of measurable variables and parameters. For example, if we were performing a mass balance for a conservative substance (that is, one that does not increase or decrease due to chemical transformations) in a reactor (Fig. 12.2), we would have to quantify the rate at which mass flows into the reactor through the two inflow pipes and out of the reactor through the outflow pipe. This can be done by taking the product of the flow rate Q (in cubic meters per minute) and the concentration c (in milligrams per cubic meter) for each pipe. For example, for pipe 1 in Fig. 12.2, $Q_1 = 2$ m^3/min and $c_1 = 25$ mg/m^3; therefore the rate at which mass flows into the reactor through pipe 1 is $Q_1c_1 = (2$ m^3/min$)(25$ mg/m$^3) = 50$ mg/min. Thus, 50 mg of chemical flows into the reactor through this pipe each minute. Similarly, for pipe 2 the mass inflow rate can be calculated as $Q_2c_2 = (1.5$ m^3/min$)(10$ mg/m$^3) = 15$ mg/min.

Notice that the concentration out of the reactor through pipe 3 is not specified by Fig. 12.2. This is because we already have sufficient information to calculate it on the basis of the conservation of mass. Because the reactor is at steady state, Eq. (12.2) holds and the inputs should be in balance with the outputs, as in

$$Q_1c_1 + Q_2c_2 = Q_3c_3$$

Substituting the given values into this equation yields

$$50 + 15 = 3.5c_3$$

which can be solved for $c_3 = 18.6$ mg/m^3. Thus, we have determined the concentration in the third pipe. However, the computation yields an additional bonus. Because the reactor is well mixed (as represented by the propeller in Fig. 12.2), the concentration will be uniform, or homogeneous, throughout the tank. Therefore the concentration in pipe 3 should be identical to the concentration throughout the reactor. Consequently, the mass balance has allowed us to compute both the concentration in the reactor and in the outflow pipe. Such

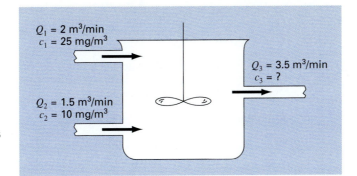

FIGURE 12.2
A steady-state, completely mixed reactor with two inflow pipes and one outflow pipe. The flows Q are in cubic meters per minute, and the concentrations c are in milligrams per cubic meter.

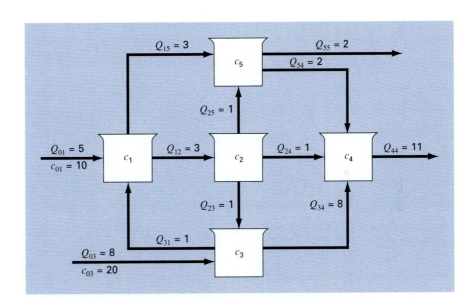

FIGURE 12.3
Five reactors linked by pipes.

information is of great utility to chemical and petroleum engineers who must design reactors to yield mixtures of a specified concentration.

Because simple algebra was used to determine the concentration for the single reactor in Fig. 12.2, it might not be obvious how computers figure in mass-balance calculations. Figure 12.3 shows a problem setting where computers are not only useful but are a practical necessity. Because there are five interconnected, or coupled, reactors, five simultaneous mass-balance equations are needed to characterize the system. For reactor 1, the rate of mass flow in is

$$5(10) + Q_{31}c_3$$

and the rate of mass flow out is

$$Q_{12}c_1 + Q_{15}c_1$$

Because the system is at steady state, the inflows and outflows must be equal:

$$5(10) + Q_{31}c_3 = Q_{12}c_1 + Q_{15}c_1$$

or, substituting the values for flow from Fig. 12.3,

$$6c_1 - c_3 = 50$$

Similar equations can be developed for the other reactors:

$$-3c_1 + 3c_2 = 0$$
$$-c_2 + 9c_3 = 160$$
$$-c_2 - 8c_3 + 11c_4 - 2c_5 = 0$$
$$-3c_1 - c_2 + 4c_5 = 0$$

A numerical method can be used to solve these five equations for the five unknown concentrations:

$$\{C\}^T = \lfloor 11.51 \quad 11.51 \quad 19.06 \quad 17.00 \quad 11.51 \rfloor$$

In addition, the matrix inverse can be computed as

$$[A]^{-1} = \begin{bmatrix} 0.16981 & 0.00629 & 0.01887 & 0 & 0 \\ 0.16981 & 0.33962 & 0.01887 & 0 & 0 \\ 0.01887 & 0.03774 & 0.11321 & 0 & 0 \\ 0.06003 & 0.07461 & 0.08748 & 0.09091 & 0.04545 \\ 0.16981 & 0.08962 & 0.01887 & 0 & 0.25000 \end{bmatrix}$$

Each of the elements a_{ij} signifies the change in concentration of reactor i due to a unit change in loading to reactor j. Thus, the zeros in column 4 indicate that a loading to reactor 4 will have no impact on reactors 1, 2, 3, and 5. This is consistent with the system configuration (Fig. 12.3), which indicates that flow out of reactor 4 does not feed back into any of the other reactors. In contrast, loadings to any of the first three reactors will affect the entire system as indicated by the lack of zeros in the first three columns. Such information is of great utility to engineers who design and manage such systems.

12.2 ANALYSIS OF A STATICALLY DETERMINATE TRUSS (CIVIL/ENVIRONMENTAL ENGINEERING)

Background.   An important problem in structural engineering is that of finding the forces and reactions associated with a statically determinate truss. Figure 12.4 shows an example of such a truss.

The forces (F) represent either tension or compression on the members of the truss. External reactions (H_2, V_2, and V_3) are forces that characterize how the truss interacts with the supporting surface. The hinge at node 2 can transmit both horizontal and vertical forces to the surface, whereas the roller at node 3 transmits only vertical forces. It is observed that the effect of the external loading of 1000 lb is distributed among the various members of the truss.

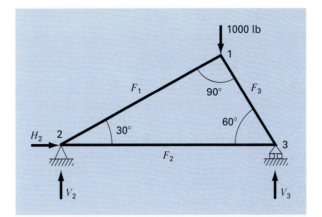

FIGURE 12.4
Forces on a statically
determinate truss.

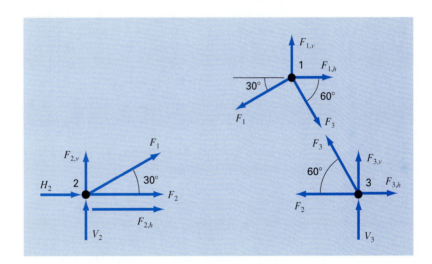

FIGURE 12.5
Free-body force diagrams for
the nodes of a statically
determinate truss.

Solution.   This type of structure can be described as a system of coupled linear algebraic equations. Free-body force diagrams are shown for each node in Fig. 12.5. The sum of the forces in both horizontal and vertical directions must be zero at each node, because the system is at rest. Therefore, for node 1,

$$\Sigma F_H = 0 = -F_1 \cos 30° + F_3 \cos 60° + F_{1,h} \tag{12.3}$$

$$\Sigma F_V = 0 = -F_1 \sin 30° - F_3 \sin 60° + F_{1,v} \tag{12.4}$$

for node 2,

$$\Sigma F_H = 0 = F_2 + F_1 \cos 30° + F_{2,h} + H_2 \tag{12.5}$$

$$\Sigma F_V = 0 = F_1 \sin 30° + F_{2,v} + V_2 \tag{12.6}$$

for node 3,

$$\Sigma F_H = 0 = -F_2 - F_3 \cos 60° + F_{3,h} \tag{12.7}$$

$$\Sigma F_V = 0 = F_3 \sin 60° + F_{3,v} + V_3 \tag{12.8}$$

where $F_{i,h}$ is the external horizontal force applied to node i (where a positive force is from left to right) and $F_{1,v}$ is the external vertical force applied to node i (where a positive force is upward). Thus, in this problem, the 1000-lb downward force on node 1 corresponds to $F_{1,v} = -1000$. For this case all other $F_{i,v}$'s and $F_{i,h}$'s are zero. Note that the directions of the internal forces and reactions are unknown. Proper application of Newton's laws requires only consistent assumptions regarding direction. Solutions are negative if the directions are assumed incorrectly. Also note that in this problem, the forces in all members are assumed to be in tension and act to pull adjoining nodes together. A negative solution therefore corresponds to compression. This problem can be written as the following system of six equations and six unknowns:

$$
\begin{bmatrix}
0.866 & 0 & -0.5 & 0 & 0 & 0 \\
0.5 & 0 & 0.866 & 0 & 0 & 0 \\
-0.866 & -1 & 0 & -1 & 0 & 0 \\
-0.5 & 0 & 0 & 0 & -1 & 0 \\
0 & 1 & 0.5 & 0 & 0 & 0 \\
0 & 0 & -0.866 & 0 & 0 & -1
\end{bmatrix}
\begin{Bmatrix}
F_1 \\ F_2 \\ F_3 \\ H_2 \\ V_2 \\ V_3
\end{Bmatrix}
=
\begin{Bmatrix}
0 \\ -1000 \\ 0 \\ 0 \\ 0 \\ 0
\end{Bmatrix}
\tag{12.9}
$$

Notice that, as formulated in Eq. (12.9); partial pivoting is required to avoid division by zero diagonal elements. Employing a pivot strategy, the system can be solved using any of the elimination techniques discussed in Chap. 9 or 10. However, because this problem is an ideal case study for demonstrating the utility of the matrix inverse, the *LU* decomposition can be used to compute

$$F_1 = -500 \qquad F_2 = 433 \qquad F_3 = -866$$
$$H_2 = 0 \qquad V_2 = 250 \qquad V_3 = 750$$

and the matrix inverse is

$$
[A]^{-1} =
\begin{bmatrix}
0.866 & 0.5 & 0 & 0 & 0 & \\
0.25 & -0.433 & 0 & 0 & 1 & 0 \\
-0.5 & 0.866 & 0 & 0 & 0 & 0 \\
-1 & 0 & -1 & 0 & -1 & 0 \\
-0.433 & -0.25 & 0 & -1 & 0 & 0 \\
0.433 & -0.75 & 0 & 0 & 0 & -1
\end{bmatrix}
$$

Now, realize that the right-hand-side vector represents the externally applied horizontal and vertical forces on each node, as in

$$\{F\}^T = \lfloor F_{1,h} \quad F_{1,v} \quad F_{2,h} \quad F_{2,v} \quad F_{3,h} \quad F_{3,v} \rfloor \tag{12.10}$$

Because the external forces have no effect on the *LU* decomposition, the method need not be implemented over and over again to study the effect of different external forces on the truss. Rather, all that we have to do is perform the forward- and backward-substitution steps for each right-hand-side vector to efficiently obtain alternative solutions. For example,

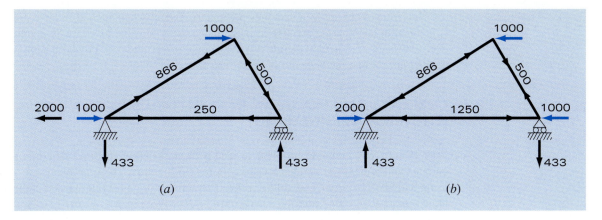

FIGURE 12.6
Two test cases showing (a) winds from the left and (b) winds from the right.

we might want to study the effect of horizontal forces induced by a wind blowing from left to right. If the wind force can be idealized as two point forces of 1000 lb on nodes 1 and 2 (Fig. 12.6a), the right-hand-side vector is

$$\{F\}^T = \lfloor -1000 \quad 0 \quad 1000 \quad 0 \quad 0 \quad 0 \rfloor$$

which can be used to compute

$$F_1 = -866 \qquad F_2 = 250 \qquad F_3 = -500$$
$$H_2 = -2000 \qquad V_2 = -433 \qquad V_3 = 433$$

For a wind from the right (Fig. 12.6b), $F_{1,h} = -1000$, $F_{3,h} = -1000$, and all other external forces are zero, with the result that

$$F_1 = -866 \qquad F_2 = -1250 \qquad F_3 = 500$$
$$H_2 = 2000 \qquad V_2 = 433 \qquad V_3 = -433$$

The results indicate that the winds have markedly different effects on the structure. Both cases are depicted in Fig. 12.6.

The individual elements of the inverted matrix also have direct utility in elucidating stimulus-response interactions for the structure. Each element represents the change of one of the unknown variables to a unit change of one of the external stimuli. For example, element a_{32}^{-1} indicates that the third unknown (F_3) will change 0.866 due to a unit change of the second external stimulus ($F_{1,v}$). Thus, if the vertical load at the first node were increased by 1, F_3 would increase by 0.866. The fact that elements are 0 indicates that certain unknowns are unaffected by some of the external stimuli. For instance $a_{13}^{-1} = 0$ means that F_1 is unaffected by changes in $F_{2,h}$. This ability to isolate interactions has a number of engineering applications, including the identification of those components that are most sensitive to external stimuli and, as a consequence, most prone to failure. In addition, it can be used to determine components that may be unnecessary (see Prob. 12.16).

The foregoing approach becomes particularly useful when applied to large complex structures. In engineering practice, it may be necessary to solve trusses with hundreds or even thousands of structural members. Linear equations provide one powerful approach for gaining insight into the behavior of these structures.

12.3 CURRENTS AND VOLTAGES IN RESISTOR CIRCUITS (ELECTRICAL ENGINEERING)

Background. A common problem in electrical engineering involves determining the currents and voltages at various locations in resistor circuits. These problems are solved using Kirchhoff's current and voltage rules. The current (or point) rule states that the algebraic sum of all currents entering a node must be zero (see Fig. 12.7*a*), or

$$\Sigma i = 0 \tag{12.11}$$

where all current entering the node is considered positive in sign. The current rule is an application of the principle of conservation of charge (recall Table 1.1).

The voltage (or loop) rule specifies that the algebraic sum of the potential differences (that is, voltage changes) in any loop must equal zero. For a resistor circuit, this is expressed as

$$\Sigma \xi - \Sigma iR = 0 \tag{12.12}$$

where ξ is the emf (electromotive force) of the voltage sources and R is the resistance of any resistors on the loop. Note that the second term derives from Ohm's law (Fig. 12.7*b*), which states that the voltage drop across an ideal resistor is equal to the product of the current and the resistance. Kirchhoff's voltage rule is an expression of the conservation of energy.

Solution. Application of these rules results in systems of simultaneous linear algebraic equations because the various loops within a circuit are coupled. For example, consider the circuit shown in Fig. 12.8. The currents associated with this circuit are unknown both in magnitude and direction. This presents no great difficulty because one simply assumes a direction for each current. If the resultant solution from Kirchhoff's laws is negative, then the assumed direction was incorrect. For example, Fig. 12.9 shows some assumed currents.

FIGURE 12.7
Schematic representations of (a) Kirchhoff's current rule and (b) Ohm's law.

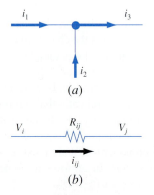

(a)

(b)

FIGURE 12.8
A resistor circuit to be solved using simultaneous linear algebraic equations.

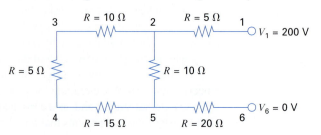

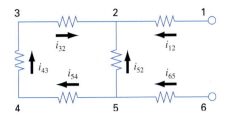

FIGURE 12.9
Assumed currents.

Given these assumptions, Kirchhoff's current rule is applied at each node to yield

$$i_{12} + i_{52} + i_{32} = 0$$
$$i_{65} - i_{52} - i_{54} = 0$$
$$i_{43} - i_{32} = 0$$
$$i_{54} - i_{43} = 0$$

Application of the voltage rule to each of the two loops gives

$$-i_{54}R_{54} - i_{43}R_{43} - i_{32}R_{32} + i_{52}R_{52} = 0$$
$$-i_{65}R_{65} - i_{52}R_{52} + i_{12}R_{12} - 200 = 0$$

or, substituting the resistances from Fig. 12.8 and bringing constants to the right-hand side,

$$-15i_{54} - 5i_{43} - 10i_{32} + 10i_{52} = 0$$
$$-20i_{65} - 10i_{52} + 5i_{12} = 200$$

Therefore the problem amounts to solving the following set of six equations with six unknown currents:

$$
\begin{bmatrix}
1 & 1 & 1 & 0 & 0 & 0 \\
0 & -1 & 0 & 1 & -1 & 0 \\
0 & 0 & -1 & 0 & 0 & 1 \\
0 & 0 & 0 & 0 & 1 & -1 \\
0 & 10 & -10 & 0 & -15 & -5 \\
5 & -10 & 0 & -20 & 0 & 0
\end{bmatrix}
\begin{Bmatrix}
i_{12} \\ i_{52} \\ i_{32} \\ i_{65} \\ i_{54} \\ i_{43}
\end{Bmatrix}
=
\begin{Bmatrix}
0 \\ 0 \\ 0 \\ 0 \\ 0 \\ 200
\end{Bmatrix}
$$

Although impractical to solve by hand, this system is easily handled using an elimination method. Proceeding in this manner, the solution is

$$i_{12} = 6.1538 \qquad i_{52} = -4.6154 \qquad i_{32} = -1.5385$$
$$i_{65} = -6.1538 \qquad i_{54} = -1.5385 \qquad i_{43} = -1.5385$$

Thus, with proper interpretation of the signs of the result, the circuit currents and voltages are as shown in Fig. 12.10. The advantages of using numerical algorithms and computers for problems of this type should be evident.

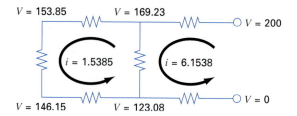

FIGURE 12.10
The solution for currents and voltages obtained using an elimination method.

12.4 SPRING-MASS SYSTEMS (MECHANICAL/AEROSPACE ENGINEERING)

Background.   Idealized spring-mass systems play an important role in mechanical and other engineering problems. Figure 12.11 shows such a system. After they are released, the masses are pulled downward by the force of gravity. Notice that the resulting displacement of each spring in Fig. 12.11b is measured along local coordinates referenced to its initial position in Fig. 12.11a.

As introduced in Chap. 1, Newton's second law can be employed in conjunction with force balances to develop a mathematical model of the system. For each mass, the second law can be expressed as

$$m\frac{d^2x}{dt^2} = F_D - F_U \tag{12.13}$$

To simplify the analysis we will assume that all the springs are identical and follow Hooke's law. A free-body diagram for the first mass is depicted in Fig. 12.12a. The upward force is merely a direct expression of Hooke's law:

$$F_U = kx_1 \tag{12.14}$$

The downward component consists of the two spring forces along with the action of gravity on the mass,

$$F_D = k(x_2 - x_1) + k(x_2 - x_1) + m_1 g \tag{12.15}$$

Note how the force component of the two springs is proportional to the displacement of the second mass, x_2, corrected for the displacement of the first mass, x_1.

Equations (12.14) and (12.15) can be substituted into Eq. (12.13) to give

$$m_1\frac{d^2x_1}{dt^2} = 2k(x_2 - x_1) + m_1 g - kx_1 \tag{12.16}$$

Thus, we have derived a second-order ordinary differential equation to describe the displacement of the first mass with respect to time. However, notice that the solution cannot be obtained because the model includes a second dependent variable, x_2. Consequently, free-body diagrams must be developed for the second and the third masses (Fig. 12.12b

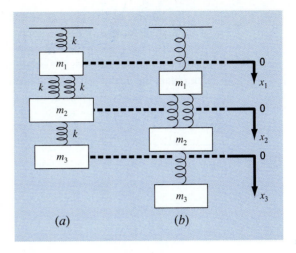

FIGURE 12.11
A system composed of three masses suspended vertically by a series of springs. (a) The system before release, that is, prior to extension or compression of the springs. (b) The system after release. Note that the positions of the masses are referenced to local coordinates with origins at their position before release.

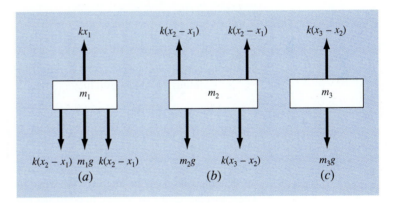

FIGURE 12.12
Free-body diagrams for the three masses from Fig. 12.11.

and c) that can be employed to derive

$$m_2 \frac{d^2 x_2}{dt^2} = k(x_3 - x_2) + m_2 g - 2k(x_2 - x_1) \tag{12.17}$$

and

$$m_3 \frac{d^2 x_3}{dt^2} = m_3 g - k(x_3 - x_2) \tag{12.18}$$

Equations (12.16), (12.17), and (12.18) form a system of three differential equations with three unknowns. With the appropriate initial conditions, they could be used to solve for the displacements of the masses as a function of time (that is, their oscillations). We will discuss numerical methods for obtaining such solutions in Part Seven. For the present, we can obtain the displacements that occur when the system eventually comes to rest, that is, to the steady state. To do this, the derivatives in Eqs. (12.16), (12.17), and (12.18) are set to zero to give

$$
\begin{aligned}
3kx_1 - 2kx_2 &= m_1 g \\
-2kx_1 + 3kx_2 - kx_3 &= m_2 g \\
- kx_2 + kx_3 &= m_3 g
\end{aligned}
$$

or, in matrix form,

$$[K]\{X\} = \{W\}$$

where $[K]$, called the *stiffness matrix*, is

$$
[K] = \begin{bmatrix}
3k & -2k & \\
-2k & 3k & -k \\
& -k & k
\end{bmatrix}
$$

and $\{X\}$ and $\{W\}$ are the column vectors of the unknowns X and the weights mg, respectively.

Solution. At this point, numerical methods can be employed to obtain a solution. If $m_1 = 2$ kg, $m_2 = 3$ kg, $m_3 = 2.5$ kg, and the k's $= 10$ kg/s^2, use LU decomposition to solve for the displacements and generate the inverse of $[K]$.

Substituting the model parameters gives

$$
[K] = \begin{bmatrix}
30 & -20 & \\
-20 & 30 & -10 \\
& -10 & 10
\end{bmatrix}
\qquad
\{W\} = \begin{Bmatrix}
19.6 \\
29.4 \\
24.5
\end{Bmatrix}
$$

LU decomposition can be employed to solve for $x_1 = 7.35$, $x_2 = 10.045$, and $x_3 = 12.495$. These displacements were used to construct Fig. 12.11b. The inverse of the stiffness matrix is computed as

$$
[K] = \begin{bmatrix}
0.1 & 0.1 & 0.1 \\
0.1 & 0.15 & 0.15 \\
0.1 & 0.15 & 0.25
\end{bmatrix}
$$

Each element of this matrix k_{ji}^{-1} tells us the displacement of mass i due to a unit force imposed on mass j. Thus, the values of 0.1 in column 1 tell us that a downward unit load to the first mass will displace all of the masses 0.1 m downward. The other elements can be interpreted in a similar fashion. Therefore, the inverse of the stiffness matrix provides a fundamental summary of how the system's components respond to externally applied forces.

PROBLEMS

Chemical/Petroleum Engineering

12.1 Perform the same computation as in Sec. 12.1, but change c_{01} to 20 and c_{03} to 6.

12.2 If the input to reactor 1 in Sec. 12.1 is decreased 25 percent, what is the percent change in the concentration of reactors 2 and 3?

12.3 Because the system shown in Fig. 12.3 is at steady state, what can be said regarding the four flows: Q_{01}, Q_{03}, Q_{44}, and Q_{55}?

12.4 Recompute the concentrations for the five reactors shown in Fig. 12.2, if the flows are changed to:

$$Q_{01} = 5 \quad Q_{31} = 2 \quad Q_{25} = 3 \quad Q_{23} = 1$$
$$Q_{15} = 3 \quad Q_{55} = 4 \quad Q_{54} = 2 \quad Q_{34} = 9$$
$$Q_{12} = 4 \quad Q_{03} = 10 \quad Q_{24} = 0$$

12.5 Solve the same system as specified in Prob. 12.4, but set Q_{12} and Q_{54} equal to zero. Use conservation of flow to recompute the values for the other flows. What does the answer indicate to you regarding the physical system?

12.6 Figure P12.6 shows three reactors linked by pipes. As indicated, the rate of transfer of chemicals through each pipe is equal to a flow rate (Q, with units of cubic meters per second) multiplied by the concentration of the reactor from which the flow originates (c, with units of milligrams per cubic meter). If the system is at a steady state, the transfer into each reactor will balance the transfer out. Develop mass-balance equations for the reactors and solve the three simultaneous linear algebraic equations for their concentrations.

12.7 Employing the same basic approach as in Sec. 12.1, determine the concentration of chloride in each of the Great Lakes using the information shown in Fig. P12.7.

12.8 A stage extraction process is depicted in Fig. P12.8. In such systems, a stream containing a weight fraction Y_{in} of a chemical enters from the left at a mass flow rate of F_1. Simultaneously, a solvent carrying a weight fraction X_{in} of the same chemical enters from the right at a flow rate of F_2. Thus, for stage i, a mass balance can be represented as

$$F_1 Y_{i-1} + F_2 X_{i+1} = F_1 Y_i + F_2 X_i \qquad (P12.8a)$$

At each stage, an equilibrium is assumed to be established between Y_i and X_i as in

$$K = \frac{X_i}{Y_i} \qquad (P12.8b)$$

where K is called a distribution coefficient. Equation (P12.8b) can be solved for X_i and substituted into Eq. (P12.8a) to yield

$$Y_{i-1} - \left(1 + \frac{F_2}{F_1} K\right) Y_i + \left(\frac{F_2}{F_1} K\right) Y_{i+1} = 0 \qquad (P12.8c)$$

If $F_1 = 400$ kg/h, $Y_{in} = 0.1$, $F_2 = 800$ kg/h, $X_{in} = 0$, and $K = 5$, determine the values of Y_{out} and X_{out} if a five-stage reactor is used. Note that Eq. (P12.8c) must be modified to account for the inflow weight fractions when applied to the first and last stages.

Civil/Environmental Engineering

12.9 A civil engineer involved in construction requires 4800, 5810, and 5690 m^3 of sand, fine gravel, and coarse gravel, respectively, for a building project. There are three pits from which these materials can be obtained. The composition of these pits is

	Sand %	Fine Gravel %	Coarse Gravel %
Pit 1	52	30	18
Pit 2	20	50	30
Pit 3	25	20	55

FIGURE P12.6
Three reactors linked by pipes. The rate of mass transfer through each pipe is equal to the product of flow Q and concentration c of the reactor from which the flow originates.

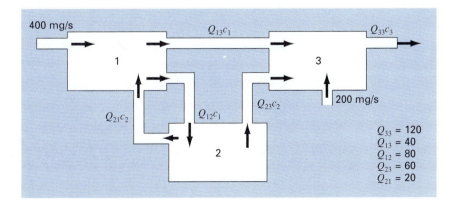

400 mg/s $Q_{13}c_1$ $Q_{33}c_3$

1   3

$Q_{21}c_2$ $Q_{12}c_1$ $Q_{23}c_2$ 200 mg/s

2

$Q_{33} = 120$
$Q_{13} = 40$
$Q_{12} = 80$
$Q_{23} = 60$
$Q_{21} = 20$

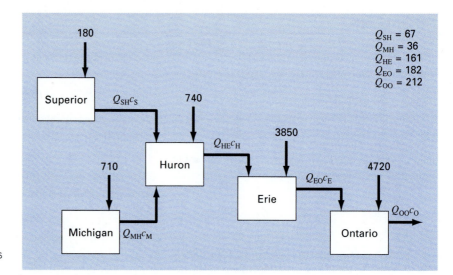

FIGURE P12.7
A chloride balance for the Great Lakes. Numbered arrows are direct inputs.

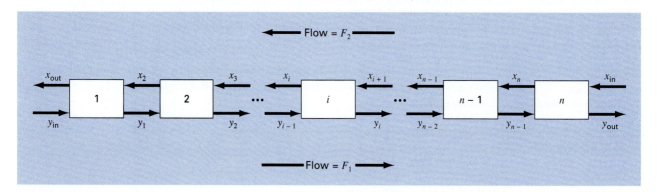

FIGURE P12.8
A state extraction process.

How many cubic meters must be hauled from each pit in order to meet the engineer's needs?

12.10 Perform the same computation as in Sec. 12.2, but change the angle at node 2 to $40°$ and at node 3 to $50°$, and the force at node 1 to 750 lbs.

12.11 Perform the same computation as in Sec. 12.2, but for the truss depicted in Fig. P12.11.

12.12 Perform the same computation as in Sec. 12.2, but for the truss depicted in Fig. P12.12.

FIGURE P12.11

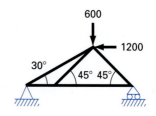

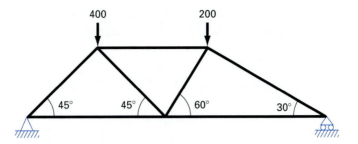

FIGURE P12.12

12.13 Calculate the forces and reactions for the truss in Fig. 12.4 if a downward force of 2200 kg and a horizontal force to the right of 1800 kg are applied at node 1.

12.14 In the example for Fig. 12.4, where a 1000-lb downward force is applied at node 1, the external reactions V_2 and V_3 were calculated. But if the lengths of the truss members had been given, we could have calculated V_2 and V_3 by utilizing the fact that $V_2 + V_3$ must equal 1000 and by summing moments around node 2. However, because we do know V_2 and V_3, we can work backward to solve for the lengths of the truss members. Note that because there are three unknown lengths and only two equations, we can solve for only the relationship between lengths. Solve for this relationship.

12.15 Employing the same methods as used to analyze Fig. 12.4, determine the forces and reactions for the truss shown in Fig. P12.15.

12.16 Solve for the forces and reaction for the truss in Fig. P12.16. Determine the matrix inverse for the system. Does the vertical-member force in the middle member seem reasonable? Why?

12.17 As the name implies, indoor air pollution deals with air contamination in enclosed spaces such as homes, offices, work areas, etc. Suppose that you are designing a ventilation system for a restaurant as shown in Fig. P12.17. The restaurant serving area consists of two square rooms and one elongated room. Room 1 and room 3 have sources of carbon monoxide from smokers and a faulty grill. Steady-state mass balances can be written for each room. For example for the smoking section (Room 1), the balance can be written as

$$0 = W_{smoker} + \quad Q_a c_a \quad - \quad Q_a c_1 \quad + E_{13}(c_3 - c_1)$$
$$\text{(load)} \quad + \text{(inflow)} - \text{(outflow)} + \quad \text{(mixing)}$$

or substituting the parameters

$$225c_1 - 25c_3 = 1400$$

Similar balances can be written for the other rooms.
(a) Solve for the steady-state concentration of carbon monoxide in each room.

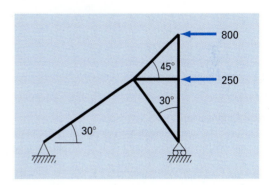

FIGURE P12.15

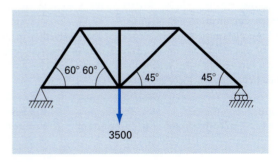

FIGURE P12.16

(b) Determine what percent of the carbon monoxide in the kids' section is due to (*i*) the smokers, (*ii*) the grill, and (*iii*) the air in the intake vents.

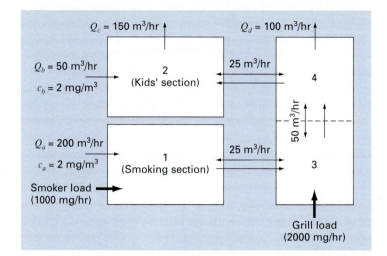

FIGURE P12.17

Overhead view of rooms in a restaurant. The one-way arrows represent volumetric airflows whereas the two-way arrows represent diffusive mixing. The smoker and grill loads add carbon monoxide mass to the system but negligible airflow.

Electrical Engineering

12.18 Perform the same computation as in Sec. 12.3, but change the resistance between nodes 3 and 4 to 25 Ω and change V_6 to 60 V.

12.19 Perform the same computation as in Sec. 12.3, but for the circuit depicted in Fig. P12.19.

12.20 Perform the same computation as in Sec. 12.3, but for the circuit depicted in Fig P12.20.

12.21 Solve the resistor circuit in Fig. 12.8, using Gauss elimination, if $V_1 = 180$ V and $R_{56} = 50$ ohms.

12.22 Solve the circuit in Fig. P12.22 for the currents in each wire. Use Gauss elimination with pivoting.

12.23 An electrical engineer supervises the production of three types of electrical components. Three kinds of material—metal, plastic, and rubber—are required for production. The amounts needed to produce each component are

Component	Metal, g/component	Plastic, g/component	Rubber, g/component
1	15	0.25	1.0
2	17	0.33	1.2
3	19	0.42	1.6

If totals of 2.12, 0.0434, and 0.164 kg of metal, plastic, and rubber, respectively, are available each day, how many components can be produced per day?

Mechanical/Aerospace Engineering

12.24 Perform the same computation as in Sec. 12.4, but triple the spring constants.

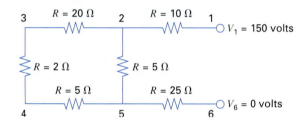

FIGURE P12.19

12.25 Perform the same computation as in Sec. 12.4, but add a third spring between masses 1 and 2 and double k.

12.26 Perform the same computation as in Sec. 12.4, but change the masses from 2, 3, and 2.5 kg to 15, 3, and 2 kg.

12.27 Idealized spring-mass systems have numerous applications throughout engineering. Figure P12.27 shows an arrangement of four springs in series being depressed with a force of 2000 kg. At equilibrium, force-balance equations can be developed defining the interrelationships between the springs;

$$k_2(x_2 - x_1) = k_1 x_1$$
$$k_3(x_3 - x_2) = k_2(x_2 - x_1)$$
$$k_4(x_4 - x_3) = k_3(x_3 - x_2)$$
$$F = k_4(x_4 - x_3)$$

where the k's are spring constants. If k_1 through k_4 are 150, 50, 75, and 225 kg/s^2, respectively, compute the x's.

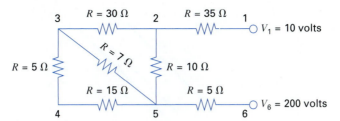

FIGURE P12.20

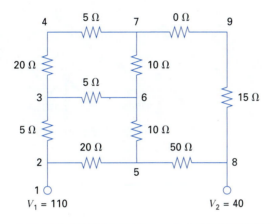

FIGURE P12.22

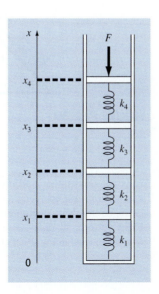

FIGURE P12.27

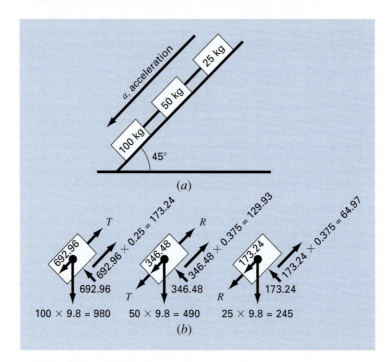

FIGURE P12.28

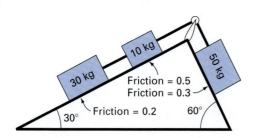

FIGURE P12.29

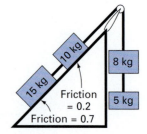

FIGURE P12.30

12.28 Three blocks are connected by a weightless cord and rest on an inclined plane (Fig. P12.28a). Employing a procedure similar to the one used in the analysis of the falling parachutists in Example 9.11 yields the following set of simultaneous equations (free-body diagrams are shown in Fig. P12.28b):

$$100a + T \qquad = 519.72$$
$$50a - T + R = 216.55$$
$$25a \qquad - R = 108.27$$

Solve for acceleration a and the tensions T and R in the two ropes.

12.29 Perform a computation similar to that called for in Prob. 12.28, but for the system shown in Fig. P12.29.

12.30 Perform the same computation as in Prob. 12.28, but for the system depicted in Fig. P12.30 (angles are $45°$).

EPILOGUE: PART THREE

PT3.4 TRADE-OFFS

Table PT3.2 provides a summary of the trade-offs involved in solving simultaneous linear algebraic equations. Two methods—graphical and Cramer's rule—are limited to small (≤ 3) numbers of equations and thus have little utility for practical problem solving. However, these techniques are useful didactic tools for understanding the behavior of linear systems in general.

The numerical methods themselves are divided into two general categories: exact and approximate methods. As the name implies, the former are intended to yield exact answers. However, because they are affected by round-off errors, they sometimes yield imprecise results. The magnitude of the round-off error varies from system to system and is dependent on a number of factors. These include the system's dimensions, its condition, and whether the matrix of coefficients is sparse or full. In addition, computer precision will affect round-off error.

It is recommended that a pivoting strategy be employed in any computer program implementing exact elimination methods. The inclusion of such a strategy minimizes round-off error and avoids problems such as division by zero. All other things being equal, *LU* decomposition–based algorithms are the methods of choice because of their efficiency and flexibility.

Although elimination methods have great utility, their use of the entire matrix of coefficients can be somewhat limiting when dealing with very large, sparse systems. This is due

TABLE PT3.2 Comparison of the characteristics of alternative methods for finding solutions of simultaneous linear algebraic equations.

Method	Stability	Precision	Breadth of Application	Programming Effort	Comments
Graphical	—	Poor	Limited	—	May take more time than the numerical method
Cramer's rule	—	Affected by round-off error	Limited	—	Excessive computational effort required for more than three equations
Gauss elimination (with partial pivoting)	—	Affected by round-off error	General	Moderate	
LU decomposition	—	Affected by round-off error	General	Moderate	Preferred elimination method; allows computation of matrix inverse
Gauss-Seidel	May not converge if not diagonally dominant	Excellent	Appropriate only for diagonally dominant systems	Easy	

to the fact that large portions of computer memory would be devoted to storage of meaningless zeros. For banded systems, techniques are available to implement elimination methods without having to store the entire coefficient matrix.

The approximate technique described in this book is called the Gauss-Seidel method. It differs from the exact techniques in that it employs an iterative scheme to obtain progressively closer estimates of the solution. Thus, the effect of round-off is a moot point with the Gauss-Seidel method, because the iterations can be continued as long as is necessary to obtain the desired precision. In addition, versions of the Gauss-Seidel method can be developed to efficiently utilize computer storage requirements for sparse systems. Consequently, the Gauss-Seidel technique has utility for large systems of equations, where storage requirements would pose significant problems for the exact techniques.

The disadvantage of the Gauss-Seidel method is that it does not always converge or sometimes converges slowly on the true solution. It is strictly reliable only for those systems that are diagonally dominant. However, relaxation methods are available that sometimes offset these disadvantages. In addition, because many sets of linear algebraic equations originating from physical systems exhibit diagonal dominance, the Gauss-Seidel method has great utility for engineering problem solving.

In summary, a variety of factors will bear on your choice of a technique for a particular problem involving linear algebraic equations. However, as outlined above, the size and sparseness of the system are particularly important factors in determining your choice.

PT3.5   IMPORTANT RELATIONSHIPS AND FORMULAS

Every part of this book includes a section that summarizes important formulas. Although Part Three does not really deal with single formulas, we have used Table PT3.3 to summarize the algorithms that were covered. The table provides an overview that should be helpful for review and in elucidating the major differences between the methods.

PT3.6   ADVANCED METHODS AND ADDITIONAL REFERENCES

General references on the solution of simultaneous linear equations can be found in Faddeev and Faddeeva (1963), Stewart (1973), Varga (1962), and Young (1971). Ralston and Rabinowitz (1978) provide a general summary.

Many advanced techniques are available to increase time and/or space savings of linear algebraic equations. Most of these focus on exploiting properties of the equations such as symmetry and bandedness. In particular, algorithms are available to operate on sparse matrices to convert them to a minimum banded format. Tewerson (1973) and Jacobs (1977) include information on this area. Once they are in a minimum banded format, there are a variety of efficient solution strategies that are employed such as the active column storage approach of Bathe and Wilson (1976).

Aside from $n \times n$ sets of equations, there are other systems where the number of equations, m, and number of unknowns, n, are not equal. Systems where $m < n$ are called *underdetermined*. In such cases there can be either no solution or else more than one. Sys-

TABLE PT3.3 Summary of important information presented in Part Three.

Method	Procedure	Potential Problems and Remedies
Gauss elimination	$\begin{bmatrix} a_{11} & a_{12} & a_{13} & \vert & c_1 \\ a_{21} & a_{22} & a_{23} & \vert & c_2 \\ a_{31} & a_{32} & a_{33} & \vert & c_3 \end{bmatrix} \Rightarrow \begin{bmatrix} a_{11} & a_{12} & a_{13} & \vert & c_1 \\ & a'_{22} & a'_{23} & \vert & c'_2 \\ & & a''_{33} & \vert & c''_3 \end{bmatrix} \Rightarrow$ $\begin{aligned} x_3 &= c''_3/a''_{33} \\ x_2 &= (c'_2 - a'_{23}x_3)/a'_{22} \\ x_1 &= (c_1 - a_{12}x_1 - a_{13}x_3)/a_{11} \end{aligned}$	Problems: Ill conditioning Round-off Division by zero Remedies: Higher precision Partial pivoting
LU decomposition	Decomposition Back Substitution $\begin{bmatrix} a_{11} & a_{12} & a_{13} \\ a_{21} & a_{22} & a_{23} \\ a_{31} & a_{32} & a_{33} \end{bmatrix} \Rightarrow \begin{bmatrix} 1 & 0 & 0 \\ l_{21} & 1 & 0 \\ l_{31} & l_{32} & 1 \end{bmatrix} \begin{Bmatrix} d_1 \\ d_2 \\ d_3 \end{Bmatrix} = \begin{Bmatrix} c_1 \\ c_2 \\ c_3 \end{Bmatrix} \Rightarrow \begin{bmatrix} u_{11} & u_{12} & u_{13} \\ 0 & u_{22} & u_{23} \\ 0 & 0 & u_{33} \end{bmatrix} \begin{Bmatrix} x_1 \\ x_2 \\ x_3 \end{Bmatrix} = \begin{Bmatrix} d_1 \\ d_2 \\ d_3 \end{Bmatrix} \Rightarrow \begin{Bmatrix} x_1 \\ x_2 \\ x_3 \end{Bmatrix}$ Forward Substitution	Problems: Ill conditioning Round-off Division by zero Remedies: Higher precision Partial pivoting
Gauss-Seidel method	$\left. \begin{aligned} x_1^j &= (c_1 - a_{12}x_2^{j-1} - a_{13}x_3^{j-1})/a_{11} \\ x_2^j &= (c_2 - a_{21}x_1^j - a_{23}x_3^{j-1})/a_{22} \\ x_3^j &= (c_3 - a_{31}x_1^j - a_{32}x_2^j)/a_{33} \end{aligned} \right\}$ continue iteratively until $\left\lvert \dfrac{x_i^j - x_i^{j-1}}{x_i^j} \right\rvert 100\% < \epsilon_s$ for all x_i's	Problems: Divergent or converges slowly Remedies: Diagonal dominance Relaxation

tems where $m > n$ are called *overdetermined*. For such situations, there is in general no exact solution. However, it is often possible to develop a compromise solution that attempts to determine answers that come "closest" to satisfying all the equations simultaneously. A common approach is to solve the equation in a "least squares" sense (Lawson and Hanson, 1974; Wilkinson and Reinsch, 1971). Alternatively, linear programming methods can be used where the equations are solved in an "optimal" sense by minimizing some objective function (Rabinowitz, 1968; Dantzig, 1963; and Luenberger, 1973). We describe this approach in detail in Chap. 15.

PART FOUR

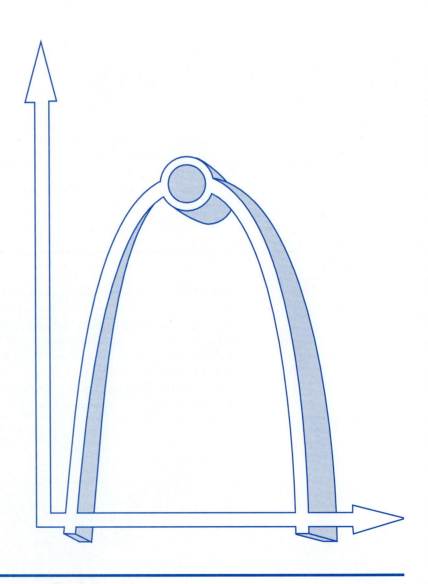

OPTIMIZATION

PT4.1 MOTIVATION

Root location (Part 2) and optimization are related in the sense that both involve guessing and searching for a point on a function,. The fundamental difference between the two types of problems is illustrated in Fig. PT4.1. Root location involves searching for zeros of a function or functions. In contrast, *optimization* involves searching for either the minimum or the maximum.

The optimum is the point where the curve is flat. In mathematical terms, this corresponds to the x value where the derivative $f'(x)$ is equal to zero. Additionally, the second derivative, $f''(x)$, indicates whether the optimum is a minimum or a maximum: if $f''(x) < 0$, the point is a maximum; if $f''(x) > 0$, the point is a minimum.

Now, understanding the relationship between roots and optima would suggest a possible strategy for finding the latter. That is, you can differentiate the function and locate the root (i.e., the zero) of the new function. In fact, some optimization methods seek to find an optima by solving the root problem: $f'(x) = 0$. It should be noted that such searches are often complicated because $f'(x)$ is not available analytically. Thus, one must sometimes use finite-difference approximations to estimate the derivative.

Beyond viewing optimization as a roots problem, it should be noted that the task of locating optima is aided by some extra mathematical structure that is not part of simple root finding. This tends to make optimization a more tractable task, particularly for multidimensional cases.

FIGURE PT4.1
A function of a single variable illustrating the difference between roots and optima.

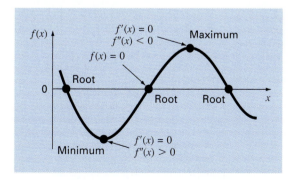

PT4.1.1 Noncomputer Methods and History

As mentioned above, differential calculus methods are still used to determine optimum solutions. All engineering and science students recall working maxima-minima problems by determining first derivatives of functions in their calculus courses. Bernoulli, Euler, Lagrange, and others laid the foundations of the calculus of variations, which deals with the minimization of functions. The Lagrange multiplier method was developed to optimize constrained problems; that is, optimization problems where the variables are bounded in some way.

The first major advances in numerical approaches occurred only with the development of digital computers after World War II. Koopmans in the United Kingdom and Kantorovich in the former Soviet Union independently worked on the general problem of least-cost distribution of supplies and products. In 1947, Koopman's student Dantzig invented the *simplex procedure* for solving linear programming problems. This approach paved the way for other methods of constrained optimization by a number of investigators, notably Charnes and his co-workers. Approaches for unconstrained optimization also developed rapidly following the widespread availability of computers.

PT4.1.2 Optimization and Engineering Practice

Most of the mathematical models we have dealt with to this point have been *descriptive* models. That is, they have been derived to simulate the behavior of an engineering device or system. In contrast, optimization typically deals with finding the "best result," or optimum solution, of a problem. Thus, in the context of modeling, they are often termed *prescriptive* models since they can be used to prescribe a course of action or the best design.

Engineers must continuously design devices and products that perform tasks in an efficient fashion. In doing so, they are constrained by the limitations of the physical world. Further, they must keep costs down. Thus, they are always confronting optimization problems that balance performance and limitations. Some common instances are listed in Table PT4.1. The following example has been developed to help you get a feel for the way in which such problems might be formulated.

TABLE PT4.1 Some common examples of optimization problems in engineering.

- Design aircraft for minimum weight and maximum strength.
- Optimal trajectories of space vehicles.
- Design civil engineering structures for minimum cost.
- Design water-resource projects like dams to mitigate flood damage while yielding maximum hydropower.
- Predict structural behavior by minimizing potential energy.
- Material-cutting strategy for minimum cost.
- Design pump and heat transfer equipment for maximum efficiency.
- Maximize power output of electrical networks and machinery while minimizing heat generation.
- Shortest route of salesperson visiting various cities during one sales trip.
- Optimal planning and scheduling.
- Statistical analysis and models with minimum error.
- Optimal pipeline networks.
- Inventory control.
- Maintenance planning to minimize cost.
- Minimize waiting and idling times.
- Design waste treatment systems to meet water-quality standards at least cost.

EXAMPLE PT4.1 Optimization of Parachute Cost

Problem Statement.   Throughout the rest of the book, we've used the falling parachutist to illustrate the basic problem areas of numerical methods. You may have noticed that none of these examples concentrate on what happens after the chute opens. In this example, we'll examine a case were the chute has opened and we are interested in predicting impact velocity at the ground.

You are an engineer working for an agency planning to airlift supplies to refugees in a war zone. The supplies will be dropped at low altitude (500 m) so that the drop is not detected and the supplies fall as close as possible to the refugee camp. The chutes open immediately upon leaving the plane. To reduce damage, the vertical velocity on impact must be below a critical value of $v_c = 20$ m/s.

The parachute used for the drop is depicted in Fig. PT4.2. The cross-sectional area of the chute is that of a half sphere,

$$A = 2\pi r^2 \tag{PT4.1}$$

The length of each of the 16 cords connecting the chute to the mass is related to the chute radius by

$$\ell = \sqrt{2}\,r \tag{PT4.2}$$

You know that the drag force for the chute is a linear function of its cross-sectional area described by the following formula

$$c = k_c A \tag{PT4.3}$$

where c = drag coefficient (kg/s) and k_c = a proportionality constant parameterizing the effect of area on drag [kg/(s·m^2)].

Also, you can divide the payload into as many parcels as you like. That is, the mass of each individual parcel can be calculated as

$$m = \frac{M_t}{n}$$

FIGURE PT4.2
A deployed parachute.

where m = mass of an individual parcel (kg), M_t = total load being dropped (kg), and n = total number of parcels.

Finally, the cost of each chute is related to chute size in a nonlinear fashion,

$$\text{Cost per chute} = c_0 + c_1 \ell + c_2 A^2 \tag{PT4.4}$$

where c_0, c_1, and c_2 = cost coefficients. The constant term, c_0, is the base price for the chutes. The nonlinear relationship with area is because larger chutes are much more difficult to construct than small chutes.

Determine the size (r) and number of chutes (n) that result in minimum cost while at the same time meeting the requirement of having a sufficiently small impact velocity.

Solution. The objective here is to determine the number and size of parachutes to minimize the cost of the airlift. The problem is constrained because the parcels must have an impact velocity less than a critical value.

The cost can be computed by multiplying the cost of the individual parachute [Eq. (PT4.4)] by the number of parachutes (n). Thus, the function you wish to minimize, which is formally called the *objective function,* is written as

$$\text{Minimize } C = n(c_0 + c_1 \ell + c_2 A^2) \tag{PT4.5}$$

where C = cost (\$) and A and ℓ are calculated by Eqs. (PT4.1) and (PT4.2), respectively.

Next, we must specify the *constraints.* For this problem there are two constraints. First, the impact velocity must be equal to or less than the critical velocity,

$$v \leq v_c \tag{PT4.6}$$

Second, the number of parcels must be an integer and greater than or equal to 1,

$$n \geq 1 \tag{PT4.7}$$

where n is an integer.

At this point, the optimization problem has been formulated. As can be seen, it is a nonlinear constrained problem.

Although the problem has been broadly formulated, one more issue must be addressed: How do we determine the impact velocity v? Recall from Chap. 1 that the velocity of a falling object can be computed with

$$v = \frac{gm}{c}(1 - e^{-(c/m)t}) \tag{1.10}$$

where v = velocity (m/s), g = acceleration of gravity (m/s^2), m = mass (kg), and t = time (s).

Although Eq. (1.10) provides a relationship between v and t, we need to know how long the mass falls. Therefore, we need a relationship between the drop distance z and the time of fall t. The drop distance can be calculated from the velocity in Eq. (1.10) by integration

$$z = \int_0^t \frac{gm}{c}(1 - e^{-(c/m)t}) \, dt \tag{PT4.8}$$

This integral can be evaluated to yield

$$z = z_0 - \frac{gm}{c}t + \frac{gm^2}{c^2}(1 - e^{-(c/m)t}) \tag{PT4.9}$$

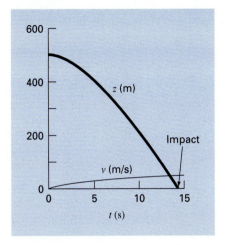

FIGURE PT4.3
The height z and velocity v of a deployed parachute as it falls to earth $(z = 0)$.

where z_0 = initial height (m). This function, as plotted in Fig. PT4.3, provides a way to predict z given knowledge of t.

However, we do not need z as a function of t to solve this problem. Rather, we need to compute the time required for the parcel to fall the distance z_0. Thus, we recognize that we must reformulate Eq. (PT4.9) as a root-finding problem. That is, we must solve for the time at which z goes to zero,

$$f(t) = 0 = z_0 - \frac{gm}{c}t + \frac{gm^2}{c^2}(1 - e^{-(c/m)t}) \tag{PT4.10}$$

Once the time to impact is computed, we can substitute it into Eq. (1.10) to solve for the impact velocity.

The final specification of the problem would be

$$\text{Minimize } C = n(c_0 + c_1 \ell + c_2 A^2) \tag{PT4.11}$$

subject to

$$v \leq v_c \tag{PT4.12}$$

$$n \geq 1 \tag{PT4.13}$$

where

$$A = 2\pi r^2 \tag{PT4.14}$$

$$\ell = \sqrt{2}r \tag{PT4.15}$$

$$c = k_c A \tag{PT4.16}$$

$$m = \frac{M_t}{n} \tag{PT4.17}$$

$$t = \text{root}\left[z_0 - \frac{gm}{c}t + \frac{gm^2}{c^2}(1 - e^{-(c/m)t})\right] \qquad \text{(PT4.18)}$$

$$v = \frac{gm}{c}(1 - e^{-(c/m)t}) \qquad \text{(PT4.19)}$$

We will solve this problem in Example 15.4 at the end of Chap. 15. For the time being recognize that it has most of the fundamental elements of other optimization problems you will routinely confront in engineering practice. These are

- The problem will involve an *objective function* that embodies your goal.
- There will be a number of *design variables*. These variables can be real numbers or they can be integers. In our example, these are r (real) and n (integer).
- The problem will include *constraints* that reflect the limitations you are working under.

We should make one more point before proceeding. Although the objective function and constraints may superficially appear to be simple equations [e.g., Eq. (PT4.12)], they may in fact be the "tip of the iceberg." That is, they may be underlain by complex dependencies and models. For instance, as in our example, they may involve other numerical methods [Eq. (PT4.18)]. This means that the functional relationships you will be using could actually represent large and complicated calculations. Thus, techniques that can find the optimal solution, while minimizing function evaluations, can be extremely valuable.

PT4.2 MATHEMATICAL BACKGROUND

There are a number of mathematical concepts and operations that underlie optimization. Because we believe that they will be more relevant to you in context, we will defer discussion of specific mathematical prerequisites until they are needed. For example, we will discuss the important concepts of the gradient and Hessians at the beginning of Chap. 14 on multivariate unconstrained optimization. In the meantime, we will limit ourselves here to the more general topic of how optimization problems are classified.

An *optimization* or *mathematical programming* problem generally can be stated as

Find x, which minimizes or maximizes $f(x)$

subject to

$$d_i(x) \le a_i \qquad i = 1, 2, \ldots, m \qquad \text{(PT4.20)}$$

$$e_i(x) = b_i \qquad i = 1, 2, \ldots, p \qquad \text{(PT4.21)}$$

where x is an n-dimensional *design vector*, $f(x)$ is the *objective function*, $d_i(x)$ are *inequality constraints*, $e_i(x)$ are *equality constraints*, and a_i and b_i are constants.

Optimization problems can be classified on the basis of the form of $f(x)$:

- If $f(x)$ and the constraints are linear, we have *linear programming*.
- If $f(x)$ is quadratic and the constraints are linear, we have *quadratic programming*.
- If $f(x)$ is not linear or quadratic and/or the constraints are nonlinear, we have *nonlinear programming*.

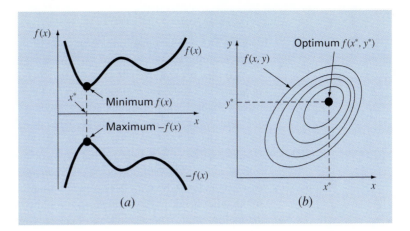

FIGURE PT4.4

(a) One-dimensional optimization. This figure also illustrates how minimization of $f(x)$ is equivalent to the maximization of $-f(x)$. (b) Two-dimensional optimization. Note that this figure can be taken to represent either a maximization (contours increase in elevation up to the maximum like a mountain) or a minimization (contours decrease in elevation down to the minimum like a valley).

Further, when Eqs. (PT4.20) and (PT4.21) are included, we have a *constrained optimization* problem; otherwise, it is an *unconstrained optimization* problem.

Note that for constrained problems, the degrees of freedom are given by $n-p$. Generally, to obtain a solution, p must be $\leq n$. If $p > n$, the problem is said to be *overconstrained*.

Another way in which optimization problems are classified is by dimensionality. This is most commonly done by dividing them into one-dimensional and multidimensional problems. As the name implies, *one-dimensional problems* involve functions that depend on a single dependent variable. As in Fig. PT4.4a, the search then consists of climbing or descending one-dimensional peaks and valleys. *Multidimensional problems* involve functions that depend on two or more dependent variables. In the same spirit, a two-dimensional optimization can again be visualized as searching out peaks and valleys (PT4.4b). However, just as in real hiking, we are not constrained to walk a single direction, instead the *topography* is examined to efficiently reach the goal.

Finally, the process of finding a maximum versus finding a minimum is essentially identical because the same value, x^*, both minimizes $f(x)$ and maximizes $-f(x)$. This equivalence is illustrated graphically for a one-dimensional function in Fig. PT4.4a.

PT4.3   ORIENTATION

Some orientation is helpful before proceeding to the numerical methods for optimization. The following is intended to provide an overview of the material in Part Four. In addition, some objectives have been included to help you focus your efforts when studying the material.

PT4.3.1 Scope and Preview

Figure PT4.5 is a schematic representation of the organization of Part Four. Examine this figure carefully, starting at the top and working clockwise.

After the present introduction, *Chap. 13* is devoted to *one-dimensional unconstrained optimization*. Methods are presented to find the minimum or maximum of a function of a single variable. Three methods are covered: *golden-section search, quadratic interpolation,* and *Newton's method.* These methods also have relevance to multidimensional optimization.

Chapter 14 covers two general types of methods to solve *multidimensional unconstrained optimization* problems. *Direct methods* such as *random searches, univariate searches,* and *pattern searches* do not require the evaluation of the function's derivatives. On the other hand, *gradient methods* use either first and sometimes second derivatives to find the optimum. The chapter introduces the *gradient* and the *Hessian,* which are multidimensional representations of the first and second derivatives. The method of *steepest ascent/descent* is then covered in some detail. This is followed by descriptions of some advanced methods: *conjugate gradient, Newton's method, Marquardt's method,* and *quasi-Newton methods.*

Chapter 15 is devoted to *constrained optimization. Linear programming* is described in detail using both a graphical representation and the *simplex method.* The detailed analysis of *nonlinear constrained optimization* is beyond this book's scope, but we provide an overview of the major approaches. In addition, we illustrate how such problems (along with the problems covered in Chaps. 13 and 14) can be obtained with software packages and libraries such as Excel, Mathcad, and IMSL.

Chapter 16 extends the above concepts to actual engineering problems. Engineering applications are used to illustrate how optimization problems are formulated and provide insight into the application of the solution techniques in professional practice.

An epilogue is included at the end of Part Four. It contains an overview of the methods discussed in Chaps. 13, 14, and 15. This overview includes a description of trade-offs related to the proper use of each technique. This section also provides references for some numerical methods that are beyond the scope of this text.

PT4.3.2 Goals and Objectives

Study Objectives.   After completing Part Four, you should have sufficient information to successfully approach a wide variety of engineering problems dealing with optimization. In general, you should have mastered the techniques, have learned to assess their reliability, and be capable of analyzing alternative methods for any particular problem. In addition to these general goals, the specific concepts in Table PT4.2 should be assimilated for a comprehensive understanding of the material in Part Four.

Computer Objectives.   You should be able to write a subprogram to implement a simple one-dimensional (like golden-section search or quadratic interpolation) and multidimensional (like the random-search method) search. In addition, program libraries like IMSL and software packages such as Excel or Mathcad have varying capabilities for optimization. You can use this part of the book to become familiar with these capabilities.

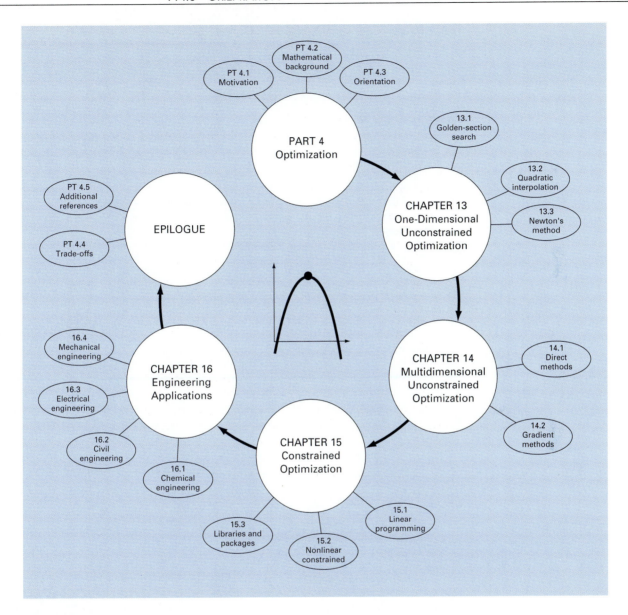

FIGURE PT4.5
Schematic of the organization of the material in Part Four: Optimization.

TABLE PT4.2 Specific study objectives for Part Four.

1. Understand why and where optimization occurs in engineering problem solving.
2. Understand the major elements of the general optimization problem: objective function, decision variables, and constraints.
3. Be able to distinguish between linear and nonlinear optimization, and between constrained and unconstrained problems.
4. Be able to define the golden ratio and understand how it makes one-dimensional optimization efficient.
5. Locate the optimum of a single variable function with the golden-section search, quadratic interpolation, and Newton's method. Also, recognize the trade-offs among these approaches, with particular attention to initial guesses and convergence.
6. Be capable of writing a program and solving for the optimum of a multivariable function using random searching.
7. Understand the ideas behind pattern searches, conjugate directions, and Powell's method.
8. Be able to define and evaluate the gradient and Hessian of a multivariable function both analytically and numerically.
9. Compute by hand the optimum of a two-variable function using the method of steepest ascent/descent.
10. Understand the basic ideas behind the conjugate gradient, Newton's, Marquardt's, and quasi-Newton methods. In particular, understand the trade-offs among the approaches and recognize how each improves on the steepest ascent/descent.
11. Be capable of recognizing and setting up a linear programming problem to represent applicable engineering problems.
12. Be able to solve a two-dimensional linear programming problem with both the graphical and simplex methods.
13. Understand the four possible outcomes of a linear programming problem.
14. Be able to set up and solve nonlinear constrained optimization problems using a software package.

CHAPTER 13

One-Dimensional Unconstrained Optimization

This section will describe techniques to find the minimum or maximum of a function of a single variable, $f(x)$. A useful image in this regard is the one-dimensional, "roller coaster"–like function depicted in Fig. 13.1. Recall from Part Two that root location was complicated by the fact that several roots can occur for a single function. Similarly, both local and global optima can occur in optimization. Such cases are called *multimodal*. In almost all instances, we will be interested in finding the absolute highest or lowest value of a function. Thus, we must take care that we do not mistake a local result for the global optimum.

Distinguishing a global from a local extremum can be a very difficult problem for the general case. There are three usual ways to approach this problem. First, insight into the behavior of low-dimensional functions can sometimes be obtained graphically. Second, finding optima based on widely varying and perhaps randomly generated starting guesses, and then selecting the largest of these as global. Finally, perturbing the starting point associated with a local optimum and seeing if the routine returns a better point, or always returns to the same point. Although all these approaches can have utility, the fact is that in some problems (usually the large ones), there may be no practical way to ensure that you've located a global optimum. However, although you should always be sensitive to the

FIGURE 13.1
A function that asymptotically approaches zero at plus and minus ∞ and has two maximum and two minimum points in the vicinity of the origin. The two points to the right are local optima, whereas the two to the left are global.

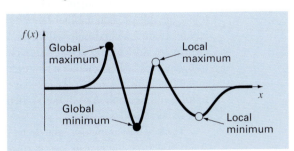

issue, it is fortunate that there are numerous engineering problems where you can locate the global optimum in an unambiguous fashion.

Just as in root location, optimization in one dimension can be divided into bracketing and open methods. As described in the next section, the golden-section search is an example of a bracketing method that depends on initial guesses that bracket a single optimum. This is followed by a somewhat more sophisticated bracketing approach—quadratic interpolation.

The final method described in this chapter is an open method based on the idea from calculus that the minimum or maximum can be found by solving $f'(x) = 0$. This reduces the optimization problem to finding the root of $f'(x)$ using techniques of the sort described in Part Two. We will demonstrate one version of this approach—Newton's method.

13.1 GOLDEN-SECTION SEARCH

In solving for the root of a single nonlinear equation, the goal was to find the value of the variable x that yields a *zero* of the function $f(x)$. *Single-variable optimization* has the goal of finding the value of x that yields an *extremum*, either a maximum or minimum of $f(x)$.

The golden-section search is a simple, general-purpose, single-variable search technique. It is similar in spirit to the bisection approach for locating roots in Chap. 5. Recall that bisection hinged on defining an interval, specified by a lower guess (x_l) and an upper guess (x_u), that bracketed a single root. The presence of a root between these bounds was verified by determining that $f(x_l)$ and $f(x_u)$ had different signs. The root was then estimated as the midpoint of this interval,

$$x_r = \frac{x_l + x_u}{2}$$

The final step in a bisection iteration involved determining a new smaller bracket. This was done by replacing whichever of the bounds x_l or x_u had a function value with the same sign as $f(x_r)$. An efficiency of this approach was that the new value x_r replaced one of the old bounds.

Now we can develop a similar approach for locating the optimum of a one-dimensional function. For simplicity, we will focus on the problem of finding a maximum. When we discuss the computer algorithm, we will describe the minor modifications needed to simulate a minimum.

As with bisection, we can start by defining an interval that contains a single answer. That is, the interval should contain a single maximum, and hence is called *unimodal*. We can adopt the same nomenclature as for bisection, where x_l and x_u defined the lower and upper bounds of such an interval. However, in contrast to bisection, we need a new strategy for finding a maximum within the interval. Rather than using only two function values (which are sufficient to detect a sign change, and hence a zero), we would need three function values to detect whether a maximum occurred. Thus, an additional point within the interval has to be chosen. Next, we have to pick a fourth point. Then the test for the maximum could be applied to discern whether the maximum occurred within the first three or the last three points.

The key to making this approach efficient is the wise choice of the intermediate points. As in bisection, the goal is to minimize function evaluations by replacing old values with

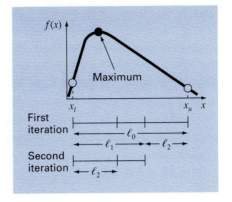

FIGURE 13.2
The initial step of the golden-section search algorithm involves choosing two interior points according to the golden ratio.

new values. This goal can be achieved by specifying that the following two conditions hold (Fig. 13.2):

$$\ell_0 = \ell_1 + \ell_2 \tag{13.1}$$

$$\frac{\ell_1}{\ell_0} = \frac{\ell_2}{\ell_1} \tag{13.2}$$

The first condition specifies that the sum of the two sublengths ℓ_1 and ℓ_2 must equal the original interval length. The second says that the ratio of the lengths must be equal. Equation (13.1) can be substituted into Eq. (13.2),

$$\frac{\ell_1}{\ell_1 + \ell_2} = \frac{\ell_2}{\ell_1} \tag{13.3}$$

If the reciprocal is taken and $R = \ell_2 / \ell_1$, we arrive at

$$1 + R = \frac{1}{R} \tag{13.4}$$

or

$$R^2 + R - 1 = 0 \tag{13.5}$$

which can be solved for the positive root

$$R = \frac{-1 + \sqrt{1 - 4(-1)}}{2} = \frac{\sqrt{5} - 1}{2} = 0.61803\ldots \tag{13.6}$$

This value, which has been know since antiquity, is called the *golden ratio* (see Box 13.1). Because it allows optima to be found efficiently, it is the key element of the golden-section method we have been developing conceptually. Now let's derive an algorithm to implement this approach on the computer.

Box 13.1   The Golden Ratio and Fibonacci Numbers

In many cultures, certain numbers are ascribed qualities. For example, we in the West are all familiar with "Lucky 7" and "Friday the 13th." Ancient Greeks called the following number the "golden ratio":

$$\frac{\sqrt{5}-1}{2} = 0.61803\ldots$$

This ratio was employed for a number of purposes, including the development of the rectangle in Fig. 13.3. These proportions were considered aesthetically pleasing by the Greeks. Among other things, many of their temples followed this shape.

The golden ratio is related to an important mathematical series known as the *Fibonacci numbers,* which are

0, 1, 1, 2, 3, 5, 8, 13, 21, 34, . . .

Thus, each number after the first two represents the sum of the preceding two. This sequence pops up in many diverse areas of science and engineering. In the context of the present discussion, an interesting property of the Fibonacci sequence relates to the ratio of consecutive numbers in the sequence; that is, $0/1 = 0$, $1/1 = 1$, $1/2 = 0.5$, $2/3 = 0.667$, $3/5 = 0.6$, $5/8 = 0.625$, $8/13 = 0.615$, and so on. As one proceeds, the ratio of consecutive numbers approaches the golden ratio!

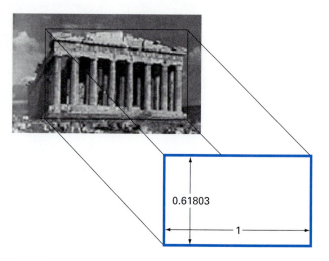

FIGURE 13.3

The Parthenon in Athens, Greece, was constructed in the 5th century B.C. Its front dimensions can be fit almost exactly within a golden rectangle.

As mentioned above and as depicted in Fig. 13.4, the method starts with two initial guesses, x_l and x_u, that bracket one local extremum of $f(x)$. Next, two interior points x_1 and x_2 are chosen according to the golden ratio,

$$d = \frac{\sqrt{5}-1}{2}(x_u - x_l)$$

$$x_1 = x_l + d$$

$$x_2 = x_u - d$$

The function is evaluated at these two interior points. Two results can occur:

1. If, as is the case in Fig. 13.4, $f(x_1) > f(x_2)$, then the domain of x to the left of x_2, from x_l to x_2, can be eliminated because it does not contain the maximum. For this case, x_2 becomes the new x_l for the next round.

2. If it had occurred that $f(x_2) > f(x_1)$, then the domain of x to the right of x_1, from x_1 to x_u would have been eliminated. In this case, x_1 becomes the new x_u for the next round.

Now, here is the real benefit from the use of the golden ratio. Because the original x_1 and x_2 were chosen using the golden ratio, we don't have to recalculate all the function

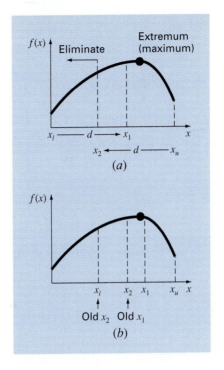

FIGURE 13.4
(a) The initial step of the golden-section search algorithm involves choosing two interior points according to the golden ratio. (b) The second step involves defining a new interval that includes the optimum.

values for the next iteration. For example, for the case illustrated in Fig. 13.4, the old x_1 becomes the new x_2. This means that we already have the value for the new $f(x_2)$, since it is the same as the function value at the old x_1.

To complete the algorithm, we now only need to determine the new x_1. This is done with the same proportionality as before,

$$x_1 = x_l + \frac{\sqrt{5} - 1}{2}(x_u - x_l)$$

A similar approach would be used for the alternate case where the optimum fell in the left subinterval.

As the iterations are repeated, the interval containing the extremum is reduced rapidly. In fact, each round the interval is reduced by a factor of the golden ratio (about 61.8%). That means that after 10 rounds, the interval is shrunk to about 0.618^{10} or 0.008 or 0.8% of its initial length. After 20 rounds, it is about 0.0066%. This is not quite as good as the reduction achieved with bisection, but this is a harder problem.

EXAMPLE 13.1   Golden-Section Search

Problem Statement.   Use the golden-section search to find the maximum of

$$f(x) = 2 \sin x - \frac{x^2}{10}$$

within the interval $x_l = 0$ and $x_u = 4$.

Solution.   First, the golden ratio is used to create the two interior points

$$d = \frac{\sqrt{5} - 1}{2}(4 - 0) = 2.472$$

$$x_1 = 0 + 2.472 = 2.472$$
$$x_2 = 4 - 2.472 = 1.528$$

The function can be evaluated at the interior points

$$f(x_2) = f(1.528) = \frac{1.528^2}{10} - 2 \sin (1.528) = 1.765$$

$$f(x_1) = f(2.472) = 0.63$$

Because $f(x_2) > f(x_1)$, the maximum is in the interval defined by x_l, x_2, and x_1. Thus, for the new interval, the lower bound remains $x_l = 0$, and x_1 becomes the upper bound; that is, $x_u = 2.472$. In addition, the former x_2 value becomes the new x_1; that is, $x_1 = 1.528$. In addition, we do not have to recalculate $f(x_1)$ because it was determined on the previous iteration as $f(1.528) = 1.765$.

All that remains is to compute the new golden ratio and x_2,

$$d = \frac{\sqrt{5} - 1}{2}(2.472 - 0) = 1.528$$

$$x_2 = 4 - 1.528 = 0.944$$

The function evaluation at x_2 is $f(0.994) = 1.531$. Since this value is less than the function value at x_1, the maximum is in the interval prescribed by x_2, x_1, and x_u.

The process can be repeated, with the results tabulated below:

i	x_l	$f(x_l)$	x_2	$f(x_2)$	x_1	$f(x_1)$	x_u	$f(x_u)$	d
1	0	0	1.5279	1.7647	2.4721	0.6300	4.0000	−3.1136	2.4721
2	0	0	0.9443	1.5310	1.5279	1.7647	2.4721	0.6300	1.5279
3	0.9443	1.5310	1.5279	1.7647	1.8885	1.5432	2.4721	0.6300	0.9443
4	0.9443	1.5310	1.3050	1.7595	1.5279	1.7647	1.8885	1.5432	0.5836
5	1.3050	1.7595	1.5279	1.7647	1.6656	1.7136	1.8885	1.5432	0.3607
6	1.3050	1.7595	1.4427	1.7755	1.5279	1.7647	1.6656	1.7136	0.2229
7	1.3050	1.7595	1.3901	1.7742	1.4427	1.7755	1.5279	1.7647	0.1378
8	1.3901	1.7742	1.4427	1.7755	1.4752	1.7732	1.5279	1.7647	0.0851

Note that the current maximum is highlighted for every iteration. After the eighth iteration, the maximum occurs at $x = 1.4427$ with a function value of 1.7755. Thus, the result is converging on the true value of 1.7757 at $x = 1.4276$.

Recall that for bisection (Sec. 5.2.1), an exact upper bound for the error can be calculated at each iteration. Using similar reasoning, an upper bound for golden-section search can be derived as follows.

Once an iteration is complete, the optimum will either fall in one of two intervals. If x_2 is the optimum function value, it will be in the lower interval (x_l, x_2, x_1). If x_1 is the optimum function value, it will be in the upper interval (x_2, x_1, x_u). Because the interior points are symmetrical, either case can be used to define the error.

Looking at the upper interval, if the true value were at the far left, the maximum distance from the estimate would be

$$
\begin{aligned}
\Delta x_a &= x_1 - x_2 \\
&= x_l + R(x_u - x_l) - x_u + R(x_u - x_l) \\
&= (x_l - x_u) + 2R(x_u - x_l) \\
&= (2R - 1)(x_u - x_l)
\end{aligned}
$$

or $0.236(x_u - x_l)$.

If the true value were at the far right, the maximum distance from the estimate would be

$$
\begin{aligned}
\Delta x_b &= x_u - x_1 \\
&= x_u - x_l - R(x_u - x_l) \\
&= (1 - R)(x_u - x_l)
\end{aligned}
$$

or $0.382(x_u - x_l)$. Therefore, this case would represent the maximum error. This result can then be normalized to the optimal value for that iteration, x_{opt} to yield

$$
\varepsilon_a = (1 - R) \left| \frac{x_u - x_l}{x_{opt}} \right| 100\%
$$

This estimate provides a basis for terminating the iterations.

Pseudocode for the golden-section-search algorithm for maximization is presented in Fig. 13.5a. The minor modifications to convert the algorithm to minimization are listed in Fig. 13.5b. In both versions the x value for the optimum is returned as the function value ($gold$). In addition, the value of $f(x)$ at the optimum is returned as the variable (fx).

You may be wondering why we have stressed the reduced function evaluations of the golden-section search. Of course, for solving a single optimization, the speed savings would be negligible. However, there are two important contexts where minimizing the number of function evaluations can be important. These are

1. *Many evaluations.* There are cases where the golden-section-search algorithm may be a part of a much larger calculation. In such cases, it may be called many times. Therefore, keeping function evaluations to a minimum could pay great dividends for such cases.

FIGURE 13.5
Algorithm for the golden-section search.

```
FUNCTION Gold (xlow, xhigh, maxit, es, fx)
R = (5^0.5 − 1)/2
xℓ = xlow; xu = xhigh
iter = 1
d  = R * (xu − xℓ)
x1  = xℓ + d; x2 = xu − d
f1  = f(x1)
f2  = f(x2)
IF f1 > f2 THEN                                        IF f1 < f2 THEN
   xopt = x1
   fx = f1
ELSE
   xopt = x2
   fx = f2
END IF
DO
   d = R*d
   IF f1 > f2 THEN                                     IF f1 < f2 THEN
      xℓ = x2
      x2= x1
      x1 = xℓ+d
      f2= f1
      f1 = f(x1)
   ELSE
      xu= x1
      x1 = x2
      x2= xu−d
      f1 = f2
      f2= f(x2)
   END IF
   iter = iter+1
   IF f1 > f2 THEN                                     IF f1 < f2 THEN
      xopt = x1
      fx = f1
   ELSE
      xopt = x2
      fx = f2
   END IF
   IF xopt ≠ 0. THEN
      ea = (1.−R) *ABS((xu − xℓ)/xopt) * 100.
   END IF
   IF ea ≤ es OR iter ≥ maxit EXIT
END DO
Gold = xopt
END Gold
```

        (a) Maximization   **(b) Minimization**

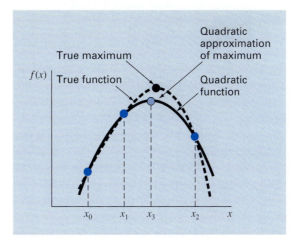

FIGURE 13.6
Graphical description of quadratic interpolation.

2. *Time-consuming evaluation.* For pedagogical reasons, we use simple functions in most of our examples. You should understand that a function can be very complex and time-consuming to evaluate. For example, in a later part of this book, we will describe how optimization can be used to estimate the parameters of a model consisting of a system of differential equations. For such cases, the "function" involves time-consuming model integration. Any method that minimizes such evaluations would be advantageous.

13.2 QUADRATIC INTERPOLATION

Quadratic interpolation takes advantage of the fact that a second-order polynomial often provides a good approximation to the shape of $f(x)$ near an optimum (Fig. 13.6).

Just as there is only one straight line connecting two points, there is only one quadratic or parabola connecting three points. Thus, if we have three points that jointly bracket an optimum, we can fit a parabola to the points. Then we can differentiate it, set the result equal to zero, and solve for an estimate of the optimal x. It can be shown through some algebraic manipulations that the result is

$$x_3 = \frac{f(x_0)\left(x_1^2 - x_2^2\right) + f(x_1)\left(x_2^2 - x_0^2\right) + f(x_2)\left(x_0^2 - x_1^2\right)}{2f(x_0)\left(x_1 - x_2\right) + 2f(x_1)\left(x_2 - x_0\right) + 2f(x_2)\left(x_0 - x_1\right)} \qquad (13.7)$$

where x_0, x_1, and x_2 are the guesses that bracket the extremum, and x_3 is the value of x that corresponds to the maximum value of the quadratic fit to the guesses.

EXAMPLE 13.2   Quadratic Interpolation

Problem Statement.   Use quadratic interpolation to approximate the maximum of

$$f(x) = 2 \sin x - \frac{x^2}{10}$$

with initial guesses of $x_0 = 0$, $x_1 = 1$, and $x_2 = 4$.

Solution.   The function values at the three guesses can be evaluated,

$$x_0 = 0 \qquad f(x_0) = 0$$
$$x_1 = 1 \qquad f(x_1) = 1.5829$$
$$x_2 = 4 \qquad f(x_2) = -3.1136$$

and substituted into Eq. (13.7) to give,

$$x_3 = \frac{0\left(1^2 - 4^2\right) + 1.5829\left(4^2 - 0^2\right) + (-3.1136)\left(0^2 - 1^2\right)}{2(0)\,(1 - 4) + 2(1.5829)\,(4 - 0) + 2(-3.1136)\,(0 - 1)} = 1.5055$$

which has a function value of $f(1.5055) = 1.7691$.

Next, a strategy similar to the golden-section search can be employed to determine which point should be discarded. Because the function value for the new point is higher than for the intermediate point (x_1) and the new x value is to the right of the intermediate point, the lower guess (x_0) is discarded. Therefore, for the next iteration,

$$x_0 = 1 \qquad\qquad f(x_0) = 1.5829$$
$$x_1 = 1.5055 \qquad f(x_1) = 1.7691$$
$$x_2 = 4 \qquad\qquad f(x_2) = -3.1136$$

which can be substituted into Eq. (13.7) to give

$$x_3 = \frac{1.5829\left(1.5055^2 - 4^2\right) + 1.7691\left(4^2 - 1^2\right) + (-3.1136)\left(1^2 - 1.5055^2\right)}{2(1.5829)(1.5055 - 4) + 2(1.7691)(4 - 1) + 2(-3.1136)\,(1 - 1.5055)}$$
$$= 1.4903$$

which has a function value of $f(1.4903) = 1.7714$.

The process can be repeated, with the results tabulated below:

i	x_0	$f(x_0)$	x_1	$f(x_1)$	x_2	$f(x_2)$	x_3	$f(x_3)$
1	0.0000	0.0000	1.0000	1.5829	4.0000	−3.1136	1.5055	1.7691
2	1.0000	1.5829	1.5055	1.7691	4.0000	−3.1136	1.4903	1.7714
3	1.0000	1.5829	1.4903	1.7714	1.5055	1.7691	1.4256	1.7757
4	1.0000	1.5829	1.4256	1.7757	1.4903	1.7714	1.4266	1.7757
5	1.4256	1.7757	1.4266	1.7757	1.4903	1.7714	1.4275	1.7757

Thus, within five iterations, the result is converging rapidly on the true value of 1.7757 at $x = 1.4276$.

We should mention that just like the false-position method, quadratic interpolation can get hung up with just one end of the interval converging. Thus, convergence can be slow. For example, notice that in our example, 1.0000 was an endpoint for most of the iterations.

This method, as well as others using third-order polynomials, can be formulated into algorithms that contain convergence tests, careful selection strategies for the points to retain on each iteration, and attempts to minimize round-off error accumulation. In particular, see Brent's method in Press et al. (1992).

13.3 NEWTON'S METHOD

Recall that the Newton-Raphson method of Chap. 6 is an open method that finds the root x of a function such that $f(x) = 0$. The method is summarized as

$$x_{i+1} = x_i - \frac{f(x_i)}{f'(x_i)}$$

A similar open approach can be used to find an optimum of $f(x)$ by defining a new function, $g(x) = f'(x)$. Thus, because the same optimal value x^* satisfies both

$$f'(x^*) = g(x^*) = 0$$

we can use the following,

$$x_{i+1} = x_i - \frac{f'(x_i)}{f''(x_i)} \tag{13.8}$$

as a technique to find the minimum or maximum of $f(x)$. It should be noted that this equation can also be derived by writing a second-order Taylor series for $f(x)$ and setting the derivative of the series equal to zero. Newton's method is an open method similar to Newton-Raphson because it does not require initial guesses that bracket the optimum. In addition, it also shares the disadvantage that it may be divergent. Finally, it's usually a good idea to check that the second derivative has the correct sign to confirm that the technique is converging on the result you desire.

EXAMPLE 13.3   Newton's Method

Problem Statement.   Use Newton's method to find the maximum of

$$f(x) = 2 \sin x - \frac{x^2}{10}$$

with an initial guess of $x_0 = 2.5$.

Solution. The first and second derivatives of the function can be evaluated as

$$f'(x) = 2\cos x - \frac{x}{5}$$

$$f''(x) = -2\sin x - \frac{1}{5}$$

which can be substituted into Eq. (13.8) to give

$$x_{i+1} = x_i - \frac{2\cos x_i - x_i/5}{-2\sin x_i - 1/5}$$

Substituting the initial guess yields

$$x_1 = 2.5 - \frac{2\cos 2.5 - 2.5/5}{-2\sin 2.5 - 1/5} = 0.99508$$

which has a function value of 1.57859. The second iteration gives

$$x_1 = 0.995 - \frac{2\cos 0.995 - 0.995/5}{-2\sin 0.995 - 1/5} = 1.46901$$

which has a function value of 1.77385.

The process can be repeated, with the results tabulated below:

i	x	f(x)	f'(x)	f''(x)
0	2.5	0.57194	−2.10229	−1.39694
1	0.99508	1.57859	0.88985	−1.87761
2	1.46901	1.77385	−0.09058	−2.18965
3	1.42764	1.77573	−0.00020	−2.17954
4	1.42755	1.77573	0.00000	−2.17952

Thus, within four iterations, the result converges rapidly on the true value.

Although Newton's method works well in some cases, it is impractical for cases where the derivatives cannot be conveniently evaluated. For these cases, other approaches that do not involve derivative evaluation are available. For example, a secant-like version of Newton's method can be developed by using finite-difference approximations for the derivative evaluations.

A bigger reservation regarding the approach is that it may diverge based on the nature of the function and the quality of the initial guess. Thus, it is usually employed only when we are close to the optimum. Hybrid techniques that use *bracketing approaches* far from the optimum and *open methods* near the optimum attempt to exploit the strong points of both approaches.

This concludes our treatment of methods to solve the optima of functions of a single variable. Some engineering examples are presented in Chap. 16. In addition, the techniques described here are an important element of some procedures to optimize multivariable functions, as discussed in the next chapter.

PROBLEMS

13.1 Given the formula

$$f(x) = x^2 - 8x + 12$$

(a) Determine the maximum and the corresponding value of x for this function analytically (i.e., using differentiation).
(b) Verify that Eq. (13.7) yields the same results based on initial guesses of $x_0 = 0$, $x_1 = 2$, and $x_2 = 6$.

13.2 Given

$$f(x) = -1.5x^6 - 2x^4 + 12x$$

(a) Plot the function.
(b) Use analytical methods to prove that the function is concave for all values of x.
(c) Differentiate the function and then use a root-location method to solve for the maximum $f(x)$ and the corresponding value of x.

13.3 Solve for the value of x that maximizes $f(x)$ in Prob. 13.2 using the golden-section search. Employ initial guesses of $x_l = 0$ and $x_u = 2$ and perform 3 iterations.

13.4 Repeat Prob. 13.3, except use quadratic interpolation. Employ initial guesses of $x_0 = 0$, $x_1 = 1$ and $x_2 = 2$ and perform 3 iterations.

13.5 Repeat Prob. 13.3 but use Newton's method. Employ an initial guess of $x_0 = 2$ and perform three iterations.

13.6 Discuss the advantages and disadvantages of golden-section search, quadratic interpolation, and Newton's method for locating an optimum value in one dimension.

13.7 Employ the following methods to find the maximum of

$$f(x) = 2x - 1.75x^2 + 1.1x^3 - 0.25x^4$$

(a) Golden-section search ($x_l = -2$, $x_u = 4$, $\varepsilon_s = 1\%$).
(b) Quadratic interpolation ($x_0 = 1.75$, $x_1 = 2$, $x_2 = 2.25$, iterations = 5).
(c) Newton's method ($x_0 = 2.5$, $\varepsilon_s = 1\%$).

13.8 Consider the following function:

$$f(x) = 6x + 7.5x^2 + 3x^3 + x^4$$

Use analytical and graphical methods to show the function has a minimum for some value of x in the range $-2 \le x \le 1$.

13.9 Employ the following methods to find the maximum of the function from Prob. 13.8:
(a) Golden-section search ($x_l = -2$, $x_u = 1$, $\varepsilon_s = 1\%$).
(b) Quadratic interpolation ($x_0 = -2$, $x_1 = -1$, $x_2 = 1$, iterations = 5).
(c) Newton's method ($x_0 = -1$, $\varepsilon_s = 1\%$).

13.10 Consider the following function:

$$f(x) = x + \frac{1}{x}$$

Perform 10 iterations of quadratic interpolation to locate the minimum. Comment on the convergence of your results. ($x_0 = 0.1$, $x_1 = 0.5$, $x_2 = 5.0$)

13.11 Consider the following function:

$$f(x) = 2 + 5x + 6x^2 + 2x^3 + 2x^4$$

Locate the minimum by finding the root of the derivative of this function. Use bisection with initial guesses of $x_l = -2$ and $x_u = 1$.

13.12 Determine the minimum of the function from Prob. 13.11 with the following methods:
(a) Newton's method ($x_0 = -1$, $\varepsilon_s = 1\%$).
(b) Newton's method, but using a finite difference approximation for the derivative estimates.

$$f'(x_i) \cong \frac{f(x_i + \delta x_i) - f(x_i - \delta x_i)}{2\delta x_i}$$

$$f''(x_i) \cong \frac{f(x_i + \delta x_i) - 2f(x_i) + f(x_i - \delta x_i)}{\delta x_i^2}$$

where $\delta = $ a perturbation fraction ($= 0.01$). Use an initial guess of $x_0 = -1$ and iterate to $\varepsilon_s = 1\%$.

13.13 Develop a subprogram using a program or macro language to implement the golden-section search algorithm. Design the subprogram so that it is expressly designed to locate a maximum. The subroutine should have the following features:

• Check whether the guesses bracket a maximum. If not, the subroutine should not implement the algorithm, but should return an error message.
• Iterate until the relative error falls below a stopping criterion or exceeds a maximum number of iterations.

- Return both the optimal x and $f(x)$.
- Minimize the number of function evaluations.

Test your program with the same problem as Example 13.1.

13.14 Develop a subprogram as described in Prob. 13.13, but make it recognize whether the problem involves minimization or maximization, and then based on this recognition implement the golden-section search properly.

13.15 Develop a subprogram using a program or macro language to implement the quadratic interpolation algorithm. Design the subprogram so that it is expressly designed to locate a maximum. The subroutine should have the following features:

- Base it on two initial guesses and have the program generate the third initial value at the midpoint of the interval.
- Check whether the guesses bracket a maximum. If not, the subroutine should not implement the algorithm, but should return an error message.

- Iterate until the relative error falls below a stopping criterion or exceeds a maximum number of iterations.
- Return both the optimal x and $f(x)$.
- Minimize the number of function evaluations.

Test your program with the same problem as Example 13.2.

13.16 Develop a subprogram using a program or macro language to implement Newton's method. The subroutine should have the following features:

- Iterate until the relative error falls below a stopping criterion or exceeds a maximum number of iterations.
- Returns both the optimal x and $f(x)$.

Test your program with the same problem as Example 13.3.

CHAPTER 14

Multidimensional Unconstrained Optimization

This chapter describes techniques to find the minimum or maximum of a function of several variables. Recall from Chap. 13 that our visual image of a one-dimensional search was like a roller coaster. For two-dimensional cases, the image becomes that of mountains and valleys (Fig. 14.1). For higher-dimensional problems, convenient images are not possible.

We have chosen to limit this chapter to the two-dimensional case. We've adopted this approach because the essential features of multidimensional searches are often best communicated visually.

Techniques for multidimensional unconstrained optimization can be classified in a number of ways. For purposes of the present discussion, we will divide them depending on whether they require derivative evaluation. The approaches that do not require derivative evaluation are called *nongradient*, or *direct*, *methods*. Those that require derivatives are called *gradient*, or *descent* (or *ascent*), *methods*.

FIGURE 14.1

The most tangible way to visualize two-dimensional searches is in the context of ascending a mountain (maximization) or descending into a valley (minimization). (*a*) A 2-D topographic map that corresponds to the 3-D mountain in (*b*).

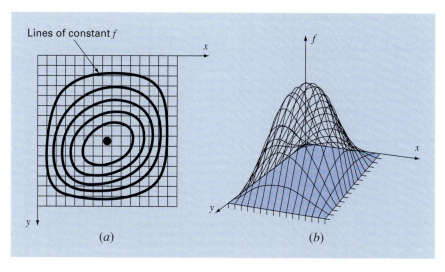

14.1 DIRECT METHODS

These methods vary from simple brute force approaches to more elegant techniques that attempt to exploit the nature of the function. We will start our discussion with a brute force approach.

14.1.1 Random Search

A simple example of a brute force approach is the *random search method*. As the name implies, this method repeatedly evaluates the function at randomly selected values of the independent variables. If a sufficient number of samples are conducted, the optimum will eventually be located.

EXAMPLE 14.1 Random Search Method

Problem Statement.   Use a random number generator to locate the maximum of

$$f(x, y) = y - x - 2x^2 - 2xy - y^2 \tag{E14.1.1}$$

in the domain bounded by $x = -2$ to 2 and $y = 1$ to 3. The domain is depicted in Fig. 14.2. Notice that a single maximum of 1.5 occurs at $x = -1$ and $y = 1.5$.

Solution.   Random number generators typically generate values between 0 and 1. If we designate such a number as r, the following formula can be used to generate x values randomly within a range between x_l to x_u:

$$x = x_l + (x_u - x_l)r$$

For the present application, $x_l = -2$ and $x_u = 2$, and the formula is

$$x = -2 + \big(2 - (-2)\big)r = -2 + 4r$$

This can be tested by substituting 0 and 1 to yield -2 and 2, respectively.

FIGURE 14.2
Equation (E14.1.1) showing the maximum at $x = -1$ and $y = 1.5$.

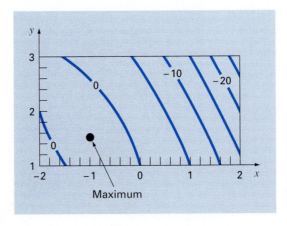

Similarly for y, a formula for the present example could be developed as

$$y = y_l + (y_u - y_l)r = 1 + (3 - 1)r = 1 + 2r$$

The following Fortran code uses the Digital Visual Fortran 90 random number subroutine RANDOM, to generate (x, y) pairs. These are then substituted into Eq. (E14.1.1). The maximum value from among these random trials is stored in the variable *maxf*, and the corresponding x and y values in *maxx* and *maxy*, respectively. A total of 300 samples are taken and the results displayed every 30 tries.

```
Program Jump

IMPLICIT NONE
INTEGER ::i,j,iter
REAL ::x,y,f,fn,maxf,maxx,maxy,rnd

f(x, y) = y - x - 2.*x**2 - 2.*x*y - y**2

maxf = -999.e9

DO i = 1, 10
  DO j = 1, 30
    iter = iter + 1
    CALL Random(rnd)
    x = -2. + 4. * rnd
    y = 1. + 2. * rnd
    fn = f(x, y)
    IF (fn.GT.maxf) THEN
      maxf = fn
      maxx = x
      maxy = y
    END IF
  END DO
  PRINT *, iter, maxx, maxy, maxf
END DO

OUTPUT:

    30    -9.483753E-01      1.525812      1.241338
    60    -9.483753E-01      1.525812      1.241338
    90    -9.799424E-01      1.510029      1.248693
   120    -9.799424E-01      1.510029      1.248693
   150    -9.799424E-01      1.510029      1.248693
   180    -9.799424E-01      1.510029      1.248693
   210    -9.799424E-01      1.510029      1.248693
   240    -9.938087E-01      1.503096      1.249875
   270      -1.003075        1.498462      1.249969
   300      -1.003075        1.498462      1.249969
```

The results indicate that the technique rapidly homes in on the true maximum.

This simple brute force approach works even for discontinuous and nondifferentiable functions. Furthermore, it always finds the global optimum rather than a local optimum. Its major shortcoming is that as the number of independent variables grows, the implementation effort required can become onerous. In addition, it is not efficient because it takes no account of the behavior of the underlying function. The remainder of the approaches described in this chapter do take function behavior into account as well as the results of previous trials to improve the speed of convergence. Thus, although the random search can certainly prove useful in specific problem contexts, the following methods have more general utility and almost always lead to more efficient convergence.

It should be noted that more sophisticated search techniques are available. These are heuristic approaches that were developed to handle either nonlinear and/or discontinuous problems that classical optimization cannot usually handle well, if at all. Simulated annealing, tabu search, artificial neural networks, and genetic algorithms are a few. The most widely applied is the *genetic algorithm,* with a number of commercial packages available. Holland (1975) pioneered the genetic algorithm approach and Goldberg (1989) and Davis (1991) provide good overviews of the theory and application of the method.

14.1.2 Univariate and Pattern Searches

It is very appealing to have an efficient optimization approach that does not require evaluation of derivatives. The random search method described above does not require derivative evaluation, but is not very efficient. This section describes an approach, the univariate search method, that is more efficient and still does not require derivative evaluation.

The basic strategy underlying the *univariate search method* is to change one variable at a time to improve the approximation while the other variables are held constant. Since only one variable is changed, the problem reduces to a sequence of one-dimensional searches that can be solved using a variety of methods (including those described in Chap. 13).

Let's perform a univariate search graphically, as shown in Fig. 14.3. Start at point 1, and move along the *x* axis with *y* constant to the maximum at point 2. You can see that point 2

FIGURE 14.3
A graphical depiction of how a univariate search is conducted.

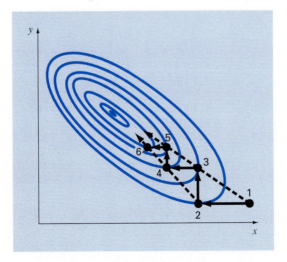

is a maximum by noticing that the trajectory along the x axis just touches a contour line at the point. Next, move along the y axis with x constant to point 3. Continue this process generating points 4, 5, 6, etc.

Although we are gradually moving toward the maximum, the search becomes less efficient as we move along the narrow ridge toward the maximum. However, also note that lines joining alternate points such as 1-3, 3-5 or 2-4, 4-6 point in the general direction of the maximum. These trajectories present an opportunity to shoot directly along the ridge toward the maximum. Such trajectories are called *pattern directions*.

Formal algorithms are available that capitalize on the idea of pattern directions to find optimum values efficiently. The best known of these algorithms is called *Powell's method*. It is based on the observation (see Fig. 14.4) that if points 1 and 2 are obtained by one-dimensional searches in the same direction but from different starting points, then the line formed by 1 and 2 will be directed toward the maximum. Such lines are called *conjugate directions*.

In fact, it can be proved that if $f(x, y)$ is a quadratic function, sequential searches along conjugate directions will converge exactly in a finite number of steps regardless of the starting point. Since a general nonlinear function can often be reasonably approximated by a quadratic function, methods based on conjugate directions are usually quite efficient and are in fact quadratically convergent as they approach the optimum.

Let's graphically implement a simplified version of Powell's method to find the maximum of

$$f(x, y) = c - (x - a)^2 - (y - b)^2$$

where a, b, and c are positive constants. This equation results in circular contours in the x, y plane, as shown in Fig. 14.5.

Initiate the search at point 0 with starting directions h_1 and h_2. Note that h_1 and h_2 are not necessarily conjugate directions. From zero, move along h_1 until a maximum is located

FIGURE 14.4
Conjugate directions.

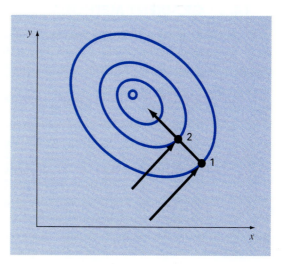

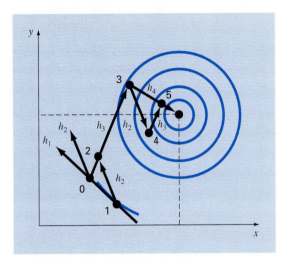

FIGURE 14.5
Powell's method.

at point 1. Then search from point 1 along direction h_2 to find point 2. Next, form a new search direction h_3 through points 0 and 2. Search along this direction until the maximum at point 3 is located. Then search from point 3 in the h_2 direction until the maximum at point 4 is located. From point 4 arrive at point 5 by again searching along h_3. Now, observe that both points 5 and 3 have been located by searching in the h_3 direction from two different points. Powell has shown that h_4 (formed by points 3 and 5) and h_3 are conjugate directions. Thus, searching from point 5 along h_4 brings us directly to the maximum.

Powell's method can be refined to make it more efficient, but the formal algorithms are beyond the scope of this text. However, it is an efficient method that is quadratically convergent without requiring derivative evaluation.

14.2   GRADIENT METHODS

As the name implies, *gradient methods* explicitly use derivative information to generate efficient algorithms to locate optima. Before describing specific approaches, we must first review some key mathematical concepts and operations.

14.2.1 Gradients and Hessians

Recall from calculus that the first derivative of a one-dimensional function provides a slope or tangent to the function being differentiated. From the standpoint of optimization, this is useful information. For example, if the slope is positive, it tells us that increasing the independent variable will lead to a higher value of the function we are exploring.

From calculus, also recall that the first derivative may tell us when we have reached an optimal value, since this is the point that the derivative goes to zero. Further, the sign of the second derivative can tell us whether we've reached a minimum (positive second derivative) or a maximum (negative second derivative).

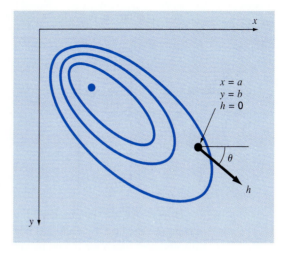

FIGURE 14.6
The directional gradient is defined along an axis *h* that forms an angle θ with the *x* axis.

These ideas were useful to us in the one-dimensional search algorithms we explored in the previous chapter. However, to fully understand multidimensional searches, we must first understand how the first and second derivatives are expressed in a multidimensional context.

The Gradient. Suppose we have a two-dimensional function $f(x, y)$. An example might be your elevation on a mountain as a function of your position. Suppose that you are at a specific location on the mountain (a, b) and you want to know the slope in an arbitrary direction. One way to define the direction is along a new axis h that forms an angle θ with the x axis (Fig. 14.6). The elevation along this new axis can be thought of as a new function $g(h)$. If you define your position as being the origin of this axis (that is, $h = 0$), the slope in this direction would be designated as $g'(0)$. This slope, which is called the *directional derivative,* can be calculated from the partial derivatives along the x and y axes by

$$g'(0) = \frac{\partial f}{\partial x} \cos \theta + \frac{\partial f}{\partial y} \sin \theta \tag{14.1}$$

where the partial derivatives are evaluated at $x = a$ and $y = b$.

Assuming that your goal is to gain the most elevation with the next step, the next logical question would be: what direction is the steepest ascent? The answer to this question is provided very neatly by what is referred to mathematically as the *gradient,* which is defined as

$$\nabla f = \frac{\partial f}{\partial x}\mathbf{i} + \frac{\partial f}{\partial y}\mathbf{j} \tag{14.2}$$

This vector is also referred to as "del f." It represents the directional derivative of $f(x, y)$ at point $x = a$ and $y = b$.

Vector notation provides a concise means to generalize the gradient to n dimensions, as

$$\nabla f(\mathbf{x}) = \begin{Bmatrix} \dfrac{\partial f}{\partial x_1}(\mathbf{x}) \\[2mm] \dfrac{\partial f}{\partial x_2}(\mathbf{x}) \\[2mm] \cdot \\ \cdot \\ \cdot \\[2mm] \dfrac{\partial f}{\partial x_n}(\mathbf{x}) \end{Bmatrix}$$

How do we use the gradient? For the mountain-climbing problem, if we're interested in gaining elevation as quickly as possible, the gradient tells us what direction to move locally and how much we'll gain by taking it. Note, however, that this strategy does not necessarily take us on a direct path to the summit! We will discuss these ideas in more depth later in this chapter.

EXAMPLE 14.2   Using the Gradient to Evaluate the Path of Steepest Ascent

Problem Statement.   Employ the gradient to evaluate the steepest ascent direction for the function

$$f(x, y) = xy^2$$

at the point (2, 2). Assume that positive x is pointed east and positive y is pointed north.

Solution: First, our elevation can be determined as

$$f(4, 2) = 2(2)^2 = 8$$

Next, the partial derivatives can be evaluated,

$$\frac{\partial f}{\partial x} = y^2 = 2^2 = 4$$

$$\frac{\partial f}{\partial y} = 2xy = 2(2)(2) = 8$$

which can be used to determine the gradient as

$$\nabla f = 4\mathbf{i} + 8\mathbf{j}$$

This vector can be sketched on a topographical map of the function, as in Fig. 14.7. This immediately tells us that the direction we must take is

$$\theta = \tan^{-1}\left(\frac{8}{4}\right) = 1.107 \text{ radians } (= 63.4°)$$

relative to the x axis. The slope in this direction, which is the magnitude of ∇f, can be calculated as

$$\sqrt{4^2 + 8^2} = 8.944$$

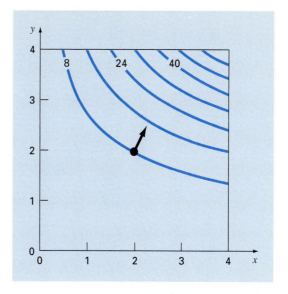

FIGURE 14.7
The arrow follows the direction of steepest ascent calculated with the gradient.

Thus, during our first step, we will initially gain 8.944 units of elevation rise for a unit distance advanced along this steepest path. Observe that Eq. (14.1) yields the same result,

$$g'(0) = 4 \cos (1.107) + 8 \sin (1.107) = 8.944$$

Note that for any other direction, say $\theta = 1.107/2 = 0.5235$, $g'(0) = 4 \cos (0.5235) + 8 \sin (0.5235) = 7.608$, which is smaller.

As we move forward, both the direction and magnitude of the steepest path will change. These changes can be quantified at each step using the gradient, and your climbing direction modified accordingly.

A final insight can be gained by inspecting Fig. 14.7. As indicated, the direction of steepest ascent is perpendicular, or *orthogonal*, to the elevation contour at the coordinate (2, 2). This is a general characteristic of the gradient.

Aside from defining a steepest path, the first derivative can also be used to discern whether an optimum has been reached. As is the case for a one-dimensional function, if the partial derivatives with respect to both x and y are zero, a two-dimensional optimum has been reached.

The Hessian. For one-dimensional problems, both the first and second derivatives provide valuable information for searching out optima. The first derivative (*a*) provides a steepest trajectory of the function and (*b*) tells us that we have reached an optimum. Once at an optimum, the second derivative tells us whether we are a maximum [negative $f''(x)$]

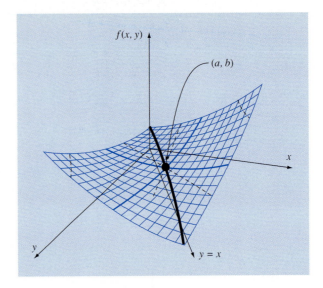

FIGURE 14.8
A saddle point ($x = a$ and $y = b$). Notice that when the curve is viewed along the x and y directions, the function appears to go through a minimum (positive second derivative), whereas when viewed along an axis $x = y$, it is concave downward (negative second derivative).

or a minimum [positive $f''(x)$]. In the previous paragraphs, we illustrated how the gradient provides best local trajectories for multidimensional problems. Now, we will examine how the second derivative is used in such contexts.

You might expect that if the partial second derivatives with respect to both x and y are both negative, then you have reached a maximum. Figure 14.8 shows a function where this is not true. The point (a, b) of this graph appears to be a minimum when observed along either the x dimension or the y dimension. In both instances, the second partial derivatives are positive. However, if the function is observed along the line $y = x$, it can be seen that a maximum occurs at the same point. This shape is called a *saddle*, and clearly, neither a maximum or a minimum occurs at the point.

Whether a maximum or a minimum occurs involves not only the partials with respect to x and y but also the second partial with respect to x and y. Assuming that the partial derivatives are continuous at and near the point being evaluated, the following quantity can be computed:

$$|H| = \frac{\partial^2 f}{\partial x^2} \frac{\partial^2 f}{\partial y^2} - \left(\frac{\partial^2 f}{\partial x \partial y} \right)^2 \tag{14.3}$$

Three cases can occur

- If $|H| > 0$ and $\partial^2 f / \partial x^2 > 0$, then $f(x, y)$ has a local minimum.
- If $|H| > 0$ and $\partial^2 f / \partial x^2 < 0$, then $f(x, y)$ has a local maximum.
- If $|H| < 0$, then $f(x, y)$ has a saddle point.

The quantity $|H|$ is equal to the determinant of a matrix made up of the second derivatives,[1]

$$
H = \begin{bmatrix} \dfrac{\partial^2 f}{\partial x^2} & \dfrac{\partial^2 f}{\partial x \partial y} \\[2ex] \dfrac{\partial^2 f}{\partial y \partial x} & \dfrac{\partial^2 f}{\partial y^2} \end{bmatrix}
\tag{14.4}
$$

where this matrix is formally referred to as the *Hessian* of f.

Besides providing a way to discern whether a multidimensional function has reached an optimum, the Hessian has other uses in optimization (e.g., for the multidimensional form of Newton's method). In particular, it allows searches to include second-order curvature to attain superior results.

Finite-Difference Approximations.   It should be mentioned that, for cases where they are difficult or inconvenient to compute analytically, both the gradient and the determinant of the Hessian can be evaluated numerically. In most cases, the approach introduced in Sec. 6.3.3 for the modified secant method is employed. That is, the independent variables can be perturbed slightly to generate the required partial derivatives. For example, if a centered-difference approach is adopted, they can be computed as

$$
\frac{\partial f}{\partial x} = \frac{f(x + \delta x, y) - f(x - \delta x, y)}{2 \delta x}
\tag{14.5}
$$

$$
\frac{\partial f}{\partial y} = \frac{f(x, y + \delta y) - f(x, y - \delta y)}{2 \delta y}
\tag{14.6}
$$

$$
\frac{\partial^2 f}{\partial x^2} = \frac{f(x + \delta x, y) - 2 f(x, y) - f(x - \delta x, y)}{\delta x^2}
\tag{14.7}
$$

$$
\frac{\partial^2 f}{\partial y^2} = \frac{f(x, y + \delta y) - 2 f(x, y) - f(x, y - \delta y)}{\delta y^2}
\tag{14.8}
$$

$$
\frac{\partial^2 f}{\partial x \partial y} =
$$

$$
\frac{f(x + \delta x, y + \delta y) - f(x + \delta x, y - \delta y) - f(x - \delta x, y + \delta y) + f(x - \delta x, y - \delta y)}{4 \delta x \delta y}
$$

$$
\tag{14.9}
$$

where δ is some small fractional value.

Note that the methods employed in commercial software packages also use forward differences. In addition, they are usually more complicated than the approximations listed in Eqs. (14.5) through (14.9). For example, the IMSL library bases the perturbation on machine epsilon. Dennis and Schnabel (1996) provide more detail on the approach.

Regardless of how the approximation is implemented, the important point is that you may have the option of evaluating the gradient and/or the Hessian analytically. This can sometimes be an arduous task, but the performance of the algorithm may benefit enough

---

[1]Note that $\partial^2 f/(\partial x \partial y) = \partial^2 f/(\partial y \partial x)$.

to make your effort worthwhile. The closed-form derivatives will be exact; but more importantly, you will reduce function evaluations. This latter point can have a critical impact on the execution time.

On the other hand, you will often exercise the option of having the quantities computed internally using numerical approaches. In many cases, the performance will be quite adequate and you will be saved the difficulty of numerous partial differentiations. Such would be the case on the optimizers used in certain spreadsheets and mathematical software packages (e.g., Excel). In such cases, you may not even be given the option of entering an analytically derived gradient and Hessian. However, for small to moderately sized problems this is usually not a major shortcoming.

14.2.2 Steepest Ascent Method

An obvious strategy for climbing a hill would be to determine the maximum slope at your starting position and then start walking in that direction. But clearly, another problem arises almost immediately. Unless you were really lucky and started on a ridge that pointed directly to the summit, as soon as you moved, your path would diverge from the steepest ascent direction.

Recognizing this fact, you might adopt the following strategy. You could walk a short distance along the gradient direction. Then you could stop, reevaluate the gradient and walk another short distance. By repeating the process you would eventually get to the top of the hill.

Although this strategy sounds superficially sound, it is not very practical. In particular, the continuous reevaluation of the gradient can be computationally demanding. A preferred approach involves moving in a fixed path along the initial gradient until $f(x, y)$ stops increasing; i.e., becomes level along your direction of travel. This stopping point becomes the starting point where ∇f is reevaluated and a new direction followed. The process is repeated until the summit is reached. This approach is called the *steepest ascent method*.[2] It is the most straightforward of the gradient search techniques. The basic idea behind the approach is depicted in Fig. 14.9.

We start at an initial point (x_0, y_0) labeled "0" in the figure. At this point, we determine the direction of steepest ascent; that is, the gradient. We then search along the direction of the gradient, h_0, until we find a maximum, which is labeled "1" in the figure. The process is then repeated.

Thus, the problem boils down to two parts: (1) determining the "best" direction to search and (2) determining the "best value" along that search direction. As we will see, the effectiveness of the various algorithms described in the coming pages depends on how clever we are at both parts.

For the time being, the steepest ascent method uses the gradient approach as its choice for the "best" direction. We have already shown how the gradient is evaluated in Example 14.1. Now, before examining how the algorithm goes about locating the maximum along the steepest direction, we must pause to explore how to transform a function of x and y into a function of h along the gradient direction.

---

[2]Because of our emphasis on maximization here, we use the terminology *steepest ascent*. The same approach can also be used for minimization, in which case the terminology *steepest descent* is used.

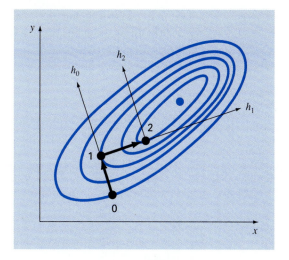

FIGURE 14.9
A graphical depiction of the method of steepest ascent.

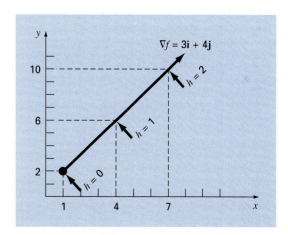

FIGURE 14.10
The relationship between an arbitrary direction h and x and y coordinates.

Starting at x_0, y_0 the coordinates of any point in the gradient direction can be expressed as

$$x = x_0 + \frac{\partial f}{\partial x} h \tag{14.10}$$

$$y = y_0 + \frac{\partial f}{\partial y} h \tag{14.11}$$

where h is distance along the h axis. For example, suppose $x_0 = 1$ and $y_0 = 2$ and $\nabla f = 3\mathbf{i} + 4\mathbf{j}$, as shown in Fig. 14.10. The coordinates of any point along the h axis are given by

$$x = 1 + 3h \tag{14.12}$$

$$y = 2 + 4h \tag{14.13}$$

The following example illustrates how we can use these transformations to convert a two-dimensional function of x and y into a one-dimensional function in h.

EXAMPLE 14.3   Developing a 1-D Function Along the Gradient Direction

Problem Statement.   Suppose we have the following two-dimensional function:

$$f(x, y) = 2xy + 2x - x^2 - 2y^2$$

Develop a one-dimensional version of this equation along the gradient direction at point $x = -1$ and $y = 1$.

Solution.   The partial derivatives can be evaluated at $(-1, 1)$,

$$\frac{\partial f}{\partial x} = 2y + 2 - 2x = 2(1) + 2 - 2(-1) = 6$$

$$\frac{\partial f}{\partial y} = 2x - 4y = 2(-1) - 4(1) = -6$$

Therefore, the gradient vector is

$$\nabla f = 6\mathbf{i} - 6\mathbf{j}$$

To find the maximum, we could search along the gradient direction; that is, along an h axis running along the direction of this vector. The function can be expressed along this axis as

$$f\left(x_0 + \frac{\partial f}{\partial x}h, y_0 + \frac{\partial f}{\partial y}h\right) = f(-1 + 6h, 1 - 6h)$$

$$= 2(-1 + 6h)(1 - 6h) + 2(-1 + 6h) - (-1 + 6h)^2 - 2(1 - 6h)^2$$

where the partial derivatives are evaluated at $x = -1$ and $y = 1$.

By combining terms, we develop a one-dimensional function $g(h)$ that maps $f(x, y)$ along the h axis,

$$g(h) = -180h^2 + 72h - 7$$

Now that we've developed a function along the path of steepest ascent, we can explore how to answer the second question. That is, how far along this path do we travel? One approach might be to move along this path until we find the maximum of this function. We will call the location of this maximum h^*. This is the value of the step that maximizes g (and hence, f) in the gradient direction. This problem is equivalent to finding the maximum of a function of a single variable h. This can be done using different one-dimensional search techniques like the ones we discussed in Chap. 13. Thus, we convert from finding the optimum of a two-dimensional function to performing a one-dimensional search along the gradient direction.

This method is called *steepest ascent* when an arbitrary step size h is used. If a value of a single step $h*$ is found that brings us directly to the maximum along the gradient direction, the method is called the *optimal steepest ascent*.

EXAMPLE 14.4   Optimal Steepest Ascent

Problem Statement.   Maximize the following function:

$$f(x, y) = 2xy + 2x - x^2 - 2y^2$$

using initial guesses, $x = -1$ and $y = 1$.

Solution.   Because this function is so simple, we can first generate an analytical solution. To do this, the partial derivatives can be evaluated as

$$\frac{\partial f}{\partial x} = 2y + 2 - 2x = 0$$

$$\frac{\partial f}{\partial y} = 2x - 4y = 0$$

This pair of equations can be solved for the optimum, $x = 2$ and $y = 1$. The second partial derivatives can also be determined and evaluated at the optimum,

$$\frac{\partial^2 f}{\partial x^2} = -2$$

$$\frac{\partial^2 f}{\partial y^2} = -4$$

$$\frac{\partial^2 f}{\partial x \partial y} = \frac{\partial^2 f}{\partial y \partial x} = 2$$

and the determinant of the Hessian is computed [Eq. (14.3)],

$$|H| = -2(-4) - 2^2 = 4$$

Therefore, because $|H| > 0$ and $\partial^2 f/\partial x^2 < 0$, function value $f(2, 1)$ is a maximum.

Now let's implement steepest ascent. Recall that, at the end of Example 14.3, we had already implemented the initial steps of the problem by generating

$$g(h) = -180h^2 + 72h - 7$$

Now, because this is a simple parabola, we can directly locate the maximum (i.e., $h = h*$) by solving the problem,

$$g'(h*) = 0$$

$$-360h* + 72 = 0$$

$$h* = 0.2$$

This means that if we travel along the h axis, $g(h)$ reaches a minimum value when $h = h* = 0.2$. This result can be placed back into Eqs. (14.10) and (14.11) to solve for the

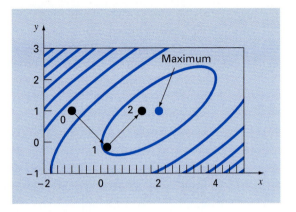

FIGURE 14.11
The method of optimal steepest ascent.

(x, y) coordinates corresponding to this point,

$$x = -1 + 6(0.2) = 0.2$$
$$y = 1 - 6(0.2) = -0.2$$

This step is depicted in Fig. 14.11 as the move from point 0 to 1.

The second step is merely implemented by repeating the procedure. First, the partial derivatives can be evaluated at the new starting point $(0.2, -0.2)$ to give

$$\frac{\partial f}{\partial x} = 2(-0.2) + 2 - 2(0.2) = 1.2$$

$$\frac{\partial f}{\partial y} = 2(0.2) - 4(-0.2) = 1.2$$

Therefore, the gradient vector is

$$\nabla f = 1.2\,\mathbf{i} + 1.2\,\mathbf{j}$$

This means that the steepest direction is now pointed up and to the right at a 45° angle with the x axis (see Fig. 14.11). The coordinates along this new h axis can now be expressed as

$$x = 0.2 + 1.2h$$
$$y = -0.2 + 1.2h$$

Substituting these values into the function yields

$$f(0.2 + 1.2h, -0.2 + 1.2h) = g(h) = -1.44h^2 + 2.88h + 0.2$$

The step h^* to take us to the maximum along the search direction can then be directly computed as

$$g'(h^*) = -2.88h^* + 2.88 = 0$$
$$h^* = 1$$

This result can be placed back into Eqs. (14.10) and (14.11) to solve for the (x, y) coordinates corresponding to this new point,

$$x = 0.2 + 1.2(1) = 1.4$$
$$y = -0.2 + 1.2(1) = 1$$

As depicted in Fig. 14.11, we move to the new coordinates, labeled point 2 in the plot, and in so doing move closer to the maximum. The approach can be repeated with the final result converging on the analytical solution, $x = 2$ and $y = 1$.

It can be shown that the method of steepest descent is linearly convergent. Further, it tends to move very slowly along long, narrow ridges. This is because the new gradient at each maximum point will be perpendicular to the original direction. Thus, the technique takes many small steps criss-crossing the direct route to the summit. Hence, although it is reliable, there are other approaches that converge much more rapidly, particularly in the vicinity of an optimum. The remainder of the section is devoted to such methods.

14.2.3 Advanced Gradient Approaches

Conjugate Gradient Method (Fletcher-Reeves).   In Sec. 14.1.2, we have seen how conjugate directions in Powell's method greatly improved the efficiency of a univariate search. In a similar manner, we can also improve the linearly convergent steepest ascent using conjugate gradients. In fact, an optimization method that makes use of conjugate gradients to define search directions can be shown to be quadratically convergent. This also ensures that the method will optimize a quadratic function exactly in a finite number of steps regardless of the starting point. Since most well-behaved functions can be approximated reasonably well by a quadratic in the vicinity of an optimum, quadratically convergent approaches are often very efficient near an optimum.

We have seen how starting with two arbitrary search directions, Powell's method produced new conjugate search directions. This method is quadratically convergent and does not require gradient information. On the other hand, if evaluation of derivatives is practical, we can devise algorithms that combine the ideas of steepest descent and conjugate directions to achieve robust initial performance and rapid convergence as the technique gravitates towards the optimum. The *Fletcher-Reeves conjugate gradient algorithm* modifies the steepest-ascent method by imposing the condition that successive gradient search directions be mutually conjugate. The proof and algorithm are beyond the scope of the text but are described by Rao (1996).

Newton's Method.   Newton's method for a single variable (recall Sec. 13.3) can be extended to multivariate cases. Write a second-order Taylor series for $f(\mathbf{x})$ near $\mathbf{x} = \mathbf{x}_i$,

$$f(\mathbf{x}) = f(\mathbf{x}_i) + \nabla f^T(\mathbf{x}_i)(\mathbf{x} - \mathbf{x}_i) + \frac{1}{2}(\mathbf{x} - \mathbf{x}_i)^T H_i(\mathbf{x} - \mathbf{x}_i)$$

where H_i is the Hessian matrix. At the minimum,

$$\frac{\partial f(\mathbf{x})}{\partial \mathbf{x}_j} = 0 \qquad \text{for } j = 1, 2, \ldots, n$$

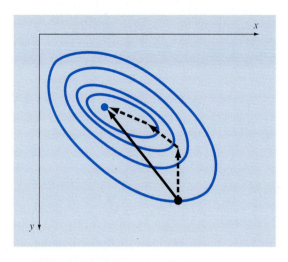

FIGURE 14.12
When the starting point is close to the optimal point, following the gradient can be inefficient. Newton methods attempt to search along a direct path to the optimum (solid line).

Thus,

$$\nabla f = \nabla f(\mathbf{x}_i) + H_i(\mathbf{x} - \mathbf{x}_i) = 0$$

If H is nonsingular,

$$\mathbf{x}_{i+1} = \mathbf{x}_i - H_i^{-1}\nabla f \tag{14.14}$$

which can be shown to converge quadratically near the optimum. This method again performs better than the steepest ascent method (see Fig. 14.12). However, note that the method requires both the computation of second derivatives and matrix inversion at each iteration. Thus, the method is not very useful in practice for functions with large numbers of variables. Furthermore, Newton's method may not converge if the starting point is not close to the optimum.

Marquardt Method.   We know that the method of steepest ascent increases the function value even if the starting point is far from an optimum. On the other hand, we have just described Newton's method, which converges rapidly near the maximum. *Marquardt's method* uses the steepest descent method when $\mathbf{x}$ is far from $\mathbf{x}^*$, and Newton's method when $\mathbf{x}$ closes in on an optimum. This is accomplished by modifying the diagonal of the Hessian in Eq. (14.14),

$$\tilde{H}_i = H_i + \alpha_i I$$

where α_i is a positive constant and I is the identity matrix. At the start of the procedure, α_i is assumed to be large and

$$\tilde{H}_i^{-1} \approx \frac{1}{\alpha_i}I$$

which reduces Eq. (14.14) to the steepest ascent method. As the iterations proceed, α_i approaches zero and the method becomes Newton's method.

Thus, Marquardt's method offers the best of both worlds: it plods along reliably from poor initial starting values yet accelerates rapidly when it approaches the optimum. Unfortunately, the method still requires Hessian evaluation and matrix inversion at each step.

It should be noted that the Marquardt method is primarily used for nonlinear least-squares problems. For example, the IMSL library contains a subroutine for this purpose.

Quasi-Newton Methods. *Quasi-Newton*, or *variable metric, methods* seek to estimate the direct path to the optimum in a manner similar to Newton's method. However, notice that the Hessian matrix in Eq. (14.14) is composed of the second derivatives of f that vary from step to step. Quasi-Newton methods attempt to avoid these difficulties by approximating H with another matrix A using only first partial derivatives of f. The approach involves starting with an initial approximation of H^{-1} and updating and improving it with each iteration. The methods are called quasi-Newton because we do not use the true Hessian, rather an approximation. Thus, we have two approximations at work simultaneously: (1) the original Taylor-series approximation and (2) the Hessian approximation.

There are two primary methods of this type: the *Davidon-Fletcher-Powell* (DFP) and the *Broyden-Fletcher-Goldfarb-Shanno* (BFGS) algorithms. They are similar except for details concerning how they handle round-off error and convergence issues. BFGS is generally recognized as being superior in most cases. Rao (1996) provides details and formal statements of both the DFP and the BFGS algorithms.

PROBLEMS

14.1 Find the directional derivative of

$$f(x, y) = 2x^2 + y^2$$

at $x = 2$ and $y = 2$ in the direction of $\boldsymbol{h} = 3\boldsymbol{i} + 2\boldsymbol{j}$.

14.2 Find the gradient vector and Hessian matrix for each of the following functions:

(a) $f(x, y) = 2xy^2 + 3e^{xy}$

(b) $f(x, y, z) = x^2 + y^2 + 2z^2$

(c) $f(x, y) = \ln(x^2 + 2xy + 3y^2)$

14.3 Given

$$f(x, y) = 2xy + 1.5y - 1.25x^2 - 2y^2$$

Construct and solve a system of linear algebraic equations that maximizes $f(x)$. Note that this is done by setting the partial derivatives of f with respect to both x and y to zero.

14.4

(a) Start with an initial guess of $x = 1$ and $y = 1$ and apply two applications of the steepest ascent method to $f(x, y)$ from Prob. 14.3.

(b) Construct a plot from the results of (a) showing the path of the search.

14.5 Find the minimum value of

$$f(x, y) = (x - 2)^2 + (y - 3)^2$$

starting at $x = 1$ and $y = 1$, using the steepest descent method with a stopping criterion of $\varepsilon_s = 1\%$

14.6 Perform one iteration of the steepest ascent method to locate the maximum of

$$f(x, y) = 3.5x + 2y + x^2 - x^4 - 2xy - y^2$$

using initial guesses $x = 0$, $y = 0$. Employ bisection to find the optimal step size in the gradient search direction.

14.7 Perform one iteration of the optimal gradient steepest descent method to locate the minimum of

$$f(x, y) = -7x + 1.2x^2 + 11y + 2y^2 - 2xy$$

using initial guesses $x = 0$ and $y = 0$.

14.8 Develop a subprogram using a program or macro language to implement the random search method. Design the subprogram so that it is expressly designed to locate a maximum. Test the program with $f(x, y)$ from Prob. 14.6. Use a range of -2 to 2 for both x and y.

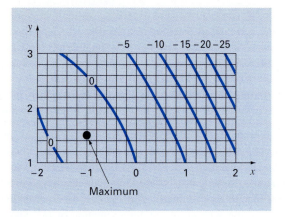

FIGURE P14.9
The grid search.

14.9 The *grid search* is another brute force approach to optimization. The two-dimensional version is depicted in Fig. P14.9. The *x* and *y* dimensions are divided into increments to create a grid. The function is then evaluated at each node of the grid. The denser the grid, the more likely it would be to locate the optimum.

Develop a subprogram using a program or macro language to implement the grid search method. Design the subprogram so that it is expressly designed to locate a maximum. Test your program with the same problem as Example 14.1.

CHAPTER 15

Constrained Optimization

This chapter deals with optimization problems where constraints come into play. We first discuss problems where both the objective function and the constraints are linear. For such cases, special methods are available that exploit the linearity of the underlying functions. Called linear programming methods, the resulting algorithms solve very large problems with thousands of variables and constraints with great efficiency. They are used in a wide range of problems in engineering and management.

Then we will turn briefly to the more general problem of nonlinear constrained optimization. Finally, we provide an overview of how packages and libraries can be employed for optimization.

15.1 LINEAR PROGRAMMING

Linear programming (or LP for short) is an optimization approach that deals with meeting a desired objective such as maximizing profit or minimizing cost in the presence of constraints such as limited resources. The term *linear* connotes that the mathematical functions representing both the objective and the constraints are linear. The term *programming* does not mean "computer programming," but rather, connotes "scheduling" or "setting an agenda" (Revelle et al. 1997).

15.1.1 Standard Form

The basic linear programming problem consists of two major parts: the objective function and a set of constraints. For a maximization problem, the objective function is generally expressed as

$$\text{Maximize } Z = c_1 x_1 + c_2 x_2 + \cdots + c_n x_n \tag{15.1}$$

where c_j = payoff of each unit of the jth activity that is undertaken and x_j = magnitude of the jth activity. Thus, the value of the objective function, Z, is the total payoff due to the total number of activities, n.

The constraints can be represented generally as

$$a_{i1} x_1 + a_{i2} x_2 + \cdots + a_{in} x_n \leq b_i \tag{15.2}$$

where a_{ij} = amount of the ith resource that is consumed for each unit of the jth activity and b_i = amount of the ith resource that is available. That is, the resources are limited.

The second general type of constraint specifies that all activities must have a positive value,

$$x_1 \geq 0 \tag{15.3}$$

In the present context, this expresses the realistic notion that, for some problems, negative activity is physically impossible (for example, we cannot produce negative goods).

Together, the objective function and the constraints specify the linear programming problem. They say that we are trying to maximize the payoff for a number of activities under the constraint that these activities utilize finite amounts of resources. Before showing how this result can be obtained, we will first develop an example.

EXAMPLE 15.1   Setting Up the LP Problem

Problem Statement. The following problem is developed from the area of chemical or petroleum engineering. However, it is relevant to all areas of engineering that deal with producing products with limited resources.

Suppose that a gas processing plant receives a fixed amount of raw gas each week. The raw gas is processed into two grades of heating gas, regular and premium quality. These grades of gas are in high demand (that is, they are guaranteed to sell) and yield different profits to the company. However, their production involves both time and on-site storage constraints. For example, only one of the grades can be produced at a time, and the facility is open for only 80 hrs/week. Further, there is limited on-site storage for each of the products. All these factors are listed below (note that a metric ton, or *tonne*, is equal to 1000 kg):

| | **Product** | | |
Resource	Regular	Premium	Resource Availability
Raw gas	7 m³/tonne	11 m³/tonne	77 m³/week
Production time	10 hr/tonne	8 hr/tonne	80 hr/week
Storage	9 tonnes	6 tonnes	
Profit	150/tonne	175/tonne	

Develop a linear programming formulation to maximize the profits for this operation.

Solution. The engineer operating this plant must decide how much of each gas to produce to maximize profits. If the amounts of regular and premium produced weekly are designated as x_1 and x_2, respectively, the total weekly profit can be calculated as

Total profit = $150x_1 + 175x_2$

or written as a linear programming objective function,

Maximize $Z = 150x_1 + 175x_2$

The constraints can be developed in a similar fashion. For example, the total raw gas used can be computed as

Total gas used = $7x_1 + 11x_2$

This total cannot exceed the available supply of 77 m³/week, so the constraint can be represented as

$$7x_1 + 11x_2 \leq 77$$

The remaining constraints can be developed in a similar fashion, with the resulting total LP formulation given by

Maximize $Z = 150x_1 + 175x_2$ (maximize profit)

subject to

$7x_1 + 11x_2 \leq 77$	(material constraint)
$10x_1 + 8x_2 \leq 80$	(time constraint)
$x_1 \leq 9$	("regular" storage constraint)
$x_2 \leq 6$	("premium" storage constraint)
$x_1, x_2 \geq 0$	(positivity constraints)

Note that the above set of equations constitute the total LP formulation. The parenthetical explanations at the right have been appended to clarify the meaning of each term.

15.1.2 Graphical Solution

Because they are limited to two or three dimensions, graphical solutions have limited practical utility. However, they are very useful for demonstrating some basic concepts that underlie the general algebraic techniques used to solve higher-dimensional problems with the computer.

For a two-dimensional problem, such as the one in Example 15.1, the solution space is defined as a plane with x_1 measured along the abscissa and x_2 along the ordinate. Because they are linear, the constraints can be plotted on this plane as straight lines. If the LP problem was formulated properly (that is, it has a solution), these constraint lines will delineate a region, called the *feasible solution space,* encompassing all possible combinations of x_1 and x_2 that obey the constraints and hence represent feasible solutions. The objective function for a particular value of Z can then be plotted as another straight line and superimposed on this space. The value of Z can then be adjusted until it is at the maximum value while still touching the feasible space. This value of Z represents the optimal solution. The corresponding values of x_1 and x_2, where Z touches the feasible solution space, represent the optimal values for the activities. The following example should help clarify the approach.

EXAMPLE 15.2   Graphical Solution

Problem Statement.   Develop a graphical solution for the gas processing problem previously derived in Example 15.1:

Maximize $Z = 150x_1 + 175x_2$

subject to

$7x_1 + 11x_2 \leq 77$	(1)
$10x_1 + 8x_2 \leq 80$	(2)
$x_1 \leq 9$	(3)

$$x_2 \leq 6 \tag{4}$$

$$x_1 \geq 0 \tag{5}$$

$$x_2 \geq 0 \tag{6}$$

We have numbered the constraints to identify them in the following graphical solution.

Solution. First, the constraints can be plotted on the solution space. For example, the first constraint can be reformulated as a line by replacing the inequality by an equal sign and solving for x_2:

$$x_2 = -\frac{7}{11}x_1 + 7$$

Thus, as in Fig. 15.1a, the possible values of x_1 and x_2 that obey this constraint fall below this line (the direction designated in the plot by the small arrow). The other constraints can be evaluated similarly, as superimposed on Fig. 15.1a. Notice how they encompass a region where they are all met. This is the feasible solution space (the area ABCDE in the plot).

Aside from defining the feasible space, Fig. 15.1a also provides additional insight. In particular, we can see that constraint 3 (storage of regular gas) is "redundant." That is, the feasible solution space is unaffected if it were deleted.

Next, the objective function can be added to the plot. To do this, a value of Z must be chosen. For example, for $Z = 0$, the objective function becomes

$$0 = 150x_1 + 175x_2$$

FIGURE 15.1
Graphical solution of a linear programming problem. (a) The constraints define a feasible solution space. (b) The objective function can be increased until it reaches the highest value that obeys all constraints. Graphically, the function moves up and to the right until it touches the feasible space at a single optimal point.

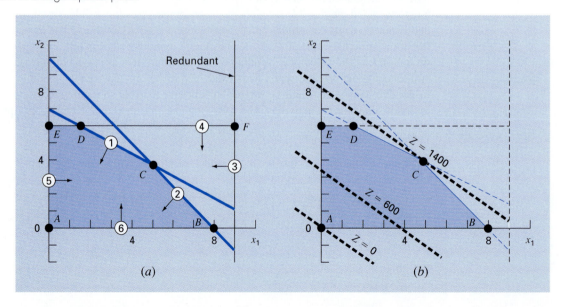

or, solving for x_2, we derive the line

$$x_2 = -\frac{150}{175}x_1$$

As displayed in Fig. 15.1b, this represents a dashed line intersecting the origin. Now, since we are interested in maximizing Z, we can increase it to say, 600, and the objective function is

$$x_2 = \frac{600}{175} - \frac{150}{175}x_1$$

Thus, increasing the value of the objective function moves the line away from the origin. Because the line still falls within the solution space, our result is still feasible. For the same reason, however, there is still room for improvement. Hence, Z can keep increasing until a further increase will take the objective beyond the feasible region. As shown in Fig. 15.1b, the maximum value of Z corresponds to approximately 1400. At this point, x_1 and x_2 are equal to approximately 4.9 and 3.9, respectively. Thus, the graphical solution tells us that if we produce these quantities of regular and premium, we will reap a maximum profit of about 1400.

Aside from determining optimal values, the graphical approach provides further insights into the problem. This can be appreciated by substituting the answers back into the constraint equations,

$$7(4.9) + 11(3.9) \cong 77$$
$$10(4.9) + 8(3.9) \cong 80$$
$$4.9 \leq 9$$
$$3.9 \leq 6$$

Consequently, as is also clear from the plot, producing at the optimal amount of each product brings us right to the point where we just meet the resource (1) and time constraints (2). Such constraints are said to be *binding*. Further, as is also evident graphically, neither of the storage constraints [(3) and (4)] acts as a limitation. Such constraints are called *nonbinding*. This leads to the practical conclusion that, for this case, we can increase profits by either increasing our resource supply (the raw gas) or increasing our production time. Further, it indicates that increasing storage would have no impact on profit.

The result obtained in the previous example is one of four possible outcomes that can be generally obtained in a linear programming problem. These are

1. *Unique solution.* As in the example, the maximum objective function intersects a single point.
2. *Alternate solutions.* Suppose that the objective function in the example had coefficients so that it was precisely parallel to one of the constraints. In our example problem, one way in which this would occur would be if the profits were changed to $140/tonne and $220/tonne. Then, rather than a single point, the problem would have an infinite number of optima corresponding to a line segment (Fig. 15.2a).

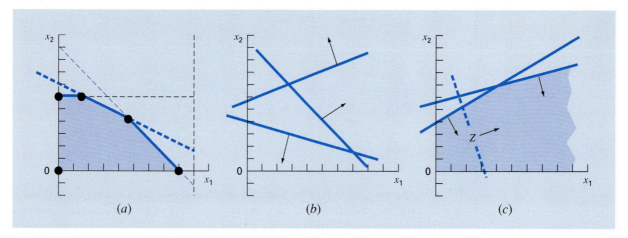

FIGURE 15.2
Aside from a single optimal solution (for example, Fig. 15.1b), there are three other possible outcomes of a linear programming problem: (a) alternative optima, (b) no feasible solution, and (c) an unbounded result.

3. *No feasible solution.* As in Fig. 15.2b, it is possible that the problem is set up so that there is no feasible solution. This can be due to dealing with an unsolvable problem or due to errors in setting up the problem. The latter can result if the problem is overconstrained to the point that no solution can satisfy all constraints.
4. *Unbounded problems.* As in Fig. 15.2c, this usually means that the problem is underconstrained and therefore open-ended. As with the no-feasible-solution case, it can often arise from errors committed during problem specification.

Now let's suppose that our problem involves a unique solution. The graphical approach might suggest an enumerative strategy for hunting down the maximum. From Fig. 15.1, it should be clear that the optimum always occurs at one of the corner points where two constraints meet. Such a point is known formally as an *extreme point*. Thus, out of the infinite number of possibilities in the decision space, focusing on extreme points clearly narrows down the possible options.

Further, we can recognize that not every extreme point is feasible; that is, satisfying all constraints. For example, notice that point *F* in Fig. 15.1a is an extreme point but is not feasible. Limiting ourselves to *feasible extreme points* narrows the field down still further.

Finally, once all feasible extreme points are identified, the one yielding the best value of the objective function represents the optimum solution. Finding this optimal solution could be done by exhaustively (and inefficiently) evaluating the value of the objective function at every feasible extreme point. The following section discusses the simplex method, which offers a preferable strategy that charts a selective course through a sequence of feasible extreme points to arrive at the optimum in an extremely efficient manner.

15.1.3 The Simplex Method

The simplex method is predicated on the assumption that the optimal solution will be an extreme point. Thus, the approach must be able to discern whether during problem solution an extreme point occurs. To do this, the constraint equations are reformulated as equalities by introducing what are called slack variables.

Slack Variables.   As the name implies, a *slack variable* measures how much of a constrained resource is available; that is, how much "slack" of the resource is available. For example, recall the resource constraint used in Examples 15.1 and 15.2,

$$7x_1 + 11x_2 \leq 77$$

We can define a slack variable S_1 as the amount of raw gas that is not used for a particular production level (x_1, x_2). If this quantity is added to the left side of the constraint, it makes the relationship exact,

$$7x_1 + 11x_2 + S_1 = 77$$

Now recognize what the slack variable tells us. If it is positive, it means that we have some "slack" for this constraint. That is, we have some surplus resource that is not being fully utilized. If it is negative, it tells us that we have exceeded the constraint. Finally, if it is zero, we exactly meet the constraint. That is, we have used up all the allowable resource. Since this is exactly the condition where constraint lines intersect, the slack variable provides a means to detect extreme points.

A different slack variable is developed for each constraint equation, resulting in what is called the *fully augmented version,*

$$\text{Maximize } Z = 150x_1 + 175x_2$$

subject to

$$7x_1 + 11x_2 + S_1 \qquad\qquad = 77 \tag{15.4a}$$
$$10x_1 + 8x_2 \quad + S_2 \qquad\quad = 80 \tag{15.4b}$$
$$x_1 \qquad\qquad + S_3 \quad = 9 \tag{15.4c}$$
$$x_2 \qquad\qquad + S_4 = 6 \tag{15.4d}$$
$$x_1, \ x_2, \ S_1, \ S_2, \ S_3, \ S_4 \ \geq \ 0$$

Notice how we have set up the four equality equations so that the unknowns are aligned in columns. We did this to underscore that we are now dealing with a system of linear algebraic equations (recall Part Three). In the following section, we will show how these equations can be used to determine extreme points algebraically.

Algebraic Solution.   In contrast to Part Three, where we had n equations with n unknowns, our example system [Eqs. (15.4)] is *underspecified;* that is, it has more unknowns than equations. In general terms, there are n *structural variables* (that is, the original unknowns), m *surplus* or *slack variables* (one per constraint), and $n + m$ total variables (structural plus surplus). For the gas production problem we have 2 structural variables, 4 slack variables, and 6 total variables. Thus, the problem involves solving 4 equations with 6 unknowns.

The difference between the number of unknowns and the number of equations (equal to 2 for our problem) is directly related to how we can distinguish a feasible extreme point. Specifically, every feasible point has 2 variables out of 6 equal to zero. For example, the five corner points of the area *ABCDE* have the following zero values:

Extreme Point	Zero Variables
A	x_1, x_2
B	x_2, S_2
C	S_1, S_2
D	S_1, S_4
E	x_1, S_4

This observation leads to the conclusion that the extreme points can be determined from the standard form by setting two of the variables equal to zero. In our example, this reduces the problem to a solvable form of 4 equations with 4 unknowns. For example, for point *E*, setting $x_1 = S_4 = 0$ reduces the standard form to

$$11x_2 + S_1 \qquad\qquad = 77$$
$$8x_2 \qquad + S_2 \qquad = 80$$
$$\qquad\qquad + S_3 = 9$$
$$x_2 \qquad\qquad = 6$$

which can be solved for $x_2 = 6$, $S_1 = 11$, $S_2 = 32$, and $S_3 = 9$. Together with $x_1 = S_4 = 0$, these values define point *E*.

To generalize, a basic solution for *m* linear equations with *n* unknowns is developed by setting $n - m$ variables to zero, and solving the *m* equations for the *m* remaining unknowns. The zero variables are formally referred to as *nonbasic variables*, whereas the remaining *m* variables are called *basic variables*. If all the basic variables are nonnegative, the result is called a *basic feasible solution*. The optimum will be one of these.

Now a direct approach to determining the optimal solution would be to calculate all the basic solutions, determine which were feasible, and among those, which had the highest value of *Z*. There are two reasons why this is not a wise approach.

First, for even moderately sized problems, the approach can involve solving a great number of equations. For *m* equations with *n* unknowns, this results in solving

$$C_m^n = \frac{n!}{m!(n-m)!}$$

simultaneous equations. For example, if there are 10 equations ($m = 10$) with 16 unknowns ($n = 16$), you would have 8008 [$= 16!/(10!\ 6!)$] 10×10 systems of equations to solve!

Second, a significant portion of these may be infeasible. For example, in the present problem, out of $C_6^4 = 15$ extreme points, only 5 are feasible. Clearly, if we could avoid solving all these unnecessary systems, a more efficient algorithm would be developed. Such an approach is described next.

Simplex Method Implementation. The simplex method avoids inefficiencies outlined in the previous section. It does this by starting with a basic feasible solution. Then it moves through a sequence of other basic feasible solutions that successively improve the value of the objective function. Eventually, the optimal value is reached and the method is terminated.

We will illustrate the approach using the gas processing problem from Examples 15.1 and 15.2. The first step is to start at a basic feasible solution (that is, at an extreme corner point of the feasible space). For cases like ours, an obvious starting point would be point A; that is, $x_1 = x_2 = 0$. The original 6 equations with 4 unknowns become

$$S_1 \qquad\qquad = 77$$
$$\quad S_2 \qquad\quad = 80$$
$$\qquad S_2 \quad\; = 9$$
$$\qquad\quad S_4 = 6$$

Thus, the starting values for the basic variables are given automatically as being equal to the right-hand sides of the constraints.

Before proceeding to the next step, the beginning information can now be summarized in a convenient tabular format called a *tableau*. As shown below, the tableau provides a concise summary of the key information constituting the linear programming problem.

Basic	Z	x_1	x_2	S_1	S_2	S_3	S_4	Solution	Intercept
Z	1	−150	−175	0	0	0	0	0	
S_1	0	7	11	1	0	0	0	77	11
S_2	0	10	8	0	1	0	0	80	8
S_3	0	1	0	0	0	1	0	9	9
S_4	0	0	1	0	0	0	1	6	∞

Notice that for the purposes of the tableau, the objective function is expressed as

$$Z - 150x_1 - 175x_2 - 0S_1 - 0S_2 - 0S_3 - 0S_4 = 0 \tag{15.5}$$

The next step involves moving to a new basic feasible solution that leads to an improvement of the objective function. This is accomplished by increasing a current nonbasic variable (at this point, x_1 or x_2) above zero so that Z increases. Recall that, for the present example, extreme points must have 2 zero values. Therefore, one of the current basic variables (S_1, S_2, S_3, or S_4) must also be set to zero.

To summarize this important step: one of the current nonbasic variables must be made basic (nonzero). This variable is called the *entering variable*. In the process, one of the current basic variables is made nonbasic (zero). This variable is called the *leaving variable*.

Now, let's develop a mathematical approach for choosing the entering and leaving variables. Because of the convention by which the objective function is written [(Eq. (15.5)], the entering variable can be any variable in the objective function having a negative coefficient (because this will make Z bigger). The variable with the largest negative value is conventionally chosen because it usually leads to the largest increase in Z. For our case, x_2

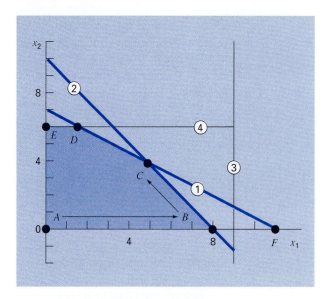

FIGURE 15.3

Graphical depiction of how the simplex method successively moves through feasible basic solutions to arrive at the optimum in an efficient manner.

would be the entering variable since its coefficient, -175, is more negative than the coefficient of x_1, -150.

At this point the graphical solution can be consulted for insight. As in Fig. 15.3, we start at the initial point A. Based on its coefficient, x_2 should be chosen to enter. However, to keep the present example brief, we choose x_1 since we can see from the graph that this will bring us to the maximum quicker.

Next, we must choose the leaving variable from among the current basic variables— $S_1, S_2, S_3,$ or S_4. Graphically, we can see that there are two possibilities. Moving to point B will drive S_2 to zero, whereas moving to point F will drive S_1 to zero. However, the graph also makes it clear that F is not possible because it lies outside the feasible solution space. Thus, we decide to move from A to B.

How is the same result detected mathematically? One way is to calculate the values at which the constraint lines intersect the axis or line corresponding to the leaving variable (in our case, the x_1 axis). We can calculate this value as the ratio of the right-hand side of the constraint (the "Solution" column of the tableau) to the corresponding coefficient of x_1. For example, for the first constraints slack variable S_1, the result is

$$\text{Intercept} = \frac{77}{7} = 11$$

The remaining intercepts can be calculated and listed as the last column of the tableau. Because 8 is the smallest positive integer, it means that the second constraint line will be reached first as x_1 is increased. Hence, S_2 should be the entering variable.

At this point, we have moved to point B ($x_2 = S_2 = 0$), and the new basic solution becomes

$$7x_1 + S_1 \qquad = 77$$
$$10x_1 \qquad\qquad = 80$$
$$x_1 \qquad + S_3 \quad = 9$$
$$S_4 = 6$$

The solution of this system of equations effectively defines the values of the basic variables at point B: $x_1 = 8$, $S_1 = 21$, $S_3 = 1$, and $S_4 = 6$.

The tableau can be used to make the same calculation by employing the Gauss-Jordan method. Recall that the basic strategy behind Gauss-Jordan involved converting the pivot element to 1 and then eliminating the coefficients in the same column above and below the pivot element (recall Sec. 9.7).

For this example, the pivot row is S_2 (the entering variable) and the pivot element is 10 (the coefficient of the leaving variable, x_1). Dividing the row by 10 and replacing S_2 by x_1 gives

Basic	Z	x_1	x_2	S_1	S_2	S_3	S_4	Solution	Intercept
Z	1	−150	−175	0	0	0	0	0	
S_1	0	7	11	1	0	0	0	77	
x_1	0	1	0.8	0	0.1	0	0	8	
S_3	0	1	0	0	0	1	0	9	
S_4	0	0	1	0	0	0	1	6	

Next, the x_1 coefficients in the other rows can be eliminated. For example, for the objective function row, the pivot row is multiplied by −150 and the result subtracted from the first row to give

Z	x_1	x_2	S_1	S_2	S_3	S_4	Solution
1	−150	−175	0	0	0	0	0
−0	−(−150)	−(−120)	−0	−(−15)	0	0	−(−1200)
1	0	−55	0	15	0	0	1200

Similar operations can be performed on the remaining rows to give the new tableau,

Basic	Z	x_1	x_2	S_1	S_2	S_3	S_4	Solution	Intercept
Z	1	0	−55	0	15	0	0	1200	
S_1	0	0	5.4	1	−0.7	0	0	21	3.889
x_1	0	1	0.8	0	0.1	0	0	8	10
S_3	0	0	−0.8	0	−0.1	1	0	1	−1.25
S_4	0	0	1	0	0	0	1	6	6

Thus, the new tableau summarizes all the information for point B. This includes the fact that the move has increased the objective function to $Z = 1200$.

This tableau can then be used to chart our next, and in this case final, step. Only one more variable, x_2, has a negative value in the objective function, and it is therefore chosen as the leaving variable. According to the intercept values (now calculated as the solution column over the coefficients in the x_2 column), the first constraint has the smallest positive value, and therefore, S_1 is selected as the entering variable. Thus, the simplex method moves us from points B to C in Fig. 15.3. Finally, the Gauss-Jordan elimination can be implemented to solve the simultaneous equations. The result is the final tableau,

Basic	Z	x_1	x_2	S_1	S_2	S_3	S_4	Solution
Z	1	0	0	10.1852	7.8704	0	0	1413.889
x_2	0	0	1	0.1852	−0.1296	0	0	3.889
x_1	0	1	0	−0.1481	0.2037	0	0	4.889
S_3	0	0	0	0.1481	−0.2037	1	0	4.111
S_4	0	0	0	−0.1852	0.1296	0	1	2.111

We know that the result is final because there are no negative coefficients remaining in the objective function row. The final solution is tabulated as $x_1 = 3.889$ and $x_2 = 4.889$, which give a maximum objective function of $Z = 1413.889$. Further, because S_3 and S_4 are still in the basis, we know that the solution is limited by the first and second constraints.

15.2   NONLINEAR CONSTRAINED OPTIMIZATION

There are a number of approaches for handling nonlinear optimization problems in the presence of constraints. These can generally be divided into indirect and direct approaches (Rao, 1996). A typical indirect approach uses so-called *penalty functions*. These involve placing additional expressions to make the objective function less optimal as the solution approaches a constraint. Thus, the solution will be discouraged from violating constraints. Although such methods can be useful in some problems, they can become arduous when the problem involves many constraints.

The *generalized reduced gradient search,* or *GRG,* method is one of the more popular of the direct methods (for details, see Lasdon et al., 1978; Lasdon and Smith, 1992). It is, in fact, the nonlinear method used within the Excel Solver.

It first "reduces" the problem to an unconstrained optimization problem. It does this by solving a set of nonlinear equations for the basic variables in terms of the nonbasic variables. Then, the unconstrained problem is solved using approaches similar to those described in Chap. 14. First, a search direction is chosen along which an improvement in the objective function is sought. The default choice is a *quasi-Newton approach* (BFGS) that, as described in Chap. 14, requires storage of an approximation of the Hessian matrix. This approach performs very well for most cases. The *conjugate gradient* approach is also available in Excel as an alternative for large problems. The Excel Solver has the nice feature that it automatically switches to the conjugate gradient method, depending on available storage. Once the search direction is established, a one-dimensional search is carried out along that direction using a variable step-size approach.

15.3   OPTIMIZATION WITH PACKAGES

Software libraries and packages have great capabilities for optimization. In this section, we will give you an introduction to some of the more useful ones.

15.3.1 Mathcad

Mathcad contains a numeric mode function called **Find** that can be used to solve up to 50 simultaneous nonlinear algebraic equations with inequality constraints. The use of this function for unconstrained applications was described in Part Two. If **Find** fails to locate a solution that satisfies the equations and constraints, it returns the error message "did not find solution." However, Mathcad also contains a similar function called **Minerr.** This function gives solution results that minimize the errors in the constraints even when exact solutions cannot be found. This function solves equations and accommodates several constraints using the Lenenberg-Marquardt method taken from the public-domain MINPACK algorithms developed and published by the Argonne National Laboratory.

Let's do an example where **Find** is used to solve a system of nonlinear equations with constraints. Initial guesses of x = −1 and y = 1 are input using the definition symbol as shown in Fig. 15.4. The word **Given** then alerts Mathcad that what follows is a system of equations. Then comes the equations and the inequality constraint. Note that for this application, Mathcad requires the use of a symbolic equal sign typed as [Ctrl]= and > to separate left and right sides of an equation. Now the vector consisting of xval and yval are computed using **Find**(x,y) and the values are shown using an equal sign.

A graph that demonstrates the equations and constraints as well as the solution can be placed anywhere on the worksheet by clicking to the desired location. This places a red cross hair at that location. Then use the Insert/Graph/X-Y Plot pull down menu to place an empty plot on the worksheet with placeholders for the expressions to be graphed and for

---

FIGURE 15.4

Mathcad screen for a nonlinear constrained optimization problem.

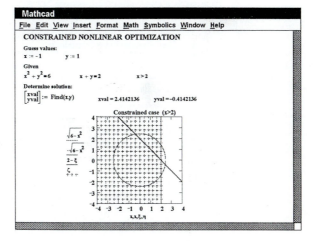

the ranges of the x and y axes. Four variables are plotted on the y axis as shown—the top and bottom halves of the equation for the circle, the linear function, and a cross pattern to represent the x > 2 constraint. The ranges for the various cases are x for the top and bottom circles, ξ for the line, and η for the constraint. Mathcad does all the rest to produce the graph. Once the graph has been created you can use the Format/Graph/X-Y Plot pull down menu to vary the type of graph; change the color, type, and weight of the trace of the function; and add titles, labels, and other features. The graph and the numerical values for xval and yval nicely portray the solution as the intersection of the circle and the line in the region where x > 2.

15.3.2 Excel for Linear Programming

There are a variety of software packages expressly designed to implement linear programming. However, because of its broad availability, we will focus on the Excel spreadsheet. This involves using the Solver option previously employed in Chap. 7 for root location.

The manner in which Solver is used for linear programming is similar to our previous applications in that the data is entered into spreadsheet cells. The basic strategy is to arrive at a single cell that is to be optimized as a function of variations of other cells on the spreadsheet. The following example illustrates how this can be done for the gas-processing problem.

EXAMPLE 15.3 Using Excel's Solver for a Linear Programming Problem

Problem Statement.   Use Excel to solve the gas-processing problem we have been examining in this chapter.

Solution.   An Excel worksheet set up to calculate the pertinent values in the gas-processing problem is shown in Fig. 15.5. The unshaded cells are those containing numeric and labeling data. The shaded cells involve quantities that are calculated based

---

FIGURE 15.5
Excel spreadsheet set up to use the Solver for linear programming.

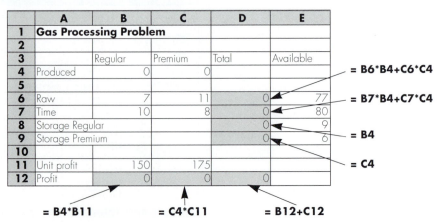

on other cells. Recognize that the cell to be maximized is D12, which contains the total profit. The cells to be varied are B4:C4, which hold the amounts of regular and premium gas produced.

Once the spreadsheet is created, **Solver** is chosen from the Tools menu. At this point a dialogue box will be displayed, querying you for pertinent information. The pertinent cells of the Solver dialogue box are filled out as

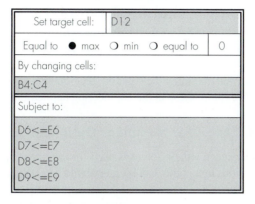

The constraints must be added one by one by selecting the "Add" button. This will open up a dialogue box that looks like

As shown, the constraint that the total raw gas (cell D6) must be less than or equal to the available supply (E6) can be added as shown. After adding each constraint, the "Add"

FIGURE 15.6
Excel spreadsheet showing solution to linear programming problem.

	A	B	C	D	E
1	**Gas Processing Problem**				
2					
3		Regular	Premium	Total	Available
4	Produced	**4.888889**	**3.888889**		
5					
6	Raw	7	11	77	77
7	Time	10	8	80	80
8	Storage Regular			4.888889	9
9	Storage Premium			3.888889	6
10					
11	Unit profit	150	175		
12	Profit	733.3333	680.5556	**1413.889**	

button can be selected. When all four constraints have been entered, the OK button is selected to return to the Solver dialogue box.

Now, before execution, the Solver options button should be selected and the box labeled "Assume linear model" should be checked off. This will make Excel employ a version of the simplex algorithm (rather than the more general nonlinear solver it usually uses) that will speed up your application.

After selecting this option, return to the Solver menu. When the OK button is selected, a dialogue box will open with a report on the success of the operation. For the present case, the Solver obtains the correct solution (Fig. 15.6)

Beyond obtaining the solution, the Solver also provides some useful summary reports. We will explore these in the engineering application described in Sec. 16.2.

15.3.3 Excel for Nonlinear Optimization

The manner in which Solver is used for nonlinear optimization is similar to our previous applications in that the data is entered into spreadsheet cells. Once again, the basic strategy is to arrive at a single cell that is to be optimized as a function of variations of other cells on the spreadsheet. The following example illustrates how this can be done for the parachutist problem we set up in the introduction to this part of the book (recall Example PT4.1).

EXAMPLE 15.4　Using Excel's Solver for Nonlinear Constrained Optimization

Problem Statement.　Recall from Example PT4.1 that we developed a nonlinear constrained optimization to minimize the cost for a parachute drop into a refugee camp. Parameters for this problem are

Parameter	Symbol	Value	Unit
Total mass	M_t	2000	kg
Acceleration of gravity	g	9.8	m/s^2
Cost coefficient (constant)	c_0	200	$
Cost coefficient (length)	c_1	56	$/m
Cost coefficient (area)	c_2	0.1	$/m^2$
Critical impact velocity	v_c	20	m/s
Area effect on drag	k_c	3	$kg/(s \cdot m^2)$
Initial drop height	z_0	500	m

Substituting these values into Eqs. (PT4.11) through (PT4.19) gives

$$\text{Minimize } C = n(200 + 56\ell + 0.1A^2)$$

subject to

$$v \leq 20$$
$$n \geq 1$$

where n is an integer and all other variables are real. In addition, the following quantities are defined as

$$A = 2\pi r^2$$
$$\ell = \sqrt{2}r$$
$$c = 3A$$

$$m = \frac{M_t}{n} \tag{15.6}$$

$$t = \text{root}\left[500 - \frac{9.8m}{c}t + \frac{9.8m^2}{c^2}(1 - e^{-(c/m)t})\right] \tag{15.7}$$

$$v = \frac{9.8m}{c}(1 - e^{-(c/m)t})$$

Use Excel to solve this problem for the design variables r and n that minimize cost C.

Solution. Before implementation of this problem on Excel, we must first deal with the problem of determining the root in the above formulation [Eq. (15.7)]. One method might be to develop a macro to implement a root-location method such as bisection or the secant method. (Note that we will illustrate how this is done in the next chapter in Sec. 16.3.)

For the time being, an easier approach is possible by developing the following fixed-point iteration solution to Eq. (15.7),

$$t_{i+1} = \left[500 + \frac{9.8m^2}{c^2}\left(1 - e^{-(c/m)t_i}\right)\right]\frac{c}{9.8m} \tag{15.8}$$

Thus, t can be adjusted until Eq. (15.8) is satisfied. It can be shown that for the range of parameters used in the present problem, this formula always converges.

Now, how can this equation be solved on a spreadsheet? As shown below, two cells can be set up to hold a value for t and for the right-hand side of Eq. (15.8) [that is, $f(t)$]. You can type in Eq. (15.8) into cell B21 so that it gets its time value from cell B20 and the other parameter values from cells elsewhere on the sheet (see below for how we set up the whole sheet). Then go to cell B20 and point its value to cell B21.

	A	**B**
20	t	0
21	f(t)	0.480856

$$= \left[500 + \frac{9.8m^2}{c^2}(1 - e^{-(c/m)t})\right]\frac{c}{9.8m}$$

Once you enter these formulations you will immediately get the error message: "Cannot resolve circular references," because B20 depends on B21 and vice versa. Now, go to the Tools/Options selections from the menu and select **calculation.** From the calculation dialogue box, check off "iteration" and hit "OK." Immediately the spreadsheet will iterate these cells and the result will come out as

	A	B
20	t	10.2551
21	f(t)	10.25595

Thus, the cells will converge on the root. If you want to make it more precise, just strike the F9 key to make it iterate some more (the default is 100 iterations, which you can change if you wish).

An Excel worksheet to calculate the pertinent values can then be set up as shown below. The unshaded cells are those containing numeric and labeling data. The shaded cells involve quantities that are calculated based on other cells. For example, the mass in B17 was computed with Eq. (15.6) based on the values for M_t (B4) and n (E5). Note also that some cells are redundant. For example, cell E11 points back to cell E5. It is repeated in cell E11 so that the structure of the constraints is evident from the sheet. Finally, recognize that the cell to be minimized is E15, which contains the total cost. The cells to be varied are E4:E5, which hold the radius and the number of parachutes.

Once the spreadsheet is created, the selection Solver is chosen from the Tools menu. At this point a dialogue box will be displayed, querying you for pertinent information. The

FIGURE 15.7
Excel spreadsheet set up for the nonlinear parachute optimization problem.

	A	B	C	D	E	F	G
1	Parachute optimization problem						
2							
3	Parameters:			Design variables:			
4	Mt	2000		r		1	
5	g	9.8		n		1	
6	cost1	200					
7	cost2	56		Constraints:			
8	cost3	0.1					
9	vc	20		variable		type	limit
10	kc	3		v	95.87867	<=	20
11	z0	500		n	1	>=	1
12							
13	Calculated values:			Objective function:			
14	A	6.283185					
15	l	1.414214		Cost	283.1438		
16	c	18.84956					
17	m	2000					
18							
19	Root: location:						
20	t	10.26439					
21	f(t)	10.26439					

pertinent cells of the Solver dialogue box would be filled out as

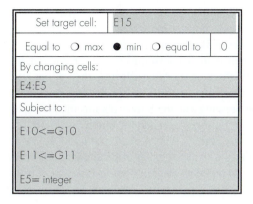

The constraints must be added one by one by selecting the "Add" button. This will open up a dialogue box that looks like

Cell reference:		Constraint
E10	<= ↓	G10

FIGURE 15.8
Excel spreadsheet showing solution for the nonlinear parachute optimization problem.

	A	B	C	D	E	F	G
1	Parachute optimization problem						
2							
3	Parameters:			Design variables:			
4	Mt	2000		r	2.943651		
5	g	9.8		n	6		
6	cost1	200					
7	cost2	56		Constraints:			
8	cost3	0.1					
9	vc	20		variable		type	limit
10	kc	3		v	20.00001	<=	20
11	z0	500		n	6	>=	1
12							
13	Calculated values:			Objective function:			
14	A	54.44432					
15	l	4.162952		Cost	4377.262		
16	c	163.333					
17	m	333.3333					
18							
19	Root: location:						
20	t	27.04076					
21	f(t)	27.04076					

As shown, the constraint that the actual impact velocity (cell E10) must be less than or equal to the required velocity (G10) can be added as shown. After adding each constraint, the "Add" button can be selected. Note that the down arrow allows you to choose among several types of constraints ($<=$, $>=$, $=$, and integer). Thus, we can force the number of parachutes (E5) to be an integer.

When all three constraints have been entered, the OK button is selected to return to the Solver dialogue box. After selecting this option return to the Solver menu. When the OK button is selected, a dialogue box will open with a report on the success of the operation. For the present case, the Solver obtains the correct solution as in Fig. 15.8.

Thus, we determine that the minimum cost of $4,377.26 will occur if we break the load up into six parcels with a chute radius of 2.944 m. Beyond obtaining the solution, the Solver also provides some useful summary reports. We will explore these in the engineering application described in Sec. 16.2.

15.3.4 IMSL

IMSL has several Fortran subroutines for optimization (Table 15.1). In the present discussion, we will focus on the UVMID routine. This routine locates the minimum point of a smooth function of a single variable using function and first-derivative evaluations.

UVMID is implemented by the following CALL statement:

```
CALL UVMID (F, G, XGUESS, ERREL, GTOL, MAXFN, A, B, X, FX, GX)
```

where F = User-supplied FUNCTION to compute the value of the function to be minimized. The form is F(X), where X = the point at which the function is evaluated. (Input). X should not be changed by F,.and F = the computed function value at the point X. (Output)

G = User-supplied FUNCTION to compute the derivative of the function, where G = The computed function value at the point X. (Output)

F and G must be declared EXTERNAL in the calling program.

XGUESS = An initial guess of the minimum point of F. (Input)

ERREL = Required relative accuracy of the final value of X. (Input)

GTOL = The derivative tolerance used to decide if the current point is a minimum. (Input)

MAXFN = Maximum number of function evaluations allowed. (Input)

A = Lower endpoint of interval in which maximum is to be located. (Input)

B = Upper endpoint of interval in which maximum is to be located. (Input)

FX = Function value at X. (Output)

GX = Derivative value at X. (Output)

TABLE 15.1 IMSL routines for optimization.

Category	Routine	Capability
Unconstrained minimization		
Univariate function		
	UVMIF	Using function values only
	UVMID	Using function and first-derivative values
	UVMGS	Nonsmooth function
Multivariate function		
	UMINF	Using finite-difference gradient
	UMING	Using analytic gradient
	UMIDH	Using finite-difference Hessian
	UMIAH	Using analytic Hessian
	UMCGF	Using conjugate gradient with finite-difference gradient
	UMCGG	Using conjugate gradient with analytic gradient
	UMPOL	Nonsmooth function
Nonlinear least squares		
	UNLSF	Using finite-difference Jacobian
	UNLSJ	Using analytic Jacobian
Minimization with simple bounds		
	BCONF	Using finite-difference gradient
	BCONG	Using analytic gradient
	BCODH	Using finite-difference Hessian
	BCOAH	Using analytic Hessian
	BCPOL	Nonsmooth function
	BCLSF	Nonlinear least squares using finite-difference Jacobian
	BCLSJ	Nonlinear least squares using analytic Jacobian
Linearly constrained minimization		
	DLPRS	Dense linear programming
	QPROG	Quadratic programming
	LCONF	General objective function with finite-difference gradient
	LCONG	General objective function with analytic gradient
Nonlinearly constrained minimization		
	NCONF	Using finite-difference gradient
	NCONG	Using analytic gradient
Service routines		
	CDGRD	Central-difference gradient
	FDGRD	Forward-difference gradient
	FDHES	Forward-difference Hessian
	GDHES	Forward-difference Hessian using analytic gradient
	FDJAC	Forward-difference Jacobian
	CHGRD	Check user-supplied gradient
	CHHES	Check user-supplied Hessian
	CHJAC	Check user-supplied Jacobian
	GGUES	Generate starting points

EXAMPLE 15.5   Using IMSL to Locate a Single Optimum

Problem Statement.   Use UVMID to determine the maximum of the one-dimensional function we solved in Chap. 13 (recall Examples 15.1 through 15.3).

$$f(x) = 2 \sin x - \frac{x^2}{10}$$

Solution.   An example of a main Fortran 90 program and function using UVMIF to solve this problem can be written as

```
PROGRAM Oned

USE mimsl

IMPLICIT NONE
INTEGER::maxfn=50
REAL::xguess=0.,errel=1.E-6,gtol=1.E-6,a=-2.,b=2.
REAL::x,f,g,fx,gx
EXTERNAL f,g
CALL UVMID(f,g,xguess,errel,gtol,maxfn,a,b,x,fx,gx)
PRINT *,x,fx,gx
END PROGRAM

FUNCTION f(x)
IMPLICIT NONE
REAL::x,f
f=-(2.*SIN(X) - x**2/10.)
END FUNCTION

FUNCTION g(x)
IMPLICIT NONE
REAL::x,g
g=-(2.*COS(x) - 2.*x/10.)
END FUNCTION
```

Notice that because the routine is set up for minimization, we enter the negative of the function. An example run is

```
1.427334    -1.775726    -4.739729E-04
```

PROBLEMS

15.1 A company makes two types of products, A and B. These products are produced during a 40-hour work week and then shipped out at the end of the week. They require 20 and 5 kg of raw material per kg of product, respectively, and the company has access to 10,000 kg of raw material per week. Only one product can be created at a time with production times for each of 0.05 and 0.15 hrs, respectively. The plant can only store 550 kg of total product per week. Finally, the company makes profits of $45 and $30 on each unit of A and B, respectively.
(a) Set up the linear programming problem to maximize profit.
(b) Solve the linear programming problem graphically.
(c) Solve the linear programming problem with the simplex method.
(d) Solve the problem with a software package.
(e) Evaluate which of the following options will raise profits the most: increasing raw material, storage, or production time.

15.2 Suppose that for Example 15.1, the gas processing plant decides to produce a third grade of product with the following characteristics:

	Supreme
Raw gas	15 m³/tonne
Production time	12 hr/tonne
Storage	5 tonnes
Profit	$250/tonne

In addition, suppose that a new source of raw gas has been discovered so that the total available is doubled to 154 m³/week.
(a) Set up the linear programming problem to maximize profit.
(b) Solve the linear programming problem with the simplex method.
(c) Solve the problem with a software package.
(d) Evaluate which of the following options will raise profits the most: increasing raw material, storage, or production time.

15.3 Consider the linear programming problem:

$$\text{Minimize } f(x, y) = \frac{5}{3}x + y$$

subject to

$$x + 2.5y \le 15$$
$$x + y \le 7$$
$$2x + y \le 9$$
$$x \ge 0$$
$$y \ge 0$$

Obtain the solution
(a) Graphically.
(b) Using the simplex method.
(c) Using an appropriate package or software library (e.g., Excel, Mathcad, IMSL).

15.4. Consider the linear programming problem:

$$\text{Maximize } f(x, y) = 12x + 10y$$

subject to

$$5x + 4y \le 1700$$
$$x + y \le 7$$
$$4.5x + 3.5y \le 1600$$
$$x + 2y \le 500$$
$$x \ge 0$$
$$y \ge 0$$

Obtain the solution
(a) Graphically.
(b) Using the simplex method.
(c) Using an appropriate package or software library (e.g., Excel, Mathcad, IMSL).

15.5 Use a package or software library (e.g., Excel, Mathcad, IMSL) to solve the following constrained nonlinear optimization problem.

$$\text{Maximize } f(x, y) = 1.1x + 1.9y - y^3$$

subject to

$$x + y \le 0.9$$
$$x \ge 0$$
$$y \ge 0$$

15.6 Use a package or software library (e.g., Excel, Mathcad, IMSL) to solve the following constrained nonlinear optimization problem:

Minimize $f(x, y) = 19x + 11y$

subject to

$$x^2 + y^2 \geq 0.95$$
$$x + 2y \leq 2$$
$$x \geq 0$$
$$y \geq 0$$

15.7 Consider the following constrained nonlinear optimization problem:

Minimize $f(x, y) = (x - 2.5)^2 + (y - 2.5)^2$

subject to

$$x + 2y = 4$$

(a) Use a graphical approach to estimate the solution.
(b) Use a package or software library (e.g., Excel, Mathcad, IMSL) to obtain a more accurate estimate.

15.8 Use a package or software library to determine the maximum of

$$f(x, y) = 2xy + 1.5y - 1.25x^2 - 2y^2$$

15.9 Use a package or software library to determine the maximum of

$$f(x, y) = 3.5x + 2y + x^2 - x^4 - 2xy - y^2$$

15.10 Given the following function,

$$f(x, y) = -7x + 1.2x^2 + 11y + 2y^2 - 2xy$$

use a package or software library to determine the minimum
(a) Graphically.
(b) Numerically.
(c) Substitute the result of **(b)** back into the function to determine the minimum $f(x, y)$.
(d) Determine the Hessian and its determinant, and substitute the result of part **(b)** back into the latter to verify that a minimum has been detected.

CHAPTER 16

Engineering Applications: Optimization

The purpose of this chapter is to use the numerical procedures discussed in Chaps. 13 through 15 to solve actual engineering problems involving optimization. These problems are important because engineers are often called upon to come up with the "best" solution to a problem. Because many of these cases involve complex systems and interactions, numerical methods and computers are often a necessity for developing optimal solutions.

The following applications are typical of those that are routinely encountered during upper-class and graduate studies. Furthermore, they are representative of problems you will address professionally. The problems are drawn from the major discipline areas of engineering: chemical/petroleum, civil/environmental, electrical, and mechanical/aerospace.

The first application, taken from *chemical/petroleum engineering,* deals with using nonlinear constrained optimization to design an optimal cylindrical tank. The Excel Solver is used to develop the solution.

Next, we use linear programming to assess a problem from *civil/environmental engineering:* minimizing the cost of waste treatment to meet water-quality objectives in a river. In this example, we introduce the notion of shadow prices and their use in assessing the sensitivity of a linear programming solution.

The third application, taken from *electrical engineering,* involves maximizing the power across a potentiometer in an electric circuit. The solution involves one-dimensional unconstrained optimization. Aside from solving the problem, we illustrate how the Visual Basic macro language allows access to the golden-section search algorithm within the context of the Excel environment.

Finally, the fourth application, taken from *mechanical/aerospace engineering,* involves finding the displacements of the strut on a mountain bike by minimizing a two-dimensional equation of potential energy.

16.1 LEAST-COST DESIGN OF A TANK (CHEMICAL/PETROLEUM ENGINEERING)

Background. Chemical and petroleum engineers (as well as other specialists such as mechanical and civil engineers) often encounter the general problem of designing containers to transport liquids and gases. Suppose that you are asked to determine the dimensions of

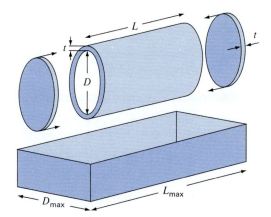

FIGURE 16.1
Parameters for determining the optimal dimensions of a cylindrical tank.

TABLE 16.1 Parameters for determining the optimal dimensions of a cylindrical tank used to transport toxic wastes.

Parameter	Symbol	Value	Units
Required volume	V_o	0.8	m^3
Thickness	t	3	cm
Density	ρ	8000	kg/m^3
Bed length	L_{max}	2	m
Bed width	D_{max}	1	m
Material cost	c_m	4.5	$/kg
Welding cost	c_w	20	$/m

a small cylindrical tank to transport toxic waste that is to be mounted on the back of a pickup truck. Your overall objective will be to minimize the cost of the tank. However, aside from cost, you must ensure that it holds the required amount of liquid and that it does not exceed the dimensions of the truck's bed. Note that because the tank will be carrying toxic waste, the tank thickness is specified by regulations.

A schematic of the tank and bed are shown in Fig. 16.1. As can be seen, the tank consists of a cylinder with two plates welded on each end.

The cost of the tank involves two components: (1) material expense, which is based on weight, and (2) welding expense based on length of weld. Note that the latter involves welding both the interior and the exterior seams where the plates connect with the cylinder. The data needed for the problem are summarized in Table 16.1.

Solution.   The objective here is to construct a tank for a minimum cost. The cost is related to the design variables (length and diameter) as they effect the mass of the tank and the welding lengths. Further, the problem is constrained because the tank must (1) fit within the truck bed and (2) carry the required volume of material.

The cost consists of tank material and welding costs. Therefore, the objective function can be formulated as minimizing

$$C = c_m m + c_w \ell_w \tag{16.1}$$

where C = cost (\$), m = mass (kg), ℓ_w = weld length (m), and c_m and c_w = cost factors for mass (\$/kg) and weld length (\$/m), respectively.

Next, we will formulate how the mass and weld lengths are related to the dimensions of the drum. First, the mass can be calculated as the volume of material times its density. The volume of the material used to create the side walls (i.e., the cylinder) can be computed as

$$V_{\text{cylinder}} = L\pi \left[\left(\frac{D}{2} + t \right)^2 - \left(\frac{D}{2} \right)^2 \right]$$

For each circular end plate, it is

$$V_{\text{plate}} = \pi \left(\frac{D}{2} + t \right)^2 t$$

Thus, the mass is computed by

$$m = \rho \left\{ L\pi \left[\left(\frac{D}{2} + t \right)^2 - \left(\frac{D}{2} \right)^2 \right] + 2\pi \left(\frac{D}{2} + t \right)^2 t \right\} \tag{16.2}$$

where ρ = density (kg/m^3).

The weld length for attaching each plate is equal to the cylinder's inside and outside circumference. For the two plates, the total weld length would be

$$\ell_w = 2 \left[2\pi \left(\frac{D}{2} + t \right) + 2\pi \frac{D}{2} \right] = 4\pi (D + t) \tag{16.3}$$

Given values for D and L (remember, thickness t is fixed by regulations), Eqs. (16.1) through (16.3) provide a means to compute cost. Also recognize that when Eqs. (16.2) and (16.3) are substituted into Eq. (16.1), the resulting objective function is nonlinear in the unknowns.

Next, we can formulate the constraints. First, we must compute how much volume can be held within the finished tank,

$$V = \frac{\pi D^2}{4} L$$

This value must be equal to the desired volume. Thus, one constraint is

$$\frac{\pi D^2 L}{4} = V_o$$

where V_o is the desired volume (m^3).

The remaining constraints deal with ensuring that the tank will fit within the dimensions of the truck bed,

$$L \leq L_{\text{max}}$$
$$D \leq D_{\text{max}}$$

The problem is now specified. Substituting the values from Table 16.1, it can be summarized as

Maximize $C = 4.5m + 20\ell_w$

subject to

$$\frac{\pi D^2 L}{4} = 0.8$$

$$L \leq 2$$

$$D \leq 1$$

where

$$m = 8000 \left\{ L\pi \left[\left(\frac{D}{2} + 0.03 \right)^2 - \left(\frac{D}{2} \right)^2 \right] + 2\pi \left(\frac{D}{2} + 0.03 \right)^2 0.03 \right\}$$

and

$$\ell_w = 4\pi(D + 0.03)$$

The problem can now be solved in a number of ways. However, the simplest approach for a problem of this magnitude is to use a tool like the Excel Solver. The spreadsheet to accomplish this is shown in Fig. 16.2.

For the case shown, we enter the upper limits for D and L. For this case, the volume is more than required ($1.57 > 0.8$).

Once the spreadsheet is created, the selection Solver is chosen from the Tools menu. At this point a dialogue box will be displayed, querying you for pertinent information. The

FIGURE 16.2
Excel spreadsheet set up to evaluate the cost of a tank subject to a volume requirement and size constraints.

	A	B	C	D	E	F	G
1	Optimum tank design						
2							
3	Parameters:			Design variables			
4							
5	VO	0.8		D	1		
6	t	0.03		L	2		
7	rho	8000					
8	Lmax	2		Constraints			
9	Dmax	1					
10	cm	4.5		D	1	< =	1
11	cw	20		L	2	< =	2
12				Vol	1.570796	=	0.8
13	Computed values:						
14				Objective function:			
15	m	1976.791					
16	lw	12.94336		C	9154.425		
17							
18	Vshell	0.19415					
19	Vends	0.052948					

pertinent cells of the Solver dialogue box would be filled out as

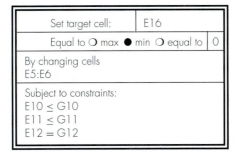

When the OK button is selected, a dialogue box will open with a report on the success of the operation. For the present case, the Solver obtains the correct solution, which is shown in Fig. 16.3. Notice that the optimal diameter is nudging up against the constraint of 1 m. Thus, if the required capacity of the tank were increased, we would run up against this constraint and the problem would reduce to a one-dimensional search for length.

	A	B	C	D	E	F	G
1	Optimum tank design						
2							
3	Parameters:			Design variables			
4							
5	V0	0.8		D	0.982949		
6	t	0.03		L	1.054235		
7	rho	8000					
8	Lmax	2		Constraints			
9	Dmax	1					
10	cm	4.5		D	0.982949	< =	1
11	cw	20		L	1.064235	< =	2
12				Vol	0.799998	=	0.8
13	Computed values:						
14				Objective function:			
15	m	1215.236					
16	lw	12.72909		C	5723.145		
17							
18	Vshell	0.100646					
19	Vends	0.061259					

FIGURE 16.3
Results of minimization. The price is reduced from $9154 to $5723 because of the smaller volume using dimensions of $D = 0.98$ m and $l = 1.05$ m.

16.2   LEAST-COST TREATMENT OF WASTEWATER (CIVIL/ENVIRONMENTAL ENGINEERING)

Background.   Wastewater discharges from big cities are often a major cause of river pollution. Figure 16.4 illustrates the type of system that an environmental engineer might confront. Several cities are located on a river and its tributary. Each generates pollution at a

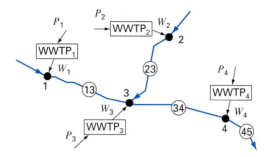

FIGURE 16.4
Four wastewater treatment plants discharging pollution to a river system. The river segments between the cities are labeled with circled numbers.

loading rate P that has units of milligrams per day (mg/d). The pollution loading is subject to waste treatment that results in a fractional removal x. Thus, the amount discharged to the river is the excess not removed by treatment,

$$W_i = (1 - x_i)P_i \tag{16.4}$$

where W_i = waste discharge from the ith city.

When the waste discharge enters the stream, it mixes with pollution from upstream sources. If complete mixing is assumed at the discharge point, the resulting concentration at the discharge point can be calculated by a simple mass balance,

$$c_i = \frac{W_i + Q_u c_u}{Q_i} \tag{16.5}$$

where Q_u = flow (L/d), c_u = concentration (mg/L) in the river immediately upstream of the discharge, and Q_i = flow downstream of the discharge point (L/d).

After the concentration at the mixing point is established, chemical and biological decomposition processes can remove some of the pollution as it flows downstream. For the present case, we will assume that this removal can be represented by a simple fractional reduction R.

Assuming that the headwaters (i.e., the river above cities 1 and 2) are pollution-free, the concentrations at the four nodes can be computed as

$$
\begin{aligned}
c_1 &= \frac{(1 - x_1)P_1}{Q_{13}} \\[1em]
c_2 &= \frac{(1 - x_2)P_2}{Q_{23}} \\[1em]
c_3 &= \frac{R_{13}Q_{13}c_1 + R_{23}Q_{23}c_2 + (1 - x_3)P_3}{Q_{34}} \\[1em]
c_4 &= \frac{R_{34}Q_{34}c_3 + (1 - x_4)P_4}{Q_{45}}
\end{aligned}
\tag{16.6}
$$

TABLE 16.2   Parameters for four wastewater treatment plants discharging pollution to a river system, along with the resulting concentration (c_i) for zero treatment. Flow, removal, and standards for the river segments are also listed.

City	P_i (mg/d)	d_i ($\$10^{-6}$/mg)	c_i (mg/L)	Segment	Q (L/d)	R	c_s (mg/L)
1	1.00×10^9	2	100	1–3	1.00×10^7	0.5	20
2	2.00×10^9	2	40	2–3	5.00×10^7	0.35	20
3	4.00×10^9	4	47.3	3–4	1.10×10^8	0.6	20
4	2.50×10^9	4	22.5	4–5	2.50×10^8		20

Next, it is recognized that the waste treatment costs a different amount, d_i ($\$1000$/mg removed), at each of the facilities. Thus, the total cost of treatment (on a daily basis) can be calculated as

$$Z = d_1 P_1 x_1 + d_2 P_2 x_2 + d_3 P_3 x_3 + d_4 P_4 x_4 \tag{16.7}$$

where Z is total daily cost of treatment ($\$1000$/d).

The final piece in the "decision puzzle" involves environmental regulations. To protect the beneficial uses of the river (e.g., boating, fisheries, bathing), regulations say that the river concentration must not exceed a water-quality standard of c_s.

Parameters for the river system in Fig. 16.4 are summarized in Table 16.2. Notice that there is a difference in treatment cost between the upstream (1 and 2) and the downstream cities (3 and 4) because of the outmoded nature of the downstream plants.

The concentration can be calculated with Eq. (16.6) and the result listed in the shaded column for the case where no waste treatment is implemented (i.e., all the x's $= 0$). Notice that the standard of 20 mg/L is being violated at all mixing points.

Use linear programming to determine the treatment levels that meet the water-quality standards for the minimum cost. Also, evaluate the impact of making the standard more stringent below city 3. That is, redo the exercise, but with the standards for segments 3-4 and 4-5 lowered to 10 mg/L.

Solution.   All the factors outlined above can be combined into the following linear programming problem:

$$\text{Minimize } Z = d_1 P_1 x_1 + d_2 P_2 x_2 + d_3 P_3 x_3 + d_4 P_4 x_4 \tag{16.8}$$

subject to the following constraints

$$\frac{(1 - x_1) P_1}{Q_{13}} \le c_{s1}$$

$$\frac{(1 - x_2) P_2}{Q_{23}} \le c_{s2}$$

$$\frac{R_{13} Q_{13} c_1 + R_{23} Q_{23} c_2 + (1 - x_3) P_3}{Q_{34}} \le c_{s3} \tag{16.9}$$

$$\frac{R_{34} Q_{34} c_3 + (1 - x_4) P_4}{Q_{45}} \le c_{s4}$$

$$0 \le x_1, x_2, x_3, x_4 \le 1 \tag{16.10}$$

	A	B	C	D	E	F	G	H
1	**Least-Cost Treatment of Wastewater**							
2		Untreated	Treatment	Discharge	Unit cost	River	WQ	Treatment
3	City	P	x	W	d	conc	standard	Cost
4	1	1.00E + 09	0	1E + 09	2.00E-06	100	20	0
5	2	2.00E + 09	0	2E + 09	2.00E-06	40.00	20	0
6	3	4.00E + 09	0	4E + 09	4.00E-06	47.27	20	0
7	4	2.50E + 09	0	2.5E + 09	4.00E-06	22.48	20	0
8		River	River					
9	Segment	Flow	Removal				Total	0
10	1-3	1.00E + 07	0.5					
11	2-3	5.00E + 07	0.35					
12	3-4	1.10E + 08	0.6					
13	4-5	2.50E + 08						

= $D$4/$B$10

= SUM($H$4:/$H$7)

= $B$4*$C$4*$E$4

FIGURE 16.5

Excel spreadsheet set up to evaluate the cost of waste treatment on a regulated river system. Column F contains the calculation of concentration according to Eq. (16.6). Cells F4 and H4 re highlighted to show the formulas used to calculate c_1 and treatment cost for City 1. In addition, highlighted cell H9 shows the formula (Eq. 16.8) for total cost that is to be minimized.

Thus, the objective function is to minimize treatment cost [Eq. (16.8)] subject to the constraint that water-quality standards must be met for all parts of the system (16.9). In addition, treatment cannot be negative or greater than 100% removal [Eq. (16.10)].

The problem can be solved using a variety of packages. For the present application, we use the Excel spreadsheet. As in Fig. 16.5, the data along with the concentration calculations can be set up nicely in the spreadsheet cells.

Once the spreadsheet is created, the selection Solver is chosen from the Tools menu. At this point a dialogue box will be displayed, querying you for pertinent information. The pertinent cells of the Solver dialogue box would be filled out as

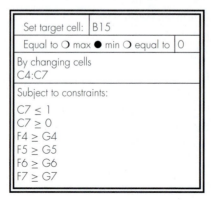

Set target cell: | B15

Equal to ○ max ● min ○ equal to | 0

By changing cells
C4:C7

Subject to constraints:

C7 ≤ 1
C7 ≥ 0
F4 ≥ G4
F5 ≥ G5
F6 ≥ G6
F7 ≥ G7

	A	B	C	D	E	F	G	H
1	**Least-Cost Treatment of Wastewater**							
2		Untreated	Treatment	Discharge	Unit cost	River	WQ	Treatment
3	City	P	x	W	d	conc	standard	Cost
4	1	1.00E + 09	0.8	2E + 08	2.00E-06	20	20	1600
5	2	2.00E + 09	0.5	1E + 09	2.00E-06	20.00	20	2000
6	3	4.00E + 09	0.5625	1.75E + 09	4.00E-06	20.00	20	9000
7	4	2.50E + 09	0	2.5E + 09	4.00E-06	15.28	20	0
8		River	River					
9	Segment	Flow	Removal				Total	12600
10	1-3	1.00E + 07	0.5					
11	2-3	5.00E + 07	0.35					
12	3-4	1.10E + 08	0.6					
13	4-5	2.50E + 08						

FIGURE 16.6

Results of minimization. The water-quality standards are met at a cost of $12,600/day. Notice that despite the fact that no treatment is required for city 4, the concentration at its mixing point actually exceeds the standard.

Microsoft Excel 5.0 Sensitivity Report
Worksheet: [CASE1502.XLS]Waste Treat
Report Created: 3/9/97 8:29

Changing Cells

Cell	Name	Final Value	Reduced Cost	Objective Coefficient	Allowable Increase	Allowable Decrease
$C$4	x	0.8	0	2000	1E + 30	0
$C$5	x	0.5	0	4000	1E + 30	1200
$C$6	x	0.5625	0	16000	0	16000
$C$7	x	0	10000	10000	1E + 30	10000

Constraints

Cell	Name	Final Value	Shadow Price	Constraint R.H. Side	Allowable Increase	Allowable Decrease
$F$4	conc	20	0	20	80	20
$F$5	conc	20.00	−30.00	20	20	20
$F$6	conc	20.00	−440.00	20	17.87878788	15.90909091
$F$7	conc	15.28	0.00	20	1E+30	4.72

FIGURE 16.7

Sensitivity Report for spreadsheet set up to evaluate the cost of waste treatment on a regulated river system.

Notice that not all the constraints are shown, because the dialogue box displays only six constraints at a time.

When the OK button is selected, a dialogue box will open with a report on the success of the operation. For the present case, the Solver obtains the correct solution, which is shown in Fig. 16.6. Before accepting the solution (by selecting the OK button on the Solver Reports box), notice that 3 reports can be generated: Answer, Sensitivity, and Limits. Select the Sensitivity Report and then hit the OK button to accept the solution. The Solver will automatically generate a Sensitivity Report, as in Fig. 16.7.

TABLE 16.3 Comparison of two scenarios involving the impact of different regulations on treatment costs.

	Scenario 1: All $c_s = 20$			Scenario 2: Downstream $c_s = 10$		
City	x	c	City	x	c
1	0.8	20	1	0.8	20
2	0.5	20	2	0.5	20
3	0.5625	20	3	0.8375	10
4	0	15.28	4	0.264	10
	Cost = $12,600			Cost = $19,640		

Now let's examine the solution (Fig. 16.6). Notice that the standard will be met at all the mixing points. In fact, the concentration at city 4 will actually be less than the standard (16.28 mg/L), even though no treatment would be required for city 4.

As a final exercise, we can lower the standards for reaches 3-4 and 4-5 to 10 mg/L. Before doing this, we can examine the Sensitivity Report. For the present case, the key column of Fig. 16.7 is the Shadow Price. The *shadow price* is a value that expresses the sensitivity of the objective function (in our case, cost) to a unit change of one of the constraints (water-quality standards). It therefore represents the additional cost that will be incurred by making the standards more stringent. For our example, it is revealing that the largest shadow price, $-\$440/\Delta c_{s3}$, occurs for one of the standard changes (i.e., downstream from City 3) that we are contemplating. This tips us off that our modification will be costly.

This is confirmed when we rerun Solver with the new standards (i.e., we lower cells G6 and G7 to 10). As in Table 16.3, the result is that treatment cost is increased from $12,600/day to $19,640/day. In addition, reducing the standard concentrations for the lower reaches means that city 4 must begin to treat its waste, and city 3 must upgrade its treatment. Notice also that the treatment of the upstream cities is unaffected.

16.3 MAXIMUM POWER TRANSFER FOR A CIRCUIT (ELECTRICAL ENGINEERING)

Background.   The simple resistor circuit in Fig. 16.8 contains three fixed resistors and one adjustable resistor. Adjustable resistors are called *potentiometers*. The values for the parameters are $V = 80$ V, $R_1 = 8$ Ω, $R_2 = 12$ Ω, and $R_3 = 10$ Ω. (*a*) Find the value of the adjustable resistance R_a that maximizes the power transfer across terminals 1 and 2. (*b*) Perform a sensitivity analysis to determine how the maximum power and the corresponding setting of the potentiometer (R_a) varies as V is varied over the range from 45 to 105 V.

Solution.   An expression for power for the circuit can be derived from Kirchhoff's laws as

$$P(R_a) = \frac{\left[\dfrac{VR_3R_a}{R_1(R_a + R_2 + R_3) + R_3R_a + R_3R_2} \right]^2}{R_a} \tag{16.11}$$

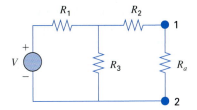

FIGURE 16.8
A resistor circuit with an adjustable resistor, or potentiometer.

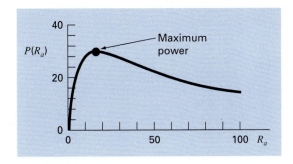

FIGURE 16.9
A plot of power transfer across terminals 1-2 from Fig. 16.8, as a function of the potentiometer resistance R_a.

	A	B	C	D
1	Maximum Power Transfer			
2				
3	V	80		
4	R1	8		
5	R2	12		
6	R3	10		
7				
8	Ra	16.44445		
9	P(Ra)	**30.03003**		

FIGURE 16.10
Excel determination of maximum power across a potentiometer using trial-and-error.

=(B3*B6*B8/(B4*(B8+B5+B6)+B6*B8+B6*B5))^2/B8

Substituting the parameter values gives the plot shown in Fig. 16.9. Notice that a maximum power transfer occurs at a resistance of about 16 Ω.

We will solve this problem in three ways with the Excel spreadsheet. First, we will employ trial-and-error and the Solver option. Then, we will develop a Visual BASIC macro program to perform the sensitivity analysis.

(*a*) An Excel spreadsheet to implement Eq. (16.11) is shown in Fig. 16.10. As indicated, Eq. (16.11) can be entered into cell B9. Then the value of R_a (cell B8) can be varied in a trial-and-error fashion until a minimum drag was determined. For this example, the result is a power of 30.03 W and a potentiometer setting of $R_a = 16.44 \ \Omega$.

A superior approach involves using the Solver option from the spreadsheet's Tools menu. At this point a dialogue box will be displayed, querying you for pertinent information.

FIGURE 16.11
Excel macro written in Visual BASIC to determine a minimum with the golden-section search.

```
Function Golden(xlow, xhigh, R1, R2, R3, V)
maxit = 50 : es = 0.01 : r = (5 ^ 0.5 - 1) / 2
xl = xlow : xu = xhigh : iter = 1
d = r * (xu - xl)
x1 = xl + d : x2 = xu - d
f1 = Power(x1, R1, R2, R3, V)
f2 = Power(x2, R1, R2, R3, V)
If f1 > f2 Then
  xopt = x1 : fx = f1
Else
  xopt = x2 : fx = f2
End If
Do
  d = r * d
  If f1 > f2 Then
    xl = x2 : x2 = x1
    x1 = xl + d : f2 = f1
    f1 = Power(x1, R1, R2, R3, V)
  Else
    xu = x1 : x1 = x2
    x2 = xu - d : f1 = f2
    f2 = Power(x2, R1, R2, R3, V)
  End If
  iter = iter + 1
  If f1 > f2 Then
    xopt = x1 : fx = f1
  Else
    xopt = x2 : fx = f2
  End If
  If xopt <> 0 then ea = (1 - r) * Abs((xu - xl) / xopt) * 100
  If ea <= es Or iter > maxit Then Exit Do
Loop
Golden = xopt
End Function

Function Power(Ra, R1, R2, R3, V)
  Num = (V * R3 * Ra / (R1 * (Ra + R2 + R3) + R3 * Ra + R3 * R2)) ^ 2
  Power = Num/Ra
End Function
```

The pertinent cells of the Solver dialogue box would be filled out as

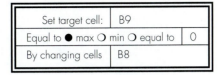

When the OK button is selected, a dialogue box will open with a report on the success of the operation. For the present case, the Solver obtains the same correct solution shown in Fig. 16.10.

(*b*) Now, although the foregoing approach is excellent for a single evaluation, it is not convenient for cases where multiple optimizations would be employed. Such would be the case for the second part of this application, where we are interested in determining how the maximum power varies for different voltage settings. Of course, the Solver could be invoked multiple times for different parameter values, but this would be inefficient. A preferable course would involve developing a macro function to come up with the optimum.

Such a function is listed in Fig. 16.11. Notice how closely it resembles the golden-section-search pseudocode previously presented in Fig. 13.5. In addition, notice that a function must also be defined to compute power according to Eq. (16.11).

An Excel spreadsheet utilizing this macro to evaluate the sensitivity of the solution to voltage is given in Fig. 16.12. A column of values is set up that spans the range of V's (that is, from 45 to 105 V). A function call to the macro is written in cell B9 that references the adjacent value of V (the 45 in A9). In addition, the other parameters in the function argument are also included. Notice that, whereas the reference to V is relative, the references to the

FIGURE 16.12
Excel spreadsheet to implement a sensitivity analysis of the maximum power to variations of voltage. This routine accesses the macro program for golden-section search from Fig. 16.11.

	A	B	C	D
1	**Maximum Power Transfer**			
2				
3	R1	8		
4	R2	12		
5	R3	10		
6	Rmin	0.1		
7	Rmax	100		
8	V	Ra	P(Ra)	
9	45	**16.44444**	9.501689	
10	60	16.44444	16.89189	
11	75	16.44444	26.39358	
12	90	16.44444	38.00676	
13	105	16.44444	51.73142	

Call to Visual BASIC macro function

= Golden($B$6,$B$7,$B$3,$B$4,$B$5,A9)

Power calculation

=(A9*$B$5*B9/($B$3*(B9+$B$4+$B$5)+$B$5*B9+$B$3*$B$4))^2/B9

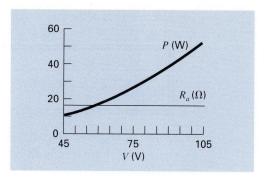

FIGURE 16.13
Results of sensitivity analysis of the effect of voltage variations on maximum power.

lower and upper guesses and the resistances are absolute (that is, including leading $). This was done so that when the formula is copied down, the absolute references stay fixed, whereas the relative reference corresponds to the voltage in the same row. A similar strategy is used to place Eq. (16.11) in cell C9.

When the formulas are copied downward, the result is as shown in Fig. 16.12. The maximum power can be plotted to visualize the impact of voltage variations. As in Fig. 16.13, the power increases with V.

The results for the corresponding potentiometer settings (R_a) are more interesting. The spreadsheet indicates that the same setting, 16.44 Ω, results in maximum power. Such a result might be difficult to intuit based on casual inspection of Eq. (16.11).

16.4 MOUNTAIN BIKE DESIGN (MECHANICAL/AEROSPACE ENGINEERING)

Background.   Because of their work in the building industry, civil engineers are most commonly associated with structural design. However, other engineering specialties also must deal with the impact of forces on the devices they design. In particular, mechanical and aerospace engineers must assess both the static and dynamic responses of a wide assortment of vehicles ranging from automobiles to space vehicles.

Recent interest in competitive and recreational cycling has meant that engineers have directed their skills toward the design and testing of mountain bikes (Fig. 16.14a). Suppose that you are given the task of predicting the horizontal and vertical displacement of a bike bracketing system in response to a force. Assume the forces you must analyze can be simplified as depicted in Fig. 16.14b. You are interested in testing the response of the truss to a force exerted in any number of directions designated by the angle θ.

The parameters for the problem are $E =$ Young's modulus $= 2 \times 10^{11}$ Pa, $A =$ cross-sectional area $= 0.0001$ m^2, $w =$ width $= 0.44$ m, $\ell =$ length $= 0.56$ m, and $h =$ height $= 0.5$ m. The displacements x and y can be solved by determining the values that yield a

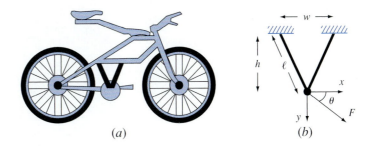

FIGURE 16.14
(a) A mountain bike along with (b) a free-body diagram for a part of the frame.

minimum potential energy. Determine the displacements for a force of 10,000 N and a range of θ's from 0° (horizontal) to 90° (vertical).

Solution.   This problem can be approached by developing the following equation for the potential energy of the bracketing system,

$$V(x, y) = \frac{EA}{\ell}\left(\frac{w}{2\ell}\right)^2 x^2 + \frac{EA}{\ell}\left(\frac{h}{\ell}\right)^2 y^2 - Fx\cos\theta - Fy\sin\theta \tag{16.12}$$

Solving for an individual angle is straightforward. For $\theta = 30°$, the given parameter values can be substituted into Eq. (16.12) to yield

$$V(x, y) = 5,512,026x^2 + 28,471,210y^2 - 5000x - 8660y$$

The minimum of this function can be determined in a number of ways. For example, using the Excel Solver, the minimum potential energy is −3.62 with deflections of $x = 0.000786$ and $y = 0.0000878$ m.

Of course, the Excel Solver could be implemented repeatedly for different values of θ to assess how the solution changes as the angle changes. Alternatively, a macro can be written in the same fashion as in Sec. 16.3 so that multiple optimizations can be implemented simultaneously. Of course, for this case a multidimensional search algorithm would have to be implemented. A third way to approach the problem would be to use a programming language such as Fortran 90, along with a numerical methods software library like IMSL.

In any event, the results are as displayed in Fig. 16.15. As might be expected (Fig. 16.15a), the x deflection is at a maximum when the load is pointed in the x direction ($\theta = 0°$) and the y deflection is at a maximum when the load is pointed in the y direction ($\theta = 90°$). However, notice that the x deflection is much more pronounced than that in the y direction. This is also manifest in Fig. 16.15b, where the potential energy is higher at low angles. Both results are due to the geometry of the strut. If w were made bigger, the deflections would be more uniform.

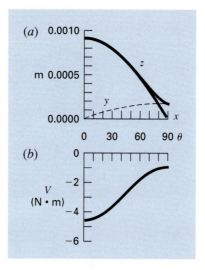

FIGURE 16.15

(a) The impact of different angles on the deflections (note that z is the resultant of the x and y components) and (b) potential energy of a part of the frame of a mountain bike subjected to a constant force.

PROBLEMS

Chemical/Petroleum Engineering

16.1 Design the optimal cylindrical container (Fig. P16.1) that is open at one end and has walls of negligible thickness. The container is to hold 0.2 m³. Design it so that the area of its bottom and sides are minimized.

16.2 Design the optimal conical container (Fig. P16.2) that has a cover and has walls of negligible thickness. The container is to hold 0.2 m³. Design it so that the area of its top and sides are minimized.

16.3 Design the optimal cylindrical tank with dished ends (Fig. P16.3). The container is to hold 0.2 m³ and has walls of negligible thickness. Note that the area and volume of each of the dished ends can be computed with

$$A = \pi(h^2 + r^2) \qquad V = \frac{2\pi h(h^2 + 3r^2)}{3}$$

(a) Design the tank so that its surface area is minimized. Interpret the result.

(b) Repeat part **(a)**, but add the constraint $L \geq 2h$.

16.4 The specific growth rate of a yeast that produces an antibiotic is a function of the food concentration c,

$$g = \frac{2c}{4 + 0.8c + c^2 + 0.2c^3}$$

FIGURE P16.1

A cylindrical container with no lid.

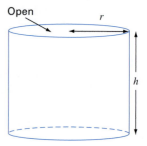

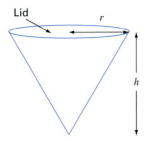

FIGURE P16.2
A conical container with a lid.

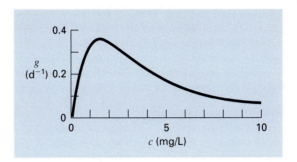

FIGURE P16.4
The specific growth rate of a yeast that produces an antibiotic versus the food concentration.

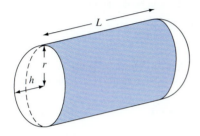

FIGURE P16.3
A cylindrical container with dished ends.

As depicted in Fig. P16.4, growth goes to zero at very low concentration due to food limitation. It also goes to zero at high concentrations due to toxicity effects. Find the value of c at which growth is a maximum.

16.5 A chemical plant makes three major products on a weekly basis. Each of these products requires a certain quantity of raw chemical, different production times, and yields different profits. The pertinent information is summarized below:

Note that there is sufficient warehouse space at the plant to store a total of 450 kg/week.

(a) Set up a linear programming problem to maximize profit.
(b) Solve the linear programming problem with the simplex method.
(c) Solve the problem with a software package.
(d) Evaluate which of the following options will raise profits the most: increasing raw chemical, production time or storage.

16.6 Recently chemical engineers have become involved in the area known as *waste minimization*. This involves the operation of a chemical plant so that impacts on the environment are minimized. Suppose a refinery develops a product, $Z1$, made from two raw materials X and Y. The production of 1 metric tonne of the product involves 1 tonne of X and 2.5 tonnes of Y and produces 1 tonne of a liquid waste, W. The engineers have come up with three alternative ways to handle the waste:

- Produce a tonne of a secondary product, $Z2$, by adding an additional tonne of X to each tonne of W.
- Produce a tonne of another secondary product, $Z3$, by adding 1 tonne of Y to each tonne of W.
- Treat the waste so that it is permissible to discharge it.

	Product 1	Product 2	Product 3	Resource Availability
Raw chemical	5 kg/kg	4 kg/kg	10 kg/kg	3000 kg
Production time	0.05 hr/kg	0.1 hr/kg	0.2 hr/kg	55 hr/week
New Profit	$30/kg	$30/kg	$35/kg	

The products yield profits of $2500, $-\$50$ and $200/tonne for Z1, Z2, and Z3, respectively. Note that producing Z2 actually creates a loss. The treatment process costs $300/tonne. In addition, the company has access to a limit of 7500 and 10,000 tonnes of X and Y during the production period. Determine how much of the products and waste must be created in order to maximize profit.

Civil/Environmental Engineering

16.7 A finite-element model of a cantilever beam subject to loading and moments (Fig. P16.7) is given by optimizing

$$f(x, y) = 5x^2 - 5xy + 2.5y^2 - x - 1.5y$$

where x = end displacement, and y = end moment. Find the values of x and y that minimize $f(x, y)$.

16.8 Suppose that you are asked to design a column to support a compressive load P as shown in Fig. P16.8a. The column has a cross-section shaped as a thin-walled pipe as shown in Fig. P16.8b.

The design variables are the pipe mean diameter, d, and the wall thickness, t. The cost of the pipe is computed by

$$\text{Cost} = f(t, d) = c_1 W + c_2 d$$

FIGURE P16.7
A cantilever beam.

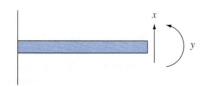

FIGURE P16.8
(a) A column supporting a compressive load P. (b) The column has a cross-section shaped as a thin-walled pipe.

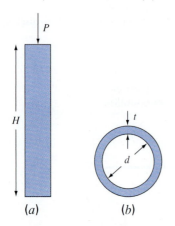

(a)   (b)

where $c_1 = 4$ and $c_2 = 2$ are cost factors and W = weight of the pipe,

$$W = \pi dt H \rho$$

where ρ = density of the pipe material = 0.0025 kg/cm^3. The column must support the load under compressive stress and not buckle. Therefore,

Actual stress $(\sigma) \leq$ maximum compressive yield stress
$$= \sigma_y = 550 \text{ kg/cm}^2$$

Actual stress $\leq$ buckling stress

The actual stress is given by

$$\sigma = \frac{P}{A} = \frac{P}{\pi dt}$$

The buckling stress can be shown to be

$$\sigma_b = \frac{\pi EI}{H^2 dt}$$

where E = modulus of elasticity and I = second moment of the area of the cross section. Calculus can be used to show that

$$I = \frac{\pi}{8} dt(d^2 + t^2)$$

Finally, diameters of available pipes are between d_1 and d_2 and thicknesses between t_1 and t_2. Develop and solve this problem by determining the values of d and t that minimize the cost. Note that $H = 275$ cm, $P = 2000$ kg, $E = 900,000$ kg/cm^2, $d_1 = 1$ cm, $d_2 = 10$ cm, $t_1 = 0.1$ cm, and $t_2 = 1$ cm.

16.9 The *Streeter-Phelps model* can be used to compute the dissolved oxygen concentration in a river below a point discharge of sewage (Fig. P16.9),

$$o = o_s - \frac{k_d L_o}{k_d + k_s - k_a}\left(e^{-k_a t} - e^{-(k_d + k_s)t}\right) - \frac{S_b}{k_a}\left(1 - e^{-k_a t}\right)$$

(P16.9)

where o = dissolved oxygen concentration [mg/L], o_s = oxygen saturation concentration [mg/L], t = travel time [d], L_o = BOD concentration at the mixing point [mg/L], k_d = rate of decomposition of biochemical oxygen demand (BOD) [d^{-1}], k_s = rate of settling of (BOD) [d^{-1}], k_a = reaeration rate [d^{-1}], and S_b = sediment oxygen demand [mg/L/d].

As indicated in Fig. P16.9, Eq. (P16.9) produces an oxygen "sag" that reaches a critical minimum level, o_c, some travel time, t_c, below the point discharge. This point is called "critical" because it represents the location where biota that depend on oxygen (like fish) would be the most stressed. Determine the critical travel time and concentration, given the following values:

$$o_s = 10 \text{ mg/L} \qquad k_d = 0.1 \text{ d}^{-1} \qquad k_a = 0.6 \text{ d}^{-1}$$
$$k_s = 0.05 \text{ d}^{-1} \qquad L_o = 50 \text{ mg/L} \qquad S_b = 1 \text{ mg/L/d}$$

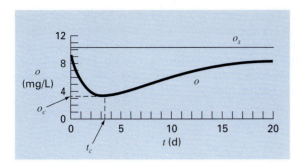

FIGURE P16.9
A dissolved oxygen "sag" below a point discharge of sewage into a river.

16.10 The two-dimensional distribution of pollutant concentration in a channel can be described by

$$c(x, y) = 7.9 + 0.13x + 0.21y - 0.05x^2$$
$$- 0.016y^2 - 0.007xy$$

Determine the exact location of the peak concentration given the function and the knowledge that the peak lies within the bounds $-10 \le x \le 10$ and $0 \le y \le 20$.

16.11 The flow Q [m³/s] in an open channel can be predicted with the Manning equation (recall Sec. 8.2)

$$Q = \frac{1}{n} A_c R^{2/3} S^{1/2}$$

where n = Manning roughness coefficient (a dimensionless number used to parameterize the channel friction), A_c = cross-sectional area of the channel (m²), S = channel slope (dimensionless, meters drop per meter length), and R = hydraulic radius (m), which is related to more fundamental parameters by

$$R = \frac{A_c}{P}$$

where P = wetted perimeter (m). As the name implies, the wetted perimeter is the length of the channel sides and bottom that is under water. For example, for a rectangular channel, it is defined as

$$P = B + 2H$$

where H = depth (m). Suppose that you are using this formula to design a lined canal (note that farmers line canals to minimize leakage losses).
(a) Given the parameters $n = 0.035$, $S = 0.003$, and $Q = 1$ m³/s, determine the values of B and H that minimize the wetted perimeter. Note that such a calculation would minimize cost if lining costs were much larger than excavation costs.
(b) Repeat part **(a)**, but this time include the cost of excavation. To do this minimize the following cost function,

$$C = c_1 A + c_2 P$$

where c_1 is a cost factor for excavation = 100 \$/m² and c_2 is a cost factor for lining \$50/m.
(c) Discuss the implications of your results.

Electrical Engineering
16.12 A model for a solenoid can be expressed as

$$L = \frac{7a^2 N/b}{8 + 6(a/b) + 10(c/b)}$$

where L = inductance, N = number of turns, and a, b, and c describe the geometry as shown in Fig. P16.12.
We would like to maximize L for a given length of wire ℓ with a cross-sectional area A. If $\ell = 2$ m and $A = 10^{-6}$ m², then we require that

$$2\pi NA = 2 \quad \text{and} \quad \frac{bc}{N} = 10^{-6}$$

Therefore,

$$N = \frac{1}{\pi A} \quad \text{and} \quad c = \frac{10^{-6}N}{b}$$

Find a and b that maximize L.
16.13 An electric circuit is designed to use a 40-volt source to charge 15 V, 6 V, and 25 V batteries connected in parallel. Find the currents that maximize the power transferred to the batteries.
This problem can be formulated as the following linear programming problem:

Maximize $P = 15I_2 + 6I_4 + 25I_5$

subject to

$$I_1 = I_2 + I_3$$
$$I_3 = I_4 + I_5$$
$$I_1 \le 5$$

FIGURE P16.12
The dimensions of a solenoid.

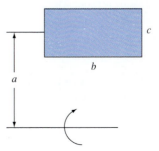

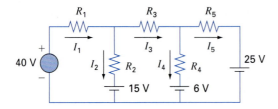

FIGURE P16.13
A parallel electric circuit to charge three batteries.

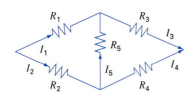

FIGURE P16.14
An electric network.

$$I_2 \leq 4$$
$$I_3 \leq 3$$
$$I_4 \leq 2$$
$$I_5 \leq 2$$
$$I_1, I_2, I_3, I_4, I_5 \geq 0$$

16.14 A network shown in Fig. P16.14 consists of five resistors and currents.

Find the resistances so that the total power dissipated by the network is a minimum. Assume that each current can vary between upper and lower limits,

$$I_{i,\min} \leq I_i \leq I_{i,\max}$$

and that the voltage drop across each resistor must be constant; i.e.,

$$V_i = R_i I_i = \text{constant}_i$$

Formulate this case as a constrained optimization problem.
16.15 A system consists of two power plants that must deliver loads over a transmission network. The costs of generating power at plants 1 and 2 are given by

$$F_1 = 2p_1 + 2$$
$$F_2 = 10p_2$$

where p_1 and p_2 = power produced by each plant. The losses of power due to transmission, L, are given by

$$L_1 = 0.2p_1 + 0.1p_2$$
$$L_2 = 0.2p_1 + 0.5p_2$$

The total demand for power is 30. Determine the power generation needed to meet demands while minimizing cost using an optimization routine such as those found on Excel, Mathcad, IMSL, etc.
16.16 The torque transmitted to an induction motor is a function of the slip between the rotation of the stator field and the rotor speed where slip is defined as

$$s = \frac{n - n_R}{n}$$

where n = revolutions per second of rotating stator speed and n_R = rotor speed. Kirchhoff's laws can be used to show that the torque (expressed in dimensionless form) and slip are related by

$$T = \frac{15s(1 - s)}{(1 - s)(4s^2 - 3s + 4)}$$

Figure P16.16 shows this function. Use a numerical method to determine the slip at which the maximum torque occurs.

Mechanical/Aerospace Engineering
16.17 The total drag on an airfoil can be estimated by

$$D = \underbrace{0.01\sigma V^2}_{\text{friction}} + \underbrace{\frac{0.95}{\sigma}\left(\frac{W}{V}\right)^2}_{\text{lift}}$$

where D = drag, σ = ratio of air density between the flight altitude and sea level, W = weight, and V = velocity. As in Fig. P16.17, the two factors contributing to drag are affected differently as velocity increases. Whereas friction drag increases with velocity, the drag due to lift decreases. The combination of the two factors leads to a minimum drag. **(a)** If $\sigma = 0.5$ and $W = 15,000$, determine the minimum drag and the velocity at which it occurs. **(b)** In addition, develop a sensitivity analysis to determine how this optimum varies in response to a range of $W = 12,000$ to $18,000$ with $\sigma = 0.5$.

FIGURE P16.16
Torque transmitted to an inductor as a function of slip.

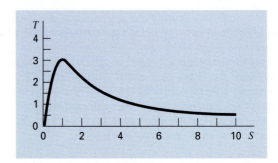

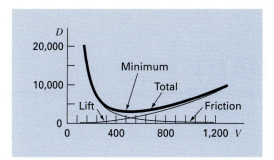

FIGURE P16.17
Plot of drag versus velocity for an airfoil.

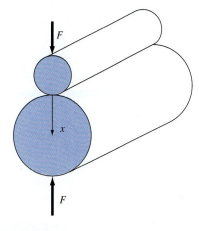

FIGURE P16.19
Roller bearings.

16.18 Four helical springs can be used to support a railroad car carrying mountain bikes to Denver. It is desirable to find the wire diameter (d), the coil diameter (D), and the number of turns (N) that minimizes the weight of the spring and limits the deflection to 0.15 inches, the shear stress to 9000 psi, and requires the shear stress to be greater than 120. This problem can be formulated as the following optimization problem:

$$\text{Minimize } F(d, D, N) = \frac{\pi d^2}{4}(\pi DNp)$$

subject to

$$\text{Deflection} = \frac{8FD^3N}{d^4G} \leq 0.15$$

$$\text{Shear stress} = K_s\frac{8FD}{\pi d^3} \leq 9000$$

$$\text{Natural frequency} = \frac{\sqrt{Gg}}{2\sqrt{2p\pi}}\frac{d}{D^2N} \geq 120$$

$$d, D, N > 0$$

If given values for the load and material parameters are substituted, we obtain

$$\text{Minimize } F(d, D, N) = 0.7d^2DN$$

subject to

$$110\frac{d^4}{D^3N} > 1$$

$$3\frac{d^3}{D} > 1$$

$$135\frac{d}{D^2N} > 1$$

$$d, D, N > 0$$

Solve for d, D, and N.

16.19 Roller bearings are subject to fatigue failure caused by large contact loads F (Fig. P16.19).

The problem of finding the location of the maximum stress along the x axis can be shown to be equivalent to maximizing the function

$$f(x) = \frac{0.4}{\sqrt{1+x^2}} - \sqrt{1+x^2}\left(1 - \frac{0.4}{1+x^2}\right) + x$$

Find the x that maximizes $f(x)$.

16.20 An aerospace company is developing a new fuel additive for commercial airliners. The additive is composed of three ingredients: X, Y, and Z. For peak performance, the total amount of additive must be at least 6 mL/L of fuel. For safety reasons, the sum of the highly flammable X and Y ingredients must not exceed 3 mL/L. In addition, the amount of the X ingredient must always be equal to or greater than the Y, and the Z must be greater than half the Y. If the cost per mL for the ingredients X, Y, and Z are 0.15, 0.025 and 0.05, respectively, determine the minimum cost mixture for each liter of fuel.

EPILOGUE: PART FOUR

The epilogues of other parts of this book contain a discussion and a tabular summary of the trade-offs among various methods as well as important formulas and relationships. Most of the methods of this part are quite complicated and, consequently, cannot be summarized with simple formulas and tabular summaries. Therefore, we deviate somewhat here by providing the following narrative discussion of trade-offs and further references.

PT4.4 TRADE-OFFS

Chapter 13 dealt with finding the optimum of an unconstrained function of a single variable. The golden-section search method is a bracketing method requiring that an interval containing a single optimum be known. It has the advantage that it minimizes function evaluations and always converges. Quadratic interpolation also works best when implemented as a bracketing method, although it can also be programmed as an open method. However, in such cases, it may diverge. Both the golden-section search and quadratic interpolation do not require derivative evaluations. Thus, they are both appropriate methods when the bracket can be readily defined and function evaluations are costly.

Newton's method is an open method not requiring that an optimum be bracketed. It can be implemented in a closed-form representation when first and second derivatives can be determined analytically. It can also be implemented in a fashion similar to the secant method with finite-difference representations of the derivatives. Although Newton's method converges rapidly near the optimum, it is often divergent for poor guesses. Convergence is also dependent on the nature of the function.

Chapter 14 covered two general types of methods to solve multidimensional unconstrained optimization problems. Direct methods such as random searches and univariate searches do not require the evaluation of the function's derivatives and are often inefficient. However they also provide a tool to find global rather than local optima. Pattern search methods like Powell's method can be very efficient and also do not require derivative evaluation.

Gradient methods use either first and sometimes second derivatives to find the optimum. The method of steepest ascent/descent provides a reliable but sometimes slow approach. In contrast, Newton's method often converges rapidly when in the vicinity of a root, but sometimes suffers from divergence. The Marquardt method uses the steepest descent method at the starting location far away from the optimum and switches to Newton's method near the optimum in an attempt to take advantage of the strengths of each method.

The Newton method can be computationally costly because it requires computation of both the gradient vector and the Hessian matrix. Quasi-Newton approaches attempt to circumvent these problems by using approximations to reduce the number of matrix evaluations (particularly the evaluation, storage, and inversion of the Hessian).

Research investigations continue today that explore the characteristics and respective strengths of various hybrid and tandem methods. Some examples are the Fletcher-Reeves conjugate gradient method and Davidon-Fletcher-Powell quasi-Newton methods.

Chapter 15 was devoted to constrained optimization. For linear problems, linear programming based on the simplex method provides an efficient means to obtain solutions. Approaches such as the GRG method are available to solve nonlinear constrained problems.

Software packages and libraries include a wide variety of optimization capabilities. The most generic is the IMSL library which contains many subroutines to implement most standard optimization algorithms. At the time of this book's printing, Excel had the most useful optimization capabilities in the form of its Solver tool. Because this tool is designed to implement the most general form of optimization—nonlinear constrained optimization—it can be used to solve problems in all the areas covered in this part of the book.

PT4.5   ADDITIONAL REFERENCES

For one-dimensional problems, Brent's method is a hybrid that attempts to account for the nature of the function to ensure both slow and steady convergence for poor initial guesses and rapid convergence near the optimum. See Press et al. (1992) for details. For multidimensional problems, additional information can be found in Fletcher (1980, 1981), Gill et al. (1981), Dennis and Schnabel (1996), and Luenberger (1984).

PART FIVE

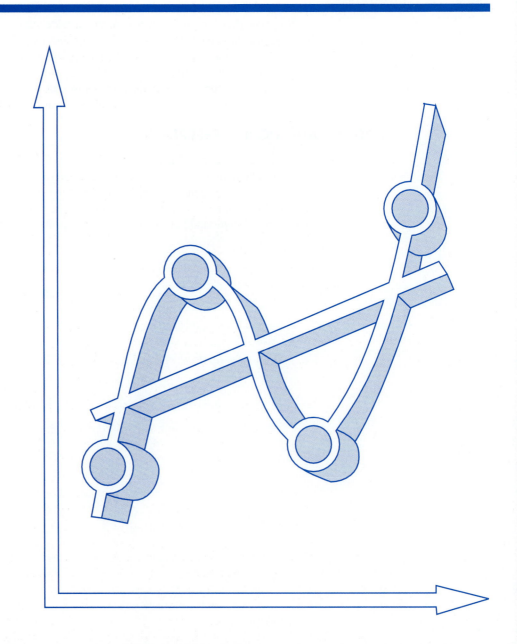

CURVE FITTING

PT5.1 MOTIVATION

Data is often given for discrete values along a continuum. However, you may require estimates at points between the discrete values. The present part of this book describes techniques to fit curves to such data to obtain intermediate estimates. In addition, you may require a simplified version of a complicated function. One way to do this is to compute values of the function at a number of discrete values along the range of interest. Then, a simpler function may be derived to fit these values. Both of these applications are known as *curve fitting*.

There are two general approaches for curve fitting that are distinguished from each other on the basis of the amount of error associated with the data. First, where the data exhibits a significant degree of error or "noise," the strategy is to derive a single curve that represents the general trend of the data. Because any individual data point may be incorrect, we make no effort to intersect every point. Rather, the curve is designed to follow the pattern of the points taken as a group. One approach of this nature is called *least-squares regression* (Fig. PT5.1*a*).

Second, where the data is known to be very precise, the basic approach is to fit a curve or a series of curves that pass directly through each of the points. Such data usually originates from tables. Examples are values for the density of water or for the heat capacity of gases as a function of temperature. The estimation of values between well-known discrete points is called *interpolation* (Fig. PT5.1*b* and *c*).

PT5.1.1 Noncomputer Methods for Curve Fitting

The simplest method for fitting a curve to data is to plot the points and then sketch a line that visually conforms to the data. Although this is a valid option when quick estimates are required, the results are dependent on the subjective viewpoint of the person sketching the curve.

For example, Fig. PT5.1 shows sketches developed from the same set of data by three engineers. The first did not attempt to connect the points, but rather, characterized the general upward trend of the data with a straight line (Fig. PT5.1*a*). The second engineer used straight-line segments or linear interpolation to connect the points (Fig. PT5.1*b*). This is a very common practice in engineering. If the values are truly close to being linear or are spaced closely, such an approximation provides estimates that are adequate for many engineering calculations. However, where the underlying relationship is highly curvilinear or the data is widely spaced, significant errors can be introduced by such linear interpolation. The third engineer used curves to try to capture the meanderings suggested by the data (Fig. PT5.1*c*). A fourth or fifth engineer would likely develop alternative fits. Obviously,

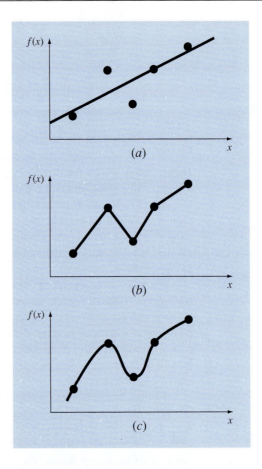

FIGURE PT5.1
Three attempts to fit a "best" curve through five data points. (a) Least-squares regression, (b) linear interpolation, and (c) curvilinear interpolation.

our goal here is to develop systematic and objective methods for the purpose of deriving such curves.

PT5.1.2 Curve Fitting and Engineering Practice

Your first exposure to curve fitting may have been to determine intermediate values from tabulated data—for instance, from interest tables for engineering economics or from steam tables for thermodynamics. Throughout the remainder of your career, you will have frequent occasion to estimate intermediate values from such tables.

Although many of the widely used engineering properties have been tabulated, there are a great many more that are not available in this convenient form. Special cases and new problem contexts often require that you measure your own data and develop your own predictive relationships. Two types of applications are generally encountered when fitting experimental data: trend analysis and hypothesis testing.

Trend analysis represents the process of using the pattern of the data to make predictions. For cases where the data is measured with high precision, you might utilize interpolating polynomials. Imprecise data is often analyzed with least-squares regression.

Trend analysis may be used to predict or forecast values of the dependent variable. This can involve extrapolation beyond the limits of the observed data or interpolation within the range of the data. All fields of engineering commonly involve problems of this type.

A second engineering application of experimental curve fitting is *hypothesis testing*. Here, an existing mathematical model is compared with measured data. If the model coefficients are unknown, it may be necessary to determine values that best fit the observed data. On the other hand, if estimates of the model coefficients are already available, it may be appropriate to compare predicted values of the model with observed values to test the adequacy of the model. Often, alternative models are compared and the "best" selected on the basis of empirical observations.

In addition to the above engineering applications, curve fitting is important in other numerical methods such as integration and the approximate solution of differential equations. Finally, curve-fitting techniques can be used to derive simple functions to approximate complicated functions.

PT5.2   MATHEMATICAL BACKGROUND

The prerequisite mathematical background for interpolation is found in the material on Taylor series expansions and finite divided differences introduced in Chap. 4. Least-squares regression requires additional information from the field of statistics. If you are familiar with the concepts of the mean, standard deviation, residual sum of the squares, normal distribution, and confidence intervals, feel free to skip the following pages and proceed directly to PT5.3. If you are unfamiliar with these concepts or are in need of a review, the following material is designed as a brief introduction to these topics.

PT5.2.1 Simple Statistics

Suppose that in the course of an engineering study, several measurements were made of a particular quantity. For example, Table PT5.1 contains 24 readings of the coefficient of thermal expansion of a structural steel. Taken at face value, the data provides a limited amount of information—that is, that the values range from a minimum of 6.395 to a maximum of 6.775. Additional insight can be gained by summarizing the data in one or more well-chosen statistics that convey as much information as possible about specific characteristics of the data set. These descriptive statistics are most often selected to represent

TABLE PT5.1   Measurements of the coefficient of thermal expansion of structural steel [$\times 10^{-6}$ in/(in $\cdot$ °F)].

6.495	6.595	6.615	6.635	6.485	6.555
6.665	6.505	6.435	6.625	6.715	6.655
6.755	6.625	6.715	6.575	6.655	6.605
6.565	6.515	6.555	6.395	6.775	6.685

(1) the location of the center of the distribution of the data and (2) the degree of spread of the data set.

The most common location statistic is the arithmetic mean. The *arithmetic mean* ($\bar{y}$) of a sample is defined as the sum of the individual data points (y_i) divided by the number of points (n), or

$$\bar{y} = \frac{\Sigma y_i}{n} \tag{PT5.1}$$

where the summation (and all the succeeding summations in this introduction) is from $i = 1$ through n.

The most common measure of spread for a sample is the *standard deviation* (s_y) about the mean,

$$s_y = \sqrt{\frac{S_t}{n-1}} \tag{PT5.2}$$

where S_t is the total sum of the squares of the residuals between the data points and the mean, or

$$S_t = \Sigma \, (y_i - \bar{y})^2 \tag{PT5.3}$$

Thus, if the individual measurements are spread out widely around the mean, S_t (and, consequently, s_y) will be large. If they are grouped tightly, the standard deviation will be small. The spread can also be represented by the square of the standard deviation, which is called the *variance*:

$$s_y^2 = \frac{S_t}{n-1} \tag{PT5.4}$$

Note that the denominator in both Eqs. (PT5.2) and (PT5.4) is $n - 1$. The quantity $n - 1$ is referred to as the degrees of freedom. Hence S_t and s_y are said to be based on $n - 1$ *degrees of freedom*. This nomenclature derives from the fact that the sum of the quantities upon which S_t is based (that is, $\bar{y} - y_1$, $\bar{y} - y_2$, ..., $\bar{y} - y_n$) is zero. Consequently, if $\bar{y}$ is known and $n - 1$ of the values are specified, the remaining value is fixed. Thus, only $n - 1$ of the values are said to be freely determined. Another justification for dividing by $n - 1$ is the fact that there is no such thing as the spread of a single data point. For the case where $n = 1$, Eqs. (PT5.2) and (PT5.4) yield a meaningless result of infinity.

It should be noted that an alternative, more convenient formula is available to compute the standard deviation,

$$s_y^2 = \frac{\Sigma y_i^2 - (\Sigma y_i)^2 / n}{n-1}$$

This version does not require precomputation of $\bar{y}$ and yields an identical result as Eq. (PT5.4).

A final statistic that has utility in quantifying the spread of data is the *coefficient of variation* (*c.v.*). This statistic is the ratio of the standard deviation to the mean. As such, it provides a normalized measure of the spread. It is often multiplied by 100 so that it can be expressed in the form of a percent:

$$c.v. = \frac{s_y}{\bar{y}} 100\% \tag{PT5.5}$$

Notice that the coefficient of variation is similar in spirit to the percent relative error (ε_t) discussed in Sec. 3.3. That is, it is the ratio of a measure of error (s_y) to an estimate of the true value ($\bar{y}$).

EXAMPLE PT5.1   Simple Statistics of a Sample

Problem Statement.   Compute the mean, variance, standard deviation, and coefficient of variation for the data in Table PT5.1.

TABLE PT5.2   Computations for statistics for the readings of the coefficient of thermal expansion. The frequencies and bounds are developed to construct the histogram in Fig. PT5.2.

				Interval	
i	y_i	$(y_i - \bar{y})^2$	Frequency	Lower Bound	Upper Bound
1	6.395	0.042025	1	6.36	6.40
2	6.435	0.027225	1	6.40	6.44
3	6.485	0.013225			
4	6.495	0.011025	4	6.48	6.52
5	6.505	0.009025			
6	6.515	0.007225			
7	6.555	0.002025	2	6.52	6.56
8	6.555	0.002025			
9	6.565	0.001225			
10	6.575	0.000625	3	6.56	6.60
11	6.595	0.000025			
12	6.605	0.000025			
13	6.615	0.000225			
14	6.625	0.000625	5	6.60	6.64
15	6.625	0.000625			
16	6.635	0.001225			
17	6.655	0.003025			
18	6.655	0.003025	3	6.64	6.68
19	6.665	0.004225			
20	6.685	0.007225			
21	6.715	0.013225	3	6.68	6.72
22	6.715	0.013225			
23	6.755	0.024025	1	6.72	6.76
24	6.775	0.030625	1	6.76	6.80
Σ	158.4	0.217000			

Solution. The data is added (Table PT5.2) and the results are used to compute [Eq. (PT5.1)]

$$\bar{y} = \frac{158.4}{24} = 6.6$$

As in Table PT5.2, the sum of the squares of the residuals is 0.217000, which can be used to compute the standard deviation [Eq. (PT5.2)]:

$$s_y = \sqrt{\frac{0.217000}{24 - 1}} = 0.097133$$

the variance [Eq. (PT5.4)]:

$$s_y^2 = 0.009435$$

and the coefficient of variation [Eq. (PT5.5)]:

$$\text{c.v.} = \frac{0.097133}{6.6} 100\% = 1.47\%$$

PT5.2.2 The Normal Distribution

Another characteristic that bears on the present discussion is the *data distribution*—that is, the shape with which the data is spread around the mean. A *histogram* provides a simple visual representation of the distribution. As in Table PT5.2, the histogram is constructed by sorting the measurements into intervals. The units of measurement are plotted on the abscissa and the frequency of occurrence of each interval is plotted on the ordinate. Thus, five of the measurements fall between 6.60 and 6.64. As in Fig. PT5.2, the histogram suggests that most of the data is grouped close to the mean value of 6.6.

If we have a very large set of data, the histogram often can be approximated by a smooth curve. The symmetric, bell-shaped curve superimposed on Fig. PT5.2 is one such characteristic shape—the *normal distribution*. Given enough additional measurements, the histogram for this particular case could eventually approach the normal distribution.

The concepts of the mean, standard deviation, residual sum of the squares, and normal distribution all have great relevance to engineering practice. A very simple example is their use to quantify the confidence that can be ascribed to a particular measurement. If a quantity is normally distributed, the range defined by $\bar{y} - s_y$ to $\bar{y} + s_y$ will encompass approximately 68 percent of the total measurements. Similarly, the range defined by $\bar{y} - 2s_y$ to $\bar{y} + 2s_y$ will encompass approximately 95 percent.

For example, for the data in Table PT5.1 ($\bar{y} = 6.6$ and $s_y = 0.097133$), we can make the statement that approximately 95 percent of the readings should fall between 6.405734 and 6.794266. If someone told us that they had measured a value of 7.35, we would suspect that the measurement might be erroneous. The following section elaborates on such evaluations.

PT5.2.3 Estimation of Confidence Intervals

As should be clear from the previous sections, one of the primary aims of statistics is to estimate the properties of a *population* based on a limited *sample* drawn from that population.

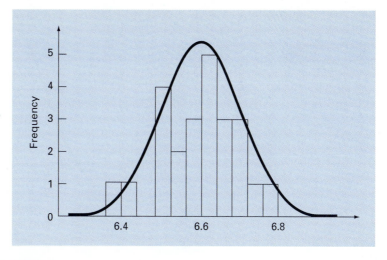

FIGURE PT5.2
A histogram used to depict the distribution of data. As the number of data points increases, the histogram could approach the smooth, bell-shaped curve called the normal distribution.

Clearly, it is impossible to measure the coefficient of thermal expansion for every piece of structural steel that has ever been produced. Consequently, as in Tables PT5.1 and PT5.2, we can randomly make a number of measurements and, on the basis of the sample, attempt to characterize the properties of the entire population.

Because we "infer" properties of the unknown population from a limited sample, the endeavor is called *statistical inference*. Because the results are often reported as estimates of the population parameters, the process is also referred to as *estimation*.

We have already shown how we estimate the central tendency (sample mean, $\bar{y}$) and spread (sample standard deviation and variance) of a limited sample. Now, we will briefly describe how we can attach probabilistic statements to the quality of these estimates. In particular, we will discuss how we can define a confidence interval around our estimate of the mean. We have chosen this particular topic because of its direct relevance to the regression models we will be describing in Chap. 17.

Note that in the following discussion, the nomenclature $\bar{y}$ and s_y refer to the sample mean and standard deviation. The nomenclature μ and σ refer to the population mean and standard deviation. The former are sometimes referred to as the "estimated" mean and standard deviation, whereas the latter are sometimes called the "true" mean and standard deviation.

An *interval estimator* gives the range of values within which the parameter is expected to lie with a given probability. Such intervals are described as being one-sided or two-sided. As the name implies, a *one-sided interval* expresses our confidence that the parameter estimate is less than or greater than the true value. In contrast, the *two-sided interval* deals with the more general proposition that the estimate agrees with the truth with no consideration to the sign of the discrepancy. Because it is more general, we will focus on the two-sided interval.

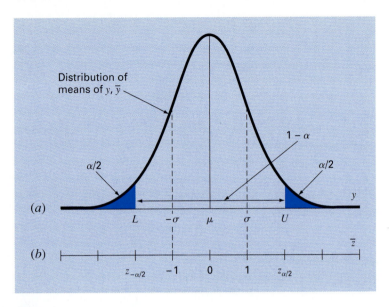

FIGURE PT5.3
A two-sided confidence interval. The abscissa scale in (a) is written in the natural units of the random variable y. (b) is a normalized version of the abscissa that places the origin at the location of the mean and scales the axis so that the standard deviation corresponds to a unit value.

A two-sided interval can be described by the statement

$$P\{L \leq \mu \leq U\} = 1 - \alpha$$

which reads, "the probability that the true mean of y, μ, falls within the bound from L to U is $1 - \alpha$." The quantity α is called the *significance level*. So the problem of defining a confidence interval reduces to estimating L and U. Although it is not absolutely necessary, it is customary to view the two-sided interval with the α probability distributed evenly as $\alpha/2$ in each tail of the distribution, as in Fig. PT5.3.

If the true variance of the distribution of y, σ^2, is known (which is not usually the case), statistical theory states that the sample mean $\bar{y}$ comes from a normal distribution with mean μ and variance σ^2/n (Box PT5.1). In the case illustrated in Fig. PT5.3, we really do not know μ. Therefore, we do not know where the normal curve is exactly located with respect to $\bar{y}$. To circumvent this dilemma, we compute a new quantity, the *standard normal estimate*

$$\bar{z} = \frac{\bar{y} - \mu}{\sigma/\sqrt{n}} \tag{PT5.6}$$

which represents the normalized distance between $\bar{y}$ and μ. According to statistical theory, this quantity should be normally distributed with a mean of 0 and a variance of σ^2/n. Furthermore, the probability that $\bar{z}$ would fall within the unshaded region of Fig. PT5.3

Box PT5.1 A Little Statistics

Most engineers take several courses to become proficient at statistics. Because you may not have taken such a course yet, we'd like to mention a few ideas that might make this present section more coherent.

As we have stated, the "game" of inferential statistics assumes that the random variable you are sampling, y, has a true mean (μ) and variance (σ^2). Further, in the present discussion, we also assume that it has a particular distribution: the normal distribution. The variance of this normal distribution has a finite value that specifies the "spread" of the normal distribution. If the variance is large, the distribution is broad. Conversely, if the variance is small, the distribution is narrow. Thus, the true variance quantifies the intrinsic uncertainty of the random variable.

In the game of statistics, we take a limited number of measurements of this quantity called a sample. From this sample, we can compute an estimated mean ($\bar{y}$) and variance (s_y^2). The more measurements we take, the better the estimates approximate the true values. That is, as $n \rightarrow \infty$, $\bar{y} \rightarrow \mu$ and $s_y^2 \rightarrow \sigma^2$.

Suppose that we take n samples and compute an estimated mean $\bar{y}_1$. Then, we take another n samples and compute another, $\bar{y}_2$. We can keep repeating this process until we have generated a sample of means: $\bar{y}_1, \bar{y}_2, \bar{y}_3, \ldots, \bar{y}_m$, where m is large. We can then develop a histogram of these means and determine a "distribution of the means," as well as a "mean of the means" and a "standard deviation of the means." Now the question arises: does this new distribution of means and its statistics behave in a predictable fashion?

There is an extremely important theorem known as the *Central Limit Theorem* that speaks directly to this question. It can be stated as

Let $y_1, y_2, \ldots, y_n$ be a random sample of size n from a distribution with mean μ and variance σ^2. Then, for large n, $\bar{y}$ is approximately normal with mean μ and variance σ^2/n. Furthermore, for large n, the random variable $(\bar{y} - \mu)/(\sigma/\sqrt{n})$ is approximately standard normal.

Thus, the theorem states the remarkable result that the distribution of means will always be normally distributed regardless of the underlying distribution of the random variables! It also yields the expected result that given a sufficiently large sample, the mean of the means should converge on the true population mean μ.

Further, the theorem says that as the sample size gets larger, the variance of the means should approach zero. This makes sense, because if n is small, our individual estimates of the mean should be poor and the variance of the means should be large. As n increases, our estimates of the mean will improve and hence their spread should shrink. The Central Limit Theorem neatly defines exactly how this shrinkage relates to both the true variance and the sample size, i.e., as σ^2/n.

Finally, the theorem states the important result that we have given as Eq. (PT5.6). As is shown in this section, this result is the basis for constructing confidence intervals for the mean.

should be $1 - \alpha$. Therefore, the statement can be made that

$$\frac{\bar{y} - \mu}{\sigma/\sqrt{n}} < -z_{\alpha/2} \qquad \text{or} \qquad \frac{\bar{y} - \mu}{\sigma/\sqrt{n}} > z_{\alpha/2}$$

with a probability of α.

The quantity $z_{\alpha/2}$ is a standard normal random variable. This is the distance measured along the normalized axis above and below the mean that encompasses $1 - \alpha$ probability (Fig. PT5.3b). Values of $z_{\alpha/2}$ are tabulated in statistics books (e.g., Milton and Arnold 1995). They can also be calculated using functions on software packages and libraries like Excel and IMSL. As an example, for $\alpha = 0.05$ (in other words, defining an interval encompassing 95%), $z_{\alpha/2}$ is equal to about 1.96. This means that an interval around the mean of width ± 1.96 times the standard deviation will encompass approximately 95% of the distribution.

These results can be rearranged to yield

$$L \leq \mu \leq U$$

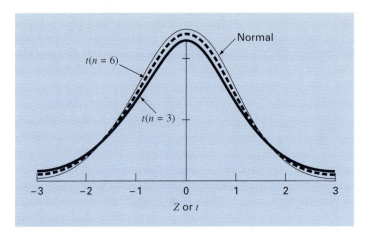

FIGURE PT5.4
Comparison of the normal distribution with the t distribution for $n = 3$ and $n = 6$. Notice how the t distribution is generally flatter.

with a probability of $1 - \alpha$, where

$$L = \bar{y} - \frac{\sigma}{\sqrt{n}} z_{\alpha/2} \qquad U = \bar{y} + \frac{\sigma}{\sqrt{n}} z_{\alpha/2} \tag{PT5.7}$$

Now, although the foregoing provides an estimate of L and U, it is based on knowledge of the true variance σ. For our case, we know only the estimated variance s_y. A straightforward alternative would be to develop a version of Eq. (PT5.6) based on s_y,

$$t = \frac{\bar{y} - \mu}{s_y/\sqrt{n}} \tag{PT5.8}$$

Even when we sample from a normal distribution, this fraction will not be normally distributed, particularly when n is small. It was found by W.S. Gossett that the random variable defined by Eq. (PT5.8) follows the so-called Student-t, or simply, t *distribution*. For this case,

$$L = \bar{y} - \frac{s_y}{\sqrt{n}} t_{\alpha/2, n-1} \qquad U = \bar{y} + \frac{s_y}{\sqrt{n}} t_{\alpha/2, n-1} \tag{PT5.9}$$

where $t_{\alpha/2, n-1}$ is the standard random variable for the t distribution for a probability of $\alpha/2$. As was the case for $z_{\alpha/2}$, values are tabulated in statistics books, and can also be calculated using software packages and libraries. For example, if $\alpha = 0.05$ and $n = 20$, $t_{\alpha/2, n-1} = 2.086$.

The t distribution can be thought of as a modification of the normal distribution that accounts for the fact that we have an imperfect estimate of the standard deviation. When n is small, it tends to be flatter than the normal (see Fig. PT5.4). Therefore, for small numbers of measurements, it yields wider and hence more conservative confidence intervals. As n grows larger, the t distribution converges on the normal.

EXAMPLE PT5.2   Confidence Interval on the Mean

Problem Statement.   Determine the mean and the corresponding 95% confidence interval for the data from Table PT5.1. Perform three estimates based on (a) the first 8, (b) the first 16, and (c) all 24 measurements.

Solution.   (a) The mean and standard deviation for the first 8 points is

$$\bar{y} = \frac{52.72}{8} = 6.59 \qquad s_y = \sqrt{\frac{347.4814 - (52.72)^2/8}{8 - 1}} = 0.089921$$

The appropriate t statistic can be calculated as

$$t_{0.05/2, 8-1} = t_{0.025, 7} = 2.364623$$

which can be used to compute the interval

$$L = 6.59 - \frac{0.089921}{\sqrt{8}} 2.364623 = 6.5148$$

$$U = 6.59 + \frac{0.089921}{\sqrt{8}} 2.364623 = 6.6652$$

or

$$6.5148 \le \mu \le 6.6652$$

Thus, based on the first eight measurements, we conclude that there is a 95% probability that the true mean falls within the range 6.5148 to 6.6652.

FIGURE PT5.5
Estimates of the mean and 95% confidence intervals for different numbers of sample size.

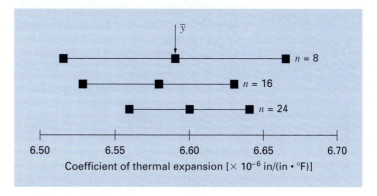

The two other cases for (*b*) 16 points and (*c*) 24 points can be calculated in a similar fashion and the results tabulated along with case (*a*) as

n	$\bar{y}$	s_y	$t_{\alpha/2,n-1}$	L	U
8	6.5900	0.089921	2.364623	6.5148	6.6652
16	6.5794	0.095845	2.131451	6.5283	6.6304
24	6.6000	0.097133	2.068655	6.5590	6.6410

These results, which are also summarized in Fig. PT5.5, indicate the expected outcome that the confidence interval becomes more narrow as *n* increases. Thus, the more measurements we take, our estimate of the true value becomes more refined.

The above is just one simple example of how statistics can be used to make judgments regarding uncertain data. These concepts will also have direct relevance to our discussion of regression models. You can consult any basic statistics book (for example, Milton and Arnold, 1995) to obtain additional information on the subject.

PT5.3   ORIENTATION

Before we proceed to numerical methods for curve fitting, some orientation might be helpful. The following is intended as an overview of the material discussed in Part Five. In addition, we have formulated some objectives to help focus your efforts when studying the material.

PT5.3.1 Scope and Preview

Figure PT5.5 provides a visual overview of the material to be covered in Part Five. *Chapter 17* is devoted to *least-squares regression*. We will first learn how to fit the "best" straight line through a set of uncertain data points. This technique is called *linear regression*. Besides discussing how to calculate the slope and intercept of this straight line, we also present quantitative and visual methods for evaluating the validity of the results.

In addition to fitting a straight line, we also present a general technique for fitting a "best" polynomial. Thus, you will learn to derive a parabolic, cubic, or higher-order polynomial that optimally fits uncertain data. Linear regression is a subset of this more general approach, which is called *polynomial regression*.

The next topic covered in Chap. 17 is *multiple linear regression*. It is designed for the case where the dependent variable *y* is a linear function of two or more independent variables $x_1, x_2, \ldots, x_m$. This approach has special utility for evaluating experimental data where the variable of interest is dependent on a number of different factors.

After multiple regression, we illustrate how polynomial and multiple regression are both subsets of a *general linear least-squares model*. Among other things, this will allow us to introduce a concise matrix representation of regression and discuss its general statistical properties.

Finally, the last sections of Chap. 17 are devoted to *nonlinear regression*. This approach is designed to compute a least-squares fit of a nonlinear equation to data.

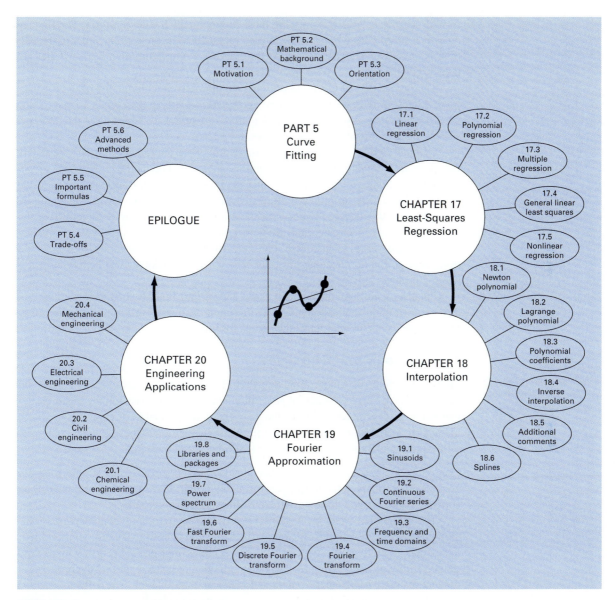

FIGURE PT5.6
Schematic of the organization of the material in Part Five: Curve Fitting.

In *Chap. 18*, the alternative curve-fitting technique called *interpolation* is described. As discussed previously, interpolation is used for estimating intermediate values between precise data points. In Chap. 18, polynomials are derived for this purpose. We introduce the basic concept of polynomial interpolation by using straight lines and parabolas to connect points. Then, we develop a generalized procedure for fitting an nth-order polynomial. Two formats are presented for expressing these polynomials in equation form. The first, called *Newton's interpolating polynomial,* is preferable when the appropriate order of the polynomial is unknown. The second, called the *Lagrange interpolating polynomial,* has advantages when the proper order is known beforehand.

The last section of Chap. 18 presents an alternative technique for fitting precise data points. This technique, called *spline interpolation,* fits polynomials to data but in a piecewise fashion. As such, it is particularly well-suited for fitting data that is generally smooth but exhibits abrupt local changes.

Chapter 19 deals with the Fourier transform approach to curve fitting where periodic functions are fit to data. Our emphasis in this section will be on the *fast Fourier transform.* At the end of this chapter, we also include an overview of several software packages and libraries that can be used for curve fitting. These are Mathcad, Excel, MATLAB, and IMSL.

Chapter 20 is devoted to engineering applications that illustrate the utility of the numerical methods in engineering problem contexts. Examples are drawn from the four major specialty areas of chemical, civil, electrical, and mechanical engineering. In addition, some of the applications illustrate how software packages can be applied for engineering problem solving.

Finally, an epilogue is included at the end of Part Five. It contains a summary of the important formulas and concepts related to curve fitting as well as a discussion of trade-offs among the techniques and suggestions for future study.

PT5.3.2 Goals and Objectives

Study Objectives.   After completing Part Five, you should have greatly enhanced your capability to fit curves to data. In general, you should have mastered the techniques, have learned to assess the reliability of the answers, and be capable of choosing the preferred method (or methods) for any particular problem. In addition to these general goals, the specific concepts in Table PT5.3 should be assimilated and mastered.

Computer Objectives.   You have been provided with software and simple computer algorithms to implement the techniques discussed in Part Five. You may also have access to software packages and libraries. All have utility as learning tools.

The Numerical Methods TOOLKIT software includes polynomial regression. The graphics associated with this software will enable you to easily visualize your problem and the associated mathematical operations. The graphics are a critical part of your assessment of the validity of a regression fit. They also provide guidance regarding the proper order of polynomial regression and the potential dangers of extrapolation. The software is very easy to apply to solve practical problems and can be used to check the results of any computer programs you may develop yourself.

In addition, pseudocode algorithms are provided for most of the other methods in Part Five. This information will allow you to expand your software library to include techniques beyond polynomial regression. For example, you may find it useful from a professional viewpoint to have software to implement multiple linear regression, Newton's interpolating polynomial, cubic spline interpolation, and the fast Fourier transform.

Finally, one of your most important goals should be to master several of the general-purpose software packages that are widely available. In particular, you should become adept at using these tools to implement numerical methods for engineering problem solving.

TABLE PT5.3   Specific study objectives for Part Five.

1. Understand the fundamental difference between regression and interpolation and realize why confusing the two could lead to serious problems
2. Understand the derivation of linear least-squares regression and be able to assess the reliability of the fit using graphical and quantitative assessments
3. Know how to linearize data by transformation
4. Understand situations where polynomial, multiple, and nonlinear regression are appropriate
5. Be able to recognize general linear models, understand the general matrix formulation of linear least squares, and know how to compute confidence intervals for parameters
6. Understand that there is one and only one polynomial of degree n or less that passes exactly through $n + 1$ points
7. Know how to derive the first-order Newton's interpolating polynomial
8. Understand the analogy between Newton's polynomial and the Taylor series expansion and how it relates to the truncation error
9. Recognize that the Newton and Lagrange equations are merely different formulations of the same interpolating polynomial and understand their respective advantages and disadvantages
10. Realize that more accurate results are generally obtained if data used for interpolation is centered around and close to the unknown point
11. Realize that data points do not have to be equally spaced nor in any particular order for either the Newton or Lagrange polynomials
12. Know why equispaced interpolation formulas have utility
13. Recognize the liabilities and risks associated with extrapolation
14. Understand why spline functions have utility for data with local areas of abrupt change
15. Recognize how the Fourier series is used to fit data with periodic functions
16. Understand the difference between the frequency and time domains

CHAPTER 17

Least-Squares Regression

Where substantial error is associated with data, polynomial interpolation is inappropriate and may yield unsatisfactory results when used to predict intermediate values. Experimental data is often of this type. For example, Fig. 17.1a shows seven experimentally derived data points exhibiting significant variability. Visual inspection of the data suggests a positive relationship between y and x. That is, the overall trend indicates that higher values of y are associated with higher values of x. Now, if a sixth-order interpolating polynomial is fitted to this data (Fig. 17.1b), it will pass exactly through all of the points. However, because of the variability in the data, the curve oscillates widely in the interval between the points. In particular, the interpolated values at $x = 1.5$ and $x = 6.5$ appear to be well beyond the range suggested by the data.

A more appropriate strategy for such cases is to derive an approximating function that fits the shape or general trend of the data without necessarily matching the individual points. Figure 17.1c illustrates how a straight line can be used to generally characterize the trend of the data without passing through any particular point.

One way to determine the line in Fig. 17.1c is to visually inspect the plotted data and then sketch a "best" line through the points. Although such "eyeball" approaches have commonsense appeal and are valid for "back-of-the-envelope" calculations, they are deficient because they are arbitrary. That is, unless the points define a perfect straight line (in which case, interpolation would be appropriate), different analysts would draw different lines.

To remove this subjectivity, some criterion must be devised to establish a basis for the fit. One way to do this is to derive a curve that minimizes the discrepancy between the data points and the curve. A technique for accomplishing this objective, called *least-squares regression,* will be discussed in the present chapter.

17.1 LINEAR REGRESSION

The simplest example of a least-squares approximation is fitting a straight line to a set of paired observations: $(x_1, y_1), (x_2, y_2), \ldots, (x_n, y_n)$. The mathematical expression for the straight line is

$$y = a_0 + a_1 x + e \tag{17.1}$$

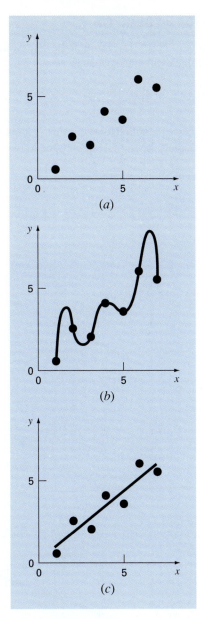

FIGURE 17.1
(a) Data exhibiting significant error. (b) Polynomial fit oscillating beyond the range of the data. (c) More satisfactory result using the least-squares fit.

where a_0 and a_1 are coefficients representing the intercept and the slope, respectively, and e is the error, or residual, between the model and the observations, which can be represented by rearranging Eq. (17.1) as

$$e = y - a_0 - a_1 x$$

Thus, the error, or *residual,* is the discrepancy between the true value of y and the approximate value, $a_0 + a_1 x$, predicted by the linear equation.

17.1.1 Criteria for a "Best" Fit

One strategy for fitting a "best?" line through the data would be to minimize the sum of the residual errors for all the available data, as in

$$\sum_{i=1}^{n} e_i = \sum_{i=1}^{n} (y_i - a_0 - a_1 x_i) \tag{17.2}$$

where n = total number of points. However, this is an inadequate criterion, as illustrated by Fig. 17.2a which depicts the fit of a straight line to two points. Obviously, the best fit is

FIGURE 17.2

Examples of some criteria for "best fit" that are inadequate for regression: (a) minimizes the sum of the residuals, (b) minimizes the sum of the absolute values of the residuals, and (c) minimizes the maximum error of any individual point.

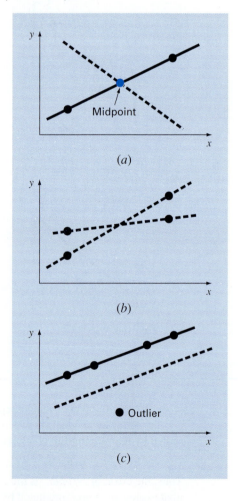

the line connecting the points. However, any straight line passing through the midpoint of the connecting line (except a perfectly vertical line) results in a minimum value of Eq. (17.2) equal to zero because the errors cancel.

Therefore, another logical criterion might be to minimize the sum of the absolute values of the discrepancies, as in

$$\sum_{i=1}^{n} |e_i| = \sum_{i=1}^{n} |y_i - a_0 - a_1 x_i|$$

Figure 17.2b demonstrates why this criterion is also inadequate. For the four points shown, any straight line falling within the dashed lines will minimize the absolute value of the sum. Thus, this criterion also does not yield a unique best fit.

A third strategy for fitting a best line is the *minimax* criterion. In this technique, the line is chosen that minimizes the maximum distance that an individual point falls from the line. As depicted in Fig. 17.2c, this strategy is ill-suited for regression because it gives undue influence to an outlier, that is, a single point with a large error. It should be noted that the minimax principle is sometimes well-suited for fitting a simple function to a complicated function (Carnahan, Luther, and Wilkes, 1969).

A strategy that overcomes the shortcomings of the aforementioned approaches is to minimize the sum of the squares of the residuals between the measured y and the y calculated with the linear model

$$S_r = \sum_{i=1}^{n} e_i^2 = \sum_{i=1}^{n} \left(y_{i,\text{measured}} - y_{i,\text{model}} \right)^2 = \sum_{i=1}^{n} \left(y_i - a_0 - a_1 x_i \right)^2 \qquad (17.3)$$

This criterion has a number of advantages, including the fact that it yields a unique line for a given set of data. Before discussing these properties, we will present a technique for determining the values of a_0 and a_1 that minimize Eq. (17.3).

17.1.2 Least-Squares Fit of a Straight Line

To determine values for a_0 and a_1, Eq. (17.3) is differentiated with respect to each coefficient:

$$\frac{\partial S_r}{\partial a_0} = -2 \sum \left(y_i - a_0 - a_1 x_i \right)$$

$$\frac{\partial S_r}{\partial a_1} = -2 \sum \left[\left(y_i - a_0 - a_1 x_i \right) x_i \right]$$

Note that we have simplified the summation symbols; unless otherwise indicated, all summations are from $i = 1$ to n. Setting these derivatives equal to zero will result in a minimum S_r. If this is done, the equations can be expressed as

$$0 = \sum y_i - \sum a_0 - \sum a_1 x_i$$

$$0 = \sum y_i x_i - \sum a_0 x_i - \sum a_1 x_i^2$$

Now, realizing that $\Sigma a_0 = na_0$, we can express the equations as a set of two simultaneous linear equations with two unknowns (a_0 and a_1):

$$na_0 + \left(\sum x_i\right) a_1 = \sum y_i \tag{17.4}$$

$$\left(\sum x_i\right) a_0 + \left(\sum x_i^2\right) a_1 = \sum x_i y_i \tag{17.5}$$

These are called the *normal equations*. They can be solved simultaneously

$$a_1 = \frac{n\sum x_i y_i - \sum x_i \sum y_i}{n\sum x_i^2 - \left(\sum x_i\right)^2} \tag{17.6}$$

This result can then be used in conjunction with Eq. (17.4) to solve for

$$a_0 = \bar{y} - a_1\bar{x} \tag{17.7}$$

where $\bar{y}$ and $\bar{x}$ are the means of y and x, respectively.

EXAMPLE 17.1   Linear Regression

Problem Statement.   Fit a straight line to the x and y values in the first two columns of Table 17.1.

Solution.   The following quantities can be computed:

$$n = 7 \qquad \sum x_i y_i = 119.5 \qquad \sum x_i^2 = 140$$

$$\sum x_i = 28 \qquad \bar{x} = \frac{28}{7} = 4$$

$$\sum y_i = 24 \qquad \bar{x} = \frac{24}{7} = 3.428571$$

Using Eqs. (17.6) and (17.7),

$$a_1 = \frac{7(119.5) - 28(24)}{7(140) - (28)^2} = 0.8392857$$

$$a_0 = 3.428571 - 0.8392857(4) = 0.07142857$$

TABLE 17.1   Computations for an error analysis of the linear fit.

x_i	y_i	$(y_i - \bar{y})^2$	$(y_i - a_0 - a_1 x_i)^2$
1	0.5	8.5765	0.1687
2	2.5	0.8622	0.5625
3	2.0	2.0408	0.3473
4	4.0	0.3265	0.3265
5	3.5	0.0051	0.5896
6	6.0	6.6122	0.7972
7	5.5	4.2908	0.1993
Σ	24.0	22.7143	2.9911

Therefore, the least-squares fit is

$$y = 0.07142857 + 0.8392857x$$

The line, along with the data, is shown in Fig. 17.1c.

17.1.3 Quantification of Error of Linear Regression

Any line other than the one computed in Example 17.1 results in a larger sum of the squares of the residuals. Thus, the line is unique and in terms of our chosen criterion is a "best" line through the points. A number of additional properties of this fit can be elucidated by examining more closely the way in which residuals were computed. Recall that the sum of the squares is defined as [Eq. (17.3)]

$$S_r = \sum_{i=1}^{n} e_i^2 = \sum_{i=1}^{n} (y_i - a_0 - a_1 x_i)^2 \tag{17.8}$$

Notice the similarity between Eqs. (PT5.3) and (17.8). In the former case, the square of the residual represented the square of the discrepancy between the data and a single estimate of the measure of central tendency—the mean. In Eq. (17.8), the square of the residual represents the square of the vertical distance between the data and another measure of central tendency—the straight line (Fig. 17.3).

The analogy can be extended further for cases where (1) the spread of the points around the line is of similar magnitude along the entire range of the data and (2) the distribution of these points about the line is normal. It can be demonstrated that if these criteria are met, least-squares regression will provide the best (that is, the most likely) estimates of a_0 and a_1 (Draper and Smith, 1981). This is called the *maximum likelihood principle* in

FIGURE 17.3
The residual in linear regression represents the vertical distance between a data point and the straight line.

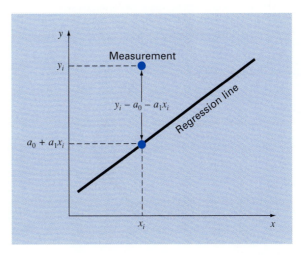

statistics. In addition, if these criteria are met, a "standard deviation" for the regression line can be determined as [compare with Eq. (PT5.2)]

$$s_{y/x} = \sqrt{\frac{S_r}{n-2}}$$

(17.9)

where $s_{y/x}$ is called the *standard error of the estimate*. The subscript notation "y/x" designates that the error is for a predicted value of y corresponding to a particular value of x. Also, notice that we now divide by $n - 2$ because two data-derived estimates—a_0 and a_1—were used to compute S_r; thus, we have lost two degrees of freedom. As with our discussion of the standard deviation in PT5.2.1, another justification for dividing by $n - 2$ is that there is no such thing as the "spread of data" around a straight line connecting two points. Thus, for the case where $n = 2$, Eq. (17.9) yields a meaningless result of infinity.

Just as was the case with the standard deviation, the standard error of the estimate quantifies the spread of the data. However, $s_{y/x}$ quantifies the spread *around the regression line* as shown in Fig. 17.4b in contrast to the original standard deviation s_y that quantified the spread *around the mean* (Fig. 17.4a).

The above concepts can be used to quantify the "goodness" of our fit. This is particularly useful for comparison of several regressions (Fig. 17.5). To do this, we return to the original data and determine the *total sum of the squares* around the mean for the dependent variable (in our case, y). As was the case for Eq. (PT5.3), this quantity is designated S_t. This is the magnitude of the residual error associated with the dependent variable prior to regression. After performing the regression, we can compute S_r, the sum of the squares of the residuals around the regression line. This characterizes the residual error that remains after the regression. It is, therefore, sometimes called the unexplained sum of the squares. The

FIGURE 17.4

Regression data showing (a) the spread of the data around the mean of the dependent variable and (b) the spread of the data around the best-fit line. The reduction in the spread in going from (a) to (b), as indicated by the bell-shaped curves at the right, represents the improvement due to linear regression.

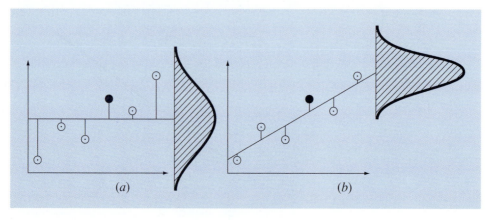

(a)   (b)

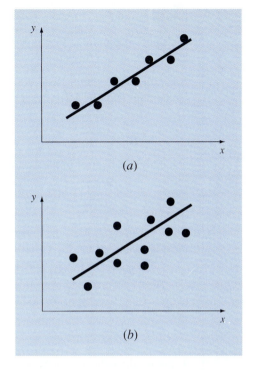

FIGURE 17.5
Examples of linear regression with (a) small and (b) large residual errors.

difference between the two quantities, $S_t - S_r$, quantifies the improvement or error reduction due to describing the data in terms of a straight line rather than as an average value. Because the magnitude of this quantity is scale-dependent, the difference is normalized to S_t to yield

$$r^2 = \frac{S_t - S_r}{S_t}$$

(17.10)

where r^2 is called the *coefficient of determination* and r is the *correlation coefficient* $(= \sqrt{r^2})$. For a perfect fit, $S_r = 0$ and $r = r^2 = 1$, signifying that the line explains 100 percent of the variability of the data. For $r = r^2 = 0$, $S_r = S_t$ and the fit represents no improvement. An alternative formulation for r that is more convenient for computer implementation is

$$r = \frac{n\sum x_i y_i - (\sum x_i)(\sum y_i)}{\sqrt{n\sum x_i^2 - (\sum x_i)^2}\sqrt{n\sum y_i^2 - (\sum y_i)^2}}$$

(17.11)

EXAMPLE 17.2   Estimation of Errors for the Linear Least-Squares Fit

Problem Statement.   Compute the total standard deviation, the standard error of the estimate, and the correlation coefficient for the data in Example 17.1.

Solution.   The summations are performed and presented in Table 17.1. The standard deviation is [Eq. (PT5.2)]

$$s_y = \sqrt{\frac{22.7143}{7-1}} = 1.9457$$

and the standard error of the estimate is [Eq. (17.9)]

$$s_{y/x} = \sqrt{\frac{2.9911}{7-2}} = 0.7735$$

Thus, because $s_{y/x} < s_y$, the linear regression model has merit. The extent of the improvement is quantified by [Eq. (17.10)]

$$r^2 = \frac{22.7143 - 2.9911}{22.7143} = 0.868$$

or

$$r = \sqrt{0.868} = 0.932$$

These results indicate that 86.8 percent of the original uncertainty has been explained by the linear model.

Before proceeding to the computer program for linear regression, a word of caution is in order. Although the correlation coefficient provides a handy measure of goodness-of-fit, you should be careful not to ascribe more meaning to it than is warranted. Just because r is "close" to 1 does not mean that the fit is necessarily "good." For example, it is possible to obtain a relatively high value of r when the underlying relationship between y and x is not even linear. Draper and Smith (1981) provide guidance and additional material regarding assessment of results for linear regression. In addition, at the minimum, you should *always* inspect a plot of the data along with your regression curve. As described in the next section, the Numerical Methods TOOLKIT software includes such a capability.

17.1.4 Computer Program for Linear Regression

It is a relatively trivial matter to develop a pseudocode for linear regression (Fig. 17.6). As mentioned above, a plotting option is critical to the effective use and interpretation of regression and is included in the supplementary Numerical Methods TOOLKIT software. In addition, popular software packages like Excel and Mathcad can implement regression and have plotting capabilities. If your computer language has plotting capabilities, we recommend that you expand your program to include a plot of y versus x showing both the data and the regression line. The inclusion of the capability will greatly enhance the utility of the program in problem-solving contexts.

SUB Regres(x, y, n, a1, a0, syx, r2)

sumx = 0: sumxy = 0: st = 0
sumy = 0: sumx2 = 0: sr = 0
DO i = 1, n
    sumx = sumx + x_i
    sumy = sumy + y_i
    sumxy = sumxy + x_i*y_i
    sumx2 = sumx2 + x_i*x_i
END DO
xm = sumx/n
ym = sumy/n
a1 = (n*sumxy − sumx*sumy)/(n*sumx2 − sumx*sumx)
a0 = ym − a1*xm
DO i = 1, n
    st = st + (y_i − ym)^2
    sr = sr + (y_i − a1*x_i − a0)^2
END DO
syx = (sr/(n − 2))^{0.5}
r2 = (st − sr)/st

END Regres

FIGURE 17.6
Algorithm for linear regression.

EXAMPLE 17.3   Linear Regression Using the Computer

Problem Statement.   A user-friendly computer program to implement linear regression is contained in the Numerical Methods TOOLKIT software package associated with this text. We can use this software to solve a hypothesis-testing problem associated with the falling parachutist discussed in Chap. 1. A theoretical mathematical model for the velocity of the parachutist was given as the following [Eq. (1.10)]:

$$v(t) = \frac{gm}{c}\left(1 - e^{(-c/m)t}\right)$$

where v = velocity (m/s), g = gravitational constant (9.8 m/s^2), m = mass of the parachutist equal to 68.1 kg, and c = drag coefficient of 12.5 kg/s. The model predicts the velocity of the parachutist as a function of time, as described in Example 1.1. A plot of the velocity variation was developed in Example 2.1.

An alternative empirical model for the velocity of the parachutist is given by

$$v(t) = \frac{gm}{c}\left(\frac{t}{3.75 + t}\right) \tag{E17.3.1}$$

TABLE 17.2 Measured and calculated velocities for the falling parachutist.

Time, s	Measured v, m/s (a)	Model-calculated v, m/s [Eq. (1.10)] (b)	Model-calculated v, m/s [Eq. (E17.3.1)] (c)
1	10.00	8.953	11.240
2	16.30	16.405	18.570
3	23.00	22.607	23.729
4	27.50	27.769	27.556
5	31.00	32.065	30.509
6	35.60	35.641	32.855
7	39.00	38.617	34.766
8	41.50	41.095	36.351
9	42.90	43.156	37.687
10	45.00	44.872	38.829
11	46.00	46.301	39.816
12	45.50	47.490	40.678
13	46.00	48.479	41.437
14	49.00	49.303	42.110
15	50.00	49.988	42.712

Suppose that you would like to test and compare the adequacy of these two mathematical models. This might be accomplished by measuring the actual velocity of the parachutist at known values of time and comparing these results with the predicted velocities according to each model.

Such an experimental-data-collection program was implemented, and the results are listed in column (a) of Table 17.2. Computed velocities for each model are listed in columns (b) and (c).

Solution.   The adequacy of the models can be tested by plotting the model-calculated velocity versus the measured velocity. Linear regression can be used to calculate the slope and the intercept of the plot. This line will have a slope of 1, an intercept of 0, and an $r^2 = 1$ if the model matches the data perfectly. A significant deviation from these values can be used as an indication of the inadequacy of the model.

Figure 17.7a and b are plots of the line and data for the regressions of columns (b) and (c), respectively, versus column (a). For the first model [Eq. (1.10) as depicted in Fig. 17.7a],

$$v_{model} = -0.859 + 1.032 v_{measure}$$

and for the second model [Eq. (E17.3.1) as depicted in Fig. 17.7b],

$$v_{model} = 5.776 + 0.752 v_{measure}$$

These plots indicate that the linear regression between the data and each of the models is highly significant. Both models match the data with a correlation coefficient of greater than 0.99.

However, the model described by Eq. (1.10) conforms to our hypothesis test criteria much better than that described by Eq. (E17.3.1) because the slope and intercept are more

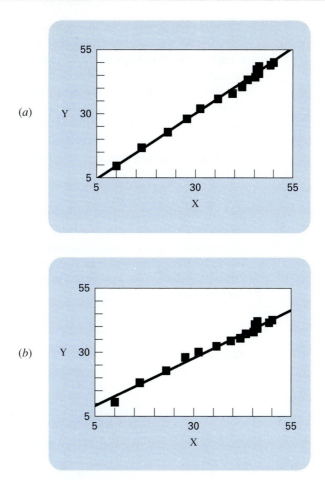

FIGURE 17.7

(a) Results using linear regression to compare predictions computed with the theoretical model [Eq. (1.10)] versus measured values. (b) Results using linear regression to compare predictions computed with the empirical model [Eq. (E17.3.1)] versus measured values.

nearly equal to 1 and 0. Thus, although each plot is well described by a straight line, Eq. (1.10) appears to be a better model than Eq. (E17.3.1).

Model testing and selection are common and extremely important activities performed in all fields of engineering. The background material provided in this chapter, together with your software, should allow you to address many practical problems of this type.

There is one shortcoming with the analysis in Example 17.3. The example was unambiguous because the empirical model [Eq. (E17.3.1)] was clearly inferior to Eq. (1.10). Thus, the slope and intercept for the former were so much closer to the desired result of 1 and 0, that it was obvious which model was superior.

However, suppose that the slope were 0.85 and the intercept were 2. Obviously this would make the conclusion that the slope and intercept were 1 and 0 open to debate. Clearly, rather than relying on a subjective judgment, it would be preferable to base such a conclusion on a quantitative criterion.

This can be done by computing confidence intervals for the model parameters in the same way that we developed confidence intervals for the mean in Sec. PT5.2.3. We will return to this topic at the end of this chapter.

17.1.5 Linearization of Nonlinear Relationships

Linear regression provides a powerful technique for fitting a "best" line to data. However, it is predicated on the fact that the relationship between the dependent and independent variables is linear. This is not always the case, and the first step in any regression analysis should be to plot and visually inspect the data to ascertain whether a linear model applies. For example, Fig. 17.8 shows some data that is obviously curvilinear. In some cases, techniques such as polynomial regression, which is described in Sec. 17.2, are appropriate. For others, transformations can be used to express the data in a form that is compatible with linear regression.

FIGURE 17.8
(a) Data that is ill-suited for linear least-squares regression. (b) Indication that a parabola is preferable.

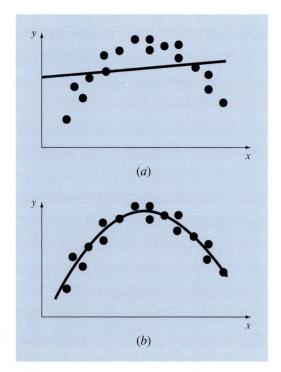

One example is the *exponential model*

$$y = a_1 e^{b_1 x} \tag{17.12}$$

where a_1 and b_1 are constants. This model is used in many fields of engineering to characterize quantities that increase (positive b_1) or decrease (negative b_1) at a rate that is directly proportional to their own magnitude. For example, population growth or radioactive decay can exhibit such behavior. As depicted in Fig. 17.9a, the equation represents a nonlinear relationship (for $b_1 \neq 0$) between y and x.

Another example of a nonlinear model is the simple power equation

$$y = a_2 x^{b_2} \tag{17.13}$$

FIGURE 17.9

(a) The exponential equation, (b) the power equation, and (c) the saturation-growth-rate equation. Parts (d), (e), and (f) are linearized versions of these equations that result from simple transformations.

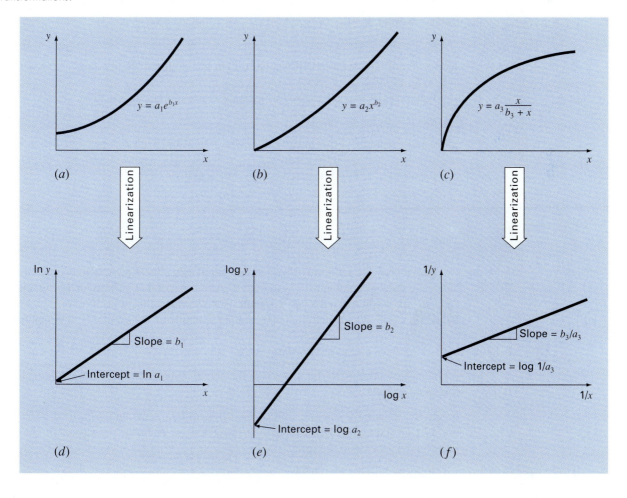

where a_2 and b_2 are constant coefficients. This model has wide applicability in all fields of engineering. As depicted in Fig. 17.9b, the equation (for $b_2 \neq 0$ or 1) is nonlinear.

A third example of a nonlinear model is the saturation-growth-rate equation [recall Eq. (E17.3.1)]

$$y = a_3 \frac{x}{b_3 + x} \tag{17.14}$$

where a_3 and b_3 are constant coefficients. This model, which is particularly well-suited for characterizing population growth rate under limiting conditions, also represents a nonlinear relationship between y and x (Fig. 17.9c) that levels off, or "saturates," as x increases.

Nonlinear regression techniques are available to fit these equations to experimental data directly. (Note that we will discuss nonlinear regression in Sec. 17.5.) However, a simpler alternative is to use mathematical manipulations to transform the equations into a linear form. Then, simple linear regression can be employed to fit the equations to data.

For example, Eq. (17.12) can be linearized by taking its natural logarithm to yield

$$\ln y = \ln a_1 + b_1 x \ln e$$

But because $\ln e = 1$,

$$\ln y = \ln a_1 + b_1 x \tag{17.15}$$

Thus a plot of $\ln y$ versus x will yield a straight line with a slope of b_1 and an intercept of $\ln a_1$ (Fig. 17.9d).

Equation (17.14) is linearized by taking its base-10 logarithm to give

$$\log y = b_2 \log x + \log a_2 \tag{17.16}$$

Thus, a plot of $\log y$ versus $\log x$ will yield a straight line with a slope of b_2 and an intercept of $\log a_2$ (Fig. 17.9e).

Equation (17.14) is linearized by inverting it to give

$$\frac{1}{y} = \frac{b_3}{a_3} \frac{1}{x} + \frac{1}{a_3} \tag{17.17}$$

Thus, a plot of $1/y$ versus $1/x$ will be linear, with a slope of b_3/a_3 and an intercept of $1/a_3$ (Fig. 17.9f).

In their transformed forms, these models are fit using linear regression in order to evaluate the constant coefficients. They could then be transformed back to their original state and used for predictive purposes. Example 17.4 illustrates this procedure for Eq. (17.13). In addition, Sec. 20.1 provides an engineering example of the same sort of computation.

EXAMPLE 17.4   Linearization of a Power Equation

Problem Statement.   Fit Eq. (17.13) to the data in Table 17.3 using a logarithmic transformation of the data.

Solution.   Figure 17.10a is a plot of the original data in its untransformed state. Figure 17.10b shows the plot of the transformed data. A linear regression of the log-transformed data yields the result

$$\log y = 1.75 \log x - 0.300$$

TABLE 17.3 Data to be fit to the power equation.

x	y	log x	log y
1	0.5	0	−0.301
2	1.7	0.301	0.226
3	3.4	0.477	0.534
4	5.7	0.602	0.753
5	8.4	0.699	0.922

FIGURE 17.10
(a) Plot of untransformed data with the power equation that fits the data. (b) Plot of transformed data used to determine the coefficients of the power equation.

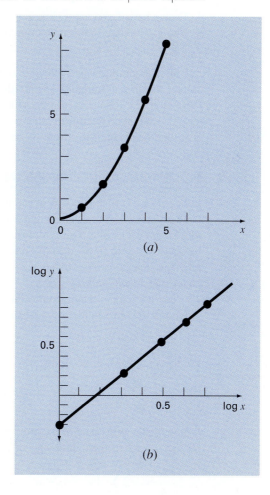

Thus, the intercept, log a_2, equals -0.300, and therefore, by taking the antilogarithm, $a_2 = 10^{-0.3} = 0.5$. The slope is $b_2 = 1.75$. Consequently, the power equation is

$$y = 0.5x^{1.75}$$

This curve, as plotted in Fig. 17.10a, indicates a good fit.

17.1.6 General Comments on Linear Regression

Before proceeding to curvilinear and multiple linear regression, we must emphasize the introductory nature of the foregoing material on linear regression. We have focused on the simple derivation and practical use of equations to fit data. You should be cognizant of the fact that there are theoretical aspects of regression that are of practical importance but are beyond the scope of this book. For example, some statistical assumptions that are inherent in the linear least-squares procedures are

1. Each x has a fixed value; it is not random and is known without error.
2. The y values are independent random variables and all have the same variance.
3. The y values for a given x must be normally distributed.

Such assumptions are relevant to the proper derivation and use of regression. For example, the first assumption means that (1) the x values must be error-free and (2) the regression of y versus x is not the same as x versus y (try Prob. 17.4 at the end of the chapter). You are urged to consult other references such as Draper and Smith (1981) to appreciate aspects and nuances of regression that are beyond the scope of this book.

17.2 POLYNOMIAL REGRESSION

In Sec. 17.1, a procedure was developed to derive the equation of a straight line using the least-squares criterion. Some engineering data, although exhibiting a marked pattern such as seen in Fig. 17.8, is poorly represented by a straight line. For these cases, a curve would be better suited to fit the data. As discussed in the previous section, one method to accomplish this objective is to use transformations. Another alternative is to fit polynomials to the data using *polynomial regression*.

The least-squares procedure can be readily extended to fit the data to a higher-order polynomial. For example, suppose that we fit a second-order polynomial or quadratic:

$$y = a_0 + a_1 x + a_2 x^2 + e$$

For this case the sum of the squares of the residuals is [compare with Eq. (17.3)]

$$S_r = \sum_{i=1}^{n} \left(y_i - a_0 - a_1 x_i - a_2 x_i^2 \right)^2 \tag{17.18}$$

Following the procedure of the previous section, we take the derivative of Eq. (17.18) with respect to each of the unknown coefficients of the polynomial, as in

$$\frac{\partial S_r}{\partial a_0} = -2 \sum \left(y_i - a_0 - a_1 x_i - a_2 x_i^2 \right)$$

$$\frac{\partial S_r}{\partial a_1} = -2 \sum x_i \left(y_i - a_0 - a_1 x_i - a_2 x_i^2 \right)$$

$$\frac{\partial S_r}{\partial a_2} = -2 \sum x_i^2 \left(y_i - a_0 - a_1 x_i - a_2 x_i^2 \right)$$

These equations can be set equal to zero and rearranged to develop the following set of normal equations:

$$(n)a_0 + \left(\sum x_i \right) a_1 + \left(\sum x_i^2 \right) a_2 = \sum y_i$$

$$\left(\sum x_i \right) a_0 + \left(\sum x_i^2 \right) a_1 + \left(\sum x_i^3 \right) a_2 = \sum x_i y_i \tag{17.19}$$

$$\left(\sum x_i^2 \right) a_0 + \left(\sum x_i^3 \right) a_1 + \left(\sum x_i^4 \right) a_2 = \sum x_i^2 y_i$$

where all summations are from $i = 1$ through n. Note that the above three equations are linear and have three unknowns: a_0, a_1, and a_2. The coefficients of the unknowns can be calculated directly from the observed data.

For this case, we see that the problem of determining a least-squares second-order polynomial is equivalent to solving a system of three simultaneous linear equations. Techniques to solve such equations were discussed in Part Three.

The two-dimensional case can be easily extended to an mth-order polynomial as

$$y = a_0 + a_1 x + a_2 x^2 + \cdots + a_m x^m + e$$

The foregoing analysis can be easily extended to this more general case. Thus, we can recognize that determining the coefficients of an mth-order polynomial is equivalent to solving a system of $m + 1$ simultaneous linear equations. For this case, the standard error is formulated as

$$s_{y/x} = \sqrt{\frac{S_r}{n - (m + 1)}} \tag{17.20}$$

This quantity is divided by $n - (m + 1)$ because $(m + 1)$ data-derived coefficients—a_0, $a_1, \ldots, a_m$—were used to compute S_r; thus, we have lost $m + 1$ degrees of freedom. In addition to the standard error, a coefficient of determination can also be computed for polynomial regression with Eq. (17.10).

EXAMPLE 17.5 Polynomial Regression

Problem Statement. Fit a second-order polynomial to the data in the first two columns of Table 17.4.

Solution. From the given data,

$$m = 2 \qquad \sum x_i = 15 \qquad \sum x_i^4 = 979$$

$$n = 6 \qquad \sum y_i = 152.6 \qquad \sum x_i y_i = 585.6$$

$$\bar{x} = 2.5 \qquad \sum x_i^2 = 55 \qquad \sum x_i^2 y_i = 2488.8$$

$$\bar{y} = 25.433 \qquad \sum x_i^3 = 225$$

TABLE 17.4 Computations for an error analysis of the quadratic least-squares fit.

x_i	y_i	$(y_i - \bar{y})^2$	$(y_i - a_0 - a_1 x_i - a_2 x_i^2)$
0	2.1	544.44	0.14332
1	7.7	314.47	1.00286
2	13.6	140.03	1.08158
3	27.2	3.12	0.80491
4	40.9	239.22	0.61951
5	61.1	1272.11	0.09439
Σ	152.6	2513.39	3.74657

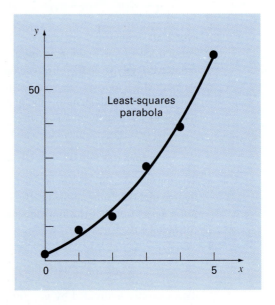

FIGURE 17.11
Fit of a second-order polynomial.

Therefore, the simultaneous linear equations are

$$\begin{bmatrix} 6 & 15 & 55 \\ 15 & 55 & 225 \\ 55 & 225 & 979 \end{bmatrix} \begin{Bmatrix} a_0 \\ a_1 \\ a_2 \end{Bmatrix} = \begin{Bmatrix} 152.6 \\ 585.6 \\ 2488.8 \end{Bmatrix}$$

Solving these equations through a technique such as Gauss elimination gives $a_0 = 2.47857$, $a_1 = 2.35929$, and $a_2 = 1.86071$. Therefore, the least-squares quadratic equation for this case is

$$y = 2.47857 + 2.35929x + 1.86071x^2$$

The standard error of the estimate based on the regression polynomial is [Eq. (17.20)]

$$s_{y/x} = \sqrt{\frac{3.74657}{6 - 3}} = 1.12$$

The coefficient of determination is

$$r^2 = \frac{2513.39 - 3.74657}{2513.39} = 0.99851$$

and the correlation coefficient is $r = 0.99925$.

These results indicate that 99.851 percent of the original uncertainty has been explained by the model. This result supports the conclusion that the quadratic equation represents an excellent fit, as is also evident from Fig. 17.11.

17.2.1 Algorithm for Polynomial Regression

An algorithm for polynomial regression is delineated in Fig. 17.12. Note that the primary task is the generation of the coefficients of the normal equations [Eq. (17.19)]. (Pseudocode for accomplishing this is presented in Fig. 17.13.) Then, techniques from Part Three can be applied to solve these simultaneous equations for the coefficients.

A potential problem associated with implementing polynomial regression on the computer is that the normal equations are sometimes ill-conditioned. This is particularly true for higher-order versions. For these cases, the computed coefficients may be highly susceptible to round-off error, and consequently, the results can be inaccurate. Among other things, this problem is related to the structure of the normal equations and to the fact that for higher-order polynomials the normal equations can have very large and very small coefficients. This is because the coefficients are summations of the data raised to powers.

Although the strategies for mitigating round-off error discussed in Part Three, such as pivoting, can help to partially remedy this problem, a simpler alternative is to use a computer with higher precision. Fortunately, most practical problems are limited to lower-order polynomials for which round-off is usually negligible. In situations where higher-order versions are required, other alternatives are available for certain types of data. However, these techniques (such as orthogonal polynomials) are beyond the scope of this book. The reader should consult texts on regression, such as Draper and Smith (1981), for additional information regarding the problem and possible alternatives.

FIGURE 17.12
Algorithm for implementation of polynomial and multiple linear regression.

Step 1: Input order of polynomial to be fit, m.
Step 2: Input number of data points, n.
Step 3: If $n < m + 1$, print out an error message that regression is impossible and terminate the process. If $n \geq m + 1$, continue.
Step 4: Compute the elements of the normal equation in the form of an augmented matrix.
Step 5: Solve the augmented matrix for the coefficients $a_0, a_1, a_2, \ldots, a_m$, using an elimination method.
Step 6: Print out the coefficients.

$$
\begin{aligned}
&DO\; i = 1,\, order + 1\\
&\quad DO\; j = 1,\, i\\
&\qquad k = i + j - 2\\
&\qquad sum = 0\\
&\qquad DO\; \ell = 1,\, n\\
&\qquad\quad sum = sum + x_\ell^k\\
&\qquad END\; DO\\
&\qquad a_{i,j} = sum\\
&\qquad a_{j,i} = sum\\
&\quad END\; DO\\
&\quad sum = 0\\
&\quad DO\; \ell = 1,\, n\\
&\qquad sum = sum + y_\ell \cdot x_\ell^{i-1}\\
&\quad END\; DO\\
&\quad a_{i,\, order+2} = sum\\
&END\; DO
\end{aligned}
$$

FIGURE 17.13
Pseudocode to assemble the elements of the normal equations for polynomial regression.

EXAMPLE 17.6   Polynomial Regression Using the Computer

Problem Statement. A user-friendly computer program to implement polynomial regression is contained in the Numerical Methods TOOLKIT software associated with the text. We can use this software to fit polynomials to the following data:

x	2	4	5	6	6	7	9	1	0.5	7.5
y	6	2	3	7	8	8	1	5	3	7

Solution.   Press the Fit Data with Curve button on the TOOLKIT main menu to obtain a blank screen similar to Fig. 17.14. This screen contains spaces for the input and output information needed to fit data with an mth-order least-squares regression polynomial.

The first step is to click the Input X vs Y Values table and enter up to 100 pairs of values for X and Y. Next you might decide to plot the data alone before making a decision concerning the order of the polynomial. This is done using a procedure similar to that described in Example 2.1. Inspection of the data shows two peaks and suggests that a polynomial of at least order 4 would be appropriate. For our example, we will first try a fifth-order polynomial. Simply enter a value of 5 for the order of the polynomial and the plot parameters in the Input Parameters table and click the red Calc and Plot buttons (in the process changing the buttons to black) to produce Fig. 17.14. The issue of determining the best order can be explored by examining how the standard error varies as a func-

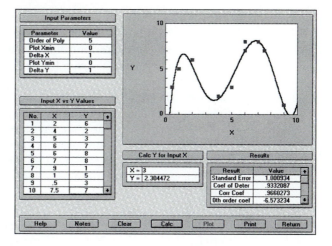

FIGURE 17.14
Screen from Numerical Methods TOOLKIT for fifth-order polynomial regression.

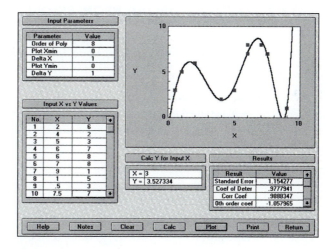

FIGURE 17.15
Plot for eighth-order polynomial regression.

tion of the order of the regression. The results for various order regression fits is tabulated below:

Order	1	2	3	4	5	6	7	8
Standard Error	2.71	2.69	2.34	1.38	1.00	1.12	1.17	1.15

Note that the standard error drops dramatically from order 3 to 4 and reaches a minimum for the order-5 polynomial. This suggests that not much is gained by expending the computational effort to perform higher than fifth-order regression.

Figure 17.15 shows the plots for the eighth-order case. For this case, overshoot begins to become a problem in a manner similar to higher-order interpolation (we'll discuss this phenomenon in detail in the next chapter). Figure 17.15 shows that the eighth-order polynomial produces negative Y values for X values between 8 and 9. Also observe from both Figs. 17.14 and 17.15 that while the regression curves follow the trend of the data, it is highly inappropriate to extrapolate for values Y beyond the range of the data for X.

Interpolation can be performed by entering a value for X in the Calc Y for Input X table. For example at X = 3, Y = 2.304472 as calculated from the fifth-order polynomial (Fig. 17.14).

Finally, take a look at the Results table on the lower right. The first three results are statistical summaries of the regression: Standard Error, Coefficient of Determination, and Correlation Coefficient. Note how these values change for different orders of regression. The scroll bar on the Results table is used to observe the actual coefficients of the regression polynomial. Again, these values change with different orders.

17.3   MULTIPLE LINEAR REGRESSION

A useful extension of linear regression is the case where y is a linear function of two or more independent variables. For example, y might be a linear function of x_1 and x_2, as in

$$y = a_0 + a_1 x_1 + a_2 x_2 + e$$

Such an equation is particularly useful when fitting experimental data where the variable being studied is often a function of two other variables. For this two-dimensional case, the regression "line" becomes a "plane" (Fig. 17.16).

FIGURE 17.16

Graphical depiction of multiple linear regression where y is a linear function of x_1 and x_2.

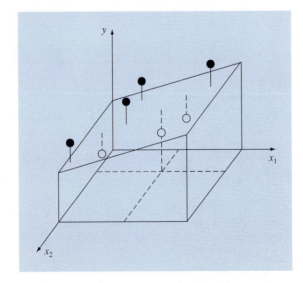

As with the previous cases, the "best" values of the coefficients are determined by setting up the sum of the squares of the residuals,

$$S_r = \sum_{i=1}^{n} (y_i - a_0 - a_1 x_{1i} - a_2 x_{2i})^2 \tag{17.21}$$

and differentiating with respect to each of the unknown coefficients,

$$\frac{\partial S_r}{\partial a_0} = -2 \sum (y_i - a_0 - a_1 x_{1i} - a_2 x_{2i})$$

$$\frac{\partial S_r}{\partial a_1} = -2 \sum x_{1i} (y_i - a_0 - a_1 x_{1i} - a_2 x_{2i})$$

$$\frac{\partial S_r}{\partial a_2} = -2 \sum x_{2i} (y_i - a_0 - a_1 x_{1i} - a_2 x_{2i})$$

The coefficients yielding the minimum sum of the squares of the residuals are obtained by setting the partial derivatives equal to zero and expressing the result in matrix form as

$$\begin{bmatrix} n & \Sigma x_{1i} & \Sigma x_{2i} \\ \Sigma x_{1i} & \Sigma x_{1i}^2 & \Sigma x_{1i} x_{2i} \\ \Sigma x_{2i} & \Sigma x_{1i} x_{2i} & \Sigma x_{2i}^2 \end{bmatrix} \begin{Bmatrix} a_0 \\ a_1 \\ a_2 \end{Bmatrix} = \begin{Bmatrix} \Sigma y_i \\ \Sigma x_{1i} y_i \\ \Sigma x_{2i} y_i \end{Bmatrix} \tag{17.22}$$

EXAMPLE 17.7   Multiple Linear Regression

Problem Statement. The following data was calculated from the equation $y = 5 + 4x_1 - 3x_2$:

x_1	x_2	y
0	0	5
2	1	10
2.5	2	9
1	3	0
4	6	3
7	2	27

Use multiple linear regression to fit this data.

Solution. The summations required to develop Eq. (17.22) are computed in Table 17.5. The result is

$$\begin{bmatrix} 6 & 16.5 & 14 \\ 16.5 & 76.25 & 48 \\ 14 & 48 & 54 \end{bmatrix} \begin{Bmatrix} a_0 \\ a_1 \\ a_2 \end{Bmatrix} = \begin{Bmatrix} 54 \\ 243.5 \\ 100 \end{Bmatrix}$$

which can be solved using a method such as Gauss elimination for

$$a_0 = 5 \qquad a_1 = 4 \qquad a_2 = -3$$

which is consistent with the original equation from which the data was derived.

TABLE 17.5 Computations required to develop the normal equations for Example 17.7.

	y	x_1	x_2	x_1^2	x_2^2	x_1x_2	x_1y	x_2y
	5	0	0	0	0	0	0	0
	10	2	1	4	1	2	20	10
	9	2.5	2	6.25	4	5	22.5	18
	0	1	3	1	9	3	0	0
	3	4	6	16	36	24	12	18
	27	7	2	49	4	14	189	54
Σ	54	16.5	14	76.25	54	48	243.5	100

The foregoing two-dimensional case can be easily extended to m dimensions, as in

$$y = a_0 + a_1x_1 + a_2x_2 + \cdots + a_mx_m + e$$

where the standard error is formulated as

$$s_{y/x} = \sqrt{\frac{S_r}{n - (m + 1)}}$$

and the coefficient of determination is computed as in Eq. (17.10). An algorithm to set up the normal equations is listed in Fig. 17.17.

Although there may be certain cases where a variable is linearly related to two or more other variables, multiple linear regression has additional utility in the derivation of power equations of the general form

$$y = a_0x_1^{a_1}x_2^{a_2}\cdots x_m^{a_m}$$

FIGURE 17.17
Pseudocode to assemble the elements of the normal equations for multiple regression. Note that aside from storing the independent variables in $x_{1,i}$, $x_{2,i}$, etc., 1's must be stored in $x_{0,i}$ for this algorithm to work.

```
DO i = 1, order + 1
    DO j = 1, i
        sum = 0
        DO ℓ = 1, n
            sum = sum + x_{i−1,ℓ} · x_{j−1,ℓ}
        END DO
        a_{i,j} = sum
        a_{j,i} = sum
    END DO
    sum = 0
    DO ℓ = 1, n
        sum = sum + y_ℓ · x_{i−1,ℓ}
    END DO
    a_{i, order+2} = sum
END DO
```

Such equations are extremely useful when fitting experimental data. To use multiple linear regression, the equation is transformed by taking its logarithm to yield

$$\log y = \log a_0 + a_1 \log x_1 + a_2 \log x_2 + \cdots + a_m \log x_m$$

This transformation is similar in spirit to the one used in Sec. 17.1.5 and Example 17.4 to fit a power equation when y was a function of a single variable x. Section 20.4 provides an example of such an application for two independent variables.

17.4 GENERAL LINEAR LEAST SQUARES

To this point, we have focused on the mechanics of obtaining least-squares fits of some simple functions to data. Before turning to nonlinear regression, there are several issues that we would like to discuss to enrich your understanding of the preceding material.

17.4.1 General Matrix Formulation for Linear Least Squares

In the preceding pages, we have introduced three types of regression: simple linear, polynomial, and multiple linear. In fact, all three belong to the following general linear least-squares model:

$$y = a_0 z_0 + a_1 z_1 + a_2 z_2 + \cdots + a_m z_m + e \tag{17.23}$$

where $z_0, z_1, \ldots, z_m$ are $m + 1$ different functions. It can easily be seen how simple and multiple linear regression fall within this model—that is, $z_0 = 1, z_1 = x_1, z_2 = x_2, \ldots, z_m = x_m$. Further, polynomial regression is also included if the z's are simple monomials as in $z_0 = x^0 = 1, z_1 = x, z_2 = x^2, \ldots, z_m = x^m$.

Note that the terminology "linear" refers only to the model's dependence on its parameters—that is, the a's. As in the case of polynomial regression, the functions themselves can be highly nonlinear. For example, the z's can be sinusoids, as in

$$y = a_0 + a_1 \cos(\omega t) + a_2 \sin(\omega t)$$

Such a format is the basis of Fourier analysis described in Chap. 19.

On the other hand, a simple looking model like

$$f(x) = a_0 \left(1 - e^{-a_1 x}\right)$$

is truly nonlinear because it cannot be manipulated into the format of Eq. (17.23). We will turn to such models at the end of this chapter.

For the time being, Eq. (17.23) can be expressed in matrix notation as

$$\{Y\} = [Z]\{A\} + \{E\} \tag{17.24}$$

where $[Z]$ is a matrix of the calculated values of the z functions at the measured values of the independent variables,

$$[Z] = \begin{bmatrix} z_{01} & z_{11} & \cdots & z_{m1} \\ z_{02} & z_{12} & \cdots & z_{m2} \\ \cdot & \cdot & \cdot & \\ \cdot & \cdot & \cdot & \\ \cdot & \cdot & \cdot & \\ z_{0n} & z_{1n} & \cdots & z_{mn} \end{bmatrix}$$

where m is the number of variables in the model and n is the number of data points. Because $n \geq m + 1$, you should recognize that most of the time $[Z]$ is not a square matrix.

The column vector $\{Y\}$ contains the observed values of the dependent variable

$$\{Y\}^T = \lfloor y_1 \quad y_2 \quad \cdots \quad y_n \rfloor$$

The column vector $\{A\}$ contains the unknown coefficients

$$\{A\}^T = \lfloor a_0 \quad a_1 \quad \cdots \quad a_m \rfloor$$

and the column vector $\{E\}$ contains the residuals

$$\{E\}^T = \lfloor e_1 \quad e_2 \quad \cdots \quad e_n \rfloor$$

As was done throughout this chapter, the sum of the squares of the residuals for this model can be defined as

$$S_r = \sum_{i=1}^{n} \left(y_i - \sum_{j=0}^{m} a_j z_{ji} \right)^2$$

This quantity can be minimized by taking its partial derivative with respect to each of the coefficients and setting the resulting equation equal to zero. The outcome of this process is the normal equations that can be expressed concisely in matrix form as

$$\left[[Z]^T [Z] \right] \{A\} = \left\{ [Z]^T \{Y\} \right\} \tag{17.25}$$

It can be shown that Eq. (17.25) is, in fact, equivalent to the normal equations developed previously for simple linear, polynomial, and multiple linear regression.

Our primary motivation for the foregoing has been to illustrate the unity among the three approaches and to show how they can all be expressed simply in the same matrix notation. It also sets the stage for the next section where we will gain some insights into the preferred strategies for solving Eq. (17.25). The matrix notation will also have relevance when we turn to nonlinear regression in the last section of this chapter.

17.4.2 Solution Techniques

In previous discussions in this chapter, we have glossed over the issue of the specific numerical techniques to solve the normal equations. Now that we have established the unity among the various models, we can explore this question in more detail.

First, it should be clear that Gauss-Seidel cannot be employed because the normal equations are not diagonally dominant. We are thus left with the elimination methods. For the present purposes, we can divide these techniques into three categories: (1) *LU* decomposition methods including Gauss elimination, (2) Cholesky's method, and (3) matrix inversion approaches. There are obviously overlaps involved in this breakdown. For example, Cholesky's method is, in fact, an *LU* decomposition, and all the approaches can be formulated so that they can generate the matrix inverse. However, this breakdown has merit in that each category offers benefits regarding the solution of the normal equations.

LU Decomposition.   If you are merely interested in applying a least-squares fit for the case where the appropriate model is known a priori, any of the *LU* decomposition approaches described in Chap. 9 is perfectly acceptable. In fact, the non-*LU*-decomposition formulation of Gauss elimination can also be employed. It is a relatively straightforward programming task to incorporate any of these into an algorithm for linear least squares. In fact, if a modular approach has been followed, it is almost trivial.

Cholesky's Method.   Cholesky's decomposition algorithm has several advantages with regard to the solution of the general linear regression problem. First, it is expressly designed for solving symmetric matrices like the normal equations. Thus, it is fast and requires less storage space to solve such systems. Second, it is ideally suited for cases where the order of the model [that is, the value of m in Eq. (17.23)] is not known beforehand (see Ralston and Rabinowitz, 1978). A case in point would be polynomial regression. For this case, we might not know a priori whether a linear, quadratic, cubic, or higher-order polynomial is the "best" model to describe our data. Because of the way in which both the normal equations are constructed and the Cholesky algorithm proceeds (Fig. 11.3), we can develop successively higher-order models in an extremely efficient manner. At each step we could examine the residual sum of the squares error (and a plot!) to examine whether the inclusion of higher-order terms significantly improves the fit.

The analogous situation for multiple linear regression occurs when independent variables are added to the model one at a time. Suppose that the dependent variable of interest is a function of a number of independent variables: say, temperature, moisture content, pressure, etc. We could first perform a linear regression with temperature and compute a residual error. Next, we could include moisture content by performing a two-variable multiple regression and see whether the additional variable results in an improved fit. Cholesky's method makes this process efficient because the decomposition of the linear model would merely be supplemented to incorporate a new variable.

Matrix Inverse Approaches.   From Eq. (PT3.6), recall that the matrix inverse can be employed to solve Eq. (17.25), as in

$$\{A\} = \left[[Z]^T [Z] \right]^{-1} \left\{ [Z]^T \{Y\} \right\} \tag{17.26}$$

Each of the elimination methods can be used to determine the inverse and, thus, can be used to implement Eq. (17.26). However, as we have learned in Part Three, this is an inefficient approach for solving a set of simultaneous equations. Thus, if we were merely interested in solving for the regression coefficients, it is preferable to employ an LU decomposition approach without inversion. However, from a statistical perspective, there are a number of reasons why we might be interested in obtaining the inverse and examining its coefficients. These reasons will be discussed next.

17.4.3 Statistical Aspects of Least-Squares Theory

In Sec. PT5.2.1, we reviewed a number of descriptive statistics that can be used to describe a sample. These included the arithmetic mean, the standard deviation, and the variance.

Aside from yielding a solution for the regression coefficients, the matrix formulation of Eq. (17.26) provides estimates of their statistics. It can be shown (Draper and Smith, 1981) that the diagonal and off-diagonal terms of the matrix $\left[[Z]^T [Z] \right]^{-1}$ give, respectively, the variances and the covariances[1] of the a's. If the diagonal elements of

---

[1]The covariance is a statistic that measures the dependency of one variable on another. Thus, cov (x, y) indicates the dependency of x and y. For example, cov$(x, y) = 0$ would indicate that x and y are totally independent.

$\left[[Z]^T[Z]\right]^{-1}$ are designated as z_{ii}^{-1}, then

$$\text{var}\ (a_{i-1}) = z_{ii}^{-1} s_{y/x}^2 \tag{17.27}$$

and

$$\text{cov}\ (a_{i-1}, a_j) = z_{i-1,j}^{-1} s_{y/x}^2 \tag{17.28}$$

These statistics have a number of important applications. For our present purposes, we will illustrate how they can be used to develop confidence intervals for the intercept and slope.

Using an approach similar to that in Sec. PT5.2.3, it can be shown that lower and upper bounds on the intercept can be formulated as (see Milton and Arnold 1995 for details)

$$L = a_0 - t_{\alpha/2,n-2}\, s(a_0) \qquad U = a_0 + t_{\alpha/2,n-2}\, s(a_0) \tag{17.29}$$

where $s(a_j)$ = the standard error of coefficient $a_j = \sqrt{\text{var}\ (a_j)}$. In a similar manner, lower and upper bounds on the slope can be formulated as

$$L = a_1 - t_{\alpha/2,n-2}\, s(a_1) \qquad U = a_1 + t_{\alpha/2,n-2}\, s(a_1) \tag{17.30}$$

The following example illustrates how these intervals can be used to make quantitative inferences related to linear regression.

EXAMPLE 17.8   Confidence Intervals for Linear Regression

Problem Statement.   In Example 17.3, we used regression to develop the following relationship between measurements and model predictions:

$$y = -0.859 + 1.032x$$

where y = the model predictions and x = the measurements. We concluded that there was a good agreement between the two because the intercept was approximately equal to 0 and the slope approximately equal to 1. Recompute the regression but use the matrix approach to estimate standard errors for the parameters. Then employ these errors to develop confidence intervals and use these to make a probabilistic statement regarding the goodness of fit.

Solution.   The data can be written in matrix format for simple linear regression as:

$$[Z] = \begin{bmatrix} 1 & 10 \\ 1 & 16.3 \\ 1 & 23 \\ \cdot & \cdot \\ \cdot & \cdot \\ \cdot & \cdot \\ 1 & 50 \end{bmatrix} \qquad \{Y\} = \begin{Bmatrix} 8.953 \\ 16.405 \\ 22.607 \\ \cdot \\ \cdot \\ \cdot \\ 49.988 \end{Bmatrix}$$

Matrix transposition and multiplication can then be used to generate the normal equations as

$$\begin{array}{ccc} \left[[Z]^T[Z]\right] & \{A\} = & \{[Z]^T\{Y\}\} \end{array}$$

$$\begin{bmatrix} 15 & 548.3 \\ 548.3 & 22191.21 \end{bmatrix} \begin{Bmatrix} a_0 \\ a_1 \end{Bmatrix} = \begin{Bmatrix} 552.741 \\ 22421.43 \end{Bmatrix}$$

Matrix inversion can be used to obtain the slope and intercept as

$$\{A\} = \quad \left[[Z]^T[Z]\right]^{-1} \quad \{[Z]^T\{Y\}\}$$

$$= \begin{bmatrix} 0.688414 & -0.01701 \\ -0.01701 & 0.000465 \end{bmatrix} \begin{Bmatrix} 552.741 \\ 22421.43 \end{Bmatrix} = \begin{Bmatrix} -0.85872 \\ 1.031592 \end{Bmatrix}$$

Thus, the intercept and the slope are determined as $a_0 = -0.85872$ and $a_1 = 1.031592$, respectively. These values in turn can be used to compute the standard error of the estimate as $s_{y/x} = 0.863403$. This value can be used along with the diagonal elements of the matrix inverse to calculate the standard errors of the coefficients,

$$s(a_0) = \sqrt{z_{11}^{-1} s_{y/x}^2} = \sqrt{0.688414(0.863403)^2} = 0.716372$$

$$s(a_1) = \sqrt{z_{22}^{-1} s_{y/x}^2} = \sqrt{0.000465(0.863403)^2} = 0.018625$$

The statistic, $t_{\alpha/2,n-1}$ needed for a 95% confidence interval with $n - 2 = 15 - 2 = 13$ degrees of freedom can be determined from a statistics table or using software. We used an Excel function, TINV, to come up with the proper value, as in

$$= \text{TINV}(0.05, 13)$$

which yielded a value of 2.160368. Equations (17.29) and (17.30) can then be used to compute the confidence intervals as

$$a_0 = -0.85872 \pm 2.160368(0.716372)$$

$$= -0.85872 \pm 1.547627 = [-2.40634, 0.688912]$$

$$a_1 = 1.031592 \pm 2.160368(0.018625)$$

$$= 1.031592 \pm 0.040237 = [0.991355, 1.071828]$$

Notice that the desired values (0 for intercept and slope and 1 for the intercept) fall within the intervals. On the basis of this analysis we could make the following statement regarding the slope: We have strong grounds for believing that the slope of the true regression line lies within the interval from 0.991355 to 1.071828. Because 1 falls within this interval, we also have strong grounds for believing that the result supports the agreement between the measurements and the model. Because zero falls within the intercept interval, a similar statement can be made regarding the intercept.

The foregoing is a limited introduction to the rich topic of statistical inference and its relationship to regression. There are many subleties that are beyond the scope of this book. Our primary motivation has been to illustrate the power of the matrix approach to general linear least squares. You should consult some of the excellent books on the subject (e.g., Draper and Smith 1981) for additional information. In addition, it should be noted that software packages and libraries can generate least-squares regression fits along with information relevant to inferential statistics. We will explore some of these capabilities when we describe these packages at the end of Chap. 19.

17.5 NONLINEAR REGRESSION

There are many cases in engineering where nonlinear models must be fit to data. In the present context, these models are defined as those that have a nonlinear dependence on their parameters. For example,

$$f(x) = a_0 \left(1 - e^{-a_1 x}\right) + e \qquad (17.31)$$

This equation cannot be manipulated so that it conforms to the general form of Eq. (17.23).

As with linear least squares, nonlinear regression is based on determining the values of the parameters that minimize the sum of the squares of the residuals. However, for the nonlinear case, the solution must proceed in an iterative fashion.

The *Gauss-Newton method* is one algorithm for minimizing the sum of the squares of the residuals between data and nonlinear equations. The key concept underlying the technique is that a Taylor series expansion is used to express the original nonlinear equation in an approximate, linear form. Then, least-squares theory can be used to obtain new estimates of the parameters that move in the direction of minimizing the residual.

To illustrate how this is done, first the relationship between the nonlinear equation and the data can be expressed generally as

$$y_i = f(x_i; a_0, a_1, \ldots, a_m) + e_i$$

where y_i = a measured value of the dependent variable, $f(x_i; a_0, a_1, \ldots, a_m)$ = the equation that is a function of the independent variable x_i and a nonlinear function of the parameters $a_0, a_1, \ldots, a_m$, and e_i = a random error. For convenience, this model can be expressed in abbreviated form by omitting the parameters,

$$y_i = f(x_i) + e_i \qquad (17.32)$$

The nonlinear model can be expanded in a Taylor series around the parameter values and curtailed after the first derivatives. For example, for a two-parameter case,

$$f(x_i)_{j+1} = f(x_i)_j + \frac{\partial f(x_i)_j}{\partial a_0} \Delta a_0 + \frac{\partial f(x_i)_j}{\partial a_1} \Delta a_1 \qquad (17.33)$$

where j = the initial guess, $j + 1$ = the prediction, $\Delta a_0 = a_{0,j+1} - a_{0,j}$, and $\Delta a_1 = a_{1,j+1} - a_{1,j}$. Thus, we have linearized the original model with respect to the parameters. Equation (17.33) can be substituted into Eq. (17.32) to yield

$$y_i - f(x_i)_j = \frac{\partial f(x_i)_j}{\partial a_0} \Delta a_0 + \frac{\partial f(x_i)_j}{\partial a_1} \Delta a_1 + e_i$$

or in matrix form [compare with Eq. (17.24)],

$$\{D\} = [Z_j]\{\Delta A\} + \{E\} \qquad (17.34)$$

where $[Z_j]$ is the matrix of partial derivatives of the function evaluated at the initial guess j,

$$[Z_j] = \begin{bmatrix} \partial f_1/\partial a_0 & \partial f_1/\partial a_1 \\ \partial f_2/\partial a_0 & \partial f_2/\partial a_1 \\ \cdot & \cdot \\ \cdot & \cdot \\ \cdot & \cdot \\ \partial f_n/\partial a_0 & \partial f_n/\partial a_1 \end{bmatrix}$$

where n = the number of data points and $\partial f_i/\partial a_k$ = the partial derivative of the function with respect to the kth parameter evaluated at the ith data point. The vector $\{D\}$ contains the differences between the measurements and the function values,

$$\{D\} = \begin{Bmatrix} y_1 - f(x_1) \\ y_2 - f(x_2) \\ \cdot \\ \cdot \\ \cdot \\ y_n - f(x_n) \end{Bmatrix}$$

and the vector $\{\Delta A\}$ contains the changes in the parameter values,

$$\{\Delta A\} = \begin{Bmatrix} \Delta a_0 \\ \Delta a_1 \\ \cdot \\ \cdot \\ \cdot \\ \Delta a_m \end{Bmatrix}$$

Applying linear least-squares theory to Eq. (17.34) results in the following normal equations [recall Eq. (17.25)]:

$$\left[[Z_j]^T[Z_j]\right]\{\Delta A\} = \left\{[Z_j]^T\{D\}\right\} \tag{17.35}$$

Thus, the approach consists of solving Eq. (17.35) for $\{\Delta A\}$ which can be employed to compute improved values for the parameters, as in

$$a_{0,j+1} = a_{0,j} + \Delta a_0$$

and

$$a_{1,j+1} = a_{1,j} + \Delta a_1$$

This procedure is repeated until the solution converges—that is, until

$$|\varepsilon_a|_k = \left| \frac{a_{k,j+1} - a_{k,j}}{a_{k,j+1}} \right| 100\% \tag{17.36}$$

falls below an acceptable stopping criterion.

EXAMPLE 17.9   Gauss-Newton Method

Problem Statement.   Fit the function $f(x; a_0, a_1) = a_0(1 - e^{-a_1 x})$ to the data:

x	0.25	0.75	1.25	1.75	2.25
y	0.28	0.57	0.68	0.74	0.79

Use initial guesses of $a_0 = 1.0$ and $a_1 = 1.0$ for the parameters. Note that for these guesses the initial sum of the squares of the residuals is 0.0248.

Solution.   The partial derivatives of the function with respect to the parameters are

$$\frac{\partial f}{\partial a_0} = 1 - e^{-a_1 x} \tag{E17.9.1}$$

and

$$\frac{\partial f}{\partial a_1} = a_0 x e^{-a_1 x}$$
(E17.9.2)

Equations (E17.9.1) and (E17.9.2) can be used to evaluate the matrix

$$[Z_0] = \begin{bmatrix} 0.2212 & 0.1947 \\ 0.5276 & 0.3543 \\ 0.7135 & 0.3581 \\ 0.8262 & 0.3041 \\ 0.8946 & 0.2371 \end{bmatrix}$$

This matrix multiplied by its transpose results in

$$[Z_0]^T [Z_0] = \begin{bmatrix} 2.3193 & 0.9489 \\ 0.9489 & 0.4404 \end{bmatrix}$$

which in turn can be inverted to yield

$$\big[[Z_0]^T [Z_0]\big]^{-1} = \begin{bmatrix} 3.6397 & -7.8421 \\ -7.8421 & 19.1678 \end{bmatrix}$$

The vector $\{D\}$ consists of the differences between the measurements and the model predictions,

$$\{D\} = \begin{Bmatrix} 0.28 - 0.2212 \\ 0.57 - 0.5276 \\ 0.68 - 0.7135 \\ 0.74 - 0.8262 \\ 0.79 - 0.8946 \end{Bmatrix} = \begin{Bmatrix} 0.0588 \\ 0.0424 \\ -0.0335 \\ -0.0862 \\ -0.1046 \end{Bmatrix}$$

It is multiplied by $[Z_0]^T$ to give

$$[Z_0]^T \{D\} = \begin{bmatrix} -0.1533 \\ -0.0365 \end{bmatrix}$$

The vector $\{\Delta A\}$ is then calculated by solving Eq. (17.35) for

$$\Delta A = \begin{Bmatrix} -0.2714 \\ 0.5019 \end{Bmatrix}$$

which can be added to the initial parameter guesses to yield

$$\begin{Bmatrix} a_0 \\ a_1 \end{Bmatrix} = \begin{Bmatrix} 1.0 \\ 1.0 \end{Bmatrix} + \begin{Bmatrix} -0.2714 \\ 0.5019 \end{Bmatrix} = \begin{Bmatrix} 0.7286 \\ 1.5019 \end{Bmatrix}$$

Thus, the improved estimates of the parameters are $a_0 = 0.7286$ and $a_1 = 1.5019$. The new parameters result in a sum of the squares of the residuals equal to 0.0242. Equation (17.36) can be used to compute ε_0 and ε_1 equal to 37 and 33 percent, respectively. The computation would then be repeated until these values fell below the prescribed stopping criterion. The final result is $a_0 = 0.79186$ and $a_1 = 1.6751$. These coefficients give a sum of the squares of the residuals of 0.000662.

A potential problem with the Gauss-Newton method as developed to this point is that the partial derivatives of the function may be difficult to evaluate. Consequently, many computer programs use difference equations to approximate the partial derivatives. One method is

$$\frac{\partial f_i}{\partial a_k} \cong \frac{f(x_i; a_0, \ldots, a_k + \delta a_k, \ldots, a_m) - f(x_i; a_0, \ldots, a_k, \ldots, a_m)}{\delta a_k} \qquad (17.37)$$

where δ = a small fractional perturbation.

The Gauss-Newton method has a number of other possible shortcomings:

1. It may converge slowly.
2. It may oscillate widely, i.e., continually change directions.
3. It may not converge at all.

Modifications of the method (Booth and Peterson, 1958; Hartley, 1961) have been developed to remedy the shortcomings.

In addition, although there are several approaches expressly designed for regression, a more general approach is to use nonlinear optimization routines as described in Part Four. To do this, a guess for the parameters is made, and the sum of the squares of the residuals is computed. For example, for Eq. (17.31) it would be computed as

$$S_r = \sum_{i=1}^{n} \left[y_i - a_0(1 - e^{-a_1 x_i}) \right]^2 \qquad (17.38)$$

Then, the parameters would be adjusted systematically to minimize S_r using search techniques of the type described previously in Chap. 14. We will illustrate how this is done when we describe software applications at the end of Chap. 19.

PROBLEMS

17.1 Given the data

0.90	1.42	1.30	1.55	1.63
1.32	1.35	1.47	1.95	1.66
1.96	1.47	1.92	1.35	1.05
1.85	1.74	1.65	1.78	1.71
2.29	1.82	2.06	2.14	1.27

determine (a) the mean, (b) the standard deviation, (c) the variance, (d) the coefficient of variation, and (e) the 95% confidence interval for the mean.

17.2 Construct a histogram for the data in Prob. 17.1. Use a range of 0.6 to 2.4 with intervals of 0.2.

17.3 Given the data

15	6	18	21	26	28	32
39	22	28	24	27	27	33
2	12	17	34	29	31	38
45	36	41	37	43	38	26

determine (a) the mean, (b) the standard deviation, (c) the variance, (d) the coefficient of variation, and (e) the 90% confidence interval for the mean. (f) Construct a histogram. Use a range from 0 to 55 with increments of 5. (g) Assuming that the distribution is normal and that your estimate of the standard deviation is valid, compute the range (that is, the lower and the upper values) that encompasses 68% of the readings. Determine whether this is a valid estimate for the data in this problem.

17.4 Use least-squares regression to fit a straight line to

x	1	3	5	7	10	12	13	16	18	20
y	4	5	6	5	8	7	6	9	12	11

Along with the slope and intercept, compute the standard error of the estimate and the correlation coefficient. Plot the data and the regression line. Then repeat the problem, but regress x versus y—that is, switch the variables. Interpret your results.

17.5 Use least-squares regression to fit a straight line to

x	5	6	10	14	16	20	22	28	28	36	38
y	30	22	28	14	22	16	8	8	14	0	4

Along with the slope and the intercept, compute the standard error of the estimate and the correlation coefficient. Plot the data and the regression line. If someone made an additional measurement of $x = 5$, $y = 5$, would you suspect, based on a visual assessment and the standard error, that the measurement was valid or faulty? Justify your conclusion.

17.6 Use least-squares regression to fit a straight line to

x	2	3	4	7	8	9	5	5
y	9	6	5	10	9	11	2	3

(a) Along with the slope and intercept, compute the standard error of the estimate and the correlation coefficient. Plot the data and the straight line. Assess the fit.
(b) Recompute (a), but use polynomial regression to fit a parabola to the data. Compare the results with those of (a).

17.7 Fit a saturation-growth-rate model to

x	0.75	2	2.5	4	6	8	8.5
y	0.8	1.3	1.2	1.6	1.7	1.8	1.7

Plot the data and the equation. Find the standard error.

17.8 Fit a power equation to the data from Prob. 17.7. Plot the data and the equation, and find the standard error.

17.9 Fit a parabola to the data from Prob. 17.7. Plot the data and the equation, and find the standard error.

17.10 Fit a power equation to

x	2.5	3.5	5	6	7.5	10	12.5	15	17.5	20
y	7	5.5	3.9	3.6	3.1	2.8	2.6	2.4	2.3	2.3

Plot y versus x along with the power equation.

17.11 Fit an exponential model to

x	0.4	0.8	1.2	1.6	2.0	2.3
y	750	1000	1400	2000	2700	3750

Plot the data and the equation on both standard and semi-logarithmic graph paper. Discuss your results.

17.12 Fit a parabola to the data in Prob. 17.11. Plot the data and the equation.

17.13 Given the data

x	5	10	15	20	25	30	35	40	45	50
y	16	25	32	33	38	36	39	40	42	42

use least-squares regression to fit (a) a straight line, (b) a power equation, (c) a saturation-growth-rate equation, and (d) a parabola. Plot the data along with all the curves. Is any one of the curves superior? If so, justify.

17.14 Use multiple linear regression to fit

x_1	0	1	1	2	2	3	3	4	4
x_2	0	1	2	1	2	1	2	1	2
y	15	18	12.8	25.7	20.6	35.0	29.8	45.5	40.3

Compute the coefficients, the standard error of the estimate, and the correlation coefficient.

17.15 Use multiple linear regression to fit

x_1	0	0	1	2	0	1	2	2	1
x_2	0	2	2	4	4	6	6	2	1
y	15	19	12	11	24	22	15	5	19

Compute the coefficients, the standard error of the estimate, and the correlation coefficient.

17.16 Use nonlinear regression to fit a parabola to the following data,

x	0.075	0.5	1	1.2	1.7	2.0	2.3
y	600	800	1200	1400	2050	2650	3750

17.17 Use nonlinear regression to fit a saturation-growth-rate equation to the data in Prob. 17.13.

17.18 Recompute the regression fits from Probs. (a) 17.4, and (b) 17.13, using the matrix approach. Estimate the standard errors and develop 90% confidence intervals for the slope and the intercept.

17.19 Develop, debug, and test a subprogram in either a high-level language or macro language of your choice to implement linear regression. Among other things: (a) Add statements to document the code, and (b) determine the standard error and the coefficient of determination.

17.20 Use the Numerical Methods TOOLKIT software to solve Probs. (a) 17.4, (b) 17.5, (c) 17.6, (d) 17.9, and (e) 17.12.

CHAPTER 18

Interpolation

You will frequently have occasion to estimate intermediate values between precise data points. The most common method used for this purpose is polynomial interpolation. Recall that the general formula for an nth-order polynomial is

$$f(x) = a_0 + a_1 x + a_2 x^2 + \cdots + a_n x^n \tag{18.1}$$

For $n + 1$ data points, there is one and only one polynomial of order n that passes through all the points. For example, there is only one straight line (that is, a first-order polynomial) that connects two points (Fig. 18.1a). Similarly, only one parabola connects a set of three points (Fig. 18.1b). *Polynomial interpolation* consists of determining the unique nth-order polynomial that fits $n + 1$ data points. This polynomial then provides a formula to compute intermediate values.

Although there is one and only one nth-order polynomial that fits $n + 1$ points, there are a variety of mathematical formats in which this polynomial can be expressed. In this chapter, we will describe two alternatives that are well-suited for computer implementation: the Newton and the Lagrange polynomials.

FIGURE 18.1
Examples of interpolating polynomials: (a) first-order (linear) connecting two points, (b) second-order (quadratic or parabolic) connecting three points, and (c) third-order (cubic) connecting four points.

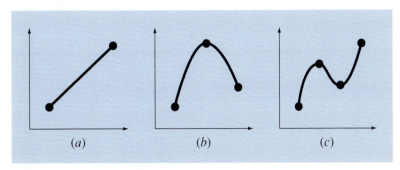

(a) (b) (c)

18.1 NEWTON'S DIVIDED-DIFFERENCE INTERPOLATING POLYNOMIALS

As stated above, there are a variety of alternative forms for expressing an interpolating polynomial. *Newton's divided-difference interpolating polynomial* is among the most popular and useful forms. Before presenting the general equation, we will introduce the first- and second-order versions because of their simple visual interpretation.

18.1.1 Linear Interpolation

The simplest form of interpolation is to connect two data points with a straight line. This technique, called *linear interpolation,* is depicted graphically in Fig. 18.2. Using similar triangles,

$$\frac{f_1(x) - f(x_0)}{x - x_0} = \frac{f(x_1) - f(x_0)}{x_1 - x_0}$$

which can be rearranged to yield

$$f_1(x) = f(x_0) + \frac{f(x_1) - f(x_0)}{x_1 - x_0}(x - x_0) \tag{18.2}$$

which is a *linear-interpolation formula*. The notation $f_1(x)$ designates that this is a first-order interpolating polynomial. Notice that besides representing the slope of the line connecting the points, the term $[f(x_1) - f(x_0)]/(x_1 - x_0)$ is a finite-divided-difference

---

FIGURE 18.2
Graphical depiction of linear interpolation. The shaded areas indicate the similar triangles used to derive the linear-interpolation formula [Eq. (18.2)].

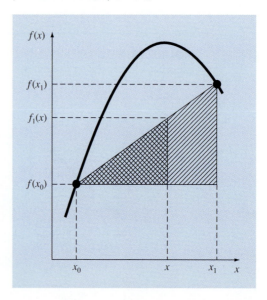

approximation of the first derivative [recall Eq. (4.17)]. In general, the smaller the interval between the data points, the better the approximation. This is due to the fact that, as the interval decreases, a continuous function will be better approximated by a straight line. This characteristic is demonstrated in the following example.

EXAMPLE 18.1 Linear Interpolation

Problem Statement. Estimate the natural logarithm of 2 using linear interpolation. First, perform the computation by interpolating between $\ln 1 = 0$ and $\ln 6 = 1.791759$. Then, repeat the procedure, but use a smaller interval from $\ln 1$ to $\ln 4$ (1.386294). Note that the true value of $\ln 2$ is 0.6931472.

Solution. We use Eq. (18.2) and a linear interpolation for $\ln(2)$ from $x_0 = 1$ to $x_1 = 6$ to give

$$f_1(2) = 0 + \frac{1.791759 - 0}{6 - 1}(2 - 1) = 0.3583519$$

which represents an error of $\varepsilon_t = 48.3\%$. Using the smaller interval from $x_0 = 1$ to $x_1 = 4$ yields

$$f_1(2) = 0 + \frac{1.386294 - 0}{4 - 1}(2 - 1) = 0.4620981$$

Thus, using the shorter interval reduces the percent relative error to $\varepsilon_t = 33.3\%$. Both interpolations are shown in Fig. 18.3, along with the true function.

FIGURE 18.3
Two linear interpolations to estimate ln 2. Note how the smaller interval provides a better estimate.

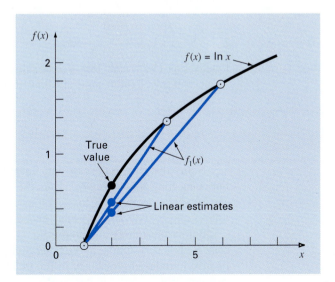

18.1.2 Quadratic Interpolation

The error in Example 18.1 resulted from our approximating a curve with a straight line. Consequently, a strategy for improving the estimate is to introduce some curvature into the line connecting the points. If three data points are available, this can be accomplished with a second-order polynomial (also called a quadratic polynomial or a *parabola*). A particularly convenient form for this purpose is

$$f_2(x) = b_0 + b_1(x - x_0) + b_2(x - x_0)(x - x_1) \tag{18.3}$$

Note that although Eq. (18.3) might seem to differ from the general polynomial [Eq. (18.1)], the two equations are equivalent. This can be shown by multiplying the terms in Eq. (18.3) to yield

$$f_2(x) = b_0 + b_1 x - b_1 x_0 + b_2 x^2 + b_2 x_0 x_1 - b_2 x x_0 - b_2 x x_1$$

or, collecting terms,

$$f_2(x) = a_0 + a_1 x + a_2 x^2$$

where

$$a_0 = b_0 - b_1 x_0 + b_2 x_0 x_1$$
$$a_1 = b_1 - b_2 x_0 - b_2 x_1$$
$$a_2 = b_2$$

Thus, Eqs. (18.1) and (18.3) are alternative, equivalent formulations of the unique second-order polynomial joining the three points.

A simple procedure can be used to determine the values of the coefficients. For b_0, Eq. (18.3) with $x = x_0$ can be used to compute

$$b_0 = f(x_0) \tag{18.4}$$

Equation (18.4) can be substituted into Eq. (18.3), which can be evaluated at $x = x_1$ for

$$b_1 = \frac{f(x_1) - f(x_0)}{x_1 - x_0} \tag{18.5}$$

Finally, Eqs. (18.4) and (18.5) can be substituted into Eq. (18.3), which can be evaluated at $x = x_2$ and solved (after some algebraic manipulations) for

$$b_2 = \frac{\dfrac{f(x_2) - f(x_1)}{x_2 - x_1} - \dfrac{f(x_1) - f(x_0)}{x_1 - x_0}}{x_2 - x_0} \tag{18.6}$$

Notice that, as was the case with linear interpolation, b_1 still represents the slope of the line connecting points x_0 and x_1. Thus, the first two terms of Eq. (18.3) are equivalent to linear interpolation from x_0 to x_1, as specified previously in Eq. (18.2). The last term, $b_2(x - x_0)(x - x_1)$, introduces the second-order curvature into the formula.

Before illustrating how to use Eq. (18.3), we should examine the form of the coefficient b_2. It is very similar to the finite-divided-difference approximation of the second derivative introduced previously in Eq. (4.24). Thus, Eq. (18.3) is beginning to manifest a structure that is very similar to the Taylor series expansion. This observation will be explored further when we relate Newton's interpolating polynomials to the Taylor series in Sec. 18.1.4. But first, we will do an example that shows how Eq. (18.3) is used to interpolate among three points.

EXAMPLE 18.2   Quadratic Interpolation

Problem Statement.   Fit a second-order polynomial to the three points used in Example 18.1:

$$x_0 = 1 \qquad f(x_0) = 0$$
$$x_1 = 4 \qquad f(x_1) = 1.386294$$
$$x_2 = 6 \qquad f(x_2) = 1.791759$$

Use the polynomial to evaluate ln 2.

Solution.   Applying Eq. (18.4) yields

$$b_0 = 0$$

Equation (18.5) yields

$$b_1 = \frac{1.386294 - 0}{4 - 1} = 0.4620981$$

and Eq. (18.6) gives

$$b_2 = \frac{\dfrac{1.791759 - 1.386294}{6 - 4} - 0.4620981}{6 - 1} = -0.0518731$$

FIGURE 18.4
The use of quadratic interpolation to estimate ln 2. The linear interpolation from $x = 1$ to 4 is also included for comparison.

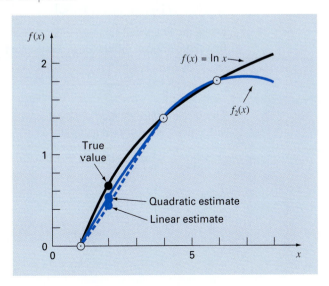

Substituting these values into Eq. (18.3) yields the quadratic formula

$$f_2(x) = 0 + 0.4620981(x - 1) - 0.0518731(x - 1)(x - 4)$$

which can be evaluated at $x = 2$ for

$$f_2(2) = 0.5658444$$

which represents a relative error of $\varepsilon_t = 18.4\%$. Thus, the curvature introduced by the quadratic formula (Fig. 18.4) improves the interpolation compared with the result obtained using straight lines in Example 18.1 and Fig. 18.3.

18.1.3 General Form of Newton's Interpolating Polynomials

The preceding analysis can be generalized to fit an nth-order polynomial to $n + 1$ data points. The nth-order polynomial is

$$f_n(x) = b_0 + b_1(x - x_0) + \cdots + b_n(x - x_0)(x - x_1) \cdots (x - x_{n-1}) \tag{18.7}$$

As was done previously with the linear and quadratic interpolations, data points can be used to evaluate the coefficients $b_0, b_1, \ldots, b_n$. For an nth-order polynomial, $n + 1$ data points are required: $[x_0, f(x_0)], [x_1, f(x_1)], \ldots, [x_n, f(x_n)]$. We use these data points and the following equations to evaluate the coefficients:

$$b_0 = f(x_0) \tag{18.8}$$
$$b_1 = f[x_1, x_0] \tag{18.9}$$
$$b_2 = f[x_2, x_1, x_0] \tag{18.10}$$

$$\cdot$$
$$\cdot$$
$$\cdot$$

$$b_n = f[x_n, x_{n-1}, \ldots, x_1, x_0] \tag{18.11}$$

where the bracketed function evaluations are finite divided differences. For example, the first finite divided difference is represented generally as

$$f[x_i, x_j] = \frac{f(x_i) - f(x_j)}{x_i - x_j} \tag{18.12}$$

The *second finite divided difference*, which represents the difference of two first divided differences, is expressed generally as

$$f[x_i, x_j, x_k] = \frac{f[x_i, x_j] - f[x_j, x_k]}{x_i - x_k} \tag{18.13}$$

Similarly, the *nth finite divided difference* is

$$f[x_n, x_{n-1}, \ldots, x_1, x_0] = \frac{f[x_n, x_{n-1}, \ldots, x_1] - f[x_{n-1}, x_{n-2}, \ldots, x_0]}{x_n - x_0} \tag{18.14}$$

i	x_i	$f(x_i)$	First	Second	Third
0	x_0	$f(x_0)$	$f[x_1, x_0]$	$f[x_2, x_1, x_0]$	$f[x_3, x_2, x_1, x_0]$
1	x_1	$f(x_1)$	$f[x_2, x_1]$	$f[x_3, x_2, x_1]$	
2	x_2	$f(x_2)$	$f[x_3, x_2]$		
3	x_3	$f(x_3)$			

FIGURE 18.5
Graphical depiction of the recursive nature of finite divided differences.

These differences can be used to evaluate the coefficients in Eqs. (18.8) through (18.11), which can then be substituted into Eq. (18.7) to yield the interpolating polynomial

$$f_n(x) = f(x_0) + (x - x_0)f[x_1, x_0] + (x - x_0)(x - x_1)f[x_2, x_1, x_0]$$
$$+ \cdots + (x - x_0)(x - x_1) \cdots (x - x_{n-1})f[x_n, x_{n-1}, \ldots, x_0] \qquad (18.15)$$

which is called *Newton's divided-difference interpolating polynomial*. It should be noted that it is not necessary that the data points used in Eq. (18.15) be equally spaced or that the abscissa values necessarily be in ascending order, as illustrated in the following example. Also, notice how Eqs. (18.12) through (18.14) are recursive—that is, higher-order differences are computed by taking differences of lower-order differences (Fig. 18.5). This property will be exploited when we develop an efficient computer program in Sec. 18.1.5 to implement the method.

EXAMPLE 18.3   Newton's Divided-Difference Interpolating Polynomials

Problem Statement.   In Example 18.2, data points at $x_0 = 1$, $x_1 = 4$, and $x_2 = 6$ were used to estimate ln 2 with a parabola. Now, adding a fourth point [$x_3 = 5$; $f(x_3) = 1.609438$], estimate ln 2 with a third-order Newton's interpolating polynomial.

Solution.   The third-order polynomial, Eq. (18.7) with $n = 3$, is

$$f_3(x) = b_0 + b_1(x - x_0) + b_2(x - x_0)(x - x_1) + b_3(x - x_0)(x - x_1)(x - x_2)$$

The first divided differences for the problem are [Eq. (18.12)]

$$f[x_1, x_0] = \frac{1.386294 - 0}{4 - 0} = 0.4620981$$

$$f[x_2, x_1] = \frac{1.791759 - 1.386294}{6 - 4} = 0.2027326$$

$$f[x_3, x_2] = \frac{1.609438 - 1.386294}{5 - 6} = 0.1823216$$

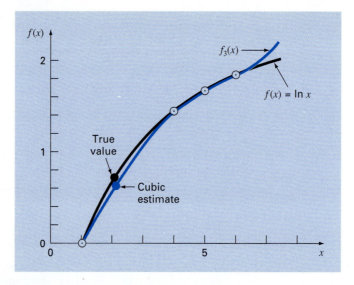

FIGURE 18.6
The use of cubic interpolation to estimate ln 2.

The second divided differences are [Eq. (18.13)]

$$f[x_2, x_1, x_0] = \frac{0.2027326 - 0.4620981}{6 - 1} = -0.05187311$$

$$f[x_3, x_2, x_1] = \frac{0.1823216 - 0.2027326}{5 - 4} = -0.02041100$$

The third divided difference is [Eq. (18.14) with $n = 3$]

$$f[x_3, x_2, x_1, x_0] = \frac{-0.02041100 - (-0.05187311)}{5 - 1} = 0.007865529$$

The results for $f[x_1, x_0]$, $f[x_2, x_1, x_0]$, and $f[x_3, x_2, x_1, x_0]$ represent the coefficients b_1, b_2, and b_3 of Eq. (18.7). Along with $b_0 = f(x_0) = 0.0$, Eq. (18.7) is

$$f_3(x) = 0 + 0.4620981(x - 1) - 0.05187311(x - 1)(x - 4)$$
$$+ 0.007865529(x - 1)(x - 4)(x - 6)$$

which can be used to evaluate $f_3(2) = 0.6287686$, which represents a relative error of $\varepsilon_t = 9.3\%$. The complete cubic polynomial is shown in Fig. 18.6.

18.1.4 Errors of Newton's Interpolating Polynomials

Notice that the structure of Eq. (18.15) is similar to the Taylor series expansion in the sense that terms are added sequentially to capture the higher-order behavior of the underlying function. These terms are finite divided differences and, thus, represent approximations of the higher-order derivatives. Consequently, as with the Taylor series, if the true underlying function is an nth-order polynomial, the nth-order interpolating polynomial based on $n + 1$ data points will yield exact results.

Also, as was the case with the Taylor series, a formulation for the truncation error can be obtained. Recall from Eq. (4.6) that the truncation error for the Taylor series could be expressed generally as

$$R_n = \frac{f^{(n+1)}(\xi)}{(n+1)!}(x_{i+1} - x_i)^{n+1} \tag{4.6}$$

where ξ is somewhere in the interval x_i to x_{i+1}. For an nth-order interpolating polynomial, an analogous relationship for the error is

$$R_n = \frac{f^{(n+1)}(\xi)}{(n+1)!}(x - x_0)(x - x_1)\cdots(x - x_n) \tag{18.16}$$

where ξ is somewhere in the interval containing the unknown and the data. For this formula to be of use, the function in question must be known and differentiable. This is not usually the case. Fortunately, an alternative formulation is available that does not require prior knowledge of the function. Rather, it uses a finite divided difference to approximate the $(n + 1)$th derivative,

$$R_n = f[x, x_n, x_{n-1}, \ldots, x_0](x - x_0)(x - x_1)\cdots(x - x_n) \tag{18.17}$$

where $f[x, x_n, x_{n-1}, \ldots, x_0)]$ is the $(n + 1)$th finite divided difference. Because Eq. (18.17) contains the unknown $f(x)$, it cannot be solved for the error. However, if an additional data point $f(x_{n+1})$ is available, Eq. (18.17) can be used to estimate the error, as in

$$R_n \cong f[x_{n+1}, x_n, x_{n-1}, \ldots, x_0](x - x_0)(x - x_1)\cdots(x - x_n) \tag{18.18}$$

EXAMPLE 18.4   Error Estimation for Newton's Polynomial

Problem Statement. Use Eq. (18.18) to estimate the error for the second-order polynomial interpolation of Example 18.2. Use the additional data point $f(x_3) = f(5) = 1.609438$ to obtain your results.

Solution. Recall that in Example 18.2, the second-order interpolating polynomial provided an estimate of $f_2(2) = 0.5658444$, which represents an error of $0.6931472 - 0.5658444 = 0.1273028$. If we had not known the true value, as is most usually the case, Eq. (18.18), along with the additional value at x_3, could have been used to estimate the error, as in

$$R_2 = f[x_3, x_2, x_1, x_0](x - x_0)(x - x_1)(x - x_2)$$

or

$$R_2 = 0.007865529(x - 1)(x - 4)(x - 6)$$

where the value for the third-order finite divided difference is as computed previously in Example 18.3. This relationship can be evaluated at $x = 2$ for

$$R_2 = 0.007865529(2 - 1)(2 - 4)(2 - 6) = 0.0629242$$

which is of the same order of magnitude as the true error.

From the previous example and from Eq. (18.18), it should be clear that the error estimate for the nth-order polynomial is equivalent to the difference between the $(n + 1)$th order and the nth-order prediction. That is,

$$R_n = f_{n+1}(x) - f_n(x) \tag{18.19}$$

In other words, the increment that is added to the nth-order case to create the $(n + 1)$th-order case [that is, Eq. (18.18)] is interpreted as an estimate of the nth-order error. This can be clearly seen by rearranging Eq. (18.19) to give

$$f_{n+1}(x) = f_n(x) + R_n$$

The validity of this approach is predicated on the fact that the series is strongly convergent. For such a situation, the $(n + 1)$th-order prediction should be much closer to the true value than the nth-order prediction. Consequently, Eq. (18.19) conforms to our standard definition of error as representing the difference between the truth and an approximation. However, note that whereas all other error estimates for iterative approaches introduced up to this point have been determined as a present prediction minus a previous one, Eq. (18.19) represents a future prediction minus a present one. This means that for a series that is converging rapidly, the error estimate of Eq. (18.19) could be less than the true error. This would represent a highly unattractive quality if the error estimate were being employed as a stopping criterion. However, as will be described in the following section, higher-order interpolating polynomials are highly sensitive to data errors—that is, they are very ill-conditioned. When employed for interpolation, they often yield predictions that diverge significantly from the true value. By "looking ahead" to sense errors, Eq. (18.19) is more sensitive to such divergence. As such, it is more valuable for the sort of exploratory data analysis for which Newton's polynomial is best-suited.

18.1.5 Computer Algorithm for Newton's Interpolating Polynomial

Three properties make Newton's interpolating polynomials extremely attractive for computer applications:

1. As in Eq. (18.7), higher-order versions can be developed sequentially by adding a single term to the next lower-order equation. This facilitates the evaluation of several different-order versions in the same program. Such a capability is especially valuable when the order of the polynomial is not known a priori. By adding new terms sequentially, we can determine when a point of diminishing returns is reached—that is, when addition of higher-order terms no longer significantly improves the estimate or in certain situations actually detracts from it. The error equations discussed below in (3) are useful in devising an objective criterion for identifying this point of diminishing terms.
2. The finite divided differences that constitute the coefficients of the polynomial [Eqs. (18.8) through (18.11)] can be computed efficiently. That is, as in Eq. (18.14) and Fig. 18.5, lower-order differences are used to compute higher-order differences. By utilizing this previously determined information, the coefficients can be computed efficiently. The algorithm in Fig. 18.7 contains such a scheme.
3. The error estimate [Eq. (18.18)] can be very simply incorporated into a computer algorithm because of the sequential way in which the prediction is built.

```
SUBROUTINE NewtInt (x, y, n, xi, yint, ea)
  LOCAL fdd_{n,n}
  DO i = 0, n
    fdd_{i,0} = y_i
  END DO
  DO j = 1, n
    DO i = 0, n − j
      fdd_{i,j} = (fdd_{i+1,j−1} − fdd_{i,j−1})/(x_{i+j} − x_i)
    END DO
  END DO
  xterm = 1
  yint_0 = fdd_{0,0}
  DO order = 1, n
    xterm = xterm * (xi − x_{order−1})
    yint2 = yint_{order−1} + fdd_{0,order} * xterm
    Ea_{order−1} = yint2 − yint_{order−1}
    yint_{order} = yint2
  END order
END NewtInt
```

FIGURE 18.7
An algorithm for Newton's interpolating polynomial written in pseudocode.

All the above characteristics can be exploited and incorporated into a general algorithm for implementing Newton's polynomial (Fig. 18.7). Note that the algorithm consists of two parts: the first determines the coefficients from Eq. (18.7); the second determines the predictions and their associated error. The utility of this algorithm is demonstrated in the following example.

EXAMPLE 18.5   Error Estimates to Determine the Appropriate Order of Interpolation

Problem Statement.   After incorporating the error [Eq. (18.18)], utilize the computer algorithm given in Fig. 18.7 and the following information to evaluate $f(x) = \ln x$ at $x = 2$:

x	f(x) = ln x
0	1
4	1.3862944
6	1.7917595
5	1.6094379
3	1.0986123
1.5	0.4054641
2.5	0.9162907
3.5	1.2527630

Solution. The results of employing the algorithm in Fig. 18.7 to obtain a solution are shown in Fig. 18.8. The error estimates, along with the true error (based on the fact that ln $2 = 0.6931472$), are depicted in Fig. 18.9. Note that the estimated error and the true error are similar and that their agreement improves as the order increases. From these results, it can be concluded that the fifth-order version yields a good estimate and that higher-order terms do not significantly enhance the prediction.

This exercise also illustrates the importance of the positioning and ordering of the points. For example, up through the third-order estimate, the rate of improvement is slow because the points that are added (at $x = 4, 6$, and 5) are distant and on one side of the point in question at $x = 2$. The fourth-order estimate shows a somewhat greater improvement because the new point at $x = 3$ is closer to the unknown. However, the most dramatic decrease in the error is associated with the inclusion of the fifth-order term using the data point at $x = 1.5$. Not only is this point close to the unknown but it is also positioned on the opposite side from most of the other points. As a consequence, the error is reduced almost an order of magnitude.

The significance of the position and sequence of the data can also be illustrated by using the same data to obtain an estimate for ln 2, but considering the points in a different sequence. Figure 18.9 shows results for the case of reversing the order of the original data, that is, $x_0 = 3.5, x_1 = 2.5, x_3 = 1.5$, and so forth. Because the initial points for this case are closer to and spaced on either side of ln 2, the error decreases much more rapidly than for the original situation. By the second-order term, the error has been reduced to less than $\varepsilon_t = 2\%$. Other combinations could be employed to obtain different rates of convergence.

FIGURE 18.8
The output of a program, based on the algorithm from Fig. 18.7, to evaluate ln 2.

```
NUMBER OF POINTS? 8
X( 0 ), y( 0 ) = ? 1,0
X( 1 ), y( 1 ) = ? 4,1.3862944
X( 2 ), y( 2 ) = ? 6,1.7917595
X( 3 ), y( 3 ) = ? 5,1.6094379
X( 4 ), y( 4 ) = ? 3,1.0986123
X( 5 ), y( 5 ) = ? 1.5,0.40546411
X( 6 ), y( 6 ) = ? 2.5,0.91629073
X( 7 ), y( 7 ) = ? 3.5,1.2527630

INTERPOLATION AT X = 2
ORDER          F(X)               ERROR
0              0.000000           0.462098
1              0.462098           0.103746
2              0.565844           0.062924
3              0.628769           0.046953
4              0.675722           0.021792
5              0.697514          -0.003616
6              0.693898          -0.000459
7              0.693439
```

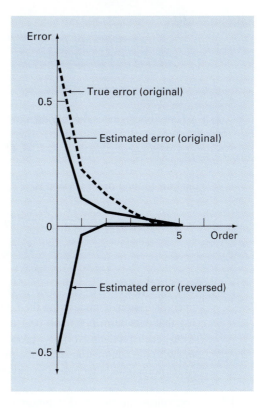

FIGURE 18.9
Percent relative errors for the prediction of ln 2 as a function of the order of the interpolating polynomial.

The foregoing example illustrates the importance of the choice of base points. As should be intuitively obvious, the points should be centered around and as close as possible to the unknown. This observation is also supported by direct examination of the error equation [Eq. (18.17)]. If we assume that the finite divided difference does not vary markedly along the range of the data, the error is proportional to the product: $(x - x_0)(x - x_1) \cdots (x - x_n)$. Obviously, the closer the base points are to x, the smaller the magnitude of this product.

18.2 LAGRANGE INTERPOLATING POLYNOMIALS

The *Lagrange interpolating polynomial* is simply a reformulation of the Newton polynomial that avoids the computation of divided differences. It can be represented concisely as

$$f_n(x) = \sum_{i=0}^{n} L_i(x) f(x_i) \qquad (18.20)$$

where

$$L_i(x) = \prod_{\substack{j=0 \\ j \neq i}}^{n} \frac{x - x_j}{x_i - x_j} \qquad (18.21)$$

where Π designates the "product of." For example, the linear version ($n = 1$) is

$$f_1(x) = \frac{x - x_1}{x_0 - x_1} f(x_0) + \frac{x - x_0}{x_1 - x_0} f(x_1) \qquad (18.22)$$

and the second-order version is

$$f_2(x) = \frac{(x - x_1)(x - x_2)}{(x_0 - x_1)(x_0 - x_2)} f(x_0) + \frac{(x - x_0)(x - x_2)}{(x_1 - x_0)(x_1 - x_2)} f(x_1)$$

$$+ \frac{(x - x_0)(x - x_1)}{(x_2 - x_0)(x_2 - x_1)} f(x_2) \qquad (18.23)$$

Equation (18.20) can be derived directly from Newton's polynomial (Box 18.1). However, the rationale underlying the Lagrange formulation can be grasped directly by realizing that each term $L_i(x)$ will be 1 at $x = x_i$ and 0 at all other sample points (Fig. 18.10). Thus, each product $L_i(x)f(x_i)$ takes on the value of $f(x_i)$ at the sample point x_i. Consequently, the summation of all the products designated by Eq. (18.20) is the unique nth-order polynomial that passes exactly through all $n + 1$ data points.

EXAMPLE 18.6   Lagrange Interpolating Polynomials

Problem Statement.   Use a Lagrange interpolating polynomial of the first and second order to evaluate ln 2 on the basis of the data given in Example 18.2:

$$x_0 = 1 \qquad f(x_0) = 0$$
$$x_1 = 4 \qquad f(x_1) = 1.386294$$
$$x_2 = 6 \qquad f(x_2) = 1.791760$$

Solution.   The first-order polynomial [Eq. (18.22)] can be used to obtain the estimate at $x = 2$,

$$f_1(2) = \frac{2 - 4}{1 - 4} 0 + \frac{2 - 1}{4 - 1} 1.386294 = 0.4620981$$

In a similar fashion, the second-order polynomial is developed as [Eq. (18.23)]

$$f_2(2) = \frac{(2 - 4)(2 - 6)}{(1 - 4)(1 - 6)} 0 + \frac{(2 - 1)(2 - 6)}{(4 - 1)(4 - 6)} 1.386294$$

$$+ \frac{(2 - 1)(2 - 4)}{(6 - 1)(6 - 4)} 1.791760 = 0.5658444$$

As expected, both these results agree with those previously obtained using Newton's interpolating polynomial.

Box 18.1 Derivation of the Lagrange Form Directly from Newton's Interpolating Polynomial

The Lagrange interpolating polynomial can be derived directly from Newton's formulation. We will do this for the first-order case only [Eq. (18.2)]. To derive the Lagrange form, we reformulate the divided differences. For example, the first divided difference,

$$f[x_1, x_0] = \frac{f(x_1) - f(x_0)}{x_1 - x_0} \qquad \text{(B18.1.1)}$$

can be reformulated as

$$f[x_1, x_0] = \frac{f(x_1)}{x_1 - x_0} + \frac{f(x_0)}{x_0 - x_1} \qquad \text{(B18.1.2)}$$

which is referred to as the *symmetric form*. Substituting Eq. (B18.1.2) into Eq. (18.2) yields

$$f_1(x) = f(x_0) + \frac{x - x_0}{x_1 - x_0} f(x_1) + \frac{x - x_0}{x_0 - x_1} f(x_0)$$

Finally, grouping similar terms and simplifying yields the Lagrange form,

$$f_1(x) = \frac{x - x_1}{x_0 - x_1} f(x_0) + \frac{x - x_0}{x_1 - x_0} f(x_1)$$

FIGURE 18.10

A visual depiction of the rationale behind the Lagrange polynomial. This figure shows a second-order case. Each of the three terms in Eq. (18.23) passes through one of the data points and is zero at the other two. The summation of the three terms must, therefore, be the unique second-order polynomial $f_2(x)$ that passes exactly through the three points.

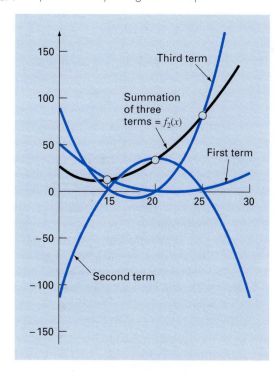

```
FUNCTION Lagrng(x, y, n, x)
  sum = 0
  DO i = 0, n
    product = y_i
    DO j = 0, n
      IF i ≠ j THEN
        product = product*(x − x_j)/(x_i − x_j)
      ENDIF
    END DO
    sum = sum + product
  END DO
  Lagrng = sum
END Lagrng
```

FIGURE 18.11

Pseudocode to implement Lagrange interpolation. This algorithm is set up to compute a single nth-order prediction, where $n + 1$ is the number of data points.

Note that, as with Newton's method, the Lagrange version has an estimated error of [Eq. (18.17)]

$$R_n = f[x, x_n, x_{n-1}, \ldots, x_0] \prod_{i=0}^{n} (x − x_i)$$

Thus, if an additional point is available at $x = x_{n+1}$, an error estimate can be obtained. However, because the finite divided differences are not employed as part of the Lagrange algorithm, this is rarely done.

Equations (18.20) and (18.21) can be very simply programmed for implementation on a computer. Figure 18.11 shows pseudocode that can be employed for this purpose.

In summary, for cases where the order of the polynomial is unknown, the Newton method has advantages because of the insight it provides into the behavior of the different-order formulas. In addition, the error estimate represented by Eq. (18.18) can usually be integrated easily into the Newton computation because the estimate employs a finite difference (Example 18.5). Thus, for exploratory computations, Newton's method is often preferable.

When only one interpolation is to be performed, the Lagrange and Newton formulations require comparable computational effort. However, the Lagrange version is somewhat easier to program. Because it does not require computation and storage of divided differences, the Lagrange form is often used when the order of the polynomial is known a priori.

EXAMPLE 18.7   Lagrange Interpolation Using the Computer

Problem Statement.   We can use the algorithm from Fig. 18.11 to study a trend analysis problem associated with our now-familiar falling parachutist. Assume that we have

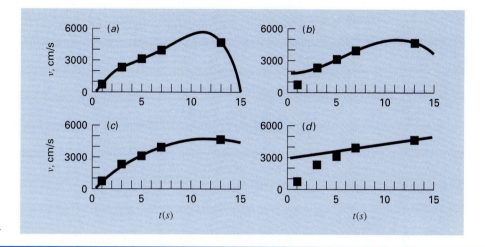

FIGURE 18.12
Plots showing (a) fourth-order, (b) third-order, (c) second-order, and (d) first-order interpolations.

developed instrumentation to measure the velocity of the parachutist. The measured data obtained for a particular test case is

Time, s	Measured Velocity v, cm/s
1	800
3	2310
5	3090
7	3940
13	4755

Our problem is to estimate the velocity of the parachutist at $t = 10$ s to fill in the large gap in the measurements between $t = 7$ and $t = 13$ s. We are aware that the behavior of interpolating polynomials can be unexpected. Therefore, we will construct polynomials of orders 4, 3, 2, and 1 and compare the results.

Solution.   The Lagrange algorithm can be used to construct fourth-, third-, second-, and first-order interpolating polynomials.

The fourth-order polynomial and the input data can be plotted as shown in Fig. 18.12a. It is evident from this plot that the estimated value of y at $x = 10$ is higher than the overall trend of the data.

Figure 18.12b through d shows plots of the results of the computations for third-, second-, and first-order interpolating polynomials. It is noted that the lower the order, the lower the estimated value of the velocity at $t = 10$ s. The plots of the interpolating polynomials indicate that the higher-order polynomials tend to overshoot the trend of the data. This suggests that the first- or second-order versions are most appropriate for this particular trend analysis. It should be remembered, however, that because we are dealing with uncertain data, regression would actually be more appropriate.

The preceding example illustrates that higher-order polynomials tend to be ill-conditioned; that is, they tend to be highly sensitive to round-off error. The same problem applies to higher-order polynomial regression. Double-precision arithmetic sometimes helps mitigate the problem. However, as the order increases, there will come a point at which round-off error will interfere with the ability to interpolate using the simple approaches covered to this point.

18.3   COEFFICIENTS OF AN INTERPOLATING POLYNOMIAL

Although both the Newton and the Lagrange polynomials are well-suited for determining intermediate values between points, they do not provide a convenient polynomial of the conventional form

$$f(x) = a_0 + a_1 x + a_2 x^2 + \cdots + a_n x^n \tag{18.24}$$

A straightforward method for computing the coefficients of this polynomial is based on the fact that $n + 1$ data points are required to determine the $n + 1$ coefficients. Thus, simultaneous linear algebraic equations can be used to calculate the a's. For example, suppose that you desired to compute the coefficients of the parabola

$$f(x) = a_0 + a_1 x + a_2 x^2 \tag{18.25}$$

Three data points are required: $[x_0, f(x_0)]$, $[x_1, f(x_1)]$, and $[x_2, f(x_2)]$. Each can be substituted into Eq. (18.25) to give

$$
\begin{aligned}
f(x_0) &= a_0 + a_1 x_0 + a_2 x_0^2 \\
f(x_1) &= a_0 + a_1 x_1 + a_2 x_1^2 \\
f(x_2) &= a_0 + a_1 x_2 + a_2 x_2^2
\end{aligned}
\tag{18.26}
$$

Thus, for this case, the x's are the knowns and the a's are the unknowns. Because there are the same number of equations as unknowns, Eq. (18.26) could be solved by an elimination method from Part Three.

It should be noted that the foregoing approach is not the most efficient method that is available to determine the coefficients of an interpolating polynomial. Press et al. (1986) provide a discussion and computer codes for more efficient approaches. Whatever technique is employed, a word of caution is in order. Systems such as Eq. (18.26) are notoriously ill-conditioned. Whether they are solved with an elimination method or with a more efficient algorithm, the resulting coefficients can be highly inaccurate, particularly for large n. When used for a subsequent interpolation, they often yield erroneous results.

In summary, if you are interested in determining an intermediate point, employ Newton or Lagrange interpolation. If you must determine an equation of the form of Eq. (18.24), limit yourself to lower-order polynomials and check your results carefully.

18.4   INVERSE INTERPOLATION

As the nomenclature implies, the $f(x)$ and x values in most interpolation contexts are the dependent and independent variables, respectively. As a consequence, the values of the x's are typically uniformly spaced. A simple example is a table of values derived for the function $f(x) = 1/x$,

x	1	2	3	4	5	6	7
$f(x)$	1	0.5	0.3333	0.25	0.2	0.1667	0.1429

Now suppose that you must use the same data, but you are given a value for $f(x)$ and must determine the corresponding value of x. For instance, for the data above, suppose that you were asked to determine the value of x that corresponded to $f(x) = 0.3$. For this case, because the function is available and easy to manipulate, the correct answer can be determined directly as $x = 1/0.3 = 3.3333$.

Such a problem is called *inverse interpolation*. For a more complicated case, you might be tempted to switch the $f(x)$ and x values [that is, merely plot x versus $f(x)$] and use an approach like Lagrange interpolation to determine the result. Unfortunately, when you reverse the variables, there is no guarantee that the values along the new abscissa [the $f(x)$'s] will be evenly spaced. In fact, in many cases, the values will be "telescoped." That is, they will have the appearance of a logarithmic scale with some adjacent points bunched together and others spread out widely. For example, for $f(x) = 1/x$ the result is

$f(x)$	0.1429	0.1667	0.2	0.25	0.3333	0.5	1
x	7	6	5	4	3	2	1

Such nonuniform spacing on the abscissa often leads to oscillations in the resulting interpolating polynomial. This can occur even for lower-order polynomials.

An alternative strategy is to fit an nth-order interpolating polynomial, $f_n(x)$, to the original data [i.e., with $f(x)$ versus x]. In most cases, because the x's are evenly spaced, this polynomial will not be ill-conditioned. The answer to your problem then amounts to finding the value of x that makes this polynomial equal to the given $f(x)$. Thus, the interpolation problem reduces to a roots problem!

For example, for the problem outlined above, a simple approach would be to fit a quadratic polynomial to the three points: $(2, 0.5)$, $(3, 0.3333)$ and $(4, 0.25)$. The result would be

$$f_2(x) = 1.08333 - 0.375x + 0.041667x^2$$

The answer to the inverse interpolation problem of finding the x corresponding to $f(x) = 0.3$ would therefore involve determining the root of

$$0.3 = 1.08333 - 0.375x + 0.041667x^2$$

For this simple case, the quadratic formula can be used to calculate

$$x = \frac{0.375 \pm \sqrt{(-0.375)^2 - 4(0.041667)0.78333}}{2(0.041667)} = \frac{5.704158}{3.295842}$$

Thus, the second root, 3.296, is a good approximation of the true value of 3.333. If additional accuracy were desired, a third- or fourth-order polynomial along with one of the root location methods from Part Two could be employed.

18.5 ADDITIONAL COMMENTS

Before proceeding to the next section, we must mention two additional topics: interpolation with equally spaced data and extrapolation.

Box 18.2 Interpolation with Equally Spaced Data

If data is equally spaced and in ascending order, then the independent variable assumes values of

$$x_1 = x_0 + h$$
$$x_2 = x_0 + 2h$$
$$\cdot$$
$$\cdot$$
$$\cdot$$
$$x_n = x_0 + nh$$

where h is the interval, or step size, between the data. On this basis, the finite divided differences can be expressed in concise form. For example, the second forward divided difference is

$$f[x_0, x_1, x_2] = \frac{\dfrac{f(x_2) - f(x_1)}{x_2 - x_1} - \dfrac{f(x_1) - f(x_0)}{x_1 - x_0}}{x_2 - x_0}$$

which can be expressed as

$$f[x_0, x_1, x_2] = \frac{f(x_2) - 2f(x_1) + f(x_0)}{2h^2} \tag{B18.2.1}$$

because $x_1 - x_0 = x_2 - x_1 = (x_2 - x_0)/2 = h$. Now recall that the second forward difference is equal to [numerator of Eq. (4.24)]

$$\Delta^2 f(x_0) = f(x_2) - 2f(x_1) + f(x_0)$$

Therefore, Eq. (B18.2.1) can be represented as

$$f[x_0, x_1, x_2] = \frac{\Delta^2 f(x_0)}{2!h^2}$$

or, in general,

$$f[x_0, x_1, \ldots, x_n] = \frac{\Delta^n f(x_0)}{n!h^n} \tag{B18.2.2}$$

Using Eq. (B18.2.2), we can express Newton's interpolating polynomial [Eq. (18.15)] for the case of equispaced data as

$$f_n(x) = f(x_0) + \frac{\Delta f(x_0)}{h}(x - x_0)$$

$$+ \frac{\Delta^2 f(x_0)}{2!h^2}(x - x_0)(x - x_0 - h)$$

$$+ \cdots + \frac{\Delta^n f(x_0)}{n!h^n}(x - x_0)(x - x_0 - h)$$

$$\cdots [x - x_0 - (n-1)h] + R_n \tag{B18.2.3}$$

where the remainder is the same as Eq. (18.16). This equation is known as *Newton's formula,* or the *Newton-Gregory forward formula.* It can be simplified further by defining a new quantity, α:

$$\alpha = \frac{x - x_0}{h}$$

This definition can be used to develop the following simplified expressions for the terms in Eq. (B18.2.3):

$$x - x_0 = \alpha h$$
$$x - x_0 - h = \alpha h - h = h(\alpha - 1)$$
$$\cdot$$
$$\cdot$$
$$\cdot$$
$$x - x_0 - (n-1)h = \alpha h - (n-1)h = h(\alpha - n + 1)$$

which can be substituted into Eq. (B18.2.3) to give

$$f_n(x) = f(x_0) + \Delta f(x_0)\alpha + \frac{\Delta^2 f(x_0)}{2!}\alpha(\alpha - 1)$$

$$+ \cdots + \frac{\Delta^n f(x_0)}{n!}\alpha(\alpha - 1)\cdots(\alpha - n + 1) + R_n \tag{B18.2.4}$$

where

$$R_n = \frac{f^{(n+1)}(\xi)}{(n+1)!}h^{n+1}\alpha(\alpha - 1)(\alpha - 2)\cdots(\alpha - n)$$

This concise notation will have utility in our derivation and error analyses of the integration formulas in Chap. 21.

In addition to the forward formula, backward and central Newton-Gregory formulas are also available. Carnahan, Luther, and Wilkes (1969) can be consulted for further information regarding interpolation for equally spaced data.

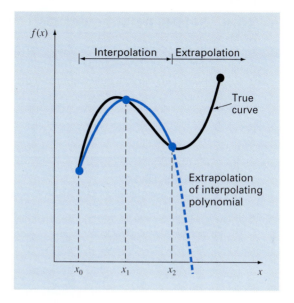

FIGURE 18.13

Illustration of the possible divergence of an extrapolated prediction. The extrapolation is based on fitting a parabola through the first three known points.

Because both the Newton and Lagrange polynomials are compatible with arbitrarily spaced data, you might wonder why we address the special case of equally spaced data (Box 18.2). Prior to the advent of digital computers, these techniques had great utility for interpolation from tables with equally spaced arguments. In fact, a computational framework known as a divided-difference table was developed to facilitate the implementation of these techniques. (Figure 18.5 is an example of such a table.)

However, because the formulas are subsets of the computer-compatible Newton and Lagrange schemes and because many tabular functions are available as library subroutines, the need for the equispaced versions has waned. In spite of this, we have included them at this point because of their relevance to later parts of this book. In particular, they are needed to derive numerical integration formulas that typically employ equispaced data (Chap. 21). Because the numerical integration formulas have relevance to the solution of ordinary differential equations, the material in Box 18.2 also has significance to Part Seven.

Extrapolation is the process of estimating a value of $f(x)$ that lies outside the range of the known base points, $x_0, x_1, \ldots, x_n$ (Fig. 18.13). In a previous section, we mentioned that the most accurate interpolation is usually obtained when the unknown lies near the center of the base points. Obviously, this is violated when the unknown lies outside the range, and consequently, the error in extrapolation can be very large. As depicted in Fig. 18.13, the open-ended nature of extrapolation represents a step into the unknown because the process extends the curve beyond the known region. As such, the true curve could easily diverge from the prediction. Extreme care should, therefore, be exercised whenever a case arises where one must extrapolate.

18.6 SPLINE INTERPOLATION

In the previous sections, nth-order polynomials were used to interpolate between $n + 1$ data points. For example, for eight points, we can derive a perfect seventh-order polynomial. This curve would capture all the meanderings (at least up to and including seventh derivatives) suggested by the points. However, there are cases where these functions can lead

FIGURE 18.14
A visual representation of a situation where the splines are superior to higher-order interpolating polynomials. The function to be fit undergoes an abrupt increase at $x = 0$. Parts (a) through (c) indicate that the abrupt change induces oscillations in interpolating polynomials. In contrast, because it is limited to third-order curves with smooth transitions, the cubic spline (d) provides a much more acceptable approximation.

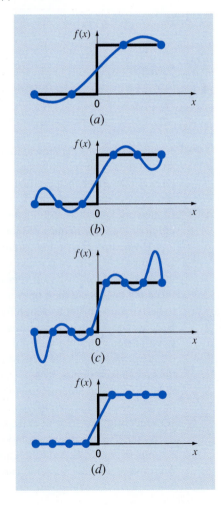

to erroneous results because of round-off error and overshoot. An alternative approach is to apply lower-order polynomials to subsets of data points. Such connecting polynomials are called *spline functions*.

For example, third-order curves employed to connect each pair of data points are called *cubic splines*. These functions can be constructed so that the connections between adjacent cubic equations are visually smooth. On the surface, it would seem that the third-order approximation of the splines would be inferior to the seventh-order expression. You might wonder why a spline would ever be preferable.

Figure 18.14 illustrates a situation where a spline performs better than a higher-order polynomial. This is the case where a function is generally smooth but undergoes an abrupt change somewhere along the region of interest. The step increase depicted in Fig. 18.14 is an extreme example of such a change and serves to illustrate the point.

Figure 18.14*a* through *c* illustrates how higher-order polynomials tend to swing through wild oscillations in the vicinity of an abrupt change. In contrast, the spline also connects the points, but because it is limited to third-order changes, the oscillations are kept to a minimum. As such, the spline usually provides a superior approximation of the behavior of functions that have local, abrupt changes.

The concept of the spline originated from the drafting technique of using a thin, flexible strip (called a *spline*) to draw smooth curves through a set of points. The process is depicted in Fig. 18.15 for a series of five pins (data points). In this technique, the drafter places paper over a wooden board and hammers nails or pins into the paper (and board) at the location of the data points. A smooth cubic curve results from interweaving the strip between the pins. Hence, the name "cubic spline" has been adopted for polynomials of this type.

In this section, simple linear functions will first be used to introduce some basic concepts and problems associated with spline interpolation. Then we derive an algorithm for fitting quadratic splines to data. Finally, we present material on the cubic spline, which is the most common and useful version in engineering practice.

FIGURE 18.15
The drafting technique of using a spline to draw smooth curves through a series of points. Notice how, at the end points, the spline straightens out. This is called a "natural" spline.

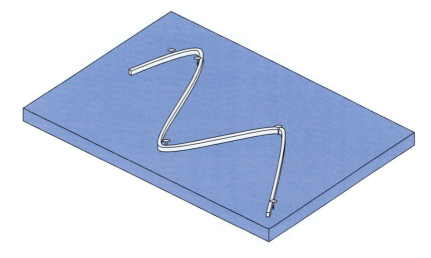

18.6.1 Linear Splines

The simplest connection between two points is a straight line. The first-order splines for a group of ordered data points can be defined as a set of linear functions,

$$f(x) = f(x_0) + m_0(x - x_0) \qquad\qquad x_0 \leq x \leq x_1$$
$$f(x) = f(x_1) + m_1(x - x_1) \qquad\qquad x_1 \leq x \leq x_2$$

$$\cdot$$
$$\cdot$$
$$\cdot$$

$$f(x) = f(x_{n-1}) + m_{n-1}(x - x_{n-1}) \qquad x_{n-1} \leq x \leq x_n$$

where m_i is the slope of the straight line connecting the points:

$$m_i = \frac{f(x_{i+1}) - f(x_i)}{x_{i+1} - x_i} \tag{18.27}$$

These equations can be used to evaluate the function at any point between x_0 and x_n by first locating the interval within which the point lies. Then the appropriate equation is used to determine the function value within the interval. The method is obviously identical to linear interpolation.

EXAMPLE 18.8   First-Order Splines

Problem Statement.   Fit the data in Table 18.1 with first-order splines. Evaluate the function at $x = 5$.

Solution.   The data can be used to determine the slopes between points. For example, for the interval $x = 4.5$ to $x = 7$ the slope can be computed using Eq. (18.27):

$$m = \frac{2.5 - 1}{7 - 4.5} = 0.60$$

The slopes for the other intervals can be computed, and the resulting first-order splines are plotted in Fig. 18.16a. The value at $x = 5$ is 1.3.

TABLE 18.1
Data to be fit with spline functions.

x	f(x)
3.0	2.5
4.5	1.0
7.0	2.5
9.0	0.5

Visual inspection of Fig. 18.16*a* indicates that the primary disadvantage of first-order splines is that they are not smooth. In essence, at the data points where two splines meet (called a *knot*), the slope changes abruptly. In formal terms, the first derivative of the function is discontinuous at these points. This deficiency is overcome by using higher-order polynomial splines that ensure smoothness at the knots by equating derivatives at these points, as discussed in the next section.

18.6.2 Quadratic Splines

To ensure that the *m*th derivatives are continuous at the knots, a spline of at least $m + 1$ order must be used. Third-order polynomials or cubic splines that ensure continuous first and second derivatives are most frequently used in practice. Although third and higher

FIGURE 18.16
Spline fits of a set of four points. (*a*) Linear spline, (*b*) quadratic spline, and (*c*) cubic spline, with a cubic interpolating polynomial also plotted.

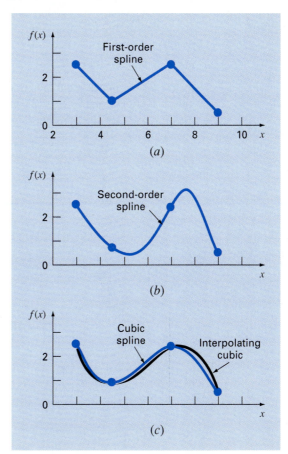

derivatives could be discontinuous when using cubic splines, they usually cannot be detected visually and consequently are ignored.

Because the derivation of cubic splines is somewhat involved, we have chosen to include them in a subsequent section. We have decided to first illustrate the concept of spline interpolation using second-order polynomials. These "quadratic splines" have continuous first derivatives at the knots. Although quadratic splines do not ensure equal second derivatives at the knots, they serve nicely to demonstrate the general procedure for developing higher-order splines.

The objective in quadratic splines is to derive a second-order polynomial for each interval between data points. The polynomial for each interval can be represented generally as

$$f_i(x) = a_i x^2 + b_i x + c_i \tag{18.28}$$

Figure 18.17 has been included to help clarify the notation. For $n + 1$ data points ($i = 0, 1, 2, \ldots, n$), there are n intervals and, consequently, $3n$ unknown constants (the a's, b's, and c's) to evaluate. Therefore, $3n$ equations or conditions are required to evaluate the unknowns. These are:

1. *The function values of adjacent polynomials must be equal at the interior knots.* This condition can be represented as

$$a_{i-1} x_{i-1}^2 + b_{i-1} x_{i-1} + c_{i-1} = f(x_{i-1}) \tag{18.29}$$

$$a_i x_{i-1}^2 + b_i x_{i-1} + c_i = f(x_{i-1}) \tag{18.30}$$

for $i = 2$ to n. Because only interior knots are used, Eqs. (18.29) and (18.30) each provide $n - 1$ for a total of $2n - 2$ conditions.

2. *The first and last functions must pass through the end points.* This adds two additional equations:

$$a_1 x_0^2 + b_1 x_0 + c_1 = f(x_0) \tag{18.31}$$

FIGURE 18.17

Notation used to derive quadratic splines. Notice that there are n intervals and $n + 1$ data points. The example shown is for $n = 3$.

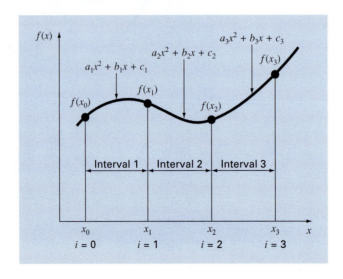

$$a_n x_n^2 + b_n x_n + c_n = f(x_n) \tag{18.32}$$

for a total of $2n - 2 + 2 = 2n$ conditions.

3. *The first derivatives at the interior knots must be equal.* The first derivative of Eq. (18.28) is

$$f'(x) = 2ax + b$$

Therefore, the condition can be represented generally as

$$2a_{i-1}x_{i-1} + b_{i-1} = 2a_i x_{i-1} + b_i \tag{18.33}$$

for $i = 2$ to n. This provides another $n - 1$ conditions for a total of $2n + n - 1 = 3n - 1$. Because we have $3n$ unknowns, we are one condition short. Unless we have some additional information regarding the functions or their derivatives, we must make an arbitrary choice to successfully compute the constants. Although there are a number of different choices that can be made, we select the following:

4. *Assume that the second derivative is zero at the first point.* Because the second derivative of Eq. (18.28) is $2a_i$, this condition can be expressed mathematically as

$$a_1 = 0 \tag{18.34}$$

The visual interpretation of this condition is that the first two points will be connected by a straight line.

EXAMPLE 18.9   Quadratic Splines

Problem Statement.   Fit quadratic splines to the same data used in Example 18.8 (Table 18.1). Use the results to estimate the value at $x = 5$.

Solution.   For the present problem, we have four data points and $n = 3$ intervals. Therefore, $3(3) = 9$ unknowns must be determined. Equations (18.29) and (18.30) yield $2(3) - 2 = 4$ conditions:

$$20.25a_1 + 4.5b_1 + c_1 = 1.0$$
$$20.25a_2 + 4.5b_2 + c_2 = 1.0$$
$$49a_2 + \quad 7b_2 + c_2 = 2.5$$
$$49a_3 + \quad 7b_3 + c_3 = 2.5$$

Passing the first and last functions through the initial and final values adds 2 more [Eq. (18.31)]:

$$9a_1 + 3b_1 + c_1 = 2.5$$

and [Eq. (18.32)]

$$81a_3 + 9b_3 + c_3 = 0.5$$

Continuity of derivatives creates an additional $3 - 1 = 2$ [Eq. (18.33)]:

$$9a_1 + b_1 = 9a_2 + b_2$$
$$14a_2 + b_2 = 14a_3 + b_3$$

Finally, Eq. (18.34) specifies that $a_1 = 0$. Because this equation specifies a_1 exactly, the problem reduces to solving eight simultaneous equations. These conditions can be expressed in matrix form as

$$\begin{bmatrix} 4.5 & 1 & 0 & 0 & 0 & 0 & 0 & 0 \\ 0 & 0 & 20.25 & 4.5 & 1 & 0 & 0 & 0 \\ 0 & 0 & 49 & 7 & 1 & 0 & 0 & 0 \\ 0 & 0 & 0 & 0 & 0 & 49 & 7 & 1 \\ 3 & 1 & 0 & 0 & 0 & 0 & 0 & 0 \\ 0 & 0 & 0 & 0 & 0 & 81 & 9 & 1 \\ 1 & 0 & -9 & -1 & 0 & 0 & 0 & 0 \\ 0 & 0 & 14 & 1 & 0 & -14 & -1 & 0 \end{bmatrix} \begin{Bmatrix} b_1 \\ c_1 \\ a_2 \\ b_2 \\ c_2 \\ a_3 \\ b_3 \\ c_3 \end{Bmatrix} = \begin{Bmatrix} 1 \\ 1 \\ 2.5 \\ 2.5 \\ 2.5 \\ 0.5 \\ 0 \\ 0 \end{Bmatrix}$$

These equations can be solved using techniques from Part Three, with the results:

$$a_1 = 0 \qquad b_1 = -1 \qquad c_1 = 5.5$$
$$a_2 = 0.64 \qquad b_2 = -6.76 \qquad c_2 = 18.46$$
$$a_3 = -1.6 \qquad b_3 = 24.6 \qquad c_3 = -91.3$$

which can be substituted into the original quadratic equations to develop the following relationships for each interval:

$$f_1(x) = -x + 5.5 \qquad\qquad 3.0 \le x \le 4.5$$
$$f_2(x) = 0.64x^2 - 6.76x + 18.46 \qquad 4.5 \le x \le 7.0$$
$$f_3(x) = -1.6x^2 + 24.6x - 91.3 \qquad 7.0 \le x \le 9.0$$

When we use f_2, the prediction for $x = 5$ is, therefore,

$$f_2(5) = 0.64(5)^2 - 6.76(5) + 18.46 = 0.66$$

The total spline fit is depicted in Fig. 18.16b. Notice that there are two shortcomings that detract from the fit: (1) the straight line connecting the first two points and (2) the spline for the last interval seems to swing too high. The cubic splines in the next section do not exhibit these shortcomings and, as a consequence, are better methods for spline interpolation.

18.6.3 Cubic Splines

The objective in cubic splines is to derive a third-order polynomial for each interval between knots, as in

$$f_i(x) = a_i x^3 + b_i x^2 + c_i x + d_i \tag{18.35}$$

Thus, for $n + 1$ data points ($i = 0, 1, 2, \ldots, n$), there are n intervals and, consequently, $4n$ unknown constants to evaluate. Just as for quadratic splines, $4n$ conditions are required to evaluate the unknowns. These are:

1. The function values must be equal at the interior knots ($2n - 2$ conditions).
2. The first and last functions must pass through the end points (2 conditions).

3. The first derivatives at the interior knots must be equal ($n - 1$ conditions).
4. The second derivatives at the interior knots must be equal ($n - 1$ conditions).
5. The second derivatives at the end knots are zero (2 conditions).

The visual interpretation of condition 5 is that the function becomes a straight line at the end knots. Specification of such an end condition leads to what is termed a "natural" spline. It is given this name because the drafting spline naturally behaves in this fashion (Fig. 18.15). If the value of the second derivative at the end knots is nonzero (that is, there is some curvature), this information can be used alternatively to supply the two final conditions.

The above five types of conditions provide the total of $4n$ equations required to solve for the $4n$ coefficients. Whereas it is certainly possible to develop cubic splines in this fashion, we will present an alternative technique that requires the solution of only $n - 1$ equations. Although the derivation of this method (Box 18.3) is somewhat less straightforward than that for quadratic splines, the gain in efficiency is well worth the effort.

Box 18.3   Derivation of Cubic Splines

The first step in the derivation (Cheney and Kincaid, 1985) is based on the observation that because each pair of knots is connected by a cubic, the second derivative within each interval is a straight line. Equation (18.35) can be differentiated twice to verify this observation. On this basis, the second derivatives can be represented by a first-order Lagrange interpolating polynomial [Eq. (18.22)]:

$$f_i''(x) = f_i''(x_{i-1}) \frac{x - x_i}{x_{i-1} - x_i} + f_i''(x_i) \frac{x - x_{i-1}}{x_i - x_{i-1}} \qquad \text{(B18.3.1)}$$

where $f_i''(x)$ is the value of the second derivative at any point x within the ith interval. Thus, this equation is a straight line connecting the second derivative at the first knot $f''(x_{i-1})$ with the second derivative at the second knot $f''(x_i)$.

Next, Eq. (B18.3.1) can be integrated twice to yield an expression for $f_i(x)$. However, this expression will contain two unknown constants of integration. These constants can be evaluated by invoking the function-equality conditions—$f(x)$ must equal $f(x_{i-1})$ at x_{i-1} and $f(x)$ must equal $f(x_i)$ at x_i. By performing these evaluations, the following cubic equation results:

$$f_i(x) = \frac{f_i''(x_{i-1})}{6(x_i - x_{i-1})} (x_i - x)^3 + \frac{f_i''(x_i)}{6(x_i - x_{i-1})} (x - x_{i-1})^3$$

$$+ \left[\frac{f(x_{i-1})}{x_i - x_{i-1}} - \frac{f''(x_{i-1})(x_i - x_{i-1})}{6} \right] (x_i - x)$$

$$+ \left[\frac{f(x_i)}{x_i - x_{i-1}} - \frac{f''(x_i)(x_i - x_{i-1})}{6} \right] (x - x_{i-1})$$

$$\text{(B18.3.2)}$$

Now, admittedly, this relationship is a much more complicated expression for the cubic spline for the ith interval than, say,

Eq. (18.35). However, notice that it contains only two unknown "coefficients," the second derivatives at the beginning and the end of the interval—$f''(x_{i-1})$ and $f''(x_i)$. Thus, if we can determine the proper second derivative at each knot, Eq. (B18.3.2) is a third-order polynomial that can be used to interpolate within the interval.

The second derivatives can be evaluated by invoking the condition that the first derivatives at the knots must be continuous:

$$f_{i-1}'(x_i) = f_i'(x_i) \qquad \text{(B18.3.3)}$$

Equation (B18.3.2) can be differentiated to give an expression for the first derivative. If this is done for both the $(i - 1)$th and the ith intervals and the two results are set equal according to Eq. (B18.3.3), the following relationship results:

$$(x_i - x_{i-1})f''(x_{i-1}) + 2(x_{i+1} - x_{i-1})f''(x_i)$$

$$+ (x_{i+1} - x_i)f''(x_{i+1})$$

$$= \frac{6}{x_{i+1} - x_i} [f(x_{i+1}) - f(x_i)]$$

$$+ \frac{6}{x_i - x_{i-1}} [f(x_{i-1}) - f(x_i)] \qquad \text{(B18.3.4)}$$

If Eq. (B18.3.4) is written for all interior knots, $n - 1$ simultaneous equations result with $n + 1$ unknown second derivatives. However, because this is a natural cubic spline, the second derivatives at the end knots are zero and the problem reduces to $n - 1$ equations with $n - 1$ unknowns. In addition, notice that the system of equations will be tridiagonal. Thus, not only have we reduced the number of equations but we have also cast them in a form that is extremely easy to solve (recall Sec. 11.1.1).

The derivation from Box 18.3 results in the following cubic equation for each interval:

$$f_i(x) = \frac{f_i''(x_{i-1})}{6(x_i - x_{i-1})}(x_i - x)^3 + \frac{f_i''(x_i)}{6(x_i - x_{i-1})}(x - x_{i-1})^3$$

$$+ \left[\frac{f(x_{i-1})}{x_i - x_{i-1}} - \frac{f''(x_{i-1})(x_i - x_{i-1})}{6}\right](x_i - x)$$

$$+ \left[\frac{f(x_i)}{x_i - x_{i-1}} - \frac{f''(x_i)(x_i - x_{i-1})}{6}\right](x - x_{i-1}) \tag{18.36}$$

This equation contains only two unknowns—the second derivatives at the end of each interval. These unknowns can be evaluated using the following equation:

$$(x_i - x_{i-1})f''(x_{i-1}) + 2(x_{i+1} - x_{i-1})f''(x_i) + (x_{i+1} - x_i)f''(x_{i+1})$$

$$= \frac{6}{x_{i+1} - x_i}[f(x_{i+1}) - f(x_i)] + \frac{6}{x_i - x_{i-1}}[f(x_{i-1}) - f(x_i)] \tag{18.37}$$

If this equation is written for all the interior knots, $n - 1$ simultaneous equations result with $n - 1$ unknowns. (Remember, the second derivatives at the end knots are zero.) The application of these equations is illustrated in the following example.

EXAMPLE 18.10   Cubic Splines

Problem Statement.   Fit cubic splines to the same data used in Examples 18.8 and 18.9 (Table 18.1). Utilize the results to estimate the value at $x = 5$.

Solution.   The first step is to employ Eq. (18.37) to generate the set of simultaneous equations that will be utilized to determine the second derivatives at the knots. For example, for the first interior knot, the following data is used:

$$x_0 = 3 \qquad f(x_0) = 2.5$$
$$x_1 = 4.5 \qquad f(x_1) = 1$$
$$x_2 = 7 \qquad f(x_2) = 2.5$$

These values can be substituted into Eq. (18.37) to yield

$$(4.5 - 3)f''(3) + 2(7 - 3)f''(4.5) + (7 - 4.5)f''(7)$$

$$= \frac{6}{7 - 4.5}(2.5 - 1) + \frac{6}{4.5 - 3}(2.5 - 1)$$

Because of the natural spline condition, $f''(3) = 0$, and the equation reduces to

$$8f''(4.5) + 2.5f''(7) = 9.6$$

In a similar fashion, Eq. (18.37) can be applied to the second interior point to give

$$2.5f''(4.5) + 9f''(7) = -9.6$$

These two equations can be solved simultaneously for

$$f''(4.5) = 1.67909$$
$$f''(7) = -1.53308$$

These values can then be substituted into Eq. (18.36), along with values for the x's and the $f(x)$'s, to yield

$$f_1(x) = \frac{1.67909}{6(4.5 - 3)}(x - 3)^3 + \frac{2.5}{4.5 - 3}(4.5 - x)$$
$$+ \left[\frac{1}{4.5 - 3} - \frac{1.67909(4.5 - 3)}{6}\right](x - 3)$$

or

$$f_1(x) = 0.186566\,(x - 3)^3 + 1.666667\,(4.5 - x) + 0.246894\,(x - 3)$$

This equation is the cubic spline for the first interval. Similar substitutions can be made to develop the equations for the second and third intervals:

$$f_2(x) = 0.111939(7 - x)^3 - 0.102205(x - 4.5)^3 - 0.299621(7 - x)$$
$$+ 1.638783(x - 4.5)$$

and

$$f_3(x) = -0.127757(9 - x)^3 + 1.761027(9 - x) + 0.25(x - 7)$$

The three equations can then be employed to compute values within each interval. For example, the value at $x = 5$, which falls within the second interval, is calculated as

$$f_2(5) = 0.111939(7 - 5)^3 - 0.102205(5 - 4.5)^3 - 0.299621(7 - 5)$$
$$+ 1.638783(5 - 4.5) = 1.102886$$

Other values are computed and the results are plotted in Fig. 18.16c.

The results of Examples 18.8 through 18.10 are summarized in Fig. 18.16. Notice the progressive improvement of the fit as we move from linear to quadratic to cubic splines. We have also superimposed a cubic interpolating polynomial on Fig. 18.16c. Although the cubic spline consists of a series of third-order curves, the resulting fit differs from that obtained using the third-order polynomial. This is due to the fact that the natural spline requires zero second derivatives at the end knots, whereas the cubic polynomial has no such constraint.

18.6.4 Computer Algorithm for Cubic Splines

The method for calculating cubic splines outlined in the previous section is ideal for computer implementation. Recall that, by some clever manipulations, the method reduced to solving $n - 1$ simultaneous equations. An added benefit of the derivation is that, as specified by Eq. (18.37), the system of equations is tridiagonal. As described in Sec. 11.1, algorithms are available to solve such systems in an extremely efficient manner. Figure 18.18 outlines a computational framework that incorporates these features.

Note that the routine in Fig. 18.18 returns a single interpolated value, yu, for a given value of the dependent variable, xu. This is but one way in which spline interpolation can be implemented. For example, you might want to determine the coefficients once, and then

the sinusoids that comprise it. The alternative is to display these sinusoids—that is, $(4/\pi) \cos (\omega_0 t)$, $-(4/3\pi) \cos (3\omega_0 t)$, $(4/5\pi) \cos (5\omega_0 t)$, etc. This alternative does not provide an adequate visualization of the structure of these harmonics. In contrast, Fig. 19.9a and b provides a graphic display of this structure. As such, the line spectra represent "fingerprints" that can help us to characterize and understand a complicated waveform. They are particularly valuable for nonidealized cases where they sometimes allow us to discern structure in otherwise obscure signals. In the next section, we will describe the Fourier transform that will allow us to extend such analyses to nonperiodic waveforms.

19.4 FOURIER INTEGRAL AND TRANSFORM

Although the Fourier series is a useful tool for investigating the spectrum of a periodic function, there are many waveforms that do not repeat themselves regularly. For example, a lightning bolt occurs only once (or at least it will be a long time until it occurs again), but it will cause interference with receivers operating on a broad range of frequencies—for example, TVs, radios, shortwave receivers, etc. Such evidence suggests that a nonrecurring signal such as that produced by lightning exhibits a continuous frequency spectrum. Because such phenomena are of great interest to engineers, an alternative to the Fourier series would be valuable for analyzing these aperiodic waveforms.

The *Fourier integral* is the primary tool available for this purpose. It can be derived from the exponential form of the Fourier series

$$f(t) = \sum_{k=-\infty}^{\infty} \tilde{c}_k e^{ik\omega_0 t} \tag{19.23}$$

where

$$\tilde{c}_k = \frac{1}{T} \int_{-T/2}^{T/2} f(t) e^{-ik\omega_0 t} \, dt \tag{19.24}$$

where $\omega_0 = 2\pi/T$ and $k = 0, 1, 2, \ldots$.

The transition from a periodic to a nonperiodic function can be effected by allowing the period to approach infinity. In other words, as T becomes infinite, the function never repeats itself and thus becomes aperiodic. If this is allowed to occur, it can be demonstrated (for example, Van Valkenburg, 1974; Hayt and Kemmerly, 1986) that the Fourier series reduces to

$$f(t) = \frac{1}{2\pi} \int_{-\infty}^{\infty} F(i\omega_0) e^{i\omega_0 t} \, d\omega_0 \tag{19.25}$$

and the coefficients become a continuous function of the frequency variable ω, as in

$$F(i\omega_0) = \int_{-\infty}^{\infty} f(t) e^{-i\omega_0 t} \, dt \tag{19.26}$$

The function $F(i\omega_0)$, as defined by Eq. (19.26), is called the *Fourier integral* of $f(t)$. In addition, Eqs. (19.25) and (19.26) are collectively referred to as the *Fourier transform pair.* Thus, along with being called the Fourier integral, $F(i\omega_0)$ is also called the *Fourier transform* of $f(t)$. In the same spirit, $f(t)$, as defined by Eq. (19.25), is referred to as the *inverse*

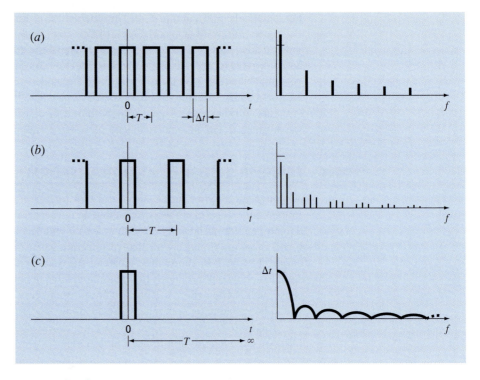

FIGURE 19.10

Illustration of how the discrete frequency spectrum of a Fourier series for a pulse train (a) approaches a continuous frequency spectrum of a Fourier integral (c) as the period is allowed to approach infinity.

Fourier transform of $F(i\omega_0)$. Thus, the pair allows us to transform back and forth between the time and the frequency domains for an aperiodic signal.

The distinction between the Fourier series and transform should now be quite clear. The major difference is that each applies to a different class of functions—the series to periodic and the transform to nonperiodic waveforms. Beyond this major distinction, the two approaches differ in how they move between the time and the frequency domains. The Fourier series converts a continuous, periodic time-domain function to frequency-domain magnitudes at discrete frequencies. In contrast, the Fourier transform converts a continuous time-domain function to a continuous frequency-domain function. Thus, the discrete frequency spectrum generated by the Fourier series is analogous to a continuous frequency spectrum generated by the Fourier transform.

The shift from a discrete to a continuous spectrum can be illustrated graphically. In Fig. 19.10a, we can see a pulse train of rectangular waves with pulse widths equal to one-half the period along with its associated discrete spectrum. This is the same function as was investigated previously in Example 19.2, with the exception that it is shifted vertically.

In Fig. 19.10b, a doubling of the pulse train's period has two effects on the spectrum. First, two additional frequency lines are added on either side of the original components. Second, the amplitudes of the components are reduced.

As the period is allowed to approach infinity, these effects continue as more and more spectral lines are packed together until the spacing between lines goes to zero. At the limit, the series converges on the continuous Fourier integral, depicted in Fig. 19.10c.

Now that we have introduced a way to analyze an aperiodic signal, we will take the final step in our development. In the next section, we will acknowledge the fact that a signal is rarely characterized as a continuous function of the sort needed to implement Eq. (19.26). Rather, the data is invariably in a discrete form. Thus, we will now show how to compute a Fourier transform for such discrete measurements.

19.5   DISCRETE FOURIER TRANSFORM (DFT)

In engineering, functions are often represented by finite sets of discrete values. Additionally, data is often collected in or converted to such a discrete format. As depicted in Fig. 19.11, an interval from 0 to t can be divided into N equispaced subintervals with widths of $\Delta t = T/N$. The subscript n is employed to designate the discrete times at which samples are taken. Thus, f_n designates a value of the continuous function $f(t)$ taken at t_n.

Note that the data points are specified at $n = 0, 1, 2, \ldots, N - 1$. A value is not included at $n = N$. (See Ramirez, 1985, for the rationale for excluding f_N.)

FIGURE 19.11
The sampling points of the discrete Fourier series.

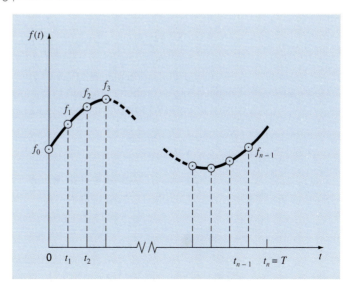

For the system in Fig. 19.11, a discrete Fourier transform can be written as

$$F_k = \sum_{n=0}^{N-1} f_n e^{-ik\omega_0 n} \qquad \text{for } k = 0 \text{ to } N - 1 \tag{19.27}$$

and the inverse Fourier transform as

$$f_n = \frac{1}{N} \sum_{k=0}^{N-1} F_k e^{ik\omega_0 n} \qquad \text{for } n = 0 \text{ to } N - 1 \tag{19.28}$$

where $\omega_0 = 2\pi/N$.

Equations (19.27) and (19.28) represent the discrete analogs of Eqs. (19.26) and (19.25), respectively. As such, they can be employed to compute both a direct and an inverse Fourier transform for discrete data. Although such calculations can be performed by hand, they are extremely arduous. As expressed by Eq. (19.27), the DFT requires N^2 complex operations. Thus, we will now develop a computer algorithm to implement the DFT.

Computer Algorithm for the DFT. Note that the factor $1/N$ in Eq. (19.28) is merely a scale factor that can be included in either Eq. (19.27) or (19.28), but not both. For our computer algorithm, we will shift it to Eq. (19.27) so that the first coefficient F_0 (which is the analog of the continuous coefficient a_0) is equal to the arithmetic mean of the samples. Also, to develop an algorithm that can be implemented in languages that do not accommodate complex variables, we can use Euler's identity,

$$e^{\pm ia} = \cos a \pm i \sin a$$

to re-express Eqs. (19.27) and (19.28) as

$$F_k = \frac{1}{N} \sum_{n=0}^{N} \left[f_n \cos (k\omega_0 n) - i f_n \sin (k\omega_0 n) \right] \tag{19.29}$$

and

$$f_n = \sum_{k=0}^{N-1} \left[F_k \cos (k\omega_0 n) + i F_k \sin (k\omega_0 n) \right] \tag{19.30}$$

Pseudocode to implement Eq. (19.29) is listed in Fig. 19.12. This algorithm can be developed into a computer program to compute the DFT. The output from such a program is listed in Fig. 19.13 for the analysis of a cosine function.

FIGURE 19.12
Pseudocode for computing the DFT.

```
DO k = 0, N − 1
    DO n = 0, N − 1
        angle = kω₀n
        realₖ = realₖ + fₙ cos(angle)/N
        imaginaryₖ = imaginaryₖ − fₙ sin(angle)/N
    END DO
END DO
```

INDEX	f(t)	REAL	IMAGINARY
0	1.000	0.0000	0.0000
1	0.707	0.0000	0.0000
2	-0.000	0.0000	-0.0000
3	-0.707	0.0000	-0.0000
4	-1.000	0.5000	0.0000
5	-0.707	0.0000	-0.0000
6	-0.000	0.0000	0.0000
7	0.707	0.0000	0.0000
8	1.000	-0.0000	0.0000
9	0.707	0.0000	0.0000
10	-0.000	-0.0000	-0.0000
11	-0.707	0.0000	0.0000
12	-1.000	-0.0000	0.0000
13	-0.707	0.0000	-0.0000
14	0.000	0.0000	0.0000
15	0.707	0.0000	-0.0000
16	1.000	-0.0000	0.0000
17	0.707	0.0000	0.0000
18	-0.000	-0.0000	0.0000
19	-0.707	-0.0000	-0.0000
20	-1.000	-0.0000	0.0000
21	-0.707	0.0000	0.0000
22	0.000	0.0000	0.0000
23	0.707	-0.0000	-0.0000
24	1.000	-0.0000	0.0000
25	0.707	-0.0000	0.0000
26	-0.000	-0.0000	0.0000
27	-0.707	-0.0000	-0.0000
28	-1.000	0.5000	-0.0000
29	-0.707	0.0000	-0.0000
30	0.000	0.0000	0.0000
31	0.707	0.0000	0.0000

FIGURE 19.13
Output of a program based on the algorithm from Fig. 19.12 for the DFT of data generated by a cosine function $f(t) = \cos [2\pi(12.5)t]$ at 32 points with $\Delta t = 0.01$ s.

19.6 FAST FOURIER TRANSFORM (FFT)

Although the algorithm described in the previous section adequately calculates the DFT, it is computationally burdensome because N^2 operations are required. Consequently, for data samples of even moderate size, the direct determination of the DFT can be extremely time-consuming.

The *fast Fourier transform*, or *FFT*, is an algorithm that has been developed to compute the DFT in an extremely economical fashion. Its speed stems from the fact that it utilizes the

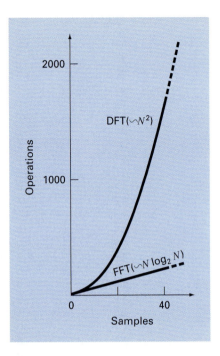

FIGURE 19.14
Plot of number of operations vs. sample size for the standard DFT and the FFT.

results of previous computations to reduce the number of operations. In particular, it exploits the periodicity and symmetry of trigonometric functions to compute the transform with approximately $N \log_2 N$ operations (Fig. 19.14). Thus, for $N = 50$ samples, the FFT is about 10 times faster than the standard DFT. For $N = 1000$, it is about 100 times faster.

The first FFT algorithm was developed by Gauss in the early nineteenth century (Heideman et al., 1984). Other major contributions were made by Runge, Danielson, Lanczos, and others in the early twentieth century. However, because discrete transforms often took days to weeks to calculate by hand, they did not attract broad interest prior to the development of the modern digital computer.

In 1965, J. W. Cooley and J. W. Tukey published a key paper in which they outlined an algorithm for calculating the FFT. This scheme, which is similar to those of Gauss and other earlier investigators, is called the *Cooley-Tukey algorithm.* Today, there are a host of other approaches that are offshoots of this method.

The basic idea behind each of these algorithms is that a DFT of length N is decomposed, or "decimated," into successively smaller DFTs. There are a variety of different ways to implement this principle. For example, the Cooley-Tukey algorithm is a member of what are called *decimation-in-time* techniques. In the present section, we will describe an alternative approach called the *Sande-Tukey algorithm.* This method is a member of another class of algorithms called *decimation-in-frequency* techniques. The distinction between the two classes will be discussed after we have elaborated on the method.

19.6.1 Sande-Tukey Algorithm

In the present case, N will be assumed to be an integral power of 2,

$$N = 2^M \tag{19.31}$$

where M is an integer. This constraint is introduced to simplify the resulting algorithm. Now, recall that the DFT can be generally represented as

$$F_k = \sum_{n=0}^{N-1} f_n e^{-i(2\pi/N)nk} \qquad \text{for } k = 0 \text{ to } N - 1 \tag{19.32}$$

where $2\pi/N = \omega_0$. Equation (19.32) can also be expressed as

$$F_k = \sum_{n=0}^{N-1} f_n W^{nk}$$

where W is a complex-valued weighting function defined as

$$W = e^{-i(2\pi/N)} \tag{19.33}$$

Suppose now that we divide the sample in half and express Eq. (19.32) in terms of the first and last $N/2$ points:

$$F_k = \sum_{n=0}^{(N/2)-1} f_n e^{-i(2\pi/N)kn} + \sum_{n=N/2}^{N-1} f_n e^{-i(2\pi/N)kn}$$

where $k = 0, 1, 2, \ldots, N - 1$. A new variable, $m = n - N/2$, can be created so that the range of the second summation is consistent with the first,

$$F_k = \sum_{n=0}^{(N/2)-1} f_n e^{-i(2\pi/N)kn} + \sum_{m=0}^{(N/2)-1} f_{m+N/2} e^{-i(2\pi/N)k(m+N/2)}$$

or

$$F_k = \sum_{n=0}^{(N/2)-1} \left(f_n + e^{-i\pi k} f_{n+N/2} \right) e^{-i2\pi kn/N} \tag{19.34}$$

Next, recognize that the factor $e^{-i\pi k} = (-1)^k$. Thus, for even points it is equal to 1 and for odd points it is equal to -1. Therefore, the next step in the method is to separate Eq. (19.34) according to even values and odd values of k. For the even values,

$$F_{2k} = \sum_{n=0}^{(N/2)-1} \left(f_n + f_{n+N/2} \right) e^{-i2\pi(2k)n/N} = \sum_{n=0}^{(N/2)-1} \left(f_n + f_{n+N/2} \right) e^{-i2\pi kn/(N/2)}$$

and for the odd values,

$$F_{2k+1} = \sum_{n=0}^{(N/2)-1} \left(f_n - f_{n+N/2} \right) e^{-i2\pi(2k+1)n/N}$$

$$= \sum_{n=0}^{(N/2)-1} \left(f_n - f_{n+N/2} \right) e^{-i2\pi n/N} e^{-i2\pi kn/(N/2)}$$

for $k = 0, 1, 2, \ldots, (N/2) - 1$.

These equations can also be expressed in terms of Eq. (19.33). For the even values,

$$F_{2k} = \sum_{n=0}^{(N/2)-1} \left(f_n + f_{n+N/2} \right) W^{2kn}$$

and for the odd values,

$$F_{2k+1} = \sum_{n=0}^{(N/2)-1} \left(f_n - f_{n+N/2} \right) W^n W^{2kn}$$

Now, a key insight can be made. These even and odd expressions can be interpreted as being equal to the transforms of the $(N/2)$-length sequences

$$g_n = f_n + f_{n+N/2} \tag{19.35}$$

and

$$h_n = (f_n - f_{n+N/2})W^n \qquad \text{for } n = 0, 1, 2, \ldots, (N/2) - 1 \tag{19.36}$$

Thus, it directly follows that

$$\left. \begin{array}{r} F_{2k} = G_k \\ F_{2k+1} = H_k \end{array} \right\} \qquad \text{for } k = 0, 1, 2, \ldots, (N/2) - 1$$

In other words, one N-point computation has been replaced by two $(N/2)$-point computations. Because each of the latter requires approximately $(N/2)^2$ complex multiplications and additions, the approach produces a factor-of-2 savings—that is, N^2 versus $2(N/2)^2 = N^2/2$.

The scheme is depicted in Fig. 19.15 for $N = 8$. The DFT is computed by first forming the sequence g^n and h^n and then computing the $N/2$ DFTs to obtain the even- and odd-numbered transforms. The weights W^n are sometimes called *twiddle factors*.

Now it is clear that this "divide-and-conquer" approach can be repeated at the second stage. Thus, we can compute the $(N/4)$-point DFTs of the four $N/4$ sequences composed of the first and last $N/4$ points of Eqs. (19.35) and (19.36).

The strategy is continued to its inevitable conclusion when $N/2$ two-point DFTs are computed (Fig. 19.16). The total number of calculations for the entire computation is on the order of $N \log_2 N$. The contrast between this level of effort and that of the standard DFT (Fig. 19.14) illustrates why the FFT is so important.

Computer Algorithm.   It is a relatively straightforward proposition to express Fig. 19.16 as an algorithm. As was the case for the DFT algorithm of Fig. 19.12, we will use Euler's identity,

$$e^{\pm ia} = \cos a \pm i \sin a$$

to allow the algorithm to be implemented in languages that do not explicitly accommodate complex variables.

Close inspection of Fig. 19.16 indicates that its fundamental computational molecule is the so-called *butterfly network* depicted in Fig. 19.17a. Pseudocode to implement one of these molecules is shown in Fig. 19.17b.

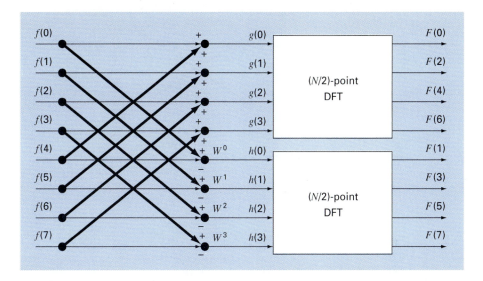

FIGURE 19.15
Flow graph of the first stage in a decimation-in-frequency decomposition of an *N*-point DFT into two (*N*/2)-point DFTs for *N* = 8.

FIGURE 19.16
Flow graph of the complete decimation-in-frequency decomposition of an eight-point DFT.

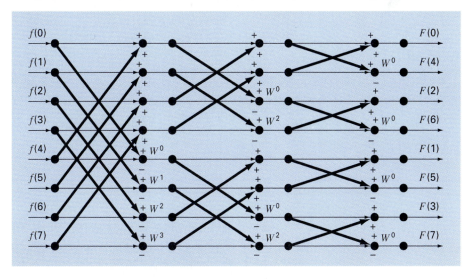

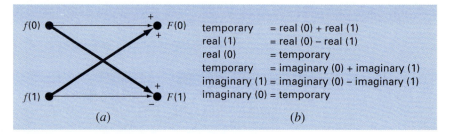

FIGURE 19.17
(a) A butterfly network that represents the fundamental computation of Fig. 19.16. (b) Pseudocode to implement (a).

FIGURE 19.18
Pseudocode to implement a decimation-in-frequency FFT. Note that the pseudocode is composed of two parts: (a) the FFT itself and (b) a bit-reversal routine to unscramble the order of the resulting Fourier coefficients.

(a)
```
m = LOG(N) / LOG(2)
N2 = N
DO k = 1, m
    N1 = N2
    N2 = N2/2
    angle = 0
    arg = 2π / N1
    DO j = 0, N2 − 1
        c = cos(angle)
        s = −sin(angle)
        DO i = j, N − 1, N1
            kk = i + N2
            xt = x(i) − x(kk)
            x(i) = x(i) + x(kk)
            yₜ = y(i) − y(kk)
            y(i) = y(i) + y(kk)
            x(kk) = xt * c − yt * s
            y(kk) = yt * c + xt * s
        END DO
        angle = (j + 1) * arg
    END DO
END DO
```

(b)
```
j = 0
DO i = 0, N-2
    IF (i < j) THEN
        xt = xⱼ
        xⱼ = xᵢ
        xᵢ = xt
        yt = yⱼ
        yⱼ = yᵢ
        yᵢ = yt
    END IF
    k = N / 2
    DO
        IF (k ≥ j + 1) EXIT
        j = j − k
        k = k / 2
    END DO
    j = j + k
END DO
DO i = 0, N − 1
    x(i) = x(i) / N
    y(i) = y(i) / N
END DO
```

Pseudocode for the FFT is listed in Fig. 19.18. The first part consists essentially of three nested loops to implement the computation embodied in Fig. 19.16. Note that the real-valued data is originally stored in the array x. Also note that the outer loop steps through the M stages [recall Eq. (19.31)] of the flow graph.

After this first part is executed, the DFT will have been computed but in a scrambled order (see the right-hand side of Fig. 19.16). These Fourier coefficients can be unscrambled

Scrambled Order (Decimal)		Scrambled Order (Binary)		Bit-Reversed Order (Binary)		Final Result (Decimal)
F(0)		F(000)		F(000)		F(0)
F(4)		F(100)		F(001)		F(1)
F(2)		F(010)		F(010)		F(2)
F(6)	$\Rightarrow$	F(110)	$\Rightarrow$	F(011)	$\Rightarrow$	F(3)
F(1)		F(001)		F(100)		F(4)
F(5)		F(101)		F(101)		F(5)
F(3)		F(011)		F(110)		F(6)
F(7)		F(111)		F(111)		F(7)

FIGURE 19.19
Depiction of the bit-reversal process.

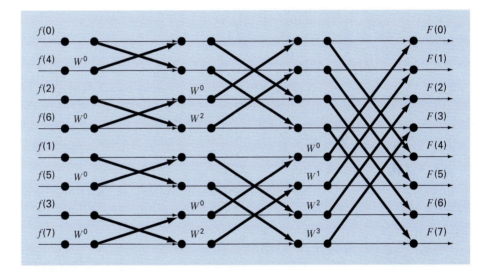

FIGURE 19.20
Flow graph of a decimation-in-time FFT of an eight-point DFT.

by a procedure called *bit reversal*. If the subscripts 0 through 7 are expressed in binary, the correct ordering can be obtained by reversing these bits (Fig. 19.19). The second part of the algorithm implements this procedure.

19.6.2 Cooley-Tukey Algorithm

Figure 19.20 shows a flow network to implement the Cooley-Tukey algorithm. For this case, the sample is initially divided into odd- and even-numbered points, and the final results are in correct order.

This approach is called a *decimation in time.* It is the reverse of the Sande-Tukey algorithm described in the previous section. Although the two classes of method differ in organization, they both exhibit the $N \log_2 N$ operations that are the strength of the FFT approach.

19.7 THE POWER SPECTRUM

The FFT has many engineering applications, ranging from vibration analysis of structures and mechanisms to signal processing. As described previously, amplitude and phase spectra provide a means to discern the underlying structure of seemingly random signals. Similarly, a useful analysis called a power spectrum can be developed from the Fourier transform.

As the name implies, the power spectrum derives from the analysis of the power output of electrical systems. In mathematical terms, the power of a periodic signal in the time domain can be defined as

$$P = \frac{1}{T} \int_{-T/2}^{T/2} f^2(t)\, dt \tag{19.37}$$

Now another way to look at this information is to express it in the frequency domain by calculating the power associated with each frequency component. This information can be then displayed as a *power spectrum,* a plot of the power versus frequency.

If the Fourier series for $f(t)$ is

$$f(t) = \sum_{k=-\infty}^{\infty} F_n e^{ik\omega_0 t} \tag{19.38}$$

the following relation holds (see Gabel and Roberts 1987) for details):

$$\frac{1}{T} \int_{-T/2}^{T/2} f^2(t)\, dt = \sum_{k=-\infty}^{\infty} |F_k|^2 \tag{19.39}$$

Thus, the power in $f(t)$ can be determined by adding together the squares of the Fourier coefficients; that is, the powers associated with the individual frequency components.

Now, remember that in this representation, the single real harmonic consists of both frequency components at $\pm k\omega_0$. We also know that the positive and negative coefficients are equal. Therefore, the power in $f_k(t)$, the kth real harmonic of $f(t)$, is

$$p_k = 2|F_k|^2 \tag{19.40}$$

The power spectrum is the plot of p_k as a function of frequency $k\omega_0$. We will devote Sec. 20.3 to an engineering application involving the FFT and the power spectrum generated with software packages.

Additional Information.   The foregoing has been a brief introduction to Fourier approximation and the FFT. Additional information on the former can be found in Van Valkenburg

(1974), Chirlian (1969), and Hayt and Kemmerly (1986). References on the FFT include Davis and Rabinowitz (1975); Cooley, Lewis, and Welch (1977); and Brigham (1974). Nice introductions to both can be found in Ramirez (1985), Oppenheim and Schafer (1975), and Gabel and Roberts (1987).

19.8 CURVE FITTING WITH LIBRARIES AND PACKAGES

Software libraries and packages have great capabilities for curve fitting. In this section, we will give you a taste of some of the more useful ones.

19.8.1 Excel

In the present context, the most useful application of Excel is for regression analysis and, to a lesser extent, polynomial interpolation. Aside from a few built-in functions (see Table 19.1), there are two primary ways in which this capability can be implemented: the Trendline command and the Data Analysis Toolpack.

The Trendline Command (Insert Menu). This command allows a number of different trend models to be added to a chart. These models include linear, polynomial, logarithmic, exponential, power, and moving average fits. The following example illustrates how the Trendline command is invoked.

TABLE 19.1 Excel built-in functions related to regression fits of data.

Function	Description
FORECAST	Returns a value along a linear trend
GROWTH	Returns values along an exponential trend
INTERCEPT	Returns the intercept of the linear regression line
LINEST	Returns the parameters of a linear trend
LOGEST	Returns the parameters of an exponential trend
SLOPE	Returns the slope of the linear regression line
TREND	Returns values along a linear trend

EXAMPLE 19.3   Using Excel's Trendline Command

Problem Statement. You may have noticed that several of the fits available on Trendline were discussed previously in Chap. 17 (e.g., linear, polynomial, exponential, and power). An additional capability is the logarithmic model

$$y = a_0 + a_1 \log x$$

Fit the following data with this model using Excel's Trendline command:

x	0.5	1	1.5	2	2.5	3	3.5	4	4.5	5	5.5
y	0.53	0.69	1.5	1.5	2	2.06	2.28	2.23	2.73	2.42	2.79

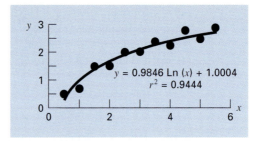

FIGURE 19.21
Fit of a logarithmic model to the data from Example 19.3.

Solution.   To invoke the Trendline command, a chart relating a series of dependent and independent variables must be created. For the present case, we use the Excel Chart Wizard to create an XY-plot of the data.

Next, we can select the chart (by double clicking on it) and the series (by positioning the mouse arrow on one of the values and single clicking). The Insert and Trendline commands are then invoked with the mouse or by the key sequence

/ **I**nsert T**r**endline

At this point, a dialogue box opens with two tabs: Options tab and the Type tab. The Options tab provides ways to customize the fit. The most important in the present context is to display both the equation and the value for the coefficient of determination (r^2) on the chart. The primary choice on the Type tab is to specify the type of trendline. For the present case, select **Logarithmic.** The resulting fit along with r^2 is displayed in Fig. 19.21.

The Trendline command provides a handy way to fit a number of commonly used models to data. In addition, its inclusion of the **Polynomial** option means that it can also be used for polynomial interpolation. However, the fact that its statistical content is limited to r^2 means that it does not allow statistical inferences to be drawn regarding the model fit. The Data Analysis Toolpack described next provides a nice alternative where such inferences are necessary.

The Data Analysis Toolpack.   This Excel Add-in Package contains a comprehensive capability for curve fitting with general linear least squares. As previously described in Sec. 17.4, such models are of the general form

$$y = a_0 z_0 + a_1 z_1 + a_2 z_2 + \cdots + a_m z_m + e \qquad (17.23)$$

where $z_0, z_1, \ldots, z_m$ are $m + 1$ different functions. The next example illustrates how such models can be fit with Excel.

EXAMPLE 19.4   Using Excel's Data Analysis Toolpack

Problem Statement.   The following data was collected for the slope, hydraulic radius, and velocity of water flowing in a canal:

S, m/m	0.0002	0.0002	0.0005	0.0005	0.001	0.001
R, m	0.2	0.5	0.2	0.5	0.2	0.5
U, m/s	0.25	0.5	0.4	0.75	0.5	1

There are theoretical reasons (recall Sec. 8.2) for believing that this data can be fit to a power model of the form

$$U = \alpha S^\sigma R^\rho$$

where α, σ, and ρ are empirically derived coefficients. There are theoretical reasons (again, see Sec. 8.2) for believing that σ and ρ should have values of approximately 0.5 and 0.667, respectively. Fit this data with Excel and evaluate whether your regression estimates contradict the expected values for the model coefficients.

Solution.   The logarithm of the power model is first used to convert it to the linear format of Eq. (17.23),

$$U = \log \alpha + \sigma \log S + \rho \log R$$

An Excel spreadsheet can be developed with both the original data along with their common logarithms, as in the following:

	A	B	C	D	E	F
	S	Rh	U	log(S)	log(Rh)	log(U)
1						
2	0.0002	0.2	0.25	−3.69897	−0.69897	−0.60206
3	0.0002	0.5	0.5	−3.69897	−0.30103	−0.30103
4	0.0005	0.2	0.4	−3.30103	−0.69897	−0.39794
5	0.0005	0.5	0.75	−3.30103	−0.30103	−0.12494
6	0.001	0.2	0.5	−3	−0.69897	−0.30103
7	0.001	0.5	1	−3	−0.30103	0

= log(A2)

As shown, an efficient way to generate the logarithms is to type the formula to compute the first log(S). This formula can then be copied to the right and down to generate the other logarithms.

Because of its status as an "Add-In" on the version of Excel available at the time of this book's printing, the Data Analysis Toolpack must sometimes be loaded into Excel. To do this, merely use the mouse or the key sequence

/**T**ools Add-I**n**s

Then select **Analysis Toolpack** and OK. If the add-in is successful, the selection Data Analysis will be added to the Tools menu.

After selecting **Data Analysis** from the Tools menu, a Data Analysis menu will appear on the screen containing a large number of statistically oriented routines. Select **Regression** and a dialogue box will appear, prompting you for information on the regression. After making sure that the default selection **New Worksheet Ply** is selected, fill in F2:F7 for the y range and D2:E7 for the x range, and select OK. The following worksheet will be created:

	A	B	C	D	E	F	G
1	SUMMARY OUTPUT						
2							
3	*Regression Statistics*						
4	Multiple R	0.998353					
5	R Square	0.996708					
6	Adjusted R Square	0.994513					
7	Standard Error	0.015559					
8	Observations	6					
9							
10	ANOVA						
11		*df*	*SS*	*MS*	*F*	*Significance F*	
12	Regression	2	0.219867	0.109933	454.1106	0.0001889	
13	Residual	3	0.000726	0.000242			
14	Total	5	0.220593				
15							
16		*Coeffs*	*Std. Error*	*t Stat*	*P-value*	*Lower 95%*	*Upper 95%*
17	Intercept	**1.522452**	0.075932	20.0501	0.0002712	1.2808009	1.7641028
18	X Variable 1	**0.433137**	0.022189	19.5203	0.0002937	**0.3625211**	**0.5037521**
19	X Variable 2	**0.732993**	0.031924	22.96038	0.000181	**0.6313953**	**0.8345899**

Thus, the resulting fit is

$$\log U = 1.522 + 0.433 \log S + 0.733 \log R$$

or by taking antilog,

$$U = 33.3 S^{0.433} R^{0.733}$$

Notice that 95% confidence intervals are generated for the coefficients. Thus, there is a 95% probability that the true slope exponent falls between 0.363 and 0.504, and the true hydraulic radius coefficient falls between 0.631 and 0.835. Thus, the fit does not contradict the theoretical exponents.

Finally, it should be noted that the Excel Solver tool can be used to perform *nonlinear regression* by directly minimizing the sum of the squares of the residuals between a nonlinear model prediction and data. We devote Sec. 20.1 to an example of how this can be done.

19.8.2 Mathcad

Mathcad can perform a wide variety of statistical, curve fitting, and data-smoothing tasks. These include relatively simple jobs like plotting histograms and calculating population statistic summaries such as mean, median, variance, standard deviations, and correlation coefficients. In addition, Mathcad can predict intermediate values by connecting known data points with either straight lines (linear interpolation) using **linterp** or with sections of cubic polynomials (cubic spline interpolation) using **cspline, pspline,** or **lspline.** These spline functions allow you to try different ways to deal with interpolation at the end points of the data. The **lspline** function generates a spline curve that is a straight line at the end points. The **pspline** function generates a spline curve that is a parabola at the end points. The **cspline** function generates a spline curve that is cubic at the end points. The **interp** function uses the curve fitting results and returns an interpolated y value given an x value. In addition, you can perform two-dimensional cubic spline interpolation by passing a surface through a grid of points.

Mathcad contains a number of functions for performing regression. The **slope** and **intercept** functions return the slope and intercept of the least-squares regression fit line. The **regress** function is used for nth-order polynomial regression of a complete data set. The **loess** function performs localized nth-order polynomial regression over spans of the data that you can specify. The **interp** function can also be used to return intermediate values of y from a regression fit for a given x point. The **regress** and **loess** functions can also perform multivariate polynomial regression. Mathcad also provides the **linfit** function that is used to model data with a linear combination of arbitrary functions. Finally, the **genfit** function is available for cases where model coefficients appear in arbitrary form. In this case, the more difficult nonlinear equations must be solved by iteration.

Let's do an example that shows how Mathcad is used to perform two-dimensional spline interpolation (Fig. 19.22). The data we will fit is

				x			
	z	0	1	2	3	4	5
	0	0.17500	0.14100	−0.13900	−0.51400	−0.29000	0.32700
	1	0.93500	0.16700	−0.76400	−0.98600	−0.30800	0.82600
y	2	0.64900	−0.00302	−0.33400	−0.65900	−0.00678	0.23900
	3	−0.55300	−0.22500	0.46700	0.73600	0.10600	−0.09200
	4	−0.97900	0.17500	0.36800	0.81400	0.39000	−0.78200
	5	−0.70700	0.12600	0.76100	0.30200	0.30300	−0.16400

FIGURE 19.22
2D spline with Mathcad.

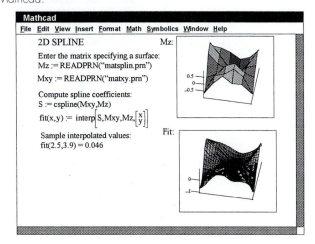

Note that the numbers along the top and left side are the x and y coordinates of the z values in the interior of the matrix.

The first step is to supply the data to Mathcad. To do this, we can create two data files called **matsplin.prn** and **matxy.prn.** The first two active lines in Fig. 19.22 use the **READ-PRN** command to read data from these files. The **matsplin.prn** file is a simple text file that contains the values of the function (z) to be interpolated at various x and y locations on a rectangular grid. The dimensions of the grid are defined by the data in the **matxy.prn** text file. The elements of this file are pairs of x and y values that characterize the diagonal elements of the region. The definition symbol is used to assign the data from the data files to the variables Mz and Mxy. Next, the definition symbol and the **csplin** function are used to define the S matrix. This is a matrix that contains values of the second derivative and other numerical results at the various grid locations. This matrix, along with Mz and Mxy, are used by the **interp** function to return values of z as the variable fit(x,y) based on the cubic spline interpolation at input values of x and y. Mathcad designed this sequence of operations in this manner so that the interpolating polynomials would not have to be recalculated every time interpolation is required at different values of x and y. With these operations in place, you can interpolate at any location using fit(x,y), as shown with x = 2.5 and y = 3.9. You can also construct a plot of the interpolated surface as shown in Fig. 19.22.

As another example of demonstrating some of Mathcad's curve fitting capabilities, let's use the **fft** function for Fourier analysis as in Fig. 19.23. The first line uses the definition symbol to create i as a range variable. Next x_i is formulated using the **rnd** Mathcad function to impart a random component to a sinusoidal signal. The graph of the signal can be placed on the worksheet by clicking to the desired location. This places a red cross hair at that location. Then use the Insert/Graph/X-Y Plot pull down menu to place an empty plot on the worksheet with placeholders for the expressions to be graphed and for the ranges of the x and y axes. Simply type x_i in the placeholder on the y axis and 0 and 80 for

FIGURE 19.23
FFT with Mathcad.

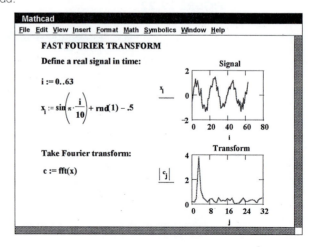

the x-axis range. Mathcad does all the rest to produce the graph shown in Fig. 19.23. Once the graph has been created you can use the Format/Graph/X-Y Plot pull down menu to vary the type of graph; change the color, type, and weight of the trace of the function; and add titles, labels, and other features. Next, c is defined as **fft**(x). This function returns the Fourier transform of x. The result is a vector, c, of complex coefficients that represent values in the frequency domain. A plot of the magnitude of c_j is constructed as above.

19.8.3 MATLAB

As summarized in Table 19.2, MATLAB has a variety of built-in functions that span the total capabilities described in this part of the book. The following example illustrates how a few of them can be used.

TABLE 19.2 Some of the MATLAB functions to implement interpolation, regression, splines, and the FFT.

Function	Description
polyfit	Fit polynomial to data
interp1	1-D interpolation (1-D table lookup)
interp2	2-D interpolation (2-D table lookup)
spline	Cubic spline data interpolation
fft	Discrete Fourier transform

EXAMPLE 19.5   Using MATLAB for Curve Fitting

Problem Statement.   Explore how MATLAB can be employed to fit curves to data. To do this, use the sine function to generate equally spaced $f(x)$ values from 0 to 10. Employ a step size of 1 so that the resulting characterization of the sine wave is sparse (Fig. 19.24). Then, fit it with (*a*) linear interpolation, (*b*) a fifth-order polynomial, and (*c*) a cubic spline.

Solution.

(a) The values for the independent and the dependent variables can be entered into vectors by

```
>> x=0:10;
>> y=sin(x);
```

A new, more finely spaced vector of independent variable values can be generated and stored in the vector **xi,**

```
>> xi=0:.25:10;
```

The MATLAB function **interp1** can then be used to generate dependent variable values yi for all the xi values using linear interpolation. Both the original data (x, y) along with

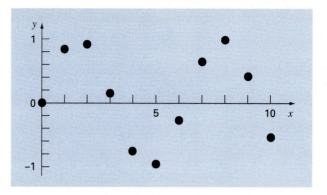

FIGURE 19.24
Eleven points sampled from a sinusoid.

the linearly interpolated values can be plotted together, as shown in the graph below:

```
>> yi=interp1(x,y,xi);
>> plot(x,y,'o',xi,yi)
```

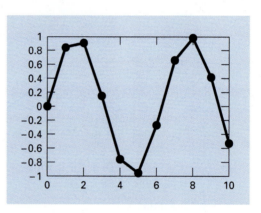

(b) Next, the MATLAB **polyfit** function can be used to generate the coefficients of a fifth-order polynomial fit of the original sparse data,

```
>> p=polyfit(x,y,5)
p =
     0.0008  -0.0290   0.3542  -1.6854   2.5860  -0.0915
```

where the vector **p** holds the polynomial's coefficients. These can, in turn, be used to generate a new set of yi values, which can again be plotted along with the original sparse sample,

```
>> yi = polyval(p,xi);
>> plot(x,y,'o',xi,yi)
```

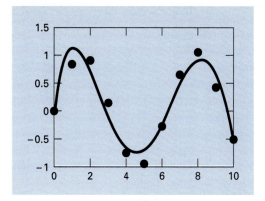

Thus, the polynomial captures the general pattern of the data, but misses most of the points.

(c) Finally, the MATLAB **spline** function can be used to fit a cubic spline to the original sparse data in the form of a new set of yi values, which can again be plotted along with the original sparse sample,

```
>> yi=spline(x,y,xi);
>> plot(x,y,'o',xi,yi)
```

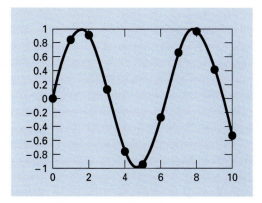

It should be noted that MATLAB also has excellent capabilities to perform Fourier analysis. We devote Sec. 20.3 to an example of how this can be done.

19.8.4 IMSL

IMSL has numerous routines for curve fitting that span all the capabilities covered in this book, and then some. A sample is presented in Table 19.3. In the present discussion, we will focus on the RCURV routine. This routine fits a least-squares polynomial to data.

RCURV is implemented by the following CALL statement:

```
CALL RCURV (NOBS, XDATA, YDATA, NDEG, B, SSPOLY, STAT)
```

TABLE 19.3 IMSL routines for curve fitting.

Category	Routines	Description
• Cubic spline interpolation	CSIEZ	Easy to use cubic spline routine
	CSINT	Not-a-knot
	CSDEC	Derivative end conditions
• Cubic spline evaluation and integration	CSVAL	Evaluation
	CSDER	Evaluation of the derivative
	CS1GD	Evaluation on a grid
	CSITG	Integration
• B-spline interpolation		
• Piecewise polynomial		
• Quadratic polynomial interpolation routines for gridded data		
• Scattered data interpolation		
• Least-squares approximation	RLINE	Linear polynomial
	RCURV	General polynomial
	FNLSQ	General functions
• Cubic spline smoothing		
• Rational weighted Chebyshev approximation		Rational weighted Chebyshev approximation
• Real trigonometric FFT	FFTRF	Forward transform
	FFTRB	Backward or inverse transform
	FFTRI	Initialization routine for FFTR
• Complex exponential FFT	FFTCF	Forward transform
	FFTCB	Backward or inverse transform
	FFTCI	Initialization routine for FFTC
• Real sine and cosine FFTs		
• Real quarter sine and quarter cosine FFTs		
• Two- and three-dimensional complex FFTs		
• Convolutions and correlations		
• Laplace transform		

where   NOBS = Number of observations. (Input)

XDATA = Vector of length NOBS containing the x values. (Input)

YDATA = Vector of length NOBS containing the y values. (Input)

NDEG = Degree of polynomial. (Input)

B = Vector of length NDEG + 1 containing the coefficients.

SSPOLY = Vector of length NDEG + 1 containing the sequential sums of squares. (Output) SSPOLY(1) contains the sum of squares due to the mean. For $i = 1, 2, \ldots$, NDEG, SSPOLY($i + 1$) contains the sum of squares due to x^i adjusted for the mean, x, $x^2, \ldots$, and x^{i-1}.

STAT = Vector of length 10 containing statistics described in Table 19.4. (Output) where 1 = Mean of x

2 = Mean of y

3 = Sample variance of x

4 = Sample variance of y

5 = R-squared (in percent)

6 = Degrees of freedom for regression

7 = Regression sum of squares

8 = Degrees of freedom for error

9 = Error sum of squares

10 = Number of data points (x, y) containing NaN (not a number) as an x or y value.

EXAMPLE 19.6 Using IMSL for Polynomial Regression

Problem Statement. Use RCURV to determine the cubic polynomial that provides a least-squares fit of the following data:

x	0.05	0.12	0.15	0.30	0.45	0.70	0.84	1.05
y	0.957	0.851	0.832	0.720	0.583	0.378	0.295	0.156

Solution. An example of a main Fortran 90 program and function using RCURV to solve this problem can be written as

```
PROGRAM Fitpoly
use msimsl
IMPLICIT NONE
INTEGER::ndeg,nobs,i,j
PARAMETER (ndeg=3, nobs=8)
REAL::b(ndeg+1),sspoly(ndeg+1),stat(10),x(nobs),y(nobs),
   ycalc(nobs)
DATA x/0.05,0.12,0.15,0.30,0.45,0.70,0.84,1.05/
DATA y/0.957,0.851,0.832,0.720,0.583,0.378,0.295,0.156/
CALL RCURV(nobs,x,y,ndeg,B,sspoly,stat)
PRINT *, 'Fitted polynomial is'
DO i = 1,ndeg+1
   PRINT '(1X, ''X^'',I1,'' TERM: '',F8.4)', i-1, b(i)
END DO
PRINT *
PRINT '(1X,''R^2: '',F5.2,''%'')',stat(5)
PRINT *
PRINT *, 'NO.   X   Y   YCALC'
DO i = 1,nobs
```

```
ycalc=0.
DO j = 1,ndeg+1
  ycalc(i)=ycalc(i)+b(j)*x(i)**(j-1)
END DO
PRINT '(1X,I8,3(5X,F8.4))', i, x(i), y(i), ycalc(i)
END DO
END
```

An example run is

```
Fitted polynomial is
X^0 TERM:    .9909
X^1 TERM:  -1.0312
X^2 TERM:    .2785
X^3 TERM:   -.0513
R^2: 99.81%
        NO.        X          Y         YCALC
         1       .0500      .9570      .9401
         2       .1200      .8510      .8711
         3       .1500      .8320      .8423
         4       .3000      .7200      .7053
         5       .4500      .5830      .5786
         6       .7000      .3780      .3880
         7       .8400      .2950      .2908
         8      1.0500      .1560      .1558
```

PROBLEMS

19.1 The pH in a reactor varies sinusoidally over the course of a day. Use least-squares regression to fit Eq. (19.11) to the following data. Use your fit to determine the mean, amplitude, and time of maximum pH.

Time, hr	0	2	4	5	7	8.5	12	15	20	22	24
pH	7.3	7	7.1	6.4	7.4	7.2	8.9	8.8	8.9	7.9	7

19.2 The solar radiation for Georgetown, South Carolina has been tabulated as

Time, mo	J	F	M	A	M	J	J	A	S	O	N	D
Radiation, W/m²	122	—	188	230	267	270	252	—	196	160	138	120

Assuming each month is 30 days long, fit a sinusoid to this data. Use the resulting equation to predict the radiation in mid-August.

19.3 The average values of a function can be determined by

$$\overline{f(x)} = \frac{\int_0^x f(x)\, dx}{x}$$

Use this relationship to verify the results of Eq. (19.13).

19.4 Use a continuous Fourier series to approximate the sawtooth wave in Fig. P19.4. Plot the first three terms along with the summation.

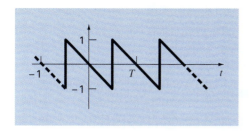

FIGURE P19.4
A sawtooth wave.

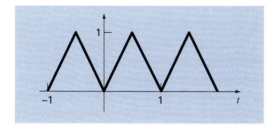

FIGURE P19.5
A triangular wave

19.5 Use a continuous Fourier series to approximate the wave form in Fig. P19.5. Plot the first three terms along with the summation.

19.6 Construct amplitude and phase line spectra for Prob. 19.4.

19.7 Construct amplitude and phase line spectra for Prob. 19.5.

19.8 A half-wave rectifier can be characterized by

$$y(t) = C_1 \left[\frac{1}{\pi} + \frac{1}{2} \sin t - \frac{2}{3\pi} \cos 2t - \frac{2}{15\pi} \cos 4t \right.$$
$$\left. - \frac{2}{35\pi} \cos 6t - \cdots \right]$$

where C_1 is the amplitude of the wave. Plot the first four terms along with the summation.

19.9 Construct amplitude and phase line spectra for Prob. 19.8.

19.10 Develop a user-friendly subprogram for the DFT based on the algorithm from Fig. 19.12. Test it by duplicating Fig. 19.13.

19.11 Use the program from Prob. 19.10 to compute a DFT for the

triangular wave from Prob. 19.8. Sample the wave from $t = 0$ to $4T$. Use 32, 64, and 128 sample points. Time each run and plot execution versus N to verify Fig. 19.14.

19.12 Develop a user-friendly subprogram for the FFT based on the algorithm from Fig. 19.18. Test it by duplicating Fig. 19.13.

19.13 Repeat Prob. 19.11 using the software you developed in Prob. 19.12.

19.14 Use Excel's Trendline command to fit a power equation to

x	2.5	3.5	5	6	7.5	10	12.5	15	17.5	20
y	7	5.5	3.9	3.6	3.1	2.8	2.6	2.4	2.3	2.3

Plot y versus x along with the power equation and r^2.

19.15 Use the Excel Data Analysis Toolpack to develop a fourth-order regression polynomial to the following data for the dissolved oxygen concentration of fresh water versus temperature at sea level.

T, °C	0	8	16	24	32	40
o, mg/L	14.621	11.843	9.870	8.418	7.305	6.413

19.16 Use the Excel Data Analysis Toolpack to fit a straight line to the following data. Determine the 90% confidence interval for the intercept. If it encompasses zero, redo the regression, but with the intercept forced to be zero (this is an option on the **Regression** dialogue box)

x	2	4	6	8	10	12	14
y	5.7	6.4	12.3	17.7	18.9	25.7	26.8

19.17 Use Mathcad to fit a cubic spline (with a straight line at the end points) to the following data:

x	0	2	4	7	10	12
y	20	20	12	7	6	5.6

Determine the value of y at $x = 3$.

19.18 Use Mathcad to generate 64 points from the function

$$f(t) = \cos(3t) + \sin(10t)$$

from $t = 0$ to 2π. As in Sec. 19.8.2 add a random component to the signal. Take an FFT of these values and plot the results.

19.19 In a fashion similar to Sec. 19.8.3, use MATLAB to fit the data from Prob. 19.17 using **(a)** linear interpolation, **(b)** a fifth-order polynomial, and **(c)** a spline.

19.20 Repeat Prob. 19.18, but use MATLAB to perform the analysis.

19.21 Repeat Prob. 19.15, but use the IMSL routine, RCURV.

CHAPTER 20

Engineering Applications: Curve Fitting

The purpose of this chapter is to use the numerical methods for curve fitting to solve some engineering problems. The first application, which is taken from chemical engineering, demonstrates how a nonlinear model can be linearized and fit to data using linear regression. The second application employs splines to study a problem that has relevance to the environmental area of civil engineering: heat and mass-transport in a stratified lake.

The third application illustrates how a fast Fourier transform can be employed in electrical engineering to analyze a signal by determining its major harmonics. The final application demonstrates how multiple linear regression is used to analyze experimental data for a fluids problem taken from mechanical and aerospace engineering.

20.1 LINEAR REGRESSION AND POPULATION MODELS (CHEMICAL/PETROLEUM ENGINEERING)

Background. Population growth models are important in many fields of engineering. Fundamental to many of the models is the assumption that the rate of change of the population (dp/dt) is proportional to the actual population (p) at any time (t), or in equation form,

$$\frac{dp}{dt} = kp \tag{20.1}$$

where k = a proportionality factor called the specific growth rate and has units of time^{-1}. If k is a constant, then the solution of Eq. (20.1) can be obtained from the theory of differential equations:

$$p(t) = p_0 e^{kt} \tag{20.2}$$

where p_0 = the population when $t = 0$. It is observed that $p(t)$ in Eq. (20.2) approaches infinity as t becomes large. This behavior is clearly impossible for real systems. Therefore, the model must be modified to make it more realistic.

Solution. First, it must be recognized that the specific growth rate k cannot be constant as the population becomes large. This is the case because, as p approaches infinity, the organism being modeled will become limited by factors such as food shortages and toxic waste production. One way to express this mathematically is to use a saturation-growth-rate model such that

$$k = k_{max} \frac{f}{K + f} \tag{20.3}$$

where k_{max} = the *maximum attainable growth rate* for large values of food (f) and K = the *half-saturation constant*. The plot of Eq. (20.3) in Fig. 20.1 shows that when $f = K$, $k = k_{max}/2$. Therefore, K is that amount of available food that supports a population growth rate equal to one-half the maximum rate.

The constants K and k_{max} are empirical values based on experimental measurements of k for various values of f. As an example, suppose the population p represents a yeast employed in the commercial production of beer and f is the concentration of the carbon source to be fermented. Measurements of k versus f for the yeast are shown in Table 20.1. It is

FIGURE 20.1

Plot of specific growth rate versus available food for the saturation-growth-rate model used to characterize microbial kinetics. The value K is called a half-saturation constant because it conforms to the concentration where the specific growth rate is half its maximum value.

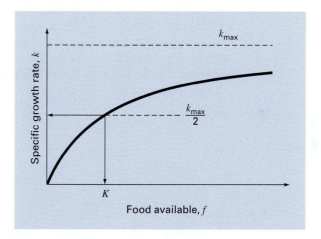

TABLE 20.1   Data used to evaluate the constants for a saturation-growth-rate model to characterize microbial kinetics.

f, mg/L	k, day^{-1}	1/f, L/mg	1/k, day
7	0.29	0.14286	3.448
9	0.37	0.11111	2.703
15	0.48	0.06666	2.083
25	0.65	0.04000	1.538
40	0.80	0.02500	1.250
75	0.97	0.01333	1.031
100	0.99	0.01000	1.010
150	1.07	0.00666	0.935

required to calculate k_{max} and K from this empirical data. This is accomplished by inverting Eq. (20.3) in a manner similar to Eq. (17.17) to yield

$$\frac{1}{k} = \frac{K + f}{k_{max} f} = \frac{K}{k_{max}} \frac{1}{f} + \frac{1}{k_{max}} \qquad (20.4)$$

By this manipulation, we have transformed Eq. (20.3) into a linear form; that is, $1/k$ is a linear function of $1/f$, with slope K/k_{max} and intercept $1/k_{max}$. These values are plotted in Fig. 20.2.

Because of this transformation, the linear least-squares procedures described in Chap. 17 can be used to determine $k_{max} = 1.23$ day^{-1} and $K = 22.18$ mg/L. These results combined with Eq. (20.3) are compared to the untransformed data in Fig. 20.3, and when substituted into the model in Eq. (20.1), give

$$\frac{dp}{dt} = 1.23 \frac{f}{22.18 + f} p \qquad (20.5)$$

Note that the fit yields a sum of the squares of the residuals (as computed for the untransformed data) of 0.001305.

Equation (20.5) can be solved using the theory of differential equations or using numerical methods discussed in Chap. 25 when $f(t)$ is known. If f approaches zero as p becomes large, then dp/dt approaches zero and the population stabilizes.

The linearization of Eq. (20.3) is one way to evaluate the constants k_{max} and K. An alternative approach, which fits the relationship in its original form, is the nonlinear regression described in Sec. 17.5. Figure 20.4 shows how the Excel Solver tool can be used to estimate the parameters with nonlinear regression. As can be seen, a column of predicted values is developed based on the model and the parameter guesses. These are used to gen-

FIGURE 20.2
Linearized version of the saturation-growth-rate model. The line is a least-squares fit that is used to evaluate the model coefficients $k_{max} = 1.23$ day^{-1} and $K = 22.18$ mg/L for a yeast that is used to produce beer.

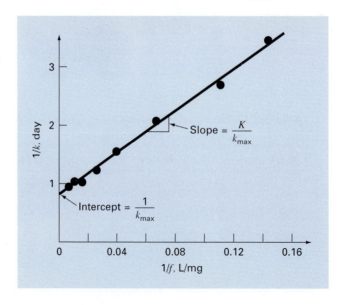

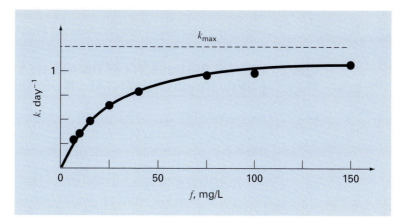

FIGURE 20.3
Fit of the saturation-growth-rate model to a yeast employed in the commercial production of beer.

	A	**B**	**C**	**D**	
1	kmax	1.2301			=$B$1*A5/($B$2 + A5)
2	K	22.1386			
3					=(B5 − C5)^2
4	f	k	k-predicted	Res^2	
5	7	0.29	**0.295508**	**0.000030**	
6	9	0.37	0.355536	0.000209	
7	15	0.48	0.496828	0.000283	
8	25	0.65	0.652384	0.000006	
9	40	0.80	0.791842	0.000067	
10	75	0.97	0.949751	0.000410	
11	100	0.99	1.007134	0.000294	
12	150	1.07	1.071897	0.000004	
13					=SUM(D5..D12)
14			SSR	**0.001302**	

FIGURE 20.4
Nonlinear regression to fit the saturation-growth-rate model to a yeast employed in the commercial production of beer.

erate a column of squared residuals that are summed, and the result is placed in cell D14. The Excel Solver is then invoked to minimize cell D14 by adjusting cells B1:B2. The result, as shown in Fig. 20.4, yields estimates of $k_{max} = 1.23$ and $K = 22.14$, with an $S_r = 0.001302$. Thus, although, as expected, the nonlinear regression yields a slightly better fit, the results are almost identical. In other applications, this may not be true (or the function may not be compatible with linearization) and nonlinear regression could represent the only feasible option for obtaining a least-squares fit.

20.2   USE OF SPLINES TO ESTIMATE HEAT TRANSFER (CIVIL/ENVIRONMENTAL ENGINEERING)

Background.   Lakes in the temperate zone can become thermally stratified during the summer. As depicted in Fig. 20.5, warm, buoyant water near the surface overlies colder, denser bottom water. Such stratification effectively divides the lake vertically into two layers: the *epilimnion* and the *hypolimnion* separated by a plane called the *thermocline*.

Thermal stratification has great significance for environmental engineers studying the pollution of such systems. In particular, the thermocline greatly diminishes mixing between the two layers. As a result, decomposition of organic matter can lead to severe depletion of oxygen in the isolated bottom waters.

The location of the thermocline can be defined as the inflection point of the temperature-depth curve—that is, the point at which $d^2T/dx^2 = 0$. It is also the point at which the absolute value of the first derivative or gradient is a maximum. Use cubic splines to determine the thermocline depth for Platte Lake (Table 20.2). Also use the splines to determine the value of the gradient at the thermocline.

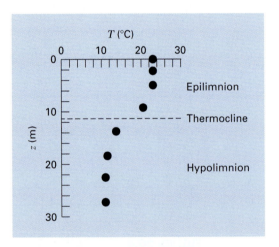

FIGURE 20.5
Temperature versus depth during summer for Platte Lake, Michigan.

TABLE 20.2   Temperature versus depth during summer for Platte Lake, Michigan.

T, °C	22.8	22.8	22.8	20.6	13.9	11.7	11.1	11.1
z, m	0	2.3	4.9	9.1	13.7	18.3	22.9	27.2

Solution. The data is analyzed with a program that was developed based on the pseudocode from Fig. 18.18. The results as displayed in Table 20.3 list the spline predictions along with first and second derivatives at intervals of 1 m down through the water column.

The results are plotted in Fig. 20.6. Notice how the thermocline is clearly located at the depth where the gradient is highest (i.e., the absolute value of the derivative is greatest) and the second derivative is zero. The depth is 11.35 m and the gradient at this point is $-1.61°C/m$.

TABLE 20.3 Output of spline program based on pseudocode from Fig. 18.18.

Depth (m)	T(C)	dT/dz	d2T/dz2	Depth (m)	T(C)	dT/dz	d2T/dz2
0.	22.8000	-.0115	.0000	15.	12.7652	-.6518	.3004
1.	22.7907	-.0050	.0130	16.	12.2483	-.3973	.2086
2.	22.7944	.0146	.0261	17.	11.9400	-.2346	.1167
3.	22.8203	.0305	-.0085	18.	11.7484	-.1638	.0248
4.	22.8374	-.0055	-.0635	19.	11.5876	-.1599	.0045
5.	22.7909	-.0966	-.1199	20.	11.4316	-.1502	.0148
6.	22.6229	-.2508	-.1884	21.	11.2905	-.1303	.0251
7.	22.2665	-.4735	-.2569	22.	11.1745	-.1001	.0354
8.	21.6531	-.7646	-.3254	23.	11.0938	-.0596	.0436
9.	20.7144	-1.1242	-.3939	24.	11.0543	-.0212	.0332
10.	19.4118	-1.4524	-.2402	25.	11.0480	.0069	.0229
11.	17.8691	-1.6034	-.0618	26.	11.0646	.0245	.0125
12.	16.2646	-1.5759	.1166	27.	11.0936	.0318	.0021
13.	14.7766	-1.3702	.2950	28.	11.1000	.0000	.0000
14.	13.5825	-.9981	.3923				

FIGURE 20.6

Plots of (a) temperature, (b) gradient, and (c) second derivative versus depth (m) generated with the cubic spline program. The thermocline is located at the inflection point of the temperature-depth curve.

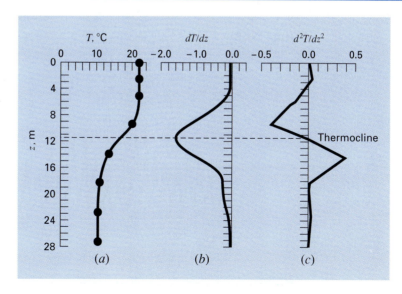

20.3   FOURIER ANALYSIS (ELECTRICAL ENGINEERING)

Background.   Fourier analysis is used in many areas of engineering. However, it is extensively employed in electrical engineering applications such as signal processing.

In 1848, Rudolph Wolf devised a method for quantifying solar activity by counting the number of individual spots and groups of spots on the sun's surface. He computed a quantity, now called a *Wolf sunspot number,* by adding 10 times the number of groups plus the total count of individual spots. As in Fig. 20.7, the record of this number extends back to 1770. On the basis of the early historical records, Wolf determined the cycle's length to be 11.1 years.

Use a Fourier analysis to confirm this result by applying an FFT to the data from Fig. 20.3. Pinpoint the period by developing a power versus period plot.

Solution.   The data for year and sunspot number was downloaded from the web[1] and stored in a tab-delimited file: sunspot.dat. The file can be loaded into MATLAB and the year and number information assigned to vectors of the same name,

```
>> load sunspot.dat
>> year=sunspot(:,1);number=sunspot(:,2);
```

Next, an FFT can be applied to the sunspot numbers

```
>> y=fft(number);
```

After getting rid of the first harmonic, the length of the FFT is determined (n) and then the power and frequency calculated,

```
>> y(1)=[ ];
>> n=length(y);
>> power=abs(y(1:n/2)).^2;
>> nyquist=1/2;
>> freq=(1:n/2)/(n/2)*nyquist;
```

FIGURE 20.7
Plot of Wolf sunspot number versus year.

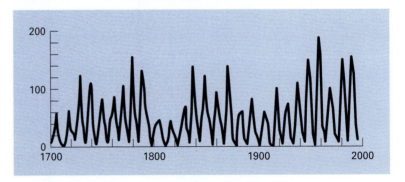

[1]At the time of this book's printing, the html was http://www.ngdc.noaa.gov//stp/SOLAR/SSN/ssn.html.

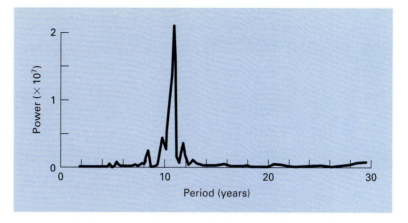

FIGURE 20.8
Power spectrum for Wolf sunspot numbers.

At this point, the power spectrum is a plot of power versus frequency. However, because period is more meaningful in the present context, we can determine the period and a power-period plot,

```
>> period=1./freq
>> plot(period,power);
```

The result, as shown in Fig. 20.8, indicates a peak at about 11 years. The exact value can be computed with

```
>> index=find(power==max(power));
>> period(index)

ans=
    10.9259
```

20.4 ANALYSIS OF EXPERIMENTAL DATA (MECHANICAL/ AEROSPACE ENGINEERING)

Background.   Engineering design variables are often dependent on several independent variables. Often this functional dependence is best characterized by multivariate power equations. As discussed in Sec. 17.3, a multiple linear regression of log-transformed data provides a means to evaluate such relationships.

For example, a mechanical engineering study indicates that fluid flow through a pipe is related to pipe diameter and slope (Table 20.4). Use multiple linear regression to analyze this data. Then use the resulting model to predict the flow for a pipe with a diameter of 2.5 ft and a slope of 0.025 ft/ft.

Solution.   The power equation to be evaluated is

$$Q = a_0 D^{a_1} S^{a_2} \tag{20.6}$$

TABLE 20.4 Experimental data for diameter, slope, and flow of concrete circular pipes.

Experiment	Diameter, ft	Slope, ft/ft	Flow, ft³/s
1	1	0.001	1.4
2	2	0.001	8.3
3	3	0.001	24.2
4	1	0.01	4.7
5	2	0.01	28.9
6	3	0.01	84.0
7	1	0.05	11.1
8	2	0.05	69.0
9	3	0.05	200.0

where Q = flow (ft³/s), S = slope (ft/ft), D = pipe diameter (ft), and a_0, a_1, and a_2 = coefficients. Taking the logarithm of this equation yields

$$\log Q = \log a_0 + a_1 \log D + a_2 \log S$$

In this form, the equation is suited for multiple linear regression because $\log Q$ is a linear function of $\log S$ and $\log D$. Using the logarithm (base 10) of the data in Table 20.4, we can generate the following normal equations expressed in matrix form [recall Eq. (17.22)]:

$$\begin{bmatrix} 9 & 2.334 & -18.903 \\ 2.334 & 0.954 & -4.903 \\ -18.903 & -4.903 & 44.079 \end{bmatrix} \begin{Bmatrix} \log a_0 \\ a_1 \\ a_2 \end{Bmatrix} = \begin{Bmatrix} 11.691 \\ 3.945 \\ -22.207 \end{Bmatrix}$$

This system can be solved using Gauss elimination for

$$\log a_0 = 1.7475$$
$$a_1 = 2.62$$
$$a_2 = 0.54$$

If $\log a_0 = 1.7475$, $a_0 = 10^{1.7475} = 55.9$, and Eq. (20.6) is

$$Q = 55.9 D^{2.62} S^{0.54} \tag{20.7}$$

Eq. (20.7) can be used to predict flow for the case of $D = 2.5$ ft and $S = 0.025$ ft/ft, as in

$$Q = 55.9(2.5)^{2.62}(0.025)S^{0.54} = 84.1 \text{ ft}^3/\text{s}$$

It should be noted that Eq. (20.7) can be used for other purposes besides computing flow. For example, the slope is related to head loss h_L and pipe length L by $S = h_L/L$. If this relationship is substituted into Eq. (20.7) and the resulting formula solved for h_L, the following equation can be developed:

$$h_L = \frac{L}{1721} Q^{1.85} D^{4.85}$$

This relationship is called the *Hazen-Williams equation*.

PROBLEMS

Chemical/Petroleum Engineering

20.1 Perform the same computation as in Sec. 20.1, but use linear regression and transformations to fit the data with a power equation. Ignore the first point when fitting the equation.

20.2 You perform experiments and determine the following values of heat capacity c at various temperatures T for a gas:

T	−40	−20	10	70	100	120
c	1250	1280	1350	1480	1580	1700

Use regression to determine a model to predict c as a function of T.

20.3 The saturation concentration of dissolved oxygen in water as a function of temperature and chloride concentration is listed in Table P20.3. Use interpolation to estimate the dissolved oxygen level for $T = 18°C$ with chloride = 10,000 mg/L.

20.4 For the data in Table P20.3, use polynomial interpolation to derive a second-order predictive equation for dissolved oxygen concentration as a function of temperature for the case where chloride concentration is equal to 20,000 mg/L. Use the equation to estimate the dissolved oxygen concentration for $T = 8°C$.

20.5 Use multiple linear regression to derive a predictive equation for dissolved oxygen concentration as a function of temperature and chloride based on the data from Table P20.3. Use the equation to estimate the concentration of dissolved oxygen for a chloride concentration of 15,000 mg/L at $T = 12°C$.

20.6 As compared to the models from Probs. 20.4 and 20.5, a somewhat more sophisticated model that accounts for the effect of both temperature and chloride on dissolved oxygen saturation can be hypothesized as being of the form,

$$o_s = f_3(T) + f_1(c)$$

That is, a third-order polynomial in temperature and a linear relationship in chloride is assumed to yield superior results. Use the general linear least-squares approach to fit this model to the data in Table P20.3. Use the resulting equation to estimate the dissolved oxygen concentration for a chloride concentration of 20,000 mg/L at $T = 30°C$.

20.7 It is known that the tensile strength of a plastic increases as a function of the time it is heat-treated. The following data is collected:

Time	10	15	20	25	40	50	55	60	75
Tensile strength	4	20	18	50	33	48	80	60	78

Fit a straight line to this data and use the equation to determine the tensile strength at a time of 30 min.

20.8 The following data was gathered to determine the relationship between pressure and temperature of a fixed volume of 1 kg of nitrogen. The volume is 10 m^3.

T, °C	−20	0	20	40	50	70	100	120
p, N/m^3	7500	8104	8700	9300	9620	10,200	11,200	11,700

Employ the ideal gas law $pV = nRT$ to determine R on the basis of this data. Note that for the law T must be expressed in kelvins.

TABLE P20.3 Dependency of dissolved oxygen concentration on temperature and chloride concentration.

Temperature, °C	Dissolved Oxygen (mg/L) for Stated Concentration of Chloride and Temperature		
	Chloride = 0 mg/L	Chloride = 10,000 mg/L	Chloride = 20,000 mg/L
5	12.8	11.6	10.5
10	11.3	10.3	9.2
15	10.0	9.1	8.2
20	9.0	8.2	7.4
25	8.2	7.4	6.7
30	7.4	6.8	6.1

20.9 The specific volume of a superheated steam is listed in steam tables for various temperatures. For example, at a pressure of 2950 lb/in^2, absolute:

T, °F	700	720	740	760	780
v	0.1058	0.1280	0.1462	0.1603	0.1703

Determine v at $T = 750$°F.

20.10 A reactor is thermally stratified as in the following table:

Depth, m	0	0.5	1.0	1.5	2.0	2.5	3.0
Temperature, °C	70	68	55	22	13	11	10

As depicted in Fig. P20.10, the tank can be idealized as two zones separated by a strong temperature gradient or *thermocline*. The depth of this gradient can be defined as the inflection point of the temperature-depth curve–that is, the point at which $d^2T/dz^2 = 0$. At this depth, the heat flux from the surface to the bottom layer can be computed with Fourier's law,

$$J = -k\frac{dT}{dz}$$

Civil/Environment Engineering

20.11 The shear stress, in kips per square foot (ksf), of nine specimens taken at various depths in a clay stratum are

Depth, m	1.9	3.1	4.2	5.1	5.8	6.9	8.1	9.3	10.0
Stress, ksf	0.3	0.6	0.4	0.9	0.7	1.1	1.5	1.3	1.6

Estimate the shear stress at a depth of 4.5 m.

20.12 A transportation engineering study was conducted to determine the proper design of bike lanes. Data was gathered on bike-lane widths and average distance between bikes and passing cars. The data from 11 streets is

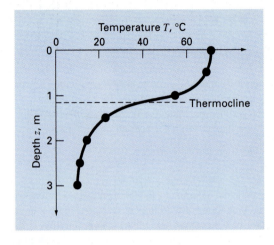

FIGURE P20.10
Use a cubic spline fit of this data to determine the thermocline depth. If $k = 0.01$ cal/(s · cm · °C) compute the flux across this interface.

Distance x, ft	3	8	5	8	6	6	10	10	4	5	7
Lane width y, ft	5	10	7	7.5	7	6	8	9	5	5.5	8

(a) Plot the data.
(b) Fit a straight line to the data with linear regression. Add this line to the plot.
(c) If the minimum safe average distance between bikes and passing cars is considered to be 7 ft, determine the corresponding minimum lane width.

20.13 In water-resources engineering the sizing of reservoirs depends on accurate estimates of water flow in the river that is being impounded. For some rivers, long-term historical records of such flow data are difficult to obtain. In contrast, meteorological data on precipitation is often available for many years past. Therefore it is often useful to determine a relationship between flow and precipitation. This relationship can then be used to estimate flows for years when only precipitation measurements were made. The following data is available for a river that is to be dammed:

Precipitation, cm	88.9	101.6	104.1	139.7	132.1	94.0	116.8	121.9	99.1
Flow, m^3/s	114.7	172.0	152.9	269.0	206.4	161.4	175.8	239.0	130.0

(a) Plot the data.
(b) Fit a straight line to the data with linear regression. Superimpose this line on your plot.
(c) Use the best-fit line to predict the annual water flow if the precipitation is 120 cm.

20.14 The concentration of total phosphorus (p in mg/m^3) and chlorophyll a (c in mg/m^3) for each of the Great Lakes is

	p	c
Lake Superior	4.5	0.8
Lake Michigan	8.0	2.0
Lake Huron	5.5	1.2
Lake Erie:		
West basin	39.0	11.0
Central basin	19.5	4.4
East basin	17.5	3.8
Lake Ontario	21.0	5.5

Chlorophyll a is a parameter that indicates how much plant life is suspended in the water. As such, it indicates how unclear and unsightly the water appears. Use the above data to determine a relationship to predict c as a function of p. Use this equation to predict the level of chlorophyll that can be expected if waste treatment is used to lower the phosphorus concentration of western Lake Erie to 15 mg/m^3.

20.15 The vertical stress σ_z under the corner of a rectangular area subjected to a uniform load of intensity q is given by the solution of Boussinesq's equation:

$$\sigma_z = \frac{q}{4\pi}\left[\frac{2mn\sqrt{m^2+n^2+1}}{m^2+n^2+1+m^2n^2}\frac{m^2+n^2+2}{m^2+n^2+1}\right.$$
$$\left. + \sin^{-1}\left(\frac{2mn\sqrt{m^2+n^2+1}}{m^2+n^2+1+m^2n^2}\right)\right]$$

Because this equation is inconvenient to solve manually, it has been reformulated as

$$\sigma_z = qf_z(m,n)$$

where $f_z(m, n)$ is called the influence value and m and n are dimensionless ratios, with $m = a/z$ and $n = b/z$ and a and b as defined in Fig. P20.36. The influence value is then listed in a table, a portion of which is given here:

m	$n = 1.2$	$n = 1.4$	$n = 1.6$
0.1	0.02926	0.03007	0.03058
0.2	0.05733	0.05894	0.05994
0.3	0.08323	0.08561	0.08709
0.4	0.10631	0.10941	0.11135
0.5	0.12626	0.13003	0.13241
0.6	0.14309	0.14749	0.15027
0.7	0.15703	0.16199	0.16515
0.8	0.16843	0.17389	0.17739

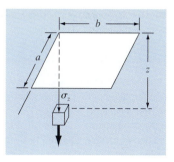

FIGURE P20.15

(a) If $a = 4.8$ and $b = 16$, use a third-order interpolating polynomial to compute σ_z at a depth 10 m below the corner of a rectangular footing that is subject to a total load of 200 t (metric tons). Express your answer in tonnes per square meter. Note that q is equal to the load per area.

(b) Solve Part **(a)** but use Mathcad's cspline function as described in Sec. 19.8.2.

20.16 Three disease-carrying organisms decay exponentially in lake water according to the following model.

$$p(t) = Ae^{-1.5t} + Be^{-0.3t} + Ce^{-0.05t}$$

Estimate the initial population of each organism (A, B, and C) given the following measurements:

t, hr	0.5	1	2	3	4	5	6	7	8	9
$p(t)$	7	5.2	3.8	3.2	2.5	2.1	1.8	1.5	1.2	1.1

20.17 The mast of a sailboat has a cross-sectional area of 10.65 cm^2 and is constructed of an experimental aluminum alloy. Tests were performed to define the relationship between stress and strain. The test results are

Strain, cm/cm	0.002	0.0045	0.0060	0.0013	0.0085	0.0005
Stress, N/cm^2	4965	5172	5517	3586	6896	1241

The stress caused by wind can be computed as F/A_c; where F = force in the mast and A_c = mast's cross-sectional area. This value can then be substituted into Hooke's law to determine the mast's deflection: $\Delta L = $ strain $\times L$; where $L = $ the mast's length. If the wind force is 25,069 N, use the data to estimate the deflection of a 9.14 m mast.

Electrical Engineering

20.18 Perform the same computations as in Sec. 20.3, but analyze data generated with $f(t) = 5\cos(7t) - 2\sin(4t) + 6$.

20.19 You measure the voltage drop V across a resistor for a number of different values of current i. The results are

I	0.25	0.75	1.25	1.5	2.0
V	−0.45	−0.6	0.70	1.88	6.0

Use polynomial interpolation to estimate the voltage drop for $i = 1.1$. Interpret your results.

20.20 Duplicate the computation for Prob. 20.19, but use polynomial regression to derive a cubic equation to fit the data. Plot and evaluate your results.

20.21 The current in a wire is measured with great precision as a function of time:

t	0	0.1250	0.2500	0.3750	0.5000
i	0	6.2402	7.7880	4.8599	0.0000

Determine i at $t = 0.22$.

20.22 The following data was taken from an experiment that measured the current in a wire for various imposed voltages:

V, V	2	3	4	5	7	10
i, A	5.2	7.8	10.7	13	19.3	27.5

On the basis of a linear regression of this data, determine current for a voltage of 6 V. Plot the line and the data and evaluate the fit. Determine whether it is a good assumption that the intercept is zero. If so, redo the regression and force the intercept to be zero.

20.23 It is known that the voltage drop across an inductor follows Faraday's law:

$$V_L = L \frac{di}{dt}$$

where V_L is the voltage drop (in volts), L is inductance (in henrys; 1 H = 1 V·s/A), and i is current (in amperes). Employ the following data to estimate L:

di/dt, A/s	1	2	4	6	8	10
V_L, V	5	12	18	31	39	50

What is the meaning, if any, of the intercept of the regression equation derived from this data?

20.24 Ohm's law states that the voltage drop V across an ideal resistor is linearly proportional to the current i flowing through the resistor as in $V = iR$, where R is the resistance. However, real resistors may not always obey Ohm's law. Suppose that you performed some very precise experiments to measure the voltage drop and corresponding current for a resistor. The results, as listed in Table P20.24, suggest a curvilinear relationship rather than the straight line represented by Ohm's law. In order to quantify this relationship, a curve must be fit to the data. Because of measurement

TABLE P20.24 Experimental data for voltage drop across a resistor subjected to various levels of current.

i	−1.00	−.50	−0.25	0.25	0.50	1.00
V	−193	−41	−13.5625	13.5625	41	193

error, regression would typically be the preferred method of curve fitting for analyzing such experimental data. However, the smoothness of the relationship, as well as the precision of the experimental methods, suggests that interpolation might be appropriate. Use a fifth-order interpolating polynomial to fit the data and compute V for $i = 0.10$.

20.25 Repeat Prob. 20.24 but determine the coefficients of the fifth-order equation (Sec. 18.4) that fit the data in Table P20.24.

20.26 An experiment is performed to determine the % elongation of electrical conducting material as a function of temperature. The resulting data is

Temperature, °C	200	250	300	375	425	475	525	600
% elongation	11	13	13	15	17	19	20	23

Predict the % elongation for a temperature of 400°F.

20.27 Bessel functions often arise in advanced engineering analyses such as the study of electric fields. These functions are usually not amenable to straightforward evaluation and, therefore, are often compiled in standard mathematical tables. For example,

x	1.8	2.0	2.2	2.4	2.6
$J_0(x)$	0.3400	0.2239	0.1104	0.0025	0.0968

Estimate $J_0(2.1)$, **(a)** using an interpolating polynomial and **(b)** using cubic splines. Note that the true value is 0.1666.

20.28 The population (p) of a small community on the outskirts of a city grows rapidly over a 20-year period:

t	0	5	10	15	20
p	100	212	448	949	2009

As an engineer working for a utility company, you must forecast the population 5 years into the future in order to anticipate the demand for power. Employ an exponential model and linear regression to make this prediction.

Mechanical/Aerospace Engineering

20.29 Based on Table 20.4, use linear and quadratic interpolation to compute Q for $D = 1.23$ ft and $S = 0.01$ ft/ft. Compare your results with the same value computed with the formula derived in Sec. 20.4.

20.30 Reproduce Sec. 20.4, but develop an equation to predict diameter as a function of slope and flow. Compare your results with the formula from Sec. 20.4 and discuss your results.

20.31 Kinematic viscosity of water, v, is related to temperature in the following manner:

T, °C	0	4	8	12	16	20	24
v, 10^{-2} cm²/s	1.7923	1.5676	1.3874	1.2396	1.1168	1.0105	0.9186

Plot this data.
(a) Use interpolation to predict v at $T = 7.5$ °C.
(b) Use polynomial regression to fit a parabola to the data in order to make the same prediction.

20.32 Hooke's law, which holds when a spring is not stretched too far, signifies that the extension of the spring and the applied force are linearly related. The proportionality is parameterized by the spring constant k. A value for this parameter can be established experimentally by placing known weights onto the spring and measuring the resulting compression. Such data is contained in Table P20.32 and plotted in Fig. P20.32. Notice that above a weight of 40×10^4 N, the linear relationship between the force and displacement breaks down. This sort of behavior is typical of what is termed a "hardening spring." Employ linear regression to determine a value of k for the linear portion of this system. In addition, fit a nonlinear relationship to the nonlinear portion.

20.33 Repeat Prob. 20.32 but fit a power curve to all the data in Table P20.32. Comment on your results.

20.34 The distance required to stop an automobile is a function of its speed. The following experimental data was collected to quantify this relationship:

Speed, mi/h	15	20	25	30	40	50	60
Stopping distance, ft	16	20	34	40	60	90	120

Estimate the stopping distance for a car traveling at 45 mi/h.

FIGURE P20.32
Plot of force (in 10^4 newtons) versus displacement (in meters) for the spring from the automobile suspension system.

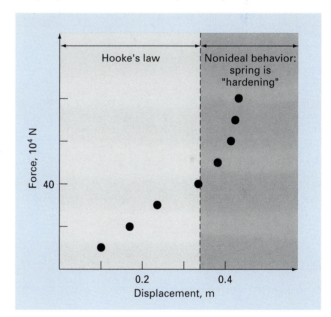

TABLE P20.32 Experimental values for elongation x and force F for the spring on an automobile suspension system.

Displacement, m	0.10	0.17	0.27	0.35	0.39	0.42	0.43	0.44
Force, 10^4 N	10	20	30	40	50	60	70	80

20.35 An experiment is performed to define the relationship between applied stress and the time to fracture for a stainless steel. Eight different values of stress are applied, and the resulting data is

Applied stress, x, kg/mm^2	5	10	15	20	25	30	35	40
Fracture time, y, h	40	30	25	40	18	20	22	15

(a) Plot the data.

(b) Fit a straight line to the data with linear regression. Superimpose this line on your plot.

(c) Use the best-fit equation to predict the fracture time for an applied stress of 17 kg/mm^2.

20.36 The acceleration due to gravity at an altitude y above the surface of the earth is given by

y, m	0	20,000	40,000	60,000	80,000
g, m/s^2	9.8100	9.7487	9.6879	9.6278	9.5682

Compute g at $y = 55{,}000$ m.

EPILOGUE: PART FIVE

PT5.4 TRADE-OFFS

Table PT5.4 provides a summary of the trade-offs involved in curve fitting. The techniques are divided into two broad categories, depending on the uncertainty of the data. For imprecise measurements, regression is used to develop a "best" curve that fits the overall trend of the data without necessarily passing through any of the individual points. For precise measurements, interpolation is used to develop a curve that passes directly through each of the points.

All the regression methods are designed to fit functions that minimize the sum of the squares of the residuals between the data and the function. Such methods are termed least-squares regression. Linear least-squares regression is used for cases where a dependent and an independent variable are related to each other in a linear fashion. For situations where a dependent and an independent variable exhibit a curvilinear relationship, several options are available. In some cases, transformations can be used to linearize the relationship. In these instances, linear regression can be applied to the transformed variables to determine the best straight line. Alternatively, polynomial regression can be employed to fit a curve directly to the data.

TABLE PT5.4 Comparison of the characteristics of alternative methods for curve fitting.

Method	Error Associated with Data	Match of Individual Data Points	Number of Points Matched Exactly	Programming Effort	Comments
Regression					
Linear regression	Large	Approximate	0	Easy	
Polynomial regression	Large	Approximate	0	Moderate	Round-off error becomes pronounced for higher-order versions
Multiple linear regression	Large	Approximate	0	Moderate	
Nonlinear regression	Large	Approximate	0	Difficult	
Interpolation					
Newton's divided-difference polynomials	Small	Exact	$n + 1$	Easy	Usually preferred for exploratory analyses
Lagrange polynomials	Small	Exact	$n + 1$	Easy	Usually preferred when order is known
Cubic splines	Small	Exact	Piecewise fit of data points	Moderate	First and second derivatives equal at knots

Multiple linear regression is utilized when a dependent variable is a linear function of two or more independent variables. Logarithmic transformations can also be applied to this type of regression for some cases where the multiple dependency is curvilinear.

Polynomial and multiple linear regression (note that simple linear regression is a member of both) belong to a more general class of linear least-squares models. They are classified in this way because they are linear with respect to their coefficients. These models are typically implemented using linear algebraic systems that are sometimes ill-conditioned. However, in many engineering applications (that is, for lower-order fits), this does not come into play. For cases where it is a problem, alternative approaches are available. For example, a technique called orthogonal polynomials is available to perform polynomial regression (see Sec. PT5.6).

Equations that are not linear with respect to their coefficients are called nonlinear. Special regression techniques are available to fit such equations. These are approximate methods that start with initial parameter estimates and then iteratively home in on values that minimize the sum of the squares.

Polynomial interpolation is designed to fit a unique nth-order polynomial that passes exactly through $n + 1$ precise data points. This polynomial is presented in two alternative formats. Newton's divided-difference interpolating polynomial is ideally suited for those cases where the proper order of the polynomial is unknown. Newton's polynomial is appropriate for such situations because it is easily programmed in a format to compare results with different orders. In addition, an error estimate can be simply incorporated into the technique. Thus, you can compare and choose from results using several different-order polynomials.

The Lagrange interpolating polynomial is an alternative formulation that is appropriate when the order is known a priori. For these situations, the Lagrange version is somewhat simpler to program and does not require the computation and storage of finite divided differences.

Another approach to curve fitting is spline interpolation. This technique fits a low-order polynomial to each interval between data points. The fit is made smooth by setting the derivatives of adjacent polynomials to the same value at their connecting points. The cubic spline is the most common version. Splines are of great utility when fitting data that is generally smooth but exhibits local areas of abrupt change. Such data tends to induce wild oscillations in higher-order interpolating polynomials. Cubic splines are less prone to these oscillations because they are limited to third-order variations.

The final method covered in this part of the book is Fourier approximation. This area deals with using trigonometric functions to approximate waveforms. In contrast to the other techniques, the major emphasis of this approach is not to fit a curve to data points. Rather, the curve fit is employed to analyze the frequency characteristics of a signal. In particular, a fast Fourier transform is available to very efficiently transform a function from the time to the frequency domain to elucidate its underlying harmonic structure.

PT5.5 IMPORTANT RELATIONSHIPS AND FORMULAS

Table PT5.5 summarizes important information that was presented in Part Five. This table can be consulted to quickly access important relationships and formulas.

TABLE PT5.5   Summary of important information presented in Part Four.

Method	Formulation	Graphical Interpretation	Errors
Linear regression	$y = a_0 + a_1 x$ where $a_1 = \dfrac{n\sum x_i y_i - \sum x_i \sum y_i}{n\sum x_i^2 - (\sum x_i)^2}$ $a_0 = \bar{y} - a_1 \bar{x}$		$s_{y/x} = \sqrt{\dfrac{S_r}{n-2}}$ $r^2 = \dfrac{S_t - S_r}{S_t}$
Polynomial regression	$y = a_0 + a_1 x + \cdots + a_m x^m$ (Evaluation of a's equivalent to solution of $m+1$ linear algebraic equations)		$s_{y/x} = \sqrt{\dfrac{S_r}{n-(m+1)}}$ $r^2 = \dfrac{S_t - S_r}{S_t}$
Multiple linear regression	$y = a_0 + a_1 x_1 + \cdots + a_m x_m$ (Evaluation of a's equivalent to solution of $m+1$ linear algebraic equations)		$s_{y/x} = \sqrt{\dfrac{S_r}{n-(m+1)}}$ $r^2 = \dfrac{S_t - S_r}{S_t}$
Newton's divided-difference interpolating polynomial*	$f_2(x) = b_0 + b_1(x-x_0) + b_2(x-x_0)(x-x_1)$ where $b_0 = f(x_0)$ $b_1 = f[x_1, x_0]$ $b_2 = f[x_2, x_1, x_0]$		$R_2 = (x-x_0)(x-x_1)(x-x_2)\dfrac{f^{(3)}(\xi)}{6}$ or $R_2 = (x-x_0)(x-x_1)(x-x_2)f[x_3, x_2, x_1, x_0]$
Lagrange interpolating polynomial*	$f_2(x) = f(x_0)\left(\dfrac{x-x_1}{x_0-x_1}\right)\left(\dfrac{x-x_2}{x_0-x_2}\right)$ $+ f(x_1)\left(\dfrac{x-x_0}{x_1-x_0}\right)\left(\dfrac{x-x_2}{x_1-x_2}\right)$ $+ f(x_2)\left(\dfrac{x-x_0}{x_2-x_0}\right)\left(\dfrac{x-x_1}{x_2-x_1}\right)$		$R_2 = (x-x_0)(x-x_1)(x-x_2)\dfrac{f^{(3)}(\xi)}{6}$ or $R_2 = (x-x_0)(x-x_1)(x-x_2)f[x_3, x_2, x_1, x_0]$
Cubic splines	A cubic: $a_i x^3 + b_i x^2 + c_i x + d_i$ is fit to each interval between knots. First and second derivatives are equal at each knot		

*Note: For simplicity, second-order versions are shown.

PT5.6   ADVANCED METHODS AND ADDITIONAL REFERENCES

Although polynomial regression with normal equations is adequate for many engineering applications, there are problem contexts where its sensitivity to round-off error can represent a serious limitation. An alternative approach based on *orthogonal polynomials* can mitigate these effects. It should be noted that this approach does not yield a best-fit equation, but rather, yields individual predictions for given values of the independent variable. Information on orthogonal polynomials can be found in Shampine and Allen (1973) and Guest (1961).

Whereas the orthogonal polynomial technique is helpful for developing a polynomial regression, it does not represent a solution to the instability problem for the general linear regression model [Eq. (17.23)]. An alternative approach based on *single-value decomposition,* called the SVD method, is available for this purpose. Forsythe et al. (1977), Lawson and Hanson (1974), and Press et al. (1992) contain information on this approach.

In addition to the Gauss-Newton algorithm, there are a number of optimization methods that can be used to directly develop a least-squares fit for a nonlinear equation. These nonlinear regression techniques include Marquardt's and the steepest-descent methods (recall Part Four). General information on regression can be found in Draper and Smith (1981).

All the methods in Part Five have been couched in terms of fitting a curve to data points. In addition, you may also desire to fit a curve to another curve. The primary motivation for such *functional approximation* is to represent a complicated function by a simpler version that is easier to manipulate. One way to do this is to use the complicated function to generate a table of values. Then the techniques discussed in this part of the book can be used to fit polynomials to these discrete values.

An alternative approach is based on the *minimax principle* (recall Fig. 17.2c). This principle specifies that the coefficients of the approximating polynomial be chosen so that the maximum discrepancy is as small as possible. Thus, although the approximation may not be as good as that given by the Taylor series at the base point, it is generally better across the entire range of the fit. *Chebyshev economization* is an example of an approach for functional approximation based on such a strategy (Ralston and Rabinowitz, 1978; Gerald and Wheatley, 1984; and Carnahan, Luther, and Wilkes, 1969).

An important area in curve fitting is the combining of splines with least-squares regression. Thus, a cubic spline is generated that does not intercept every point, but rather, minimizes the sum of the squares of the residuals between the data points and the spline curves. The approach involves using the so-called *B splines* as basis functions. These are so named because of their use as *basis* function, but also because of their characteristic bell shape. Such curves are consistent with a spline approach in that their value and their first and second derivatives would have continuity at their extremes. Thus, continuity of $f(x)$ and its lower derivatives at the knots is ensured. Wold (1974), Prenter (1974), and Cheney and Kincaid (1994) present discussions of this approach.

In summary, the foregoing is intended to provide you with avenues for deeper exploration of the subject. Additionally, all the above references provide descriptions of the basic techniques covered in Part Five. We urge you to consult these alternative sources to broaden your understanding of numerical methods for curve fitting.

PART SIX

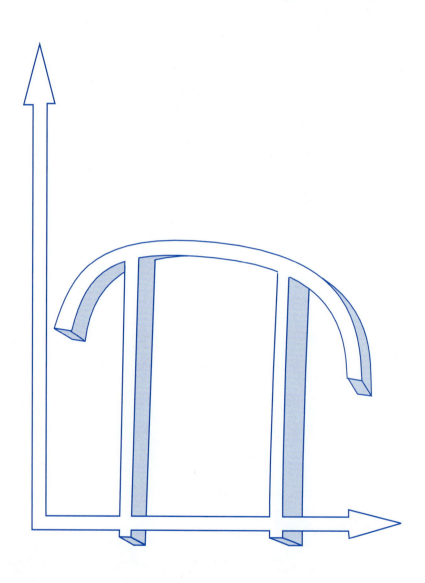

NUMERICAL DIFFERENTIATION AND INTEGRATION

PT6.1 MOTIVATION

Calculus is the mathematics of change. Because engineers must continuously deal with systems and processes that change, calculus is an essential tool of our profession. Standing at the heart of calculus are the related mathematical concepts of differentiation and integration.

According to the dictionary definition, to *differentiate* means "to mark off by differences; distinguish; . . . to perceive the difference in or between." Mathematically, the *derivative,* which serves as the fundamental vehicle for differentiation, represents the rate of change of a dependent variable with respect to an independent variable. As depicted in Fig. PT6.1, the mathematical definition of the derivative begins with a difference approximation:

$$\frac{\Delta y}{\Delta x} = \frac{f(x_i + \Delta x) - f(x_i)}{\Delta x} \tag{PT6.1}$$

where y and $f(x)$ are alternative representatives for the dependent variable and x is the independent variable. If Δx is allowed to approach zero, as occurs in moving from Fig. PT6.1a to c, the difference becomes a derivative

$$\frac{dy}{dx} = \lim_{\Delta x \to 0} \frac{f(x_i + \Delta x) - f(x_i)}{\Delta x}$$

FIGURE PT6.1
The graphical definition of a derivative: as Δx approaches zero in going from (a) to (c), the difference approximation becomes a derivative.

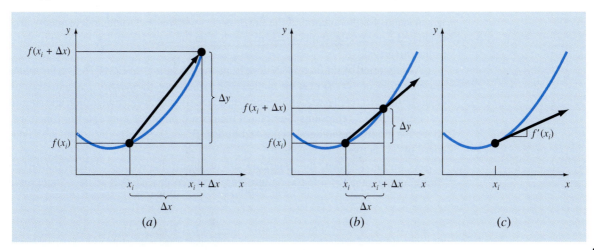

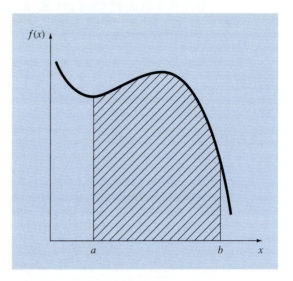

FIGURE PT6.2
Graphical representation of the integral of $f(x)$ between the limits $x = a$ to b. The integral is equivalent to the area under the curve.

where dy/dx [which can also be designated as y' or $f'(x_i)$] is the first derivative of y with respect to x evaluated at x_i. As seen in the visual depiction of Fig. PT6.1c, the derivative is the slope of the tangent to the curve at x_i.

The inverse process to differentiation in calculus is integration. According to the dictionary definition, to *integrate* means "to bring together, as parts, into a whole; to unite; to indicate the total amount . . . ." Mathematically, integration is represented by

$$I = \int_a^b f(x)\,dx \tag{PT6.2}$$

which stands for the integral of the function $f(x)$ with respect to the independent variable x, evaluated between the limits $x = a$ to $x = b$. The function $f(x)$ in Eq. (PT6.2) is referred to as the *integrand*.

As suggested by the dictionary definition, the "meaning" of Eq. (PT6.2) is the *total value*, or *summation*, of $f(x)\,dx$ over the range $x = a$ to b. In fact, the symbol $\int$ is actually a stylized capital S that is intended to signify the close connection between integration and summation.

Figure PT6.2 represents a graphical manifestation of the concept. For functions lying above the x axis, the integral expressed by Eq. (PT6.2) corresponds to the area under the curve of $f(x)$ between $x = a$ and b.[1]

As outlined above, the "marking off or discrimination" of differentiation and the "bringing together" of integration are closely linked processes that are, in fact, inversely

---

[1] It should be noted that the process represented by Eq. (PT6.2) and Fig. PT6.2 is called *definite integration*. There is another type called indefinite integration in which the limits a and b are unspecified. As will be discussed in Part Seven, *indefinite integration* deals with determining a function whose derivative is given.

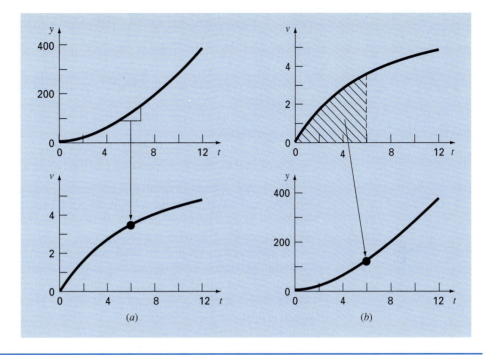

FIGURE PT6.3

The contrast between (a) differentiation and (b) integration.

related (Fig. PT6.3). For example, if we are given a function $y(t)$ that specifies an object's position as a function of time, differentiation provides a means to determine its velocity, as in (Fig. PT6.3a).

$$v(t) = \frac{d}{dt} y(t)$$

Conversely, if we are provided with velocity as a function of time, integration can be used to determine its position (Fig. PT6.3b),

$$y(t) = \int_0^t v(t)\, dt$$

Thus, we can make the general claim that the evaluation of the integral

$$I = \int_a^b f(x)\, dx$$

is equivalent to solving the differential equation

$$\frac{dy}{dx} = f(x)$$

for $y(b)$ given the initial condition $y(a) = 0$.

Because of this close relationship, we have opted to devote this part of the book to both processes. Among other things, this will provide the opportunity to highlight their similarities and differences from a numerical perspective. In addition, our discussion will have relevance to the next parts of the book, where we will cover differential equations.

PT6.1.1 Noncomputer Methods for Differentiation and Integration

The function to be differentiated or integrated will typically be in one of the following three forms:

1. A simple continuous function such as a polynomial, an exponential, or a trigonometric function.
2. A complicated continuous function that is difficult or impossible to differentiate or integrate directly.
3. A tabulated function where values of x and $f(x)$ are given at a number of discrete points, as is often the case with experimental or field data.

In the first case, the derivative or integral of a simple function may be evaluated analytically using calculus. For the second case, analytical solutions are often impractical, and sometimes impossible, to obtain. In these instances, as well as in the third case of discrete data, approximate methods must be employed.

A noncomputer method for determining derivatives from data is called *equal-area graphical differentiation*. In this method, the (x, y) data are tabulated and, for each interval, a simple divided difference $\Delta y / \Delta x$ is employed to estimate the slope. Then these values are plotted as a stepped curve versus x (Fig. PT6.4). Next, a smooth curve is drawn that attempts to approximate the area under the stepped curve. That is, it is drawn so that visually, the positive and negative areas are balanced. The rates at given values of x can then be read from the curve.

In the same spirit, visually oriented approaches were employed to integrate tabulated data and complicated functions in the precomputer era. A simple intuitive approach is to plot the function on a grid (Fig. PT6.5) and count the number of boxes that approximate the area. This number multiplied by the area of each box provides a rough estimate of the total

FIGURE PT6.4
Equal-area differentiation.
(a) Centered finite divided differences are used to estimate the derivative for each interval between the data points.
(b) The derivative estimates are plotted as a bar graph. A smooth curve is superimposed on this plot to approximate the area under the bar graph. This is accomplished by drawing the curve so that equal positive and negative areas are balanced.
(c) Values of dy/dx can then be read off the smooth curve.

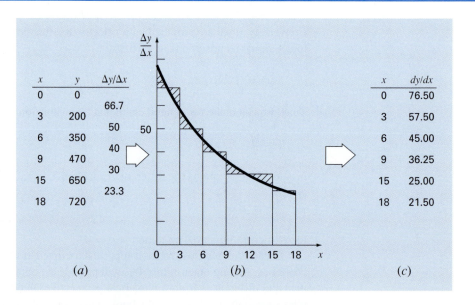

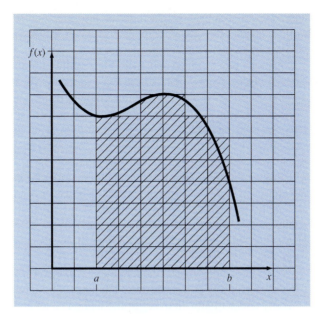

FIGURE PT6.5
The use of a grid to approximate an integral.

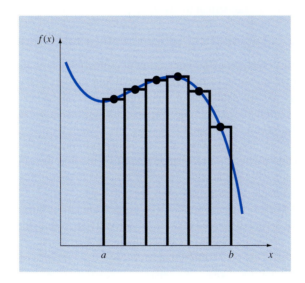

FIGURE PT6.6
The use of rectangles, or strips, to approximate the integral.

area under the curve. This estimate can be refined, at the expense of additional effort, by using a finer grid.

Another commonsense approach is to divide the area into vertical segments, or strips, with a height equal to the function value at the midpoint of each strip (Fig. PT6.6). The area of the rectangles can then be calculated and summed to estimate the total area. In this approach, it is assumed that the value at the midpoint provides a valid approximation of the average height of the function for each strip. As with the grid method, refined estimates are possible by using more (and thinner) strips to approximate the integral.

Although such simple approaches have utility for quick estimates, alternative numerical techniques are available for the same purpose. Not surprisingly, the simplest of these methods is similar in spirit to the noncomputer techniques.

For differentiation, the most fundamental numerical techniques use finite divided differences to estimate derivatives. For data with error, an alternative approach is to fit a smooth curve to the data with a technique such as least-squares regression and then differentiate this curve to obtain derivative estimates.

In a similar spirit, numerical integration or *quadrature* methods are available to obtain integrals. These methods, which are actually easier to implement than the grid approach, are similar in spirit to the strip method. That is, function heights are multiplied by strip widths and summed to estimate the integral. However, through clever choices of weighting factors, the resulting estimate can be made more accurate than that from the simple "strip method."

As in the simple strip method, numerical integration and differentiation techniques utilize data at discrete points. Because tabulated information is already in such a form, it is naturally compatible with many of the numerical approaches. Although continuous functions are not originally in discrete form, it is usually a simple proposition to use the given equation to generate a table of values. As depicted in Fig. PT6.7, this table can then be evaluated with a numerical method.

PT6.1.2 Numerical Differentiation and Integration in Engineering

The differentiation and integration of a function has so many engineering applications that you were required to take differential and integral calculus in your first year at college. Many specific examples of such applications could be given in all fields of engineering.

Differentiation is commonplace in engineering because so much of our work involves characterizing the changes of variables in both time and space. In fact, many of the laws and other generalizations that figure so prominently in our work are based on the predictable ways in which change manifests itself in the physical world. A prime example is Newton's second law, which is not couched in terms of the position of an object but rather in its change of position with respect to time.

Aside from such temporal examples, numerous laws governing the spatial behavior of variables are expressed in terms of derivatives. Among the most common of these are those laws involving potentials or gradients. For example, *Fourier's law* of heat conduction quantifies the observation that heat flows from regions of high to low temperature. For the one-dimensional case, this can be expressed mathematically as

$$\text{Heat flux} = -k' \frac{dT}{dx}$$

Thus, the derivative provides a measure of the intensity of the temperature change, or *gradient*, that drives the transfer of heat. Similar laws provide workable models in many other areas of engineering, including the modeling of fluid dynamics, mass transfer, chemical reaction kinetics, and electromagnetic flux. The ability to accurately estimate derivatives is an important facet of our capability to work effectively in these areas.

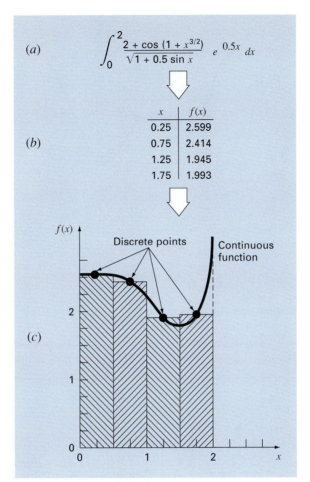

FIGURE PT6.7

Application of a numerical integration method: (a) A complicated, continuous function. (b) Table of discrete values of $f(x)$ generated from the function. (c) Use of a numerical method (the strip method here) to estimate the integral on the basis of the discrete points. For a tabulated function, the data is already in tabular form (b); therefore, step (a) is unnecessary.

Just as accurate estimates of derivatives are important in engineering, the calculation of integrals is equally valuable. A number of examples relate directly to the idea of the integral as the area under a curve. Figure PT6.8 depicts a few cases where integration is used for this purpose.

Other common applications relate to the analogy between integration and summation. For example, a common application is to determine the mean of continuous functions. In Part Five, you were introduced to the concept of the mean of n discrete data points [recall Eq. (PT6.1)]:

$$\text{Mean} = \frac{\displaystyle\sum_{i=1}^{n} y_i}{n} \tag{PT6.3}$$

where y_i are individual measurements. The determination of the mean of discrete points is depicted in Fig. PT6.9a.

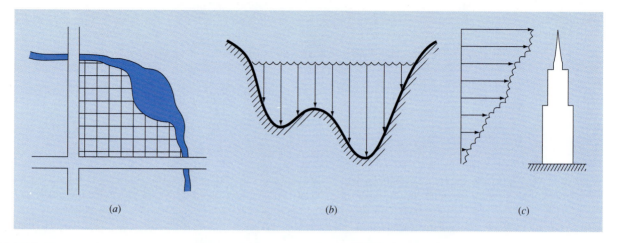

FIGURE PT6.8

Examples of how integration is used to evaluate areas in engineering applications. (a) A surveyor might need to know the area of a field bounded by a meandering stream and two roads. (b) A water-resource engineer might need to know the cross-sectional area of a river. (c) A structural engineer might need to determine the net force due to a nonuniform wind blowing against the side of a skyscraper.

FIGURE PT6.9

An illustration of the mean for (a) discrete and (b) continuous data.

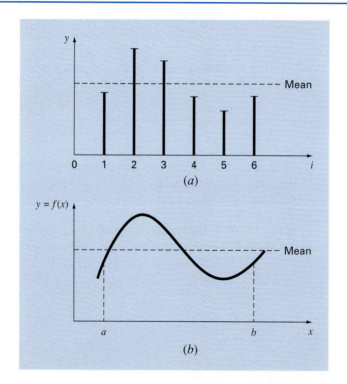

In contrast, suppose that y is a continuous function of an independent variable x, as depicted in Fig. PT6.9b. For this case, there are an infinite number of values between a and b. Just as Eq. (PT6.3) can be applied to determine the mean of the discrete readings, you might also be interested in computing the mean or average of the continuous function $y = f(x)$ for the interval from a to b. Integration is used for this purpose, as specified by the formula

$$\text{Mean} = \frac{\displaystyle\int_a^b f(x)\,dx}{b - a} \tag{PT6.4}$$

This formula has hundreds of engineering applications. For example, it is used to calculate the center of gravity of irregular objects in mechanical and civil engineering and to determine the root-mean-square current in electrical engineering.

Integrals are also employed by engineers to evaluate the total amount or quantity of a given physical variable. The integral may be evaluated over a line, an area, or a volume. For example, the total mass of chemical contained in a reactor is given as the product of the concentration of chemical and the reactor volume, or

$$\text{Mass} = \text{concentration} \times \text{volume}$$

where concentration has units of mass per volume. However, suppose that concentration varies from location to location within the reactor. In this case, it is necessary to sum the products of local concentrations c_i and corresponding elemental volumes ΔV_i:

$$\text{Mass} = \sum_{i=1}^{n} c_i\,\Delta V_i$$

where n is the number of discrete volumes. For the continuous case, where $c(x, y, z)$ is a known function and x, y, and z are independent variables designating position in Cartesian coordinates, integration can be used for the same purpose:

$$\text{Mass} = \iiint c(x, y, z)\,dx\,dy\,dz$$

or

$$\text{Mass} = \iiint_V c(V)\,dV$$

which is referred to as a *volume integral*. Notice the strong analogy between summation and integration.

Similar examples could be given in other fields of engineering. For example, the total rate of energy transfer across a plane where the flux (in calories per square centimeter per second) is a function of position is given by

$$\text{Heat transfer} = \iint_A \text{flux}\,dA$$

which is referred to as an *areal integral,* where $A = $ area.

Similarly, for the one-dimensional case, the total mass of a variable-density rod with constant cross-sectional area is given by

$$m = A \int_0^L \rho(x)\, dx$$

where m = total weight (kg), L = length of the rod (m), $\rho(x)$ = known density (kg/m^3) as a function of length x (m), and A = cross-sectional area of the rod (m^2).

Finally, integrals are used to evaluate differential or rate equations. Suppose the velocity of a particle is a known continuous function of time $v(t)$,

$$\frac{dy}{dt} = v(t)$$

The total distance y traveled by this particle over a time t is given by (Fig. PT6.3b)

$$y = \int_0^t v(t)\, dt \tag{PT6.5}$$

These are just a few of the applications of differentiation and integration that you might face regularly in the pursuit of your profession. When the functions to be analyzed are simple, you will normally choose to evaluate them analytically. For example, in the falling parachutist problem, we determined the solution for velocity as a function of time [Eq. (1.10)]. This relationship could be substituted into Eq. (PT6.5), which could then be integrated easily to determine how far the parachutist fell over a time period t. For this case, the integral is simple to evaluate. However, it is difficult or impossible when the function is complicated, as is typically the case in more realistic examples. In addition, the underlying function is often unknown and defined only by measurement at discrete points. For both these cases, you must have the ability to obtain approximate values for derivatives and integrals using numerical techniques. Several such techniques will be discussed in this part of the book.

PT6.2 MATHEMATICAL BACKGROUND

In high school or during your first year of college, you were introduced to *differential* and *integral calculus*. There you learned techniques to obtain analytical or exact derivatives and integrals.

When we differentiate a function analytically, we generate a second function that can be used to compute the derivative for different values of the independent variable. General rules are available for this purpose. For example, in the case of the monomial

$$y = x^n$$

the following simple rule applies ($n \neq 0$):

$$\frac{dy}{dx} = nx^{n-1}$$

which is the expression of the more general rule for

$$y = u^n$$

where $u =$ a function of x. For this equation, the derivative is computed via

$$\frac{dy}{dx} = nu^{n-1}\frac{du}{dx}$$

Two other useful formulas apply to the products and quotients of functions. For example, if the product of two functions of x (u and v) is represented as $y = uv$, then the derivative can be computed as

$$\frac{dy}{dx} = u\frac{dv}{dx} + v\frac{du}{dx}$$

For the division, $y = u/v$, the derivative can be computed as

$$\frac{dy}{dx} = \frac{v\dfrac{du}{dx} - u\dfrac{dv}{dx}}{v^2}$$

Other useful formulas are summarized in Table PT6.1.

Similar formulas are available for definite integration, which deals with determining an integral between specified limits, as in

$$I = \int_a^b f(x)\, dx \qquad\qquad (\text{PT6.6})$$

According to the *fundamental theorem* of integral calculus, Eq. (PT6.6) is evaluated as

$$\int_a^b f(x)\, dx = F(x)\Big|_a^b$$

where $F(x) =$ the integral of $f(x)$—that is, any function such that $F'(x) = f(x)$. The nomenclature on the right-hand side stands for

$$F(x)\Big|_a^b = F(b) - F(a) \qquad\qquad (\text{PT6.7})$$

TABLE PT6.1 Some commonly used derivatives.

$\dfrac{d}{dx}\sin x = \cos x$	$\dfrac{d}{dx}\cot x = -\csc^2 x$
$\dfrac{d}{dx}\cos x = -\sin x$	$\dfrac{d}{dx}\sec x = \sec x \tan x$
$\dfrac{d}{dx}\tan x = \sec^2 x$	$\dfrac{d}{dx}\csc x = -\csc x \cot x$
$\dfrac{d}{dx}\ln x = \dfrac{1}{x}$	$\dfrac{d}{dx}\log_a x = \dfrac{1}{x \ln a}$
$\dfrac{d}{dx}e^x = e^x$	$\dfrac{d}{dx}a^x = a^x \ln a$

An example of a definite integral is

$$I = \int_0^{0.8} (0.2 + 25x - 200x^2 + 675x^3 - 900x^4 + 400x^5) \, dx \tag{PT6.8}$$

For this case, the function is a simple polynomial that can be integrated analytically by evaluating each term according to the rule

$$\int_a^b x^n \, dx = \frac{x^{n+1}}{n+1} \Big|_a^b \tag{PT6.9}$$

where n cannot equal -1. Applying this rule to each term in Eq. (PT6.8) yields

$$I = 0.2x + 12.5x^2 - \frac{200}{3}x^3 + 168.75x^4 - 180x^5 + \frac{400}{6}x^6 \Big|_0^{0.8} \tag{PT6.10}$$

which can be evaluated according to Eq. (PT6.7) as $I = 1.6405333$. This value is equal to the area under the original polynomial [Eq. (PT6.8)] between $x = 0$ and 0.8.

The foregoing integration depends on knowledge of the rule expressed by Eq. (PT6.9). Other functions follow different rules. These "rules" are all merely instances of *antidifferentiation*, that is, finding $F(x)$ so that $F'(x) = f(x)$. Consequently, analytical integration depends on prior knowledge of the answer. Such knowledge is acquired by training and

TABLE PT6.2   Some simple integrals that are used in Part Six. The a and b in this table are constants and should not be confused with the limits of integration discussed in the text.

$$\int u \, dv = uv - \int v \, du$$

$$\int u^n \, du = \frac{u^{n+1}}{n+1} + C \qquad n \neq -1$$

$$\int a^{bx} \, dx = \frac{a^{bx}}{b \ln a} + C \qquad a > 0, a \neq 1$$

$$\int \frac{dx}{x} = \ln |x| + C \qquad x \neq 0$$

$$\int \sin (ax + b) \, dx = -\frac{1}{a} \cos (ax + b) + C$$

$$\int \cos (ax + b) \, dx = \frac{1}{a} \sin (ax + b) + C$$

$$\int \ln|x| \, dx = x \ln|x| - x + C$$

$$\int e^{ax} \, dx = \frac{e^{ax}}{a} + C$$

$$\int x e^{ax} \, dx = \frac{e^{ax}}{a^2} (ax - 1) + C$$

$$\int \frac{dx}{a + bx^2} = \frac{1}{\sqrt{ab}} \tan^{-1} \frac{\sqrt{ab}}{a} x + C$$

experience. Many of the rules are summarized in handbooks and in tables of integrals. We list some commonly encountered integrals in Table PT6.2. However, many functions of practical importance are too complicated to be contained in such tables. One reason why the techniques in the present part of the book are so valuable is that they provide a means to evaluate relationships such as Eq. (PT6.8) without knowledge of the rules.

PT6.3   ORIENTATION

Before proceeding to the numerical methods for integration, some further orientation might be helpful. The following is intended as an overview of the material discussed in Part Six. In addition, we have formulated some objectives to help focus your efforts when studying the material.

PT6.3.1 Scope and Preview

Figure PT6.10 provides an overview of Part Six. *Chapter 21* is devoted to the most common approaches for numerical integration—the *Newton-Cotes formulas*. These relationships are based on replacing a complicated function or tabulated data with a simple polynomial that is easy to integrate. Three of the most widely used Newton-Cotes formulas are discussed in detail: the *trapezoidal rule, Simpson's 1/3 rule,* and *Simpson's 3/8 rule.* All these formulas are designed for cases where the data to be integrated is evenly spaced. In addition, we also include a discussion of numerical integration of unequally spaced data. This is a very important topic because many real-world applications deal with data that is in this form.

All the above material relates to closed integration, where the function values at the ends of the limits of integration are known. At the end of Chap. 21, we present *open integration formulas,* where the integration limits extend beyond the range of the known data. Although they are not commonly used for definite integration, open integration formulas are presented here because they are utilized extensively in the solution of ordinary differential equations in Part Seven.

The formulations covered in Chap. 21 can be employed to analyze both tabulated data and equations. *Chapter 22* deals with two techniques that are expressly designed to integrate equations and functions: *Romberg integration* and *Gauss quadrature.* Computer algorithms are provided for both of these methods. In addition, methods for evaluating *improper integrals* are discussed.

In *Chap. 23,* we present additional information on *numerical differentiation* to supplement the introductory material from Chap. 4. Topics include high-accuracy finite-difference formulas, Richardson's extrapolation, and the differentiation of unequally spaced data. The effect of errors on both numerical differentiation and integration is discussed. Finally, the chapter is concluded with a description of the application of several software packages and libraries for integration and differentiation.

Chapter 24 demonstrates how the methods can be applied for problem solving. As with other parts of the book, applications are drawn from all fields of engineering.

A review section, or *epilogue,* is included at the end of Part Six. This review includes a discussion of trade-offs that are relevant to implementation in engineering practice. In addition, important formulas are summarized. Finally, we present a short review of advanced

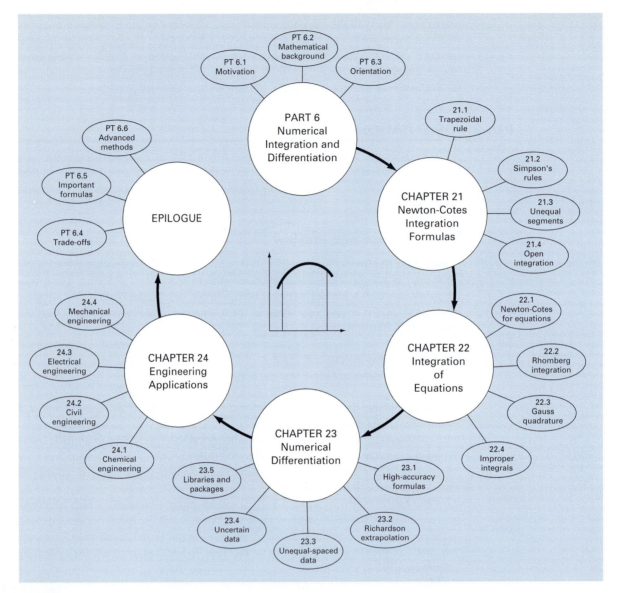

FIGURE PT6.10
Schematic of the organization of material in Part Six: Numerical Integration and Differentiation.

methods and alternative references that will facilitate your further studies of numerical differentiation and integration.

PT6.3.2 Goals and Objectives

Study Objectives.   After completing Part Six you should be able to solve many numerical integration and differentiation problems and appreciate their application for engineer-

TABLE PT6.3 Specific study objectives for Part Six.

1. Understand the derivation of the Newton-Cotes formulas; know how to derive the trapezoidal rule and how to set up the derivation of both of Simpson's rules; recognize that the trapezoidal and Simpson's 1/3 and 3/8 rules represent the areas under first-, second-, and third-order polynomials, respectively.
2. Know the formulas and error equations for (a) the trapezoidal rule, (b) the multiple-application trapezoidal rule, (c) Simpson's 1/3 rule, (d) Simpson's 3/8 rule, and (e) the multiple-application Simpson's rule. Be able to choose the "best" among these formulas for any particular problem context.
3. Recognize that Simpson's 1/3 rule is fourth-order accurate even though it is based on only three points; realize that all the even-segment–odd-point Newton-Cotes formulas have similar enhanced accuracy.
4. Know how to evaluate the integral and derivative of unequally spaced data.
5. Recognize the difference between open and closed integration formulas.
6. Understand the theoretical basis of Richardson extrapolation and how it is applied in the Romberg integration algorithm and for numerical differentiation.
7. Understand the fundamental difference between Newton-Cotes and Gauss quadrature formulas.
8. Recognize why both Romberg integration and Gauss quadrature have utility when integrating equations (as opposed to tabular or discrete data).
9. Know how open integration formulas are employed to evaluate improper integrals.
10. Understand the application of high-accuracy numerical-differentiation formulas.
11. Know how to differentiate unequally spaced data.
12. Recognize the differing effects of data error on the processes of numerical integration and differentiation.

ing problem solving. You should strive to master several techniques and assess their reliability. You should understand the trade-offs involved in selecting the "best" method (or methods) for any particular problem. In addition to these general objectives, the specific concepts listed in Table PT6.3 should be assimilated and mastered.

Computer Objectives.   You have been provided with software and simple computer algorithms to implement the techniques discussed in Part Six. All have utility as learning tools.

The Numerical Methods TOOLKIT personal computer software is user-friendly. It employs the trapezoidal rule to evaluate the integral of either continuous functions or tabular data. The graphics associated with this software will enable you to easily visualize your problem and the associated mathematical operations as the area between the curve and the x axis. The software is very easy to apply to solve many practical problems and can be used to check the results of any computer programs you may develop yourself.

In addition, algorithms are provided for most of the other methods in Part Five. This information will allow you to expand your software library to include techniques beyond the trapezoidal rule. For example, you may find it useful from a professional viewpoint to have software to implement numerical integration and differentiation of unequally spaced data. You may also want to develop your own software for Simpson's rules, Romberg integration, and Gauss quadrature, which are usually more efficient and accurate than the trapezoidal rule.

Finally, one of your most important goals should be to master several of the general-purpose software packages that are widely available. In particular, you should become adept at using these tools to implement numerical methods for engineering problem solving.

CHAPTER 21

Newton-Cotes Integration Formulas

The *Newton-Cotes formulas* are the most common numerical integration schemes. They are based on the strategy of replacing a complicated function or tabulated data with an approximating function that is easy to integrate:

$$I = \int_a^b f(x)\,dx \cong \int_a^b f_n(x)\,dx \tag{21.1}$$

where $f_n(x) = $ a polynomial of the form

$$f_n(x) = a_0 + a_1 x + \cdots + a_{n-1} x^{n-1} + a_n x^n$$

where n is the order of the polynomial. For example, in Fig. 21.1a, a first-order polynomial (a straight line) is used as an approximation. In Fig. 21.1b, a parabola is employed for the same purpose.

The integral can also be approximated using a series of polynomials applied piecewise to the function or data over segments of constant length. For example, in Fig. 21.2, three

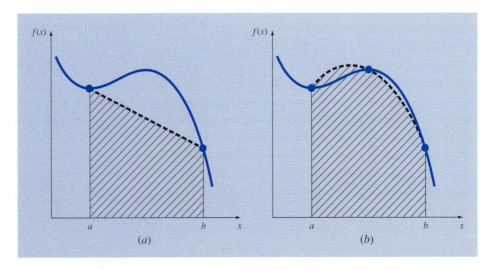

straight-line segments are used to approximate the integral. Higher-order polynomials can be utilized for the same purpose. With this background, we now recognize that the "strip method" in Fig. PT6.6 employed a series of zero-order polynomials (that is, constants) to approximate the integral.

Closed and open forms of the Newton-Cotes formulas are available. The *closed forms* are those where the data points at the beginning and end of the limits of integration are known (Fig. 21.3*a*). The *open forms* have integration limits that extend beyond the range of the data (Fig. 21.3*b*). In this sense, they are akin to extrapolation as discussed in Sec. 18.5. Open Newton-Cotes formulas are not generally used for definite integration.

FIGURE 21.2
The approximation of an integral by the area under three straight-line segments.

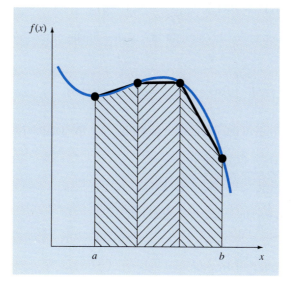

FIGURE 21.3
The difference between (*a*) closed and (*b*) open integration formulas.

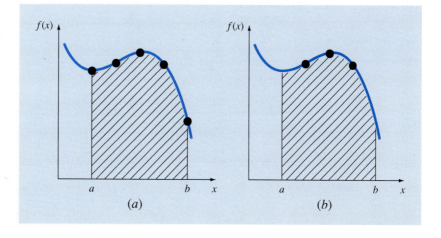

However, they are utilized for evaluating improper integrals and for the solution of ordinary differential equations. This chapter emphasizes the closed forms. However, material on open Newton-Cotes formulas is briefly introduced at the end of this chapter.

21.1 THE TRAPEZOIDAL RULE

The *trapezoidal rule* is the first of the Newton-Cotes closed integration formulas. It corresponds to the case where the polynomial in Eq. (21.1) is first-order:

$$I = \int_a^b f(x)\, dx \cong \int_a^b f_1(x)\, dx$$

Recall from Chap. 18 that a straight line can be represented as [Eq. (18.2)]

$$f_1(x) = f(a) + \frac{f(b) - f(a)}{b - a}(x - a) \tag{21.2}$$

The area under this straight line is an estimate of the integral of $f(x)$ between the limits a and b:

$$I = \int_a^b \left[f(a) + \frac{f(b) - f(a)}{b - a}(x - a) \right] dx$$

The result of the integration (see Box 21.1 for details) is

$$I = (b - a)\frac{f(a) + f(b)}{2} \tag{21.3}$$

which is called the *trapezoidal rule*.

Box 21.1   Derivation of Trapezoidal Rule

Before integration, Eq. (21.2) can be expressed as

$$f_1(x) = \frac{f(b) - f(a)}{b - a}x + f(a) - \frac{af(b) - af(a)}{b - a}$$

Grouping the last two terms gives

$$f_1(x) = \frac{f(b) - f(a)}{b - a}x + \frac{bf(a) - af(a) - af(b) + af(a)}{b - a}$$

or

$$f_1(x) = \frac{f(b) - f(a)}{b - a}x + \frac{bf(a) - af(b)}{b - a}$$

which can be integrated between $x = a$ and $x = b$ to yield

$$I = \frac{f(b) - f(a)}{b - a}\frac{x^2}{2} + \frac{bf(a) - af(b)}{b - a}x \bigg|_a^b$$

This result can be evaluated to give

$$I = \frac{f(b) - f(a)}{b - a}\frac{(b^2 - a^2)}{2} + \frac{bf(a) - af(b)}{b - a}(b - a)$$

Now, since $b^2 - a^2 = (b - a)(b + a)$,

$$I = [f(b) - f(a)]\frac{b + a}{2} + bf(a) - af(b)$$

Multiplying and collecting terms yields

$$I = (b - a)\frac{f(a) + f(b)}{2}$$

which is the formula for the trapezoidal rule.

Geometrically, the trapezoidal rule is equivalent to approximating the area of the trapezoid under the straight line connecting $f(a)$ and $f(b)$ in Fig. 21.4. Recall from geometry that the formula for computing the area of a trapezoid is the height times the average of the bases (Fig. 21.5a). In our case, the concept is the same but the trapezoid is on its side (Fig. 21.5b). Therefore, the integral estimate can be represented as

$$I \cong \text{width} \times \text{average height} \tag{21.4}$$

FIGURE 21.4
Graphical depiction of the trapezoidal rule.

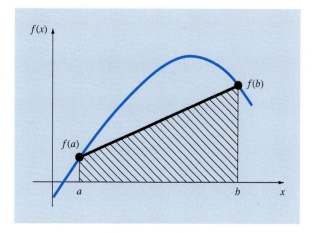

FIGURE 21.5
(a) The formula for computing the area of a trapezoid: height times the average of the bases.
(b) For the trapezoidal rule, the concept is the same but the trapezoid is on its side.

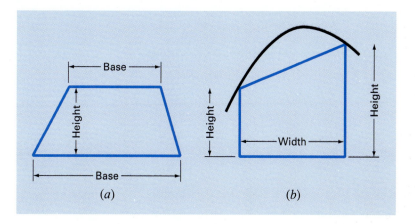

or

$$I \cong (b - a) \times \text{average height} \tag{21.5}$$

where, for the trapezoidal rule, the average height is the average of the function values at the end points, or $[f(a) + f(b)]/2$.

All the Newton-Cotes closed formulas can be expressed in the general format of Eq. (21.5). In fact, they differ only with respect to the formulation of the average height.

21.1.1 Error of the Trapezoidal Rule

When we employ the integral under a straight-line segment to approximate the integral under a curve, we obviously can incur an error that may be substantial (Fig. 21.6). An estimate for the local truncation error of a single application of the trapezoidal rule is (Box. 21.2)

$$E_t = -\frac{1}{12} f''(\xi)(b - a)^3 \tag{21.6}$$

where ξ lies somewhere in the interval from a to b. Equation (21.6) indicates that if the function being integrated is linear, the trapezoidal rule will be exact. Otherwise, for functions with second- and higher-order derivatives (that is, with curvature), some error can occur.

FIGURE 21.6
Graphical depiction of the use of a single application of the trapezoidal rule to approximate the integral of $f(x) = 0.2 + 25x - 200x^2 + 675x^3 - 900x^4 + 400x^5$ from $x = 0$ to 0.8.

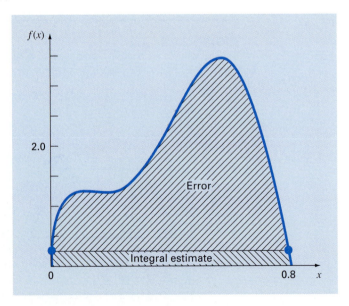

Box 21.2   Derivation and Error Estimate of the Trapezoidal Rule

An alternative derivation of the trapezoidal rule is possible by integrating the forward Newton-Gregory interpolating polynomial. Recall that for the first-order version with error term, the integral would be (Box 18.2)

$$I = \int_a^b \left[f(a) + \Delta f(a)\alpha + \frac{f''(\xi)}{2}\alpha(\alpha - 1)h^2 \right] dx \quad \text{(B21.2.1)}$$

To simplify the analysis, realize that because $\alpha = (x - a)/h$,

$$dx = h \, d\alpha$$

Inasmuch as $h = b - a$ (for the one-segment trapezoidal rule), the limits of integration a and b correspond to 0 and 1, respectively. Therefore, Eq. (B21.2.1) can be expressed as

$$I = h \int_0^1 \left[f(a) + \Delta f(a)\alpha + \frac{f''(\xi)}{2}\alpha(\alpha - 1)h^2 \right] d\alpha$$

If it is assumed that, for small h, the term $f''(\xi)$ is approximately

constant, this equation can be integrated:

$$I = h \left[\alpha f(a) + \frac{\alpha^2}{2}\Delta f(a) + \left(\frac{\alpha^3}{6} - \frac{\alpha^2}{4} \right) f''(\xi)h^2 \right]_0^1$$

and evaluated as

$$I = h \left[f(a) + \frac{\Delta f(a)}{2} \right] - \frac{1}{12}f''(\xi)h^3$$

Because $\Delta f(a) = f(b) - f(a)$, the result can be written as

$$I = h \underbrace{\frac{f(a) + f(b)}{2}}_{\text{Trapezoidal rule}} - \underbrace{\frac{1}{12}f''(\xi)h^3}_{\text{Truncation error}}$$

Thus, the first term is the trapezoidal rule and the second is an approximation for the error.

EXAMPLE 21.1   Single Application of the Trapezoidal Rule

Problem Statement.   Use Eq. (21.3) to numerically integrate

$$f(x) = 0.2 + 25x - 200x^2 + 675x^3 - 900x^4 + 400x^5$$

from $a = 0$ to $b = 0.8$. Recall from PT6.2 that the exact value of the integral can be determined analytically to be 1.640533.

Solution.   The function values

$$f(0) = 0.2$$
$$f(0.8) = 0.232$$

can be substituted into Eq. (21.3) to yield

$$I \cong 0.8 \frac{0.2 + 0.232}{2} = 0.1728$$

which represents an error of

$$E_t = 1.640533 - 0.1728 = 1.467733$$

which corresponds to a percent relative error of $\varepsilon_t = 89.5\%$. The reason for this large error is evident from the graphical depiction in Fig. 21.6. Notice that the area under the straight line neglects a significant portion of the integral lying above the line.

In actual situations, we would have no foreknowledge of the true value. Therefore, an approximate error estimate is required. To obtain this estimate, the function's second derivative over the interval can be computed by differentiating the original function twice to give

$$f''(x) = -400 + 4050x - 10,800x^2 + 8000x^3$$

The average value of the second derivative can be computed using Eq. (PT6.4):

$$\bar{f}''(x) = \frac{\displaystyle\int_0^{0.8} \left(-400 + 4050x - 10,800x^2 + 8000x^3 \right) dx}{0.8 - 0} = -60$$

which can be substituted into Eq. (21.6) to yield

$$E_a = -\frac{1}{12}(-60)(0.8)^3 = 2.56$$

which is of the same order of magnitude and sign as the true error. A discrepancy does exist, however, because of the fact that for an interval of this size, the average second derivative is not necessarily an accurate approximation of $f''(\xi)$. Thus, we denote that the error is approximate by using the notation E_a, rather than exact by using E_t.

21.1.2 The Multiple-Application Trapezoidal Rule

One way to improve the accuracy of the trapezoidal rule is to divide the integration interval from a to b into a number of segments and apply the method to each segment (Fig. 21.7). The areas of individual segments can then be added to yield the integral for the entire interval. The resulting equations are called *multiple-application,* or *composite, integration formulas.*

Figure 21.8 shows the general format and nomenclature we will use to characterize multiple-application integrals. There are $n + 1$ equally spaced base points ($x_0, x_1, x_2, \ldots, x_n$). Consequently, there are n segments of equal width:

$$h = \frac{b - a}{n} \tag{21.7}$$

If a and b are designated as x_0 and x_n, respectively, the total integral can be represented as

$$I = \int_{x_0}^{x_1} f(x)\,dx + \int_{x_1}^{x_2} f(x)\,dx + \cdots + \int_{x_{n-1}}^{x_n} f(x)\,dx$$

Substituting the trapezoidal rule for each integral yields

$$I = h\frac{f(x_0) + f(x_1)}{2} + h\frac{f(x_1) + f(x_2)}{2} + \cdots + h\frac{f(x_{n-1}) + f(x_n)}{2} \tag{21.8}$$

or, grouping terms,

$$I = \frac{h}{2}\left[f(x_0) + 2\sum_{i=1}^{n-1} f(x_i) + f(x_n) \right] \tag{21.9}$$

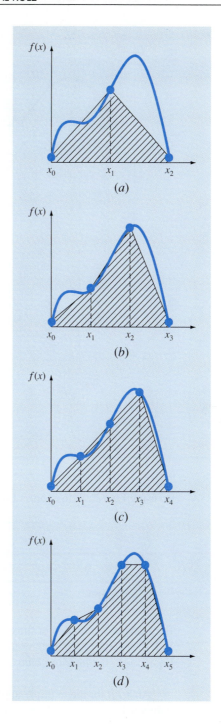

FIGURE 21.7
Illustration of the multiple-application trapezoidal rule. (*a*) Two segments, (*b*) three segments, (*c*) four segments, and (*d*) five segments.

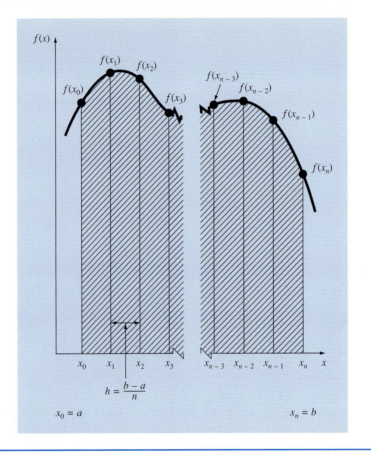

FIGURE 21.8
The general format and nomenclature for multiple-application integrals.

or, using Eq. (21.7) to express Eq. (21.9) in the general form of Eq. (21.5),

$$I = (b - a) \underbrace{\frac{f(x_0) + 2 \sum\limits_{i=1}^{n-1} f(x_i) + f(x_n)}{2n}}_{\text{Average height}} \tag{21.10}$$

Width

Because the summation of the coefficients of $f(x)$ in the numerator divided by $2n$ is equal to 1, the average height represents a weighted average of the function values. According to Eq. (21.10), the interior points are given twice the weight of the two end points $f(x_0)$ and $f(x_n)$.

An error for the multiple-application trapezoidal rule can be obtained by summing the individual errors for each segment to give

$$E_t = -\frac{(b-a)^3}{12n^3} \sum_{i=1}^{n} f''(\xi_i) \tag{21.11}$$

where $f''(\xi_i)$ is the second derivative at a point ξ_i located in segment i. This result can be simplified by estimating the mean or average value of the second derivative for the entire interval as [Eq. (PT6.3)]

$$\bar{f}'' \cong \frac{\displaystyle\sum_{i=1}^{n} f''(\xi_i)}{n} \tag{21.12}$$

Therefore $\Sigma f''(\xi_i) \cong n\bar{f}''$ and Eq. (21.11) can be rewritten as

$$E_a = -\frac{(b-a)^3}{12n^2}\bar{f}'' \tag{21.13}$$

Thus, if the number of segments is doubled, the truncation error will be quartered. Note that Eq. (21.13) is an approximate error because of the approximate nature of Eq. (21.12).

EXAMPLE 21.2   Multiple-Application Trapezoidal Rule

Problem Statement.   Use the two-segment trapezoidal rule to estimate the integral of

$$f(x) = 0.2 + 25x - 200x^2 + 675x^3 - 900x^4 + 400x^5$$

from $a = 0$ to $b = 0.8$. Employ Eq. (21.13) to estimate the error. Recall that the correct value for the integral is 1.640533.

Solution. $n = 2(h = 0.4)$:

$$f(0) = 0.2 \qquad f(0.4) = 2.456 \qquad f(0.8) = 0.232$$

$$I = 0.8\frac{0.2 + 2(2.456) + 0.232}{4} = 1.0688$$

$$E_t = 1.640533 - 1.0688 = 0.57173 \qquad \varepsilon_t = 34.9\%$$

$$E_a = -\frac{0.8^3}{12(2)^2}(-60) = 0.64$$

where -60 is the average second derivative determined previously in Example 21.1.

The results of the previous example, along with three- through ten-segment applications of the trapezoidal rule, are summarized in Table 21.1. Notice how the error decreases as the number of segments increases. However, also notice that the rate of decrease is gradual. This is because the error is inversely related to the square of n [Eq. (21.13)]. Therefore, doubling the number of segments quarters the error. In subsequent sections we develop higher-order formulas that are more accurate and that converge more quickly on the true integral as the segments are increased. However, before investigating these formulas, we will first discuss computer algorithms to implement the trapezoidal rule.

21.1.3 Computer Algorithms for the Trapezoidal Rule

Two simple algorithms for the trapezoidal rule are listed in Fig. 21.9. The first (Fig. 21.9a) is for the single-segment version. The second (Fig. 21.9b) is for the multiple-segment version with a constant segment width. Note that both are designed for data that is in tabu-

TABLE 21.1 Results for multiple-application trapezoidal rule to estimate the integral of $f(x) = 0.2 + 25x - 200x^2 + 675x^3 - 900x^4 + 400x^5$ from $x = 0$ to 0.8. The exact value is 1.640533.

n	h	I	$\varepsilon_t(\%)$
2	0.4	1.0688	34.9
3	0.2667	1.3695	16.5
4	0.2	1.4848	9.5
5	0.16	1.5399	6.1
6	0.1333	1.5703	4.3
7	0.1143	1.5887	3.2
8	0.1	1.6008	2.4
9	0.0889	1.6091	1.9
10	0.08	1.6150	1.6

(a) Single-segment

```
FUNCTION Trap (h, f0, f1)
  Trap = h * (f0 + f1) / 2
END Trap
```

(b) Multiple-segment

```
FUNCTION Trapm (h, n, f)
  sum = f0
  DO i = 1, n − 1
    sum = sum + 2 * fi
  END DO
  sum = sum + fn
  Trapm = h * sum / 2
END Trapm
```

FIGURE 21.9
Algorithms for the (a) single-segment and (b) multiple-segment trapezoidal rule.

lated form. A general program should have the capability to evaluate known functions or equations as well. We will illustrate how functions are handled in the next chapter.

An example of a user-friendly program for the multiple-segment trapezoidal rule is included in the supplementary Numerical Methods TOOLKIT software associated with this text. This software can evaluate the integrals of either tabulated data or user-defined functions. The following example demonstrates its utility for evaluating integrals. It also provides a good reference for assessing and testing your own software.

EXAMPLE 21.3   Evaluating Integrals with the Computer

Problem Statement.   Use the Numerical Methods TOOLKIT software to solve a problem related to our friend the falling parachutist. As you recall from Example 1.1, the velocity of the parachutist is given as the following function of time:

$$v(t) = \frac{gm}{c}\left(1 - e^{-(c/m)t}\right)$$ (E21.3.1)

where v = velocity (m/s), g = the gravitational constant of 9.8 m/s², m = mass of the parachutist equal to 68.1 kg, and c = the drag coefficient of 12.5 kg/s. The model predicts the velocity of the parachutist as a function of time as described in Example 1.1.

Suppose we would like to know how far the parachutist has fallen after a certain time t. This distance is given by [Eq. (PT6.5)]

$$d = \int_0^t v(t)\, dt$$

where d is the distance in meters. Substituting Eq. (E21.3.1),

$$d = \frac{gm}{c} \int_0^t \left(1 - e^{-(c/m)t}\right) dt$$

Use the Numerical Methods TOOLKIT software and your own software to determine this integral with the multiple-segment trapezoidal rule using different numbers of segments. Note that performing the integration analytically and substituting known parameter values results in an exact value of $d = 289.43515$ m.

Solution.   Click the Integrate Function button on the TOOLKIT main menu to obtain a blank screen similar to Fig. 21.10. This screen contains the input and output information needed to integrate an analytical function or tabular data.

First, we can click the Input Function table and enter the function,

$$v(t) = \frac{9.8(68.1)}{12.5} \left(1 - e^{-(12.5/68.1)t}\right)$$

Next click the Input Parameters block and enter values for the lower and upper limits of integration of 0 and 10. Next, enter the value 10 for the number of segments along with plot dimensions, as in Fig. 21.10.

FIGURE 21.10
Computer screen showing the integral of a function using the multiple-segment trapezoidal rule program from the Numerical Methods TOOLKIT.

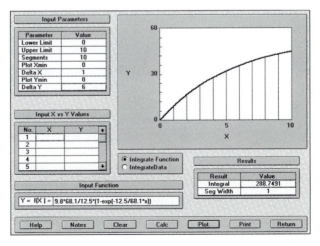

After clicking the Calc and Plot buttons, a calculated integral of 288.7491 is displayed. Thus, we have attained the integral to three significant digits of accuracy. The integral is equivalent to the area under the $v(t)$ versus t curve, as shown in Fig. 21.10. Visual inspection confirms that the integral is the width of the interval (10 s) times the average height (about 29 m/s). A plot of the function showing the segments is shown in Fig. 21.10. We can conveniently try other numbers of segments. By $n = 500$, the result is accurate to six significant digits.

At this point, it is important to recognize that the Numerical Methods TOOLKIT uses double precision to obtain its integral estimate. We therefore repeated this problem using a program based on the pseudocode from Fig. 21.9b and employed single-precision numbers (i.e., about seven significant decimal digits of precision). The results are

Segments	Segment size	Estimated d, m	ε_t (%)
10	1.0	288.7491	0.237
20	0.5	289.2636	0.0593
50	0.2	289.4076	9.5×10^{-3}
100	0.1	289.4282	2.4×10^{-3}
200	0.05	289.4336	5.4×10^{-4}
500	0.02	289.4348	1.2×10^{-4}
1,000	0.01	289.4360	-3.0×10^{-4}
2,000	0.005	289.4369	-5.9×10^{-4}
5,000	0.002	289.4337	5.2×10^{-4}
10,000	0.001	289.4317	1.2×10^{-3}

Thus, up to about 500 segments, the multiple-application trapezoidal rule attains excellent accuracy. However, notice how the error changes sign and begins to increase in absolute value beyond the 500-segment case. The 10,000-segment case actually seems to be diverging from the true value. This is due to the intrusion of round-off error because of the great number of computations for this many segments. Thus, the level of precision is limited, and we would never reach the exact result of 289.4351 obtained analytically. This limitation will be discussed in further detail in Chap. 22.

Three major conclusions can be drawn from the Example 21.3:

- For individual applications with nicely behaved functions, the multiple-segment trapezoidal rule is just fine for attaining the type of accuracy required in many engineering applications.
- If high accuracy is required, the multiple-segment trapezoidal rule demands a great deal of computational effort. Although this effort may be negligible for a single application, it could be very important when (a) numerous integrals are being evaluated or (b) where the function itself is time consuming to evaluate. For such cases, more efficient approaches (like those in the remainder of this chapter and the next) may be necessary.
- Finally, round-off error can represent a limitation on our ability to determine integrals. This is due both to the machine precision as well as to the numerous computations involved in simple techniques like the multiple-segment trapezoidal rule.

We now turn to one way in which efficiency is improved. That is, by using higher-order polynomials to approximate the integral.

21.2 SIMPSON'S RULES

Aside from applying the trapezoidal rule with finer segmentation, another way to obtain a more accurate estimate of an integral is to use higher-order polynomials to connect the points. For example, if there is an extra point midway between $f(a)$ and $f(b)$, the three points can be connected with a parabola (Fig. 21.11a). If there are two points equally spaced between $f(a)$ and $f(b)$, the four points can be connected with a third-order polynomial (Fig. 21.11b). The formulas that result from taking the integrals under these polynomials are called *Simpson's rules*.

21.2.1 Simpson's 1/3 Rule

Simpson's 1/3 rule results when a second-order interpolating polynomial is substituted into Eq. (21.1):

$$I = \int_a^b f(x)\,dx \cong \int_a^b f_2(x)\,dx$$

If a and b are designated as x_0 and x_2 and $f_2(x)$ is represented by a second-order Lagrange polynomial [Eq. (18.23)], the integral becomes

$$I = \int_{x_0}^{x_2} \left[\frac{(x-x_1)(x-x_2)}{(x_0-x_1)(x_0-x_2)} f(x_0) + \frac{(x-x_0)(x-x_2)}{(x_1-x_0)(x_1-x_2)} f(x_1) \right. $$
$$\left. + \frac{(x-x_0)(x-x_1)}{(x_2-x_0)(x_2-x_1)} f(x_2) \right] dx$$

FIGURE 21.11
(a) Graphical depiction of Simpson's 1/3 rule: it consists of taking the area under a parabola connecting three points. (b) Graphical depiction of Simpson's 3/8 rule: it consists of taking the area under a cubic equation connecting four points.

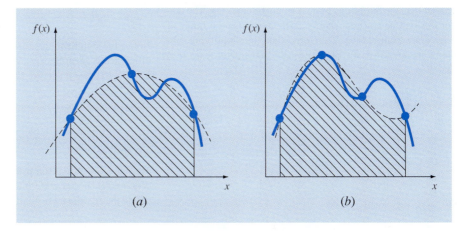

After integration and algebraic manipulation, the following formula results:

$$I \cong \frac{h}{3}\left[f(x_0) + 4f(x_1) + f(x_2)\right] \tag{21.14}$$

where, for this case, $h = (b - a)/2$. This equation is known as *Simpson's 1/3 rule*. It is the second Newton-Cotes closed integration formula. The label "1/3" stems from the fact that h is divided by 3 in Eq. (21.14). An alternative derivation is shown in Box 21.3 where the Newton-Gregory polynomial is integrated to obtain the same formula.

Simpson's 1/3 rule can also be expressed using the format of Eq. (21.5):

$$I \cong \underbrace{(b - a)}_{\text{Width}} \underbrace{\frac{f(x_0) + 4f(x_1) + f(x_2)}{6}}_{\text{Average height}} \tag{21.15}$$

Box 21.3   Derivation and Error Estimate of Simpson's 1/3 Rule

As was done in Box 21.2 for the trapezoidal rule, Simpson's 1/3 rule can be derived by integrating the forward Newton-Gregory interpolating polynomial (Box 18.2):

$$I = \int_{x_0}^{x_2} \left[f(x_0) + \Delta f(x_0)\alpha + \frac{\Delta^2 f(x_0)}{2}\alpha(\alpha - 1) \right.$$
$$+ \frac{\Delta^3 f(x_0)}{6}\alpha(\alpha - 1)(\alpha - 2)$$
$$\left. + \frac{f^{(4)}(\xi)}{24}\alpha(\alpha - 1)(\alpha - 2)(\alpha - 3)h^4 \right] dx$$

Notice that we have written the polynomial up to the fourth-order term rather than the third-order term as would be expected. The reason for this will be apparent shortly. Also notice that the limits of integration are from x_0 to x_2. Therefore, when the simplifying substitutions are made (recall Box 21.2), the integral is from $\alpha = 0$ to 2:

$$I = h \int_0^2 \left[f(x_0) + \Delta f(x_0)\alpha + \frac{\Delta^2 f(x_0)}{2}\alpha(\alpha - 1) \right.$$
$$+ \frac{\Delta^3 f(x_0)}{6}\alpha(\alpha - 1)(\alpha - 2)$$
$$\left. + \frac{f^{(4)}(\xi)}{24}\alpha(\alpha - 1)(\alpha - 2)(\alpha - 3)h^4 \right] d\alpha$$

which can be integrated to yield

$$I = h \left[\alpha f(x_0) + \frac{\alpha^2}{2}\Delta f(x_0) + \left(\frac{\alpha^3}{6} - \frac{\alpha^2}{4}\right)\Delta^2 f(x_0) \right.$$
$$+ \left(\frac{\alpha^4}{24} - \frac{\alpha^3}{6} + \frac{\alpha^2}{6}\right)\Delta^3 f(x_0)$$
$$\left. + \left(\frac{\alpha^5}{120} - \frac{\alpha^4}{16} + \frac{11\alpha^3}{72} - \frac{\alpha^2}{8}\right)f^{(4)}(\xi)h^4 \right]_0^2$$

and evaluated for the limits to give

$$I = h \left[2f(x_0) + 2\Delta f(x_0) + \frac{\Delta^2 f(x_0)}{3} \right.$$
$$\left. + (0)\Delta^3 f(x_0) - \frac{1}{90}f^{(4)}(\xi)h^4 \right] \tag{B21.3.1}$$

Notice the significant result that the coefficient of the third divided difference is zero. Because $\Delta f(x_0) = f(x_1) - f(x_0)$ and $\Delta^2 f(x_0) = f(x_2) - 2f(x_1) + f(x_0)$, Eq. (B21.3.1) can be rewritten as

$$I = \underbrace{\frac{h}{3}[f(x_0) + 4f(x_1) + f(x_2)]}_{\text{Simpson's 1/3 rule}} - \underbrace{\frac{1}{90}f^{(4)}(\xi)h^5}_{\text{Truncation error}}$$

Thus, the first term is Simpson's 1/3 rule and the second is the truncation error. Because the third divided difference dropped out, we obtain the significant result that the formula is third-order accurate.

where $a = x_0$, $b = x_2$, and $x_1 =$ the point midway between a and b, which is given by $(b + a)/2$. Notice that, according to Eq. (21.15), the middle point is weighted by two-thirds and the two end points by one-sixth.

It can be shown that a single-segment application of Simpson's 1/3 rule has a truncation error of (Box 21.3)

$$E_t = -\frac{1}{90}h^5 f^{(4)}(\xi)$$

or, because $h = (b - a)/2$,

$$E_t = -\frac{(b - a)^5}{2880} f^{(4)}(\xi) \qquad (21.16)$$

where ξ lies somewhere in the interval from a to b. Thus, Simpson's 1/3 rule is more accurate than the trapezoidal rule. However, comparison with Eq. (21.6) indicates that it is more accurate than expected. Rather than being proportional to the third derivative, the error is proportional to the fourth derivative. This is because, as shown in Box 21.3, the coefficient of the third-order term goes to zero during the integration of the interpolating polynomial. Consequently, Simpson's 1/3 rule is third-order accurate even though it is based on only three points. In other words, it yields exact results for cubic polynomials even though it is derived from a parabola!

EXAMPLE 21.4   Single Application of Simpson's 1/3 Rule

Problem Statement.   Use Eq. (21.15) to integrate

$$f(x) = 0.2 + 25x - 200x^2 + 675x^3 - 900x^4 + 400x^5$$

from $a = 0$ to $b = 0.8$. Recall that the exact integral is 1.640533.

Solution.

$$f(0) = 0.2 \qquad f(0.4) = 2.456 \qquad f(0.8) = 0.232$$

Therefore, Eq. (21.15) can be used to compute

$$I \cong 0.8 \frac{0.2 + 4(2.456) + 0.232}{6} = 1.367467$$

which represents an exact error of

$$E_t = 1.640533 - 1.367467 = 0.2730667 \qquad \varepsilon_t = 16.6\%$$

which is approximately 5 times more accurate than for a single application of the trapezoidal rule (Example 21.1).

The estimated error is [Eq . (21.16)]

$$E_a = -\frac{(0.8)^5}{2880}(-2400) = 0.2730667$$

where -2400 is the average fourth derivative for the interval as obtained using Eq. (PT6.4). As was the case in Example 21.1, the error is approximate (E_a) because the average fourth derivative is not an exact estimate of $f^{(4)}(\xi)$. However, because this case deals with a fifth-order polynomial, the result matches.

21.2.2 The Multiple-Application Simpson's 1/3 Rule

Just as with the trapezoidal rule, Simpson's rule can be improved by dividing the integration interval into a number of segments of equal width (Fig. 21.12):

$$h = \frac{b - a}{n} \tag{21.17}$$

The total integral can be represented as

$$I = \int_{x_0}^{x_2} f(x)\,dx + \int_{x_2}^{x_4} f(x)\,dx + \cdots + \int_{x_{n-2}}^{x_n} f(x)\,dx$$

Substituting Simpson's 1/3 rule for the individual integral yields

$$I \cong 2h\frac{f(x_0) + 4f(x_1) + f(x_2)}{6} + 2h\frac{f(x_2) + 4f(x_3) + f(x_4)}{6}$$

$$+ \cdots + 2h\frac{f(x_{n-2}) + 4f(x_{n-1}) + f(x_n)}{6}$$

or, combining terms and using Eq. (21.17),

$$I \cong \underbrace{(b - a)}_{\text{Width}} \underbrace{\frac{f(x_0) + 4\sum_{i=1,3,5}^{n-1} f(x_i) + 2\sum_{j=2,4,6}^{n-2} f(x_j) + f(x_n)}{3n}}_{\text{Average height}} \tag{21.18}$$

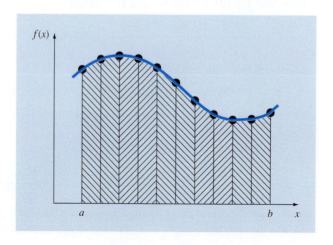

FIGURE 21.12
Graphical representation of the multiple application of Simpson's 1/3 rule. Note that the method can be employed only if the number of segments is even.

Notice that, as illustrated in Fig. 21.12, an even number of segments must be utilized to implement the method. In addition, the coefficients "4" and "2" in Eq. (21.18) might seem peculiar at first glance. However, they follow naturally from Simpson's 1/3 rule. The odd points represent the middle term for each application and hence carry the weight of 4 from Eq. (21.15). The even points are common to adjacent applications and hence are counted twice.

An error estimate for the multiple-application Simpson's rule is obtained in the same fashion as for the trapezoidal rule by summing the individual errors for the segments and averaging the derivative to yield

$$E_a = -\frac{(b-a)^5}{180n^4} \bar{f}^{(4)}$$

(21.19)

where $\bar{f}^{(4)}$ is the average fourth derivative for the interval.

EXAMPLE 21.5   Multiple-Application Version of Simpson's 1/3 Rule

Problem Statement.   Use Eq. (21.18) with $n = 4$ to estimate the integral of

$$f(x) = 0.2 + 25x - 200x^2 + 675x^3 - 900x^4 + 400x^5$$

from $a = 0$ to $b = 0.8$. Recall that the exact integral is 1.640533.

Solution. $n = 4$ ($h = 0.2$):

$$f(0) = 0.2 \qquad f(0.2) = 1.288$$
$$f(0.4) = 2.456 \qquad f(0.6) = 3.464$$
$$f(0.8) = 0.232$$

From Eq. (21.18),

$$I = 0.8 \frac{0.2 + 4(1.288 + 3.464) + 2(2.456) + 0.232}{12} = 1.623467$$

$$E_t = 1.640533 - 1.623467 = 0.017067 \qquad \varepsilon_t = 1.04\%$$

The estimated error [Eq. (21.19)] is

$$E_a = -\frac{(0.8)^5}{180(4)^4}(-2400) = 0.017067$$

The previous example illustrates that the multiple-application version of Simpson's 1/3 rule yields very accurate results. For this reason, it is considered superior to the trapezoidal rule for most applications. However, as mentioned previously, it is limited to cases where the values are equispaced. Further, it is limited to situations where there are an even number of segments and an odd number of points. Consequently, as discussed in the next section, an odd-segment–even-point formula known as Simpson's 3/8 rule is used in conjunction with the 1/3 rule to permit evaluation of both even and odd numbers of segments.

21.2.3 Simpson's 3/8 Rule

In a similar manner to the derivation of the trapezoidal and Simpson's 1/3 rule, a third-order Lagrange polynomial can be fit to four points and integrated:

$$I = \int_a^b f(x)\,dx \cong \int_a^b f_3(x)\,dx$$

to yield

$$I \cong \frac{3h}{8}\left[f(x_0) + 3f(x_1) + 3f(x_2) + f(x_3)\right]$$

where $h = (b - a)/3$. This equation is called *Simpson's 3/8 rule* because h is multiplied by 3/8. It is the third Newton-Cotes closed integration formula. The 3/8 rule can also be expressed in the form of Eq. (21.5):

$$I \cong \underbrace{(b - a)}_{\text{Width}} \underbrace{\frac{f(x_0) + 3f(x_1) + 3f(x_2) + f(x_3)}{8}}_{\text{Average height}} \tag{21.20}$$

FIGURE 21.13

Illustration of how Simpson's 1/3 and 3/8 rules can be applied in tandem to handle multiple applications with odd numbers of intervals.

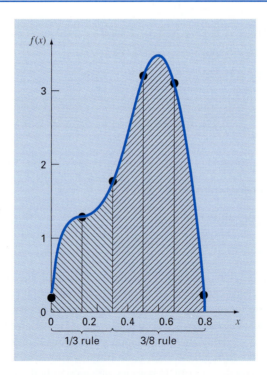

Thus, the two interior points are given weights of three-eighths, whereas the end points are weighted with one-eighth. Simpson's 3/8 rule has an error of

$$E_t = -\frac{3}{80} h^5 f^{(4)}(\xi)$$

or, because $h = (b - a)/3$,

$$E_t = -\frac{(b - a)^5}{6480} f^{(4)}(\xi) \tag{21.21}$$

Because the denominator of Eq. (21.21) is larger than for Eq. (21.16), the 3/8 rule is somewhat more accurate than the 1/3 rule.

Simpson's 1/3 rule is usually the method of preference because it attains third-order accuracy with three points rather than the four points required for the 3/8 version. However, the 3/8 rule has utility when the number of segments is odd. For instance, in Example 21.5 we used Simpson's rule to integrate the function for four segments. Suppose that you desired an estimate for five segments. One option would be to use a multiple-application version of the trapezoidal rule as was done in Examples 21.2 and 21.3. This may not be advisable, however, because of the large truncation error associated with this method. An alternative would be to apply Simpson's 1/3 rule to the first two segments and Simpson's 3/8 rule to the last three (Fig. 21.13). In this way, we could obtain an estimate with third-order accuracy across the entire interval.

EXAMPLE 21.6 Simpson's 3/8 Rule

Problem Statement.

(a) Use Simpson's 3/8 rule to integrate

$$f(x) = 0.2 + 25x - 200x^2 + 675x^3 - 900x^4 + 400x^5$$

from $a = 0$ to $b = 0.8$.
(b) Use it in conjunction with Simpson's 1/3 rule to integrate the same function for five segments.

Solution.

(a) A single application of Simpson's 3/8 rule requires four equally spaced points:

$$f(0) = 0.2 \qquad\qquad f(0.2667) = 1.432724$$
$$f(0.5333) = 3.487177 \qquad\qquad f(0.8) = 0.232$$

Using Eq. (21.20),

$$I \cong 0.8 \frac{0.2 + 3(1.432724 + 3.487177) + 0.232}{8} = 1.519170$$

$$E_t = 1.640533 - 1.519170 = 0.1213630 \qquad \varepsilon_t = 7.4\%$$

$$E_a = -\frac{(0.8)^5}{6480}(-2400) = 0.1213630$$

(b) The data needed for a five-segment application ($h = 0.16$) is

$$f(0) = 0.2 \qquad\qquad f(0.16) = 1.296919$$

$$f(0.32) = 1.743393 \qquad f(0.48) = 3.186015$$

$$f(0.64) = 3.181929 \qquad f(0.80) = 0.232$$

The integral for the first two segments is obtained using Simpson's 1/3 rule:

$$I \cong 0.32 \frac{0.2 + 4(1.296919) + 1.743393}{6} = 0.3803237$$

For the last three segments, the 3/8 rule can be used to obtain

$$I \cong 0.48 \frac{1.743393 + 3(3.186015 + 3.181929) + 0.232}{8} = 1.264754$$

The total integral is computed by summing the two results:

$$I = 0.3803237 + 1.264753 = 1.645077$$

$$E_t = 1.640533 - 1.645077 = -0.00454383 \qquad \varepsilon_t = -0.28\%$$

21.2.4 Computer Algorithms for Simpson's Rules

Pseudocode for a number of forms of Simpson's rule are outlined in Fig. 21.14. Note that all are designed for data that is in tabulated form. A general program should have the capability to evaluate known functions or equations as well. We will illustrate how functions are handled in the next chapter.

Notice that the program in Fig. 21.14d is set up so that either an even or odd number of segments may be used. For the even case, Simpson's 1/3 rule is applied to each pair of segments, and the results are summed to compute the total integral. For the odd case, Simpson's 3/8 rule is applied to the last three segments, and the 1/3 rule is applied to all the previous segments.

21.2.5 Higher-Order Newton-Cotes Closed Formulas

As noted previously, the trapezoidal rule and both of Simpson's rules are members of a family of integrating equations known as the Newton-Cotes closed integration formulas. Some of the formulas are summarized in Table 21.2, along with their truncation-error estimates.

Notice that, as was the case with Simpson's 1/3 and 3/8 rules, the five- and six-point formulas have the same order error. This general characteristic holds for the higher-point formulas and leads to the result that the even-segment–odd-point formulas (for example, 1/3 rule and Boole's rule) are usually the methods of preference.

However, it must also be stressed that, in engineering practice, the higher-order (that is, greater than four-point) formulas are rarely used. Simpson's rules are sufficient for most applications. Accuracy can be improved by using the multiple-application version. Furthermore, when the function is known and high accuracy is required, methods such as

(a)
```
FUNCTION Simp13 (h, f0, f1, f2)
   Simp13 = 2*h* (f0+4*f1+f2) / 6
END Simp13
```

(b)
```
FUNCTION Simp38 (h, f0, f1, f2, f3)
   Simp38 = 3*h* (f0+3*(f1+f2)+f3) / 8
END Simp38
```

(c)
```
FUNCTION Simp13m (h, n, f)
  sum = f(0)
  DO i = 1, n − 2, 2
    sum = sum + 4 * f_i + 2 * f_{i+1}
  END DO
  sum = sum + 4 * f_{n−1} + f_n
  Simp13m = h * sum / 3
END Simp13m
```

(d)
```
FUNCTION SimpInt(a, b, n, f)
  h = (b − a) / n
  IF n = 1 THEN
    sum = Trap(h, f_{n−1}, f_n)
  ELSE
    m = n
    odd = n / 2 − INT(n / 2)
    IF odd > 0 AND n > 1 THEN
      sum = sum+Simp38(h,f_{n−3},f_{n−2},f_{n−1},f_n)
      m = n − 3
    END IF
    IF m > 1 THEN
      sum = sum + Simp13m(h, m, f)
    END IF
  END IF
  SimpInt = sum
END SimpInt
```

FIGURE 21.14
Pseudocode for Simpson's rules. (a) Single-application Simpson's 1/3 rule, (b) single-application Simpson's 3/8 rule, (c) multiple-application Simpson's 1/3 rule, and (d) multiple-application Simpson's rule for both odd and even number of segments. Note that for all cases n must be ≥ 1.

TABLE 21.2 Newton-Cotes closed integration formulas. The formulas are presented in the format of Eq. (21.5) so that the weighting of the data points to estimate the average height is apparent. The step size is given by $h = (b − a)/n$.

Segments (n)	Points	Name	Formula	Truncation Error
1	2	Trapezoidal rule	$(b − a) \dfrac{f(x_0) + f(x_1)}{2}$	$-(1/12)h^3 f''(\xi)$
2	3	Simpson's 1/3 rule	$(b − a) \dfrac{f(x_0) + 4f(x_1) + f(x_2)}{6}$	$-(1/90)h^5 f^{(4)}(\xi)$
3	4	Simpson's 3/8 rule	$(b − a) \dfrac{f(x_0) + 3f(x_1) + 3f(x_2) + f(x_3)}{8}$	$-(3/80)h^5 f^{(4)}(\xi)$
4	5	Boole's rule	$(b − a) \dfrac{7f(x_0) + 32f(x_1) + 12f(x_2) + 32f(x_3) + 7f(x_4)}{90}$	$-(8/945)h^7 f^{(6)}(\xi)$
5	6		$(b − a) \dfrac{19f(x_0) + 75f(x_1) + 50f(x_2) + 50f(x_3) + 75f(x_4) + 19f(x_5)}{288}$	$-(275/12{,}096)h^7 f^{(6)}(\xi)$

Romberg integration or Gauss quadrature, described in Chap. 22, offer viable and attractive alternatives.

21.3 INTEGRATION WITH UNEQUAL SEGMENTS

To this point, all formulas for numerical integration have been based on equally spaced data points. In practice, there are many situations where this assumption does not hold and we must deal with unequal-sized segments. For example, experimentally derived data is often of this type. For these cases, one method is to apply the trapezoidal rule to each segment and sum the results:

$$I = h_1 \frac{f(x_0) + f(x_1)}{2} + h_2 \frac{f(x_1) + f(x_2)}{2} + \cdots + h_n \frac{f(x_{n-1}) + f(x_n)}{2} \qquad (21.22)$$

where h_i = the width of segment i. Note that this was the same approach used for the multiple-application trapezoidal rule. The only difference between Eqs. (21.8) and (21.22) is that the h's in the former are constant. Consequently, Eq. (21.8) could be simplified by grouping terms to yield Eq. (21.9). Although this simplification cannot be applied to Eq. (21.22), a computer program can be easily developed to accommodate unequal-sized segments. Before describing such an algorithm, we will illustrate in the following example how Eq. (21.22) is applied to evaluate an integral.

EXAMPLE 21.7   Trapezoidal Rule with Unequal Segments

Problem Statement.   The information in Table 21.3 was generated using the same polynomial employed in Example 21.1. Use Eq. (21.22) to determine the integral for this data. Recall that the correct answer is 1.640533.

Solution.   Applying Eq. (21.22) to the data in Table 21.3 yields

$$I = 0.12 \frac{1.309729 + 0.2}{2} + 0.10 \frac{1.305241 + 1.309729}{2} + \cdots + 0.10 \frac{0.232 + 2.363}{2}$$

$$= 0.090584 + 0.130749 + \cdots + 0.12975 = 1.594801$$

which represents an absolute percent relative error of $\varepsilon_t = 2.8\%$

TABLE 21.3   Data for $f(x) = 0.2 + 25x - 200x^2 + 675x^3 - 900x^4 + 400x^5$, with unequally spaced values of x.

x	f(x)	x	f(x)
0.0	0.200000	0.44	2.842985
0.12	1.309729	0.54	3.507297
0.22	1.305241	0.64	3.181929
0.32	1.743393	0.70	2.363000
0.36	2.074903	0.80	0.232000
0.40	2.456000		

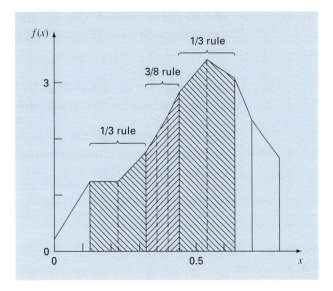

FIGURE 21.15
Use of the trapezoidal rule to determine the integral of unevenly spaced data. Notice how the shaded segments could be evaluated with Simpson's rule to attain higher accuracy.

The data from Example 21.7 is depicted in Fig. 21.15. Notice that some adjacent segments are of equal width and, consequently, could have been evaluated using Simpson's rules. This usually leads to more accurate results, as illustrated by the following example.

EXAMPLE 21.8   Inclusion of Simpson's Rules in the Evaluation of Uneven Data

Problem Statement.   Recompute the integral for the data in Table 21.3, but use Simpson's rules for those segments where they are appropriate.

Solution.   The first segment is evaluated with the trapezoidal rule:

$$I = 0.12\frac{1.309729 + 0.2}{2} = 0.09058376$$

Because the next two segments from $x = 0.12$ to 0.32 are of equal length, their integral can be computed with Simpson's 1/3 rule:

$$I = 0.2\frac{1.743393 + 4(1.305241) + 1.309729}{6} = 0.2758029$$

The next three segments are also equal and, as such, may be evaluated with the 3/8 rule to give $I = 0.2726863$. Similarly, the 1/3 rule can be applied to the two segments from $x = 0.44$ to 0.64 to yield $I = 0.6684701$. Finally, the last two segments, which are of unequal length, can be evaluated with the trapezoidal rule to give values of 0.1663479 and 0.1297500, respectively. The area of these individual segments can be summed to yield a total integral of 1.603641. This represents an error of $\varepsilon_t = 2.2\%$, which is superior to the result using the trapezoidal rule in Example 21.7.

Computer Program for Unequally Spaced Data. It is a fairly simple proposition to program Eq. (21.22). Such an algorithm is listed in Fig. 21.16a.

However, as demonstrated in Example 21.8, the approach is enhanced if it implements Simpson's rules wherever possible. For this reason, we have developed a second algorithm that incorporates this capability. As depicted in Fig 21.16b, the algorithm checks the length of adjacent segments. If two consecutive segments are of equal length, Simpson's 1/3 rule is applied. If three are equal, the 3/8 rule is used. When adjacent segments are of unequal length, the trapezoidal rule is implemented.

FIGURE 21.16

Pseudocode for integrating unequally spaced data. (a) Trapezoidal rule and (b) combination Simpson's and trapezoidal rules.

(a)

```
FUNCTION Trapun (x, y, n)
   LOCAL i, sum
   sum = 0
   DO i = 1, n
      sum = sum + (xᵢ − xᵢ₋₁)*(yᵢ₋₁ + yᵢ)/2
   END DO
   Trapun = sum
END Trapun
```

(b)

```
FUNCTION Uneven (n, x, f)
   h = x₁ − x₀
   k = 1
   sum = 0.
   DO j = 1, n
      hf = xⱼ₊₁ − xⱼ
      IF ABS (h − hf)< .000001 THEN
         IF k = 3 THEN
            sum = sum + Simp13 (h, fⱼ₋₃,fⱼ₋₂,fⱼ₋₁)
            k = k − 1
         ELSE
            k = k + 1
         END IF
      ELSE
         IF k = 1 THEN
            sum = sum + Trap (h, fⱼ₋₁,fⱼ)
         ELSE
            IF k = 2 THEN
               sum = sum + Simp13 (h, fⱼ₋₂, fⱼ₋₁, fⱼ)
            ELSE
               sum = sum + Simp38 (h, fⱼ₋₃, fⱼ₋₂, fⱼ₋₁, fⱼ)
            END IF
            k = 1
         END IF
      END IF
      h = hf
   END DO
   Uneven = sum
END Uneven
```

TABLE 21.4 Newton-Cotes open integration formulas. The formulas are presented in the format of Eq. (21.5) so that the weighting of the data points to estimate the average height is apparent. The step size is given by $h = (b - a)/n$.

Segments (n)	Points	Name	Formula	Truncation Error
2	1	Midpoint method	$(b - a) f(x_1)$	$(1/3)h^3 f''(\xi)$
3	2		$(b - a) \dfrac{f(x_1) + f(x_2)}{2}$	$(3/4)h^3 f''(\xi)$
4	3		$(b - a) \dfrac{2f(x_1) - f(x_2) + 2f(x_3)}{3}$	$(14/45)h^5 f^{(4)}(\xi)$
5	4		$(b - a) \dfrac{11f(x_1) + f(x_2) + f(x_3) + 11f(x_4)}{24}$	$(95/144)h^5 f^{(4)}(\xi)$
6	5		$(b - a) \dfrac{11f(x_1) - 14f(x_2) + 26f(x_3) - 14f(x_4) + 11f(x_5)}{20}$	$(41/140)h^7 f^{(6)}(\xi)$

Thus, not only does it allow evaluation of unequal segment data, but if equally spaced information is used, it reduces to Simpson's rules. As such, it represents a basic, all-purpose algorithm for the determination of the integral of tabulated data.

21.4 OPEN INTEGRATION FORMULAS

Recall from Fig 21.3b that open integration formulas have limits that extend beyond the range of the data. Table 21.4 summarizes the Newton-Cotes open integration formulas. The formulas are expressed in the form of Eq. (21.5) so that the weighting factors are evident. As with the closed versions, successive pairs of the formulas have the same order error. The even-segment–odd-point formulas are usually the methods of preference because they require fewer points to attain the same accuracy as the odd-segment–even-point formulas.

The open formulas are not often used for definite integration. However, as discussed in Chap. 22, they have utility for analyzing improper integrals. In addition, they will have relevance to our discussion of multistep methods for solving ordinary differential equations in Chap. 26.

PROBLEMS

21.1 Use analytical means to evaluate

(a) $\displaystyle\int_0^3 (1 - e^{-x})\, dx$

(b) $\displaystyle\int_{-2}^4 (1 - x - 4x^3 + x^5)\, dx$

(c) $\displaystyle\int_0^{\pi/2} (8 + 4\sin x)\, dx$

21.2 Use a single application of the trapezoidal rule to evaluate the integrals from Prob. 21.1.

21.3 Evaluate the integrals from Prob. 21.1 with a multiple-application trapezoidal rule, with $n = 2, 4$, and 6.

21.4 Evaluate the integrals from Prob. 21.1 with a single application of Simpson's 1/3 rule.

21.5 Evaluate the integrals from Prob. 21.1 with a multiple-application Simpson's 1/3 rule, with $n = 4$ and 6.

21.6 Evaluate the integrals from Prob. 21.1 with a single application of Simpson's 3/8 rule.

21.7 Evaluate the integrals from Prob. 21.1, but use a multiple-application Simpson's rule, with $n = 5$.

21.8 Integrate the following function using the trapezoidal rule, with $n = 1, 2, 3$, and 4:

$$\int_1^2 (x + 1/x)^2 \, dx$$

Compute percent relative errors with respect to the true value of 4.8333 to evaluate the accuracy of the trapezoidal approximations.

21.9 Integrate the following function both analytically and using Simpson's rules, with $n = 4$ and 5:

$$\int_{-3}^5 (4x + 5)^3 \, dx$$

Discuss the results.

21.10 Integrate the following function both analytically and numerically. Use both the trapezoidal and Simpson's 1/3 rules to numerically integrate the function. For both cases, use the multiple-application version, with $n = 4$.

$$\int_0^3 x e^{2x} \, dx$$

Compute percent relative errors for the numerical results.

21.11 Integrate the following function both analytically and numerically. For the numerical evaluations use (a) a single application of the trapezoidal rule, (b) Simpson's 1/3 rule, (c) Simpson's 3/8 rule, (d) Boole's rule, (e) the midpoint method, (f) the 3 segment/2-point open integration formula, and (g) the 4-segment/3-point open integration formula.

$$\int_0^1 15^{2x} \, dx$$

Compute percent relative errors for the numerical results.

21.12 Integrate the following function both analytically and numerically. For the numerical evaluations use (a) single application of the trapezoidal rule, (b) Simpson's 1/3 rule, (c) Simpson's 3/8 rule, (d) multiple application of Simpson's rules ($n = 5$), (e) Boole's rule, (f) the midpoint method, (g) the 3-segment/2-point open integration formula, and (h) the 4-segment/3-point open integration formula.

$$\int_0^\pi (5 + 3 \sin x) \, dx$$

Compute percent relative errors for the numerical results.

21.13 Integrate the following function,

$$\int_0^1 x^{0.1}(1.2 - x)(1 - e^{20(x-1)}) \, dx$$

Note that the true value is 0.602297. Evaluate this integral with the multiple-segment trapezoidal rule. Use a sufficiently high n that you get 4 significant digits of accuracy. Discuss your results.

21.14 Evaluate the integral of the following tabular data with the trapezoidal rule:

x	0	0.1	0.2	0.3	0.4	0.5
$f(x)$	1	7	4	3	5	2

21.15 Perform the same evaluation as in Prob. 21.14, but use Simpson's rules.

21.16 Evaluate the integral of the following tabular data using the trapezoidal rule:

x	−3	−1	1	3	5	7	9	11
$f(x)$	1	−4	−9	2	4	2	6	−3

21.17 Perform the same evaluation as in Prob. 21.16, but use Simpson's rules.

21.18 Determine the mean value of the function

$$f(x) = -46 + 45.4x - 13.8x^2 + 1.71x^3 - 0.0729x^4$$

between $x = 2$ and 10 by (a) graphing the function and visually estimating the mean value, (b) using Eq. (PT6.4) and the analytical evaluation of the integral, and (c) using Eq. (PT6.4) and a five-segment version of Simpson's rule to estimate the integral. Calculate the relative percent error.

21.19 The function

$$f(x) = e^{-x}$$

can be used to generate the following table of unequally spaced data:

x	0	0.1	0.3	0.5	0.7	0.95	1.2
$f(x)$	1	0.9048	0.7408	0.6065	0.4966	0.3867	0.3012

Evaluate the integral from $a = 0$ to $b = 1.2$ using (a) analytical means, (b) the trapezoidal rule, and (c) a combination of the trapezoidal and Simpson's rules; employ Simpson's rules wherever possible to obtain the highest accuracy. For (b) and (c), compute the percent relative error (ε_t).

21.20 Evaluate the following double integral:

$$\int_{-2}^2 \int_0^4 (x^2 - 3y^2 + xy^3) \, dx \, dy$$

(a) analytically, (b) using a multiple-application trapezoidal rule ($n = 2$), (c) using single applications of Simpson's 1/3 rule. For (b) and (c), compute the percent relative error (ε_t).

21.21 Evaluate the triple integral

$$\int_{-4}^{4} \int_{0}^{6} \int_{-1}^{3} (x^3 - 2yz) \, dx \, dy \, dz$$

(a) analytically and (b) using single applications of Simpson's 1/3 rule. For (b) compute the percent relative error (ε_t).

21.22 Develop a user-friendly computer program for the multiple-application trapezoidal rule based on Fig. 21.9. Among other things, (a) add documentation statements to the code, (b) make the input and output user-oriented, and (c) modify the program so that it is capable of evaluating given functions in addition to tabular data. Test your program by duplicating the computation from Example 21.2.

21.23 Develop a user-friendly computer program for the multiple-application version of Simpson's rule based on Fig. 21.14c. Test it by duplicating the computations from Example 21.5.

21.24 Develop a user-friendly computer program for integrating unequally spaced data based on Fig. 21.16b. Test it by duplicating the computation from Example 21.8.

21.25 Use the **Integrate Function** program on the Numerical Methods TOOLKIT disk (or your own program from Prob. 21.22) to repeat (a) Prob. 21.2, (b) Prob. 21.3, (c) Prob. 21.8, (d) Prob. 21.10, and (e) Prob. 21.12. Use the graphical option to help you visualize the concept that $I = \int_{a}^{b} f(x) \, dx$ is the area between the $f(x)$ curve and the axis. Try several different step sizes for each problem.

CHAPTER 22

Integration of Equations

In the introduction to Part Six, we noted that functions to be integrated numerically will typically be of two forms: a table of values or a function. The form of the data has an important influence on the approaches that can be used to evaluate the integral. For tabulated information, you are limited by the number of points that are given. In contrast, if the function is available, you can generate as many values of $f(x)$ as are required to attain acceptable accuracy (recall Fig. PT6.7).

This chapter is devoted to two techniques that are expressly designed to analyze cases where the function is given. Both capitalize on the ability to generate function values to develop efficient schemes for numerical integration. The first is based on *Richardson's extrapolation,* which is a method for combining two numerical integral estimates to obtain a third, more accurate value. The computational algorithm for implementing Richardson's extrapolation in a highly efficient manner is called *Romberg integration.* This technique is recursive and can be used to generate an integral estimate within a prespecified error tolerance.

The second method is called *Gauss quadrature.* Recall that, in the last chapter, values of $f(x)$ for the Newton-Cotes formulas were determined at specified values of x. For example, if we used the trapezoidal rule to determine an integral, we were constrained to take the weighted average of $f(x)$ at the ends of the interval. Gauss-quadrature formulas employ x values that are positioned between a and b in such a manner that a much more accurate integral estimate results.

In addition to these two standard techniques, we devote a final section to the evaluation of *improper integrals.* In this discussion, we focus on integrals with infinite limits and show how a change of variable and open integration formulas prove useful for such cases.

22.1 NEWTON-COTES ALGORITHMS FOR EQUATIONS

In Chap. 21, we presented algorithms for multiple-application versions of the trapezoidal rule and Simpson's rules. Although these pseudocodes can certainly be used to analyze equations, in our effort to make them compatible with either data or functions, they could not exploit the convenience of the latter.

Figure 22.1 shows pseudocodes that are expressly designed for cases where the function is analytical. In particular, notice that neither the independent nor the dependent

(a)
```
FUNCTION TrapEq (n, a, b)
   h = (b − a) / n
   x = a
   sum = f(x)
   DO i = 1, n − 1
      x = x + h
      sum = sum + 2 * f(x)
   END DO
   sum = sum + f(b)
   TrapEq = (b − a) * sum / (2 * n)
END TrapEq
```

(b)
```
FUNCTION SimpEq (n, a, b)
   h = (b − a) / n
   x = a
   sum = f(x)
   DO i = 1, n − 2, 2
      x = x + h
      sum = sum + 4 * f(x)
      x = x + h
      sum = sum + 2 * f(x)
   END DO
   x = x + h
   sum = sum + 4 * f(x)
   sum = sum + f(b)
   SimpEq = (b − a) * sum / (3 * n)
END SimpEq
```

FIGURE 22.1
Algorithms for multiple applications of the (a) trapezoidal and (b) Simpson's 1/3 rules, where the function is available.

FIGURE 22.2
Absolute value of the true percent relative error versus number of segments for the determination of the integral of $f(x) = 0.2 + 25x − 200x^2 + 675x^3 − 900x^4 + 400x^5$, evaluated from $a = 0$ to $b = 0.8$ using the multiple-application trapezoidal rule and the multiple-application Simpson's 1/3 rule. Note that both results indicate that for a large number of segments, round-off errors limit precision.

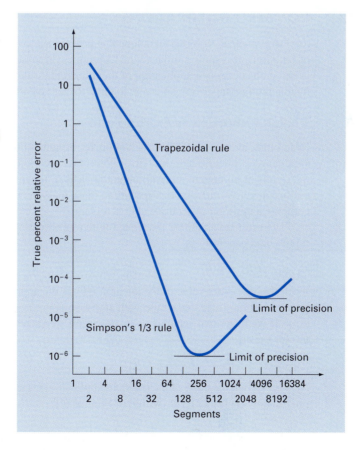

variable values are passed into the function via its argument as was the case for the codes in Chap. 21. For the independent variable x, the integration interval (a, b) and the number of segments are passed. This information is then employed to generate equispaced values of x within the function. For the dependent variable, the function values in Fig. 22.1 are computed using calls to the function being analyzed, $f(x)$.

We developed single-precision programs based on these pseudocodes to analyze the effort involved and the errors incurred as we progressively used more segments to estimate the integral of a simple function. For an analytical function, the error equations [Eqs. (21.13) and (21.19)] indicate that increasing the number of segments n will result in more accurate integral estimates. This observation is borne out by Fig. 22.2, which is a plot of true error versus n for the integral of $f(x) = 0.2 + 25x - 200x^2 + 675x^3 - 900x^4 + 400x^5$. Notice how the error drops as n increases. However, also notice that at large values of n, the error starts to increase as round-off errors begin to dominate. Also observe that a very large number of function evaluations (and, hence, computational effort) is required to attain high levels of accuracy. As a consequence of these shortcomings, the multiple-application trapezoidal rule and Simpson's rules are sometimes inadequate for problem contexts where high efficiency and low errors are needed.

22.2 ROMBERG INTEGRATION

Romberg integration is one technique that is designed to attain efficient numerical integrals of functions. It is quite similar to the techniques discussed in Chap. 21, in the sense that it is based on successive application of the trapezoidal rule. However, through mathematical manipulations, superior results are attained for less effort.

22.2.1 Richardson's Extrapolation

Recall that, in Sec. 10.3.3, we used iterative refinement to improve the solution of a set of simultaneous linear equations. Error-correction techniques are also available to improve the results of numerical integration on the basis of the integral estimates themselves. Generally called *Richardson's extrapolation,* these methods use two estimates of an integral to compute a third, more accurate approximation.

The estimate and error associated with a multiple-application trapezoidal rule can be represented generally as

$$I = I(h) + E(h)$$

where I = the exact value of the integral, $I(h)$ = the approximation from an n-segment application of the trapezoidal rule with step size $h = (b - a)/n$, and $E(h)$ = the truncation error. If we make two separate estimates using step sizes of h_1 and h_2 and have exact values for the error,

$$I(h_1) + E(h_1) = I(h_2) + E(h_2) \tag{22.1}$$

Now recall that the error of the multiple-application trapezoidal rule can be represented approximately by Eq. (21.13) [with $n = (b - a)/h$]:

$$E \cong -\frac{b - a}{12} h^2 \bar{f}'' \tag{22.2}$$

If it is assumed that $\bar{f}''$ is constant regardless of step size, Eq. (22.2) can be used to determine that the ratio of the two errors will be

$$\frac{E(h_1)}{E(h_2)} \cong \frac{h_1^2}{h_1^2} \tag{22.3}$$

This calculation has the important effect of removing the term $\bar{f}''$ from the computation. In so doing, we have made it possible to utilize the information embodied by Eq. (22.2) without prior knowledge of the function's second derivative. To do this, we rearrange Eq. (22.3) to give

$$E(h_1) \cong E(h_2)\left(\frac{h_1}{h_2}\right)^2$$

which can be substituted into Eq. (22.1):

$$I(h_1) + E(h_2)\left(\frac{h_1}{h_2}\right)^2 \cong I(h_2) + E(h_2)$$

which can be solved for

$$E(h_2) \cong \frac{I(h_1) - I(h_2)}{1 - (h_1/h_2)^2}$$

Thus, we have developed an estimate of the truncation error in terms of the integral estimates and their step sizes. This estimate can then be substituted into

$$I = I(h_2) + E(h_2)$$

to yield an improved estimate of the integral:

$$I \cong I(h_2) + \frac{1}{(h_1/h_2)^2 - 1}\left[I(h_2) - I(h_1)\right] \tag{22.4}$$

It can be shown (Ralston and Rabinowitz, 1978) that the error of this estimate is $O(h^4)$. Thus, we have combined two trapezoidal rule estimates of $O(h^2)$ to yield a new estimate of $O(h^4)$. For the special case where the interval is halved ($h_2 = h_1/2$), this equation becomes

$$I \cong I(h_2) + \frac{1}{2^2 - 1}\left[I(h_2) - I(h_1)\right]$$

or, collecting terms,

$$I \cong \frac{4}{3}I(h_2) - \frac{1}{3}I(h_1) \tag{22.5}$$

EXAMPLE 22.1   Error Corrections of the Trapezoidal Rule

Problem Statement.   In the previous chapter (Example 21.1 and Table 21.1), we used a variety of numerical integration methods to evaluate the integral of $f(x) = 0.2 + 25x - 200x^2 + 675x^3 - 900x^4 + 400x^5$ from $a = 0$ to $b = 0.8$. For example, single and multiple applications of the trapezoidal rule yielded the following results:

Segments	h	Integral	$\varepsilon_t,$ %
1	0.8	0.1728	89.5
2	0.4	1.0688	34.9
4	0.2	1.4848	9.5

Use this information along with Eq. (22.5) to compute improved estimates of the integral.

Solution. The estimates for one and two segments can be combined to yield

$$I \cong \frac{4}{3}(1.0688) - \frac{1}{3}(0.1728) = 1.367467$$

The error of the improved integral is $E_t = 1.640533 - 1.367467 = 0.273067$ ($\varepsilon_t = 16.6\%$), which is superior to the estimates upon which it was based.

In the same manner, the estimates for two and four segments can be combined to give

$$I \cong \frac{4}{3}(1.4848) - \frac{1}{3}(1.0688) = 1.623467$$

which represents an error of $E_t = 1.640533 - 1.623467 = 0.017067$ ($\varepsilon_t = 1.0\%$).

Equation (22.4) provides a way to combine two applications of the trapezoidal rule with error $O(h^2)$ to compute a third estimate with error $O(h^4)$. This approach is a subset of a more general method for combining integrals to obtain improved estimates. For instance, in Example 22.1, we computed two improved integrals of $O(h^4)$ on the basis of three trapezoidal rule estimates. These two improved estimates can, in turn, be combined to yield an even better value with $O(h^6)$. For the special case where the original trapezoidal estimates are based on successive halving of the step size, the equation used for $O(h^6)$ accuracy is

$$I \cong \frac{16}{15}I_m - \frac{1}{15}I_l \tag{22.6}$$

where I_m and I_l are the more and less accurate estimates, respectively. Similarly, two $O(h^6)$ results can be combined to compute an integral that is $O(h^8)$ using

$$I \cong \frac{64}{63}I_m - \frac{1}{63}I_l \tag{22.7}$$

EXAMPLE 22.2 Higher-Order Error Correction of Integral Estimates

Problem Statement. In Example 22.1, we used Richardson's extrapolation to compute two integral estimates of $O(h^4)$. Utilize Eq. (22.6) to combine these estimates to compute an integral with $O(h^6)$.

Solution. The two integral estimates of $O(h^4)$ obtained in Example 22.1 were 1.367467 and 1.623467. These values can be substituted into Eq. (22.6) to yield

$$I = \frac{16}{15}(1.623467) - \frac{1}{15}(1.367467) = 1.640533$$

which is the correct answer to the seven significant figures that are carried in this example.

22.2.2 The Romberg Integration Algorithm

Notice that the coefficients in each of the extrapolation equations [Eqs. (22.5), (22.6), and (22.7)] add up to 1. Thus, they represent weighting factors that, as accuracy increases, place relatively greater weight on the superior integral estimate. These formulations can be expressed in a general form that is well-suited for computer implementation:

$$I_{j,k} \cong \frac{4^{k-1}I_{j+1,k-1} - I_{j,k-1}}{4^{k-1} - 1} \tag{22.8}$$

where $I_{j+1,k-1}$ and $I_{j,k-1}$ = the more and less accurate integrals, respectively, and $I_{j,k}$ = the improved integral. The index k signifies the level of the integration, where $k = 1$ corresponds to the original trapezoidal rule estimates, $k = 2$ corresponds to $O(h^4)$, $k = 3$ to $O(h^6)$, and so forth. The index j is used to distinguish between the more ($j + 1$) and the less (j) accurate estimates. For example, for $k = 2$ and $j = 1$, Eq. (22.8) becomes

$$I_{1,2} \cong \frac{4I_{2,1} - I_{1,1}}{3}$$

which is equivalent to Eq. (22.5).

The general form represented by Eq. (22.8) is attributed to Romberg, and its systematic application to evaluate integrals is known as *Romberg integration*. Figure 22.3 is a graphical depiction of the sequence of integral estimates generated using this approach. Each matrix corresponds to a single iteration. The first column contains the trapezoidal rule

FIGURE 22.3
Graphical depiction of the sequence of integral estimates generated using Romberg integration.

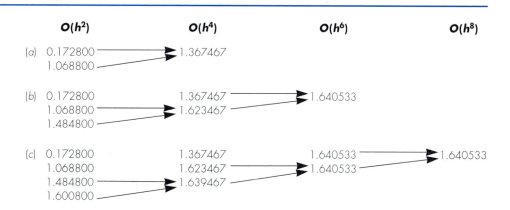

evaluations that are designated $I_{j,1}$, where $j = 1$ is for a single-segment application (step size is $b - a$), $j = 2$ is for a two-segment application [step size is $(b - a)/2$], $j = 3$ is for a four-segment application [step size is $(b - a)/4$], and so forth. The other columns of the matrix are generated by systematically applying Eq. (22.8) to obtain successively better estimates of the integral.

For example, the first iteration (Fig. 22.3a) involves computing the one- and two-segment trapezoidal rule estimates ($I_{1,1}$ and $I_{2,1}$). Equation (22.8) is then used to compute the element $I_{1,2} = 1.367467$, which has an error of $O(h^4)$.

Now, we must check to determine whether this result is adequate for our needs. As in other approximate methods in this book, a termination, or stopping, criterion is required to assess the accuracy of the results. One method that can be employed for the present purposes is [Eq. (3.5)]

$$|\varepsilon_a| = \left| \frac{I_{1,k} - I_{1,k-1}}{I_{1,k}} \right| 100\% \tag{22.9}$$

where $\varepsilon_a =$ an estimate of the percent relative error. Thus, as was done previously in other iterative processes, we compare the new estimate with a previous value. When the change between the old and new values as represented by ε_a is below a prespecified error criterion ε_s, the computation is terminated. For Fig. 22.3a, this evaluation indicates an 87.4 percent change over the course of the first iteration.

The object of the second iteration (Fig. 22.3b) is to obtain the $O(h^6)$ estimate—$I_{1,3}$. To do this, an additional trapezoidal rule estimate, $I_{3,1} = 1.4848$, is determined. Then it is combined with $I_{2,1}$ using Eq. (22.8) to generate $I_{2,2} = 1.623467$. The result is, in turn, combined

---

FIGURE 22.4
Pseudocode for Romberg integration that uses the equal-size-segment version of the trapezoidal rule from Fig. 22.1.

```
FUNCTION Rhomberg (a, b, maxit, es)
    LOCAL I(10, 10)
    n = 1
    I₁,₁ = TrapEq(n, a, b)
    iter = 0
    DO
        iter = iter + 1
        n = 2^iter
        I_iter+1,1 = TrapEq(n, a, b)
        DO k = 2, iter + 1
            j = 2 + iter − k
            I_j,k = (4^(k−1) * I_j+1,k−1 − I_j,k−1) / (4^(k−1) − 1)
        END DO
        ea = ABS((I₁,iter+1 − I₁,iter) / I₁,iter+1) * 100
        IF (iter ≥ maxit OR ea ≤ es) EXIT
    END DO
    Rhomberg = I₁,iter+1
END Rhomberg
```

with $I_{1,2}$ to yield $I_{1,3} = 1.640533$. Equation (22.9) can be applied to determine that this result represents a change of 22.6 percent when compared with the previous result $I_{1,2}$.

The third iteration (Fig. 22.3c) continues the process in the same fashion. In this case, a trapezoidal estimate is added to the first column, and then Eq. (22.8) is applied to compute successively more accurate integrals along the lower diagonal. After only three iterations, because we are evaluating a fifth-order polynomial, the result ($I_{1,4} = 1.640533$) is exact.

Romberg integration is more efficient than the trapezoidal rule and Simpson's rules discussed in Chap. 21. For example, for determination of the integral as shown in Fig. 22.1, Simpson's 1/3 rule would require a 256-segment application to yield an estimate of 1.640533. Finer approximations would not be possible because of round-off error. In contrast, Romberg integration yields an exact result (to seven significant figures) based on combining one-, two-, four-, and eight-segment trapezoidal rules; that is, with only 15 function evaluations!

Figure 22.4 presents pseudocode for Romberg integration. By using loops, this algorithm implements the method in an efficient manner. Romberg integration is designed for cases where the function to be integrated is known. This is because knowledge of the function permits the evaluations required for the initial implementations of the trapezoidal rule. Tabulated data is rarely in the form needed to make the necessary successive halvings.

22.3 GAUSS QUADRATURE

In Chap. 21, we studied the group of numerical integration or quadrature formulas known as the Newton-Cotes equations. A characteristic of these formulas (with the exception of the special case of Sec. 21.3) was that the integral estimate was based on evenly spaced function values. Consequently, the location of the base points used in these equations was predetermined or fixed.

For example, as depicted in Fig. 22.5a, the trapezoidal rule is based on taking the area under the straight line connecting the function values at the ends of the integration interval. The formula that is used to compute this area is

$$I \cong (b - a) \frac{f(a) + f(b)}{2} \tag{22.10}$$

where a and b = the limits of integration and $b - a$ = the width of the integration interval. Because the trapezoidal rule must pass through the end points, there are cases such as Fig. 22.5a where the formula results in a large error.

Now, suppose that the constraint of fixed base points was removed and we were free to evaluate the area under a straight line joining any two points on the curve. By positioning these points wisely, we could define a straight line that would balance the positive and negative errors. Hence, as in Fig. 22.5b, we would arrive at an improved estimate of the integral.

Gauss quadrature is the name for one class of techniques to implement such a strategy. The particular Gauss quadrature formulas described in this section are called *Gauss-Legendre* formulas. Before describing the approach, we will show how numerical integration formulas such as the trapezoidal rule can be derived using the method of

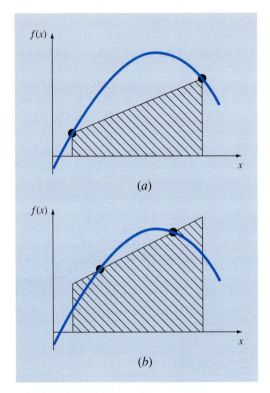

FIGURE 22.5
(a) Graphical depiction of the trapezoidal rule as the area under the straight line joining fixed end points. (b) An improved integral estimate obtained by taking the area under the straight line passing through two intermediate points. By positioning these points wisely, the positive and negative errors are balanced, and an improved integral estimate results.

undetermined coefficients. This method will then be employed to develop the Gauss-Legendre formulas.

22.3.1 Method of Undetermined Coefficients

In Chap. 21, we derived the trapezoidal rule by integrating a linear interpolating polynomial and by geometrical reasoning. The *method of undetermined coefficients* offers a third approach that also has utility in deriving other integration techniques such as Gauss quadrature.

To illustrate the approach, Eq. (22.10) is expressed as

$$I \cong c_0 f(a) + c_1 f(b) \tag{22.11}$$

where the c's = constants. Now realize that the trapezoidal rule should yield exact results when the function being integrated is a constant or a straight line. Two simple equations that represent these cases are $y = 1$ and $y = x$. Both are illustrated in Fig. 22.6. Thus, the following equalities should hold:

$$c_0 + c_1 = \int_{-(b-a)/2}^{(b-a)/2} 1 \, dx$$

and

$$-c_0 \frac{b-a}{2} + c_1 \frac{b-a}{2} = \int_{-(b-a)/2}^{(b-a)/2} x \, dx$$

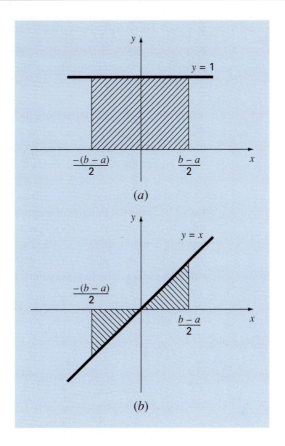

FIGURE 22.6
Two integrals that should be evaluated exactly by the trapezoidal rule: (a) a constant and (b) a straight line.

or, evaluating the integrals,

$$c_0 + c_1 = b - a$$

and

$$-c_0 \frac{b-a}{2} + c_1 \frac{b-a}{2} = 0$$

These are two equations with two unknowns that can be solved for

$$c_0 = c_1 = \frac{b-a}{2}$$

which, when substituted back into Eq. (22.11), gives

$$I = \frac{b-a}{2} f(a) + \frac{b-a}{2} f(b)$$

which is equivalent to the trapezoidal rule.

22.3.2 Derivation of the Two-Point Gauss-Legendre Formula

Just as was the case for the above derivation of the trapezoidal rule, the object of Gauss quadrature is to determine the coefficients of an equation of the form

$$I \cong c_0 f(x_0) + c_1 f(x_1) \tag{22.12}$$

where the c's = the unknown coefficients. However, in contrast to the trapezoidal rule that used fixed end points a and b, the function arguments x_0 and x_1 are not fixed at the end points, but are unknowns (Fig. 22.7). Thus, we now have a total of four unknowns that must be evaluated, and consequently, we require four conditions to determine them exactly.

Just as for the trapezoidal rule, we can obtain two of these conditions by assuming that Eq. (22.12) fits the integral of a constant and a linear function exactly. Then, to arrive at the other two conditions, we merely extend this reasoning by assuming that it also fits the integral of a parabolic ($y = x^2$) and a cubic ($y = x^3$) function. By doing this, we determine all four unknowns and in the bargain derive a linear two-point integration formula that is exact for cubics. The four equations to be solved are

$$c_0 f(x_0) + c_1 f(x_1) = \int_{-1}^{1} 1 \, dx = 2 \tag{22.13}$$

$$c_0 f(x_0) + c_1 f(x_1) = \int_{-1}^{1} x \, dx = 0 \tag{22.14}$$

$$c_0 f(x_0) + c_1 f(x_1) = \int_{-1}^{1} x^2 \, dx = \frac{2}{3} \tag{22.15}$$

$$c_0 f(x_0) + c_1 f(x_1) = \int_{-1}^{1} x^3 \, dx = 0 \tag{22.16}$$

FIGURE 22.7
Graphical depiction of the unknown variables x_0 and x_1 for integration by Gauss quadrature.

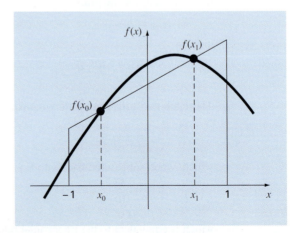

Equations (22.13) through (22.16) can be solved simultaneously for

$$c_0 = c_1 = 1$$

$$x_0 = -\frac{1}{\sqrt{3}} = -0.5773503\ldots$$

$$x_1 = \frac{1}{\sqrt{3}} = 0.5773503\ldots$$

which can be substituted into Eq. (22.12) to yield the two-point Gauss-Legendre formula

$$I \cong f\left(\frac{-1}{\sqrt{3}}\right) + f\left(\frac{1}{\sqrt{3}}\right) \tag{22.17}$$

Thus, we arrive at the interesting result that the simple addition of the function values at $x = 1/\sqrt{3}$ and $-1/\sqrt{3}$ yields an integral estimate that is third-order accurate.

Notice that the integration limits in Eqs. (22.13) through (22.16) are from -1 to 1. This was done to simplify the mathematics and to make the formulation as general as possible. A simple change of variable can be used to translate other limits of integration into this form. This is accomplished by assuming that a new variable x_d is related to the original variable x in a linear fashion, as in

$$x = a_0 + a_1 x_d \tag{22.18}$$

If the lower limit, $x = a$, corresponds to $x_d = -1$, these values can be substituted into Eq. (22.18) to yield

$$a = a_0 + a_1(-1) \tag{22.19}$$

Similarly, the upper limit, $x = b$, corresponds to $x_d = 1$, to give

$$b = a_0 + a_1(1) \tag{22.20}$$

Equations (22.19) and (22.20) can be solved simultaneously for

$$a_0 = \frac{b+a}{2} \tag{22.21}$$

and

$$a_1 = \frac{b-a}{2} \tag{22.22}$$

which can be substituted into Eq. (22.18) to yield

$$x = \frac{(b+a) + (b-a)x_d}{2} \tag{22.23}$$

This equation can be differentiated to give

$$dx = \frac{b-a}{2} dx_d \tag{22.24}$$

Equations (22.23) and (22.24) can be substituted for x and dx, respectively, in the equation to be integrated. These substitutions effectively transform the integration interval without changing the value of the integral. The following example illustrates how this is done in practice.

EXAMPLE 22.3   Two-Point Gauss-Legendre Formula

Problem Statement.   Use Eq. (22.17) to evaluate the integral of

$$f(x) = 0.2 + 25x - 200x^2 + 675x^3 - 900x^4 + 400x^5$$

between the limits $x = 0$ to 0.8. Recall that this was the same problem that we solved in Chap. 21 using a variety of Newton-Cotes formulations. The exact value of the integral is 1.640533.

Solution.   Before integrating the function, we must perform a change of variable so that the limits are from -1 to $+1$. To do this, we substitute $a = 0$ and $b = 0.8$ into Eq. (22.23) to yield

$$x = 0.4 + 0.4x_d$$

The derivative of this relationship is [Eq. (22.24)]

$$dx = 0.4\, dx_d$$

Both of these can be substituted into the original equation to yield

$$\int_0^{0.8} (0.2 + 25x - 200x^2 + 675x^3 - 900x^4 + 400x^5)\, dx$$

$$= \int_{-1}^{1} [0.2 + 25(0.4 + 0.4x_d) - 200(0.4 + 0.4x_d)^2 + 675(0.4 + 0.4x_d)^3$$

$$- 900(0.4 + 0.4x_d)^4 + 400(0.4 + 0.4x_d)^5]0.4\, dx_d$$

Therefore, the right-hand side is in the form that is suitable for evaluation using Gauss quadrature. The transformed function can be evaluated at $-1/\sqrt{3}$ to be equal to 0.516741 and at $1/\sqrt{3}$ to be equal to 1.305837. Therefore, the integral according to Eq. (22.17) is

$$I \cong 0.516741 + 1.305837 = 1.822578$$

which represents a percent relative error of -11.1 percent. This result is comparable in magnitude to a four-segment application of the trapezoidal rule (Table 21.1) or a single application of Simpson's 1/3 and 3/8 rules (Examples 21.4 and 21.6). This latter result is to be expected because Simpson's rules are also third-order accurate. However, because of the clever choice of base points, Gauss quadrature attains this accuracy on the basis of only two function evaluations.

22.3.3 Higher-Point Formulas

Beyond the two-point formula described in the previous section, higher-point versions can be developed in the general form

$$I \cong c_0 f(x_0) + c_1 f(x_1) + \cdots + c_{n-1} f(x_{n-1}) \tag{22.25}$$

TABLE 22.1   Weighting factors c and function arguments x used in Gauss-Legendre formulas.

Points	Weighting Factors	Function Arguments	Truncation Error
2	$c_0 = 1.0000000$ $c_1 = 1.0000000$	$x_0 = -0.577350269$ $x_1 = 0.577350269$	$\cong f^{(4)}(\xi)$
3	$c_0 = 0.5555556$ $c_1 = 0.8888889$ $c_2 = 0.5555556$	$x_0 = -0.774596669$ $x_1 = 0.0$ $x_2 = 0.774596669$	$\cong f^{(6)}(\xi)$
4	$c_0 = 0.3478548$ $c_1 = 0.6521452$ $c_2 = 0.6521452$ $c_3 = 0.3478548$	$x_0 = -0.861136312$ $x_1 = -0.339981044$ $x_2 = 0.339981044$ $x_3 = 0.861136312$	$\cong f^{(8)}(\xi)$
5	$c_0 = 0.2369269$ $c_1 = 0.4786287$ $c_2 = 0.5688889$ $c_3 = 0.4786287$ $c_4 = 0.2369269$	$x_0 = -0.906179846$ $x_1 = -0.538469310$ $x_2 = 0.0$ $x_3 = 0.538469310$ $x_4 = 0.906179846$	$\cong f^{(10)}(\xi)$
6	$c_0 = 0.1713245$ $c_1 = 0.3607616$ $c_2 = 0.4679139$ $c_3 = 0.4679139$ $c_4 = 0.3607616$ $c_5 = 0.1713245$	$x_0 = -0.932469514$ $x_1 = -0.661209386$ $x_2 = -0.238619186$ $x_3 = 0.238619186$ $x_4 = 0.661209386$ $x_5 = 0.932469514$	$\cong f^{(12)}(\xi)$

where $n =$ the number of points. Values for c's and x's for up to and including the six-point formula are summarized in Table 22.1.

EXAMPLE 22.4   Three-Point Gauss-Legendre Formula

Problem Statement.   Use the three-point formula from Table 22.1 to estimate the integral for the same function as in Example 22.3.

Solution.   According to Table 22.1, the three-point formula is

$$I = 0.5555556 f(-0.7745967) + 0.8888889 f(0) + 0.5555556 f(0.7745967)$$

which is equal to

$$I = 0.2813013 + 0.8732444 + 0.4859876 = 1.640533$$

which is exact.

Because Gauss quadrature requires function evaluations at nonuniformly spaced points within the integration interval, it is not appropriate for cases where the function is unknown. Thus, it is not suited for engineering problems that deal with tabulated data. However, where the function is known, its efficiency can be a decided advantage. This is particularly true when numerous integral evaluations must be performed.

EXAMPLE 22.5   Applying Gauss Quadrature to the Falling Parachutist Problem

Problem Statement.   In Example 21.3, we used the multiple-application trapezoidal rule to evaluate

$$d = \frac{gm}{c} \int_0^{10} \left[1 - e^{-(c/m)t} \right] dt$$

where $g = 9.8$, $c = 12.5$, and $m = 68.1$. The exact value of the integral was determined by calculus to be 289.4351. Recall that the best estimate obtained using a 500-segment trapezoidal rule was 289.4348 with an $|\varepsilon_t| \cong 1.15 \times 10^{-4}$ percent. Repeat this computation using Gauss quadrature.

Solution.   After modifying the function, the following results are obtained:

    Two-point estimate = 290.0145

    Three-point estimate = 289.4393

    Four-point estimate = 289.4352

    Five-point estimate = 289.4351

    Six-point estimate = 289.4351

Thus, the five- and six-point estimates yield results that are exact to seven significant figures.

22.3.4 Error Analysis for Gauss Quadrature

The error for the Gauss-Legendre formulas is specified generally by (Carnahan et al., 1969)

$$E_t = \frac{2^{2n+3} \left[(n+1)! \right]^4}{(2n+3)\left[(2n+2)! \right]^3} f^{(2n+2)}(\xi) \tag{22.26}$$

where $n =$ the number of points minus one and $f^{(2n+2)}(\xi) =$ the $(2n+2)$th derivative of the function after the change of variable with ξ located somewhere on the interval from -1 to 1. Comparison of Eq. (22.26) with Table 21.2 indicates the superiority of Gauss quadrature to Newton-Cotes formulas, provided the higher-order derivatives do not increase substantially with increasing n. Problem 22.8 at the end of this chapter illustrates a case where the Gauss-Legendre formulas perform poorly. In these situations, the multiple-application Simpson's rule or Romberg integration would be preferable. However, for many functions confronted in engineering practice, Gauss quadrature provides an efficient means for evaluating integrals.

22.4   IMPROPER INTEGRALS

To this point, we have dealt exclusively with integrals having finite limits and bounded integrands. Although these types are commonplace in engineering, there will be times when improper integrals must be evaluated. In this section we will focus on one type of improper integral—that is, one with a lower limit of $-\infty$ or an upper limit of $+\infty$.

    Such integrals usually can be evaluated by making a change of variable that transforms the infinite range to one that is finite. The following identity serves this purpose and works for any function that decreases toward zero at least as fast as $1/x^2$ as x approaches infinity:

$$\int_a^b f(x)\,dx = \int_{1/b}^{1/a} \frac{1}{t^2} f\!\left(\frac{1}{t}\right) dt \tag{22.27}$$

for $ab > 0$. Therefore, it can be used only when a is positive and b is ∞ or when a is $-\infty$ and b is negative. For cases where the limits are from $-\infty$ to a positive value or from a negative value to ∞, the integral can be implemented in two steps. For example,

$$\int_{-\infty}^b f(x)\,dx = \int_{-\infty}^{-A} f(x)\,dx + \int_{-A}^b f(x)\,dx \tag{22.28}$$

where $-A$ is chosen as a sufficiently large negative value so that the function has begun to approach zero asymptotically at least as fast as $1/x^2$. After the integral has been divided into two parts, the first can be evaluated with Eq. (22.27) and the second with a Newton-Cotes closed formula such as Simpson's 1/3 rule.

One problem with using Eq. (22.27) to evaluate an integral is that the transformed function will be singular at one of the limits. The open integration formulas can be used to circumvent this dilemma as they allow evaluation of the integral without employing data at the end points of the integration interval. To allow the maximum flexibility, a multiple-application version of one of the open formulas from Table 21.4 is required.

Multiple-application versions of the open formulas can be concocted by using closed formulas for the interior segments and open formulas for the ends. For example, the multiple-segment trapezoidal rule and the midpoint rule can be combined to give

$$\int_{x_0}^{x_n} f(x)\,dx = h\left[\frac{3}{2} f(x_1) + \sum_{i=2}^{n-2} f(x_i) + \frac{3}{2} f(x_{n-1})\right]$$

In addition, semiopen formulas can be developed for cases where one or the other end of the interval is closed. For example, a formula that is open at the lower limit and closed at the upper limit is given as

$$\int_{x_0}^{x_n} f(x)\,dx = h\left[\frac{3}{2} f(x_1) + \sum_{i=2}^{n-1} f(x_i) + \frac{1}{2} f(x_n)\right]$$

Although these relationships can be used, a preferred formula is (Press et al., 1986)

$$\int_{x_0}^{x_n} f(x)\,dx = h\left[f(x_{1/2}) + f(x_{3/2}) + \cdots + f(x_{n-3/2}) + f(x_{n-1/2})\right] \tag{22.29}$$

which is called the *extended midpoint rule*. Notice that this formula is based on limits of integration that are $h/2$ after and before the first and last data points (Fig. 22.8).

FIGURE 22.8
Placement of data points relative to integration limits for the extended midpoint rule.

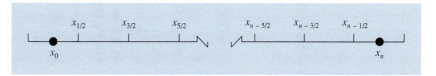

EXAMPLE 22.6 Evaluation of an Improper Integral

Problem Statement. The *cumulative normal distribution* is an important formula in statistics (see Fig. 22.9):

$$N(x) = \int_{-\infty}^{x} \frac{1}{\sqrt{2\pi}} e^{-x^2/2} \, dx \qquad (E22.6.1)$$

FIGURE 22.9
(a) The normal distribution, (b) the transformed abscissa in terms of the standardized normal deviate, and (c) the cumulative normal distribution. The shaded area in (a) and the point in (c) represent the probability that a random event will be less than the mean plus one standard deviation.

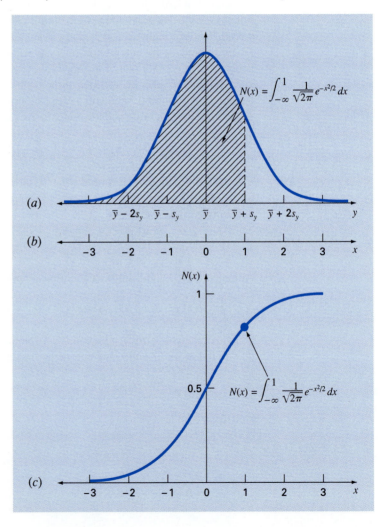

where $x = (y - \bar{y})/s_y$ is called the *normalized standard deviate*. It represents a change of variable to scale the normal distribution so that it is centered on zero and the distance along the abscissa is measured in multiples of the standard deviation (Fig. 22.9*b*).

Equation (E22.6.1) represents the probability that an event will be less than *x*. For example, if $x = 1$, Eq. (E22.6.1) can be used to determine that the probability that an event will occur that is less than one standard deviation above the mean is $N(1) = 0.8413$. In other words, if 100 events occur, approximately 84 will be less than the mean plus one standard deviation. Because Eq. (E22.6.1) cannot be evaluated in a simple functional form, it is solved numerically and listed in statistical tables. Use Eq. (22.28) in conjunction with Simpson's 1/3 rule and the extended midpoint rule to determine $N(1)$ numerically.

Solution. Equation (E22.6.1) can be re-expressed in terms of Eq. (22.28) as

$$N(x) = \frac{1}{\sqrt{2\pi}} \left(\int_{-\infty}^{-2} e^{-x^2/2}\, dx + \int_{-2}^{1} e^{-x^2/2}\, dx \right)$$

The first integral can be evaluated by applying Eq. (22.27) to give

$$\int_{-\infty}^{-2} e^{-x^2/2}\, dx = \int_{-1/2}^{0} \frac{1}{t^2} e^{-1/(2t^2)}\, dt$$

Then the extended midpoint rule with $h = 1/8$ can be employed to estimate

$$\int_{-1/2}^{0} \frac{1}{t^2} e^{-1/(2t^2)}\, dt \cong \frac{1}{8}\left[f(x_{-7/16}) + f(x_{-5/16}) + f(x_{-3/16}) + f(x_{-1/16}) \right]$$

$$= \frac{1}{8}\left[0.3833 + 0.0612 + 0 + 0 \right] = 0.0556$$

Simpson's 1/3 rule with $h = 0.5$ can be used to estimate the second integral as

$$\int_{-2}^{1} e^{-x^2/2}\, dx$$

$$= [1 - (-2)]\frac{0.1353 + 4(0.3247 + 0.8825 + 0.8825) + 2(0.6065 + 1) + 0.6065}{3(6)}$$

$$= 2.0523$$

Therefore, the final result can be computed as

$$N(1) \cong \frac{1}{\sqrt{2\pi}}(0.0556 + 2.0523) = 0.8409$$

which represents an error of $\varepsilon_t = 0.046$ percent.

The foregoing computation can be improved in a number of ways. First, higher-order formulas could be used. For example, a Romberg integration could be employed. Second, more points could be used. Press et al. (1986) explore both options in depth.

Aside from infinite limits, there are other ways in which an integral can be improper. Common examples include cases where the integral is singular at either the limits or at a point within the integral. Press et al. (1986) provide a nice discussion of ways to handle these situations.

PROBLEMS

22.1 Use Romberg integration to evaluate

$$\int_1^2 (x + 1/x)^2 \, dx$$

to an accuracy of $\varepsilon_s = 0.5\%$. Your results should be presented in the form of Fig. 22.3. Use the true value of 4.8333 to determine the true error ε_t of the result obtained with Romberg integration. Check that ε_t is less than the stopping criterion ε_s.

22.2 Use order of h^8 Romberg integration to evaluate

$$\int_0^3 x e^{2x} \, dx$$

Compare ε_a and ε_t.

22.3 Use Romberg integration to evaluate

$$\int_0^3 \frac{e^x \sin x}{1 + x^2} \, dx$$

to an accuracy of the order of h^8. Your results should be presented in the form of Fig. 22.3.

22.4 Obtain an estimate of the integral from Prob. 22.1, but using two-, three-, and four-point Gauss-Legendre formulas. Compute ε_t for each case on the basis of the analytical solution.

22.5 Obtain an estimate of the integral from Prob. 22.2, but using two-, three-, and four-point Gauss-Legendre formulas. Compute ε_t for each case on the basis of the analytical solution.

22.6 Obtain an estimate of the integral from Prob. 22.3 using six-point Gauss-Legendre formulas.

22.7 Perform the computation in Examples 21.3 and 22.5 for the falling parachutist, but use Romberg integration ($\varepsilon_s = 0.01\%$).

22.8 Employ two- through six-point Gauss-Legendre formulas to solve

$$\int_{-3}^3 \frac{2}{1 + 2x^2} \, dx$$

Interpret your results in light of Eq. (22.26).

22.9 Use numerical integration to evaluate the following:

(a) $\displaystyle\int_1^\infty \frac{dx}{x(x + 2)}$

(b) $\displaystyle\int_0^\infty e^{-y} \sin^2 y \, dy$

(c) $\displaystyle\int_0^\infty \frac{1}{(1 + y^2)(1 + y^2/2)} \, dy$

(d) $\displaystyle\int_{-2}^\infty y e^{-y} \, dy$

(e) $\displaystyle\int_{-\infty}^\infty \frac{1}{\sqrt{2\pi}} e^{-x^2} \, dx$

Note that **(d)** is the normal distribution (recall Fig. 22.9).

22.10 Develop a user-friendly computer program for the multiple-segment trapezoidal and Simpson's 1/3 rule based on Fig. 22.1. Test it by integrating

$$\int_0^1 x^{0.1}(1.2 - x)(1 - e^{20(x-1)}) \, dx$$

Use the true value of 0.602297 to compute ε_t for $n = 4$.

22.11 Develop a user-friendly computer program for Romberg integration based on Fig. 22.4. Test it by duplicating the results of Examples 22.3 and 22.4 and the function in Prob. 22.10.

22.12 Develop a user-friendly computer program for Gauss quadrature. Test it by duplicating the results of Examples 22.3 and 22.4 and the function in Prob. 22.10.

22.13 Use the program developed in Prob. 22.11 to solve Probs. **(a)** 22.1, **(b)** 22.2, and **(c)** 22.3.

22.14 Use the program developed in Prob. 22.12 to solve Probs. **(a)** 22.4, **(b)** 22.5, and **(c)** 22.6.

22.15 Develop a program to implement the extended midpoint rule iteratively. Start the iterations with an initial estimate based on a single point and the midpoint rule from Table 21.4. Then successively apply Eq. (22.29) with the interval divided by 3 at each stage; that is, $h = (b - a)/3, (b - a)/9, (b - a)/27$, etc. Note that this means that one-third of the function estimates will have been determined in the previous iteration. Develop your algorithm so that it capitalizes on this property. Perform iterations until an approximate error estimate ε_a falls below a prespecified stopping criteria ε_s. Test the program by evaluating Example 22.6.

CHAPTER 23

Numerical Differentiation

We have already introduced the notion of numerical differentiation in Chap. 4. Recall that we employed Taylor series expansions to derive finite-divided-difference approximations of derivatives. In Chap. 4, we developed forward, backward, and centered difference approximations of first and higher derivatives. Recall that, at best, these estimates had errors that were $O(h^2)$—that is, their errors were proportional to the square of the step size. This level of accuracy is due to the number of terms of the Taylor series that were retained during the derivation of these formulas. We will now illustrate how to develop more accurate formulas by retaining more terms.

23.1 HIGH-ACCURACY DIFFERENTIATION FORMULAS

As noted above, high-accuracy divided-difference formulas can be generated by including additional terms from the Taylor series expansion. For example, the forward Taylor series expansion can be written as [Eq. (4.21)]

$$f(x_{i+1}) = f(x_i) + f'(x_i)h + \frac{f''(x_i)}{2}h^2 + \cdots \tag{23.1}$$

which can be solved for

$$f'(x_i) = \frac{f(x_{i+1}) - f(x_i)}{h} - \frac{f''(x_i)}{2}h + O(h^2) \tag{23.2}$$

In Chap. 4, we truncated this result by excluding the second- and higher-derivative terms and were thus left with a final result of

$$f'(x_i) = \frac{f(x_{i+1}) - f(x_i)}{h} + O(h) \tag{23.3}$$

In contrast to this approach, we now retain the second-derivative term by substituting the following approximation of the second derivative [recall Eq. (4.24)]

$$f''(x_i) = \frac{f(x_{i+2}) - 2f(x_{i+1}) + f(x_i)}{h^2} + O(h) \tag{23.4}$$

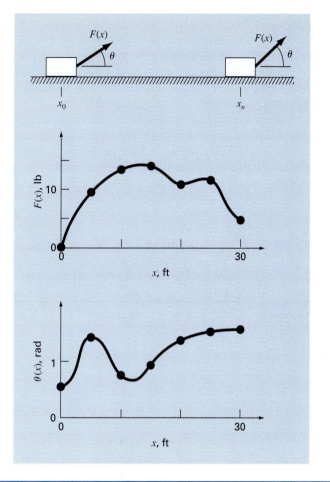

FIGURE 24.5
The case of a variable force acting on a block. For this case, the angle, as well as the magnitude, of the force varies.

TABLE 24.6 Data for force $F(x)$ and angle $\theta(x)$ as a function of position x.

x, ft	$F(x)$, lb	θ, rad	$F(x) \cos \theta$
0	0.0	0.50	0.0000
5	9.0	1.40	1.5297
10	13.0	0.75	9.5120
15	14.0	0.90	8.7025
20	10.5	1.30	2.8087
25	12.0	1.48	1.0881
30	5.0	1.50	0.3537

Suppose that you have to perform the computation for the situation depicted in Fig. 24.5. Although the figure shows the continuous values for $F(x)$ and $\theta(x)$, assume that, because of experimental constraints, you are provided with only discrete measurements at $x = 5$-ft intervals (Table 24.6). Use single- and multiple-application versions of the trapezoidal rule and Simpson's 1/3 and 3/8 rules to compute work for this data.

Solution.   The results of the analysis are summarized in Table 24.7. A percent relative error ε_t was computed in reference to a true value of the integral of 129.52 that was estimated on the basis of values taken from Fig. 24.5 at 1-ft intervals.

The results are interesting because the most accurate outcome occurs for the simple two-segment trapezoidal rule. More refined estimates using more segments, as well as Simpson's rules, yield less accurate results.

The reason for this apparently counterintuitive result is that the coarse spacing of the points is not adequate to capture the variations of the forces and angles. This is particularly evident in Fig. 24.6, where we have plotted the continuous curve for the product of $F(x)$ and $\cos[\theta(x)]$. Notice how the use of seven points to characterize the continuously varying function misses the two peaks at $x = 2.5$ and 12.5 ft. The omission of these two points

TABLE 24.7   Estimates of work calculated using the trapezoidal rule and Simpson's rules. The percent relative error ε_t as computed in reference to a true value of the integral (129.52 ft·lb) that was estimated on the basis of values at 1-ft intervals.

Technique	Segments	Work	ε_t, %
Trapezoidal	1	5.31	95.9
	2	133.19	2.84
	3	124.98	3.51
	6	119.09	8.05
Simpson's 1/3 rule	2	175.82	−35.75
	6	117.13	9.57
Simpson's 3/8 rule	3	139.93	−8.04

FIGURE 24.6

A continuous plot of $F(x) \cos[\theta(x)]$ versus position with the seven discrete points used to develop the numerical integration estimates in Table 24.7. Notice how the use of seven points to characterize this continuously varying function misses two peaks at $x = 2.5$ and 12.5 ft.

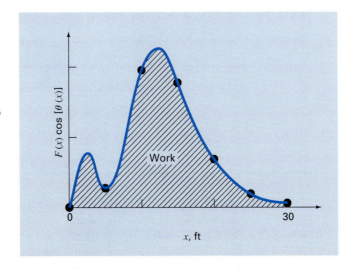

effectively limits the accuracy of the numerical integration estimates in Table 24.7. The fact that the two-segment trapezoidal rule yields the most accurate result is due to the chance positioning of the points for this particular problem (Fig. 24.7).

The conclusion to be drawn from Fig. 24.6 is that an adequate number of measurements must be made to accurately compute integrals. For the present case, if data were

FIGURE 24.7
Graphical depiction of why the two-segment trapezoidal rule yields a good estimate of the integral for this particular case. By chance, the use of two trapezoids happens to lead to an even balance between positive and negative errors.

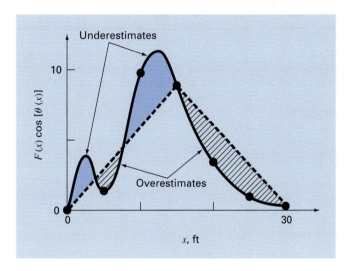

FIGURE 24.8
The unequal segmentation scheme that results from the inclusion of two additional points at $x = 2.5$ and 12.5 in the data in Table 24.6. The numerical integration formulas applied to each set of segments are shown.

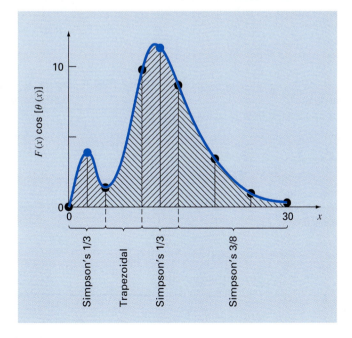

available at $F(2.5) \cos [\theta(2.5)] = 4.3500$ and $F(12.5) \cos [\theta(12.5)] = 11.3600$, we could determine an integral estimate using the algorithm for unequally spaced data described previously in Sec. 21.3. Figure 24.8 illustrates the unequal segmentation for this case. Including the two additional points yields an improved integral estimate of 126.9 ($\varepsilon_t = 2.02\%$). Thus, the inclusion of the additional data would incorporate the peaks that were missed previously and, as a consequence, lead to better results.

PROBLEMS

Chemical/Petroleum Engineering

24.1 Perform the same computation as Sec. 24.1, but compute the amount of heat required to raise the temperature of 1500 g of the material from -150 to $50°C$. Use Simpson's rule for your computation, with values of T at $50°C$ increments.

24.2 Repeat Prob. 24.1, but use Romberg integration to $\varepsilon_s = 0.01\%$.

24.3 Repeat Prob. 24.1, but use a two- and a three-point Gauss-Legendre formula. Interpret your results.

24.4 Integration provides a means to compute how much mass enters or leaves a reactor over a specified time period, as in

$$M = \int_{t_1}^{t_2} Qc \, dt$$

where t_1 and t_2 = the initial and final times. This formula makes intuitive sense if you recall the analogy between integration and summation. Thus, the integral represents the summation of the product of flow times concentration to give the total mass entering or leaving from t_1 to t_2. If the flow rate is constant, Q can be moved outside the integral:

$$M = Q \int_{t_1}^{t_2} c \, dt \qquad \text{(P24.4)}$$

Use numerical integration to evaluate this equation for the data in Table P24.4. Note that $Q = 4 \text{ m}^3/\text{min}$.

24.5 The outflow chemical concentration from a completely mixed reactor is measured as

t, min	0	2	4	6	8	12	16	20
c, mg/m³	10	20	30	40	60	72	70	50

For an outflow of $Q = 12 \text{ m}^3/\text{min}$, estimate the mass of chemical that exits the reactor from $t = 0$ to 20 min.

24.6 *Fick's first diffusion law* states that

$$\text{Mass flux} = -D \frac{dc}{dx} \qquad \text{(P24.6)}$$

where mass flux = the quantity of mass that passes across a unit area per unit time (g/cm²/s), D = a diffusion coefficient (cm²/s),

c = concentration, and x = distance (cm). An environmental engineer measures the following concentration of a pollutant in the sediments underlying a lake ($x = 0$ at the sediment-water interface and increases downward):

x, cm	0	1	3
c, 10^{-6} g/cm³	0.1	0.4	0.9

Use the best numerical differentiation technique available to estimate the derivative at $x = 0$. Employ this estimate in conjunction with Eq. (P24.6) to compute the mass flux of pollutant out of the sediments and into the overlying waters ($D = 2 \times 10^{-6} \text{ cm}^2/\text{s}$). For a lake with $3 \times 10^6 \text{ m}^2$ of sediments, how much pollutant would be transported into the lake over a year's time?

24.7 The following data were collected when a large oil tanker was loading:

t, min	0	15	30	45	60	90	120
V, 10^6 barrels	0.5	0.65	0.73	0.88	1.03	1.14	1.30

Calculate the flow rate Q (that is, dV/dt) for each time to the order of h^2.

TABLE P24.4 Values of concentration measured in the exit pipe of a reactor.

t, min	c, mg/m³
0	10
5	22
10	35
15	47
20	55
25	58
30	52
35	40
40	37
45	32
50	34

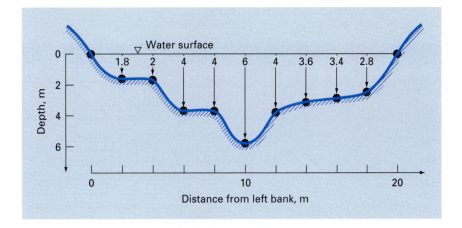

FIGURE P24.12
A stream cross section.

24.8 You are interested in measuring the fluid velocity in a narrow rectangular open channel carrying petroleum waste between locations in an oil refinery. You know that, because of bottom friction, the velocity varies with depth in the channel. If your technician has time to perform only two velocity measurements, at what depths would you take them to obtain the best estimate of the average velocity? State your recommendation in terms of the percent of total depth d measured from the fluid surface. For example, measuring at the top, would be $0\%d$, whereas at the very bottom would be $100\%d$.

Civil/Environmental Engineering

24.9 Perform the same computation in Sec. 24.2, but use order of h^8 Romberg integration to evaluate the integral.

24.10 Perform the same computation as in Sec. 24.2, but use Gauss quadrature to evaluate the integral.

24.11 Compute F using the Trapezoidal rule, Simpson's 1/3 and Simpson's 3/8 rule. Divide the mast into five-foot intervals.

$$F = \int_0^{30} \frac{250z}{6+z} e^{-z/10}\, dz$$

24.12 Stream cross-sectional areas (A) are required for a number of tasks in water resources engineering, including flood forecasting and reservoir design. Unless electronic sounding devices are available to obtain continuous profiles of the channel bottom, the engineer must rely on discrete depth measurements to compute A. An example of a typical stream cross section is shown in Fig. P24.12. The data points represent locations where a boat was anchored and depth readings taken. Use two trapezoidal rule applications ($h = 4$ and 2 m) and Simpson's 1/3 rule to estimate the cross-sectional area from this data.

24.13 During a survey, you are required to compute the area of the field shown in Fig. P24.13. Use Simpson's rules to determine the area.

24.14 A transportation engineering study requires the calculation of the total number of cars that pass through an intersection over a 24-h period. An individual visits the intersection at various times during the course of a day and counts the number of cars that pass through the intersection in a minute. Utilize this data, summarized in Table P24.14, to estimate the total number of cars that pass through the intersection per day. (Be careful of units.)

24.15 A wind force distributed against the side of a skyscraper is measured as

Height l, m	0	30	60	90	120	150	180	210	240
Force, $F(l)$, N/m	0	350	1000	1500	2600	3000	3300	3500	3600

Compute the net force and the line of action due to this distributed wind.

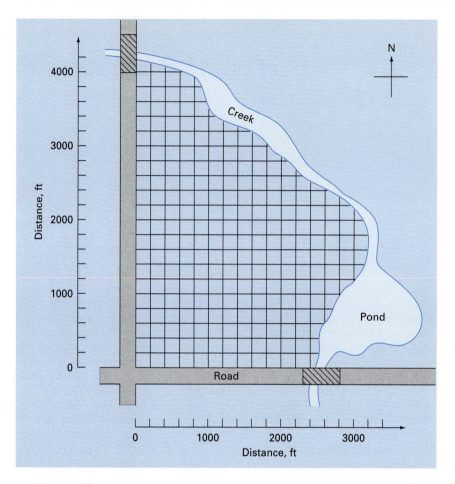

FIGURE P24.13
A field bounded by two roads and a creek.

TABLE P24.14 Traffic flow rate (cars/min) for an intersection measured at various times over a 24-h period.

Time	Rate	Time	Rate	Time	Rate
12:00 midnight	2	9:00 A.M.	12	6:00 P.M.	22
2:00 A.M.	2	10:30 A.M.	5	7:00 P.M.	10
4:00 A.M.	0	11:30 A.M.	10	8:00 P.M.	9
5:00 A.M.	2	12:30 P.M.	12	9:00 P.M.	11
6:00 A.M.	5	2:00 P.M.	7	10:00 P.M.	8
7:00 A.M.	8	4:00 P.M.	9	11:00 P.M.	9
8:00 A.M.	25	5:00 P.M.	28	12:00 midnight	3

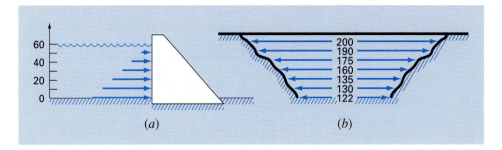

FIGURE P24.16
Water exerting pressure on the upstream face of a dam: (a) side view showing force increasing
linearly with depth; (b) front view showing width of dam in meters.

24.16 Water exerts pressure on the upstream face of a dam as shown
in Fig. P24.16. The pressure can be characterized by

$$p(z) = \rho g(D - z) \qquad\qquad\qquad (P24.16)$$

where $p(z)$ = pressure in pascals (or N/m^2) exerted at an elevation
z meters above the reservoir bottom; ρ = density of water, which
for this problem is assumed to be a constant 10^3 kg/m^3; g = acceleration due to gravity (9.8 m/s^2); and D = elevation (in m) of the
water surface above the reservoir bottom. According to Eq.
(P24.16). pressure increases linearly with depth, as depicted in
Fig. P24.16a. Omitting atmospheric pressure (because it works
against both sides of the dam face and essentially cancels out), the
total force f can be determined by multiplying pressure times the
area of the dam face (as shown in Fig. P24.16b). Because both
pressure and area vary with elevation, the total force is obtained by
evaluating

$$f_t = \int_0^D \rho g w(z)(D - z)\, dz$$

where $w(z)$ = width of the dam face (m) at elevation z (Fig.
P24.16b). The line of action can also be obtained by evaluating

$$d = \frac{\displaystyle\int_0^D \rho g z w(z)(D - z)\, dz}{\displaystyle\int_0^D \rho g w(z)(D - z)\, dz}$$

Use Simpson's rule to compute f_t and d. Check the results with your
computer program for the trapezoidal rule.

24.17 To estimate the size of a new dam, you have to determine the
total volume of water (m^3) that flows down a river in a year's time.
You have available the following long-term average data for the
river:

Date	Mid-Jan.	Mid-Feb.	Mid-Mar.	Mid-Apr.	Mid-June	Mid-Sept.	Mid-Oct.	Mid-Nov.	Mid-Dec.
Flow, m^3/s	31	37	80	119	102	20	21	23	27

Determine the volume. Be careful of units, and take care to make a
proper estimate of flow at the end points.

24.18 The data listed in the following table give hourly measurements of heat flux q (cal/cm^2/h) at the surface of a solar collector.
As an architectural engineer, you must estimate the total heat
absorbed by a 150,000-cm^2 collector panel during a 14-h period.
The panel has an absorption efficiency e_{ab} of 45%. The total heat
absorbed is given by

$$H = e_{ab} \int_0^t qA\, dt$$

where A = area and q = heat flux.

t	0	1	2	3	4	5	6	7	8	9	10	11	12	13	14
q	0.10	1.62	5.32	6.29	7.80	8.81	8.00	8.57	8.03	7.04	6.27	5.56	3.54	1.00	0.20

24.19 The heat flux q is the quantity of heat flowing through a unit area of a material per unit time. It can be computed with Fourier's law,

$$j = -k\frac{dT}{dx}$$

where J has units of $J/m^2/s$ or W/m^2 and k is a coefficient of thermal conductivity that parameterizes the heat-conducting properties of the material and has units of $W/(°C \cdot m)$. T = temperature (°C); and x = distance (m) along the path of heat flow. Fourier's law is used routinely by architectural engineers to determine heat flow through walls. The following temperatures are measured into a stone wall:

x, m	0	0.1	0.2
T, °C	20	17	15

If the flux at $x = 0$ is 60 W/m^2, compute k.

Electrical Engineering

24.20 Perform the same computation in Sec. 24.3, but for the current as specified by

$$i(t) = 4e^{-1.5t} \sin 2\pi t \qquad \text{for } 0 \le t \le T/2$$
$$i(t) = 0 \qquad \text{for } T/2 < t \le T$$

where $T = 1$ s. Use five point Gauss Quadrature to estimate the integral.

24.21 Repeat Prob. 24.20, but use a five segment Simpson's 1/3 Rule.

24.22 Repeat Prob. 24.20, but use Romberg integration to $\varepsilon_s = 1\%$.

24.23 Faraday's law characterizes the voltage drop across an inductor as

$$V_L = L\frac{di}{dt}$$

where V_L = voltage drop (V), L = inductance (in henrys; 1 H = 1 V·s/A), i = current (A), and t = time (s). Determine the voltage drop as a function of time from the following data for an inductance of 4 H.

t	0	0.1	0.2	0.3	0.5	0.7
i	0	0.15	0.3	0.55	0.8	1.9

24.24 Suppose that the current through a resistor is described by the function

$$i(t) = (60 - t)^2 + (60 - t) \sin\left(\sqrt{t}\right)$$

and the resistance is a function of the current,

$$R = 10i + 2i^{2/3}$$

Compute the average voltage over $t = 0$ to 60 using the multiple-segment Simpson's 1/3 rule.

Mechanical/Aerospace Engineering

24.25 Perform the same computation as in Sec. 24.4, but use the following equation to compute:

$$F(x) = 1.5x - 0.04x^2$$

Employ the values of θ from Table 24.6.

24.26 Perform the same computation as in Sec. 24.4, but use the following equation to compute:

$$\theta(x) = 0.8 + 0.125x - 0.009x^2 + (2 \times 10^{-4})x^3$$

Employ the equation from Prob. 24.25 for $F(x)$. Use 4-, 8-, and 16-segment trapezoidal rules to compute the integral.

24.27 Repeat Prob. 24.26, but use Simpson's 1/3 rule.

24.28 Repeat Prob. 24.26, but use Romberg integration to $\varepsilon_s = 0.5\%$.

24.29 Repeat Prob. 24.26, but use Gauss quadrature.

24.30 The work done on an object is equal to the force times the distance moved in the direction of the force. The velocity of an object in the direction of a force is given by

$$v = 4t \qquad 0 \le t \le 6$$
$$v = 24 + (6 - t)^2 \qquad 6 \le t \le 14$$

where v = m/s. Employ the multiple-application trapezoidal rule to determine the work if a constant force of 200 N is applied for all t.

24.31 The rate of cooling of a body (Fig. P24.31) can be expressed as

$$\frac{dT}{dt} = -k(T - T_a)$$

where T = temperature of the body (°C), T_a = temperature of the surrounding medium (°C), and k = a proportionality constant (per minute). Thus, this equation (called *Newton's law of cooling*) specifies that the rate of cooling is proportional to the difference in the temperatures of the body and of the surrounding medium. If a metal

FIGURE P24.31

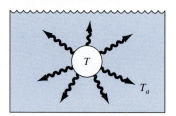

ball heated to 90°C is dropped into water that is held constant at $T_a = 25°C$, the temperature of the ball changes, as in

Time, min	0	5	10	15	20	25
T, °C	90	49.9	33.8	28.4	26.2	25.4

Utilize numerical differentiation to determine dT/dt at each value of time. Plot dT/dt versus $T - T_a$ and employ linear regression to evaluate k.

24.32 A rod subject to an axial load (Fig. P24.32a) will be deformed, as shown in the stress-strain curve in Fig. P24.32b. The area under the curve from zero stress out to the point of rupture is called the *modulus of toughness* of the material. It provides a measure of the energy per unit volume required to cause the material to rupture. As such, it is representative of the material's ability to withstand an impact load. Use numerical integration to compute the modulus of toughness for the stress-strain curve seen in Fig. P24.32b.

24.33 If the velocity distribution of a fluid flowing through a pipe is known (Fig. P24.33), the flow rate Q (that is, the volume of water passing through the pipe per unit time) can be computed by $Q = \int v \, dA$, where v is the velocity and A is the pipe's cross-sectional area. (To grasp the meaning of this relationship physically, recall the close connection between summation and integration.) For a circular pipe, $A = \pi r^2$ and $dA = 2\pi r \, dr$. Therefore,

$$Q = \int_0^r v(2\pi r) \, dr$$

where r is the radial distance measured outward from the center of the pipe. If the velocity distribution is given by

$$v = 2.0\left(1 - \frac{r}{r_0}\right)^{1/6}$$

where r_0 is the total radius (in this case, 2 cm), compute Q using the multiple-application trapezoidal rule. Discuss the results.

24.34 Using the following data, calculate the work done by stretching a spring that has a spring constant of $k \cong 3 \times 10^2$ N/m to $x = 0.45$ m.

F, 10³ N	0	0.01	0.028	0.046	0.063	0.082	0.11	0.13
x, m	0	0.05	0.10	0.15	0.20	0.25	0.30	0.35

FIGURE P24.32
(a) A rod under axial loading and (b) the resulting stress-strain curve where stress is in kips per square inch (10^3 lb/in²) and strain is dimensionless.

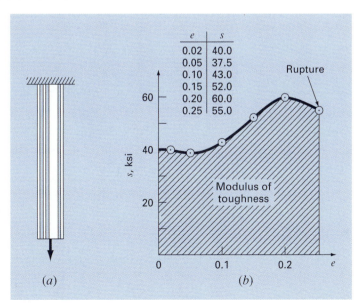

FIGURE P24.33

24.35 A jet fighter's position on an aircraft carrier's runway was timed during landing:

t, s	0	0.51	1.03	1.74	2.36	3.24	3.82
x, m	154	186	209	250	262	272	274

where x is the distance from the end of the carrier. Estimate (a) velocity (dx/dt) and (b) acceleration (dv/dt) using numerical differentiation.

24.36 Employ the multiple-application trapezoidal rule to evaluate the vertical distance traveled by a rocket if the vertical velocity is given by

$$v = 10t^2 - 5t \qquad\qquad 0 \le t \le 10$$
$$v = 1000 - 5t \qquad\qquad 10 \le t \le 20$$
$$v = 45t + 2(t - 20)^2 \qquad 20 \le t \le 30$$

24.37 The upward velocity of a rocket can be computed by the following formula:

$$v = u \ln \left(\frac{m_0}{m_0 - qt} \right) - gt$$

where $v =$ upward velocity, $u =$ velocity at which fuel is expelled relative to the rocket, $m_0 =$ initial mass of the rocket at time $t = 0$, $q =$ fuel consumption rate, and $g =$ downward acceleration of gravity (assumed constant $= 9.8$ m/s^2). If $u = 2000$ m/s, $m_0 = 150,000$ kg, and $q = 2600$ kg/s, use six-segment trapezoidal and Simpson's 1/3 rule, six-point Gauss quadrature, and order of h^8 Romberg methods to determine how high the rocket will fly in 30 s.

EPILOGUE: PART SIX

PT6.4 TRADE-OFFS

Table PT6.4 provides a summary of the trade-offs involved in numerical integration or quadrature. Most of these methods are based on the simple physical interpretation of an integral as the area under a curve. These techniques are designed to evaluate the integral of two different cases: (1) a mathematical function and (2) discrete data in tabular form.

The Newton-Cotes formulas are the primary methods discussed in Chap. 21. They are applicable to both continuous and discrete functions. Both closed and open versions of these formulas are available. The open forms, which have integration limits that extend beyond the range of the data, are rarely used for the evaluation of definite integrals. However, they have utility for the solution of ordinary differential equations and for evaluating improper integrals.

The closed Newton-Cotes formulas are based on replacing a mathematical function or tabulated data by an interpolating polynomial that is easy to integrate. The simplest version is the trapezoidal rule, which is based on taking the area below a straight line joining adjacent values of the function. One way to improve the accuracy of the trapezoidal rule is to divide the integration interval from a to b into a number of segments and apply the method to each segment.

Aside from applying the trapezoidal rule with finer segmentation, another way to obtain a more accurate estimate of the integral is to use higher-order polynomials to connect the points. If a quadratic equation is employed, the result is Simpson's 1/3 rule. If a cubic

TABLE PT6.4 Comparison of the characteristics of alternative methods for numerical integration. The comparisons are based on general experience and do not account for the behavior of special functions.

Method	Data Points Required for One Application	Data Points Required for n Applications	Truncation Error	Application	Programming Effort	Comments
Trapezoidal rule	2	$n + 1$	$\simeq h^3 f''(\xi)$	Wide	Easy	
Simpson's 1/3 rule	3	$2n + 1$	$\simeq h^5 f^{(4)}(\xi)$	Wide	Easy	
Simpson's rule (1/3 and 3/8)	3 or 4	≥ 3	$\simeq h^5 f^{(4)}(\xi)$	Wide	Easy	
Higher-order Newton-Cotes	≥ 5	N/A	$\simeq h^7 f^{(6)}(\xi)$	Rare	Easy	
Romberg integration	3			Requires $f(x)$ be known	Moderate	Inappropriate for tabular data
Gauss quadrature	≥ 2	N/A		Requires $f(x)$ be known	Easy	Inappropriate for tabular data

equation is used, the result is Simpson's 3/8 rule. Because they are much more accurate than the trapezoidal rule, these formulas are usually preferred and multiple-application versions are available. For situations with an even number of segments, the multiple application of the 1/3 rule is recommended. For an odd number of segments, the 3/8 rule can be applied to the last three segments and the 1/3 rule to the remaining segments.

Higher-order Newton-Cotes formulas are also available. However, they are rarely used in practice. Where high accuracy is required, Romberg integration and Gauss quadrature formulas are available. It should be noted that both Romberg integration and Gauss quadrature are of practical value only in cases where the function is available. These techniques are ill-suited for tabulated data.

PT6.5   IMPORTANT RELATIONSHIPS AND FORMULAS

Table PT6.5 summarizes important information that was presented in Part Six. This table can be consulted to quickly access important relationships and formulas.

PT6.6   ADVANCED METHODS AND ADDITIONAL REFERENCES

Although we have reviewed a number of numerical integration techniques, there are other methods that have utility in engineering practice. For example, *adaptive Simpson's integration* is based on dividing the integration interval into a series of subintervals of width h. Then Simpson's 1/3 rule is used to evaluate the integral of each subinterval by halving the step size in an iterative fashion, that is, with a step size of h, $h/2$, $h/4$, $h/8$, and so forth. The iterations are continued for each subinterval until an approximate error estimate falls below a prespecified stopping criterion ε_s. The total integral is then computed as the summation of the integral estimates for the subintervals. This technique is especially valuable for complicated functions that have regions exhibiting both lower- and higher-order variations. Discussions of adaptive integration may be found in Gerald and Wheatley (1984) and Rice (1983). In addition, adaptive schemes for solving ordinary differential equations can be used to evaluate complicated integrals, as was mentioned in PT6.1 and as will be discussed in Chap. 25.

Another method for obtaining integrals is to fit *cubic splines* to the data. The resulting cubic equations can be integrated easily (Forsythe et al., 1977). A similar approach is also sometimes used for differentiation. Finally, aside from the Gauss-Legendre formulas discussed in Sec. 22.2, there are a variety of other quadrature formulas. Carnahan, Luther, and Wilkes (1969) and Ralston and Rabinowitz (1978) summarize many of these approaches.

In summary, the foregoing is intended to provide you with avenues for deeper exploration of the subject. Additionally, all the above references describe basic techniques covered in Part Six. We urge you to consult these alternative sources to broaden your understanding of numerical methods for integration.

TABLE PT6.5   Summary of important information presented in Part Six.

Method	Formulation	Graphic Interpretations	Error
Trapezoidal rule	$I \simeq (b - a) \dfrac{f(a) + f(b)}{2}$		$-\dfrac{(b - a)^3}{12} f''(\xi)$
Multiple-application trapezoidal rule	$I \simeq (b - a) \dfrac{f(x_0) + 2 \sum_{i=1}^{n-1} f(x_i) + f(x_n)}{2n}$		$-\dfrac{(b - a)^3}{12n^2} \bar{f}''$
Simpson's 1/3 rule	$I \simeq (b - a) \dfrac{f(x_0) + 4f(x_1) + f(x_2)}{6}$		$-\dfrac{(b - a)^5}{2880} f^{(4)}(\xi)$
Multiple-application Simpson's 1/3 rule	$I \simeq (b - a) \dfrac{f(x_0) + 4 \sum_{i=1,3}^{n-1} f(x_i) + 2 \sum_{j=2,4}^{n-2} f(x_j) + f(x_n)}{3n}$		$-\dfrac{(b - a)^5}{180n^4} \bar{f}^{(4)}$
Simpson's 3/8 rule	$I \simeq (b - a) \dfrac{f(x_0) + 3f(x_1) + 3f(x_2) + f(x_3)}{8}$		$-\dfrac{(b - a)^5}{6480} f^{(4)}(\xi)$
Romberg integration	$I_{j,k} = \dfrac{4^{k-1} I_{j+1,k-1} - I_{j,k-1}}{4^{k-1} - 1}$		$O(h^{2k})$
Gauss quadrature	$I \simeq c_0 f(x_0) + c_1 f(x_1) + \cdots + c_{n-1} f(x_{n-1})$		$\simeq f^{(2n+2)}(\xi)$

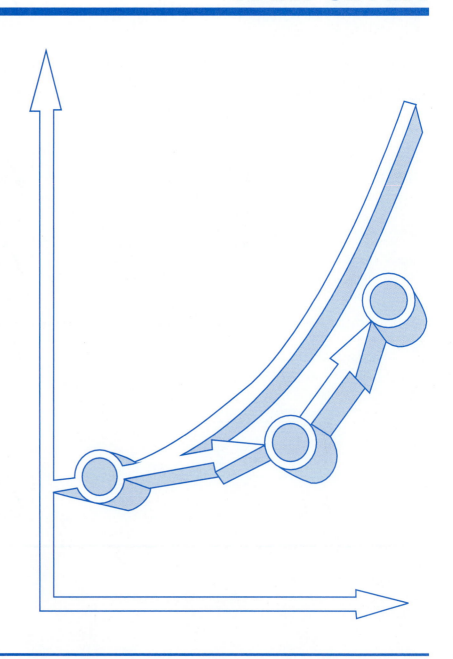

PART SEVEN

ORDINARY DIFFERENTIAL EQUATIONS

PT7.1 MOTIVATION

In the first chapter of this book, we derived the following equation based on Newton's second law to compute the velocity v of a falling parachutist as a function of time t [recall Eq. (1.9)]:

$$\frac{dv}{dt} = g - \frac{c}{m} v \tag{PT7.1}$$

where g is the gravitational constant, m is the mass, and c is a drag coefficient. Such equations, which are composed of an unknown function and its derivatives, are called *differential equations*. Equation (PT7.1) is sometimes referred to as a *rate equation* because it expresses the rate of change of a variable as a function of variables and parameters. Such equations play a fundamental role in engineering because many physical phenomena are best formulated mathematically in terms of their rate of change.

In Eq. (PT7.1), the quantity being differentiated, v, is called the *dependent variable*. The quantity with respect to which v is differentiated, t, is called the *independent variable*. When the function involves one independent variable, the equation is called an *ordinary differential equation* (or *ODE*). This is in contrast to a *partial differential equation* (or *PDE*) that involves two or more independent variables.

Differential equations are also classified as to their order. For example, Eq. (PT7.1) is called a *first-order equation* because the highest derivative is a first derivative. A *second-order equation* would include a second derivative. For example, the equation describing the position x of a mass-spring system with damping is the second-order equation (recall Sec. 8.4),

$$m\frac{d^2x}{dt^2} + c\frac{dx}{dt} + kx = 0 \tag{PT7.2}$$

where c is a damping coefficient and k is a spring constant. Similarly, an nth-order equation would include an nth derivative.

Higher-order equations can be reduced to a system of first-order equations. For Eq. (PT7.2), this is done by defining a new variable y, where

$$y = \frac{dx}{dt} \tag{PT7.3}$$

which itself can be differentiated to yield

$$\frac{dy}{dt} = \frac{d^2x}{dt^2} \tag{PT7.4}$$

Equations (PT7.3) and (PT7.4) can then be substituted into Eq. (PT7.2) to give

$$m\frac{dy}{dt} + cy + kx = 0 \tag{PT7.5}$$

or

$$\frac{dy}{dt} = -\frac{cy + kx}{m} \tag{PT7.6}$$

Thus, Eqs. (PT7.3) and (PT7.6) are a pair of first-order equations that are equivalent to the original second-order equation. Because other nth-order differential equations can be similarly reduced, this part of our book focuses on the solution of first-order equations. Some of the engineering applications in Chap. 28 deal with the solution of second-order ODEs by reduction to a pair of first-order equations.

PT7.1.1 Noncomputer Methods for Solving ODEs

Without computers, ODEs are usually solved with analytical integration techniques. For example, Eq. (PT7.1) could be multiplied by dt and integrated to yield

$$v = \int \left(g - \frac{c}{m}v\right) dt \tag{PT7.7}$$

The right-hand side of this equation is called an *indefinite integral* because the limits of integration are unspecified. This is in contrast to the definite integrals discussed previously in Part Six [compare Eq. (PT7.7) with Eq. (PT6.6)].

An analytical solution for Eq. (PT7.7) is obtained if the indefinite integral can be evaluated exactly in equation form. For example, recall that for the falling parachutist problem, Eq. (PT7.7) was solved analytically by Eq. (1.10) (assuming $v = 0$ at $t = 0$):

$$v(t) = \frac{gm}{c}\left(1 - e^{-(c/m)t}\right) \tag{1.10}$$

The mechanics of deriving such analytical solutions will be discussed in Sec. PT7.2. For the time being, the important fact is that exact solutions for many ODEs of practical importance are not available. As is true for most situations discussed in other parts of this book, numerical methods offer the only viable alternative for these cases. Because these numerical methods usually require computers, engineers in the precomputer era were somewhat limited in the scope of their investigations.

One very important method that engineers and applied mathematicians developed to overcome this dilemma was *linearization*. A linear ordinary differential equation is one that fits the general form

$$a_n(x)y^{(n)} + \cdots + a_1(x)y' + a_0(x)y = f(x) \tag{PT7.8}$$

where $y^{(n)}$ is the nth derivative of y with respect to x and the a's and f's are specified functions of x. This equation is called *linear* because there are no products or nonlinear functions of the dependent variable y and its derivatives. The practical importance of linear ODEs is that they can be solved analytically. In contrast, most nonlinear equations cannot

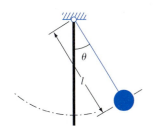

FIGURE PT7.1
The swinging pendulum.

be solved exactly. Thus, in the precomputer era, one tactic for solving nonlinear equations was to linearize them.

A simple example is the application of ODEs to predict the motion of a swinging pendulum (Fig. PT7.1). In a manner similar to the derivation of the falling parachutist problem, Newton's second law can be used to develop the following differential equation (see Sec. 28.4 for the complete derivation):

$$\frac{d^2\theta}{dt^2} + \frac{g}{l}\,\sin\,\theta = 0 \qquad \text{(PT7.9)}$$

where θ is the angle of displacement of the pendulum, g is the gravitational constant, and l is the pendulum length. This equation is nonlinear because of the term $\sin\,\theta$. One way to obtain an analytical solution is to realize that for small displacements of the pendulum from equilibrium (that is, for small values of θ),

$$\sin\,\theta \cong \theta \qquad \text{(PT7.10)}$$

Thus, if it is assumed that we are interested only in cases where θ is small, Eq. (PT7.10) can be substituted into Eq. (PT7.9) to give

$$\frac{d^2\theta}{dt^2} + \frac{g}{l}\,\theta = 0 \qquad \text{(PT7.11)}$$

We have, therefore, transformed Eq. (PT7.9) into a linear form that is easy to solve analytically.

Although linearization remains a very valuable tool for engineering problem solving, there are cases where it cannot be invoked. For example, suppose that we were interested in studying the behavior of the pendulum for large displacements from equilibrium. In such instances, numerical methods offer a viable option for obtaining solutions. Today, the widespread availability of computers places this option within reach of all practicing engineers.

PT7.1.2 ODEs and Engineering Practice

The fundamental laws of physics, mechanics, electricity, and thermodynamics are usually based on empirical observations that explain variations in physical properties and states of systems. Rather than describing the state of physical systems directly, the laws are usually couched in terms of spatial and temporal changes.

Several examples are listed in Table PT7.1. These laws define mechanisms of change. When combined with continuity laws for energy, mass, or momentum, differential equations result. Subsequent integration of these differential equations results in mathematical functions that describe the spatial and temporal state of a system in terms of energy, mass, or velocity variations.

The falling parachutist problem introduced in Chap. 1 is an example of the derivation of an ordinary differential equation from a fundamental law. Recall that Newton's second law was used to develop an ODE describing the rate of change of velocity of a falling parachutist. By integrating this relationship, we obtained an equation to predict fall velocity as a function of time (Fig. PT7.2). This equation could be utilized in a number of different ways, including design purposes.

TABLE PT7.1   Examples of fundamental laws that are written in terms of the rate of change of variables (t = time and x = position).

Law	Mathematical Expression	Variables and Parameters
Newton's second law of motion	$\dfrac{dv}{dt} = \dfrac{F}{m}$	Velocity (v), force (F), and mass (m)
Fourier's heat law	$q = -k' \dfrac{dT}{dx}$	Heat flux (q), thermal conductivity (k') and temperature (T)
Fick's law of diffusion	$J = -D \dfrac{dc}{dx}$	Mass flux (J), diffusion coefficient (D), and concentration (c)
Faraday's law (voltage drop across an inductor)	$\Delta V_L = L \dfrac{di}{dt}$	Voltage drop (ΔV_L), inductance (L), and current (i)

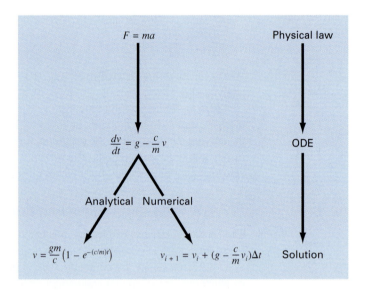

FIGURE PT7.2
The sequence of events in the application of ODEs for engineering problem solving. The example shown is the velocity of a falling parachutist.

In fact, such mathematical relationships are the basis of the solution for a great number of engineering problems. However, as described in the previous section, many of the differential equations of practical significance cannot be solved using the analytical methods of calculus. Thus, the methods discussed in the following chapters are extremely important in all fields of engineering.

PT7.2   MATHEMATICAL BACKGROUND

A solution of an ordinary differential equation is a specific function of the independent variable and parameters that satisfies the original differential equation. To illustrate this concept, let us start with a given function

$$y = -0.5x^4 + 4x^3 - 10x^2 + 8.5x + 1 \qquad \text{(PT7.12)}$$

which is a fourth-order polynomial (Fig. PT7.3a). Now, if we differentiate Eq. (PT7.12), we obtain an ODE:

$$\frac{dy}{dx} = -2x^3 + 12x^2 - 20x + 8.5 \qquad \text{(PT7.13)}$$

This equation also describes the behavior of the polynomial, but in a manner different from Eq. (PT7.12). Rather than explicitly representing the values of y for each value of x, Eq. (PT7.13) gives the rate of change of y with respect to x (that is, the slope) at every value of x. Figure PT7.3 shows both the function and the derivative plotted versus x. Notice how the

FIGURE PT7.3
Plots of (a) y versus x and (b) dy/dx versus x for the function
$y = -0.5x^4 + 4x^3 - 10x^2 + 8.5x + 1$.

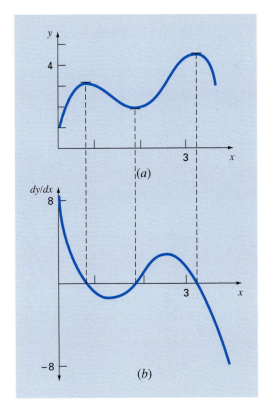

zero values of the derivatives correspond to the point at which the original function is flat—that is, has a zero slope. Also, the maximum absolute values of the derivatives are at the ends of the interval where the slopes of the function are greatest.

Although, as just demonstrated, we can determine a differential equation given the original function, the object here is to determine the original function given the differential equation. The original function then represents the solution. For the present case, we can determine this solution analytically by integrating Eq. (PT7.13):

$$y = \int \left(-2x^3 + 12x^2 - 20x + 8.5\right) dx$$

Applying the integration rule (recall Table PT6.2)

$$\int u^n \, du = \frac{u^{n+1}}{n+1} + C \qquad n \neq -1$$

to each term of the equation gives the solution

$$y = -0.5x^4 + 4x^3 - 10x^2 + 8.5x + C \tag{PT7.14}$$

which is identical to the original function with one notable exception. In the course of differentiating and then integrating, we lost the constant value of 1 in the original equation and gained the value C. This C is called a *constant of integration*. The fact that such an arbitrary constant appears indicates that the solution is not unique. In fact, it is but one of an infinite number of possible functions (corresponding to an infinite number of possible values of C) that satisfy the differential equation. For example, Fig. PT7.4 shows six possible functions that satisfy Eq. (PT7.14).

FIGURE PT7.4
Six possible solutions for the integral of $-2x^3 + 12x^2 - 20x + 8.5$. Each conforms to a different value of the constant of integration C.

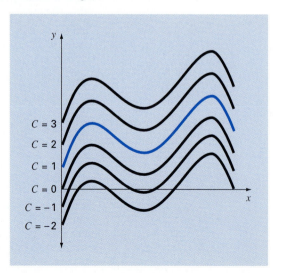

Therefore, to specify the solution completely, a differential equation is usually accompanied by *auxiliary conditions*. For first-order ODEs, a type of auxiliary condition called an *initial value* is required to determine the constant and obtain a unique solution. For example, Eq. (PT7.13) could be accompanied by the initial condition that at $x = 0$, $y = 1$. These values could be substituted into Eq. (PT7.14):

$$1 = -0.5(0)^4 + 4(0)^3 - 10(0)^2 + 8.5(0) + C \tag{PT7.15}$$

to determine $C = 1$. Therefore, the unique solution that satisfies both the differential equation and the specified initial condition is obtained by substituting $C = 1$ into Eq. (PT7.14) to yield

$$y = -0.5x^4 + 4x^3 - 10x^2 + 8.5x + 1 \tag{PT7.16}$$

Thus, we have "pinned down" Eq. (PT7.14) by forcing it to pass through the initial condition, and in so doing, we have developed a unique solution to the ODE and have come full circle to the original function [Eq. (PT7.12)].

Initial conditions usually have very tangible interpretations for differential equations derived from physical problem settings. For example, in the falling parachutist problem, the initial condition was reflective of the physical fact that at time zero the vertical velocity was zero. If the parachutist had already been in vertical motion at time zero, the solution would have been modified to account for this initial velocity.

When dealing with an nth-order differential equation, n conditions are required to obtain a unique solution. If all conditions are specified at the same value of the independent variable (for example, at x or $t = 0$), then the problem is called an *initial-value problem*. This is in contrast to *boundary-value problems* where specification of conditions occurs at different values of the independent variable. Chapters 25 and 26 will focus on initial-value problems. Boundary-value problems are covered in Chap. 27 along with eigenvalues.

PT7.3   ORIENTATION

Before proceeding to numerical methods for solving ordinary differential equations, some orientation might be helpful. The following material is intended to provide you with an overview of the material discussed in Part Seven. In addition, we have formulated objectives to focus your studies of the subject area.

PT7.3.1 Scope and Preview

Figure PT7.5 provides an overview of Part Seven. Two broad categories of numerical methods for initial-value problems will be discussed in this part of this book. One-step methods, which are covered in Chap. 25, permit the calculation of y_{i+1}, given the differential equation and y_i. Multistep methods, which are covered in Chap. 26, require additional values of y other than at i.

With all but a minor exception, the *one-step methods* in *Chap. 25* belong to what are called Runge-Kutta techniques. Although the chapter might have been organized around this theoretical notion, we have opted for a more graphical, intuitive approach to introduce the methods. Thus, we begin the chapter with *Euler's method* which has a very straightforward graphical interpretation. Then, we use visually oriented arguments to develop two

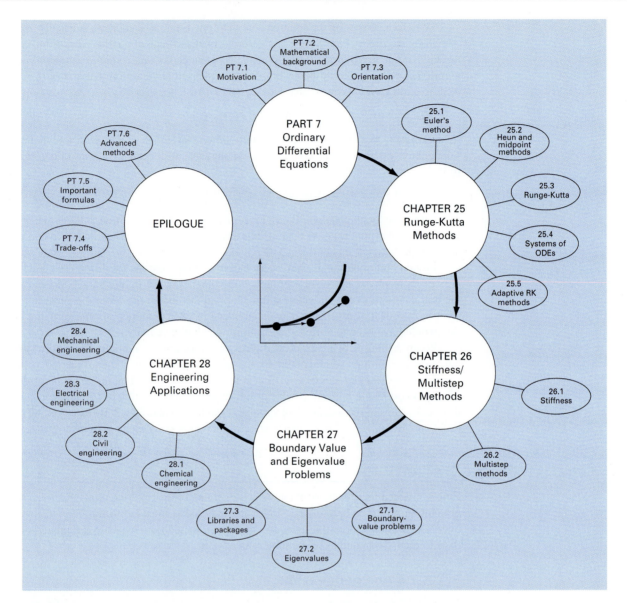

FIGURE PT7.5
Schematic representation of the organization of Part Seven: Ordinary Differential Equations.

improved versions of Euler's method—the *Heun* and the *midpoint* techniques. After this introduction, we formally develop the concept of *Runge-Kutta* (or *RK*) approaches and demonstrate how the foregoing techniques are actually first- and second-order RK methods. This is followed by a discussion of the higher-order RK formulations that are frequently used for engineering problem solving. In addition, we cover the application of one-step methods to *systems of ODEs*. Finally, the chapter ends with a discussion of *adaptive RK methods* that automatically adjust the step size in response to the truncation error of the computation.

Chapter 26 starts with a description of *stiff ODEs*. These are both individual and systems of ODEs that have both fast and slow components to their solution. We introduce the idea of an *implicit solution* technique as one commonly used remedy for this problem.

Next, we discuss *multistep methods*. These algorithms retain information of previous steps to more effectively capture the trajectory of the solution. They also yield the truncation error estimates that can be used to implement step-size control. In this section, we initially take a visual, intuitive approach by using a simple method—the *non-self-starting Heun*—to introduce all the essential features of the multistep approaches.

In *Chap. 27* we turn to *boundary-value* and *eigenvalue* problems. For the former, we introduce both *shooting* and *finite-difference methods*. For the latter, we discuss several approaches, including the *polynomial* and the *power methods*. Finally, the chapter concludes with a description of the application of several *software packages* and *libraries* for solution of ODEs and eigenvalues.

Chapter 28 is devoted to applications from all the fields of engineering. Finally, a short review section is included at the end of Part Seven. This epilogue summarizes and compares the important formulas and concepts related to ODEs. The comparison includes a discussion of trade-offs that are relevant to their implementation in engineering practice. The epilogue also summarizes important formulas and includes references for advanced topics.

PT7.3.2 Goals and Objectives

Study Objectives.   After completing Part Seven, you should have greatly enhanced your capability to confront and solve ordinary differential equations and eigenvalue problems. General study goals should include mastering the techniques, having the capability to assess the reliability of the answers, and being able to choose the "best" method (or methods) for any particular problem. In addition to these general objectives, the specific study objectives in Table PT7.2 should be mastered.

Computer Objectives.   You have been provided with software, and algorithms to implement the techniques discussed in Part Seven. All have utility as learning tools.

The Numerical Methods TOOLKIT personal computer software is user-friendly. It employs the fourth-order Runge-Kutta method for solving up to five simultaneous ODEs. The graphics associated with this software will enable you to easily visualize the solution as *xy* plots of the dependent variable(s) versus the independent variable. The software is very easy to apply to solve many practical problems and can be used to check the results of any computer programs you may develop yourself.

In addition, algorithms are provided for many of the other methods in Part Seven. This information will allow you to expand your software library. For example, you may find it useful from a professional viewpoint to have software that employs the fourth-order Runge-Kutta method for more than five equations and to solve ODEs with an adaptive step-size approach.

TABLE PT7.2 Specific study objectives for Part Seven.

1. Understand the visual representations of Euler's, Heun's, and the midpoint methods
2. Know the relationship of Euler's method to the Taylor series expansion and the insight it provides regarding the error of the method
3. Understand the difference between local and global truncation errors and how they relate to the choice of a numerical method for a particular problem
4. Know the order and the step-size dependency of the global truncation errors for all the methods described in Part Seven; understand how these errors bear on the accuracy of the techniques
5. Understand the basis of predictor-corrector methods. In particular, realize that the efficiency of the corrector is highly dependent on the accuracy of the predictor
6. Know the general form of the Runge-Kutta methods. Understand the derivation of the second-order RK method and how it relates to the Taylor series expansion; realize that there are an infinite number of possible versions for second- and higher-order RK methods
7. Know how to apply any of the RK methods to systems of equations; be able to reduce an nth-order ODE to a system of n first-order ODEs
8. Recognize the type of problem context where step size adjustment is important
9. Understand how adaptive step size control is integrated into a fourth-order RK method
10. Recognize how the combination of slow and fast components makes an equation or a system of equations stiff
11. Understand the distinction between explicit and implicit solution schemes for ODEs. In particular, recognize how the latter (1) ameliorates the stiffness problem and (2) complicates the solution mechanics.
12. Understand the difference between initial-value and boundary-value problems
13. Know the difference between multistep and one-step methods; realize that all multistep methods are predictor-correctors but that not all predictor-correctors are multistep methods
14. Understand the connection between integration formulas and predictor-corrector methods
15. Recognize the fundamental difference between Newton-Cotes and Adams integration formulas
16. Know the rationale behind the polynomial and the power methods for determining eigenvalues; in particular, recognize their strengths and limitations
17. Understand how Hoteller's deflation allows the power method to be used to compute intermediate eigenvalues
18. Know how to use software packages and/or libraries to integrate ODEs and evaluate eigenvalues.

Finally, one of your most important goals should be to master several of the general-purpose software packages that are widely available. In particular, you should become adept at using these tools to implement numerical methods for engineering problem solving.

CHAPTER 25

Runge-Kutta Methods

This chapter is devoted to solving ordinary differential equations of the form

$$\frac{dy}{dx} = f(x, y)$$

In Chap. 1, we used a numerical method to solve such an equation for the velocity of the falling parachutist. Recall that the method was of the general form

New value = old value + slope × step size

or, in mathematical terms,

$$y_{i+1} = y_i + \phi h \tag{25.1}$$

According to this equation, the slope estimate of ϕ is used to extrapolate from an old value y_i to a new value y_{i+1} over a distance h (Fig. 25.1). This formula can be applied step by step to compute out into the future and, hence, trace out the trajectory of the solution.

FIGURE 25.1
Graphical depiction of a one-step method.

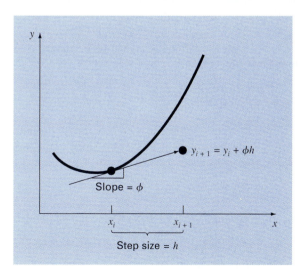

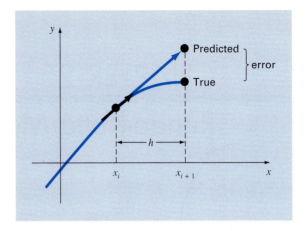

FIGURE 25.2
Euler's method.

All one-step methods can be expressed in this general form, with the only difference being the manner in which the slope is estimated. As in the falling parachutist problem, the simplest approach is to use the differential equation to estimate the slope in the form of the first derivative at x_i. In other words, the slope at the beginning of the interval is taken as an approximation of the average slope over the whole interval. This approach, called *Euler's method,* is discussed in the first part of this chapter. This is followed by other one-step methods that employ alternative slope estimates that result in more accurate predictions. All these techniques are generally called *Runge-Kutta* methods.

25.1 EULER'S METHOD

The first derivative provides a direct estimate of the slope at x_i (Fig. 25.2):

$$\phi = f(x_i, y_i)$$

where $f(x_i, y_i)$ is the differential equation evaluated at x_i and y_i. This estimate can be substituted into Eq. (25.1):

$$y_{i+1} = y_i + f(x_i, y_i)h \tag{25.2}$$

This formula is referred to as *Euler's* (or the *Euler-Cauchy* or the *point-slope*) *method.* A new value of y is predicted using the slope (equal to the first derivative at the original value of x) to extrapolate linearly over the step size h (Fig. 25.2).

EXAMPLE 25.1   Euler's Method

Problem Statement.   Use Euler's method to numerically integrate Eq. (PT7.13):

$$\frac{dy}{dx} = -2x^3 + 12x^2 - 20x + 8.5$$

from $x = 0$ to $x = 4$ with a step size of 0.5. The initial condition at $x = 0$ is $y = 1$. Recall that the exact solution is given by Eq. (PT7.16):

$$y = -0.5x^4 + 4x^3 - 10x^2 + 8.5x + 1$$

Solution. Equation (25.2) can be used to implement Euler's method:

$$y(0.5) = y(0) + f(0, 1)0.5$$

where $y(0) = 1$ and the slope estimate at $x = 0$ is

$$f(0, 1) = -2(0)^3 + 12(0)^2 - 20(0) + 8.5 = 8.5$$

Therefore,

$$y(0.5) = 1.0 + 8.5(0.5) = 5.25$$

The true solution at $x = 0.5$ is

$$y = -0.5(0.5)^4 + 4(0.5)^3 - 10(0.5)^2 + 8.5(0.5) + 1 = 3.21875$$

Thus, the error is

$$E_t = \text{true} - \text{approximate} = 3.21875 - 5.25 = -2.03125$$

or, expressed as percent relative error, $\varepsilon_t = -63.1\%$. For the second step,

$$\begin{aligned} y(1) &= y(0.5) + f(0.5, 5.25)0.5 \\ &= 5.25 + \left[-2(0.5)^3 + 12(0.5)^2 - 20(0.5) + 8.5\right]0.5 \\ &= 5.875 \end{aligned}$$

The true solution at $x = 1.0$ is 3.0, and therefore, the percent relative error is -95.8%. The computation is repeated, and the results are compiled in Table 25.1 and Fig. 25.3. Note that,

TABLE 25.1 Comparison of true and approximate values of the integral of $y' = -2x^3 + 12x^2 - 20x + 8.5$, with the initial condition that $y = 1$ at $x = 0$. The approximate values were computed using Euler's method with a step size of 0.5. The local error refers to the error incurred over a single step. It is calculated with a Taylor series expansion as in Example 25.2. The global error is the total discrepancy due to past as well as present steps.

			Percent Relative Error	
x	**y_{true}**	**y_{Euler}**	**Global**	**Local**
0.0	1.00000	1.00000		
0.5	3.21875	5.25000	−63.1	−63.1
1.0	3.00000	5.87500	−95.8	−28.0
1.5	2.21875	5.12500	131.0	−1.41
2.0	2.00000	4.50000	−125.0	20.5
2.5	2.71875	4.75000	−74.7	17.3
3.0	4.00000	5.87500	46.9	4.0
3.5	4.71875	7.12500	−51.0	−11.3
4.0	3.00000	7.00000	−133.3	−53.0

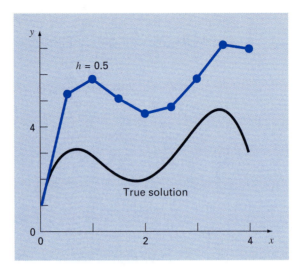

FIGURE 25.3
Comparison of the true solution with a numerical solution using Euler's method for the integral of $y' = -2x^3 + 12x^2 - 20x + 8.5$ from $x = 0$ to $x = 4$ with a step size of 0.5. The initial condition at $x = 0$ is $y = 1$.

although the computation captures the general trend of the true solution, the error is considerable. As discussed in the next section, this error can be reduced by using a smaller step size.

The preceding example uses a simple polynomial for the differential equation to facilitate the error analyses that follow. Thus,

$$\frac{dy}{dx} = f(x)$$

Obviously, a more general (and more common) case involves ODEs that depend on both x and y,

$$\frac{dy}{dx} = f(x, y)$$

As we progress through this part of the text, our examples will increasingly involve ODEs that depend on both the independent and the dependent variables.

25.1.1 Error Analysis for Euler's Method

The numerical solution of ODEs involves two types of error (recall Chaps. 3 and 4):

1. *Truncation,* or discretization, errors caused by the nature of the techniques employed to approximate values of y.

2. *Round-off* errors caused by the limited numbers of significant digits that can be retained by a computer.

The truncation errors are composed of two parts. The first is a *local truncation error* that results from an application of the method in question over a single step. The second is a *propagated truncation error* that results from the approximations produced during the previous steps. The sum of the two is the total, or *global truncation, error.*

Insight into the magnitude and properties of the truncation error can be gained by deriving Euler's method directly from the Taylor series expansion. To do this, realize that the differential equation being integrated will be of the general form

$$y' = f(x, y) \tag{25.3}$$

where $y' = dy/dx$ and x and y are the independent and the dependent variables, respectively. If the solution—that is, the function describing the behavior of y—has continuous derivatives, it can be represented by a Taylor series expansion about a starting value (x_i, y_i), as in [recall Eq. (4.7)]

$$y_{i+1} = y_i + y_i'h + \frac{y_i''}{2!}h^2 + \cdots + \frac{y_i^{(n)}}{n!}h^n + R_n \tag{25.4}$$

where $h = x_{i+1} - x_i$ and $R_n =$ the remainder term, defined as

$$R_n = \frac{y^{(n+1)}(\xi)}{(n+1)!}h^{n+1} \tag{25.5}$$

where ξ lies somewhere in the interval from x_i to x_{i+1}. An alternative form can be developed by substituting Eq. (25.3) into Eqs. (25.4) and (25.5) to yield

$$y_{i+1} = y_i + f(x_i, y_i)h + \frac{f'(x_i, y_i)}{2!}h^2 + \cdots + \frac{f^{(n-1)}(x_i, y_i)}{n!}h^n + O(h^{n+1}) \tag{25.6}$$

where $O(h^{n+1})$ specifies that the local truncation error is proportional to the step size raised to the $(n + 1)$th power.

By comparing Eqs. (25.2) and (25.6), it can be seen that Euler's method corresponds to the Taylor series up to and including the term $f(x_i, y_i)h$. Additionally, the comparison indicates that a truncation error occurs because we approximate the true solution using a finite number of terms from the Taylor series. We thus truncate, or leave out, a part of the true solution. For example, the truncation error in Euler's method is attributable to the remaining terms in the Taylor series expansion that were not included in Eq. (25.2). Subtracting Eq. (25.2) from Eq. (25.6) yields

$$E_t = \frac{f'(x_i, y_i)}{2!}h^2 + \cdots + O(h^{n+1}) \tag{25.7}$$

where $E_t =$ the true local truncation error. For sufficiently small h, the errors in the terms in Eq. (25.7) usually decrease as the order increases (recall Example 4.2 and the accompanying discussion), and the result is often represented as

$$E_a = \frac{f'(x_i, y_i)}{2!}h^2 \tag{25.8}$$

or

$$E_a = O(h^2) \tag{25.9}$$

where $E_a =$ the approximate local truncation error.

EXAMPLE 25.2   Taylor Series Estimate for the Error of Euler's Method

Problem Statement.   Use Eq. (25.7) to estimate the error of the first step of Example 25.1. Also use it to determine the error due to each higher-order term of the Taylor series expansion.

Solution.   Because we are dealing with a polynomial, we can use the Taylor series to obtain exact estimates of the errors in Euler's method. Equation (25.7) can be written as

$$E_t = \frac{f'(x_i, y_i)}{2!}h^2 + \frac{f''(x_i, y_i)}{3!}h^3 + \frac{f^{(3)}(x_i, y_i)}{4!}h^4 \tag{E25.2.1}$$

where $f'(x_i, y_i) =$ the first derivative of the differential equation (that is, the second derivative of the solution). For the present case, this is

$$f'(x_i, y_i) = -6x^2 + 24x - 20 \tag{E25.2.2}$$

and $f''(x_i, y_i)$ is the second derivative of the ODE

$$f''(x_i, y_i) = -12x + 24 \tag{E25.2.3}$$

and $f^{(3)}(x_i, y_i)$ is the third derivative of the ODE

$$f^{(3)}(x_i, y_i) = -12 \tag{E25.2.4}$$

We can omit additional terms (that is, fourth derivatives and higher) from Eq. (E25.2.1) because for this particular case they equal zero. It should be noted that for other functions (for example, transcendental functions such as sinusoids or exponentials) this would not necessarily be true, and higher-order terms would have nonzero values. However, for the present case, Eqs. (E25.2.1) through (E25.2.4) completely define the truncation error for a single application of Euler's method.

For example, the error due to truncation of the second-order term can be calculated as

$$E_{t,2} = \frac{-6(0.0)^2 + 24(0.0) - 20}{2}(0.5)^2 = -2.5 \tag{E25.2.5}$$

For the third-order term:

$$E_{t,3} = \frac{-12(0.0) + 24}{6}(0.5)^3 = 0.5$$

and the fourth-order term:

$$E_{t,4} = \frac{-12}{24}(0.5)^4 = -0.03125$$

These three results can be added to yield the total truncation error:

$$E_t = E_{t,2} + E_{t,3} + E_{t,4} = -2.5 + 0.5 - 0.03125 = -2.03125$$

which is exactly the error that was incurred in the initial step of Example 25.1. Note how $E_{t,2} > E_{t,3} > E_{t,4}$, which supports the approximation represented by Eq. (25.8).

As illustrated in Example 25.2, the Taylor series provides a means of quantifying the error in Euler's method. However, there are limitations associated with its use for this purpose:

1. The Taylor series provides only an estimate of the local truncation error—that is, the error created during a single step of the method. It does not provide a measure of the propagated and, hence, the global truncation error. In Table 25.1, we have included the local and global truncation errors for Example 25.1. The local error was computed for each time step with Eq. (25.2) but using the true value of y_i (the second column of the table) to compute each y_{i+1} rather than the approximate value (the third column), as is done in the Euler method. As expected, the average absolute local truncation error (25 percent) is less than the average global error (90 percent). The only reason that we can make these exact error calculations is that we know the true value a priori. Such would not be the case in an actual problem. Consequently, as discussed below, you must usually apply techniques such as Euler's method using a number of different step sizes to obtain an indirect estimate of the errors involved.
2. As mentioned above, in actual problems we usually deal with functions that are more complicated than simple polynomials. Consequently, the derivatives that are needed to evaluate the Taylor series expansion would not always be easy to obtain.

Although these limitations preclude exact error analysis for most practical problems, the Taylor series still provides valuable insight into the behavior of Euler's method. According to Eq. (25.9), we see that the local error is proportional to the square of the step size and the first derivative of the differential equation. It can also be demonstrated that the global truncation error is $O(h)$; that is, it is proportional to the step size (Carnahan et al. 1969). These observations lead to some useful conclusions:

1. The error can be reduced by decreasing the step size.
2. The method will provide error-free predictions if the underlying function (that is, the solution of the differential equation) is linear, because for a straight line the second derivative would be zero.

This latter conclusion makes intuitive sense because Euler's method uses straight-line segments to approximate the solution. Hence, Euler's method is referred to as a *first-order method.*

It should also be noted that this general pattern holds for the higher-order one-step methods described in the following pages. That is, an nth-order method will yield perfect results if the underlying solution is an nth-order polynomial. Further, the local truncation error will be $O(h^{n+1})$ and the global error $O(h^n)$.

EXAMPLE 25.3   Effect of Reduced Step Size on Euler's Method

Problem Statement.   Repeat the computation of Example 25.1, but use a step size of 0.25.

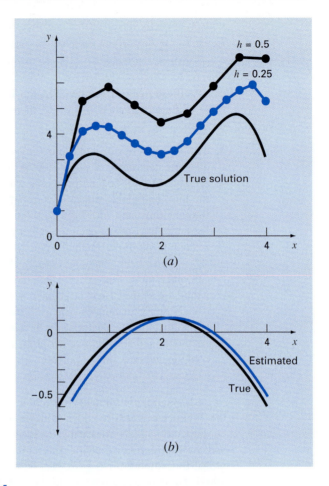

FIGURE 25.4
(a) Comparison of two numerical solutions with Euler's method using step sizes of 0.5 and 0.25.
(b) Comparison of true and estimated local truncation error for the case where the step size is
0.5. Note that the "estimated" error is based on Eq. (E25.2.5).

Solution. The computation is repeated, and the results are compiled in Fig. 25.4a.
Halving the step size reduces the absolute value of the average global error to 40 percent
and the absolute value of the local error to 6.4 percent. This is compared to global and
local errors for Example 25.1 of 90 percent and 24.8 percent. Thus, as expected, the local
error is quartered and the global error is halved.

Also, notice how the local error changes sign for intermediate values along the range.
This is due primarily to the fact that the first derivative of the differential equation is a
parabola that changes sign [recall Eq. (E25.2.2) and see Fig. 25.4b]. Because the local error
is proportional to this function, the net effect of the oscillation in sign is to keep the global
error from continuously growing as the calculation proceeds. Thus, from $x = 0$ to
$x = 1.25$, the local errors are all negative, and consequently, the global error increases over

this interval. In the intermediate section of the range, positive local errors begin to reduce the global error. Near the end, the process is reversed and the global error again inflates. If the local error continuously changes sign over the computation interval, the net effect is usually to reduce the global error. However, where the local errors are of the same sign, the numerical solution may diverge farther and farther from the true solution as the computation proceeds. Such results are said to be *unstable*.

The effect of further step-size reductions on the global truncation error of Euler's method is illustrated in Fig. 25.5. This plot shows the absolute percent relative error at $x = 5$ as a function of step size for the problem we have been examining in Examples 25.1 through 25.3. Notice that even when h is reduced to 0.001, the error still exceeds 0.1 percent. Because this step size corresponds to 5000 steps to proceed from $x = 0$ to $x = 5$, the plot suggests that a first-order technique such as Euler's method demands great computational effort to obtain acceptable error levels. Later in this chapter, we present higher-order techniques that attain much better accuracy for the same computational effort. However, it should be noted that, despite its inefficiency, the simplicity of Euler's method makes it an extremely attractive option for many engineering problems. Because it is very easy to program, the technique is particularly useful for quick analyses. In the next section, a computer algorithm for Euler's method is developed.

FIGURE 25.5
Effect of step size on the global truncation error of Euler's method for the integral of $y' = -2x^3 + 12x^2 - 20x + 8.5$. The plot shows the absolute percent relative global error at $x = 5$ as a function of step size.

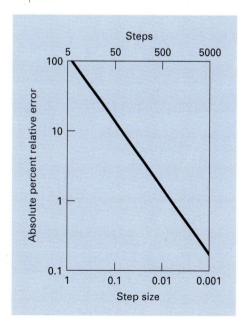

25.1.2 Algorithm for Euler's Method

Algorithms for one-step techniques such as Euler's method are extremely simple to program. As specified previously at the beginning of this chapter, all one-step methods have the general form

$$\text{New value} = \text{old value} + \text{slope} \times \text{step size} \tag{25.10}$$

The only way in which the methods differ is in the calculation of the slope.

Suppose that you want to perform the simple calculation outlined in Table 25.1. That is, you would like to use Euler's method to integrate $y' = -2x^3 + 12x^2 - 20x + 8.5$, with the initial condition that $y = 1$ at $x = 0$. You would like to integrate out to $x = 4$ using a step size of 0.5, and display all the results. A simple pseudocode to accomplish this task could be written as in Fig. 25.6.

Although this program will "do the job" of duplicating the results of Table 25.1, it is not very well designed. First, and foremost, it is not very modular. Although this is not very important for such a small program, it would be critical if we desired to modify and improve the algorithm.

Further, there are a number of issues related to the way we have set up the iterations. For example, suppose that the step size were to be made very small to obtain better accuracy. In such cases, because every computed value is displayed, the number of output values might be very large. Further, the algorithm is predicated on the assumption that the calculation interval is evenly divisible by the step size. Finally, the accumulation of x in the

FIGURE 25.6
Pseudocode for a "dumb" version of Euler's method.

```
'set integration range
xi = 0
xf = 4
'initialize variables
x = xi
y = 1
'set step size and determine
'number of calculation steps
dx = 0.5
nc = (xf − xi)/dx
'output initial condition
PRINT x, y
'loop to implement Euler's method
'and display results
DO i = 1, nc
    dydx = −2x³ + 12x² − 20x + 8.5
    y = y + dydx · dx
    x = x + dx
    PRINT x, y
END DO
```

line $x = x + dx$ can be subject to quantizing errors of the sort previously discussed in Sec. 3.4.1. For example, if dx were changed to 0.01 and standard IEEE floating point representation were used (about seven significant digits), the result at the end of the calculation would be 3.999997 rather than 4. For $dx = 0.001$, it would be 3.999892!

A much more modular algorithm that avoids these difficulties is displayed in Fig. 25.7. The algorithm does not output all calculated values. Rather, the user specifies an output interval, *xout*, that dictates the interval at which calculated results are stored in arrays, xp_m and yp_m. These values are stored in arrays so that they can be output in a variety of ways after the computation is completed (e.g., printed, graphed, written to a file).

The Driver Program takes big output steps and calls an Integrator routine that takes finer calculation steps. Note that the loops controlling both large and small steps exit on logical conditions. Thus, the intervals do not have to be evenly divisible by the step sizes.

The Integrator routine then calls an Euler routine that takes a single step with Euler's method. The Euler routine calls a Derivative routine that calculates the derivative value.

Whereas such modularization might seem like overkill for the present case, it will greatly facilitate modifying the program in later sections. For example, although the program in Fig. 25.7 is specifically designed to implement Euler's method, the Euler module

FIGURE 25.7
Pseudocode for an "improved" modular version of Euler's method.

(a) Main or "Driver" Program

Assign values for
y = initial value dependent variable
xi = initial value independent variable
xf = final value independent variable
dx = calculation step size
$xout$ = output interval

$x = xi$
$m = 0$
$xp_m = x$
$yp_m = y$
DO
  $xend = x + xout$
  IF $(xend > xf)$ THEN $xend = xf$
  $h = dx$
  CALL Integrator $(x, y, h, xend)$
  $m = m + 1$
  $xp_m = x$
  $yp_m = y$
  IF $(x \geq xf)$ EXIT
END DO
DISPLAY RESULTS
END

(b) Routine to take one output step

SUB Integrator $(x, y, h, xend)$
  DO
    IF $(xend - x < h)$ THEN $h = xend - x$
    CALL Euler $(x, y, h, ynew)$
    $y = ynew$
    IF $(x \geq xend)$ EXIT
  END DO
END SUB

(c) Euler's method for a single ODE

SUB Euler $(x, y, h, ynew)$
  CALL Derivs$(x, y, dydx)$
  $ynew = y + dydx * h$
  $x = x + h$
END SUB

(d) Routine to determine derivative

SUB Derivs $(x, y, dydx)$
  $dydx = \ldots$
END SUB

is the only part that is method-specific. Thus, all that is required to apply this algorithm to the other one-step methods is to modify this routine.

EXAMPLE 25.4   Solving ODEs with the Computer

Problem Statement.   A computer program can be developed from the pseudocode in Fig. 25.7. We can use this software to solve another problem associated with the falling parachutist. You recall from Part One that our mathematical model for the velocity was based on Newton's second law in the form

$$\frac{dv}{dt} = g - \frac{c}{m}v \tag{E25.4.1}$$

This differential equation was solved both analytically (Example 1.1) and numerically using Euler's method (Example 1.2). These solutions were for the case where $g = 9.8$, $c = 12.5$, $m = 68.1$, and $v = 0$ at $t = 0$.

The objective of the present example is to repeat these numerical computations employing a more complicated model for the velocity based on a more complete mathematical description of the drag force caused by wind resistance. This model is given by

$$\frac{dv}{dt} = g - \frac{c}{m}\left[v + a\left(\frac{v}{v_{\max}}\right)^b\right] \tag{E25.4.2}$$

where g, m, and c are the same as for Eq. (E25.4.1), and a, b, and $v_{\max}$ are empirical constants, which for this case are equal to 8.3, 2.2, and 46, respectively. Note that this model is more capable of accurately fitting empirical measurements of drag forces versus velocity than is the simple linear model of Example 1.1. However, this increased flexibility is gained at the expense of evaluating three coefficients rather than one. Furthermore, the resulting mathematical model is more difficult to solve analytically. In this case, Euler's method provides a convenient alternative to obtain an approximate numerical solution.

FIGURE 25.8
Graphical results for the solution of the nonlinear ODE [Eq. (E25.4.2)]. Notice that the plot also shows the solution for the linear model [Eq. (E25.4.1)] for comparative purposes.

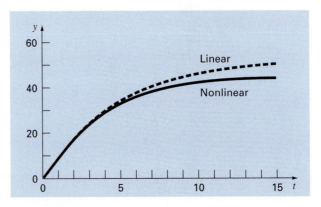

Solution.   The results for both the linear and nonlinear model are displayed in Fig. 25.8 with an integration step size of 0.1 s. The plot in Fig. 25.8 also shows an overlay of the solution of the linear model for comparison purposes.

The results of the two simulations indicate how increasing the complexity of the formulation of the drag force affects the velocity of the parachutist. In this case, the terminal velocity is lowered because of resistance caused by the higher-order terms in Eq. (E25.4.2).

Alternative models could be tested in a similar fashion. The combination of a computer-generated solution makes this an easy and efficient task. This convenience should allow you to devote more of your time to considering creative alternatives and holistic aspects of the problem rather than to tedious manual computations.

25.1.3 Higher-Order Taylor Series Methods

One way to reduce the error of Euler's method would be to include higher-order terms of the Taylor series expansion in the solution. For example, including the second-order term from Eq. (25.6) yields

$$y_{i+1} = y_i + f(x_i, y_i)h + \frac{f'(x_i, y_i)}{2!}h^2 \tag{25.11}$$

with a local truncation error of

$$E_a = \frac{f''(x_i, y_i)}{6}h^3$$

Although the incorporation of higher-order terms is simple enough to implement for polynomials, their inclusion is not so trivial when the ODE is more complicated. In particular, ODEs that are a function of both the dependent and independent variable require chain-rule differentiation. For example, the first derivative of $f(x, y)$ is

$$f'(x_i, y_i) = \frac{\partial f(x, y)}{\partial x} + \frac{\partial f(x, y)}{\partial y}\frac{dy}{dx}$$

The second derivative is

$$f''(x_i, y_i) = \frac{\partial\left[\partial f/\partial x + (\partial f/\partial y)(dy/dx)\right]}{\partial x} + \frac{\partial\left[\partial f/\partial x + (\partial f/\partial y)(dy/dx)\right]}{\partial y}\frac{dy}{dx}$$

Higher-order derivatives become increasingly more complicated.

Consequently, as described in the following sections, alternative one-step methods have been developed. These schemes are comparable in performance to the higher-order Taylor-series approaches but require only the calculation of first derivatives.

25.2   IMPROVEMENTS OF EULER'S METHOD

A fundamental source of error in Euler's method is that the derivative at the beginning of the interval is assumed to apply across the entire interval. Two simple modifications are available to help circumvent this shortcoming. As will be demonstrated in Sec. 25.3, both modifications actually belong to a larger class of solution techniques called Runge-Kutta methods. However, because they have a very straightforward graphical interpretation, we will present them prior to their formal derivation as Runge-Kutta methods.

25.2.1 Heun's Method

One method to improve the estimate of the slope involves the determination of two derivatives for the interval—one at the initial point and another at the end point. The two derivatives are then averaged to obtain an improved estimate of the slope for the entire interval. This approach, called *Heun's method,* is depicted graphically in Fig. 25.9.

Recall that in Euler's method, the slope at the beginning of an interval

$$y_i' = f(x_i, y_i) \tag{25.12}$$

is used to extrapolate linearly to y_{i+1}:

$$y_{i+1}^0 = y_i + f(x_i, y_i)h \tag{25.13}$$

For the standard Euler method we would stop at this point. However, in Heun's method the y_{i+1}^0 calculated in Eq. (25.13) is not the final answer, but an intermediate prediction. This is why we have distinguished it with a superscript 0. Equation (25.13) is called a *predictor*

FIGURE 25.9
Graphical depiction of Heun's method. (a) Predictor and (b) corrector.

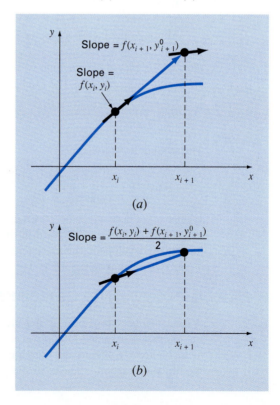

equation. It provides an estimate of y_{i+1} that allows the calculation of an estimated slope at the end of the interval:

$$y'_{i+1} = f(x_{i+1}, y^0_{i+1}) \qquad (25.14)$$

Thus, the two slopes [Eqs. (25.12) and (25.14)] can be combined to obtain an average slope for the interval:

$$\bar{y}' = \frac{y'_i + y'_{i+1}}{2} = \frac{f(x_i, y_i) + f(x_{i+1}, y^0_{i+1})}{2}$$

This average slope is then used to extrapolate linearly from y_i to y_{i+1} using Euler's method:

$$y_{i+1} = y_i + \frac{f(x_i, y_i) + f(x_{i+1}, y^0_{i+1})}{2}h$$

which is called a *corrector equation*.

The Heun method is a *predictor-corrector approach*. All the multistep methods to be discussed subsequently in Chap. 26 are of this type. The Heun method is the only one-step predictor-corrector method described in this book. As derived above, it can be expressed concisely as

Predictor (Fig. 25.9a):	$y^0_{i+1} = y_i + f(x_i, y_i)h$	(25.15)
Corrector (Fig. 25.9b):	$y_{i+1} = y_i + \dfrac{f(x_i, y_i) + f(x_{i+1}, y^0_{i+1})}{2}h$	(25.16)

Note that because Eq. (25.16) has y_{i+1} on both sides of the equal sign, it can be applied in an iterative fashion. That is, an old estimate can be used repeatedly to provide an improved estimate of y_{i+1}. The process is depicted in Fig. 25.10. It should be understood that

FIGURE 25.10
Graphical representation of iterating the corrector of Heun's method to obtain an improved estimate.

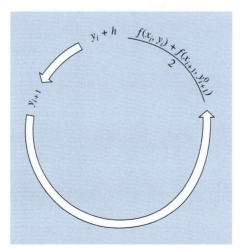

this iterative process does not necessarily converge on the true answer but will converge on an estimate with a finite truncation error, as demonstrated in the following example.

As with similar iterative methods discussed in previous sections of the book, a termination criterion for convergence of the corrector is provided by [recall Eq. (3.5)]

$$|\varepsilon_a| = \left| \frac{y_{i+1}^j - y_{i+1}^{j-1}}{y_{i+1}^j} \right| 100\% \tag{25.17}$$

where y_{i+1}^{j-1} and y_{i+1}^j are the result from the prior and the present iteration of the corrector, respectively.

EXAMPLE 25.5   Heun's Method

Problem Statement.   Use Heun's method to integrate $y' = 4e^{0.8x} - 0.5y$ from $x = 0$ to $x = 4$ with a step size of 1. The initial condition at $x = 0$ is $y = 2$.

Solution.   Before solving the problem numerically, we can use calculus to determine the following analytical solution:

$$y = \frac{4}{1.3} \left(e^{0.8x} - e^{-0.5x} \right) + 2e^{-0.5x} \tag{E25.5.1}$$

This formula can be used to generate the true solution values in Table 25.2.

First, the slope at (x_0, y_0) is calculated as

$$y_0' = 4e^0 - 0.5(2) = 3$$

This result is quite different from the actual average slope for the interval from 0 to 1.0, which is equal to 4.1946, as calculated from the differential equation using Eq. (PT6.4).

The numerical solution is obtained by using the predictor [Eq. (25.15)] to obtain an estimate of y at 1.0:

$$y_1^0 = 2 + 3(1) = 5$$

TABLE 25.2   Comparison of true and approximate values of the integral of $y' = 4e^{0.8x} - 0.5y$, with the initial condition that $y = 2$ at $x = 0$. The approximate values were computed using the Heun method with a step size of 1. Two cases, corresponding to different numbers of corrector iterations, are shown, along with the absolute percent relative error.

| | | Iterations of Heun's Method | | | |
| | | 1 | | 15 | |
x	y_{true}	y_{Heun}	$|\varepsilon_t|$ (%)	y_{Heun}	$|\varepsilon_t|$ (%)
0	2.0000000	2.0000000	0.00	2.0000000	0.00
1	6.1946314	6.7010819	8.18	6.3608655	2.68
2	14.8439219	16.3197819	9.94	15.3022367	3.09
3	33.6771718	37.1992489	10.46	34.7432761	3.17
4	75.3389626	83.3377674	10.62	77.7350962	3.18

Note that this is the result that would be obtained by the standard Euler method. The true value in Table 25.2 shows that it corresponds to a percent relative error of 25.3 percent.

Now, to improve the estimate for y_{i+1}, we use the value y_1^0 to predict the slope at the end of the interval:

$$y_1' = f(x_1, y_1^0) = 4e^{0.8(1)} - 0.5(5) = 6.402164$$

which can be combined with the initial slope to yield an average slope over the interval from $x = 0$ to 1:

$$y' = \frac{3 + 6.402164}{2} = 4.701082$$

which is closer to the true average slope of 4.1946. This result can then be substituted into the corrector [Eq. (25.16)] to give the prediction at $x = 1$:

$$y_1 = 2 + 4.701082(1) = 6.701082$$

which represents a percent relative error of -8.18 percent. Thus, the Heun method without iteration of the corrector reduces the absolute value of the error by a factor of 2.4 as compared with Euler's method.

Now this estimate can be used to refine or correct the prediction of y_1 by substituting the new result back into the right-hand side of Eq. (25.16):

$$y_1 = 2 + \frac{[3 + 4e^{0.8(1)} - 0.5(6.701082)]}{2} 1 = 6.275811$$

which represents an absolute percent relative error of 1.31 percent. This result, in turn, can be substituted back into Eq. (25.16) to further correct:

$$y_1 = 2 + \frac{[3 + 4e^{0.8(1)} - 0.5(6.275811)]}{2} 1 = 6.382129$$

which represents an $|\varepsilon_t|$ of 3.03%. Notice how the errors sometimes grow as the iterations proceed. Such increases can occur, especially for large step sizes, and they prevent us from drawing the general conclusion that an additional iteration will always improve the result. However, for a sufficiently small step size, the iterations should eventually converge on a single value. For our case, 6.360865, which represents a relative error of 2.68 percent, is attained after 15 iterations. Table 25.2 shows results for the remainder of the computation using the method with 1 and 15 iterations per step.

In the previous example, the derivative is a function of both the dependent variable y and the independent variable x. For cases such as polynomials, where the ODE is solely a function of the independent variable, the predictor step [Eq. (25.16)] is not required and the corrector is applied only once for each iteration. For such cases, the technique is expressed concisely as

$$y_{i+1} = y_i + \frac{f(x_i) + f(x_{i+1})}{2} h \qquad (25.18)$$

Notice the similarity between the right-hand side of Eq. (25.18) and the trapezoidal rule [Eq. (21.3)]. The connection between the two methods can be formally demonstrated by starting with the ordinary differential equation

$$\frac{dy}{dx} = f(x)$$

This equation can be solved for y by integration:

$$\int_{y_i}^{y_{i+1}} dy = \int_{x_i}^{x_{i+1}} f(x)\,dx \tag{25.19}$$

which yields

$$y_{i+1} - y_i = \int_{x_i}^{x_{i+1}} f(x)\,dx \tag{25.20}$$

or

$$y_{i+1} = y_i + \int_{x_i}^{x_{i+1}} f(x)\,dx \tag{25.21}$$

Now, recall from Chap. 21 that the trapezoidal rule [Eq. (21.3)] is defined as

$$\int_{x_i}^{x_{i+1}} f(x)\,dx \cong \frac{f(x_i) + f(x_{i+1})}{2} h \tag{25.22}$$

where $h = x_{i+1} - x_i$. Substituting Eq. (25.22) into Eq. (25.21) yields

$$y_{i+1} = y_i + \frac{f(x_i) + f(x_{i+1})}{2} h \tag{25.23}$$

which is equivalent to Eq. (25.18).

Because Eq. (25.23) is a direct expression of the trapezoidal rule, the local truncation error is given by [recall Eq. (21.6)]

$$E_t = -\frac{f''(\xi)}{12} h^3 \tag{25.24}$$

where ξ is between x_i and x_{i+1}. Thus, the method is second order because the second derivative of the ODE is zero when the true solution is a quadratic. In addition, the local and global errors are $O(h^3)$ and $O(h^2)$, respectively. Therefore, decreasing the step size decreases the error at a faster rate than for Euler's method. Figure 25.11, which shows the result of using Heun's method to solve the polynomial from Example 25.1, demonstrates this behavior.

25.2.2 The Midpoint (or Improved Polygon) Method

Figure 25.12 illustrates another simple modification of Euler's method. Called the *midpoint method* (or the *improved polygon* or the *modified Euler*), this technique uses Euler's method to predict a value of y at the midpoint of the interval (Fig. 25.12*a*):

$$y_{i+1/2} = y_i + f(x_i, y_i)\frac{h}{2} \tag{25.25}$$

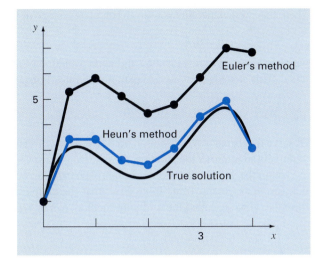

FIGURE 25.11
Comparison of the true solution with a numerical solution using Euler's and Heun's method for the integral of $y' = -2x^3 + 12x^2 - 20x + 8.5$.

FIGURE 25.12
Graphical depiction of the midpoint method.
(a) Eq. (25.25) and
(b) Eq. (25.27).

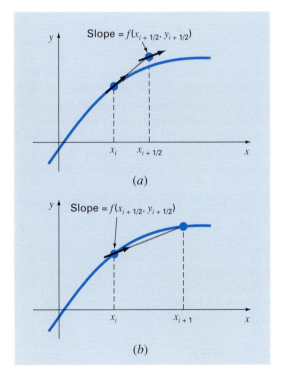

Then, this predicted value is used to calculate a slope at the midpoint:

$$y_{i+1/2} = f(x_{i+1/2}, y_{i+1/2}) \tag{25.26}$$

which is assumed to represent a valid approximation of the average slope for the entire interval. This slope is then used to extrapolate linearly from x_i to x_{i+1} (Fig. 25.12b):

$$y_{i+1} = y_i + f(x_{i+1/2}, y_{i+1/2})h \tag{25.27}$$

Observe that because y_{i+1} is not on both sides, the corrector [Eq. (25.27)] cannot be applied iteratively to improve the solution.

As in the previous section, this approach can also be linked to Newton-Cotes integration formulas. Recall from Table 21.4, that the simplest Newton-Cotes open integration formula, which is called the midpoint method, can be represented as

$$\int_a^b f(x)\, dx \cong (b-a) f(x_1)$$

where x_1 is the midpoint of the interval (a, b). Using the nomenclature for the present case, it can be expressed as

$$\int_{x_i}^{x_{i+1}} f(x)\, dx \cong h f(x_{i+1/2})$$

Substitution of this formula into Eq. (25.21) yields Eq. (25.27). Thus, just as the Heun method can be called the trapezoidal rule, the *midpoint method* gets its name from the underlying integration formula upon which it is based.

The midpoint method is superior to Euler's method because it utilizes a slope estimate at the midpoint of the prediction interval. Recall from our discussion of numerical differentiation in Sec. 4.1.3 that centered finite divided differences are better approximations of derivatives than either forward or backward versions. In the same sense, a centered approximation such as Eq. (25.26) has a local truncation error of $O(h^2)$ in comparison with the forward approximation of Euler's method, which has an error of $O(h)$. Consequently, the local and global errors of the midpoint method are $O(h^3)$ and $O(h^2)$, respectively.

25.2.3 Computer Algorithms for Heun and Midpoint Methods

Both the Heun method with a single corrector and the midpoint method can be easily programmed using the general structure depicted in Fig. 25.7. As in Fig. 25.13a and b, simple routines can be written to take the place of the Euler routine in Fig. 25.7.

However, when the iterative version of the Heun method is to be implemented, the modifications are a bit more involved (although they are still localized within a single module). We have developed pseudocode for this purpose in Fig. 25.13c. This algorithm can be combined with Fig. 25.7 to develop software for the iterative Heun method.

25.2.4 Summary

By tinkering with Euler's method, we have derived two new second-order techniques. Even though these versions require more computational effort to determine the slope, the accompanying reduction in error will allow us to conclude in a subsequent section

(a) Simple Heun without corrector

```
SUB Heun (x, y, h, ynew)
   CALL Derivs (x, y, dy1dx)
   ye = y + dy1dx · h
   CALL Derivs(x + h, ye, dy2dx)
   Slope = (dy1dx + dy2dx)/2
   ynew = y + Slope · h
   x = x + h
END SUB
```

(b) Midpoint method

```
SUB Midpoint (x, y, h, ynew)
   CALL Derivs(x, y, dydx)
   ym = y + dydx · h/2
   CALL Derivs (x + h/2, ym, dymdx)
   ynew = y + dymdx · h
   x = x + h
END SUB
```

(c) Heun with corrector

```
SUB HeunIter (x, y, h, ynew)
   es = 0.01
   maxit = 20
   CALL Derivs(x, y, dy1dx)
   ye = y + dy1dx · h
   iter = 0
   DO
      yeold = ye
      CALL Derivs(x + h, ye, dy2dx)
      slope = (dy1dx + dy2dx)/2
      ye = y + slope · h
      iter = iter + 1
```

$$ea = \left| \frac{ye - yeold}{ye} \right| 100\%$$

```
      IF (ea ≤ es OR iter > maxit) EXIT
   END DO
   ynew = ye
   x = x + h
END SUB
```

FIGURE 25.13
Pseudocode to implement the (a) simple Heun, (b) midpoint, and (c) iterative Heun methods.

(Sec. 25.3.4) that the improved accuracy is usually worth the effort. Although there are certain cases where easily programmable techniques such as Euler's method can be applied to advantage, the Heun and midpoint methods are generally superior and should be implemented if consistent with the problem objectives.

As noted at the beginning of this section, the Heun (without iterations), the midpoint method, and in fact, the Euler technique itself are versions of a broader class of one-step approaches called Runge-Kutta methods. We now turn to a formal derivation of these techniques.

25.3 RUNGE-KUTTA METHODS

Runge-Kutta (*RK*) methods achieve the accuracy of a Taylor series approach without requiring the calculation of higher derivatives. Many variations exist but all can be cast in the generalized form of Eq. (25.1):

$$y_{i+1} = y_i + \phi(x_i, y_i, h)h \tag{25.28}$$

where $\phi(x_i, y_i, h)$ is called an *increment function,* which can be interpreted as a representative slope over the interval. The increment function can be written in general form as

$$\phi = a_1 k_1 + a_2 k_2 + \cdots + a_n k_n \tag{25.29}$$

where the a's are constants and the k's are

$$k_1 = f(x_i, y_i) \tag{25.29a}$$

$$k_2 = f(x_i + p_1 h, y_i + q_{11} k_1 h) \tag{25.29b}$$

$$k_3 = f(x_i + p_2 h, y_i + q_{21} k_1 h + q_{22} k_2 h) \tag{25.29c}$$

$$\vdots$$

$$k_n = f(x_i + p_{n-1} h, y_i + q_{n-1,1} k_1 h + q_{n-1,2} k_2 h + \cdots + q_{n-1,n-1} k_{n-1} h) \tag{25.29d}$$

Notice that the k's are recurrence relationships. That is, k_1 appears in the equation for k_2, which appears in the equation for k_3, and so forth. Because each k is a functional evaluation, this recurrence makes RK methods efficient for computer calculations.

Various types of Runge-Kutta methods can be devised by employing different numbers of terms in the increment function as specified by n. Note that the first-order RK method with $n = 1$ is, in fact, Euler's method. Once n is chosen, values for the a's, p's, and q's are evaluated by setting Eq. (25.28) equal to terms in a Taylor series expansion (Box 25.1). Thus, at least for the lower-order versions, the number of terms n usually represents the order of the approach. For example, in the next section, second-order RK methods use an increment function with two terms ($n = 2$). These second-order methods will be exact if the solution to the differential equation is quadratic. In addition, because terms with h^3 and higher are dropped during the derivation, the local truncation error is $O(h^3)$ and the global error is $O(h^2)$. In subsequent sections, the third- and fourth-order RK methods ($n = 3$ and 4) are developed. For these cases, the global truncation errors are $O(h^3)$ and $O(h^4)$, respectively.

25.3.1 Second-Order Runge-Kutta Methods

The second-order version of Eq. (25.28) is

$$y_{i+1} = y_i + (a_1 k_1 + a_2 k_2) h \tag{25.30}$$

where

$$k_1 = f(x_i, y_i) \tag{25.30a}$$

$$k_2 = f(x_i + p_1 h, y_i + q_{11} k_1 h) \tag{25.30b}$$

As described in Box 25.1, values for a_1, a_2, p_1, and q_{11} are evaluated by setting Eq. (25.30) equal to a Taylor series expansion to the second-order term. By doing this, we derive three equations to evaluate the four unknown constants. The three equations are

$$a_1 + a_2 = 1 \tag{25.31}$$

$$a_1 p_2 = \frac{1}{2} \tag{25.32}$$

$$a_2 q_{11} = \frac{1}{2} \tag{25.33}$$

Box 25.1   Derivation of the Second-Order Runge-Kutta Methods

The second-order version of Eq. (25.28) is

$$y_{i+1} = y_i + (a_1 k_1 + a_2 k_2)h \tag{B25.1.1}$$

where

$$k_1 = f(x_i, y_i) \tag{B25.1.2}$$

and

$$k_2 = f(x_i + p_1 h, y_i + q_{11} k_1 h) \tag{B25.1.3}$$

To use Eq. (B25.1.1) we have to determine values for the constants a_1, a_2, p_1, and q_{11}. To do this, we recall that the second-order Taylor series for y_{i+1} in terms of y_i and $f(x_i, y_i)$ is written as [Eq. (25.11)]

$$y_{i+1} = y_i + f(x_i, y_i)h + \frac{f'(x_i, y_i)}{2!}h^2 \tag{B25.1.4}$$

where $f'(x_i, y_i)$ must be determined by chain-rule differentiation (Sec. 25.1.3):

$$f'(x_i, y_i) = \frac{\partial f(x, y)}{\partial x} + \frac{\partial f(x, y)}{\partial y}\frac{dy}{dx} \tag{B25.1.5}$$

Substituting Eq. (B25.1.5) into (B25.1.4) gives

$$y_{i+1} = y_i + f(x_i, y_i)h + \left(\frac{\partial f}{\partial x} + \frac{\partial f}{\partial y}\frac{dy}{dx}\right)\frac{h^2}{2!} \tag{B25.1.6}$$

The basic strategy underlying Runge-Kutta methods is to use algebraic manipulations to solve for values of a_1, a_2, p_1, and q_{11} that make Eqs. (B25.1.1) and (B25.1.6) equivalent.

To do this, we first use a Taylor series to expand Eq. (25.1.3). The Taylor series for a two-variable function is defined as [recall Eq. (4.26)]

$$g(x + r, y + s) = g(x, y) + r\frac{\partial g}{\partial x} + s\frac{\partial g}{\partial y} + \cdots$$

Applying this method to expand Eq. (B25.1.3) gives

$$f(x_i + p_1 h, y_i + q_{11} k_1 h) = f(x_i, y_i) + p_1 h\frac{\partial f}{\partial x}$$
$$+ q_{11} k_1 h\frac{\partial f}{\partial y} + O(h^2)$$

This result can be substituted along with Eq. (B25.1.2) into Eq. (B25.1.1) to yield

$$y_{i+1} = y_i + a_1 hf(x_i, y_i) + a_2 hf(x_i, y_i) + a_2 p_1 h^2\frac{\partial f}{\partial x}$$
$$+ a_2 q_{11} h^2 f(x_i, y_i)\frac{\partial f}{\partial y} + O(h^3)$$

or, by collecting terms,

$$y_{i+1} = y_i + [a_1 f(x_i, y_i) + a_2 f(x_i, y_i)]h$$
$$+ \left[a_2 p_1\frac{\partial f}{\partial x} + a_2 q_{11} f(x_i, y_i)\frac{\partial f}{\partial y}\right]h^2 + O(h^3) \tag{B25.1.7}$$

Now, comparing like terms in Eqs. (B25.1.6) and (B25.1.7), we determine that for the two equations to be equivalent, the following must hold:

$$a_1 + a_2 = 1$$

$$a_1 p_2 = \frac{1}{2}$$

$$a_2 q_{11} = \frac{1}{2}$$

These three simultaneous equations contain the four unknown constants. Because there is one more unknown than the number of equations, there is no unique set of constants that satisfy the equations. However, by assuming a value for one of the constants, we can determine the other three. Consequently, there is a family of second-order methods rather than a single version.

Because we have three equations with four unknowns, we must assume a value of one of the unknowns to determine the other three. Suppose that we specify a value for a_2. Then Eqs. (25.31) through (25.33) can be solved simultaneously for

$$a_1 = 1 - a_2 \tag{25.34}$$

$$p_1 = q_{11} = \frac{1}{2a_2} \tag{25.35}$$

Because we can choose an infinite number of values for a_2, there are an infinite number of second-order RK methods. Every version would yield exactly the same results if the solution to the ODE were quadratic, linear, or a constant. However, they yield different results when (as is typically the case) the solution is more complicated. We present three of the most commonly used and preferred versions:

Heun Method with a Single Corrector ($a_2 = 1/2$). If a_2 is assumed to be 1/2, Eqs. (25.34) and (25.35) can be solved for $a_1 = 1/2$ and $p_1 = q_{11} = 1$. These parameters, when substituted into Eq. (25.30), yield

$$y_{i+1} = y_i + \left(\frac{1}{2}k_1 + \frac{1}{2}k_2 \right) h \tag{25.36}$$

where

$$k_1 = f(x_i, y_i) \tag{25.36a}$$

$$k_2 = f(x_i + h, y_i + k_1 h) \tag{25.36b}$$

Note that k_1 is the slope at the beginning of the interval and k_2 is the slope at the end of the interval. Consequently, this second-order Runge-Kutta method is actually Heun's technique without iteration.

The Midpoint Method ($a_2 = 1$). If a_2 is assumed to be 1, then $a_1 = 0, p_1 = q_{11} = 1/2$, and Eq. (25.30) becomes

$$y_{i+1} = y_i + k_2 h \tag{25.37}$$

where

$$k_1 = f(x_i, y_i) \tag{25.37a}$$

$$k_2 = f\left(x_i + \frac{1}{2}h, y_i + \frac{1}{2}k_1 h \right) \tag{25.37b}$$

This is the midpoint method.

Ralston's Method ($a_2 = 2/3$). Ralston (1962) and Ralston and Rabinowitz (1978) determined that choosing $a_2 = 2/3$ provides a minimum bound on the truncation error for the second-order RK algorithms. For this version, $a_1 = 1/3$ and $p_1 = q_{11} = 3/4$:

$$y_{i+1} = y_i + \left(\frac{1}{3}k_1 + \frac{2}{3}k_2 \right) h \tag{25.38}$$

where

$$k_1 = f(x_i, y_i) \tag{25.38a}$$

$$k_2 = f\left(x_i + \frac{3}{4}h, y_i + \frac{3}{4}k_1 h \right) \tag{25.38b}$$

EXAMPLE 25.6   Comparison of Various Second-Order RK Schemes

Problem Statement.   Use the midpoint method [Eq. (25.37)] and Ralston's method [Eq. (25.38)] to numerically integrate Eq. (PT7.13):

$$f(x, y) = -2x^3 + 12x^2 - 20x + 8.5$$

from $x = 0$ to $x = 4$ using a step size of 0.5. The initial condition at $x = 0$ is $y = 1$. Compare the results with the values obtained using another second-order RK algorithm: the Heun method without corrector iteration (Table 25.3).

Solution.   The first step in the midpoint method is to use Eq. (25.37a) to compute

$$k_1 = -2(0)^3 + 12(0)^2 - 20(0) + 8.5 = 8.5$$

However, because the ODE is a function of x only, this result has no bearing on the second step—the use of Eq. (25.37b) to compute

$$k_2 = -2(0.25)^3 + 12(0.25)^2 - 20(0.25) + 8.5 = 4.21875$$

Notice that this estimate of the slope is much closer to the average value for the interval (4.4375) than the slope at the beginning of the interval (8.5) that would have been used for Euler's approach. The slope at the midpoint can then be substituted into Eq. (25.37) to predict

$$y(0.5) = 1 + 4.21875(0.5) = 3.109375 \qquad \varepsilon_t = 3.4\%$$

The computation is repeated, and the results are summarized in Fig. 25.14 and Table 25.3.

FIGURE 25.14
Comparison of the true solution with numerical solutions using three second-order RK methods and Euler's method.

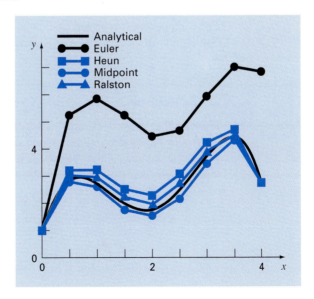

TABLE 25.3 Comparison of true and approximate values of the integral of $y' = -2x^3 + 12x^2 - 20x + 8.5$, with the initial condition that $y = 1$ at $x = 0$. The approximate values were computed using three versions of second-order RK methods with a step size of 0.5.

		Heun		Midpoint		Second-Order Ralston RK	
x	**y_{true}**	**y**	**$\|\varepsilon_t\|$ (%)**	**y**	**$\|\varepsilon_t\|$ (%)**	**y**	**$\|\varepsilon_t\|$ (%)**
0.0	1.00000	1.00000	0	1.00000	0	1.00000	0
0.5	3.21875	3.43750	6.8	3.109375	3.4	3.277344	1.8
1.0	3.00000	3.37500	12.5	2.81250	6.3	3.101563	3.4
1.5	2.21875	2.68750	21.1	1.984375	10.6	2.347656	5.8
2.0	2.00000	2.50000	25.0	1.75	12.5	2.140625	7.0
2.5	2.71875	3.18750	17.2	2.484375	8.6	2.855469	5.0
3.0	4.00000	4.37500	9.4	3.81250	4.7	4.117188	2.9
3.5	4.71875	4.93750	4.6	4.609375	2.3	4.800781	1.7
4.0	3.00000	3.00000	0	3	0	3.031250	1.0

For Ralston's method, k_1 for the first interval also equals 8.5 and [Eq. (25.38b)]

$$k_2 = -2(0.375)^3 + 12(0.375)^2 - 20(0.375) + 8.5 = 2.58203125$$

The average slope is computed by

$$\phi = \frac{1}{3}(8.5) + \frac{2}{3}(2.58203125) = 4.5546875$$

which can be used to predict

$$y(0.5) = 1 + 4.5546875(0.5) = 3.27734375 \qquad \varepsilon_t = -1.82\%$$

The computation is repeated, and the results are summarized in Fig. 25.14 and Table 25.3. Notice how all the second-order RK methods are superior to Euler's method.

25.3.2 Third-Order Runge-Kutta Methods

For $n = 3$, a derivation similar to the one for the second-order method can be performed. The result of this derivation is six equations with eight unknowns. Therefore, values for two of the unknowns must be specified a priori in order to determine the remaining parameters. One common version that results is

$$y_{i+1} = y_i + \frac{1}{6}(k_1 + 4k_2 + k_3)h \qquad (25.39)$$

where

$$k_1 = f(x_i, y_i) \qquad (25.39a)$$

$$k_2 = f\left(x_i + \frac{1}{2}h,\, y_i + \frac{1}{2}k_1 h\right) \tag{25.39b}$$

$$k_3 = f\left(x_i + h,\, y_i - k_1 h + 2k_2 h\right) \tag{25.39c}$$

Note that if the derivative is a function of x only, this third-order method reduces to Simpson's 1/3 rule. Ralston (1962) and Ralston and Rabinowitz (1978) have developed an alternative version that provides a minimum bound on the truncation error. In any case, the third-order RK methods have local and global errors of $O(h^4)$ and $O(h^3)$, respectively, and yield exact results when the solution is a cubic. When dealing with polynomials, Eq. (25.39) will also be exact when the differential equation is cubic and the solution is quartic. This is because Simpson's 1/3 rule provides exact integral estimates for cubics (recall Box 21.3).

25.3.3 Fourth-Order Runge-Kutta Methods

The most popular RK methods are fourth order. As with the second-order approaches, there are an infinite number of versions. The following is the most commonly used form, and we therefore call it the *classical fourth-order RK method:*

$$y_{i+1} = y_i + \frac{1}{6}\left(k_1 + 2k_2 + 2k_3 + k_4\right)h \tag{25.40}$$

where

$$k_1 = f(x_i, y_i) \tag{25.40a}$$

$$k_2 = f\left(x_i + \frac{1}{2}h,\, y_i + \frac{1}{2}k_1 h\right) \tag{25.40b}$$

FIGURE 25.15
Graphical depiction of the slope estimates comprising the fourth-order RK method.

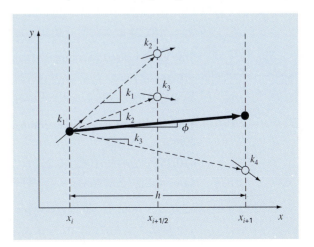

$$k_3 = f\left(x_i + \frac{1}{2}h, y_i + \frac{1}{2}k_2h\right) \tag{25.40c}$$

$$k_4 = f\left(x_i + h, y_i + k_3h\right) \tag{25.40d}$$

Notice that for ODEs that are a function of x alone, the classical fourth-order RK method is similar to Simpson's 1/3 rule. In addition, the fourth-order RK method is similar to the Heun approach in that multiple estimates of the slope are developed in order to come up with an improved average slope for the interval. As depicted in Fig. 25.15, each of the k's represents a slope. Equation (25.40) then represents a weighted average of these to arrive at the improved slope.

EXAMPLE 25.7 Classical Fourth-Order RK Method

Problem Statement. Use the classical fourth-order RK method [Eq. (25.40)] to integrate (a)

$$f(x, y) = -2x^3 + 12x^2 - 20x + 8.5$$

using a step size of $h = 0.5$ and an initial condition of $y = 1$ at $x = 0$; and (b)

$$f(x, y) = 4e^{0.8x} - 0.5y$$

using $h = 0.5$ with $y(0) = 2$ from $x = 0$ to 0.5.

Solution.

(a) Equation (25.40a) through (25.40d) are used to compute $k_1 = 8.5$, $k_2 = 4.21875$, $k_3 = 4.21875$ and $k_4 = 1.25$; which are substituted into Eq. (25.40) to yield

$$y(0.5) = 1 + \left\{\frac{1}{6}[8.5 + 2(4.21875) + 2(4.21875) + 1.25]\right\}0.5$$

$$= 3.21875$$

which is exact. Thus, because the true solution is a quartic [Eq. (PT7.16)], the fourth-order method gives an exact result.

(b) For this case, the slope at the beginning of the interval is computed as

$$k_1 = f(0, 2) = 4e^{0.8(0)} - 0.5(2) = 3$$

This value is used to compute a value of y and a slope at the midpoint,

$$y(0.25) = 2 + 3(0.25) = 2.75$$
$$k_2 = f(0.25, 2.75) = 4e^{0.8(0.25)} - 0.5(2.75) = 3.510611$$

This slope in turn is used to compute another value of y and another slope at the midpoint,

$$y(0.25) = 2 + 3.510611(0.25) = 2.877653$$
$$k_3 = f(0.25, 2.877653) = 4e^{0.8(0.25)} - 0.5(2.877653) = 3.446785$$

Next, this slope is used to compute a value of y and a slope at the end of the interval,

$$y(0.5) = 2 + 3.071785(0.5) = 3.723392$$
$$k_4 = f(0.5, 3.723392) = 4e^{0.8(0.5)} - 0.5(3.723392) = 4.105603$$

Finally, the four slope estimates are combined to yield an average slope. This average slope is then used to make the final prediction at the end of the interval.

$$\phi = \frac{1}{6}[3 + 2(3.510611) + 2(3.446785) + 4.105603] = 3.503399$$

$$y(0.5) = 2 + 3.503399(0.5) = 3.751669$$

which compares favorably with the true solution of 3.751521.

25.3.4 Higher-Order Runge-Kutta Methods

Where more accurate results are required, *Butcher's (1964) fifth-order RK method* is recommended:

$$y_{i+1} = y_i + \frac{1}{90}(7k_1 + 32k_3 + 12k_4 + 32k_5 + 7k_6)h \tag{25.41}$$

where

$$k_1 = f(x_i, y_i) \tag{25.41a}$$

$$k_2 = f\left(x_i + \frac{1}{4}h, y_i + \frac{1}{4}k_1h\right) \tag{25.41b}$$

$$k_3 = f\left(x_i + \frac{1}{4}h, y_i + \frac{1}{8}k_1h + \frac{1}{8}k_2h\right) \tag{25.41c}$$

$$k_4 = f\left(x_i + \frac{1}{2}h, y_i - \frac{1}{2}k_2h + k_3h\right) \tag{25.41d}$$

$$k_5 = f\left(x_i + \frac{3}{4}h, y_i + \frac{3}{16}k_1h + \frac{9}{16}k_4h\right) \tag{25.41e}$$

$$k_6 = f\left(x_i + h, y_i - \frac{3}{7}k_1h + \frac{2}{7}k_2h + \frac{12}{7}k_3h - \frac{12}{7}k_4h + \frac{8}{7}k_5h\right) \tag{25.41f}$$

Note the similarity between Butcher's method and Boole's Rule in Table 21.2. Higher-order RK formulas such as Butcher's method are available, but in general, beyond fourth-order methods the gain in accuracy is offset by the added computational effort and complexity.

EXAMPLE 25.8 Comparison of Runge-Kutta Methods

Problem Statement.   Use first- through fifth-order RK methods to solve

$$f(x, y) = 4e^{0.8x} - 0.5y$$

with $y(0) = 2$ from $x = 0$ to $x = 4$ with various step sizes. Compare the accuracy of the various methods for the result at $x = 4$ based on the exact answer of $y(4) = 75.33896$.

Solution.   The computation is performed using Euler's, the noniterative Heun, the third-order RK [Eq. (25.39)], the classical fourth-order RK, and Butcher's fifth-order RK

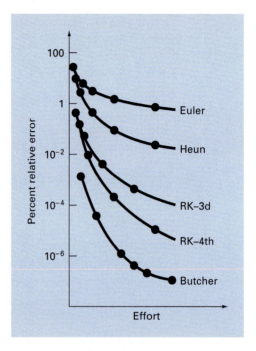

FIGURE 25.16
Comparison of percent relative error versus computational effort for first- through fifth-order RK methods.

methods. The results are presented in Fig. 25.16, where we have plotted the absolute value of the percent relative error versus the computational effort. This latter quantity is equivalent to the number of function evaluations required to attain the result, as in

$$\text{Effort} = n_f \frac{b-a}{h} \tag{E25.8.1}$$

where $n_f =$ the number of function evaluations involved in the particular RK computation. For orders ≤ 4, n_f is equal to the order of the method. However, note that Butcher's fifth-order technique requires six function evaluations [Eq. (25.41a) through (25.41f)]. The quantity $(b-a)/h$ is the total integration interval divided by the step size—that is, it is the number of applications of the RK technique required to obtain the result. Thus, because the function evaluations are usually the primary time-consuming steps, Eq. (E25.8.1) provides a rough measure of the run time required to attain the answer.

Inspection of Fig. 25.16 leads to a number of conclusions: first, that the higher-order methods attain better accuracy for the same computational effort and, second, that the gain in accuracy for the additional effort tends to diminish after a point. (Notice that the curves drop rapidly at first and then tend to level off.)

Example 25.9 and Fig. 25.16 might lead one to conclude that higher-order RK techniques are always the methods of preference. However, other factors such as programming

```
SUB RK4 (x, y, h, ynew)
    CALL Derivs(x, y, k1)
    ym = y + k1 · h/2
    CALL Derivs(x + h/2, ym, k2)
    ym = y + k2 · h/2
    CALL Derivs(x + h/2, ym, k3)
    ye = y + k3 · h
    CALL Derivs(x + h, ye, k4)
    slope = (k1 + 2(k2 + k3) + k4)/6
    ynew = y + slope · h
    x = x + h
END SUB
```

FIGURE 25.17
Pseudocode to determine a single step of the fourth-order RK method.

---

costs and the accuracy requirements of the problem also must be considered when choosing a solution technique. Such trade-offs will be explored in detail in the engineering applications in Chap. 28 and in the epilogue for Part Seven.

25.3.5 Computer Algorithms for Runge-Kutta Methods

As with all the methods covered in this chapter, the RK techniques fit nicely into the general algorithm embodied in Fig. 25.7. Figure 25.17 presents pseudocode to determine the slope of the classic fourth-order RK method [Eq. (25.40)]. Subroutines to compute slopes for all the other versions can be easily programmed in a similar fashion.

25.4 SYSTEMS OF EQUATIONS

Many practical problems in engineering and science require the solution of a system of simultaneous ordinary differential equations rather than a single equation. Such systems may be represented generally as

$$\frac{dy_1}{dx} = f_1(x, y_1, y_2, \ldots, y_n)$$

$$\frac{dy_2}{dx} = f_2(x, y_1, y_2, \ldots, y_n)$$

.

.

.

$$\frac{dy_n}{dx} = f_n(x, y_1, y_2, \ldots, y_n) \qquad (25.42)$$

The solution of such a system requires that n initial conditions be known at the starting value of x.

25.4.1 Euler Method

All the methods discussed in this chapter for single equations can be extended to the system shown above. Engineering applications can involve thousands of simultaneous equations. In each case, the procedure for solving a system of equations simply involves applying the one-step technique for every equation at each step before proceeding to the next step. This is best illustrated by the following example for the simple Euler's method.

EXAMPLE 25.9   Solving Systems of ODEs Using Euler's method

Problem Statement.   Solve the following set of differential equations using Euler's method, assuming that at $x = 0$, $y_1 = 4$, and $y_2 = 6$. Integrate to $x = 2$ with a step size of 0.5.

$$\frac{dy_1}{dx} = -0.5y_1 \qquad \frac{dy_2}{dx} = 4 - 0.3y_2 - 0.1y_1$$

Solution.   Euler's method is implemented for each variable as in Eq. (25.2):

$$y_1(0.5) = 4 + [-0.5(4)]0.5 = 3$$
$$y_2(0.5) = 6 + [4 - 0.3(6) - 0.1(4)]0.5 = 6.9$$

Note that $y_1(0) = 4$ is used in the second equation rather than the $y_1(0.5) = 3$ computed with the first equation. Proceeding in a like manner gives

x	y_1	y_2
0	4	6
0.5	3	6.9
1.0	2.25	7.715
1.5	1.6875	8.44525
2.0	1.265625	9.094087

25.4.2 Runge-Kutta Methods

Note that any of the higher-order RK methods in this chapter can be applied to systems of equations. However, care must be taken in determining the slopes. Figure 25.15 is helpful in visualizing the proper way to do this for the fourth-order method. That is, we first develop slopes for all variables at the initial value. These slopes (a set of k_1's) are then used to make predictions of the dependent variable at the midpoint of the interval. These midpoint values are in turn used to compute a set of slopes at the midpoint (the k_2's). These new slopes are then taken back to the starting point to make another set of midpoint predictions that lead to new slope predictions at the midpoint (the k_3's). These are then employed to make predictions at the end of the interval that are used to develop slopes at the end of the interval (the k_4's). Finally, the k's are combined into a set of increment functions [as in Eq. (25.40)] and brought back to the beginning to make the final prediction. The following example illustrates the approach.

EXAMPLE 25.10   Solving Systems of ODEs Using the Fourth-Order RK Method

Problem Statement.   Use the fourth-order RK method to solve the ODEs from Example 25.9.

Solution.   First, we must solve for all the slopes at the beginning of the interval:

$$k_{1,1} = f(0, 4, 6) = -0.5(4) = -2$$
$$k_{1,2} = f(0, 4, 6) = 4 - 0.3(6) - 0.1(4) = 1.8$$

where $k_{i,j}$ is the ith value of k for the jth dependent variable. Next, we must calculate the first values of y_1 and y_2 at the midpoint:

$$y_1 + k_{1,1}\frac{h}{2} = 4 + (-2)\frac{0.5}{2} = 3.5$$

$$y_2 + k_{1,2}\frac{h}{2} = 6 + (1.8)\frac{0.5}{2} = 6.45$$

which can be used to compute the first set of midpoint slopes,

$$k_{2,1} = f(0.25, 3.5, 6.45) = -1.75$$
$$k_{2,2} = f(0.25, 3.5, 6.45) = 1.715$$

These are used to determine the second set of midpoint predictions,

$$y_1 + k_{2,1}\frac{h}{2} = 4 + (-1.75)\frac{0.5}{2} = 3.5625$$

$$y_2 + k_{2,2}\frac{h}{2} = 6 + (1.715)\frac{0.5}{2} = 6.42875$$

which can be used to compute the second set of midpoint slopes,

$$k_{3,1} = f(0.25, 3.5625, 6.42875) = -1.78125$$
$$k_{3,2} = f(0.25, 3.5625, 6.42875) = 1.715125$$

These are used to determine the predictions at the end of the interval

$$y_1 + k_{3,1}h = 4 + (-1.78125)(0.5) = 3.109375$$
$$y_2 + k_{3,2}h = 6 + (1.715125)(0.5) = 6.857563$$

which can be used to compute the endpoint slopes,

$$k_{4,1} = f(0.5, 3.109375, 6.857563) = -1.554688$$
$$k_{4,2} = f(0.5, 3.109375, 6.857563) = 1.631794$$

The values of k can then be used to compute [Eq. (25.40)]:

$$y_1(0.5) = 4 + \frac{1}{6}[-2 + 2(-1.75 - 1.78125) - 1.554688]0.5 = 3.115234$$

$$y_2(0.5) = 6 + \frac{1}{6}[1.8 + 2(1.715 + 1.715125) + 1.631794]0.5 = 6.857670$$

Proceeding in a like manner for the remaining steps yields

x	y_1	y_2
0	4	6
0.5	3.115234	6.857670
1.0	2.426171	7.632106
1.5	1.889523	8.326886
2.0	1.471577	8.946865

25.4.3 Computer Algorithm for Solving Systems of ODEs

The computer code for solving a single ODE with Euler's method (Fig. 25.7) can be easily extended to systems of equations. The modifications include:

1. Inputting the number of equations, n
2. Inputting the initial values for each of the n dependent variables
3. Modifying the algorithm so that it computes slopes for each of the dependent variables
4. Including additional equations to compute derivative values for each of the ODEs
5. Including loops to compute a new value for each dependent variable

Such an algorithm is outlined in Fig. 25.18 for the fourth-order RK method. Notice how similar it is in structure and organization to Fig. 25.7. Most of the differences relate to the fact that (1) it deals with n equations and (2) the added detail of the fourth-order RK method.

EXAMPLE 25.11   Solving Systems of ODEs with the Computer

Problem Statement.   A user-friendly computer program to implement the fourth-order RK method for systems is contained in the Numerical Methods TOOLKIT software associated with the text. This software makes it convenient to compare different models of a physical system. For example, a linear model for a swinging pendulum is given by [recall Eq. (PT7.11)]

$$\frac{dy_1}{dx} = y_2 \qquad \frac{dy_2}{dx} = -16.1 y_1$$

where y_1 and y_2 = angular displacement and velocity. A nonlinear model of the same system is [recall Eq. (PT7.9)]

$$\frac{dy_3}{dx} = y_4 \qquad \frac{dy_4}{dx} = -16.1 \sin (y_3)$$

where y_3 and y_4 = angular displacement and velocity for the nonlinear case. Use the Numerical Methods TOOLKIT to solve these systems for two cases: (a) a small initial displacement ($y_1 = y_3 = 0.1$ radians; $y_2 = y_4 = 0$) and (b) a large displacement ($y_1 = y_3 = \pi/4 = 0.785398$ radians; $y_2 = y_4 = 0$).

(a) Main or "Driver" Program

Assign values for
n = number of equations
y_i = initial values of n dependent
   variables
xi = initial value independent
   variable
xf = final value independent variable
dx = calculation step size
xout = output interval

x = xi
m = 0
xp_m = x
DO i = 1, n
  $yp_{i,m}$ = y_i
END DO
DO
  xend = x + xout
  IF (xend > xf) THEN xend = xf
  h = dx
  CALL Integrator (x, y, n, h, xend)
  m = m + 1
  xp_m = x
  DO i = 1, n
    $yp_{i, m}$ = y_i
  END DO
  IF (x ≥ xf) EXIT
LOOP
DISPLAY RESULTS
END

(b) Routine to take one output step

SUB Integrator (x, y, n, h, xend)
  DO
    IF (xend − x < h) THEN h = xend − x
    CALL RK4 (x, y, n, h)
    IF (x ≥ xend) EXIT
  END DO
END SUB

(c) Fourth-order RK method for a system of ODEs

SUB RK4 (x, y, n, h)
  CALL Derivs (x, y, k1)
  DO i = 1, n
    *ym_i = y_i + $k1_i$ * h / 2*
  END DO
  CALL Derivs (x + h / 2, ym, k2)
  DO i = 1, n
    *ym_i = y_i + $k2_i$ * h / 2*
  END DO
  CALL Derivs (x + h / 2, ym, k3)
  DO i = 1, n
    *ye_i = y_i + $k3_i$ * h*
  END DO
  CALL Derivs (x + h, ye, k4)
  DO i = 1, n
    $slope_i$ = $(k1_i + 2*(k2_i+k3_i)+k4_i)/6$
    *y_i = y_i + $slope_i$ * h*
  END DO
  x = x + h
END SUB

(d) Routine to determine derivatives

SUB Derivs (x, y, dy)
  dy_1 = . . .
  dy_2 = . . .
END SUB

FIGURE 25.18
Pseudocode for the fourth-order RK method for systems.

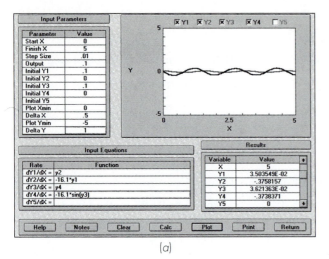
(a)

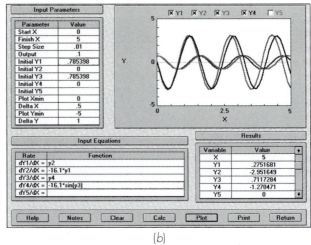
(b)

FIGURE 25.19
Two screens for the "solve ODEs" option on the Numerical Methods TOOLKIT for linear and non-linear pendulums with (a) small and (b) large initial displacement.

Solution. Press the Solve ODEs button on the TOOLKIT main menu to obtain a screen similar to Fig. 25.19a. This screen contains spaces for the input and output information associated with the solution of up to 5 simultaneous first-order ODEs.

(a) Input appropriate values for Start X, Finish X, Step Size, Output Interval, and the plot parameters as shown in Fig. 25.19a. Input the equations and initial values and then click on the Calc button. Then click on the four variables you want to display at the top of the plot (Y1, Y2, Y3, and Y4) and click the Plot button. The calculated results for the linear and nonlinear models are almost identical. This is as expected because when the initial displacement is small, $\sin(\theta) \cong \theta$.

(b) When the initial displacement is $\pi/4 = 0.785398$, the solutions are much different and the difference is magnified as time becomes larger and larger (Fig. 25.19b). This is expected because the assumption that $\sin(\theta) = \theta$ is poor when theta is large.

25.5 ADAPTIVE RUNGE-KUTTA METHODS

To this point, we have presented methods for solving ODEs that employ a constant step size. For a significant number of problems, this can represent a serious limitation. For example, suppose that we are integrating an ODE with a solution of the type depicted in Fig. 25.20. For most of the range, the solution changes gradually. Such behavior suggests that a fairly large step size could be employed to obtain adequate results. However, for a localized region from $x = 1.75$ to $x = 2.25$, the solution undergoes an abrupt change. The practical consequence of dealing with such functions is that a very small step size would be required to accurately capture the impulsive behavior. If a constant step-size algorithm were employed, the smaller step size required for the region of abrupt change would have to be applied to the entire computation. As a consequence, a much smaller step size than necessary—and, therefore, many more calculations—would be wasted on the regions of gradual change.

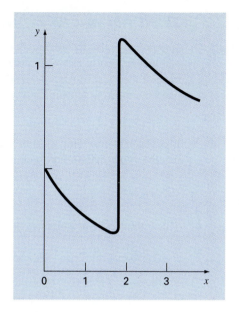

FIGURE 25.20
An example of a solution of an ODE that exhibits an abrupt change. Automatic step-size adjustment has great advantages for such cases.

Algorithms that automatically adjust the step size can avoid such overkill and hence be of great advantage. Because they "adapt" to the solution's trajectory, they are said to have *adaptive step-size control*. Implementation of such approaches requires that an estimate of the local truncation error be obtained at each step. This error estimate can then serve as a basis for either lengthening or decreasing the step size.

Before proceeding, we should mention that aside from solving ODEs, the methods described in this chapter can also be used to evaluate definite integrals. As mentioned previously in the introduction to Part Six, the evaluation of the integral

$$I = \int_a^b f(x)\,dx$$

is equivalent to solving the differential equation

$$\frac{dy}{dx} = f(x)$$

for $y(b)$ given the initial condition $y(a) = 0$. Thus, the following techniques can be employed to efficiently evaluate definite integrals involving functions that are generally smooth but exhibit regions of abrupt change.

There are two primary approaches to incorporate adaptive step-size control into one-step methods. In the first, the error is estimated as the difference between two predictions using the same-order RK method but with different step sizes. In the second, the local truncation error is estimated as the difference between two predictions using different-order RK methods.

25.5.1 Adaptive RK or Step-Halving Method

Step halving (also called *adaptive RK*) involves taking each step twice, once as a full step and independently as two half steps. The difference in the two results represents an estimate of the local truncation error. If y_1 designates the single-step prediction and y_2 designates the prediction using the two half steps, the error Δ can be represented as

$$\Delta = y_2 - y_1 \tag{25.43}$$

In addition to providing a criterion for step-size control, Eq. (25.43) can also be used to correct the y_2 prediction. For the fourth-order RK version, the correction is

$$y_2 \leftarrow y_2 + \frac{\Delta}{15} \tag{25.44}$$

This estimate is fifth-order accurate.

EXAMPLE 25.12   Adaptive Fourth-Order RK Method

Problem Statement.   Use the adaptive fourth-order RK method to integrate $y' = 4e^{0.8x} - 0.5y$ from $x = 0$ to 2 using $h = 2$ and an initial condition of $y(0) = 2$. This is the same differential equation that was solved previously in Example 25.5. Recall that the true solutions is $y(2) = 14.84392$.

Solution.   The single prediction with a step of h is computed as

$$y(2) = 2 + \frac{1}{6}[3 + 2(6.40216 + 4.70108) + 14.11105]2 = 15.10584$$

The two half-step predictions are

$$y(1) = 2 + \frac{1}{6}[3 + 2(4.21730 + 3.91297) + 5.945681]1 = 6.20104$$

and

$$y(2) = 6.20104 + \frac{1}{6}[5.80164 + 2(8.72954 + 7.99756) + 12.71283]1 = 14.86249$$

Therefore, the approximate error is

$$E_a = \frac{14.86249 - 15.10584}{15} = -0.01622$$

which compares favorably with the true error of

$$E_t = 14.84392 - 14.86249 = -0.01857$$

The error estimate can also be used to correct the prediction

$$y(2) = 14.86249 - 0.01622 = 14.84627$$

which has an $E_t = -0.00235$.

25.5.2 Runge-Kutta Fehlberg

Aside from step halving as a strategy to adjust step size, an alternative approach for obtaining an error estimate involves computing two RK predictions of different order. The results can then be subtracted to obtain an estimate of the local truncation error. One shortcoming of this approach is that it greatly increases the computational overhead. For example, a fourth- and fifth-order prediction amount to a total of 10 function evaluations per step. The *Runge-Kutta Fehlberg* or *embedded RK* method cleverly circumvents this problem by using a fifth-order RK method that employs the function evaluations from the accompanying fourth-order RK method. Thus, the approach yields the error estimate on the basis of only six function evaluations!

For the present case, we use the following fourth-order estimate

$$y_{i+1} = y_i + \left(\frac{37}{378}k_1 + \frac{250}{621}k_3 + \frac{125}{594}k_4 + \frac{512}{1771}k_6\right)h \tag{25.45}$$

along with the fifth-order formula:

$$y_{i+1} = y_i + \left(\frac{2825}{27,648}k_1 + \frac{18,575}{48,384}k_3 + \frac{13,525}{55,296}k_4 + \frac{277}{14,336}k_5 + \frac{1}{4}k_6\right)h \tag{25.46}$$

where

$$k_1 = f(x_i, y_i)$$

$$k_2 = f\left(x_i + \frac{1}{5}h, y_i + \frac{1}{5}k_1h\right)$$

$$k_3 = f\left(x_i + \frac{3}{10}h, y_i + \frac{3}{40}k_1h + \frac{9}{40}k_2h\right)$$

$$k_4 = f\left(x_i + \frac{3}{5}h, y_i + \frac{3}{10}k_1h - \frac{9}{10}k_2h + \frac{6}{5}k_3h\right)$$

$$k_5 = f\left(x_i + h, y_i - \frac{11}{54}k_1h + \frac{5}{2}k_2h - \frac{70}{27}k_3h + \frac{35}{27}k_4h\right)$$

$$k_6 = f\left(x_i + \frac{7}{8}h, y_i + \frac{1631}{55,296}k_1h + \frac{175}{512}k_2h + \frac{575}{13,824}k_3h + \frac{44,275}{110,592}k_4h \right.$$
$$\left. + \frac{253}{4096}k_5h\right)$$

Thus, the ODE can be solved with Eq. (25.46) and the error estimated as the difference of the fifth- and fourth-order estimates. It should be noted that the particular coefficients used above were developed by Cash and Karp (1990). Therefore, it is sometimes called the *Cash-Karp* RK method.

EXAMPLE 25.13   Runge-Kutta Fehlberg Method

Problem Statement.   Use the Cash-Karp version of the Runge-Kutta Fehlberg approach to perform the same calculation as in Example 25.12 from $x = 0$ to 2 using $h = 2$.

Solution. The calculation of the k's can be summarized in the following table:

	x	**y**	**f(x, y)**
k_1	0	2	3
k_2	0.4	3.2	3.908511
k_3	0.6	4.20883	4.359883
k_4	1.2	7.228398	6.832587
k_5	2	15.42765	12.09831
k_6	1.75	12.17686	10.13237

These can then be used to compute the fourth-order prediction

$$y_1 = 2 + \left(\frac{37}{378}3 + \frac{250}{621}4.359883 + \frac{125}{594}6.832587 + \frac{512}{1771}10.13237 \right)2 = 14.83192$$

along with a fifth-order formula:

$$y_1 = 2 + \left(\frac{2825}{27,648}3 + \frac{18,575}{48,384}4.359883 + \frac{13,525}{55,296}6.832587 \right.$$

$$\left. + \frac{277}{14,336}12.09831 + \frac{1}{4}10.13237 \right)2 = 14.38677$$

The error estimate is obtained by subtracting these two equations to give

$$E_a = 14.83677 - 14.83192 = 0.004842$$

25.5.3 Step-Size Control

Now that we have developed ways to estimate the local truncation error, it can be used to adjust the step size. In general, the strategy is to increase the step size if the error is too small and decrease it if the error is too large. Press et al. (1992) have suggested the following criterion to accomplish this:

$$h_{new} = h_{present} \left| \frac{\Delta_{new}}{\Delta_{present}} \right|^{\alpha} \tag{25.47}$$

where $h_{present}$ and h_{new} = the present and the new step sizes, $\Delta_{present}$ = the computed present accuracy, Δ_{new} = the desired accuracy, and α = a constant power that is equal to 0.2 when the step size is increased (i.e., when $\Delta_{present} \leq \Delta_{new}$) and 0.25 when the step size is decreased ($\Delta_{present} > \Delta_{new}$).

The key parameter in Eq. (25.47) is obviously Δ_{new} because it is your vehicle for specifying the desired accuracy. One way to do this would be to relate Δ_{new} to a relative error level. Although this works well when only positive values occur, it can cause problems for solutions that pass through zero. For example, you might be simulating an oscillating function that repeatedly passes through zero but is bounded by maximum absolute values. For such a case, you might want these maximum values to figure in the desired accuracy.

A more general way to handle such cases is to determine Δ_{new} as

$$\Delta_{new} = \varepsilon y_{scale}$$

where ε = an overall tolerance level. Your choice of y_{scale} will then determine how the error is scaled. For example, if $y_{scale} = y$, the accuracy will be couched in terms of fractional relative errors. If you are dealing with a case where you desire constant errors relative to a prescribed maximum bound, set y_{scale} equal to that bound. A trick suggested by Press et al. (1992) to obtain the constant relative errors except very near zero crossings is

$$y_{scale} = |y| + \left| h\frac{dy}{dx} \right|$$

This is the version we will use in our algorithm.

25.5.4 Computer Algorithm

Figure 25.21 and 25.22 outlines pseudocode to implement the Cash-Karp version of the Runge-Kutta Fehlberg algorithm. This algorithm is patterned after a more detailed implementation by Press et al. (1992) for systems of ODEs.

FIGURE 25.21
Pseudocode for a single step for the Cash-Karp RK method.

```
SUBROUTINE RKkc (y, dy,x,h,yout, yerr)
PARAMETER (a2=0.2,a3=0.3,a4=0.6,a5=1.,a6=0.875,
    b21=0.2,b31=3./40.,b32=9./40.,b41=0.3,b42=−0.9,
    b43=1.2,b51=−11./54.,b52=2.5,b53=−70./27.,
    b54=35./27.,b61=1631./55296.,b62=175./512.,
    b63=575./13824.,b64=44275./110592., b65=253./4096.,
    c1=37./378.,c3=250./621.,c4=125./594.,
    c6=512./1771.,dc1=c1−2825./27648.,
    dc3=c3−18575./48384.,dc4=c4−13525./55296.,
    dc5=−277./14336.,dc6=c6−0.25)
ytemp=y+b21*h*dy
CALL Derivs (x+a2*h,ytemp,k2)
ytemp=y+h*(b31*dy+b32*k2)
CALL Derivs(x+a3*h,ytemp,k3)
ytemp=y+h*(b41*dy+b42*k2+b43*k3)
CALL Derivs (x+a4*h,ytemp,k4)
ytemp=y+h*(b51*dy+b52*k2+b53*k3+b54*k4)
CALL Derivs(x+a5*h,ytemp,k5)
ytemp=y+h*(b61*dy+b62*k2+b63*k3+b64*k4+b65*k5)
CALL Derivs(x+a6*h, ytemp,k6)
yout=y+h*(c1*dy+c3*k3+c4*k4+c6*k6)
yerr=h*(dc1*dy+dc3*k3+dc4*k4+dc5*k5+dc6*k6)
END RKkc
```

(a) Driver Program

```
INPUT xi, xf, yi
maxstep=100
hi=.5; tiny = 1. × 10⁻³
eps=0.00005
print *, xi,yi
x=xi
y=yi
h=hi
istep=0
DO
   IF (istep > maxstep AND x ≤ xf) EXIT
   istep=istep+1
   CALL Derivs(x,y,dy)
   yscal=ABS(y)+ABS(h*dy)+tiny
   IF (x+h>xf) THEN h=xf−x
   CALL Adapt (x,y,dy,h,yscal,eps,hnxt)
   PRINT x,y
   h=hnxt
END DO
END
```

(b) Adaptive Step Routine

```
SUB Adapt (x,y,dy,htry,yscal,eps,hnxt)
PARAMETER (safety=0.9,econ=1.89e−4)
h=htry
DO
   CALL RKkc(y,dy,x,h,ytemp,yerr)
   emax=abs(yerr/yscal/eps)
   IF emax ≤ 1 EXIT
   htemp=safety*h*emax⁻⁰·²⁵
   h=max(abs(htemp),0.25*abs(h))
   xnew=x+h
   IF xnew = x THEN pause
END DO
IF emax > econ THEN
   hnxt=safety*emax⁻·²*h
ELSE
   hnxt=4. *h
END IF
x=x+h
y=ytemp
END Adapt
```

FIGURE 25.22
Pseudocode for a (a) driver program and an (b) adaptive step routine to solve a single ODE.

Figure 25.21 implements a single step of the Cash-Karp routine (that is Eqs. 25.45 and 25.46). Figure 25.22 outlines a general driver program along with a subroutine that actually adapts the step size.

EXAMPLE 25.14   Computer Application of an Adaptive Fourth-Order RK Scheme

Problem Statement.   The adaptive RK method is well-suited for the following ordinary differential equation

$$\frac{dy}{dx} + 0.6y = 10e^{-(x-2)^2/[2(0.075)^2]} \tag{E25.14.1}$$

Notice for the initial condition, $y(0) = 0.5$, the general solution is

$$y = 0.5e^{-0.6x} \tag{E25.14.2}$$

which is a smooth curve that gradually approaches zero as x increases. In contrast, the particular solution undergoes an abrupt transition in the vicinity of $x = 2$ due to the nature of the forcing function (Fig. 25.23a). Use a standard fourth-order RK scheme to solve Eq. (E25.14.1) from $x = 0$ to 4. Then employ the adaptive scheme described in this section to perform the same computation.

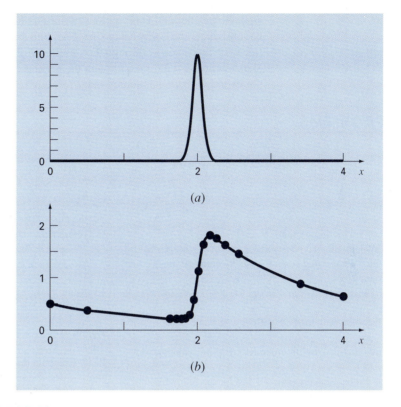

FIGURE 25.23
(a) A bell-shaped forcing function that induces an abrupt change in the solution of an ODE [Eq. (E25.14.1)]: (b) The solution. The points indicate the predictions of an adaptive step-size routine.

Solution. First, the classical fourth-order scheme is used to compute the solid curve in Fig. 25.23b. For this computation, a step size of 0.1 is used so that $4/(0.1) = 40$ applications of the technique are made. Then, the calculation is repeated with a step size of 0.05 for a total of 80 applications. The major discrepancy between the two results occurs in the region from 1.8 to 2.0. The magnitude of the discrepancy is about 0.1 to 0.2 percent.

Next, the algorithm in Figs. 25.21 and 25.22 is developed into a computer program and used to solve the same problem. An initial step size of 0.5 and an $\varepsilon = 0.00005$ were chosen. The results were superimposed on Fig. 25.23b. Notice how large steps are taken in the regions of gradual change. Then, in the vicinity of $x = 2$ the steps are decreased to accommodate the abrupt nature of the forcing function.

The utility of an adaptive integration scheme obviously depends on the nature of the functions being modeled. It is particularly advantageous for those solutions with long smooth stretches and short regions of abrupt change. In addition, it has utility in those situations where the correct step size is not known a priori. For these cases, an adaptive routine will "feel" its way through the solution while keeping the results within the desired tolerance. Thus, it will tiptoe through the regions of abrupt change and step out briskly when the variations become more gradual.

PROBLEMS

25.1 Solve the following initial-value problem analytically over the interval from $x = 0$ to 2:

$$\frac{dy}{dx} = yx^2 - 1.2y$$

where $y(0) = 1$. Plot the solution.

25.2 Use Euler's method with $h = 0.5$ and 0.25 to solve Prob. 25.1. Plot the results on the same graph to visually compare the accuracy for the two step sizes.

25.3 Use Heun's method with $h = 0.5$ to solve Prob. 25.1. Iterate the corrector to $\varepsilon_s = 1\%$.

25.4 Use the midpoint method with $h = 0.5$ and 0.25 to solve Prob. 25.1.

25.5 Use the classical fourth-order RK method with $h = 0.5$ to solve Prob. 25.1.

25.6 Repeat Probs. 25.1 through 25.5 but for the following initial-value problem over the interval from $x = 0$ to 1:

$$\frac{dy}{dx} = (1 + x)\sqrt{y} \qquad y(0) = 1$$

25.7 Use the (**a**) Euler and (**b**) Heun (without iteration) methods to solve

$$\frac{d^2 y}{dt^2} - t + y = 0$$

where $y(0) = 2$ and $y'(0) = 0$. Solve from $x = 0$ to 4 using $h = 0.1$. Compare the methods by plotting the solutions.

25.8 Solve the following problem with the fourth-order RK method:

$$\frac{d^2 y}{dx^2} + 0.5\frac{dy}{dx} + 7y = 0$$

where $y(0) = 4$ and $y'(0) = 0$. Solve from $x = 0$ to 5 with $h = 0.5$. Plot your results.

25.9 Solve from $t = 0$ to 3 with $h = 0.1$ using (**a**) Heun (without corrector) and (**b**) Ralston's 2nd-order RK method:

$$\frac{dy}{dt} = y \sin^2(t) \qquad y(0) = 1$$

25.10 Solve the following problem numerically from $t = 0$ to 3:

$$\frac{dy}{dt} = -y + t \qquad y(0) = 1$$

Use the third-order RK method with a step size of 0.5.

25.11 Use (**a**) Euler's and (**b**) the fourth-order RK method to solve

$$\frac{dy}{dx} = -2y + 5e^{-x}$$

$$\frac{dz}{dx} = -\frac{yz^2}{2}$$

over the range $x = 0$ to 1 using a step size of 0.2 with $y(0) = 2$ and $z(0) = 4$.

25.12 Compute the first step of Example 25.14 using the adaptive fourth-order RK method with $h = 0.5$. Verify whether step-size adjustment is in order.

25.13 If $\varepsilon = 0.001$, determine whether step size adjustment is required for Example 25.12.

25.14 Use the RK-Fehlberg approach to perform the same calculation as in Example 25.12 from $x = 0$ to 1 with $h = 1$.

25.15 Write a computer program based on Fig. 25.7. Among other things, place documentation statements throughout the program to identify what each section is intended to accomplish.

25.16 Test the program you developed in Prob. 25.15 by duplicating the computations from Examples 25.1, and 25.4.

25.17 Develop a user-friendly program for the Heun method with an iterative corrector. Test the program by duplicating the results in Table 25.2.

25.18 Develop a user-friendly computer program for the classical fourth-order RK method. Test the program by duplicating Example 25.7 and Prob. 25.5.

25.19 Develop a user-friendly computer program for systems of equations using the fourth-order RK method. Use this program to duplicate the computation in Example 25.10.

CHAPTER 26
Stiffness and Multistep Methods

This chapter covers two areas. First, we describe *stiff ODEs*. These are both individual and systems of ODEs that have both fast and slow components to their solution. We introduce the idea of an *implicit solution* technique as one commonly used remedy for this problem. Then we discuss *multistep methods*. These algorithms retain information of previous steps to more effectively capture the trajectory of the solution. They also yield the truncation error estimates that can be used to implement adaptive step-size control.

26.1 STIFFNESS

Stiffness is a special problem that can arise in the solution of ordinary differential equations. A *stiff system* is one involving rapidly changing components together with slowly changing ones. In many cases, the rapidly varying components are ephemeral transients that die away quickly, after which the solution becomes dominated by the slowly varying components. Although the transient phenomena exist for only a short part of the integration interval, they can dictate the time step for the entire solution.

Both individual and systems of ODEs can be stiff. An example of a single stiff ODE is

$$\frac{dy}{dt} = -1000y + 3000 - 2000e^{-t} \tag{26.1}$$

If $y(0) = 0$, the analytical solution can be developed as

$$y = 3 - 0.998e^{-1000t} - 2.002e^{-t} \tag{26.2}$$

As in Fig. 26.1, the solution is initially dominated by the fast exponential term (e^{-1000t}). After a very short period $t < 0.005$, this transient dies out and the solution becomes dictated by the slow exponential (e^{-t}).

Insight into the step size required for stability of such a solution can be gained by examining the homogeneous part of Eq. (26.1),

$$\frac{dy}{dt} = -ay \tag{26.3}$$

719

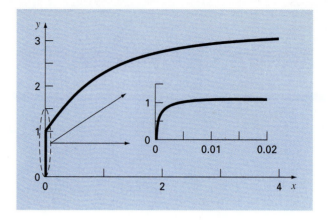

FIGURE 26.1
Plot of a stiff solution of a single ODE. Although the solution appears to start at 1, there is actually a fast transient from $y = 0$ to 1 that occurs in less than 0.005 time unit. This transient is perceptible only when the response is viewed on the finer time scale in the inset.

If $y(0) = y_0$, calculus can be used to determine the solution as

$$y = y_0 e^{-at}$$

Thus, the solution starts at y_0 and asymptotically approaches zero.

Euler's method can be used to solve the same problem numerically:

$$y_{i+1} = y_i + \frac{dy_i}{dt} h$$

Substituting Eq. (26.3) gives

$$y_{i+1} = y_i - a y_i h$$

or

$$y_{i+1} = y_i \left(1 - ah\right) \tag{26.4}$$

The stability of this formula clearly depends on the step size h. That is, $|1 - ah|$ must be less than 1. Thus, if $h > 2/a$, $|y_i| \to \infty$ as $i \to \infty$.

For the fast transient part of Eq. (26.2), this criterion can be used to show that the step size to maintain stability must be $< 2/1000 = 0.002$. In addition, it should be noted that, whereas this criterion maintains stability (i.e., a bounded solution), an even smaller step size would be required to obtain an accurate solution. Thus, although the transient occurs for only a small fraction of the integration interval, it controls the maximum allowable step size.

Superficially, you might suppose that the adaptive step-size routines described at the end of the last chapter might offer a solution for this dilemma. You might think that they would use small steps during the rapid transients and large steps otherwise. However, this is not the case, because the stability requirement will still necessitate very small steps throughout the entire solution.

Rather than using explicit approaches, implicit methods offer an alternative remedy. Such representations are called *implicit* because the unknown appears on both sides of the equation. An implicit form of Euler's method can be developed by evaluating the derivative at the future time,

$$y_{i+1} = y_i + \frac{dy_{i+1}}{dt} h$$

This is called the *backward,* or *implicit, Euler method.* Substituting Eq. (26.3) yields

$$y_{i+1} = y_i - ay_{i+1}h$$

which can be solved for

$$y_{i+1} = \frac{y_i}{1 + ah} \qquad (26.5)$$

For this case, regardless of the size of the step, $|y_i| \to 0$ as $i \to \infty$. Hence, the approach is called *unconditionally stable.*

EXAMPLE 26.1 | Explicit and Implicit Euler

Problem Statement. Use both the explicit and implicit Euler methods to solve

$$\frac{dy}{dt} = -1000y + 3000 - 2000e^{-t}$$

where $y(0) = 0$. (*a*) Use the explicit Euler with step sizes of 0.0005 and 0.0015 to solve for y between $t = 0$ and 0.006. (*b*) Use the implicit Euler with a step size of 0.05 to solve for y between 0 and 0.4.

Solution.

(**a**) For this problem, the explicit Euler's method is

$$y_{i+1} = y_i + \left(-1000y_i + 3000 - 2000e^{-t_i}\right)h$$

The result for $h = 0.005$ is displayed in Fig. 26.2*a* along with the analytical solution. Although it exhibits some truncation error, the result captures the general shape of the analytical solution. In contrast, when the step size is increased to a value just below the stability limit ($h = 0.0015$) the solution manifests oscillations. Using $h > 0.002$ would result in a totally unstable solution; i.e., it would go infinite as the solution progressed.

(**b**) The implicit Euler's method is

$$y_{i+1} = y_i + \left(-1000y_{i+1} + 3000 - 2000e^{-t_{i+1}}\right)h$$

Now because the ODE is linear, we can rearrange this equation so that y_{i+1} is isolated on the left-hand side,

$$y_{i+1} = \frac{y_i + 3000h - 2000he^{-t_{i+1}}}{1 + 1000h}$$

The result for $h = 0.05$ is displayed in Fig. 26.2*b* along with the analytical solution. Notice that even though we have used a much bigger step size than the one that

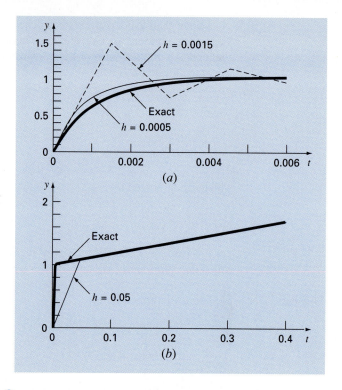

FIGURE 26.2
Solution of a "stiff" ODE with (a) the explicit and (b) implicit Euler methods.

induced instability for the explicit Euler, the numerical solution tracks nicely on the analytical result.

Systems of ODEs can also be stiff. An example is

$$\frac{dy_1}{dt} = -5y_1 + 3y_2 \tag{26.6a}$$

$$\frac{dy_2}{dt} = 100y_1 - 301y_2 \tag{26.6b}$$

For the initial conditions $y_1(0) = 52.29$ and $y_2(0) = 83.82$, the exact solution is

$$y_1 = 52.96e^{-3.9899t} - 0.67e^{-302.0101t} \tag{26.7a}$$

$$y_2 = 17.83e^{-3.9899t} + 65.99e^{-302.0101t} \tag{26.7b}$$

Note that the exponents are negative and differ by about 2 orders of magnitude. As with the single equation, it is the large exponents that respond rapidly and are at the heart of the system's stiffness.

An implicit Euler's method for systems can be formulated for the present example as

$$y_{1,i+1} = y_{1,i} + \left(-5y_{1,i+1} + 3y_{2,i+1}\right)h \tag{26.8a}$$

$$y_{2,i+1} = y_{2,i} + \left(100y_{1,i+1} - 301y_{2,i+1}\right)h \tag{26.8b}$$

Collecting terms gives

$$(1 + 5h)y_{1,i+1} - 3hy_{2,i+1} = y_{1,i} \tag{26.9a}$$

$$-100hy_{1,i+1} + (1 + 301h)y_{2,i+1} = y_{2,i} \tag{26.9b}$$

Thus, we can see that the problem consists of solving a set of simultaneous equations for each time step.

For nonlinear ODEs, the solution becomes even more difficult since it involves solving a system of nonlinear simultaneous equations (recall Sec. 6.5). Thus, although stability is gained through implicit approaches, a price is paid in the form of added solution complexity.

The implicit Euler method is unconditionally stable and only first-order accurate. It is also possible to develop in a similar manner a second-order accurate implicit trapezoidal rule integration scheme for stiff systems. It is usually desirable to have higher-order methods. The Adams-Moulton formulas described later in this chapter can also be used to devise higher-order implicit methods. However, the stability limits of such approaches is very limited when applied to stiff systems. Gear (1971) developed a special series of implicit schemes that have much larger stability limits based on backward difference formulas. Extensive efforts have been made to develop software to efficiently implement Gear's methods. As a result, this is probably the most widely used method to solve stiff systems. In addition, Rosenbrock and others have proposed implicit Runge-Kutta algorithms where the k terms appear implicitly. These methods have good stability characteristics and are quite suitable for solving systems of stiff ordinary differential equations.

26.2   MULTISTEP METHODS

The one-step methods described in the previous sections utilize information at a single point x_i to predict a value of the dependent variable y_{i+1} at a future point x_{i+1} (Fig. 26.3a). Alternative approaches, called *multistep methods* (Fig. 26.3b), are based on the insight that, once the computation has begun, valuable information from previous points is at our command. The curvature of the lines connecting these previous values provides information regarding the trajectory of the solution. The multistep methods explored in this chapter exploit this information to solve ODEs. Before describing the higher-order versions, we will present a simple second-order method that serves to demonstrate the general characteristics of multistep approaches.

26.2.1 The Non-Self-Starting Heun Method

Recall that the Heun approach uses *Euler's method* as a *predictor* [Eq. (25.15)]:

$$y_{i+1}^0 = y_i + f(x_i, y_i)h \tag{26.10}$$

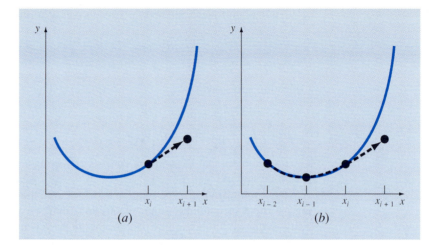

FIGURE 26.3
Graphical depiction of the fundamental difference between (a) one-step and (b) multistep methods for solving ODEs.

and the *trapezoidal rule* as a *corrector* [Eq. (25.16)]:

$$y_{i+1} = y_i + \frac{f(x_i, y_i) + f(x_{i+1}, y_{i+1}^0)}{2} h \tag{26.11}$$

Thus, the predictor and the corrector have local truncation errors of $O(h^2)$ and $O(h^3)$, respectively. This suggests that the predictor is the weak link in the method because it has the greatest error. This weakness is significant because the efficiency of the iterative corrector step depends on the accuracy of the initial prediction. Consequently, one way to improve Heun's method is to develop a predictor that has a local error of $O(h^3)$. This can be accomplished by using Euler's method and the slope at y_i, and extra information from a previous point y_{i-1}, as in

$$y_{i+1}^0 = y_{i-1} + f(x_i, y_i) 2h \tag{26.12}$$

Notice that Eq. (26.12) attains $O(h^3)$ at the expense of employing a larger step size, $2h$. In addition, note that Eq. (26.12) is not self-starting because it involves a previous value of the dependent variable y_{i-1}. Such a value would not be available in a typical initial-value problem. Because of this fact, Eqs. (26.11) and (26.12) are called the *non-self-starting Heun method*.

As depicted in Fig. 26.4, the derivative estimate in Eq. (26.12) is now located at the midpoint rather than at the beginning of the interval over which the prediction is made. As demonstrated subsequently, this centering improves the error of the predictor to $O(h^3)$. However, before proceeding to a formal derivation of the non-self-starting Heun, we will summarize the method and express it using a slightly modified nomenclature:

$$\text{Predictor:} \quad y_{i+1}^0 = y_{i-1}^m + f(x_i, y_i^m) h \tag{25.13}$$

$$\text{Corrector:} \quad y_{i+1}^j = y_i^m + \frac{f(x_i, y_i^m) + f(x_{i+1}, y_{i+1}^{j-1})}{2} h \tag{25.14}$$

$$\text{(for } j = 1, 2, \dots, m)$$

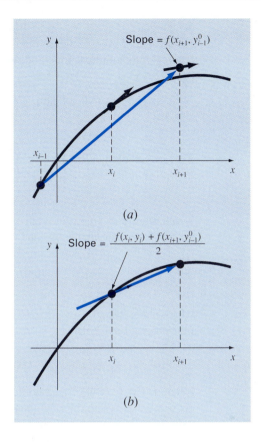

FIGURE 26.4
A graphical depiction of the non-self-starting Heun method. (a) The midpoint method that is used as a predictor. (b) The trapezoidal rule that is employed as a corrector.

where the superscripts have been added to denote that the corrector is applied iteratively from $j = 1$ to m to obtain refined solutions. Note that y_i^m and y_{i-1}^m are the final results of the corrector iterations at the previous time steps. The iterations are terminated at any time step on the basis of the stopping criterion

$$|\varepsilon_a| = \left| \frac{y_{i+1}^j - y_{i+1}^{j-1}}{y_{i+1}^j} \right| 100\% \tag{26.15}$$

When ε_a is less than a prespecified error tolerance ε_s, the iterations are terminated. At this point, $j = m$. The use of Eqs. (26.13) through (26.15) to solve an ODE is demonstrated in the following example.

EXAMPLE 26.2   Non-Self-Starting Heun Method

Problem Statement. Use the non-self-starting Heun method to perform the same computations as were performed previously in Example 25.5 using Heun's method. That is, integrate $y' = 4e^{0.8x} - 0.5y$ from $x = 0$ to $x = 4$ using a step size of 1.0. As with Example 25.5, the initial condition at $x = 0$ is $y = 2$. However, because we are now dealing with a multistep method, we require the additional information that y is equal to -0.3929953 at $x = -1$.

Solution.   The predictor [Eq. (26.13)] is used to extrapolate linearly from $x = -1$ to $x = 1$.

$$y_1^0 = -0.3929953 + \left[4e^{0.8(0)} - 0.5(2)\right]2 = 5.607005$$

The corrector [Eq. (26.14)] is then used to compute the value:

$$y_1^1 = 2 + \frac{4e^{0.8(0)} - 0.5(2) + 4e^{0.8(1)} - 0.5(5.607005)}{2}1 = 6.549331$$

which represents a percent relative error of -5.73 percent (true value $= 6.194631$). This error is somewhat smaller than the value of -8.18 percent incurred in the self-starting Heun. Now, Eq. (26.14) can be applied iteratively to improve the solution:

$$y_1^2 = 2 + \frac{3 + 4e^{0.8(1)} - 0.5(6.549331)}{2}1 = 6.313749$$

which represents an ε_t of -1.92%. An approximate estimate of the error can also be determined using Eq. (26.15):

$$|\varepsilon_a| = \left|\frac{6.313749 - 6.549331}{6.313749}\right|100\% = 3.7\%$$

Equation (26.14) can be applied iteratively until ε_a falls below a prespecified value of ε_s. As was the case with the Heun method (recall Example 25.5), the iterations converge on a value of 6.360865 ($\varepsilon_t = -2.68\%$). However, because the initial predictor value is more accurate, the multistep method converges at a somewhat faster rate.

For the second step, the predictor is

$$y_2^0 = 2 + \left[4e^{0.8(1)} - 0.5(6.360865)\right]2 = 13.44346 \qquad \varepsilon_t = 9.43\%$$

which is superior to the prediction of 12.08260 ($\varepsilon_t = 18\%$) that was computed with the original Heun method. The first corrector yields 15.76693 ($\varepsilon_t = 6.8\%$), and subsequent iterations converge on the same result as was obtained with the self-starting Heun method: 15.30224 ($\varepsilon_t = -3.1\%$). As with the previous step, the rate of convergence of the corrector is somewhat improved because of the better initial prediction.

Derivation and Error Analysis of Predictor-Corrector Formulas.   We have just employed graphical concepts to derive the non-self-starting Heun. We will now show how the same equations can be derived mathematically. This derivation is particularly interesting because it ties together ideas from curve fitting, numerical integration, and ODEs. The exercise is also useful because it provides a simple procedure for developing higher-order multistep methods and estimating their errors.

The derivation is based on solving the general ODE

$$\frac{dy}{dx} = f(x, y)$$

This equation can be solved by multiplying both sides by dx and integrating between limits at i and $i + 1$:

$$\int_{y_i}^{y_{i+1}} dy = \int_{x_i}^{x_{i+1}} f(x, y) \, dx$$

The left side can be integrated and evaluated using the fundamental theorem [recall Eq. (25.21)]:

$$y_{i+1} = y_i + \int_{x_i}^{x_{i+1}} f(x, y) \, dx \tag{26.16}$$

Equation (26.16) represents a solution to the ODE if the integral can be evaluated. That is, it provides a means to compute a new value of the dependent variable y_{i+1} on the basis of a prior value y_i and the differential equation.

Numerical integration formulas such as those developed in Chap. 21 provide one way to make this evaluation. For example, the trapezoidal rule [Eq. (21.3)] can be used to evaluate the integral, as in

$$\int_{x_i}^{x_{i+1}} f(x, y) \, dx = \frac{f(x_i, y_i) + f(x_{i+1}, y_{i+1})}{2} h \tag{26.17}$$

where $h = x_{i+1} - x_i$ is the step size. Substituting Eq. (26.17) into Eq. (26.16) yields

$$y_{i+1} = y_i + \frac{f(x_i, y_i) + f(x_{i+1}, y_{i+1})}{2} h$$

which is the corrector equation for the Heun method. Because this equation is based on the trapezoidal rule, the truncation error can be taken directly from Table 21.2,

$$E_c = -\frac{1}{12} h^3 y^{(3)}(\xi_c) = -\frac{1}{12} h^3 f''(\xi_c) \tag{26.18}$$

where the subscript c designates that this is the error of the corrector.

A similar approach can be used to derive the predictor. For this case, the integration limits are from $i - 1$ to $i + 1$:

$$\int_{y_{i-1}}^{y_{i+1}} dy = \int_{x_{i-1}}^{x_{i+1}} f(x, y) \, dx$$

which can be integrated and rearranged to yield

$$y_{i+1} = y_{i-1} + \int_{x_{i-1}}^{x_{i+1}} f(x, y) \, dx \tag{26.19}$$

Now, rather than using a closed formula from Table 21.2, the first Newton-Cotes open integration formula (see Table 21.4) can be used to evaluate the integral, as in

$$\int_{x_{i-1}}^{x_{i+1}} f(x, y) \, dx = 2h f(x_i, y_i) \tag{26.20}$$

which is called the *midpoint method*. Substituting Eq. (26.20) into Eq. (26.19) yields

$$y_{i+1} = y_{i-1} + 2h f(x_i, y_i)$$

which is the predictor for the non-self-starting Heun. As with the corrector, the local truncation error can be taken directly from Table 21.4:

$$E_p = \frac{1}{3}h^3 y^{(3)}(\xi_p) = \frac{1}{3}h^3 f''(\xi_p) \tag{26.21}$$

where the subscript p designates that this is the error of the predictor.

Thus, the predictor and the corrector for the non-self-starting Heun method have truncation errors of the same order. Aside from upgrading the accuracy of the predictor, this fact has additional benefits related to error analysis, as elaborated in the next section.

Error Estimates. If the predictor and the corrector of a multistep method are of the same order, the local truncation error may be estimated during the course of a computation. This is a tremendous advantage because it establishes a criterion for adjustment of the step size.

The local truncation error for the predictor is estimated by Eq. (26.21). This error estimate can be combined with the estimate of y_{i+1} from the predictor step to yield [recall our basic definition of Eq. (3.1)]

$$\text{True value} = y_{i+1}^0 + \frac{1}{3}h^3 y^{(3)}(\xi_p) \tag{26.22}$$

Using a similar approach, the error estimate for the corrector [Eq. (26.18)] can be combined with the corrector result y_{i+1} to give

$$\text{True value} = y_{i+1}^m - \frac{1}{12}h^3 y^{(3)}(\xi_c) \tag{26.23}$$

Equation (26.22) can be subtracted from Eq. (26.23) to yield

$$0 = y_{i+1}^m - y_{i+1}^0 - \frac{5}{12}h^3 y^{(3)}(\xi) \tag{26.24}$$

where ξ is now between x_{i-1} and x_{i+1}. Now, dividing Eq. (26.24) by 5 and rearranging the result gives

$$\frac{y_{i+1}^0 - y_{i+1}^m}{5} = -\frac{1}{12}h^3 y^{(3)}(\xi) \tag{26.25}$$

Notice that the right-hand sides of Eqs. (26.18) and (26.25) are identical, with the exception of the argument of the third derivative. If the third derivative does not vary appreciably over the interval in question, we can assume that the right-hand sides are equal, and therefore, the left-hand sides should also be equivalent, as in

$$E_c = -\frac{y_{i+1}^0 - y_{i+1}^m}{5} \tag{26.26}$$

Thus, we have arrived at a relationship that can be used to estimate the per-step truncation error on the basis of two quantities—the predictor (y_{i+1}^0) and the corrector (y_{i+1}^m)—that are routine by-products of the computation.

EXAMPLE 26.3   Estimate of Per-Step Truncation Error

Problem Statement.   Use Eq. (26.26) to estimate the per-step truncation error of Example 26.2. Note that the true values at $x = 1$ and 2 are 6.194631 and 14.84392, respectively.

Solution.   At $x_{i+1} = 1$, the predictor gives 5.607005 and the corrector yields 6.360865. These values can be substituted into Eq. (26.26) to give

$$E_c = -\frac{6.360865 - 5.607005}{5} = -0.1507722$$

which compares well with the exact error,

$$E_t = 6.194631 - 6.360865 = -0.1662341$$

At $x_{i+1} = 2$, the predictor gives 13.44346 and the corrector yields 15.30224, which can be used to compute

$$E_c = -\frac{15.30224 - 13.44346}{5} = -0.3717550$$

which also compares favorably with the exact error, $E_t = 14.84392 - 15.30224 = -0.4583148$.

The ease with which the error can be estimated using Eq. (26.26) provides a rational basis for step-size adjustment during the course of a computation. For example, if Eq. (26.26) indicates that the error is greater than an acceptable level, the step size could be decreased.

Modifiers.   Before discussing computer algorithms, we must note two other ways in which the non-self-starting Heun method can be made more accurate and efficient. First, you should realize that besides providing a criterion for step-size adjustment, Eq. (26.26) represents a numerical estimate of the discrepancy between the final corrected value at each step y_{i+1} and the true value. Thus, it can be added directly to y_{i+1} to refine the estimate further:

$$y_{i+1}^m \leftarrow y_{i+1}^m - \frac{y_{i+1}^m - y_{i+1}^0}{5} \tag{26.27}$$

Equation (26.27) is called a *corrector modifier*. (The symbol $\leftarrow$ is read "is replaced by.") The left-hand side is the modified value of y_{i+1}^m.

   A second improvement, one that relates more to program efficiency, is a *predictor modifier*, which is designed to adjust the predictor result so that it is closer to the final convergent value of the corrector. This is advantageous because, as noted previously at the beginning of this section, the number of iterations of the corrector is highly dependent on the accuracy of the initial prediction. Consequently, if the prediction is modified properly, we might reduce the number of iterations required to converge on the ultimate value of the corrector.

Predictor:

$$y_{i+1}^0 = y_{i-1}^m + f(x_i, y_i^m)2h$$

(Save result as $y_{i+1,u}^0 = y_{i+1}^0$ where the subscript u designates that the variable is unmodified.)

Predictor Modifier:

$$y_{i+1}^0 \leftarrow y_{i+1,u}^0 + \frac{4}{5}(y_{i,u}^m - y_{i,u}^0)$$

Corrector:

$$y_{i+1}^j = y_i^m + \frac{f(x, y_i^m) + f(x_{i+1}, y_{i+1}^{j-1})}{2} h \qquad \text{(for } j = 1 \text{ to maximum iterations } m\text{)}$$

Error Check:

$$|\varepsilon_a| = \left| \frac{y_{i+1}^j - y_{i+1}^{j-1}}{y_{i+1}^j} \right| 100\%$$

(If $|\varepsilon_a| >$ error criterion, set $j = j + 1$ and repeat corrector; if $\varepsilon_a \leq$ error criterion, save result as $y_{i+1,u}^m = y_{i+1}^m$.)

Corrector Error Estimate:

$$E_c = -\frac{1}{5}(y_{i+1,u}^m - y_{i+1,u}^0)$$

(If computation is to continue, set $i = i + 1$ and return to predictor.)

FIGURE 26.5
The sequence of formulas used to implement the non-self-starting Heun method. Note that the corrector error estimates can be used to modify the corrector. However, because this can affect the corrector's stability, the modifier is not included in this algorithm. The corrector error estimate is included because of its utility for step-size adjustment.

Such a modifier can be derived simply by assuming that the third derivative is relatively constant from step to step. Therefore, using the result of the previous step at i, Eq. (26.25) can be solved for

$$h^3 y^{(3)}(\xi) = -\frac{12}{5}(y_i^0 - y_i^m) \tag{26.28}$$

which, assuming that $y^{(3)}(\xi) \cong y^{(3)}(\xi_p)$, can be substituted into Eq. (26.21) to give

$$E_p = \frac{4}{5}(y_i^m - y_i^0) \tag{26.29}$$

which can then be used to modify the predictor result:

$$y_{i+1}^0 \leftarrow y_{i+1}^0 + \frac{4}{5}(y_i^m - y_i^0) \tag{26.30}$$

EXAMPLE 26.4   Effect of Modifiers on Predictor-Corrector Results

Problem Statement.   Recompute Example 26.3 using the modifiers as specified in Fig. 26.5.

Solution. As in Example 26.3, the initial predictor result is 5.607005. Because the predictor modifier [Eq. (26.30)] requires values from a previous iteration, it cannot be employed to improve this initial result. However, Eq. (26.27) can be used to modify the corrected value of 6.360865 ($\varepsilon_t = -2.684\%$), as in

$$y_1^m = 6.360865 - \frac{6.360865 - 5.607005}{5} = 6.210093$$

which represents an $\varepsilon_t = -0.25\%$. Thus, the error is reduced over an order of magnitude.

For the next iteration, the predictor [Eq. (26.13)] is used to compute

$$y_2^0 = 2 + \left[4e^{0.8(0)} - 0.5(6.210093)\right]2 = 13.59423 \qquad \varepsilon_t = 8.42\%$$

which is about half the error of the predictor for the second iteration of Example 26.3, which was $\varepsilon_t = 18.6\%$. This improvement occurs because we are using a superior estimate of y (6.210093 as opposed to 6.360865) in the predictor. In other words, the propagated and global errors are reduced by the inclusion of the corrector modifier.

Now because we have information from the prior iteration, Eq. (26.30) can be employed to modify the predictor, as in

$$y_2^0 = 13.59423 + \frac{4}{5}(6.360865 - 5.607005) = 14.19732 \qquad \varepsilon_t = -4.36\%$$

which, again, halves the error.

This modification has no effect on the final outcome of the subsequent corrector step. Regardless of whether the unmodified or modified predictors are used, the corrector will ultimately converge on the same answer. However, because the rate or efficiency of convergence depends on the accuracy of the initial prediction, the modification can reduce the number of iterations required for convergence.

Implementing the corrector yields a result of 15.21178 ($\varepsilon_t = -2.48\%$), which represents an improvement over Example 26.3 because of the reduction of global error. Finally, this result can be modified using Eq. (26.27):

$$y_2^m = 15.21178 - \frac{15.21178 - 13.59423}{5} = 14.88827 \qquad \varepsilon_t = -0.30\%$$

Again, the error has been reduced an order of magnitude.

As in the previous example, the addition of the modifiers increases both the efficiency and accuracy of multistep methods. In particular, the corrector modifier effectively increases the order of the technique. Thus, the non-self-starting Heun with modifiers is third order rather than second order as is the case for the unmodified version. However, it should be noted that there are situations where the corrector modifier will affect the stability of the corrector iteration process. As a consequence, the modifier is not included in the algorithm for the non-self-starting Heun delineated in Fig. 26.5. Nevertheless, the corrector modifier can still have utility for step-size control, as discussed next.

26.2.2 Step-Size Control and Computer Programs

Constant Step Size.   It is relatively simple to develop a constant step-size version of the non-self-starting Heun method. About the only complication is that a one-step method is required to generate the extra point required to start the computation.

Additionally, because a constant step size is employed, a value for h must be chosen prior to the computation. In general, experience indicates that an optimal step size should be small enough to ensure convergence within two iterations of the corrector (Hull and Creemer, 1963). In addition, it must be small enough to yield a sufficiently small truncation error. At the same time, the step size should be as large as possible to minimize run-time cost and round-off error. As with other methods for ODEs, the only practical way to assess the magnitude of the global error is to compare the results for the same problem but with a halved step size.

Variable Step Size.   Two criteria are typically used to decide whether a change in step size is warranted. First, if Eq. (26.26) is greater than some prespecified error criterion, the step size is decreased. Second, the step size is chosen so that the convergence criterion of the corrector is satisfied in two iterations. This criterion is intended to account for the trade-off between the rate of convergence and the total number of steps in the calculation. For smaller values of h, convergence will be more rapid but more steps are required. For larger h, convergence is slower but fewer steps result. Experience (Hull and Creemer, 1963) suggests that the total steps will be minimized if h is chosen so that the corrector converges within two iterations. Therefore, if over two iterations are required, the step size is decreased, and if less than two iterations are required, the step size is increased.

Although the above strategy specifies when step size modifications are in order, it does not indicate how they should be changed. This is a critical question because multistep methods by definition require several points to compute a new point. Once the step size is changed, a new set of points must be determined. One approach is to restart the computation and use the one-step method to generate a new set of starting points.

A more efficient strategy that makes use of existing information is to increase and decrease by doubling and halving the step size. As depicted in Fig. 26.6b, if a sufficient number of previous values have been generated, increasing the step size by doubling is a relatively straightforward task (Fig. 26.6c). All that is necessary is to keep track of subscripts so that old values of x and y become the appropriate new values. Halving the step size is somewhat more complicated because some of the new values will be unavailable (Fig. 26.6a). However, interpolating polynomials of the type developed in Chap. 18 can be used to determine these intermediate values.

In any event, the decision to incorporate step-size control represents a trade-off between initial investment in program complexity versus the long-term return because of increased efficiency. Obviously, the magnitude and importance of the problem itself will have a strong bearing on this trade-off. Fortunately, several software packages and libraries have multistep routines that you can use to obtain solutions without having to program them from scratch. We will mention some of these when we review packages and libraries at the end of Chap. 27.

26.2.3 Integration Formulas

The non-self-starting Heun method is characteristic of most multistep methods. It employs an open integration formula (the midpoint method) to make an initial estimate. This pre-

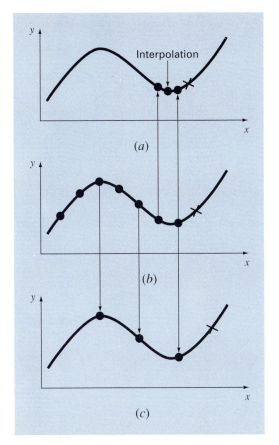

FIGURE 26.6
A plot indicating how a halving-doubling strategy allows the use of (b) previously calculated values for a third-order multistep method. (a) Halving; (c) doubling.

dictor step requires a previous data point. Then, a closed integration formula (the trapezoidal rule) is applied iteratively to improve the solution.

It should be obvious that a strategy for improving multistep methods would be to use higher-order integration formulas as predictors and correctors. For example, the higher-order Newton-Cotes formulas developed in Chap. 21 could be used for this purpose.

Before describing these higher-order methods, we will review the most common integration formulas upon which they are based. As mentioned above, the first of these are the Newton-Cotes formulas. However, there is a second class called the Adams formulas that we will also review and that are often preferred. As depicted in Fig. 26.7, the fundamental difference between the Newton-Cotes and Adams formulas relates to the manner in which the integral is applied to obtain the solution. As depicted in Fig. 26.7a, the Newton-Cotes formulas estimate the integral over an interval spanning several points. This integral is then used to project from the beginning of the interval to the end. In contrast, the Adams formulas (Fig. 26.7b) use a set of points from an interval to estimate the integral solely for the last segment in the interval. This integral is then used to project across this last segment.

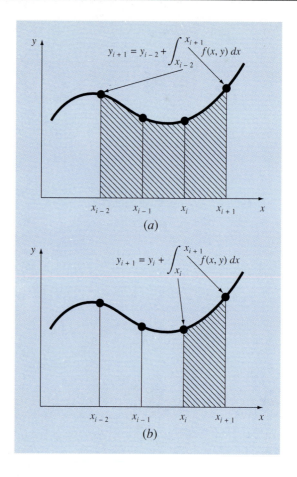

FIGURE 26.7
Illustration of the fundamental difference between the Newton-Cotes and Adams integration formulas. (a) The Newton-Cotes formulas use a series of points to obtain an integral estimate over a number of segments. The estimate is then used to project across the entire range. (b) The Adams formulas use a series of points to obtain an integral estimate for a single segment. The estimate is then used to project across the segment.

Newton-Cotes Formulas. Some of the most common formulas for solving ordinary differential equations are based on fitting an nth-degree interpolating polynomial to $n + 1$ known values of y and then using this equation to compute the integral. As discussed previously in Chap. 21, the Newton-Cotes integration formulas are based on such an approach. These formulas are of two types: open and closed forms.

Open Formulas. For n equally spaced data points, the open formulas can be expressed in the form of a solution of an ODE, as was done previously for Eq. (26.19). The general equation for this purpose is

$$y_{i+1} = y_{i-n} + \int_{x_{i-n}}^{x_{i+1}} f_n(x)\, dx \qquad (26.31)$$

where $f_n(x)$ is an nth-order interpolating polynomial. The evaluation of the integral employs the nth-order Newton-Cotes open integration formula (Table 21.4). For example, if $n = 1$,

$$y_{i+1} = y_{i-1} + 2hf_i \qquad (26.32)$$

where f_i is an abbreviation for $f(x_i, y_i)$—that is, the differential equation evaluated at x_i and y_i. Equation (26.32) is referred to as the midpoint method and was used previously as the predictor in the non-self-starting Heun method. For $n = 2$,

$$y_{i+1} = y_{i-2} + \frac{3h}{2}(f_i + f_{i-1})$$

and for $n = 3$,

$$y_{i+1} = y_{i-3} + \frac{4h}{3}(2f_i - f_{i-1} + 2f_{i-2}) \qquad (26.33)$$

Equation (26.33) is depicted graphically in Fig. 26.8a.

Closed Formulas. The closed form can be expressed generally as

$$y_{i+1} = y_{i-n+1} + \int_{x_{i-n+1}}^{x_{i+1}} f_n(x)\,dx \qquad (26.34)$$

where the integral is approximated by an nth-order Newton-Cotes closed integration formula (Table 21.2). For example, for $n = 1$,

$$y_{i+1} = y_i + \frac{h}{2}(f_i + f_{i+1})$$

which is equivalent to the trapezoidal rule. For $n = 2$,

$$y_{i+1} = y_{i-1} + \frac{h}{3}(f_{i-1} + 4f_i + f_{i+1}) \qquad (26.35)$$

which is equivalent to Simpson's 1/3 rule. Equation (26.35) is depicted in Fig. 26.8b.

Adams Formulas. The other type of integration formulas that can be used to solve ODEs are the Adams formulas. Many popular computer algorithms for multistep solution of ODEs are based on these methods.

Open Formulas (Adams-Bashforth). The Adams formulas can be derived in a variety of ways. One technique is to write a forward Taylor series expansion around x_i:

$$y_{i+1} = y_i + f_i h + \frac{f_i'}{2}h^2 + \frac{f_i''}{6}h^3 + \cdots$$

which can also be written as

$$y_{i+1} = y_i + h\left(f_i + \frac{h}{2}f_i' + \frac{h^2}{3!}f_i'' + \cdots\right) \qquad (26.36)$$

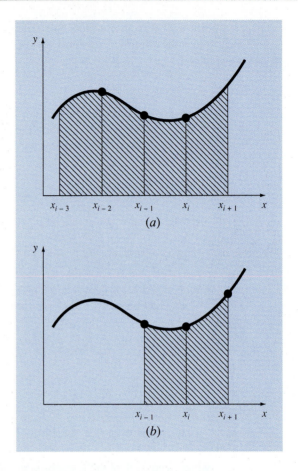

FIGURE 26.8
Graphical depiction of open and closed Newton-Cotes integration formulas. (a) The third open formula [Eq. (26.33)] and (b) Simpson's 1/3 rule [Eq. (26.35)].

Recall from Sec. 4.1.3 that a backward difference can be used to approximate the derivative:

$$f_i' = \frac{f_i - f_{i-1}}{h} + \frac{f_i''}{2}h + O(h^2)$$

which can be substituted into Eq. (26.36),

$$y_{i+1} = y_i + h\left\{f_i + \frac{h}{2}\left[\frac{f_i - f_{i-1}}{h} + \frac{f_i''}{2}h + O(h^2)\right] + \frac{h^2}{6}f_i'' + \cdots\right\}$$

or, collecting terms,

$$y_{i+1} = y_i + h\left(\frac{3}{2}f_i - \frac{1}{2}f_{i-1}\right) + \frac{5}{12}h^3 f_i'' + O(h^4) \tag{26.37}$$

TABLE 26.1   Coefficients and truncation error for Adams-Bashforth predictors.

Order	β_0	β_1	β_2	β_3	β_4	β_5	Local Truncation Error
1	1						$\frac{1}{2}h^2 f'(\xi)$
2	3/2	−1/2					$\frac{5}{12}h^3 f''(\xi)$
3	23/12	−16/12	5/12				$\frac{9}{24}h^4 f^{(3)}(\xi)$
4	55/24	−59/24	37/24	−9/24			$\frac{251}{720}h^5 f^{(4)}(\xi)$
5	1901/720	−2774/720	2616/720	−1274/720	251/720		$\frac{475}{1440}h^6 f^{(5)}(\xi)$
6	4277/720	−7923/720	9982/720	−7298/720	2877/720	−475/720	$\frac{19,087}{60,480}h^7 f^{(6)}(\xi)$

This formula is called the *second-order open Adams formula.* Open Adams formulas are also referred to as *Adams-Bashforth formulas.* Consequently, Eq. (26.37) is sometimes called the second Adams-Bashforth formula.

Higher-order Adams-Bashforth formulas can be developed by substituting higher-difference approximations into Eq. (26.36). The nth-order open Adams formula can be represented generally as

$$y_{i+1} = y_i + h \sum_{k=0}^{n-1} \beta_k f_{i-k} + O(h^{n+1}) \tag{26.38}$$

The coefficients β_k are compiled in Table 26.1. The fourth-order version is depicted in Fig. 26.9a. Notice that the first-order version is Euler's method.

Closed Formulas (Adams-Moulton). A backward Taylor series around x_{i+1} can be written as

$$y_i = y_{i+1} - f_{i+1}h + \frac{f'_{i+1}}{2}h^2 - \frac{f''_{i+1}}{3!}h^3 + \cdots$$

Solving for y_{i+1} yields

$$y_{i+1} = y_i + h\left(f_{i+1} - \frac{h}{2}f'_{i+1} + \frac{h^2}{6}f''_{i+1} + \cdots \right) \tag{26.39}$$

A difference can be used to approximate the first derivative:

$$f'_{i+1} = \frac{f_{i+1} - f_i}{h} + \frac{f''_{i+1}}{2}h + O(h^2)$$

TABLE 26.2 Coefficients and truncation error for Adams-Moulton correctors.

Order	β_0	β_1	β_2	β_3	β_4	β_5	Local Truncation Error
2	1/2	1/2					$-\dfrac{1}{12}h^3 f''(\xi)$
3	5/12	8/12	−1/12				$-\dfrac{1}{24}h^4 f^{(3)}(\xi)$
4	9/24	19/24	−5/24	1/24			$-\dfrac{19}{720}h^5 f^{(4)}(\xi)$
5	251/720	646/720	−264/720	106/720	−19/720		$-\dfrac{27}{1440}h^6 f^{(5)}(\xi)$
6	475/1440	1427/1440	−798/1440	482/1440	−173/1440	27/1440	$-\dfrac{863}{60,480}h^7 f^{(6)}(\xi)$

which can be substituted into Eq. (26.39), and collecting terms gives

$$y_{i+1} = y_i + h\left(\frac{1}{2}f_{i+1} + \frac{1}{2}f_i\right) - \frac{1}{12}h^3 f''_{i+1} - O(h^4)$$

This formula is called the *second-order closed Adams formula* or the *second Adams-Moulton formula*. Also, notice that it is the trapezoidal rule.

The *n*th-order closed Adams formula can be written generally as

$$y_{i+1} = y_i + h\sum_{k=0}^{n-1} \beta_k f_{i+1-k} + O(h^{n+1})$$

The coefficients β_k are listed in Table 26.2. The fourth-order method is depicted in Fig. 26.9b.

26.2.4 Higher-Order Multistep Methods

Now that we have formally developed the Newton-Cotes and Adams integration formulas, we can use them to derive higher-order multistep methods. As was the case with the non-self-starting Heun method, the integration formulas are applied in tandem as predictor-corrector methods. In addition, if the open and closed formulas have local truncation errors of the same order, modifiers of the type listed in Fig. 26.5 can be incorporated to improve accuracy and allow step size control. Box 26.1 provides general equations for these modifiers. In the following section, we present two of the most common higher-order multistep approaches: Milne's method and the fourth-order Adams method.

Milne's Method.   Milne's method is the most common multistep method based on Newton-Cotes integration formulas. It uses the three-point Newton-Cotes open formula as a predictor:

$$y_{i+1}^0 = y_{i-3}^m + \frac{4h}{3}\left(2f_i^m - f_{i-1}^m + 2f_{i-2}^m\right) \qquad (26.40)$$

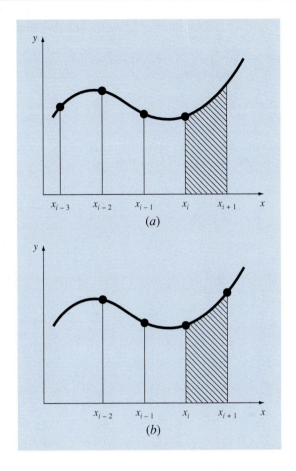

FIGURE 26.9
Graphical depiction of open and closed Adams integration formulas. (a) The fourth Adams-Bashforth open formula and (b) the fourth Adams-Moulton closed formula.

and the three-point Newton-Cotes closed formula (Simpson's 1/3 rule) as a corrector:

$$y_{i+1}^{j} = y_{i-1}^{m} + \frac{h}{3}\left(f_{i-1}^{m} + 4f_{i}^{m} + f_{i+1}^{j-1}\right) \tag{26.41}$$

The predictor and corrector modifiers for Milne's method can be developed from the formulas in Box 26.1 and the error coefficients in Tables 21.2 and 21.4:

$$E_p = \frac{28}{29}\left(y_i^m - y_i^0\right) \tag{26.42}$$

$$E_c \cong -\frac{1}{29}\left(y_{i+1}^m - y_{i+1}^0\right) \tag{26.43}$$

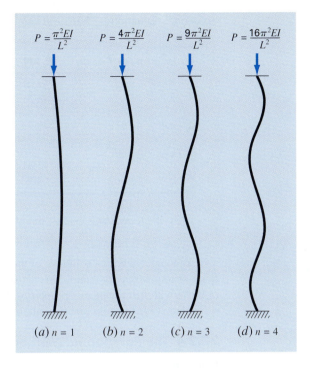

FIGURE 27.8
The first four eigenvalues for the slender rod from Fig. 27.7.

Figure 27.8, which shows the solution for the first four eigenvalues, can provide insight into the physical significance of the results. Each eigenvalue corresponds to a way in which the column buckles. Combining Eqs. (27.12) and (27.16) gives

$$P = \frac{n^2\pi^2 EI}{L^2} \qquad \text{for } n = 1, 2, 3, \ldots \tag{27.17}$$

These can be thought of as *buckling loads* because they represent the levels at which the column moves into each succeeding buckling configuration. In a practical sense, it is usually the first value that is of interest because failure will usually occur when the column first buckles. Thus, a critical load can be defined as

$$P = \frac{\pi^2 EI}{L^2}$$

which is formally known as *Euler's formula*.

EXAMPLE 27.5   Eigenvalue Analysis of an Axially Loaded Column

Problem Statement.   An axially loaded wooden column has the following characteristics: $E = 10 \times 10^9$ Pa, $I = 1.25 \times 10^{-5}$ m^4, and $L = 3$ m. Determine the first eight eigenvalues and the corresponding buckling loads.

Solution.   Equations (27.16) and (27.17) can be used to compute

n	p, m⁻²	P, kN
1	1.0472	137.078
2	2.0944	548.311
3	3.1416	1233.701
4	4.1888	2193.245
5	5.2360	3426.946
6	6.2832	4934.802
7	7.3304	6716.814
8	8.3776	8772.982

The critical buckling load is, therefore, 137.078 kN.

Although analytical solutions of the sort obtained above are useful, they are often difficult or impossible to obtain. This is usually true when dealing with complicated systems or those with heterogeneous properties. In such cases, numerical methods of the sort described next are the only practical alternative.

27.2.4 The Polynomial Method

Equation (27.11) can be solved numerically by substituting a central finite-divided-difference approximation (Table 23.3) for the second derivative to give

$$\frac{y_{i+1} - 2y_i + y_{i-1}}{h^2} + p^2 y_i = 0$$

which can be expressed as

$$y_{i-1} - (2 - h^2 p^2)y_i + y_{i+1} = 0 \qquad (27.18)$$

Writing this equation for a series of nodes along the axis of the column yields a homogeneous system of equations. For example, if the column is divided into five segments (that is, four interior nodes), the result is

$$\begin{bmatrix} (2 - h^2 p^2) & -1 & 0 & 0 \\ -1 & (2 - h^2 p^2) & -1 & 0 \\ 0 & -1 & (2 - h^2 p^2) & -1 \\ 0 & 0 & -1 & (2 - h^2 p^2) \end{bmatrix} \begin{Bmatrix} y_1 \\ y_2 \\ y_3 \\ y_4 \end{Bmatrix} = 0 \qquad (27.19)$$

Expansion of the determinant of the system yields a polynomial, the roots of which are the eigenvalues. This approach, called the *polynomial method,* is performed in the following example.

EXAMPLE 27.6   The Polynomial Method

Problem Statement.   Employ the polynomial method to determine the eigenvalues for the axially loaded column from Example 27.5 using (a) one, (b) two, (c) three, and (d) four interior nodes.

Solution.

(a) Writing Eq. (27.18) for one interior node yields ($h = 3/2$)

$$-(2 - 2.25p^2)y_1 = 0$$

Thus, for this simple case, the eigenvalue is analyzed by setting the determinant equal to zero

$$2 - 2.25p^2 = 0$$

and solving for $p = \pm 0.9428$, which is about 10 percent less than the exact value of 1.0472 obtained in Example 27.4.

(b) For two interior nodes ($h = 3/3$), Eq. (27.18) is written as

$$\begin{bmatrix} (2 - p^2) & -1 \\ -1 & (2 - p^2) \end{bmatrix} \begin{Bmatrix} y_1 \\ y_2 \end{Bmatrix} = 0$$

Expansion of the determinant gives

$$(2 - p^2)^2 - 1 = 0$$

which can be solved for $p = \pm 1$ and ± 1.73205. Thus, the first eigenvalue is now about 4.5 percent low and a second eigenvalue is obtained that is about 17 percent low.

(c) For three interior points ($h = 3/4$), Eq. (27.18) yields

$$\begin{bmatrix} 2 - 0.5625p^2 & -1 & 0 \\ -1 & 2 - 0.5625p^2 & -1 \\ 0 & -1 & 2 - 0.5625p^2 \end{bmatrix} \begin{Bmatrix} y_1 \\ y_2 \\ y_3 \end{Bmatrix} = 0 \qquad \text{(E27.6.1)}$$

The determinant can be set equal to zero and expanded to give

$$(2 - 0.5625p^2)^3 - 2(2 - 0.5625p^2) = 0$$

For this equation to hold, $2 - 0.5625p^2 = 0$ and $2 - 0.5625p^2 = \sqrt{2}$. Therefore, the first three eigenvalues can be determined as

$$p = \pm 1.0205 \qquad |\varepsilon_t| = 2.5\%$$
$$p = \pm 1.8856 \qquad |\varepsilon_t| = 10\%$$
$$p = \pm 2.4637 \qquad |\varepsilon_t| = 22\%$$

(d) For four interior points ($h = 3/5$), the result is Eq. (27.19) with $2 - 0.36p^2$ on the diagonal. Setting the determinant equal to zero and expanding it gives

$$(2 - 0.36p^2)^4 - 3(2 - 0.36p^2)^2 + 1 = 0$$

which can be solved for the first four eigenvalues

$$p = \pm 1.0301 \qquad |\varepsilon_t| = 1.6\%$$
$$p = \pm 1.9593 \qquad |\varepsilon_t| = 6.5\%$$
$$p = \pm 2.6967 \qquad |\varepsilon_t| = 14\%$$
$$p = \pm 3.1702 \qquad |\varepsilon_t| = 24\%$$

TABLE 27.2 The results of applying the polynomial method to an axially loaded column. The numbers in parentheses represent the absolute value of the true percent relative error.

		Polynomial Method			
Eigenvalue	True	h = 3/2	h = 3/3	h = 3/4	h = 3/5
1	1.0472	0.9428 (10%)	1.0000 (4.5%)	1.0205 (2.5%)	1.0301 (1.6%)
2	2.0944		1.7321 (21%)	1.8856 (10%)	1.9593 (65%)
3	3.1416			2.4637 (22%)	2.6967 (14%)
4	4.1888				3.1702 (24%)

Table 27.2, which summarizes the results of this example, illustrates some fundamental aspects of the polynomial method. As the segmentation is made more refined, additional eigenvalues are determined and the previously determined values become progressively more accurate. Thus, the approach is best suited for cases where the lower eigenvalues are required.

As in Example 27.6, the polynomial method consists of two major steps: (1) expansion of the determinant to yield a polynomial and (2) calculation of the roots of the polynomial. We now turn to a computer method for generating the polynomial.

The Fadeev-Leverrier Method. The Fadeev-Leverrier method is an efficient approach for generating the coefficients p_i of the polynomial

$$(-1)^n (\lambda^n - p_{n-1}\lambda^{n-1} - \cdots - p_1\lambda^n - p_0) = 0 \qquad (27.20)$$

which result from the expansion of the determinant of the system

$$\big[[A] - \lambda[I]\big]\{X\} = 0 \qquad (27.4)$$

Note that the term $(-1)^n$ and the negative signs are included so that the terms of the polynomial have the same signs as they would have if they were generated by expanding the determinant with minors. The method has the additional advantage that the matrix inverse $[A]^{-1}$ can be generated efficiently in the process.

The method consists of generating a sequence of matrices $[B]$ that can be employed to determine the p_i values. For example,

$$[B]_{n-1} = [A] \qquad (27.21)$$

and

$$p_{n-1} = \text{tr } [B]_{n-1} \qquad (27.22)$$

where tr $[B]_{n-1}$ is the trace of the matrix $[B]_{n-1}$, which is (recall PT3.2.2) the sum of the diagonal coefficients.

The method can be continued by generating the matrix

$$[B]_{n-2} = [A]\big[[B]_{n-1} - p_{n-1}[I]\big]$$ (27.23)

which can be used to calculate

$$p_{n-2} = \frac{1}{2}\,\text{tr}\,[B]_{n-2}$$ (27.24)

and

$$[B]_{n-3} = [A]\big[[B]_{n-2} - p_{n-2}[I]\big]$$ (27.25)

which can be used to calculate

$$p_{n-3} = \frac{1}{3}\,\text{tr}\,[B]_{n-3}$$ (27.26)

and so on until p_0 is determined from

$$[B]_0 = [A]\big[[B]_1 - p_1[I]\big]$$ (27.27)

and

$$p_0 = \frac{1}{n}\,\text{tr}\,[B]_0$$ (27.28)

The matrix inverse can then be simply computed as

$$[A]^{-1} = \frac{1}{p_0}\big[[B]_1 - p_1[I]\big]$$ (27.29)

EXAMPLE 27.7   The Fadeev-Leverrier Method

Problem Statement.   Employ the Fadeev-Leverrier method to determine the coefficients for the matrix in part (c) of Example 27.6.

Solution.   The matrix can be expressed in the general form of Eq. (27.4) by dividing each row of Eq. (E27.6.1) by h^2. For the present case $[h^2 = (3/4)^2 = 0.5625]$, this gives

$$[A] - \lambda[I] = \begin{bmatrix} 3.556 - \lambda & -1.778 & 0 \\ -1.778 & 3.556 - \lambda & -1.778 \\ 0 & -1.778 & 3.556 - \lambda \end{bmatrix}$$

where $\lambda = p^2$. Before proceeding, the determinant can be expanded by minors to give

$$-\lambda^3 + 10.667\lambda^2 - 31.607\lambda + 22.487$$

The same evaluation can be implemented with the Fadeev-Leverrier method. According to Eq. (27.21),

$$[B]_2 = \begin{bmatrix} 3.556 & -1.778 & 0 \\ -1.778 & 3.556 & -1.778 \\ 0 & -1.778 & 3.556 \end{bmatrix}$$

Therefore, according to Eq. (27.22),

$$p_2 = 3.556 + 3.556 + 3.556 = 10.667$$ (E27.7.1)

This value can be substituted into Eq. (27.23) to give

$$[B]_1 = \begin{bmatrix} 3.556 & -1.778 & 0 \\ -1.778 & 3.556 & -1.778 \\ 0 & -1.778 & 3.556 \end{bmatrix} \begin{bmatrix} -7.111 & -1.778 & 0 \\ -1.778 & -7.111 & -1.778 \\ 0 & -1.778 & -7.111 \end{bmatrix}$$

which can be evaluated to yield

$$[B]_1 = \begin{bmatrix} -22.123 & 6.321 & 3.160 \\ 6.321 & -18.963 & 6.321 \\ 3.160 & 6.321 & -22.123 \end{bmatrix}$$

Therefore, Eq. (27.24) can be employed to compute

$$p_1 = \frac{1}{2}(-22.123 - 18.963 - 22.123) = -31.605 \qquad \text{(E27.7.2)}$$

This result in turn can be substituted into Eq. (27.27) to give

$$[B]_0 = \begin{bmatrix} 3.556 & -1.778 & 0 \\ -1.778 & 3.556 & -1.778 \\ 0 & -1.778 & 3.556 \end{bmatrix} \begin{bmatrix} 9.481 & 6.321 & 3.160 \\ 6.321 & 12.642 & 6.321 \\ 3.160 & 6.321 & 9.481 \end{bmatrix}$$

which can be evaluated to yield

$$[B]_0 = \begin{bmatrix} 22.475 & 0 & 0 \\ 0 & 22.475 & 0 \\ 0 & 0 & 22.475 \end{bmatrix}$$

Therefore, Eq. (27.28) can be employed to compute

$$p_0 = \frac{1}{3}(22.475 + 22.475 + 22.475) = 22.475 \qquad \text{(E27.7.3)}$$

Substituting Eqs. (E27.7.1) through (E27.7.3) into Eq. (27.20) gives the resulting polynomial

$$-\lambda^3 + 10.667\lambda^2 - 31.605\lambda + 22.475 = 0$$

which, aside from slight differences due to round-off error, is identical to the result obtained by expanding the determinant using minors.

The matrix inverse can also be computed with Eq. (27.29),

$$[A]^{-1} = \frac{1}{22.475}$$

$$\times \begin{bmatrix} -22.123 - (-31.605) & 6.321 & 3.160 \\ 6.321 & -18.963 - (-31.605) & 6.321 \\ 3.160 & 6.321 & -22.123 - (-31.605) \end{bmatrix}$$

$$= \begin{bmatrix} 0.422 & 0.281 & 0.141 \\ 0.281 & 0.562 & 0.281 \\ 0.141 & 0.281 & 0.422 \end{bmatrix}$$

which when multiplied by $[A]$ yields $[I]$.

```
SUB Fadeev (a, n, p)
  DIM b_{n,n}
  p_n = −1
  [B] = [A]
  P_{n−1} = Trace(b, n)
  DO ii = n − 2, 0, −1
    DO i = 1, n
      b_{i,i} = b_{i,i} − p_{ii+1}
    END DO
    [B] = [A] • [B]
    p_{ii} = Trace(b, n) / (n − ii)
  END DO
END Fadeev


FUNCTION Trace(a, n)
  sum = 0!
  DO i = 1, n
    sum = sum + a_{i,i}
  END DO
  Trace = sum
END Trace
```

FIGURE 27.9
A subroutine for the Fadeev-Leverrier method. Notice that a function is also included to compute the trace.

The Fadeev-Leverrier method can be programmed concisely. Figure 27.9 lists pseudocode for this purpose. Notice that a subroutine is required to compute the trace.

Determination of the Eigenvalues as the Roots of the Characteristic Polynomial.
After applying the Fadeev-Leverrier technique, the roots of the resulting polynomial must be evaluated. The approaches presented in Chap. 7 are designed for this purpose. In the following example, Bairstow's method is used.

EXAMPLE 27.8   Roots of the Characteristic Polynomial

Problem Statement.   Determine the roots of the characteristic polynomial that was generated in Example 27.7 and compare the resulting eigenvalues with those calculated in part (c) of Example 27.6.

Solution.   The polynomial

$$-\lambda^3 + 10.667\lambda^2 - 31.605\lambda + 22.475 = 0$$

can be evaluated with Bairstow's method (or with packages like MATLAB or Mathcad) to yield

$$\lambda = 1.041, 3.555, 6.071$$

Because $\lambda = p^2$, the square root of these values can be taken to give

$$p = 1.020, 1.885, 2.464$$

which are the same results as in Example 27.6c (with small discrepancies due to round-off error).

Computer Program for the Polynomial Method.   A computer program for the polynomial method can be simply produced by combining the code for the Fadeev-Leverrier method (Fig. 27.9) with that for Bairstow's method (Fig. 7.5). This will be left as a homework exercise.

27.2.5 The Power Method

The *power method* is an iterative approach that can be employed to determine the largest eigenvalue. With slight modification, it can also be employed to determine the smallest and the intermediate values. It has the additional benefit that the corresponding eigenvector is obtained as a by-product of the method.

Determination of the Largest Eigenvalue.   To implement the power method, the system being analyzed must be expressed in the form

$$[A]\{X\} = \lambda\{X\} \tag{27.30}$$

As illustrated by the following example, Eq. (27.30) forms the basis for an iterative solution technique that eventually yields the highest eigenvalue and its associated eigenvector.

EXAMPLE 27.9   Power Method for Highest Eigenvalue

Problem Statement.   Employ the power method to determine the highest eigenvalue for part (c) of Example 27.6.

Solution.   The system is first written in the form of Eq. (27.30),

$$\begin{array}{rcl}
3.556x_1 - 1.778x_2 & = & \lambda x_1 \\
-1.778x_1 + 3.556x_2 - 1.778x_3 & = & \lambda x_2 \\
- 1.778x_2 + 3.556x_3 & = & \lambda x_3
\end{array}$$

Then, assuming the x's on the left-hand side of the equation are equal to 1,

$$\begin{array}{rcl}
3.556(1) - 1.778(1) & = & 1.778 \\
-1.778(1) + 3.556(1) - 1.778(1) & = & 0 \\
-1.778(1) + 3.556(1) & = & 1.778
\end{array}$$

Next, the right-hand side is normalized by 1.778 to make the largest element equal to

$$
\left\{ \begin{array}{c} 1.778 \\ 0 \\ 1.778 \end{array} \right\} = 1.778 \left\{ \begin{array}{c} 1 \\ 0 \\ 1 \end{array} \right\}
$$

Thus, the first estimate of the eigenvalue is 1.778. This iteration can be expressed concisely in matrix form as

$$
\left[\begin{array}{ccc} 3.556 & -1.778 & 0 \\ -1.778 & 3.556 & -1.778 \\ 0 & -1.778 & 3.556 \end{array} \right] \left\{ \begin{array}{c} 1 \\ 1 \\ 1 \end{array} \right\} = \left\{ \begin{array}{c} 1.778 \\ 0 \\ 1.778 \end{array} \right\} = 1.778 \left\{ \begin{array}{c} 1 \\ 0 \\ 1 \end{array} \right\}
$$

The next iteration consists of multiplying $[A]$ by $\lfloor 1 \quad 0 \quad 1 \rfloor^T$ to give

$$
\left[\begin{array}{ccc} 3.556 & -1.778 & 0 \\ -1.778 & 3.556 & -1.778 \\ 0 & -1.778 & 3.556 \end{array} \right] \left\{ \begin{array}{c} 1 \\ 0 \\ 1 \end{array} \right\} = \left\{ \begin{array}{c} 3.556 \\ -3.556 \\ 3.556 \end{array} \right\} = 3.556 \left\{ \begin{array}{c} 1 \\ -1 \\ 1 \end{array} \right\}
$$

Therefore, the eigenvalue estimate for the second iteration is 3.556, which can be employed to determine the error estimate

$$
|\varepsilon_a| = \left| \frac{3.556 - 1.778}{3.556} \right| 100\% = 50\%
$$

The process can then be repeated.

Third iteration:

$$
\left[\begin{array}{ccc} 3.556 & -1.778 & 0 \\ -1.778 & 3.556 & -1.778 \\ 0 & -1.778 & 3.556 \end{array} \right] \left\{ \begin{array}{c} 1 \\ -1 \\ 1 \end{array} \right\} = \left\{ \begin{array}{c} 5.334 \\ -7.112 \\ 5.334 \end{array} \right\} = -7.112 \left\{ \begin{array}{c} -0.75 \\ 1 \\ -0.75 \end{array} \right\}
$$

where $|\varepsilon_a| = 150\%$ (which is high because of the sign change).

Fourth iteration:

$$
\left[\begin{array}{ccc} 3.556 & -1.778 & 0 \\ -1.778 & 3.556 & -1.778 \\ 0 & -1.778 & 3.556 \end{array} \right] \left\{ \begin{array}{c} -0.75 \\ 1 \\ -0.75 \end{array} \right\} = \left\{ \begin{array}{c} -4.445 \\ 6.223 \\ -4.445 \end{array} \right\} = 6.223 \left\{ \begin{array}{c} -0.714 \\ 1 \\ -0.714 \end{array} \right\}
$$

where $|\varepsilon_a| = 214\%$ (again inflated because of sign change).

Fifth iteration:

$$
\left[\begin{array}{ccc} 3.556 & -1.778 & 0 \\ -1.778 & 3.556 & -1.778 \\ 0 & -1.778 & 3.556 \end{array} \right] \left\{ \begin{array}{c} -0.714 \\ 1 \\ -0.714 \end{array} \right\} = \left\{ \begin{array}{c} -4.317 \\ 6.095 \\ -4.317 \end{array} \right\} = 6.095 \left\{ \begin{array}{c} -0.708 \\ 1 \\ -0.708 \end{array} \right\}
$$

Thus, the normalizing factor is converging on the value of 6.070 ($= 2.4637^2$) obtained in part (*c*) of Example 27.6.

Note that there are some instances where the power method will converge to the second-largest eigenvalue instead of to the largest. James, Smith, and Wolford (1985) provide an illustration of such a case. Other special cases are discussed in Fadeev and Fadeeva (1963).

Determination of the Smallest Eigenvalue. There are often cases in engineering where we are interested in determining the smallest eigenvalue. Such was the case for the rod in Fig. 27.7, where the smallest eigenvalue could be used to identify a critical buckling load. This can be done by applying the power method to the matrix inverse of $[A]$. For this case, the power method will converge on the largest value of $1/\lambda$—in other words, the smallest value of λ.

EXAMPLE 27.10   **Power Method for Lowest Eigenvalue**

Problem Statement. Employ the power method to determine the lowest eigenvalue for part (c) of Example 27.6.

Solution. Recall from Example 27.7 that the matrix inverse is

$$[A]^{-1} = \begin{bmatrix} 0.422 & 0.281 & 0.141 \\ 0.281 & 0.562 & 0.281 \\ 0.141 & 0.281 & 0.422 \end{bmatrix}$$

Using the same format as in Example 27.9, the power method can be applied to this matrix.
  First iteration:

$$\begin{bmatrix} 0.422 & 0.281 & 0.141 \\ 0.281 & 0.562 & 0.281 \\ 0.141 & 0.281 & 0.422 \end{bmatrix} \begin{Bmatrix} 1 \\ 1 \\ 1 \end{Bmatrix} = \begin{Bmatrix} 0.884 \\ 1.124 \\ 0.884 \end{Bmatrix} = 1.124 \begin{Bmatrix} 0.751 \\ 1 \\ 0.751 \end{Bmatrix}$$

  Second iteration:

$$\begin{bmatrix} 0.422 & 0.281 & 0.141 \\ 0.281 & 0.562 & 0.281 \\ 0.141 & 0.281 & 0.422 \end{bmatrix} \begin{Bmatrix} 0.751 \\ 1 \\ 0.751 \end{Bmatrix} = \begin{Bmatrix} 0.704 \\ 0.984 \\ 0.704 \end{Bmatrix} = 0.984 \begin{Bmatrix} 0.715 \\ 1 \\ 0.715 \end{Bmatrix}$$

where $|\varepsilon_a| = 14.6\%$.
  Third iteration:

$$\begin{bmatrix} 0.422 & 0.281 & 0.141 \\ 0.281 & 0.562 & 0.281 \\ 0.141 & 0.281 & 0.422 \end{bmatrix} \begin{Bmatrix} 0.715 \\ 1 \\ 0.715 \end{Bmatrix} = \begin{Bmatrix} 0.684 \\ 0.964 \\ 0.684 \end{Bmatrix} = 0.964 \begin{Bmatrix} 0.709 \\ 1 \\ 0.709 \end{Bmatrix}$$

where $|\varepsilon_a| = 4\%$.
  Thus, after only three iterations, the result is converging on the value of 0.955, which is the reciprocal of the smallest eigenvalue, 1.0472, obtained in Example 27.5.

Determination of Intermediate Eigenvalues. After finding the largest eigenvalue, it is possible to determine the next highest by replacing the original matrix by one that includes only the remaining eigenvalues. The process of removing the largest known eigenvalue is called *deflation.* The technique outlined here, *Hotelling's method,* is designed for symmetric matrices. This is because it exploits the orthogonality of the eigenvectors of such matrices, which can be expressed as

$$\{X\}_i^T \{X\}_j = \begin{cases} 0 & \text{for } i \neq j \\ 1 & \text{for } i = j \end{cases} \tag{27.31}$$

where the components of the eigenvector $\{X\}$ have been normalized so that $\{X\}^T\{X\} = 1$; that is, so that the sum of the squares of the components equals 1. This can be accomplished by dividing each of the elements by the normalizing factor

$$\sqrt{\sum_{k=1}^{n} x_k^2}$$

Now, a new matrix $[A]_2$ can be computed as

$$[A]_2 = [A]_1 - \lambda_1 \{X\}_1 \{X\}_1^T \tag{27.32}$$

where $[A]_1$ = the original matrix and λ_1 = the largest eigenvalue. If the power method is applied to this matrix, the iteration process will converge to the second largest eigenvalue, λ_2. To show this, first postmultiply Eq. (27.32) by $\{X\}_1$,

$$[A]_2\{X\}_1 = [A]_1\{X\}_1 - \lambda_1 \{X\}_1 \{X\}_1^T \{X\}_1$$

Invoking the orthogonality principle converts this equation to

$$[A]_2\{X\}_1 = [A]_1\{X\}_1 - \lambda_1 \{X\}_1$$

where the right-hand side is equal to zero according to Eq. (27.30). Thus, $[A]_2\{X\}_1 = 0$. Consequently, $\lambda = 0$ and $\{X\} = \{X\}_1$ is a solution to $[A]_2\{X\} = \lambda\{X\}$. In other words, the $[A]_2$ has eigenvalues of 0, λ_2, λ_3, ..., λ_n. The largest eigenvalue, λ_1, has been replaced by a 0 and, therefore, the power method will converge on the next biggest λ_2.

The above process can be repeated by generating a new matrix $[A]_3$, etc. Although in theory this process could be continued to determine the remaining eigenvalues, it is limited by the fact that errors in the eigenvectors are passed along at each step. Thus, it is only of value in determining several of the highest eigenvalues. Although this is somewhat of a shortcoming, such information is precisely what is required in many engineering problems.

27.2.6 Other Methods

A wide variety of additional methods are available for solving eigenvalue problems. Most are based on a two-step process. The first step involves transforming the original matrix to a simpler form (e.g., tridiagonal) that retains all the original eigenvalues. Then, iterative methods are used to determine these eigenvalues.

Many of these approaches are designed for special types of matrices. In particular, a variety of techniques are devoted to symmetric systems. For example, *Jacobi's method* transforms a symmetric matrix to a diagonal matrix by eliminating off-diagonal terms in a systematic fashion. Unfortunately, the method requires an infinite number of operations because the removal of each nonzero element often creates a new nonzero value at a previous zero element. Although an infinite time is required to create all nonzero off-diagonal elements, the matrix will eventually tend toward a diagonal form. Thus, the approach is iterative in that it is repeated until the off-diagonal terms are "sufficiently" small.

Given's method also involves transforming a symmetric matrix into a simpler form. However, in contrast to the Jacobi method, the simpler form is tridiagonal. In addition, it differs in that the zeros that are created in off-diagonal positions are retained. Consequently, it is finite and, thus, more efficient than Jacobi.

Householder's method also transforms a symmetric matrix into a tridiagonal form. It is a finite method and is more efficient than Given's approach in that it reduces whole rows and columns of off-diagonal elements to zero.

Once a tridiagonal system is obtained from Given's or Householder's method, the remaining step involves finding the eigenvalues. A direct way to do this is to expand the determinant. The result is a sequence of polynomials that can be evaluated iteratively for the eigenvalues.

Aside from symmetric matrices, there are also techniques that are available when all eigenvalues of a general matrix are required. These include the *LR method* of Rutishauser and the *QR method* of Francis. Although the QR method is less efficient, it is usually the preferred approach because it is more stable. As such, it is considered to be the best general-purpose solution method.

Finally, it should be mentioned that the aforementioned techniques are often used in tandem to capitalize on their respective strengths. For example, it should be noted that Given's and Householder's methods can also be applied to nonsymmetric systems. The result will not be tridiagonal but rather a special type called the *Hessenberg form*. One approach is to exploit the speed of Householder's approach by employing it to transform the matrix to this form and then use the stable QR algorithm to find the eigenvalues. Additional information on these and other issues related to eigenvalues can be found in Ralston and Rabinowitz (1978), Wilkinson (1965), Fadeev and Fadeeva (1963), and Householder (1953). Computer codes can be found in a number of sources including Press et al. (1992). Rice (1983) discusses available software packages.

27.3 ODES AND EIGENVALUES WITH LIBRARIES AND PACKAGES

Software libraries and packages have great capabilities for solving ODEs and determining eigenvalues. This section outlines some of the ways in which they can be applied for this purpose.

27.3.1 Excel

Excel's direct capabilities for solving eigenvalue problems and ODEs is limited. However, if some programming is done (e.g., macros), they can be combined with Excel's visualization and optimization tools to implement some interesting applications. Section 28.1 provides an example of how the Excel Solver can be used for parameter estimation of an ODE.

27.3.2 Mathcad

Mathcad has a number of different functions that determine eigenvalues and eigenvectors and solve differential equations. Let's do an example by solving a system of stiff differential equations so we can compare the performance of some of the Mathcad functions. The

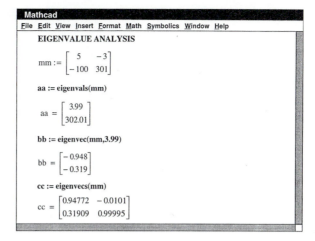

FIGURE 27.10
Mathcad screen to solve for the eigenvalues of a system of ODEs.

system is given by [recall Eq. (26.6)]

$$\frac{dy_1}{dt} = -5y_1 + 3y_2 \tag{27.33a}$$

$$\frac{dy_2}{dt} = 100y_1 - 301y_2 \tag{27.33b}$$

First, let's use Mathcad to analyze the behavior of this system by examining the eigenvalues. The function **eigenvecs(M)** returns a matrix containing normalized eigenvectors corresponding to the eigenvectors of the square matrix **M**. The functions **genvals(M,N)** and **genvecs(M,N)** produce eigenvalues and normalized eigenvectors, respectively.

The results of applying these functions to the ODEs is shown in Fig. 27.10. Because the eigenvalues (aa) are of different magnitudes, the system is stiff. Note that bb finds the specific eigenvector associated with the smaller eigenvalue, 3.99. The result cc is a matrix containing both eigenvectors as its columns.

The most basic technique employed by Mathcad to solve systems of first-order differential equations is a fixed step-size fourth-order Runge Kutta algorithm. This is provided by the **rkfixed** function. Although this is a good all-purpose integrator, it is not always efficient. Therefore, Mathcad supplies **Rkadapt,** which is a variable step sized version of **rkfixed.** It is well suited for functions that change rapidly in some regions and slowly in others. Similarly, if you know your solution is a smooth function, then you may find that the Mathcad **Bulstoer** function works well. This function employs the Bulirsch-Stoer method and is often both efficient and highly accurate for smooth functions.

Stiff differential equations are at the opposite end of the spectrum. Under these conditions the **rkfixed** function may be very inefficient or unstable. Therefore Mathcad provides two special methods specifically designed to handle stiff systems. These functions are called **Stiffb** and **Stiffr** and are based on a modified Bulirsch-Stoer method for stiff systems

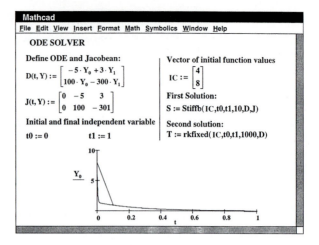

FIGURE 27.11
Mathcad screen to solve a system of ODEs.

and the Rosenbrock method. The above methods return solution values over a number of uniformly spaced intervals bounded by the integration range.

The Mathcad solution of the differential equations is shown in Fig. 27.11. First, the definition symbol is used to define the vector D(t,Y). This is the right-hand sides of the ODEs for input to **rkfixed** and **Stiffb.** Note that y_1 and y_2 in Eq. (27.33a and b) are changed to Y_0 and Y_1 to comply with Mathcad requirements. In addition, we define the J matrix for **Stiffb.** The first column of J is the derivative of the right-hand sides of Eq. (27.33) with respect to t and the Jacobian occupies the remaining four elements. The next inputs are the range for t and the initial conditions for Y. The solutions for **rkfixed** with 1000 steps between t0 and t1 and **Stiffb** with only 10 steps are stored in the T and S matrices.

The solutions for y_1 (Y_0) for both integration methods are compared in Fig. 27.11. Note that we have simplified the presentation of the plot in Fig. 27.11. The plot of the **Stiffb** solution is generated using the first and second columns of the S matrix, while the **rkfixed** solution plot is created using the first and second columns of the T matrix.

The **Stiffb** solution is stable and jumps over the details of the highly transient portion of the solution with a large step size and continues the solution efficiently to the end of the range. The **rkfixed** function solution with the same 10 steps is unstable. To maintain stability, **rkfixed** must employ a smaller step size to comply with the requirements of the larger eigenvalue. The solution in Fig. 27.11 uses 1000 steps and follows the details of the solution throughout the range of t. This is very inefficient because once past the initial transient, the solution changes gradually.

27.3.3 MATLAB

As might be expected the standard MATLAB package has excellent capabilities for determining eigenvalues and eigenvectors. However, it also has built-in functions for solving ODEs. The standard ODE solvers include two functions to implement the adaptive step-size Runge-Kutta Fehlberg method (recall Sec. 25.5.2). These are **ODE23,** which uses second- and third-order formulas to attain medium accuracy, and **ODE45,** which uses fourth- and fifth-order formulas to attain higher accuracy. The following example illustrates how they can be used to solve a system of ODEs.

EXAMPLE 27.11   Using MATLAB for Eigenvalues and ODEs

Problem Statement.   Explore how MATLAB can be used to solve the following set of nonlinear ODEs from $t = 0$ to 20:

$$\frac{dx}{dt} = 1.2x - 0.6xy \qquad \frac{dy}{dt} = -0.8y + 0.3xy$$

where $x = 2$ and $y = 1$ at $t = 0$. As we will see in the next chapter (Sec. 28.2), such equations are referred to as *predator-prey equations.*

Solution.   Before obtaining a solution with MATLAB, you must use a text processor to create an M-file containing the right-hand side of the ODEs. This M-file will then be accessed by the ODE solver [where $x = y(1)$ and $y = y(2)$]:

```
function yp = predprey(t,y)
yp = [1.2*y(1)-0.6*y(1)*y(2);-0.8*y(1)+0.3*y(1)*y(2)];
```

We stored this M-file under the name: predprey.m.

Next, start up MATLAB, and enter the following commands to specify the integration range and the initial conditions:

```
>> tspan = [0,20]';
>> y0=[2,1]';
```

The solver can then be invoked by

```
>> [t,y]=ode23('predprey',tspan,y0);
```

This command will then solve the differential equations in predprey.m over the range defined by tspan using the initial conditions found in y0. The results can be displayed by simply typing

```
>> plot(t,y)
```

which yields Fig. 27.12.

FIGURE 27.12
Solution of predator-prey model with MATLAB.

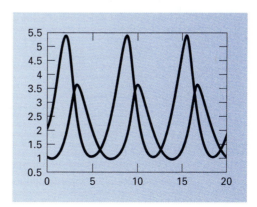

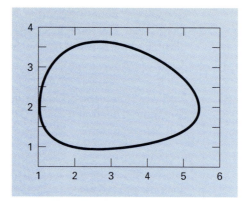

FIGURE 27.13
State-space plot of predator-prey model with MATLAB.

In addition, it is also instructive to generate a state-space plot; i.e., a plot of the dependent variables versus each other by

```
>> plot(y(:,1),y(:,2))
```

which yields Fig. 27.13.

For eigenvalues, the capabilities are also very easy to apply. Recall that, in our discussion of stiff systems in Chap. 26, we presented the stiff system defined by Eq. (27.33). Such linear ODEs can be written as an eigenvalue problem of the form

$$\begin{bmatrix} 5-\lambda & -3 \\ -100 & 301-\lambda \end{bmatrix} \begin{Bmatrix} e_1 \\ e_2 \end{Bmatrix} = \{0\}$$

where λ and $\{e\}$ = the eigenvalue and eigenvector, respectively.

MATLAB can then be employed to evaluate both the eigenvalues (d) and eigenvectors (v) with the following simple commands:

```
>> a=[5 -3;-100 301];
>> [v,d]=eig(a)

v =
    -0.9477     0.0101
    -0.3191    -0.9999

d =
     3.9899            0
          0     302.0101
```

Thus, we see that the eigenvalues are of quite different magnitudes, which is typical of a stiff system.

The eigenvalues can be interpreted by recognizing that the general solution for a system of ODEs can be represented as the sum of exponentials. For example, the solution for the present case would be of the form

$$y_1 = c_{11}e^{-3.9899t} + c_{12}e^{-302.0101t}$$
$$y_2 = c_{21}e^{-3.9899t} + c_{22}e^{-302.0101t}$$

where $c_{ij} =$ the part of the initial condition for y_i that is associated with the jth eigenvalue. It should be noted that the c's can be evaluated from the initial conditions and the eigenvectors. Any good book on differential equations, e.g., Boyce and DiPrima (1992), will provide an explanation of how this can be done.

Because, for the present case, all the eigenvalues are positive (and hence negative in the exponential function), the solution consists of a series of decaying exponentials. The one with the largest eigenvalue (in this case, 302.0101) would dictate the step size if an explicit solution technique were used.

27.3.4 IMSL

IMSL has a variety of routines for solving ODEs and determining eigenvalues (Table 27.3). In the present discussion, we will focus on the IVPRK routine. This routine integrates a system of ODEs using the Runge-Kutta method.

IVPRK is implemented by the following CALL statement:

```
CALL IVPRK (IDO, N, FCN, T, TEND, TOL, PARAM, Y)
```

TABLE 27.3 IMSL routines to solve ODEs and to determine eigenvalues.

Category	Routines	Capability
First-Order Ordinary Differential Equations		
Solution of initial-value problem for ODEs	IVPRK	Runge-Kutta method
	IVPAG	Adams or Gear method
Solution of boundary-value problem for ODEs	BVPFD	Finite-difference method
	BVPMS	Multiple-shooting method
Solution of differential-algebraic systems		
Eigenvalues and (Optionally) Eigenvectors of $Ax = \lambda x$		
Real general problem $Ax = \lambda x$	EVLRG	All eigenvalues
	EVCRG	All eigenvalues and eigenvectors
	EPIRG	Performance index
Complex general problem $Ax = \lambda x$		
Real symmetric problem $Ax = \lambda x$		
Real band symmetric matrices in band storage mode		
Complex Hermitian matrices		
Real upper Hessenberg matrices		
Complex upper Hessenberg matrices		
Eigenvalues and (optionally) eigenvectors of $Ax = \lambda Bx$		
Real general problem $Ax = \lambda Bx$		
Complex general problem $Ax = \lambda Bx$		
Real symmetric problem $Ax = \lambda Bx$		

where IDO = Flag indicating the state of the computation. Normally, the initial call is made with IDO = 1. The routine then sets IDO = 2, and this value is used for all but the last call that is made with IDO = 3. This final call is used to release workspace, which was automatically allocated by the initial call with IDO = 1. No integration is performed on this final call.

N = Number of differential equations. (Input)

FCN = User-supplied SUBROUTINE to evaluate functions.

T = Independent variable. (Input/Output) On input, T contains the initial value. On output, T is replaced by TEND unless error conditions have occurred.

TEND = Value of t where the solution is required. (Input) The value TEND may be less than the initial value of t.

TOL = Tolerance for error control. (Input) An attempt is made to control the norm of the local error such that the global error is proportional to TOL.

PARAM = A floating-point array of size 50 containing optional parameters.

Y = Array of size N of dependent variables. (Input/Output) On input, Y contains the initial values. On output, Y contains the approximate solution.

The subroutine FCN must be written that holds the differential equations. It should be of the general form,

```
subroutine fcn (n, t, y, yprime)
integer     n
real        t, y(n), yprime(n)
yprime(1) = . . .
yprime(2) = . . .
return
end
```

where the line "yprime(i) = . . ." is where the ith ODE is written. FCN must be declared EXTERNAL in the calling program.

EXAMPLE 27.12   Using IMSL to Solve ODEs

Problem Statement.   Use IVPRK to solve the same predator-prey ODEs as in Example 27.11.

Solution.   An example of a main Fortran 90 program and function using IVPRK to solve this problem can be written as

```
Program PredPrey
USE msimsl
INTEGER :: mxparm, n
PARAMETER (mxparm=50, n=2)
INTEGER :: ido, istep, nout
REAL :: param(mxparm), t, tend, tol, y(n)
EXTERNAL fcn
CALL UMACH (2, nout)
t = 0.0
y(1) = 2.0
y(2) = 1.0
tol = 0.0005
```

```
CALL SSET (mxparm, 0.0, param, 1)
param(10) = 1.0
PRINT '(4X, ''ISTEP'', 5X, ''Time'', 9X, ''Y1'', 11X, ''Y2'')'
ido = 1
istep = 0
WRITE (nout,'(I6,3F12.3)') istep, t, y
DO
  istep = istep + 1
  tend = istep
  CALL IVPRK (ido, n, fcn, t, tend, tol, param, y)
  IF (istep .LE. 10) EXIT
  WRITE (nout,'(I6,3F12.3)') istep, t, y
  IF (istep .EQ. 10) ido = 3
END DO
END PROGRAM

SUBROUTINE fcn (n, t, y, yprime)
IMPLICIT NONE
INTEGER :: n
REAL :: t, y(n), yprime(n)
yprime(1) = 1.2*y(1) - 0.6*y(1)*y(2)
yprime(2) = -0.8*y(2) + 0.3*y(1)*y(2)
END SUBROUTINE
```

An example run is

```
istep       time          y1          y2
    0        .000       2.000       1.000
    1       1.000       3.703       1.031
    2       2.000       5.433       1.905
    3       3.000       3.390       3.533
    4       4.000       1.407       3.073
    5       5.000       1.048       1.951
    6       6.000       1.367       1.241
    7       7.000       2.393        .959
    8       8.000       4.344       1.161
    9       9.000       5.287       2.421
   10      10.000       2.561       3.624
```

PROBLEMS

27.1 A steady-state heat balance for a rod can be represented as

$$\frac{d^2T}{dx^2} - 0.1T = 0$$

Obtain an analytical solution for a 10-m rod with $T(0) = 200$, and $T(10) = 100$.

27.2 Use the shooting method to solve Prob. 27.1.

27.3 Use the finite-difference approach with $\Delta x = 1$ to solve Prob. 27.1.

27.4 Use the shooting method to solve

$$8\frac{d^2y}{dx^2} - 2\frac{dy}{dx} - y + x = 0$$

with the boundary conditions $y(0) = 5$ and $y(20) = 8$.

27.5 Solve Prob. 27.4 with the finite-difference approach using $\Delta x = 2$.

27.6 Use the shooting method to solve

$$\frac{d^2T}{dx^2} - 1.2 \times 10^{-7}(T + 273)^4 + 5(150 - T) = 0 \qquad \text{(P27.6)}$$

Obtain a solution for boundary conditions: $T(0) = 200$ and $T(0.5) = 100$.

27.7 Differential equations like the one solved in Prob. 27.6 can often be simplified by linearizing their nonlinear terms. For example, a first-order Taylor series expansion can be used to linearize the quartic term in Eq. (P27.6) as

$$1.2 \times 10^{-7}(T + 273)^4 = 1.2 \times 10^{-7}(T_b + 273)^4 + 4.8$$
$$\times 10^{-7}(T_b + 273)^3(T - T_b)$$

where T_b is a base temperature about which the term is linearized. Substitute this relationship into Eq. (P27.6) and then solve the resulting linear equation with the finite-difference approach. Employ $T_b = 150$ and $\Delta x = 0.01$ to obtain your solution.

27.8 Repeat Example 27.4, but for three masses. Produce a plot like Fig. 27.6 to identify the principle modes of vibration. Change all the k's to 240.

27.9 Repeat Example 27.6, but for five interior points ($h = 3/6$).

27.10 Use minors to expand the determinant of

$$\begin{bmatrix} 2 - \lambda & 2 & 10 \\ 8 & 3 - \lambda & 4 \\ 10 & 4 & 5 - \lambda \end{bmatrix}$$

Employ the Fadeev-Leverrier method to perform the same computation. Also, compute the matrix inverse and verify that it is correct.

27.11 Use the power method to determine the highest eigenvalue and corresponding eigenvector for Prob. 27.10.

27.12 Use the power method to determine the lowest eigenvalue and corresponding eigenvector for Prob. 27.10.

27.13 Develop a user-friendly computer program to implement the shooting method for a linear second-order ODE. Test the program by duplicating Example 27.1.

27.14 Use the program developed in Prob. 27.13 to solve Probs. 27.2 and 27.4.

27.15 Develop a user-friendly computer program to implement the finite-difference approach for solving a linear second-order ODE. Test it by duplicating Example 27.2.

27.16 Use the program developed in Prob. 27.15 to solve Probs. 27.3 and 27.5.

27.17 Develop a subroutine to implement the Fadeev-Leverrier approach. Test it by duplicating Example 27.7.

27.18 Use the subroutine developed in Prob. 27.17 to solve Prob. 27.10.

27.19 Develop a user-friendly program to implement the polynomial method. Test it by duplicating Example 27.6c.

27.20 Develop a user-friendly program to solve for the largest eigenvalue with the power method. Test it by duplicating Example 27.9.

27.21 Develop a user-friendly program to solve for the smallest eigenvalue with the power method. Test it by duplicating Example 27.10.

27.22 Use the Excel Solver to directly solve (that is, without linearization) Prob. 27.6 using the finite-difference approach. As in Prob. 27.7, employ $\Delta x = 0.1$ to obtain your solution.

27.23 Use Mathcad to duplicate Example 26.1.

27.24 Use Mathcad to determine the eigenvalues and eigenvectors for the matrix in Prob. 27.10.

27.25 Use MATLAB to integrate the following pair of ODEs:

$$\frac{dy_1}{dt} = 0.3y_1 - 1.5y_1y_2 \qquad \frac{dy_2}{dt} = 0.036y_1y_2 - 0.1y_2$$

where $y_1 = 1$ and $y_2 = 0.05$ at $t = 0$. Develop a state-space plot of your results.

27.26 The following differential equation was used in Sec. 8.4 to analyze the vibrations of an automobile shock absorber:

$$1.2 \times 10^6 \frac{d^2x}{dt^2} + 1 \times 10^7 \frac{dx}{dt} + 1.4 \times 10^9 x = 0$$

Transform this equation into a pair of ODEs. **(a)** Use MATLAB to solve these equations from $t = 0$ to 0.4 for the case where $x = 0.3$ and $dx/dt = 0$ at $t = 0$. **(b)** Use MATLAB to determine the eigenvalues and eigenvectors for the system.

27.27 Use IMSL to integrate

(a) $\dfrac{dx}{dt} = ax - bxy \qquad \dfrac{dy}{dt} = -cy + dxy$

where $a = 1.2$, $b = 0.6$, $c = 0.8$, and $d = 0.3$. Employ initial conditions of $x = 2$ and $y = 1$ and integrate from $t = 0$ to 30.

(b) $\dfrac{dx}{dt} = -\sigma x + \sigma y \qquad \dfrac{dy}{dt} = rx - y - xz \qquad \dfrac{dz}{dt} = -bz + xy$

where $\sigma = 10$, $b = 2.666667$, and $r = 28$. Employ initial conditions of $x = y = z = 5$ and integrate from $t = 0$ to 20.

CHAPTER 28

Engineering Applications: Ordinary Differential Equations

The purpose of this chapter is to solve some ordinary differential equations using the numerical methods presented in Part Seven. The equations originate from practical engineering applications. Many of these applications result in nonlinear differential equations that cannot be solved using analytic techniques. Therefore, numerical methods are usually required. Thus, the techniques for the numerical solution of ordinary differential equations are fundamental capabilities that characterize good engineering practice. The problems in this chapter illustrate some of the trade-offs associated with various methods developed in Part Seven.

Section 28.1 derives from a chemical engineering problem context. It demonstrates how the transient behavior of chemical reactors can be simulated. It also illustrates how optimization can be used to estimate parameters for ODEs.

Sections 28.2 and 28.3, which are taken from civil and electrical engineering, respectively, deal with the solution of systems of equations. In both cases, high accuracy is demanded, and as a consequence, a fourth-order RK scheme is used. In addition, the electrical engineering application also deals with determining eigenvalues.

Section 28.4 employs a variety of different approaches to investigate the behavior of a swinging pendulum. This problem also utilizes two simultaneous equations. An important aspect of this example is that it illustrates how numerical methods allow nonlinear effects to be incorporated easily into an engineering analysis.

28.1 USING ODES TO ANALYZE THE TRANSIENT RESPONSE OF A REACTOR (CHEMICAL/PETROLEUM ENGINEERING)

Background.   In Sec. 12.2, we analyzed the steady state of a series of reactors. In addition to steady-state computations, we might also be interested in the transient response of a completely mixed reactor. To do this, we have to develop a mathematical expression for the accumulation term in Eq. (12.1).

Accumulation represents the change in mass in the reactor per change in time. For a constant-volume system, it can be simply formulated as

$$\text{Accumulation} = V\frac{dc}{dt} \tag{28.1}$$

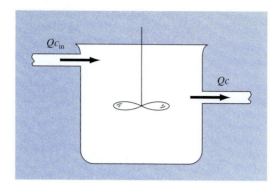

FIGURE 28.1
A single, completely mixed reactor with an inflow and an outflow.

where V = volume and c = concentration. Thus, a mathematical formulation for accumulation is volume times the derivative of c with respect to t.

In this application we will incorporate the accumulation term into the general mass-balance framework we developed in Sec. 12.1. We will then use it to simulate the dynamics of a single reactor and a system of reactors. In the latter case, we will show how the system's eigenvalues can be determined and provide insight into its dynamics. Finally, we will illustrate how optimization can be used to estimate the parameters of mass-balance models.

Solution.   Equations (28.1) and (12.1) can be used to represent the mass balance for a single reactor such as the one shown in Fig. 28.1:

$$V \frac{dc}{dt} = Q c_{in} - Q c \tag{28.2}$$

Accumulation = inputs − outputs

Equation (28.2) can be used to determine transient or time-variable solutions for the reactor. For example, if $c = c_0$ at $t = 0$, calculus can be employed to analytically solve Eq. (28.2) for

$$c = c_{in}\left(1 - e^{-(Q/V)t}\right) + c_0 e^{-(Q/V)t}$$

If $c_{in} = 50$ mg/m^3, $Q = 5$ m^3/min, $V = 100$ m^3, and $c_0 = 10$ mg/m^3, the equation is

$$c = 50\left(1 - e^{-0.05t}\right) + 10 e^{-0.05t}$$

Figure 28.2 shows this exact, analytical solution.

Euler's method provides an alternative approach for solving Eq. (28.2). Figure 28.2 includes two solutions with different step sizes. As the step size is decreased, the numerical solution converges on the analytical solution. Thus, for this case, the numerical method can be used to check the analytical result.

Besides checking the results of an analytical solution, numerical solutions have added value in those situations where analytical solutions are impossible or so difficult that they are impractical. For example, aside from a single reactor, numerical methods have utility

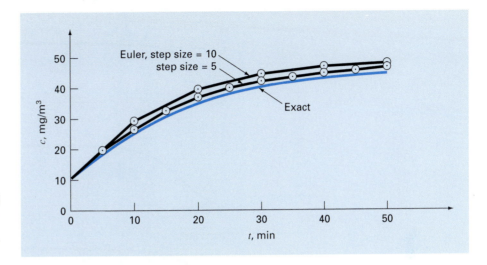

FIGURE 28.2
Plot of analytical and numerical solutions of Eq. (28.2). The numerical solutions are obtained with Euler's method using different step sizes.

when simulating the dynamics of systems of reactors. For example, ODEs can be written for the five coupled reactors in Fig. 12.3. The mass balance for the first reactor can be written as

$$V_1 \frac{dc_1}{dt} = Q_{01}c_{01} + Q_{31}c_3 - Q_{12}c_1 - Q_{15}c_1$$

or, substituting parameters (note that $Q_{01}c_{01} = 50$ mg/min, $Q_{03}c_{03} = 160$ mg/min, $V_1 = 50$ m^3, $V_2 = 20$ m^3, $V_3 = 40$ m^3, $V_4 = 80$ m^3, and $V_5 = 100$ m^3),

$$\frac{dc_1}{dt} = -0.12c_1 + 0.02c_3 + 1$$

Similarly, balances can be developed for the other reactors as

$$\frac{dc_2}{dt} = 0.15c_1 - 0.15c_2$$

$$\frac{dc_3}{dt} = 0.025c_2 - 0.225c_3 + 4$$

$$\frac{dc_4}{dt} = 0.1c_3 - 0.1375c_4 + 0.025c_5$$

$$\frac{dc_5}{dt} = 0.03c_1 + 0.01c_2 - 0.04c_5$$

Suppose that at $t = 0$ all the concentrations in the reactors are at zero. Compute how their concentrations will increase over the next hour.

The equations can be integrated with the fourth-order RK method for systems of equations and the results are depicted in Fig. 28.3. Notice that each of the reactors shows a different transient response to the introduction of chemical. These responses can be parameterized

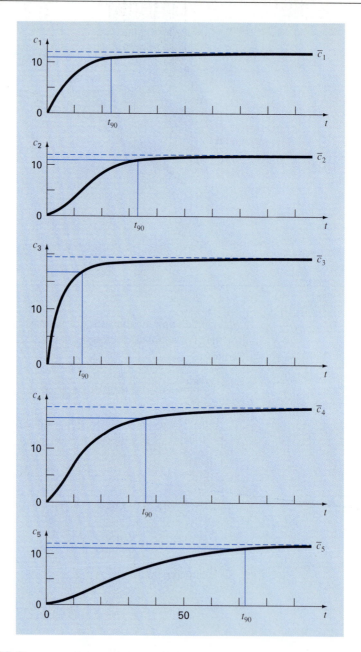

FIGURE 28.3

Plots of transient or dynamic response of the network of reactors from Fig. 12.3. Note that all the reactors eventually approach their steady-state concentrations previously computed in Sec. 12.1. In addition, the time to steady state is parameterized by the 90 percent response time t_{90}.

by a 90 percent response time t_{90}, which measures the time required for each reactor to reach 90 percent of its ultimate steady-state level. The times range from about 10 min for reactor 3 to about 70 min for reactor 5. The response times of reactors 4 and 5 are of particular concern because the two outflow streams for the system exit these tanks. Thus, a chemical engineer designing the system might change the flows or volumes of the reactors to speed up the response of these tanks while still maintaining the desired outputs. Numerical methods of the sort described in this part of the book can prove useful in these design calculations.

Further insight into the system's response characteristics can be developed by computing its eigenvalues. First, the system of ODEs can be written as an eigenvalue problem as

$$
\begin{bmatrix}
0.12 - \lambda & 0 & -0.02 & 0 & 0 \\
-0.15 & 0.15 - \lambda & 0 & 0 & 0 \\
0 & -0.025 & 0.225 - \lambda & 0 & 0 \\
0 & 0 & -0.1 & 0.1375 - \lambda & -0.025 \\
-0.03 & -0.01 & 0 & 0 & 0.04 - \lambda
\end{bmatrix}
\begin{Bmatrix}
e_1 \\ e_2 \\ e_3 \\ e_4 \\ e_5
\end{Bmatrix} = \{0\}
$$

where λ and $\{e\}$ = the eigenvalue and the eigenvector, respectively.

A package like MATLAB can be used to very conveniently generate the eigenvalues and eigenvectors,

```
>> a=[0.12 0.0 -0.02 0.0 0.0;-.15 0.15 0.0 0.0 0.0;0.0
-0.025 0.225 0.0 0.0; 0.0 0.0 -.1 0.1375 -0.025;-0.03 -0.01
0.0 0.0 0.04];

>> [e,l]=eig(a)
e =
              0           0    -0.1228    -0.1059     0.2490
              0           0     0.2983     0.5784     0.8444
              0           0     0.5637     0.3041     0.1771
         1.0000      0.2484    -0.7604    -0.7493     0.3675
              0      0.9687     0.0041    -0.0190    -0.2419

l =
         0.1375           0          0          0          0
              0      0.0400          0          0          0
              0           0     0.2118          0          0
              0           0          0     0.1775          0
              0           0          0          0     0.1058
```

The eigenvalues can be interpreted by recognizing that the general solution for a system of ODEs can be represented as the sum of exponentials. For example, for reactor 1, the general solution would be of the form

$$
c_1 = c_{11}e^{-\lambda_1 t} + c_{12}e^{-\lambda_2 t} + c_{13}e^{-\lambda_3 t} + c_{14}e^{-\lambda_4 t} + c_{15}e^{-\lambda_5 t}
$$

where c_{ij} = the part of the initial condition for reactor i that is associated with the jth eigenvalue. Thus, because, for the present case, all the eigenvalues are positive (and hence negative in the exponential function), the solution consists of a series of decaying exponentials. The one with the smallest eigenvalue (in our case, 0.04) will be the slowest. In some cases, the engineer performing this analysis could be able to relate this eigenvalue

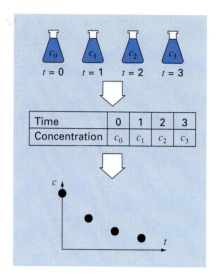

FIGURE 28.4
A simple experiment to collect rate data for a chemical compound that decays with time (reprinted from Chapra 1997).

back to the system parameters. For example, the ratio of the outflow from reactor 5 to its volume is $(Q_{55} + Q_{54})/V_5 = 4/100 = 0.04$. Such information can then be used to modify the system's dynamic performance.

The final topic we'd like to review within the present context is *parameter estimation*. One area where this occurs frequently is in *reaction kinetics;* that is, the quantification of chemical reaction rates.

A simple example is depicted in Fig. 28.4. A series of beakers are set up containing a chemical compound that decays over time. At time intervals, the concentration in one of the beakers is measured and recorded. Thus, the result is a table of times and concentrations.

One model that is commonly used to describe such data is

$$\frac{dc}{dt} = -kc^n \tag{28.3}$$

where $k = $ a reaction rate and $n = $ the order of the reaction. Chemical engineers use concentration-time data of the sort depicted in Fig. 28.4 to estimate k and n. One way to do this is to guess values of the parameters and then solve Eq. (28.3) numerically. The predicted values of concentration can be compared with the measured concentrations and an assessment of the fit made. If the fit is deemed inadequate (for example, by examining a plot or a statistical measure like the sum of the squares of the residuals), the guesses are adjusted and the procedure repeated until a decent fit is attained.

The following data can be fit in this fashion:

t, d	0	1	3	5	10	15	20
c, mg/L	12	10.7	9	7.1	4.6	2.5	1.8

	A	B	C	D	E	F	G	H
1	Fitting of reaction rate							
2	data with the integral/least-squares approach							
3	k	0.091528						
4	n	1.044425						
5	dt	1						
6	t	k1	k2	k3	k4	cp	cm	(cp-cm)^2
7	0	−1.22653	−1.16114	−1.16462	−1.10248	12	12	0
8	1	−1.10261	−1.04409	−1.04719	−0.99157	10.83658	10.7	0.018653
9	2	−0.99169	−0.93929	−0.94206	−0.89225	9.790448		
10	3	−0.89235	−0.84541	−0.84788	−0.80325	8.849344	9	0.022697
11	4	−0.80334	−0.76127	−0.76347	−0.72346	8.002317		
12	5	−0.72354	−0.68582	−0.68779	−0.65191	7.239604	7.1	0.019489
13	6	−0.65198	−0.61814	−0.61989	−0.5877	6.552494		
14	7	−0.58776	−0.55739	−0.55895	−0.53005	5.933207		
15	8	−0.53011	−0.50283	−0.50424	−0.47828	5.374791		
16	9	−0.47833	−0.45383	−0.45508	−0.43175	4.871037		
17	10	−0.4318	−0.40978	−0.4109	−0.38993	4.416389	4.6	0.033713
18	11	−0.38997	−0.37016	−0.37117	−0.35231	4.005877		
19	12	−0.35234	−0.33453	−0.33543	−0.31846	3.635053		
20	13	−0.31849	−0.30246	−0.30326	−0.28798	3.299934		
21	14	−0.28801	−0.27357	−0.2743	−0.26054	2.996949		
22	15	−0.26056	−0.24756	−0.24821	−0.23581	2.7229	2.5	0.049684
23	16	−0.23583	−0.22411	−0.22469	−0.21352	2.474917		
24	17	−0.21354	−0.20297	−0.20349	−0.19341	2.250426		
25	18	−0.19343	−0.18389	−0.18436	−0.17527	2.047117		
26	19	−0.17529	−0.16668	−0.16711	−0.1589	1.862914		
27	20	−0.15891	−0.15115	−0.15153	−0.14412	1.695953	1.8	0.010826
28								
29							SSR =	0.155062

FIGURE 28.5
The application of a spreadsheet and numerical methods to determine the order and rate coefficient of reaction data. This application was performed with the Excel spreadsheet.

The solution to this problem is shown in Fig. 28.5. The Excel spreadsheet was used to perform the computation.

Initial guesses for the reaction rate and order are entered into cells B3 and B4, respectively, and the time step for the numerical calculation is typed into cell B5. For this case, a column of calculation times is entered into column A starting at 0 (cell A7) and ending at 20 (cell A27). The k_1 through k_4 coefficients of the fourth-order RK method are then calculated in the block B7..E27. These are then used to determine the predicted concentrations (the c_p values) in column F. The measured values (c_m) are entered in column G adjacent to the corresponding predicted values. These are then used in conjunction with the predicted values to compute the squared residual in column H. These values are then summed in cell H29.

At this point, the Excel Solver can be used to determine the best parameter values. Once you have accessed the Solver, you are prompted for a target or solution cell (H29), queried whether you want to maximize or minimize the target cell (minimize), and prompted for the

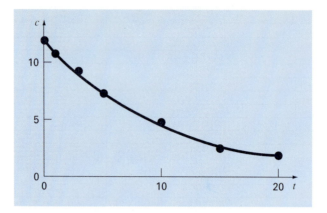

FIGURE 28.6
Plot of fit generated with the integral/least-squares approach.

cells that are to be varied (B3..B4). You then activate the algorithm [s(olve)], and the results are as in Fig. 28.5. As shown, the values in cells B3..B4 ($k = 0.0915$ and $n = 1.044$) minimize the sum of the squares of the residuals (SSR = 0.155) between the predicted and measured data. A plot of the fit along with the data is shown in Fig. 28.6.

28.2 PREDATOR-PREY MODELS AND CHAOS (CIVIL/ENVIRONMENTAL ENGINEERING)

Background. Environmental engineers deal with a variety of problems involving systems of nonlinear ordinary differential equations. In this section, we will focus on two of these applications. The first relates to the so-called predator-prey models that are used to study the cycling of nutrient and toxic pollutants in aquatic food chains and biological treatment systems. The second are equations derived from fluid dynamics that are used to simulate the atmosphere. Aside from their obvious application to weather prediction, such equations have also been used to study air pollution and global climate change.

Predator-prey models were developed independently in the early part of the twentieth century by the Italian mathematician Vito Volterra and the American biologist Alfred J. Lotka. These equations are commonly called *Lotka-Volterra equations*. The simplest example is the following pair of ODEs:

$$\frac{dx}{dt} = ax - bxy \tag{28.4}$$

$$\frac{dy}{dt} = -cy + dxy \tag{28.5}$$

where x and y = the number of prey and predators, respectively, a = the prey growth rate, c = the predator death rate, and b and d = the rate characterizing the effect of the predator-prey interaction on prey death and predator growth, respectively. The multiplicative terms (i.e., those involving xy) are what make such equations nonlinear.

An example of a simple model based on atmospheric fluid dynamics is the *Lorenz equations* developed by the American meteorologist Edward Lorenz,

$$\frac{dx}{dt} = -\sigma x + \sigma y \tag{28.6}$$

$$\frac{dy}{dt} = rx - y - xz \tag{28.7}$$

$$\frac{dz}{dt} = -bz + xy \tag{28.8}$$

Lorenz developed these equations to relate the intensity of atmospheric fluid motion, x, to temperature variations y and z in the horizontal and vertical directions, respectively. As with the predator-prey model, we see that the nonlinearity is localized in simple multiplicative terms (xz and xy).

Use numerical methods to obtain solutions for these equations. Plot the results to visualize how the dependent variables change temporally. In addition, plot the dependent variables versus each other to see whether any interesting patterns emerge.

Solution.   Use the following parameter values for the predator-prey simulation: $a = 1.2$, $b = 0.6$, $c = 0.8$, and $d = 0.3$. Employ initial conditions of $x = 2$ and $y = 1$ and integrate from $t = 0$ to 30. We will use the fourth-order RK method with double precision to obtain solutions.

The results using a step size of 0.1 are shown in Fig. 28.7. Note that a cyclical pattern emerges. Thus, because predator population is initially small, the prey grows exponentially. At a certain point, the prey become so numerous, that the predator population begins to grow. Eventually, the increased predators cause the prey to decline. This decrease, in turn, leads to a decrease of the predators. Eventually, the process repeats. Notice that, as expected, the predator peak lags the prey. Also, observe that the process has a fixed period; that is, it repeats in a set time.

Now, if the parameters used to simulate Fig. 28.7 were changed, although the general pattern would remain the same, the magnitudes of the peaks, lags, and period would change. Thus, there are an infinite number of cycles that could occur.

A state-space representation is useful in discerning the underlying structure of the model. Rather than plotting x and y versus t, we can plot x versus y. Because such a plot

FIGURE 28.7

Time-domain representation of numbers of prey and predators for the Lotka-Volterra model.

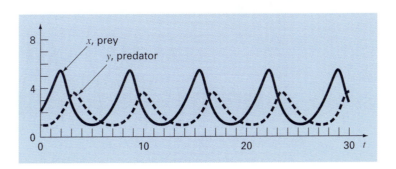

illustrates the way that the state variables (x and y) interact, it is referred to as a *state-space representation*.

Figure 28.8 shows the state-space representation for the case we are studying. Thus, the interaction between the predator and the prey defines a closed counterclockwise orbit. Notice that there is a critical or rest point at the center of the orbit. The exact location of this point can be determined by setting Eqs. (28.4) and (28.5) to steady state ($dy/dt = dx/dt = 0$) and solving for $(x, y) = (0, 0)$ and $(c/d, a/b)$. The former is the trivial result that if we start with neither predators nor prey, nothing will happen. The latter is the more interesting outcome that if the initial conditions are set at $x = c/d$ and $y = a/b$, the derivative will be zero and the populations will remain constant.

Now, let's use the same approach to investigate the trajectories of the Lorenz equations with the following parameter values: $\sigma = 10$, $b = 2.666667$, and $r = 28$. Employ initial conditions of $x = y = z = 5$ and integrate from $t = 0$ to 20. Again, we will use the fourth-order RK method with double precision to obtain solutions.

The results shown in Fig. 28.9 are quite different from the behavior of the Lotka-Volterra equations. The variable x seems to be undergoing an almost random pattern of oscillations, bouncing around from negative values to positive values. However, even though

FIGURE 28.8

State-space representation for the Lotka-Volterra model.

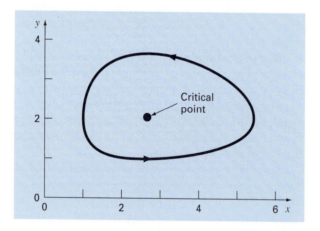

FIGURE 28.9

Time-domain representation of x versus t for the Lorenz equations. The solid time series is for the initial conditions (5, 5, 5). The dotted line is where the initial condition for x is perturbed slightly (5.001, 5, 5).

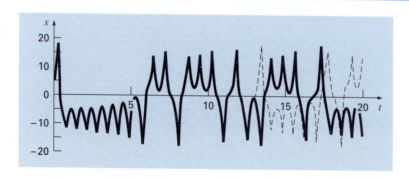

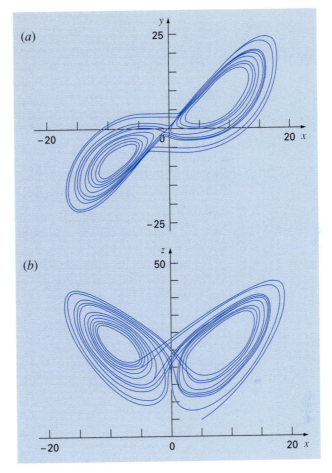

FIGURE 28.10
State-space representation for the Lorenz equations. (a) xy projection and (b) xz projection.

the patterns seem random, the frequency of the oscillation and the amplitudes seem fairly consistent.

Another interesting feature can be illustrated by changing the initial condition for x slightly (from 5 to 5.001). The results are superimposed as the dotted line in Fig. 28.9. Although the solutions track on each other for a time, after about $t = 12.5$ they diverge significantly. Thus, we can see that the Lorenz equations are quite sensitive to their initial conditions. In his original study, this led Lorenz to the conclusion that long-range weather forecasts might be impossible!

Finally, let's examine the state-space plots. Because we are dealing with three independent variables, we are limited to projections. Figure 28.10 shows projections in the xy and the xz planes. Notice how a structure is manifest when perceived from the state-space perspective. The solution forms orbits around what appear to be critical points. These points are called *strange attractors* in the jargon of mathematicians who study such non-linear systems.

Solutions such as the type we have explored for the Lorenz equations are referred to as *chaotic* solutions. The study of chaos and nonlinear systems presently represents an exciting area of analysis that has implications to mathematics as well as to science and engineering.

From a numerical perspective, the primary point is the sensitivity of such solutions to initial conditions. Thus, different numerical algorithms, computer precision, and integration time steps can all have an impact on the resulting numerical solution.

28.3 SIMULATING TRANSIENT CURRENT FOR AN ELECTRIC CIRCUIT (ELECTRICAL ENGINEERING)

Background.   Electric circuits where the current is time-variable rather than constant are common. A transient current is established in the right-hand loop of the circuit shown in Fig. 28.11 when the switch is suddenly closed.

Equations that describe the transient behavior of the circuit in Fig. 28.11 are based on Kirchhoff's law, which states that the algebraic sum of the voltage drops around a closed loop is zero (recall Sec. 8.3). Thus,

$$L\frac{di}{dt} + Ri + \frac{q}{c} - E(t) = 0 \tag{28.9}$$

where $L(di/dt)$ = voltage drop across the inductor, L = inductance (H), R = resistance (Ω), q = charge on the capacitor (C), C = capacitance (F), $E(t)$ = time-variable voltage source (V), and

$$i = \frac{dq}{dt} \tag{28.10}$$

Equations (28.9) and (28.10) are a pair of first-order linear differential equations that can be solved analytically. For example, if $E(t) = E_0 \sin \omega t$ and $R = 0$,

$$q(t) = \frac{-E_0}{L(p^2 - \omega^2)} \frac{\omega}{p} \sin \, pt + \frac{E_0}{L(p^2 - \omega^2)} \sin \omega t \tag{28.11}$$

FIGURE 28.11
An electric circuit where the current varies with time.

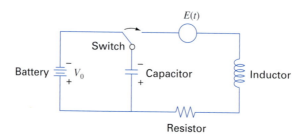

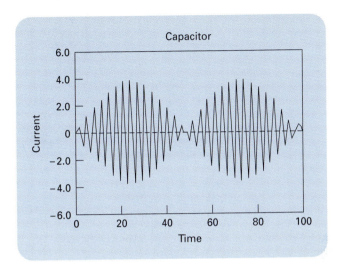

FIGURE 28.12
Computer screen showing the plot of the function represented by Eq. (28.32).

where $p = 1/\sqrt{LC}$. The values of q and dq/dt are zero for $t = 0$. Use a numerical approach to solve Eqs. (28.9) and (28.10) and compare the results with Eq. (28.11).

Solution.   This problem involves a rather long integration range and demands the use of a highly accurate scheme to solve the differential equation if good results are expected. Let's assume that $L = 1\,H$, $E_0 = 1$ V, $C = 0.25$ C, and $\omega^2 = 3.5\,s^2$. This gives $p = 2$, and Eq. (28.11) becomes

$$q(t) = -1.8708 \sin (2t) + 2 \sin (1.8708t)$$

for the analytical solution. This function is plotted in Fig. 28.12. The rapidly changing nature of the function places a severe requirement on any numerical procedure to find $q(t)$. Furthermore, because the function exhibits a slowly varying periodic nature as well as a rapidly varying component, long integration ranges are necessary to portray the solution. Thus, we expect that a high-order method is preferred for this problem.

However, we can try both Euler and fourth-order RK methods and compare the results. Using a step size of 0.1 s gives a value for q at $t = 10$ s of -6.638 with Euler's method and a value of -1.9897 with the fourth-order RK method. These results compare to an exact solution of -1.996 C.

Figure 28.13 shows the results of Euler integration every 1.0 s compared to the exact solution. Note that only every tenth output point is plotted. It is seen that the global error increases as t increases. This divergent behavior intensifies as t approaches infinity.

In addition to directly simulating a network's transient response, numerical methods can also be used to determine its eigenvalues. For example, Fig. 28.14 shows an LC network for which Kirchhoff's voltage law can be employed to develop the following system

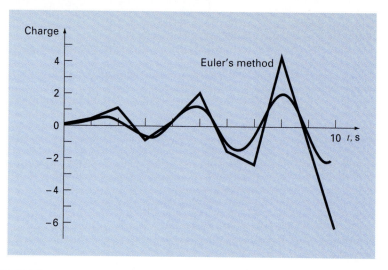

FIGURE 28.13
Results of Euler integration versus exact solution. Note that only every tenth output point is plotted.

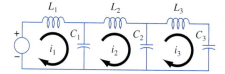

FIGURE 28.14
An *LC* network.

of ODEs:

$$-L_1 \frac{di_1}{dt} - \frac{1}{C_1} \int_{-\infty}^{t} (i_1 - i_2) \, dt = 0$$

$$-L_2 \frac{di_2}{dt} - \frac{1}{C_2} \int_{-\infty}^{t} (i_2 - i_3) \, dt + \frac{1}{C_1} \int_{-\infty}^{t} (i_1 - i_2) \, dt = 0$$

$$-L_3 \frac{di_3}{dt} - \frac{1}{C_3} \int_{-\infty}^{t} i_3 \, dt + \frac{1}{C_2} \int_{-\infty}^{t} (i_2 - i_3) \, dt = 0$$

Notice that we have represented the voltage drop across the capacitor as

$$V_C = \frac{1}{C} \int_{-\infty}^{t} i \, dt$$

This is an alternative and equivalent expression to the relationship used in Eq. (28.9) and introduced in Sec. 8.3.

The system of ODEs can be differentiated and rearranged to yield

$$L_1 \frac{d^2 i_1}{dt^2} + \frac{1}{C_1}(i_1 - i_2) = 0$$

$$L_2 \frac{d^2 i_2}{dt^2} + \frac{1}{C_2}(i_2 - i_3) - \frac{1}{C_1}(i_1 - i_2) = 0$$

$$L_3 \frac{d^2 i_3}{dt^2} + \frac{1}{C_3}i_3 - \frac{1}{C_2}(i_2 - i_3) = 0$$

Comparison of this system with the one in Eq. (27.5) indicates an analogy between a spring-mass system and an LC circuit. As was done with Eq. (27.5), the solution can be assumed to be of the form

$$i_j = A_j \sin(\omega t)$$

This solution along with its second derivative can be substituted into the simultaneous ODEs. After simplification, the result is

$$\left(\frac{1}{C_1} - L_1\omega^2\right) A_1 \qquad\qquad -\frac{1}{C_2}A_2 \qquad\qquad\qquad\qquad = 0$$

$$-\frac{1}{C_1}A_1 \qquad + \left(\frac{1}{C_1} + \frac{1}{C_2} - L_2\omega^2\right) A_2 \qquad -\frac{1}{C_2}A_3 \qquad = 0$$

$$-\frac{1}{C_2}A_2 \qquad + \left(\frac{1}{C_2} + \frac{1}{C_3} - L_3\omega^2\right) A_3 = 0$$

Thus, we have formulated an eigenvalue problem. Further simplification results for the special case where the C's and L's are constant. For this situation, the system can be expressed in matrix form as

$$\begin{bmatrix} 1-\lambda & -1 & 0 \\ -1 & 2-\lambda & -1 \\ 0 & -1 & 2-\lambda \end{bmatrix} \begin{Bmatrix} i_1 \\ i_2 \\ i_3 \end{Bmatrix} = \{0\} \tag{28.12}$$

where

$$\lambda = LC\omega^2 \tag{28.13}$$

Numerical methods can be employed to determine values for the eigenvalues and eigenvectors. MATLAB is particularly convenient in this regard. The following MATLAB session has been developed to do this:

```
>>a=[1 -1 0; -1 2 -1; 0 -1 2]

a =

        1    -1     0
       -1     2    -1
        0    -1     2
```

```
>>[v,d]=eig(a)

v =

        0.7370      0.5910      0.3280
        0.5910     -0.3280     -0.7370
        0.3280     -0.7370      0.5910

d =

        0.1981           0           0
             0      1.5550           0
             0           0      3.2470
```

The matrix v consists of the system's three eigenvectors (arranged as columns), and d is a matrix with the corresponding eigenvalues on the diagonal. Thus, the package computes that the eigenvalues are $\lambda = 0.1981, 1.555$, and 3.247. These values in turn can be substituted into Eq. (28.13) to solve for the natural circular frequencies of the system

$$
\omega = \left\{
\begin{array}{c}
\dfrac{0.4451}{\sqrt{LC}} \\[2mm]
\dfrac{1.2470}{\sqrt{LC}} \\[2mm]
\dfrac{1.8019}{\sqrt{LC}}
\end{array}
\right.
$$

Aside from providing the natural frequencies, the eigenvalues can be substituted into Eq. (28.12) to gain further insight into the circuit's physical behavior. For example, substituting $\lambda = 0.1981$ yields

$$
\begin{bmatrix}
0.8019 & -1 & 0 \\
-1 & 1.8019 & -1 \\
0 & -1 & 1.8019
\end{bmatrix}
\begin{Bmatrix} i_1 \\ i_2 \\ i_3 \end{Bmatrix} = \{0\}
$$

Although this system does not have a unque solution, it will be satisfied if the currents are in fixed ratios, as in

$$0.8019i_1 = i_2 = 1.8019i_3 \tag{28.14}$$

Thus, as depicted in Fig. 28.15a, they oscillate in the same direction with different

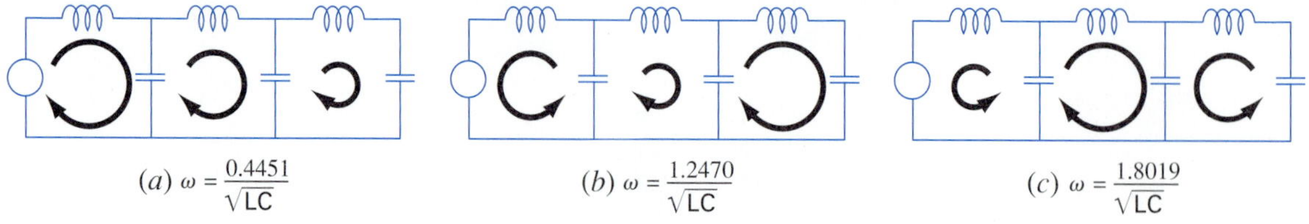

FIGURE 28.15
A visual representation of the natural modes of oscillation of the LC network for Fig. 28.14. Note that the diameters of the circular arrows are proportional to the magnitudes of the currents for each loop.

$(a)\ \omega = \dfrac{0.4451}{\sqrt{LC}}$ $(b)\ \omega = \dfrac{1.2470}{\sqrt{LC}}$ $(c)\ \omega = \dfrac{1.8019}{\sqrt{LC}}$

magnitudes. Observe that if we assume that $i_1 = 0.737$, we can use Eq. (28.14) to compute the other currents with the result

$$\{i\} = \begin{Bmatrix} 0.737 \\ 0.591 \\ 0.328 \end{Bmatrix}$$

which is the first column of the v matrix calculated with MATLAB.

In a similar fashion, the second eigenvalue of $\lambda = 1.555$ can be substituted and the result evaluated to yield

$$-1.8018 i_1 = i_2 = 2.247 i_3$$

As depicted in Fig. 28.15b, the first loop oscillates in the opposite direction from the second and third. Finally, the third mode can be determined as

$$-0.445 i_1 = i_2 = -0.8718 i_3$$

Consequently, as in Fig. 28.15c, the first and third loops oscillate in the opposite direction from the second.

28.4 THE SWINGING PENDULUM (MECHANICAL/AEROSPACE ENGINEERING)

Background. Mechanical engineers (as well as all other engineers) are frequently faced with problems concerning the periodic motion of free bodies. The engineering approach to such problems ultimately requires that the position and velocity of the body be known as a function of time. These functions of time invariably are the solution of ordinary differential equations. The differential equations are usually based on Newton's laws of motion.

As an example, consider the simple pendulum shown previously in Fig. PT7.1. The particle of weight W is suspended on a weightless rod of length l. The only forces acting on the particle are its weight and the tension R in the rod. The position of the particle at any time is completely specified in terms of the angle θ and l.

The free-body diagram in Fig. 28.16 shows the forces on the particle and the acceleration. It is convenient to apply Newton's laws of motion in the x direction tangent to the path of the particle:

$$\Sigma F = -W \sin\theta = \frac{W}{g} a$$

where g = the gravitational constant (32.2 ft/s^2) and a = the acceleration in the x direction. The angular acceleration of the particle (α) becomes

$$\alpha = \frac{a}{l}$$

Therefore, in polar coordinates ($\alpha = d^2\theta/dt^2$),

$$-W \sin\theta = \frac{Wl}{g}\alpha = \frac{Wl}{g}\frac{d^2\theta}{dt^2}$$

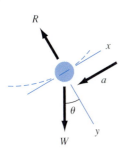

FIGURE 28.16
A free-body diagram of the swinging pendulum showing the forces on the particle and the acceleration.

or

$$\frac{d^2\theta}{dt^2} + \frac{g}{l}\sin\theta = 0$$ (28.15)

This apparently simple equation is a second-order nonlinear differential equation. In general, such equations are difficult or impossible to solve analytically. You have two choices regarding further progress. First, the differential equation might be reduced to a form that can be solved analytically (recall Sec. PT7.1.1), or second, a numerical approximation technique can be used to solve the differential equation directly. We will examine both of these alternatives in this example.

Solution.   Proceeding with the first approach, we note that the series expansion for $\sin\theta$ is given by

$$\sin\theta = \theta - \frac{\theta^3}{3!} + \frac{\theta^5}{5!} - \frac{\theta^7}{7!} + \cdots$$ (28.16)

For small angular displacements, $\sin\theta$ is approximately equal to θ when expressed in radians. Therefore, for small displacements, Eq. (28.15) becomes

$$\frac{d^2\theta}{dt^2} + \frac{g}{l}\theta = 0$$ (28.17)

which is a second-order linear differential equation. This approximation is very important because Eq. (28.17) is easy to solve analytically. The solution, based on the theory of differential equations, is given by

$$\theta(t) = \theta_0 \cos\sqrt{\frac{g}{l}}t$$ (28.18)

where θ_0 = the displacement at $t = 0$ and where it is assumed that the velocity ($v = d\theta/dt$) is zero at $t = 0$. The time required for the pendulum to complete one cycle of oscillation is called the period and is given by

$$T = 2\pi\sqrt{\frac{l}{g}}$$ (28.19)

Figure 28.17 shows a plot of the displacement θ and velocity $d\theta/dt$ as a function of time, as calculated from Eq. (28.18) with $\theta_0 = \pi/4$ and $l = 2$ ft. The period, as calculated from Eq. (28.19), is 1.5659 s.

The above calculations essentially are a complete solution of the motion of the pendulum. However, you must also consider the accuracy of the results because of the assumptions inherent in Eq. (28.17). To evaluate the accuracy, it is necessary to obtain a numerical solution for Eq. (28.15), which is a more complete physical representation of the motion. Any of the methods discussed in Chaps. 25 and 26 could be used for this purpose—for example, the Euler and fourth-order RK methods. Equation (28.15) must be transformed into two first-order equations to be compatible with the above methods. This is accomplished as follows. The velocity v is defined by

$$\frac{d\theta}{dt} = v$$ (28.20)

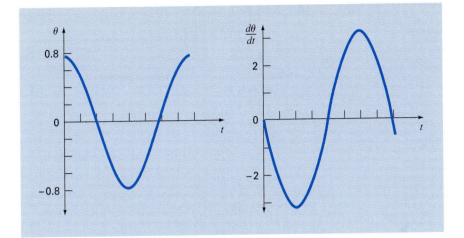

FIGURE 28.17
Plot of displacement θ and velocity $d\theta/dt$ as a function of time t, as calculated from Eq. (28.18). θ_0 is $\pi/4$ and the length is 2 ft.

TABLE 28.1 Comparison of a linear analytical solution of the swinging pendulum problem with three nonlinear numerical solutions.

		Nonlinear Numerical Solutions		
Time, s	Linear Analytical Solution (a)	Euler (h = 0.05) (b)	4th-Order RK (h = 0.05) (c)	4th-Order RK (h = 0.01) (d)
0.0	0.785398	0.785398	0.785398	0.785398
0.2	0.545784	0.615453	0.566582	0.566579
0.4	−0.026852	0.050228	0.021895	0.021882
0.6	−0.583104	−0.639652	−0.535802	−0.535820
0.8	−0.783562	−1.050679	−0.784236	−0.784242
1.0	−0.505912	−0.940622	−0.595598	−0.595583
1.2	0.080431	−0.299819	−0.065611	−0.065575
1.4	0.617698	0.621700	0.503352	0.503392
1.6	0.778062	1.316795	0.780762	0.780777

and, therefore, Eq. (28.15) can be expressed as

$$\frac{dv}{dt} = -\frac{g}{l} \sin \theta \tag{28.21}$$

Equations (28.20) and (28.21) are a coupled system of two ordinary differential equations. The numerical solutions by the Euler method and the fourth-order RK method give the results shown in Table 28.1, which compares the analytic solution for the linear equation of motion [Eq. (28.18)] in column (a) with the numerical solutions in columns (b), (c), and (d).

The Euler and fourth-order RK methods yield different results and both disagree with the analytic solution, although the fourth-order RK method for the nonlinear case is closer to the analytic solution than is the Euler method. To properly evaluate the difference

TABLE 28.2 Comparison of the period of an oscillating body calculated from linear and nonlinear models.

Initial Displacement, θ_0	Period, s	
	Linear Model $(T = 2\pi\sqrt{l/g})$	Nonlinear Model [Numerical Solution of Eq. (28.15)]
$\pi/16$	1.5659	1.57
$\pi/4$	1.5659	1.63
$\pi/2$	1.5659	1.85

between the linear and nonlinear models, it is important to determine the accuracy of the numerical results. This is accomplished in three ways. First, the Euler numerical solution is easily recognized as inadequate because it overshoots the initial condition at $t = 0.8$ s. This clearly violates conservation of energy. Second, column (c) and (d) in Table 28.1 show the solution of the fourth-order RK method for step sizes of 0.05 and 0.01. Because these vary in the fourth decimal place, it is reasonable to assume that the solution with a step size of 0.01 is also accurate with this degree of certainty. Third, for the 0.01-s step-size case, θ obtains a local maximum value of 0.785385 at $t = 1.63$ s (not shown in Table 28.1). This indicates that the pendulum returns to its original position with four-place accuracy with a period of 1.63 s. These considerations allow you to safely assume that the difference between columns (a) and (d) in Table 28.1 truly represents the difference between the linear and nonlinear model.

Another way to characterize the difference between the linear and the nonlinear model is on the basis of period. Table 28.2 shows the period of oscillation as calculated by the linear model and nonlinear model for three different initial displacements. It is seen that the calculated periods agree closely when θ is small because θ is a good approximation for sin θ in Eq. (28.16). This approximation deteriorates when θ becomes large.

These analyses are typical of cases you will routinely encounter as an engineer. The utility of the numerical techniques becomes particularly significant in nonlinear problems, and in many cases real-world problems are nonlinear.

PROBLEMS

Chemical/Petroleum Engineering

28.1 Perform the first computation in Sec. 28.1, but for the case where $h = 10$. Use the Heun (without iteration) and the fourth-order RK methods to obtain solutions.

28.2 Perform the second computation in Sec. 28.1, but for the system described in Prob. 12.4.

28.3 A mass balance for a chemical in a completely mixed reactor can be written as

$$V\frac{dc}{dt} = F - Qc - kVc^2$$

where V = volume (10 m^3), c = concentration, F = feed rate (200 g/min), Q = flow rate (1 m^3/min), and k = a second-order reaction rate (0.1 m^3/g/min). If $c(0) = 0$, solve the ODE until concentration reaches a stable level. Use the midpoint method ($h = 0.5$) and plot your results.

28.4 If $c_{in} = c_b(1 - e^{-0.1t})$, calculate the outflow concentration of a single, completely mixed reactor as a function of time. Use Heun's method (without iteration) to perform the computation. Employ values of $c_b = 50$ mg/m^3, $Q = 5$ m^3/min, $V = 100$ m^3, and $c_0 = 10$ mg/m^3. Perform the computation from $t = 0$ to 100 min using $h = 2$. Plot your results along with the inflow concentration versus time.

28.5 *Desalinization* plants are used to purify sea water so it is suitable for drinking. Sea water containing 8 g salt/kg solution is pumped into a well-mixed tank at a rate of 0.5 kg/min. Assume that the balance of the solution is pure water. Because of faulty design work, water is evaporating from the tank at a rate of 0.5 kg/min. The salt solution leaves the tank at a rate of 10 kg/min.

(a) If the tank is filled initially with 1000 kg of the inlet solution, how long after the outlet pump is turned on will the tank run dry?

(b) Use numerical methods to determine the salt concentration in the tank as a function of time?

28.6 A spherical ice cube (an "ice sphere") that is 5 cm in diameter is removed from a 0 °C freezer and placed on a mesh screen at room temperature $T_a = 20$ °C. What will be the diameter of the ice cube as a function of time out of the freezer (assuming that all the water that has melted immediately drips through the screen)? The heat transfer coefficient h for a sphere in a still room is about 3 W/(m$^2 \cdot$ K). The heat flux from the ice sphere to the air is given by

$$\text{Flux} = \frac{q}{A} = h(T_a - T)$$

where q = heat and A = surface area of the sphere. Use a numerical method to make your calculation.

28.7 The following equations define the concentrations of three reactants:

$$\frac{dc_a}{dt} = -20c_a c_c + 2c_b$$

$$\frac{dc_b}{dt} = 20c_a c_c - 2c_b$$

$$\frac{dc_c}{dt} = -20c_a c_c + 2c_b - 0.2c_c$$

If the initial conditions are $c_a = 500$, $c_b = 0$, and $c_c = 500$, find the concentrations for the times from 0 to 30 s.

28.8 Compound A diffuses through a 4-cm-long tube and reacts as it diffuses. The equation governing diffusion with reaction is

$$D\frac{d^2 A}{dx^2} - kA = 0$$

At one end of the tube, there is a large source of A at a concentration of 0.1 M. At the other end of the tube there is an adsorbent material that quickly absorbs any A, making the concentration 0 M. If $D = 1 \times 10^{-6}$ cm^2/s and $k = 4 \times 10^{-6}$ s^{-1}, what is the concentration of A as a function of distance in the tube?

Civil/Environmental Engineering

28.9 Perform the same computation for the Lotka-Volterra system in Sec. 28.2, but use **(a)** Euler's method, **(b)** Heun's method (without iterating the corrector), and **(c)** the fourth-order RK method. In all cases use single-precision variables, a step size of 0.1, and simulate from $t = 0$ to 20.

28.10 Perform the same computation for the Lorenz equations in Sec. 28.2, but use **(a)** Euler's method, **(b)** Heun's method (without iterating the corrector), and **(c)** the fourth-order RK method. In all cases use single-precision variables and a step size of 0.1 and simulate from $t = 0$ to 20.

28.11 The following equation can be used to model the deflection of a sailboat mast subject to a wind force:

$$\frac{d^2 y}{dz^2} = \frac{f}{2EI}(L - z)^2$$

where f = wind force, E = modulus of elasticity, L = mast length, and I = moment of inertia. Calculate the deflection if $y = 0$ and $dy/dz = 0$ at $z = 0$. Use parameter values of $f = 50$, $L = 30$, $E = 1.2 \times 10^8$, and $I = 0.05$ for your computation.

28.12 Perform the same computation as in Prob. 28.11, but rather than using a constant wind force, employ a force that varies with height according to (recall Sec. 24.2)

$$f(z) = \frac{200z}{5 + z}e^{-2z/30}$$

28.13 An environmental engineer is interested in estimating the mixing that occurs between a stratified lake and an adjacent embayment (Fig. P28.13).

A conservative tracer is instantaneously mixed with the bay water, and then the tracer concentration is monitored over the ensuing period in all three segments. The values are

t	0	2	4	6	8	12	16	20
c_1	0	16	12	8	5	3	2	1
c_2	0	2	5	6	6	5	4	3
c_3	100	48	28	17	11	5	3	1

FIGURE P28.13

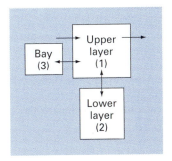

Using mass balances, the system can be modeled as the following simultaneous ODEs:

$$V_1 \frac{dc_1}{dt} = -Qc_1 + E_{12}(c_2 - c_1) + E_{13}(c_3 - c_1)$$

$$V_2 \frac{dc_2}{dt} = E_{12}(c_1 - c_2)$$

$$V_3 \frac{dc_3}{dt} = E_{13}(c_1 - c_3)$$

where V_i = volume of segment i, Q = flow, and E_{ij} = diffusive mixing rate between segments i and j. Use the data and the differential equations to estimate the E's if $V_1 = 1 \times 10^7$, $V_2 = 8 \times 10^6$, $V_3 = 5 \times 10^6$, and $Q = 4 \times 10^6$.

28.14 Population-growth dynamics are important in a variety of planning studies for areas such as transportation and water-resource engineering. One of the simplest models of such growth incorporates the assumption that the rate of change of the population p is proportional to the existing population at any time t:

$$\frac{dp}{dt} = Gp \qquad\qquad\qquad (\text{P28.14})$$

where G = a growth rate (per year). This model makes intuitive sense because the greater the population, the greater the number of potential parents. At time $t = 0$, an island has a population of 5000 people. If $G = 0.07$ per year, employ Heun's method (without iteration) to predict the population at $t = 20$ years, using a step size of 0.5 year. Plot p versus t on standard and semilog graph paper. Determine the slope of the line on the semilog plot. Discuss your results.

28.15 Although the model in Prob. 28.14 works adequately when population growth is unlimited, it breaks down when factors such as food shortages, pollution, and lack of space inhibit growth. In such cases, the growth rate itself can be thought of as being inversely proportional to population. One model of this relationship is

$$G = G'(p_{max} - p) \qquad\qquad (\text{P28.15})$$

where G' = a population-dependent growth rate (per people-year) and p_{max} = the maximum sustainable population. Thus, when population is small ($p \ll p_{max}$), the growth rate will be at a high, constant rate of $G'p_{max}$. For such cases, growth is unlimited and Eq. (P28.15) is essentially identical to Eq. (P28.14). However, as population grows (that is, p approaches p_{max}), G decreases until at $p = p_{max}$ it is zero. Thus, the model predicts that, when the population reaches the maximum sustainable level, growth is nonexistent, and the system is at a steady state. Substituting Eq. (P28.15) into Eq. (P28.14) yields

$$\frac{dp}{dt} = G'(p_{max} - p)p$$

For the same island studied in Prob. 28.14, employ Heun's method (without iteration) to predict the population at $t = 20$ years, using a step size of 0.5 year. Employ values of $G' = 10^{-5}$ per people-year and $p_{max} = 20{,}000$ people. At time $t = 0$, the island has a population of 5000 people. Plot p versus t and interpret the shape of the curve.

28.16 Isle Royale National Park is a 210-square-mile archipelago composed of a single large island and many small islands in Lake Superior. Moose arrived around 1900 and by 1930, their population approached 3000, ravaging vegetation. In 1949, wolves crossed an ice bridge from Ontario. Since the late 1950s, the numbers of the moose and wolves have been tracked.

Year	Moose	Wolves	Year	Moose	Wolves
1960	700	22	1972	836	23
1961	—	22	1973	802	24
1962	—	23	1974	815	30
1963	—	20	1975	778	41
1964	—	25	1976	641	43
1965	—	28	1977	507	33
1966	881	24	1978	543	40
1967	—	22	1979	675	42
1968	1000	22	1980	577	50
1969	1150	17	1981	570	30
1970	966	18	1982	590	13
1971	674	20	1983	811	23

(a) Integrate the Lotka-Volterra equations from 1960 through 2020. Determine the coefficient values that yield an optimal fit. Compare your simulation with the data using a time-series approach, and comment on the results.

(b) Plot the simulation of (a), but use a state-space approach.

(c) After 1993, suppose that the wildlife managers trap one wolf per year and transport it off the island. Predict how the populations of both the wolves and moose would evolve to the year 2020. Present your results as both time series and state-space plots. For this case [as well as for (d)], use the following coefficients: $a = 0.3$, $b = 0.01111$, $c = 0.2106$, $d = 0.0002632$.

(d) Suppose that in 1993, some poachers snuck onto the island and killed 50% of the moose. Predict how the populations of both the wolves and moose would evolve to the year 2020. Present your results as both time series and state-space plots.

Electrical Engineering

28.17 Perform the same computation as in the first part of Sec. 28.3, but with $R = 0.05 \ \Omega$.

28.18 Solve the ODE in Sec. 8.3 from $t = 0$ to 0.5 using numerical techniques if $q = 0.1$ and $i = -3.281515$ at $t = 0$. Use an $R = 100$ along with the other parameters from Sec. 8.3.

28.19 For a simple RL circuit, Kirchhoff's voltage law requires that (if Ohm's law holds)

$$L\frac{di}{dt} + Ri = 0$$

where i = current, L = inductance, and R = resistance. Solve for i, if $L = 1$, $R = 2$, and $i(0) = 0.6$. Solve this problem analytically and with a numerical method.

28.20 In contrast to Prob. 28.19, real resistors may not always obey Ohm's law. For example, the voltage drop may be nonlinear and the circuit dynamics described by a relationship such as

$$L\frac{di}{dt} + R\left[\frac{i}{I} - \left(\frac{i}{I}\right)^3\right] = 0$$

where all other parameters are as defined in Prob. 28.19 and I is a known reference current equal to 1. Solve for i as a function of time under the same conditions as specified in Prob. 28.19.

28.21 Develop an eigenvalue problem for an LC network similar to the one in Fig. 28.14, but with only two loops. That is, omit the i_3 loop. Draw the network, illustrating how the currents oscillate in their primary modes.

Mechanical/Aerospace Engineering

28.22 Perform the same computation as in Sec. 28.4, but for a 4-ft-long pendulum.

28.23 Use a numerical method to duplicate the computation of the position of the shock absorber versus time after hitting a pothole as described in Sec. 8.4 (recall Table 8.3).

28.24 The rate of cooling a body can be expressed as

$$\frac{dT}{dt} = -k(T - T_a)$$

where T = temperature of the body (°C), T_a = temperature of the surrounding medium (°C), and k = the proportionality constant

(min^{-1}). Thus, this equation specifies that the rate of cooling is proportional to the difference in temperature between the body and the surrounding medium. If a metal ball heated to 90°C is dropped into water that is held at a constant value of $T_a = 20$°C, use a numerical method to compute how long it takes the ball to cool to 40°C if $k = 0.2$ min^{-1}.

28.25 The rate of heat flow (conduction) between two points on a cylinder heated at one end is given by

$$\frac{dQ}{dt} = \lambda A \frac{dT}{dx}$$

where λ = a constant, A = the cylinder's cross-sectional area, Q = heat flow, T = temperature, t = time, and r = distance from the heated end. Because the equation involves two derivatives, we will simplify this equation by letting

$$\frac{dT}{dx} = \frac{100(L - x)(20 - t)}{100 - xt}$$

where L is the length of the rod. Combine the two equations and compute the heat flow for $t = 0$ to 25 s. The initial condition is $Q(0) = 0$ and the parameters are $\lambda = 0.4$ cal · cm/s, $A = 10$ cm^2, $L = 20$ cm, and $x = 2.5$ cm. Plot your results.

28.26 Repeat the falling parachutist problem (Example 1.2), but with the upward force due to drag as a second-order rate:

$$F_u = -cv^2$$

where $c = 0.235$ kg/m. Solve for $t = 0$ to 30, plot your results, and compare with those of Example 1.2.

28.27 Suppose that, after falling for 15 s, the parachutist from Examples 1.1 and 1.2 pulls the rip cord. At this point, assume that the drag coefficient is instantaneously increased to a constant value of 50 kg/s. Compute the parachutist's velocity from $t = 0$ to 30 s with Euler's method. Plot v versus t for $t = 0$ to 30 s.

EPILOGUE: PART SEVEN

PT7.4 TRADE-OFFS

Table PT7.3 contains trade-offs associated with numerical methods for the solution of initial-value ordinary differential equations. The factors in this table must be evaluated by the engineer when selecting a method for each particular applied problem.

Simple self-starting techniques such as Euler's method can be used if the problem requirements involve a short range of integration. In this case, adequate accuracy may be obtained using small step sizes to avoid large truncation errors, and the round-off errors may be acceptable. Euler's method may also be appropriate for cases where the mathematical model has an inherently high level of uncertainty or has coefficients or forcing functions with significant errors as might arise during a measurement process. In this case, the accuracy of the model itself simply does not justify the effort involved to employ a more complicated numerical method. Finally, the simpler techniques may be best when the problem

TABLE PT7.3 Comparison of the characteristics of alternative methods for the numerical solution of ODEs. The comparisons are based on general experience and do not account for the behavior of special functions.

Method	Starting Values	Iterations Required	Global Error	Ease of Changing Step Size	Programming Effort	Comments
One step						
Euler's	1	No	$O(h)$	Easy	Easy	Good for quick estimates
Heun's	1	Yes	$O(h^2)$	Easy	Moderate	—
Midpoint	1	No	$O(h^2)$	Easy	Moderate	—
Second-order Ralston	1	No	$O(h^2)$	Easy	Moderate	The second-order RK method that minimizes truncation error
Fourth-order RK	1	No	$O(h^4)$	Easy	Moderate	Widely used
Adaptive fourth-order RK or RK-Fehlberg	1	No	$O(h^5)^*$	Easy	Moderate to difficult	Error estimate allows step-size adjustment
Multistep						
Non-self-starting Heun	2	Yes	$O(h^3)^*$	Difficult	Moderate to difficult†	Simple multistep method
Milne's	4	Yes	$O(h^5)^*$	Difficult	Moderate to difficult†	Sometimes unstable
Fourth-order Adams	4	Yes	$O(h^5)^*$	Difficult	Moderate to difficult†	

*Provided error estimate is used to modify the solution.
†With variable step size.

or simulation need only be performed a few times. In these applications, it is probably best to use a simple method that is easy to program and understand, despite the fact that the method may be computationally inefficient and relatively time-consuming to run on the computer.

If the range of integration of the problem is long enough to involve a large number of steps, then it may be necessary and appropriate to use a more accurate technique than Euler's method. The fourth-order RK method is popular and reliable for many engineering problems. In these cases, it may also be advisable to estimate the truncation error for each step as a guide to selecting the best step size. This can be accomplished with the adaptive RK or fourth-order Adams approaches. If the truncation errors are extremely small, it may be wise to increase the step size to save computer time. On the other hand, if the truncation error is large, the step size should be decreased to avoid accumulation of error. Milne's method should be avoided if significant stability problems are expected. The Runge-Kutta method is simple to program and convenient to use but may be less efficient than the multistep methods. However, the Runge-Kutta method is usually employed in any event to obtain starting values for the multistep methods.

A large number of engineering problems may fall into an intermediate range of integration interval and accuracy requirement. In these cases, the second-order RK and the non-self-starting Heun methods are simple to use and are relatively efficient and accurate.

Stiff systems involve equations with slowly and rapidly varying components. Special techniques are usually required for the adequate solution of stiff equations. For example, implicit approaches are often used. You can consult Enright et al. (1975), Gear (1971), and Shampine and Gear (1979) for additional information regarding these techniques.

A variety of techniques are available for solving eigenvalue problems. For small systems or where only a few of the smallest or largest eigenvalues are required, simple approaches such as the polynomial and the power methods are available. For symmetric systems, Jacobi's, Given's, or Householder's method can be employed. Finally, the QR method represents a general approach for finding all the eigenvalues of symmetric and nonsymmetric matrices.

PT7.5 IMPORTANT RELATIONSHIPS AND FORMULAS

Table PT7.4 summarizes important information that was presented in Part Seven. This table can be consulted to quickly access important relationships and formulas.

PT7.6 ADVANCED METHODS AND ADDITIONAL REFERENCES

Although we have reviewed a number of techniques for solving ordinary differential equations there is additional information that is important in engineering practice. The question of *stability* was introduced in Sec. 26.2.4. This topic has general relevance to all methods for solving ODEs. Further discussion of the topic can be pursued in Carnahan, Luther, and Wilkes (1969), Gear (1971), and Hildebrand (1974).

In Chap. 27, we introduced methods for solving *boundary-value* problems. Isaacson and Keller (1966), Keller (1968), Na (1979), and Scott and Watts (1976) can be consulted for additional information on standard boundary-value problems. Additional material on

TABLE PT7.4 Summary of important information presented in Part Seven.

Method	Formulation	Graphic Interpretation	Errors
Euler (First-order RK)	$y_{i+1} = y_i + hk_1$ $k_1 = f(x_i, y_i)$		Local error $\simeq O(h^2)$ Global error $\simeq O(h)$
Ralston's second-order RK	$y_{i+1} = y_i + h\left(\frac{1}{3}k_1 + \frac{2}{3}k_2\right)$ $k_1 = f(x_i, y_i)$ $k_2 = f\left(x_i + \frac{3}{4}h,\ y_i + \frac{3}{4}hk_1\right)$		Local error $\simeq O(h^3)$ Global error $\simeq O(h^2)$
Classic fourth-order RK	$y_{i+1} = y_i + h\left(\frac{1}{6}k_1 + \frac{1}{3}k_2 + \frac{1}{3}k_3 + \frac{1}{6}k_4\right)$ $k_1 = f(x_i, y_i)$ $k_2 = f\left(x_i + \frac{1}{2}h,\ y_i + \frac{1}{2}hk_1\right)$ $k_3 = f\left(x_i + \frac{1}{2}h,\ y_i + \frac{1}{2}hk_2\right)$ $k_4 = f(x_i + h,\ y_i + hk_3)$		Local error $\simeq O(h^5)$ Global error $\simeq O(h^4)$
Non-self-starting Heun	Predictor: (midpoint method) $y_{i+1}^0 = y_{i-1}^m + 2hf(x_i, y_i^m)$ Corrector: (trapezoidal rule) $y_{i+1}^j = y_i^m + h\,\dfrac{f(x_i, y_i^m) + f(x_{i+1}, y_{i+1}^{j-1})}{2}$	 	Predictor modifier: $E_p \simeq \frac{4}{5}\left(y_{i,u}^m - y_{i,u}^0\right)$ Corrector modifier: $E_c \simeq -\dfrac{y_{i+1,u}^m - y_{i+1,u}^0}{5}$
Fourth-order Adams	Predictor: (fourth Adams-Bashforth) $y_{i+1}^0 = y_i^m + h\left(\frac{55}{24}f_i^m - \frac{59}{24}f_{i-1}^m + \frac{37}{24}f_{i-2}^m - \frac{9}{24}f_{i-3}^m\right)$ Corrector: (fourth Adams-Moulton) $y_{i+1}^j = y_i^m + h\left(\frac{9}{24}f_{i+1}^{j-1} + \frac{19}{24}f_i^m - \frac{5}{24}f_{i-1}^m + \frac{1}{24}f_{i-2}^m\right)$	 	Predictor modifier: $E_p \simeq \frac{251}{270}\left(y_{i,u}^m - y_{i,u}^0\right)$ Corrector modifier:

eigenvalues can be found in Ralston and Rabinowitz (1978), Wilkinson (1965), Fadeev and Fadeeva (1963), and Householder (1953).

In summary, the foregoing is intended to provide you with avenues for deeper exploration of the subject. Additionally, all the above references provide descriptions of the basic techniques covered in Part Seven. We urge you to consult these alternative sources to broaden your understanding of numerical methods for the solution of differential equations.

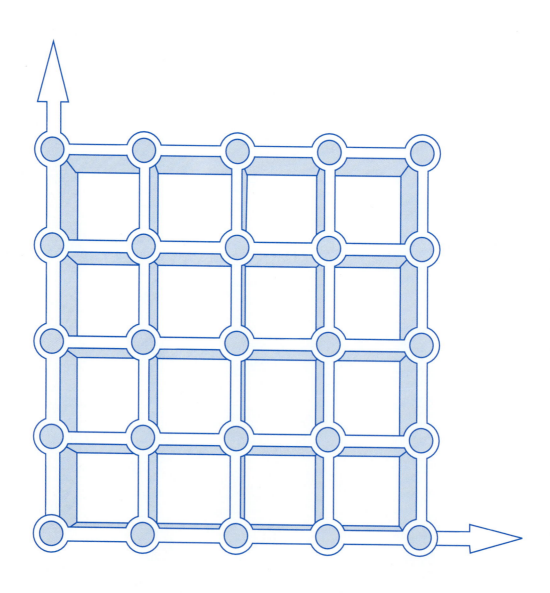

PART EIGHT

PARTIAL DIFFERENTIAL EQUATIONS

PT8.1 MOTIVATION

Given a function u that depends on both x and y, the partial derivative of u with respect to x at an arbitrary point (x, y) is defined as

$$\frac{\partial u}{\partial x} = \lim_{\Delta x \to 0} \frac{u(x + \Delta x, y) - u(x, y)}{\Delta x} \tag{PT8.1}$$

Similarly, the partial derivative with respect to y is defined as

$$\frac{\partial u}{\partial y} = \lim_{\Delta y \to 0} \frac{u(x, y + \Delta y) - u(x, y)}{\Delta y} \tag{PT8.2}$$

An equation involving partial derivatives of an unknown function of two or more independent variables is called a *partial differential equation,* or *PDE.* For example,

$$\frac{\partial^2 u}{\partial x^2} + 2xy \frac{\partial^2 u}{\partial y^2} + u = 1 \tag{PT8.3}$$

$$\frac{\partial^3 u}{\partial x^2 \partial y} + x \frac{\partial^2 u}{\partial y^2} + 8u = 5y \tag{PT8.4}$$

$$\left(\frac{\partial^2 u}{\partial x^2} \right)^3 + 6 \frac{\partial^3 u}{\partial x \partial y^2} = x \tag{PT8.5}$$

$$\frac{\partial^2 u}{\partial x^2} + xu \frac{\partial u}{\partial y} = x \tag{PT8.6}$$

The *order* of a PDE is that of the highest-order partial derivative appearing in the equation. For example, Eqs. (PT8.3) and (PT8.4) are second- and third-order, respectively.

A partial differential equation is said to be *linear* if it is linear in the unknown function and all its derivatives, with coefficients depending only on the independent variables. For example, Eq. (PT8.3) is linear, whereas Eqs. (PT8.5) and (PT8.6) are not.

Because of their widespread application in engineering, our treatment of PDEs will focus on linear, second-order equations. For two independent variables, such equations can be expressed in the following general form:

$$A \frac{\partial^2 u}{\partial x^2} + B \frac{\partial^2 u}{\partial x \partial y} + C \frac{\partial^2 u}{\partial y^2} + D = 0 \tag{PT8.7}$$

where A, B, and C are functions of x and y and D is a function of x, y, u, $\partial u/\partial x$, and $\partial u/\partial y$. Depending on the values of the coefficients of the second-derivative terms—A, B, C—

TABLE PT8.1 Categories into which linear, second-order partial differential equations in two variables can be classified.

$B^2 - 4AC$	Category	Example
< 0	Elliptic	Laplace equation (steady state with two spatial dimensions) $$\frac{\partial^2 T}{\partial x^2} + \frac{\partial^2 T}{\partial y^2} = 0$$
$= 0$	Parabolic	Heat conduction equation (time variable with one spatial dimension) $$\frac{\partial T}{\partial t} = k' \frac{\partial^2 T}{\partial x^2}$$
> 0	Hyperbolic	Wave equation (time variable with one spatial dimension) $$\frac{\partial^2 y}{\partial x^2} = \frac{1}{c^2} \frac{\partial^2 y}{\partial t^2}$$

Eq. (PT8.7) can be classified into one of three categories (Table PT8.1). This classification, which is based on the method of characteristics (for example, see Vichnevetsky, 1981, or Lapidus and Pinder, 1982), is useful because each category relates to specific and distinct engineering problem contexts that demand special solution techniques. It should be noted that for cases where A, B, and C depend on x and y, the equation may actually fall into a different category, depending on the location in the domain for which the equation holds. For simplicity, we will limit the present discussion to PDEs that remain exclusively in one of the categories.

PT8.1.1 PDEs and Engineering Practice

Each of the categories of partial differential equations in Table PT8.1 conforms to specific kinds of engineering problems. The initial sections of the following chapters will be devoted to deriving each type of equation for a particular engineering problem context. For the time being, we will discuss their general properties and applications and show how they can be employed in different physical contexts.

Elliptic equations are typically used to characterize *steady-state* systems. As in the *Laplace equation* in Table PT8.1, this is indicated by the absence of a time derivative. Thus, these equations are typically employed to determine the steady-state distribution of an unknown in two spatial dimensions.

A simple example is the heated plate in Fig. PT8.1a. For this case, the boundaries of the plate are held at different temperatures. Because heat flows from regions of high to low temperature, the boundary conditions set up a potential that leads to heat flow from the hot to the cool boundaries. If sufficient time elapses, such a system will eventually reach the stable or steady-state distribution of temperature depicted in Fig. PT8.1a. The Laplace equation, along with appropriate boundary conditions, provides a means to determine this distribution. By analogy, the same approach can be employed to tackle other problems involving potentials, such as seepage of water under a dam (Fig. PT8.1b) or the distribution of an electric field (Fig. PT8.1c).

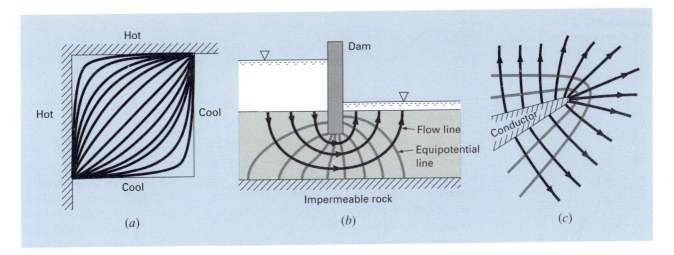

FIGURE PT8.1
Three steady-state distribution problems that can be characterized by elliptic PDEs. (*a*) Temperature distribution on a heated plate, (*b*) seepage of water under a dam, and (*c*) the electric field near the point of a conductor.

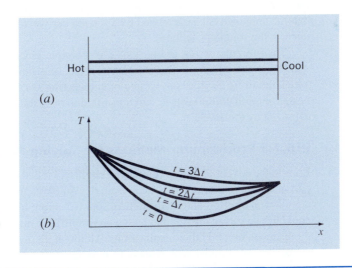

FIGURE PT8.2
(*a*) A long, thin rod that is insulated everywhere but at its end. The dynamics of the one-dimensional distribution of temperature along the rod's length can be described by a parabolic PDE. (*b*) The solution, consisting of distributions corresponding to the state of the rod at various times.

In contrast to the elliptic category, *parabolic equations* determine how an unknown varies in both space and time. This is manifested by the presence of both spatial and temporal derivatives in the *heat conduction equation* from Table PT8.1. Such cases are referred to as *propagation problems* because the solution "propagates," or changes, in time.

A simple example is a long, thin rod that is insulated everywhere except at its end (Fig. PT8.2*a*). The insulation is employed to avoid complications due to heat loss along the rod's

FIGURE PT8.3
A taut string vibrating at a low amplitude is a simple physical system that can be characterized by a hyperbolic PDE.

length. As was the case for the heated plate in Fig. PT8.1*a*, the ends of the rod are set at fixed temperatures. However, in contrast to Fig. PT8.1*a*, the rod's thinness allows us to assume that heat is distributed evenly over its cross section—that is, laterally. Consequently, lateral heat flow is not an issue and the problem reduces to studying the conduction of heat along the rod's longitudinal axis. Rather than focusing on the steady-state distribution in two spatial dimensions, the problem shifts to determining how the one-dimensional spatial distribution changes as a function of time (Fig. PT8.2*b*). Thus, the solution consists of a series of spatial distributions corresponding to the state of the rod at various times. Using an analogy from photography, the elliptic case yields a portrait of a system's stable state, whereas the parabolic case provides a motion picture of how it changes from one state to another. As with the other types of PDEs described herein, parabolic equations can be used to characterize a wide variety of other engineering problem contexts by analogy.

The final class of PDEs, the *hyperbolic* category, also deals with *propagation problems*. However, an important distinction manifested by the wave equation in Table PT8.1 is that the unknown is characterized by a second derivative with respect to time. As a consequence, the solution oscillates.

The vibrating string in Fig. PT8.3 is a simple physical model that can be described with the wave equation. The solution consists of a number of characteristic states with which the string oscillates. A variety of engineering systems such as vibrations of rods and beams, motion of fluid waves, and transmission of sound and electrical signals can be characterized by this model.

PT8.1.2 Precomputer Methods for Solving PDEs

Prior to the advent of digital computers, engineers relied on analytical or exact solutions of partial differential equations. Aside from the simplest cases, these solutions often required a great deal of effort and mathematical sophistication. In addition, many physical systems could not be solved directly but had to be simplified using linearizations, simple geometric representations, and other idealizations. Although these solutions are elegant and yield insight, they are limited with respect to how faithfully they represent real systems—especially those that are highly nonlinear and irregularly shaped.

PT8.2   ORIENTATION

Before we proceed to the numerical methods for solving partial differential equations, some orientation might be helpful. The following material is intended to provide you with an overview of the material discussed in Part Eight. In addition, we have formulated objectives to focus your studies in the subject area.

PT8.2.1 Scope and Preview

Figure PT8.4 provides an overview of Part Eight. Two broad categories of numerical methods will be discussed in this part of this book. Finite-difference approaches, which are covered in Chaps. 29 and 30, are based on approximating the solution at a finite number of points. In contrast, finite-element methods, which are covered in Chap. 31, approximate

FIGURE PT8.4
Schematic representation of the organization of material in Part Eight: Partial Differential Equations.

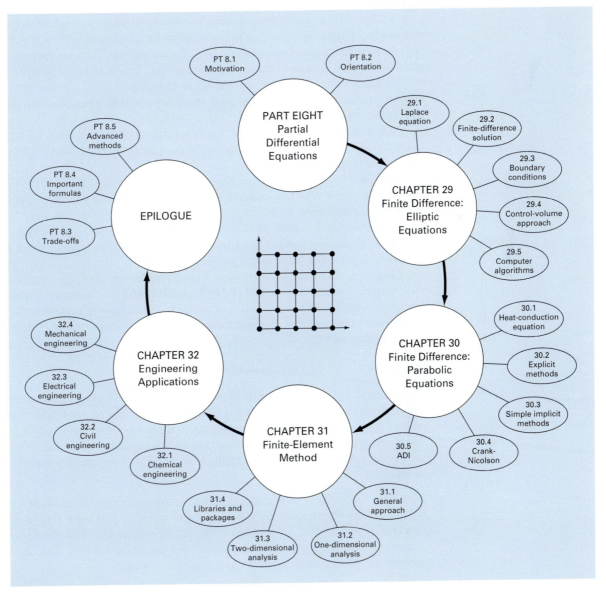

the solution in pieces, or "elements." Various parameters are adjusted until these approximations conform to the underlying differential equation in an optimal sense.

Chapter 29 is devoted to *finite-difference* solutions of *elliptic equations*. Before launching into the methods, we derive the Laplace equation for the physical problem context of the temperature distribution for a heated plate. Then, a standard solution approach, the *Liebmann method,* is described. We will illustrate how this approach is used to compute the distribution of the primary scalar variable, temperature, as well as a secondary vector variable, heat flux. The final section of the chapter deals with *boundary conditions*. This material includes procedures to handle different types of conditions as well as irregular boundaries.

In *Chap. 30,* we turn to *finite-difference* solutions of *parabolic equations*. As with the discussion of elliptic equations, we first provide an introduction to a physical problem context, the heat-conduction equation for a one-dimensional rod. Then we introduce both explicit and implicit algorithms for solving this equation. This is followed by an efficient and reliable implicit method—the *Crank-Nicolson technique*. Finally, we describe a particularly effective approach for solving two-dimensional parabolic equations—the alternating-direction implicit, or *A.D.I., method*.

Note that, because they are somewhat beyond the scope of this book, we have chosen to omit hyperbolic equations. The epilogue of this part of the book contains references related to this type of PDE.

In *Chap. 31* we turn to the other major approach for solving PDEs—the *finite-element method*. Because it is so fundamentally different from the finite-difference approach, we devote the initial section of the chapter to a general overview. Then we show how the finite-element method is used to compute the steady-state temperature distribution of a heated rod. Finally, we provide an introduction to some of the issues involved in extending such an analysis to two-dimensional problem contexts.

Chapter 32 is devoted to applications from all fields of engineering. Finally, a short review section is included at the end of Part Eight. This epilogue summarizes important information related to PDEs. This material includes a discussion of trade-offs that are relevant to their implementation in engineering practice. The epilogue also includes references for advanced topics.

PT8.2.2 Goals and Objectives

Study Objectives.   After completing Part Eight, you should have greatly enhanced your capability to confront and solve partial differential equations. General study goals should include mastering the techniques, having the capability to assess the reliability of the answers, and being able to choose the "best" method (or methods) for any particular problem. In addition to these general objectives, the specific study objectives in Table PT8.2 should be mastered.

Computer Objectives.   Computer algorithms can be developed for many of the methods in Part Eight. For example, you may find it instructive to develop a general program to simulate the steady-state distribution of temperature on a heated plate. Further, you might want to develop programs to implement both the simple explicit and the Crank-Nicolson methods for solving parabolic PDEs in one spatial dimension.

TABLE PT8.2 Specific study objectives for Part Eight.

1. Recognize the difference between elliptic, parabolic, and hyperbolic PDEs.
2. Understand the fundamental difference between finite-difference and finite-element approaches.
3. Recognize that the Liebmann method is equivalent to the Gauss-Seidel approach for solving simultaneous linear algebraic equations.
4. Know how to determine secondary variables for two-dimensional field problems.
5. Recognize the distinction between Dirichlet and derivative boundary conditions.
6. Understand how to use weighting factors to incorporate irregular boundaries into a finite-difference scheme for PDEs.
7. Understand how to implement the control-volume approach for implementing numerical solutions of PDEs.
8. Know the difference between convergence and stability of parabolic PDEs.
9. Understand the difference between explicit and implicit schemes for solving parabolic PDEs.
10. Recognize how the stability criteria for explicit methods detract from their utility for solving parabolic PDEs.
11. Know how to interpret computational molecules.
12. Recognize how the A.D.I. approach achieves high efficiency in solving parabolic equations in two spatial dimensions.
13. Understand the difference between the direct method and the method of weighted residuals for deriving element equations.
14. Know how to implement Galerkin's method.
15. Understand the benefits of integration by parts during the derivation of element equations; in particular, recognize the implications of lowering the highest derivative from a second to a first derivative.

Finally, one of your most important goals should be to master several of the general-purpose software packages that are widely available. In particular, you should become adept at using these tools to implement numerical methods for engineering problem solving.

CHAPTER 29

Finite Difference: Elliptic Equations

Elliptic equations in engineering are typically used to characterize steady-state, boundary-value problems. Before demonstrating how they can be solved, we will illustrate how a simple case—the Laplace equation—is derived from a physical problem context.

29.1 THE LAPLACE EQUATION

As mentioned in the introduction to this part of the book, the Laplace equation can be used to model a variety of problems involving the potential of an unknown variable. Because of its simplicity and general relevance to most areas of engineering, we will use a heated plate as our fundamental context for deriving and solving this elliptic PDE. Homework problems and engineering applications (Chap. 32) will be employed to illustrate the applicability of the model to other engineering problem contexts.

Figure 29.1 shows an element on the face of a thin rectangular plate of thickness Δz. The plate is insulated everywhere but at its edges, where the temperature can be set at a prescribed level. The insulation and the thinness of the plate mean that heat transfer is limited to the x and y dimensions. At steady state, the flow of heat into the element over a unit time period Δt must equal the flow out, as in

$$q(x)\,\Delta y\,\Delta z\,\Delta t + q(y)\,\Delta x\,\Delta z\,\Delta t = q(x + \Delta x)\,\Delta y\,\Delta z\,\Delta t \\ + q(y + \Delta y)\Delta x\,\Delta z\,\Delta t \tag{29.1}$$

where $q(x)$ and $q(y)$ = the heat fluxes at x and y, respectively [cal/(cm$^2 \cdot$ s)]. Dividing by Δz and Δt and collecting terms yields

$$\left[q(x) - q(x + \Delta x)\right]\Delta y + \left[q(y) - q(y + \Delta y)\right]\Delta x = 0$$

Multiplying the first term by $\Delta x/\Delta x$ and the second by $\Delta y/\Delta y$ gives

$$\frac{q(x) - q(x + \Delta x)}{\Delta x}\,\Delta x\,\Delta y + \frac{q(y) - q(y + \Delta y)}{\Delta y}\,\Delta y\,\Delta x = 0 \tag{29.2}$$

Dividing by $\Delta x\,\Delta y$ and taking the limit results in

$$-\frac{\partial q}{\partial x} - \frac{\partial q}{\partial y} = 0 \tag{29.3}$$

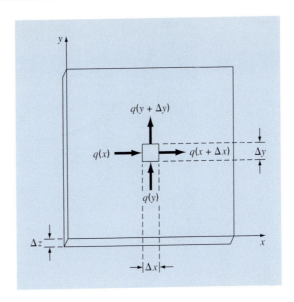

FIGURE 29.1
A thin plate of thickness Δz. An element is shown about which a heat balance is taken.

where the partial derivatives result from the definitions in Eqs. (PT7.1) and (PT7.2).

Equation (29.3) is a partial differential equation that is an expression of the conservation of energy for the plate. However, unless heat fluxes are specified at the plate's edges, it cannot be solved. Because temperature boundary conditions are given, Eq. (29.3) must be reformulated in terms of temperature. The link between flux and temperature is provided by *Fourier's law* of heat conduction, which can be represented as

$$q_i = -k\rho C \frac{\partial T}{\partial i} \tag{29.4}$$

where q_i = heat flux in the direction of the i dimension [cal/(cm$^2 \cdot$ s)], k = coefficient of *thermal diffusivity* (cm^2/s), ρ = density of the material (g/cm^3), C = heat capacity of the material [cal/(g $\cdot$ °C)], and T = temperature (°C), which is defined as

$$T = \frac{H}{\rho C V}$$

where H = heat (cal) and V = volume (cm^3). Sometimes the term in front of the differential in Eq. (29.3) is treated as a single term,

$$k' = k\rho C \tag{29.5}$$

where k' is referred to as the *coefficient of thermal conductivity* [cal/(s $\cdot$ cm $\cdot$ °C)]. In either case, both k and k' are parameters that reflect how well the material conducts heat.

Fourier's law is sometimes referred to as a *constitutive equation*. It is given this label because it provides a mechanism that defines the system's internal interactions. Inspection

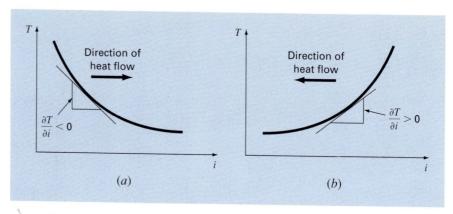

FIGURE 29.2
Graphical depiction of a temperature gradient. Because heat moves "downhill" from high to low temperature, the flow in (a) is from left to right in the positive *i* direction. However, due to the orientation of Cartesian coordinates, the slope is negative for this case. Thus, a negative gradient leads to a positive flow. This is the origin of the minus sign in Fourier's law of heat conduction. The reverse case is depicted in (b), where the positive gradient leads to a negative heat flow from right to left.

of Eq. (29.4) indicates that Fourier's law specifies that heat flux perpendicular to the *i* axis is proportional to the gradient or slope of temperature in the *i* direction. The negative sign ensures that a positive flux in the direction of *i* results from a negative slope from high to low temperature (Fig. 29.2). Substituting Eq. (29.4) into Eq. (29.3) results in

$$\frac{\partial^2 T}{\partial x^2} + \frac{\partial^2 T}{\partial y^2} = 0 \tag{29.6}$$

which is the *Laplace equation*. Note that for the case where there are sources or sinks of heat within the two-dimensional domain, the equation can be represented as

$$\frac{\partial^2 T}{\partial x^2} + \frac{\partial^2 T}{\partial y^2} = f(x, y) \tag{29.7}$$

where $f(x, y)$ is a function describing the sources or sinks of heat. Equation (29.7) is referred to as the *Poisson equation*.

29.2 SOLUTION TECHNIQUE

The numerical solution of elliptic PDEs such as the Laplace equation proceeds in the reverse manner of the derivation of Eq. (29.6) from the preceding section. Recall that the derivation of Eq. (29.6) employed a balance around a discrete element to yield an algebraic difference equation characterizing heat flux for a plate. Taking the limit turned this difference equation into a differential equation [Eq. (29.3)].

For the numerical solution, finite-difference representations based on treating the plate as a grid of discrete points (Fig. 29.3) are substituted for the partial derivatives in Eq. (29.6). As described next, the PDE is transformed into an algebraic difference equation.

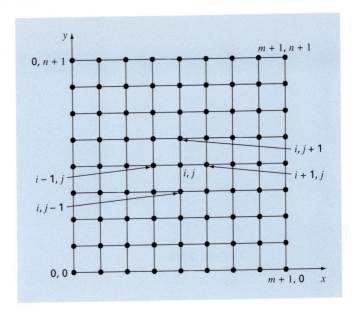

FIGURE 29.3
A grid used for the finite-difference solution of elliptic PDEs in two independent variables such as the Laplace equation.

29.2.1 The Laplacian Difference Equation

Central differences based on the grid scheme from Fig. 29.3 are (recall Fig. 23.3)

$$\frac{\partial^2 T}{\partial x^2} = \frac{T_{i+1,j} - 2T_{i,j} + T_{i-1,j}}{\Delta x^2}$$

and

$$\frac{\partial^2 T}{\partial y^2} = \frac{T_{i,j+1} - 2T_{i,j} + T_{i,j-1}}{\Delta y^2}$$

which have errors of $O[\Delta(x)^2]$ and $O[\Delta(y)^2]$, respectively. Substituting these expressions into Eq. (29.6) gives

$$\frac{T_{i+1,j} - 2T_{i,j} + T_{i-1,j}}{\Delta x^2} + \frac{T_{i,j+1} - 2T_{i,j} + T_{i,j-1}}{\Delta y^2} = 0$$

For the square grid in Fig. 29.3, $\Delta x = \Delta y$, and by collection of terms, the equation becomes

$$T_{i+1,j} + T_{i-1,j} + T_{i,j+1} + T_{i,j-1} - 4T_{i,j} = 0 \tag{29.8}$$

This relationship, which holds for all interior points on the plate, is referred to as the *Laplacian difference equation*.

In addition, boundary conditions along the edges of the plate must be specified to obtain a unique solution. The simplest case is where the temperature at the boundary is set at a fixed value. This is called a *Dirichlet boundary condition*. Such is the case for Fig. 29.4,

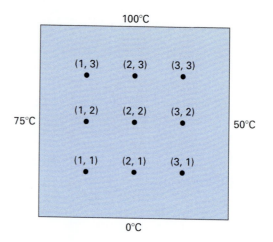

FIGURE 29.4
A heated plate where boundary temperatures are held at constant levels. This case is called a Dirichlet boundary condition.

where the edges are held at constant temperatures. For the case illustrated in Fig. 29.4, a balance for node (1, 1) is, according to Eq. (29.8),

$$T_{21} + T_{01} + T_{12} + T_{10} - 4T_{11} = 0 \qquad (29.9)$$

However, $T_{01} = 75$ and $T_{10} = 0$, and therefore, Eq. (29.9) can be expressed as

$$-4T_{11} + T_{12} + T_{21} = -75$$

Similar equations can be developed for the other interior points. The result is the following set of nine simultaneous equations with nine unknowns:

$$
\begin{array}{l}
4T_{11} \quad -T_{21} \qquad\quad -T_{12} \qquad\qquad\qquad\qquad\qquad = 75 \\
-T_{11} \ +4T_{21} \ -T_{31} \qquad -T_{22} \qquad\qquad\qquad\qquad = 0 \\
\qquad\quad -T_{21} \ +4T_{31} \qquad\qquad -T_{32} \qquad\qquad\qquad = 50 \\
-T_{11} \qquad\qquad\qquad +4T_{12} \ -T_{22} \qquad -T_{13} \qquad\qquad = 75 \\
\qquad\quad -T_{21} \qquad\qquad -T_{12} \ +4T_{22} \ -T_{32} \qquad -T_{23} \qquad = 0 \\
\qquad\qquad\qquad -T_{31} \qquad\qquad -T_{22} \ +4T_{32} \qquad\qquad -T_{33} = 50 \\
\qquad\qquad\qquad\qquad\quad -T_{12} \qquad\qquad +4T_{13} \ -T_{23} \qquad = 175 \\
\qquad\qquad\qquad\qquad\qquad\qquad -T_{22} \qquad -T_{13} \ +4T_{23} \ -T_{33} = 100 \\
\qquad\qquad\qquad\qquad\qquad\qquad\qquad -T_{32} \qquad -T_{23} \ +4T_{33} = 150
\end{array}
$$

$$(29.10)$$

29.2.2 The Liebmann Method

Most numerical solutions of the Laplace equation involve systems that are much larger than Eq. (29.10). For example, a 10-by-10 grid involves 100 linear algebraic equations. Solution techniques for these types of equations were discussed in Part Three.

Notice that there are a maximum of five unknown terms per line in Eq. (29.10). For larger-sized grids, this means that a significant number of the terms will be zero. When applied to such sparse systems, full-matrix elimination methods waste great amounts of computer memory storing these zeros. For this reason, approximate methods provide a viable approach for obtaining solutions for elliptical equations. The most commonly employed approach is *Gauss-Seidel*, which when applied to PDEs is also referred to as *Liebmann's method*. In this technique, Eq. (29.8) is expressed as

$$T_{i,j} = \frac{T_{i+1,j} + T_{i-1,j} + T_{i,j+1} + T_{i,j-1}}{4} \tag{29.11}$$

and solved iteratively for $j = 1$ to n and $i = 1$ to m. Because Eq. (29.8) is diagonally dominant, this procedure will eventually converge on a stable solution (recall Sec. 11.2.1). Overrelaxation is sometimes employed to accelerate the rate of convergence by applying the following formula after each iteration:

$$T_{i,j}^{new} = \lambda T_{i,j}^{new} + (1 - \lambda)T_{i,j}^{old} \tag{29.12}$$

where $T_{i,j}^{new}$ and $T_{i,j}^{old}$ are the values of $T_{i,j}$ from the present and the previous iteration, respectively, and λ is a weighting factor that is set between 1 and 2.

As with the conventional Gauss-Seidel method, the iterations are repeated until the absolute values of all the percent relative errors $(\varepsilon_a)_{i,j}$ fall below a prespecified stopping criterion ε_s. These percent relative errors are estimated by

$$\left|(\varepsilon_a)_{i,j}\right| = \left|\frac{T_{i,j}^{new} - T_{i,j}^{old}}{T_{i,j}^{new}}\right| 100\% \tag{29.13}$$

EXAMPLE 29.1   Temperature of a Heated Plate with Fixed Boundary Conditions

Problem Statement.   Use Liebmann's method (Gauss-Seidel) to solve for the temperature of the heated plate in Fig. 29.4. Employ overrelaxation with a value of 1.5 for the weighting factor and iterate to $\varepsilon_s = 1\%$.

Solution.   Equation (29.11) at $i = 1, j = 1$ is

$$T_{11} = \frac{0 + 75 + 0 + 0}{4} = 18.75$$

and applying overrelaxation yields

$$T_{11} = 1.5(18.75) + (1 - 1.5)0 = 28.125$$

For $i = 2, j = 1$,

$$T_{21} = \frac{0 + 28.125 + 0 + 0}{4} = 7.03125$$

$$T_{21} = 1.5(7.03125) + (1 - 1.5)0 = 10.54688$$

For $i = 3, j = 1$,

$$T_{31} = \frac{50 + 10.54688 + 0 + 0}{4} = 15.13672$$

$$T_{31} = 1.5(15.13672) + (1 - 1.5)0 = 22.70508$$

The computation is repeated for the other rows to give

$$T_{12} = 38.67188 \qquad T_{22} = 18.45703 \qquad T_{32} = 34.18579$$
$$T_{13} = 80.12696 \qquad T_{23} = 74.46900 \qquad T_{33} = 96.99554$$

Because all the $T_{i,j}$'s are initially zero, all ε_a's for the first iteration will be 100%.
For the second iteration the results are

$$T_{11} = 32.51953 \qquad T_{21} = 22.35718 \qquad T_{31} = 28.60108$$
$$T_{12} = 57.95288 \qquad T_{22} = 61.63333 \qquad T_{32} = 71.86833$$
$$T_{13} = 75.21973 \qquad T_{23} = 87.95872 \qquad T_{33} = 67.68736$$

The error for $T_{1,1}$ can be estimated as [Eq. (29.13)]

$$|(\varepsilon_a)_{11}| = \left| \frac{32.51953 - 28.12500}{32.51953} \right| 100\% = 13.5\%$$

Because this value is above the stopping criterion of 1%, the computation is continued. The ninth iteration gives the result

$$T_{11} = 43.00061 \qquad T_{21} = 33.29755 \qquad T_{31} = 33.88506$$
$$T_{12} = 63.21152 \qquad T_{22} = 56.11238 \qquad T_{32} = 52.33999$$
$$T_{13} = 78.58718 \qquad T_{23} = 76.06402 \qquad T_{33} = 69.71050$$

where the maximum error is 0.71%.

Figure 29.5 shows the results. As expected, a gradient is established as heat flows from high to low temperatures.

FIGURE 29.5
Temperature distribution for a heated plate subject to fixed boundary conditions.

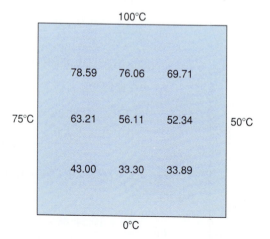

100°C

| | 78.59 | 76.06 | 69.71 |

75°C   63.21   56.11   52.34   50°C

| | 43.00 | 33.30 | 33.89 |

0°C

29.2.3 Secondary Variables

Because its distribution is described by the Laplace equation, temperature is considered to be the primary variable in the heated-plate problem. For this case, as well as for other problems involving PDEs, secondary variables may also be of interest. As a matter of fact, in certain engineering contexts, the secondary variable may actually be more important.

For the heated plate, a secondary variable is the rate of heat flux across the plate's surface. This quantity can be computed from Fourier's law. Central finite-difference approximations for the first derivatives (recall Fig. 23.3) can be substituted into Eq. (29.4) to give the following values for heat flux in the x and y dimensions:

$$q_x = -k' \frac{T_{i+1,j} - T_{i-1,j}}{2\,\Delta x} \tag{29.14}$$

and

$$q_x = -k' \frac{T_{i,j+1} - T_{i,j-1}}{2\,\Delta y} \tag{29.15}$$

The resultant heat flux can be computed from these two quantities by

$$q_n = \sqrt{q_x^2 + q_y^2} \tag{29.16}$$

where the direction of q_n is given by

$$\theta = \tan^{-1}\left(\frac{q_y}{q_x}\right) \tag{29.17}$$

for $q_x > 0$ and

$$\theta = \tan^{-1}\left(\frac{q_y}{q_x}\right) + \pi \tag{29.18}$$

for $q_x < 0$. Recall that the angle can be expressed in degrees by multiplying it by $180°/\pi$. If $q_x = 0$, θ is $\pi/2$ (90°) or $3\pi/2$ (270°), depending on whether q_y is positive or negative, respectively.

EXAMPLE 29.2   Flux Distribution for a Heated Plate

Problem Statement.   Employ the results of Example 29.1 to determine the distribution of heat flux for the heated plate from Fig. 29.4. Assume that the plate is 40 × 40 cm and is made out of aluminum [$k' = 0.49$ cal/(s·cm·°C)].

Solution.   For $i = j = 1$, Eq. (29.14) can be used to compute

$$q_x = -0.49 \frac{\text{cal}}{\text{s} \cdot \text{cm} \cdot \text{°C}} \frac{(33.29755 - 75)\text{°C}}{2(10 \text{ cm})} = 1.022 \text{ cal/(cm}^2 \cdot \text{s)}$$

and [Eq. (29.15)]

$$q_y = -0.49 \frac{\text{cal}}{\text{s} \cdot \text{cm} \cdot \text{°C}} \frac{(63.21152 - 0)\text{°C}}{2(10 \text{ cm})} = -1.549 \text{ cal/(cm}^2 \cdot \text{s)}$$

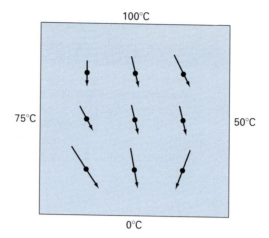

FIGURE 29.6
Heat flux for a plate subject to fixed boundary temperatures. Note that the lengths of the arrows are proportional to the magnitude of the flux.

The resultant flux can be computed with Eq. (29.16):

$$q_n = \sqrt{(1.022)^2 + (-1.549)^2} = 1.856 \text{ cal/(cm}^2 \cdot \text{s)}$$

and the angle of its trajectory by Eq. (29.17)

$$\theta = \tan^{-1}\left(\frac{-1.549}{1.022}\right) = -0.98758 \times \frac{180°}{\pi} = -56.584°$$

Thus, at this point, the heat flux is directed down and to the right. Values at the other grid points can be computed; the results are displayed in Fig. 29.6.

29.3   BOUNDARY CONDITIONS

Because it is free of complicating factors, the rectangular plate with fixed boundary conditions has been an ideal context for showing how elliptic PDEs can be solved numerically. We will now elaborate on other issues that will expand our capabilities to address more realistic problems. These involve boundaries at which the derivative is specified and boundaries that are irregularly shaped.

29.3.1 Derivative Boundary Conditions

The fixed or Dirichlet boundary condition discussed to this point is but one of several types that are used with partial differential equations. A common alternative is the case where the

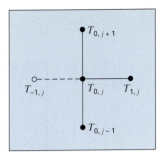

FIGURE 29.7
A boundary node (0, j) on the left edge of a heated plate. To approximate the derivative normal to the edge (that is, the x derivative), an imaginary point (−1, j) is located a distance Δx beyond the edge.

derivative is given. This is commonly referred to as a *Neumann boundary condition*. For the heated-plate problem, this amounts to specifying the heat flux rather than the temperature at the boundary. One example is the situation where the edge is insulated. In this case, which is referred to as a *natural boundary condition*, the derivative is zero. This conclusion is drawn directly from Eq. (29.4) because insulating a boundary means that the heat flux (and consequently the gradient) must be zero. Another example would be where heat is lost across the edge by predictable mechanisms such as radiation and conduction.

Figure 29.7 depicts a node $(0, j)$ at the left edge of a heated plate. Applying Eq. (29.8) at the point gives

$$T_{1,j} + T_{-1,j} + T_{0,j+1} + T_{0,j-1} - 4T_{0,j} = 0 \qquad (29.19)$$

Notice that an imaginary point $(-1, j)$ lying outside the plate is required for this equation. Although this exterior fictitious point might seem to represent a problem, it actually serves as the vehicle for incorporating the derivative boundary condition into the problem. This is done by representing the first derivative in the x dimension at $(0, j)$ by the finite divided difference

$$\frac{\partial T}{\partial x} \cong \frac{T_{1,j} - T_{-1,j}}{2\,\Delta x}$$

which can be solved for

$$T_{-1,j} = T_{1,j} - 2\,\Delta x\,\frac{\partial T}{\partial x}$$

Now we have a relationship for $T_{-1,j}$ that actually includes the derivative. It can be substituted into Eq. (29.19) to give

$$2T_{1,j} - 2\,\Delta x\,\frac{\partial T}{\partial x} + T_{0,j+1} + T_{0,j-1} - 4T_{0,j} = 0 \qquad (29.20)$$

Thus, we have incorporated the derivative into the balance.

Similar relationships can be developed for derivative boundary conditions at the other edges. The following example shows how this is done for the heated plate.

EXAMPLE 29.3   Heated Plate with an Insulated Edge

Problem Statement.   Repeat the same problem as in Example 29.1, but with the lower edge insulated.

Solution.   The general equation to characterize a derivative at the lower edge (that is, at $j = 0$) of a heated plate is

$$T_{i+1,0} + T_{i-1,0} + 2T_{i,1} - 2\,\Delta y\,\frac{\partial T}{\partial y} - 4T_{i,0} = 0$$

For an insulated edge, the derivative is zero and the equation becomes

$$T_{i+1,0} + T_{i-1,0} + 2T_{i,1} - 4T_{i,0} = 0$$

The simultaneous equations for temperature distribution on the plate in Fig. 29.4 with an insulated lower edge can be written in matrix form as

$$
\begin{bmatrix}
4 & -1 & & -2 & & & & & & & & \\
-1 & 4 & -1 & & -2 & & & & & & & \\
& -1 & 4 & & & -2 & & & & & & \\
-1 & & & 4 & -1 & & -1 & & & & & \\
& -1 & & -1 & 4 & -1 & & -1 & & & & \\
& & -1 & & -1 & 4 & & & -1 & & & \\
& & & -1 & & & 4 & -1 & & -1 & & \\
& & & & -1 & & -1 & 4 & -1 & & -1 & \\
& & & & & -1 & & -1 & 4 & & & -1 \\
& & & & & & -1 & & & 4 & -1 & \\
& & & & & & & -1 & & -1 & 4 & -1 \\
& & & & & & & & -1 & & -1 & 4 \\
\end{bmatrix}
\begin{Bmatrix}
T_{10} \\ T_{20} \\ T_{30} \\ T_{11} \\ T_{21} \\ T_{31} \\ T_{12} \\ T_{22} \\ T_{32} \\ T_{13} \\ T_{23} \\ T_{33}
\end{Bmatrix}
=
\begin{Bmatrix}
75 \\ 0 \\ 50 \\ 75 \\ 0 \\ 50 \\ 75 \\ 0 \\ 50 \\ 175 \\ 100 \\ 150
\end{Bmatrix}
$$

Note that because of the derivative boundary condition, the matrix is increased to 12×12 in contrast to the 9×9 system in Eq. (29.10) to account for the three unknown temperatures along the plate's lower edge. These equations can be solved for

$$T_{10} = 71.91 \qquad T_{20} = 67.01 \qquad T_{30} = 59.54$$
$$T_{11} = 72.81 \qquad T_{21} = 68.31 \qquad T_{31} = 60.57$$
$$T_{12} = 76.01 \qquad T_{22} = 72.84 \qquad T_{32} = 64.42$$
$$T_{13} = 83.41 \qquad T_{23} = 82.63 \qquad T_{33} = 74.26$$

FIGURE 29.8
Temperature and flux distribution for a heated plate subject to fixed boundary conditions except for an insulated lower edge.

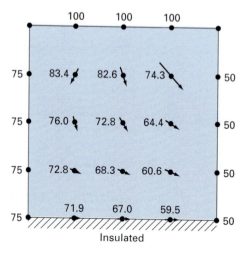

These results and computed fluxes (for the same parameters as in Example 29.2) are displayed in Fig. 29.8. Note that, because the lower edge is insulated, the plate's temperature is higher than for Fig. 29.5, where the bottom edge temperature is fixed at zero. In addition, the heat flow (in contrast to Fig. 29.6) is now deflected to the right and moves parallel to the insulated wall.

29.3.2 Irregular Boundaries

Although the rectangular plate from Fig. 29.4 has served well to illustrate the fundamental aspects of solving elliptic PDEs, many engineering problems do not exhibit such an idealized geometry. For example, a great many systems have irregular boundaries (Fig. 29.9).

Figure 29.9 is a system that can serve to illustrate how nonrectangular boundaries can be handled. As depicted, the plate's lower left boundary is circular. Notice that we have affixed parameters—α_1, α_2, β_1, β_2—to each of the lengths surrounding the node. Of course, for the plate depicted in Fig. 29.9, $\alpha_2 = \beta_2 = 1$. However, we will retain these parameters throughout the following derivation so that the resulting equation is generally applicable to any irregular boundary—not just one on the lower left-hand corner of a heated plate. The first derivatives in the x dimension can be approximated as

$$\left(\frac{\partial T}{\partial x}\right)_{i-1,i} \cong \frac{T_{i,j} - T_{i-1,j}}{\alpha_1 \, \Delta x} \tag{29.21}$$

and

$$\left(\frac{\partial T}{\partial x}\right)_{i,i+1} \cong \frac{T_{i+1,j} - T_{i,j}}{\alpha_2 \, \Delta x} \tag{29.22}$$

FIGURE 29.9
A grid for a heated plate with an irregularly shaped boundary. Note how weighting coefficients are used to account for the nonuniform spacing in the vicinity of the nonrectangular boundary.

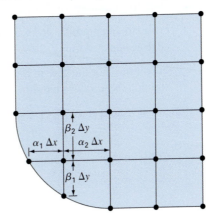

The second derivatives can be developed from these first derivatives. For the x dimension, the second derivative is

$$\frac{\partial^2 T}{\partial x^2} = \frac{\partial}{\partial x}\left(\frac{\partial T}{\partial x}\right) = \frac{\left(\dfrac{\partial T}{\partial x}\right)_{i,i+1} - \left(\dfrac{\partial T}{\partial x}\right)_{i-1,i}}{\dfrac{\alpha_1 \, \Delta x + \alpha_2 \, \Delta x}{2}} \tag{29.23}$$

Substituting Eqs. (29.21) and (29.22) into (29.23) gives

$$\frac{\partial^2 T}{\partial x^2} = 2\frac{\dfrac{T_{i-1,j} - T_{i,j}}{\alpha_1 \, \Delta x} - \dfrac{T_{i+1,j} - T_{i,j}}{\alpha_2 \, \Delta x}}{\alpha_1 \, \Delta x + \alpha_2 \, \Delta x}$$

Collecting terms yields

$$\frac{\partial^2 T}{\partial x^2} = \frac{2}{\Delta x^2}\left[\frac{T_{i-1,j} - T_{i,j}}{\alpha_1 \, (\alpha_1 + \alpha_2)} + \frac{T_{i+1,j} - T_{i,j}}{\alpha_2 \, (\alpha_1 + \alpha_2)}\right]$$

A similar equation can be developed in the y dimension:

$$\frac{\partial^2 T}{\partial y^2} = \frac{2}{\Delta y^2}\left[\frac{T_{i,j-1} - T_{i,j}}{\beta_1 \, (\beta_1 + \beta_2)} + \frac{T_{i,j+1} - T_{i,j}}{\beta_2 \, (\beta_1 + \beta_2)}\right]$$

Substituting these equations in Eq. (29.6) yields

$$\begin{aligned}\frac{2}{\Delta x^2}&\left[\frac{T_{i-1,j} - T_{i,j}}{\alpha_1 \, (\alpha_1 + \alpha_2)} + \frac{T_{i+1,j} - T_{i,j}}{\alpha_2 \, (\alpha_1 + \alpha_2)}\right] \\ &+ \frac{2}{\Delta y^2}\left[\frac{T_{i,j-1} - T_{i,j}}{\beta_1 \, (\beta_1 + \beta_2)} + \frac{T_{i,j+1} - T_{i,j}}{\beta_2 \, (\beta_1 + \beta_2)}\right] = 0\end{aligned} \tag{29.24}$$

As illustrated in the following example, Eq. (29.24) can be applied to any node that lies adjacent to an irregular, Dirichlet-type boundary.

EXAMPLE 29.4   Heated Plate with an Irregular Boundary

Problem Statement.   Repeat the same problem as in Example 29.1, but with the lower edge as depicted in Fig. 29.9.

Solution.   For the case in Fig. 29.9, $\Delta x = \Delta y$, $\alpha_1 = \beta_1 = 0.732$, and $\alpha_2 = \beta_2 = 1$. Substituting these values into Eq. (29.24) yields the following balance for node (1, 1):

$$0.788675(T_{01} - T_{11}) + 0.57735(T_{21} - T_{11})$$
$$+\, 0.788675(T_{10} - T_{11}) + 0.57735(T_{12} - T_{11}) = 0$$

Collecting terms, we can express this equation as

$$-4T_{11} + 0.8453T_{21} + 0.8453T_{12} = -1.1547T_{01} - 1.1547T_{10}$$

The simultaneous equations for temperature distribution on the plate in Fig. 29.9 with a lower-edge boundary temperature of 75 can be written in matrix form as

$$\begin{bmatrix} 4 & -0.845 & & -0.845 & & & & & \\ -1 & 4 & -1 & & -1 & & & & \\ & -1 & 4 & & & -1 & & & \\ -1 & & & 4 & -1 & & -1 & & \\ & -1 & & -1 & 4 & -1 & & -1 & \\ & & -1 & & -1 & 4 & & & -1 \\ & & & -1 & & & 4 & -1 & \\ & & & & -1 & & -1 & 4 & -1 \\ & & & & & -1 & & -1 & -4 \end{bmatrix} \begin{Bmatrix} T_{11} \\ T_{21} \\ T_{31} \\ T_{12} \\ T_{22} \\ T_{32} \\ T_{13} \\ T_{23} \\ T_{33} \end{Bmatrix} = \begin{Bmatrix} 173.2 \\ 75 \\ 125 \\ 75 \\ 0 \\ 50 \\ 175 \\ 100 \\ 150 \end{Bmatrix}$$

These equations can be solved for

$$T_{11} = 74.98 \qquad T_{21} = 72.76 \qquad T_{31} = 66.07$$
$$T_{12} = 77.23 \qquad T_{22} = 75.00 \qquad T_{32} = 66.52$$
$$T_{13} = 83.93 \qquad T_{23} = 83.48 \qquad T_{33} = 75.00$$

These results along with the computed fluxes are displayed in Fig. 29.10. Note that the fluxes are computed in the same fashion as in Sec. 29.2.3, with the exception that $(\alpha_1 + \alpha_2)$ and $(\beta_1 + \beta_2)$ are substituted for the 2's in the denominators of Eqs. (29.14) and (29.15), respectively. Section 32.3 illustrates how this is done.

FIGURE 29.10
Temperature and flux distribution for a heated plate with a circular boundary.

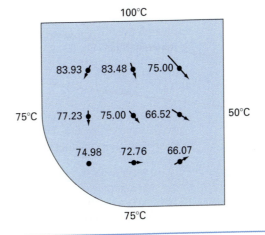

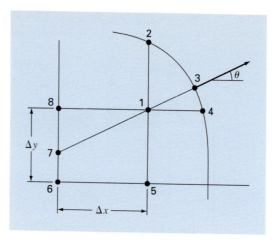

FIGURE 29.11

A curved boundary where the normal gradient is specified.

Derivative conditions for irregularly shaped boundaries are more difficult to formulate. Figure 29.11 shows a point near an irregular boundary where the normal derivative is specified.

The normal derivative at node 3 can be approximated by the gradient between nodes 1 and 7,

$$\frac{\partial T}{\partial \eta}\bigg|_3 = \frac{T_1 - T_7}{L_{17}} \tag{29.25}$$

When θ is less than 45° as shown, the distance from node 7 to 8 is $\Delta x \tan \theta$, and linear interpolation can be used to estimate

$$T_7 = T_8 + (T_6 - T_8)\frac{\Delta x \tan \theta}{\Delta y}$$

The length L_{17} is equal to $\Delta x/\cos \theta$. This length, along with the approximation for T_7, can be substituted into Eq. (29.25) to give

$$T_1 = \left(\frac{\Delta x}{\cos \theta}\right)\frac{\partial T}{\partial \eta}\bigg|_3 + T_6\frac{\Delta x \tan \theta}{\Delta y} + T_8\left(1 - \frac{\Delta x \tan \theta}{\Delta y}\right) \tag{29.26}$$

Such an equation provides a means for incorporating the normal gradient into the finite-difference approach. For cases where θ is greater than 45°, a different equation would be used. The determination of this formula will be left as a homework exercise.

29.4　THE CONTROL-VOLUME APPROACH

To summarize, the finite-difference or Taylor series approach divides the continuum into nodes (Fig. 29.12a). The underlying partial differential equation is written for each of these nodes. Finite-difference approximations are then substituted for the derivatives to convert the equations to an algebraic form.

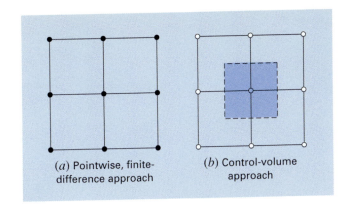

(a) Pointwise, finite-difference approach

(b) Control-volume approach

FIGURE 29.12
Two different perspectives for developing approximate solutions of PDEs: (a) finite-difference and (b) control volume.

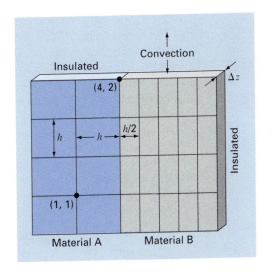

FIGURE 29.13
A heated plate with unequal grid spacing, two materials, and mixed boundary conditions.

Such an approach is quite simple and straightforward for orthogonal (i.e., rectangular) grids and constant coefficients. However, the approach becomes a more difficult endeavor for derivative conditions on irregularly shaped boundaries.

Figure 29.13 is an example of a system where additional difficulties arise. This plate is made of two different materials and has unequal grid spacing. In addition, half of its top edge is subject to convective heat transfer, whereas half is insulated. Developing equations for node (4, 2) would require some additional derivation, beyond the approaches developed to this point.

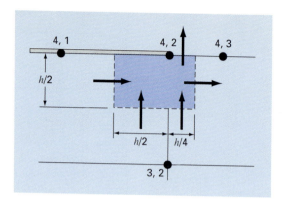

FIGURE 29.14
A control volume for node (4, 2) with arrows indicating heat transfer through the boundaries.

The *control-volume approach* (also called the *volume-integral approach*) offers an alternative way to numerically approximate PDEs that is especially useful for cases such as Fig. 29.13. As in Fig. 29.12b, the approach resembles the point-wise approach in that points are determined across the domain. However, rather than approximating the PDE at a point, the approximation is applied to a volume surrounding the point. For an orthogonal grid, the volume is formed by the perpendicular lines through the midpoint of each line joining adjacent nodes. A heat balance can then be developed for each volume in a fashion similar to Eq. (29.1).

As an example, we will apply the control-volume approach to node (4, 2). First, the volume is defined by bisecting the lines joining the nodes. As in Fig. 29.14, the volume has conductive heat transfer through its left, right, and lower boundaries and convective heat transfer through half of its upper boundary. Notice that the transfer through the lower boundary involves both materials.

A steady-state heat balance for the volume can be written in qualitative terms as

$$0 = \begin{pmatrix} \text{left-side} \\ \text{conduction} \end{pmatrix} - \begin{pmatrix} \text{right-side} \\ \text{conduction} \end{pmatrix} + \begin{pmatrix} \text{lower conduction} \\ \text{material "a"} \end{pmatrix}$$
$$+ \begin{pmatrix} \text{lower conduction} \\ \text{material "b"} \end{pmatrix} - \begin{pmatrix} \text{upper} \\ \text{convection} \end{pmatrix} \qquad (29.27)$$

Now the conduction flux rate can be represented by the finite-difference version of Fourier's law. For example, for the left-side conduction gain, it would be

$$q = -k_a' \frac{T_{42} - T_{41}}{h}$$

where q has units of cal/cm²/s. This flux rate must be then multiplied by the area across which it enters ($\Delta z \times h/2$) to give the rate of heat entering the volume per unit time,

$$Q = -k_a' \frac{T_{42} - T_{41}}{h} \frac{h}{2} \Delta z$$

where Q has units of cal/s.

The heat flux due to convection can be formulated as

$$q = h_c \, (T_a - T_{42})$$

where h_c = a heat convection coefficient [cal/(s·cm^2·°C)] and T_a = the air temperature (°C). Again, multiplication by the proper area yields the rate of heat flow per time,

$$Q = h_c \, (T_a - T_{42}) \frac{h}{4} \, \Delta z$$

The other transfers can be developed in a similar fashion and substituted into Eq. (29.27) to yield

$$0 = -k_a' \frac{T_{42} - T_{41}}{h} \frac{h}{2} \Delta z + k_b' \frac{T_{43} - T_{42}}{h/2} \frac{h}{2} \Delta z$$

(Left-side conduction) (Right-side conduction)

$$-k_a' \frac{T_{42} - T_{32}}{h} \frac{h}{2} \, \Delta z - k_b' \frac{T_{42} - T_{32}}{h} \frac{h}{4} \, \Delta z + h_c (T_a - T_{42}) \frac{h}{4} \, \Delta z$$

$$\left(\begin{array}{c} \text{Lower conduction} \\ \text{material ``a''} \end{array} \right) \left( \begin{array}{c} \text{Lower conduction} \\ \text{material ``b''} \end{array} \right) \text{(Upper convection)}$$

Parameter values can then be substituted to yield the final heat balance equation. For example, if $\Delta z = 0.5$ cm, $h = 10$ cm, $k_a' = 0.3$ cal/(s·cm·°C), $k_b' = 0.5$ cal/(s·cm·°C), and $h_c = 0.1$ cal/(s·cm^2·°C), the equation becomes

$$0.5875 T_{42} - 0.075 T_{41} - 0.25 T_{43} - 0.1375 T_{32} = 2.5$$

To make the equation comparable to the standard Laplacian, this equation can be multiplied by 4/0.5875 so that the coefficient of the base node has a coefficient of 4,

$$4 T_{42} - 0.510638 T_{41} - 1.702128 T_{43} - 0.93617 T_{32} = 17.02128$$

For the standard cases covered to this point, the control-volume and pointwise finite-difference approaches yield identical results. For example, for node (1, 1) in Fig. 29.13, the balance would be

$$0 = -k_a' \frac{T_{11} - T_{01}}{h} h \, \Delta z + k_a' \frac{T_{21} - T_{11}}{h} h \, \Delta z - k_a' \frac{T_{11} - T_{10}}{h} h \, \Delta z + k_a' \frac{T_{12} - T_{11}}{h} h \, \Delta z$$

which simplifies to the standard Laplacian,

$$0 = 4 T_{11} - T_{01} - T_{21} - T_{12} - T_{10}$$

We will look at other standard cases (e.g., the derivative boundary condition) and explore the control-volume approach in additional detail in the problems at the end of this chapter.

29.5 SOFTWARE TO SOLVE ELLIPTIC EQUATIONS

Modifying a computer program to include derivative boundary conditions for rectangular systems is a relatively straightforward task. It merely involves ensuring that additional equations are generated to characterize the boundary nodes at which the derivatives are specified. In addition, the code must be modified so that these equations incorporate the derivative in the fashion of Eq. (29.20).

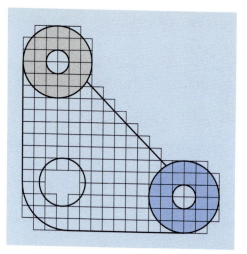

FIGURE 29.15
A finite-difference grid superimposed on an irregularly shaped gasket.

Developing general software to characterize systems with irregular boundaries is a much more difficult proposition. For example, a fairly involved algorithm would be required to model the simple gasket depicted in Fig. 29.15. This would involve two major modifications. First, a scheme would have to be developed to conveniently input the configuration of the nodes and to identify which were at the boundary. Second, an algorithm would be required to generate the proper simultaneous equations on the basis of the input information. The net result is that general software for solving elliptic (and for that matter, all) PDEs is relatively complicated.

One method used to simplify such efforts is to require a very fine grid. For such cases, it is often assumed that the closest node serves as the boundary point. In this way, the analysis does not have to consider the weighting parameters from Sec. 29.3.2. Although this introduces some error, the use of a sufficiently fine mesh can make the resulting discrepancy negligible. However, this involves a trade-off due to the computational burden introduced by the increased number of simultaneous equations.

As a consequence of these considerations, numerical analysts have developed alternative approaches that differ radically from finite-difference methods. Although these finite-element methods are more conceptually difficult, they can much more easily accommodate irregular boundaries. We will turn to these methods in Chap. 31. Before doing this, however, we will first describe finite-difference approaches for another category of PDEs—parabolic equations.

PROBLEMS

29.1 Use Liebmann's method to solve for the temperature of the square heated plate in Fig. 29.4, but with the upper boundary condition increased to 150°C and the left boundary decreased to 25°C. Use a relaxation factor of 1.2 and iterate to $\varepsilon_s = 1\%$.

29.2 Compute the fluxes for Prob. 29.1 using the parameters from Example 29.3.

29.3 Repeat Example 29.1, but use 49 interior nodes (that is, $\Delta x = \Delta y = 5$ cm).

29.4 Repeat Prob. 29.3, but for the case where the upper edge is insulated.

29.5 Repeat Examples 29.1 and 29.3, but for the case where the flux at the lower edge is directed downward with a value of 2 cal/(cm^2·s).

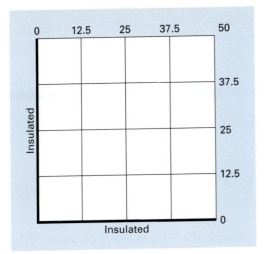

FIGURE P29.7

29.6 Repeat Example 29.5 for the case where both the lower left and the upper right corners are rounded in the same fashion as the lower left corner of Fig. 29.9. Note that all boundary temperatures on the upper and right sides are fixed at 50°C and all on the lower and left sides are fixed at 100°C.

29.7 With the exception of the boundary conditions, the plate in Fig. P29.7 has the exact same characteristics as the plate used in Examples 23.1 through 23.4. Simulate both the temperatures and fluxes for the plate.

29.8 Write equations for the darkened nodes in the grid in Fig. P29.8. Note that all units are cgs. The coefficient of thermal conductivity for the plate is 0.5 cal/(s·cm·°C), the convection coefficient is $h_c = 0.01$ cal/(cm^2·°C·s), and the thickness of the plate is 1 cm.

29.9 Write equations for the darkened nodes in the grid in Fig. P29.9. Note that all units are cgs. The convection coefficient is $h_c = 0.01$ cal/(cm^2·°C·s) and the thickness of the plate is 2 cm.

29.10 Apply the control volume approach to develop the equation for node $(0, j)$ in Fig. 29.7.

29.11 Derive an equation like Eq. (29.26) for the case where θ is greater than 45° for Fig. 29.11.

29.12 Develop a user-friendly computer program to implement Liebmann's method for a rectangular plate. Design the program so

that it can compute both temperature and flux. Test the program by duplicating the results of Examples 29.1 and 29.2.

29.13 Employ the program from Prob. 29.12 to solve Probs. 29.1 and 29.2.

29.14 Employ the program from Prob 29.12 to solve Prob. 29.3.

FIGURE P29.8

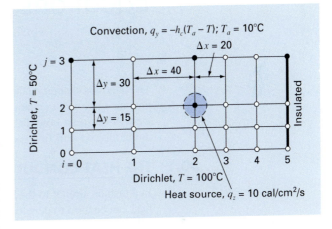

FIGURE P29.9

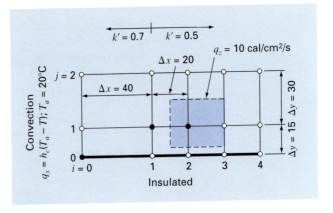

CHAPTER 30

Finite Difference: Parabolic Equations

The previous chapter dealt with steady-state PDEs. We now turn to the parabolic equations that are employed to characterize time-variable problems. In the latter part of this chapter, we will illustrate how this is done in two spatial dimensions for the heated plate. Before doing this, we will first show how the simpler one-dimensional case is approached.

30.1 THE HEAT CONDUCTION EQUATION

In a fashion similar to the derivation of the Laplace equation [Eq. (29.6)], conservation of heat can be used to develop a heat balance for the differential element in the long, thin insulated rod shown in Fig. 30.1. However, rather than examine the steady-state case, the present balance also considers the amount of heat stored in the element over a unit time period Δt. Thus, the balance is in the form, inputs − outputs = storage, or

$$q(x)\,\Delta y\,\Delta z\,\Delta t - q(x + \Delta x)\,\Delta y\,\Delta z\,\Delta t = \Delta x\,\Delta y\,\Delta z \rho C\,\Delta T$$

Dividing by the volume of the element ($= \Delta x\,\Delta y\,\Delta z$) and Δt gives

$$\frac{q(x) - q(x + \Delta x)}{\Delta x} = \rho C \frac{\Delta T}{\Delta t}$$

Taking the limit yields

$$-\frac{\partial q}{\partial x} = \rho C \frac{\partial T}{\partial t}$$

FIGURE 30.1
A thin rod, insulated at all points except at its ends.

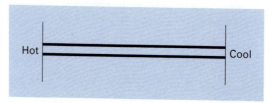

832

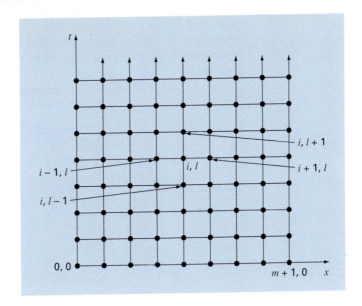

FIGURE 30.2
A grid used for the finite-difference solution of parabolic PDEs in two independent variables such as the heat-conduction equation. Note how, in contrast to Fig. 29.3, this grid is open-ended in the temporal dimension.

Substituting Fourier's law of heat conduction [Eq. (29.4)] results in

$$k\frac{\partial^2 T}{\partial x^2} = \frac{\partial T}{\partial t} \tag{30.1}$$

which is the *heat-conduction equation*.

Just as with elliptic PDEs, parabolic equations can be solved by substituting finite divided differences for the partial derivatives. However, in contrast to elliptic PDEs, we must now consider changes in time as well as in space. Whereas elliptic equations were bounded in all relevant dimensions, parabolic PDEs are temporally open-ended (Fig. 30.2). Because of their time-variable nature, solutions to these equations involve a number of new issues, notably stability. This, as well as other aspects of parabolic PDEs, will be examined in the following sections as we present two fundamental solution approaches—explicit and implicit schemes.

30.2 EXPLICIT METHODS

The heat-conduction equation requires approximations for the second derivative in space and the first derivative in time. The former is represented in the same fashion as for the Laplace equation by a centered finite divided difference:

$$\frac{\partial^2 T}{\partial x^2} = \frac{T_{i+1}^l - 2T_i^l + T_{i-1}^l}{\Delta x^2} \tag{30.2}$$

which has an error (recall Fig. 23.3) of $O[(\Delta x)^2]$. Notice the slight change in notation that superscripts are used to denote time. This is done so that a second subscript can be used to designate a second spatial dimension when the approach is expanded to two spatial dimensions.

A forward finite divided difference is used to approximate the time derivative

$$\frac{\partial T}{\partial t} = \frac{T_i^{l+1} - T_i^l}{\Delta t} \tag{30.3}$$

which has an error (recall Fig. 23.1) of $O(\Delta t)$.

Substituting Eqs. (30.2) and (30.3) into Eq. (30.1) yields

$$k\frac{T_{i+1}^l - 2T_i^l + T_{i-1}^l}{(\Delta x)^2} = \frac{T_i^{l+1} - T_i^l}{\Delta t} \tag{30.4}$$

which can be solved for

$$T_i^{l+1} = T_i^l + \lambda(T_{i+1}^l - 2T_i^l + T_{i-1}^l) \tag{30.5}$$

where $\lambda = k\,\Delta t/(\Delta x)^2$.

This equation can be written for all the interior nodes on the rod. It then provides an explicit means to compute values at each node for a future time based on the present values at the node and its neighbors. Notice that this approach is actually a manifestation of Euler's method for solving systems of ODEs. That is, if we know the temperature distribution as a function of position at an initial time, we can compute the distribution at a future time based on Eq. (30.5).

A computational molecule for the explicit method is depicted in Fig. 30.3, showing the nodes that constitute the spatial and temporal approximations. This molecule can be contrasted with others in this chapter to illustrate the differences between approaches.

FIGURE 30.3
A computational molecule for the explicit form.

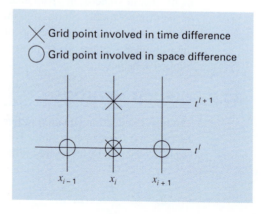

EXAMPLE 30.1   Explicit Solution of the One-Dimensional Heat-Conduction Equation

Problem Statement.   Use the explicit method to solve for the temperature distribution of a long, thin rod with a length of 10 cm and the following values: $k' = 0.49$ cal/(s·cm·°C), $\Delta x = 2$ cm, and $\Delta t = 0.1$ s. At $t = 0$, the temperature of the rod is zero and the boundary conditions are fixed for all times at $T(0) = 100\,°C$ and $T(10) = 50\,°C$. Note that the rod is aluminum with $C = 0.2174$ cal/(g·°C) and $\rho = 2.7$ g/cm³. Therefore, $k = 0.49/(2.7 \cdot 0.2174) = 0.835$ cm²/s and $\lambda = 0.835(0.1)/(2)^2 = 0.020875$.

Solution.   Applying Eq. (30.5) gives the following value at $t = 0.1$ s for the node at $x = 2$ cm:

$$T_1^1 = 0 + 0.020875[0 - 2(0) + 100] = 2.0875$$

At the other interior points, $x = 4, 6,$ and 8 cm, the results are

$$T_2^1 = 0 + 0.020875[0 - 2(0) + 0] = 0$$
$$T_3^1 = 0 + 0.020875[0 - 2(0) + 0] = 0$$
$$T_4^1 = 0 + 0.020875[50 - 2(0) + 0] = 1.0438$$

At $t = 0.2$ s, the values at the four interior nodes are computed as

$$T_1^2 = 2.0875 + 0.020875[0 - 2(2.0875) + 100] = 4.0878$$
$$T_2^2 = 0 + 0.020875[0 - 2(0) + 2.0875] = 0.043577$$
$$T_3^2 = 0 + 0.020875[1.0438 - 2(0) + 0] = 0.021788$$
$$T_4^2 = 1.0438 + 0.020875[50 - 2(1.0438) + 0] = 2.0439$$

The computation is continued, and the results at 3-s intervals are depicted in Fig. 30.4. The general rise in temperature with time indicates that the computation captures the diffusion of heat from the boundaries into the bar.

FIGURE 30.4
Temperature distribution in a long, thin rod as computed with the explicit method described in Sec. 30.2.

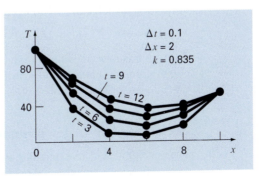

30.2.1 Convergence and Stability

Convergence means that as Δx and Δt approach zero, the results of the finite-difference technique approach the true solution. *Stability* means that errors at any stage of the computation are not amplified but are attenuated as the computation progresses. It can be shown (Carnahan et al., 1969) that the explicit method is both convergent and stable if $\lambda \leq 1/2$, or

$$\Delta t \leq \frac{1}{2} \frac{\Delta x^2}{k} \tag{30.6}$$

In addition, it should be noted that setting $\lambda \leq 1/2$ could result in a solution in which errors do not grow, but oscillate. Setting $\lambda \leq 1/4$ ensures that the solution will not oscillate. It is also known that setting $\lambda = 1/6$ tends to minimize truncation error (Carnahan et al., 1969).

Figure 30.5 is an example of instability caused by violating Eq. (30.6). This plot is for the same case as in Example 30.1 but with $\lambda = 0.735$, which is considerably greater than 0.5. As in Fig. 30.5, the solution undergoes progressively increasing oscillations. This situation will continue to deteriorate as the computation continues.

Although satisfaction of Eq. (30.6) will alleviate the instabilities of the sort manifested in Fig. 30.5, it also places a strong limitation on the explicit method. For example, suppose that Δx is halved to improve the approximation of the spatial second derivative. According to Eq. (30.6), the time step must be quartered to maintain convergence and stability. Thus, to perform comparable computations, the time steps must be increased by a factor of 4. Furthermore, the computation for each of these time steps will take twice as long because halving Δx doubles the total number of nodes for which equations must be written. Consequently, for the one-dimensional case, halving Δx results in an eightfold increase in the number of calculations. Thus, the computational burden may be large to attain acceptable accuracy. As will be described shortly, other techniques are available that do not suffer from such severe limitations.

30.2.2 Derivative Boundary Conditions

As was the case for elliptic PDEs (recall Sec. 29.3.1), derivative boundary conditions can be readily incorporated into parabolic equations. For a one-dimensional rod, this necessitates adding two equations to characterize the heat balance at the end nodes. For example, the node at the left end ($i = 0$) would be represented by

$$T_0^{l+1} = T_0^l + \lambda(T_1^l - 2T_0^l + T_{-1}^l)$$

Thus, an imaginary point is introduced at $i = -1$ (recall Fig. 29.7). However, as with the elliptic case, this point provides a vehicle for incorporating the derivative boundary condition into the analysis. Problem 30.2 at the end of the chapter deals with this exercise.

30.2.3 Higher-Order Temporal Approximations

The general idea of re-expressing the PDE as a system of ODEs is sometimes called the *method of lines*. Obviously, one way to improve on the Euler approach used above would be to employ a more accurate integration scheme for solving the ODEs. For example, the Heun method can be employed to obtain second-order temporal accuracy. An example of

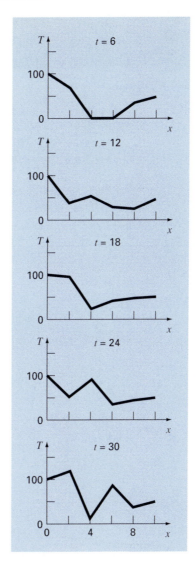

FIGURE 30.5
An illustration of instability. Solution of Example 30.1 but with $\lambda = 0.735$.

this approach is called *MacCormack's method*. This and other improved explicit methods are discussed elsewhere (e.g., Hoffman, 1992).

30.3   A SIMPLE IMPLICIT METHOD

As noted previously, explicit finite-difference formulations have problems related to stability. In addition, as depicted in Fig. 30.6, they exclude information that has a bearing on the solution. Implicit methods overcome both these difficulties at the expense of somewhat more complicated algorithms.

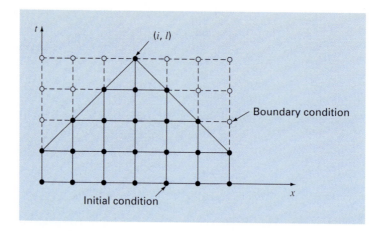

FIGURE 30.6
Representation of the effect of other nodes on the finite-difference approximation at node (i, l) using an explicit finite-difference scheme. The shaded nodes have an influence on (i, l), whereas the unshaded nodes, which in reality affect (i, l), are excluded.

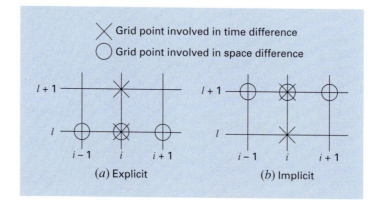

FIGURE 30.7
Computational molecules demonstrating the fundamental differences between (a) explicit and (b) implicit methods.

The fundamental difference between explicit and implicit approximations is depicted in Fig. 30.7. For the explicit form, we approximate the spatial derivative at time level l (Fig. 30.7a). Recall that when we substituted this approximation into the partial differential equation, we obtained a difference equation (30.4) with a single unknown T_i^{l+1}. Thus, we can solve "explicitly" for this unknown as in Eq. (30.5).

In implicit methods, the spatial derivative is approximated at an advanced time level $l + 1$. For example, the second derivative would be approximated by (Fig. 30.7b)

$$\frac{\partial^2 T}{\partial x^2} \cong \frac{T_{i+1}^{l+1} - 2T_i^{l+1} + T_{i-1}^{l+1}}{(\Delta x)^2} \qquad (30.7)$$

which is second-order accurate. When this relationship is substituted into the original PDE, the resulting difference equation contains several unknowns. Thus, it cannot be solved explicitly by simple algebraic rearrangement as was done in going from Eq. (30.4) to (30.5). Instead, the entire system of equations must be solved simultaneously. This is possible because, along with the boundary conditions, the implicit formulations result in a set of linear algebraic equations with the same number of unknowns. Thus, the method reduces to the solution of a set of simultaneous equations at each point in time.

To illustrate how this is done, substitute Eqs. (30.3) and (30.7) into Eq. (30.1) to give

$$k\frac{T_{i+1}^{l+1} - 2T_i^{l+1} + T_{i-1}^{l+1}}{(\Delta x)^2} = \frac{T_i^{l+1} - T_i^l}{\Delta t}$$

which can be expressed as

$$-\lambda T_{i-1}^{l+1} + (1 + 2\lambda)T_i^{l+1} - \lambda T_{i+1}^{l+1} = T_i^l \qquad (30.8)$$

where $\lambda = k\,\Delta t/(\Delta x)^2$. This equation applies to all but the first and the last interior nodes, which must be modified to reflect the boundary conditions. For the case where the temperature levels at the ends of the rod are given, the boundary condition at the left end of the rod ($i = 0$) can be expressed as

$$T_0^{l+1} = f_0(t^{l+1}) \qquad (30.9)$$

where $f_0(t^{l+1}) =$ a function describing how the boundary temperature changes with time. Substituting Eq. (30.9) into Eq. (30.8) gives the difference equation for the first interior node ($i = 1$):

$$(1 + 2\lambda)T_1^{l+1} - \lambda T_2^{l+1} = T_1^l + \lambda f_0(t^{l+1}) \qquad (30.10)$$

Similarly, for the last interior node ($i = m$),

$$-\lambda T_{m-1}^{l+1} + (1 + 2\lambda)T_m^{l+1} = T_m^l + \lambda f_{m+1}(t^{l+1}) \qquad (30.11)$$

where $f_{m+1}(t^{l+1})$ describes the specified temperature changes at the right end of the rod ($i = m + 1$).

When Eqs. (30.8), (30.10), and (30.11) are written for all the interior nodes, the resulting set of m linear algebraic equations has m unknowns. In addition, the method has the added bonus that the system is tridiagonal. Thus, we can utilize the extremely efficient solution algorithms (recall Sec. 11.1.1) that are available for tridiagonal systems.

EXAMPLE 30.2 | Simple Implicit Solution of the Heat-Conduction Equation

Problem Statement. Use the simple implicit finite-difference approximation to solve the same problem as Example 30.1.

Solution. For the rod from Example 30.1, $\lambda = 0.020875$. Therefore, at $t = 0$, Eq. (30.10) can be written for the first interior node as

$$1.04175T_1^1 - 0.020875T_2^1 = 0 + 0.020875(100)$$

or

$$1.04175T_1^1 - 0.020875T_2^1 = 2.0875$$

In a similar fashion, Eqs. (30.8) and (30.11) can be applied to the other interior nodes. This leads to the following set of simultaneous equations:

$$
\begin{bmatrix}
1.04175 & -0.020875 & & \\
-0.020875 & 1.04175 & -0.020875 & \\
& -0.020875 & 1.04175 & -0.020875 \\
& & -0.020875 & 1.04175
\end{bmatrix}
\begin{Bmatrix}
T_1^1 \\ T_2^1 \\ T_3^1 \\ T_4^1
\end{Bmatrix}
=
\begin{Bmatrix}
2.0875 \\ 0 \\ 0 \\ 1.04375
\end{Bmatrix}
$$

which can be solved for the temperature at $t = 0.1$ s:

$T_1^1 = 2.0047$

$T_2^1 = 0.0406$

$T_3^1 = 0.0209$

$T_4^1 = 1.0023$

Notice how in contrast to Example 30.1, all the points have changed from the initial condition during the first time step.

To solve for the temperatures at $t = 0.2$, the right-hand-side vector must be modified to account for the results of the first step, as in

$$
\begin{Bmatrix}
4.09215 \\
0.04059 \\
0.02090 \\
2.04069
\end{Bmatrix}
$$

The simultaneous equations can then be solved for the temperatures at $t = 0.2$ s:

$T_1^2 = 3.9305$

$T_2^2 = 0.1190$

$T_3^2 = 0.0618$

$T_4^2 = 1.9653$

Whereas the implicit method described is stable and convergent, it has the defect that the temporal difference approximation is first-order accurate, whereas the spatial difference approximation is second-order accurate (Fig. 30.8). In the next section we present an alternative implicit method that remedies the situation.

Before proceeding, it should be mentioned that, although the simple implicit method is unconditionally stable, there is an accuracy limit to the use of large time steps. Consequently, it is not that much more efficient than the explicit approaches for most time-variable problems.

Where it does shine is for steady-state problems. Recall from Chap. 29 that a form of Gauss-Seidel (Liebmann's method) can be used to obtain steady-state solutions for elliptic equations. An alternative approach would be to run a time-variable solution until it reached a steady state. In such cases, because inaccurate intermediate results are not an issue, implicit methods allow you to employ larger time steps, and hence, can generate steady-state results in an efficient manner.

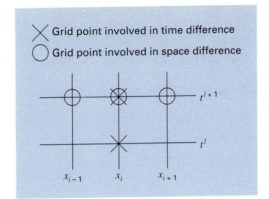

FIGURE 30.8
A computational molecule for
the simple implicit method.

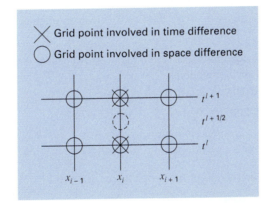

FIGURE 30.9
A computational molecule for
the Crank-Nicolson method.

30.4 THE CRANK-NICOLSON METHOD

The *Crank-Nicolson method* provides an alternative implicit scheme that is second-order accurate in both space and time. To provide this accuracy, difference approximations are developed at the midpoint of the time increment (Fig. 30.9). To do this, the temporal first derivative can be approximated at $t^{l+1/2}$ by

$$\frac{\partial T}{\partial t} \cong \frac{T_i^{l+1} - T_i^l}{\Delta t} \tag{30.12}$$

The second derivative in space can be determined at the midpoint by averaging the difference approximations at the beginning (t^l) and at the end (t^{l+1}) of the time increment

$$\frac{\partial^2 T}{\partial x^2} \cong \frac{1}{2}\left[\frac{T_{i+1}^l - 2T_i^l + T_{i-1}^l}{(\Delta x)^2} + \frac{T_{i+1}^{l+1} - 2T_i^{l+1} + T_{i-1}^{l+1}}{(\Delta x)^2}\right] \tag{30.13}$$

Substituting Eqs. (30.12) and (30.13) into Eq. (30.1) and collecting terms gives

$$-\lambda T_{i-1}^{l+1} + 2(1+\lambda)T_i^{l+1} - \lambda T_{i+1}^{l+1} = \lambda T_{i-1}^l + 2(1-\lambda)T_i^l + \lambda T_{i+1}^l \tag{30.14}$$

where $\lambda = k\,\Delta t/(\Delta x)^2$. As was the case with the simple implicit approach, boundary conditions of $T_0^{l+1} = f_0(t^{l+1})$ and $T_{m+1}^{l+1} = f_{m+1}(t^{l+1})$ can be prescribed to derive versions of Eq. (30.14) for the first and the last interior nodes. For the first interior node

$$2(1+\lambda)T_1^{l+1} - \lambda T_2^{l+1} = \lambda f_0(t^l) + 2(1-\lambda)T_1^l + \lambda T_2^l + \lambda f_0(t^{l+1}) \tag{30.15}$$

and for the last interior node,

$$-\lambda T_{m-1}^{l+1} + 2(1+\lambda)T_m^{l+1} = \lambda f_{m+1}(t^l) + 2(1-\lambda)T_m^l + \lambda T_{m-1}^l + \lambda f_{m+1}(t^{l+1}) \tag{30.16}$$

Although Eqs. (30.14) through (30.16) are slightly more complicated than Eqs. (30.8), (30.10), and (30.11), they are also tridiagonal and, therefore, efficient to solve.

EXAMPLE 30.3   Crank-Nicolson Solution to the Heat-Conduction Equation

Problem Statement.   Use the Crank-Nicolson method to solve the same problem as in Examples 30.1 and 30.2.

Solution.   Equations (30.14) through (30.16) can be employed to generate the following tridiagonal set of equations:

$$\begin{bmatrix} 2.01475 & -0.020875 & & \\ -0.020875 & 2.01475 & -0.020875 & \\ & -0.020875 & 2.01475 & -0.020875 \\ & & -0.020875 & 2.01475 \end{bmatrix}\begin{Bmatrix} T_1^1 \\ T_2^1 \\ T_3^1 \\ T_4^1 \end{Bmatrix} = \begin{Bmatrix} 4.175 \\ 0 \\ 0 \\ 2.0875 \end{Bmatrix}$$

which can be solved for the temperatures at $t = 0.1$ s:

$$T_1^1 = 2.0450$$
$$T_2^1 = 0.0210$$
$$T_3^1 = 0.0107$$
$$T_4^1 = 1.0225$$

To solve for the temperatures at $t = 0.2$ s, the right-hand-side vector must be changed to

$$\begin{Bmatrix} 8.1801 \\ 0.0841 \\ 0.0427 \\ 4.0901 \end{Bmatrix}$$

The simultaneous equations can then be solved for

$$T_1^2 = 4.0073$$
$$T_2^2 = 0.0826$$
$$T_3^2 = 0.0422$$
$$T_4^2 = 2.0036$$

30.4.1 Comparison of One-Dimensional Methods

Equation (30.1) can be solved analytically. For example, a solution is available for the case where the rod's temperature is initially at zero. At $t = 0$, the boundary condition at $x = L$ is instantaneously increased to a constant level of T while $T(0)$ is held at zero. For this case, the temperature can be computed by (Jenson and Jeffreys, 1977)

$$T = \overline{T}\left[\frac{x}{L} + \sum_{n=0}^{\infty} \frac{2}{n\pi}(-1)^n \sin\left(\frac{nx}{L}\right) \exp\left(\frac{-n^2\pi^2 kt}{L^2}\right)\right] \tag{30.17}$$

where L = total length of the rod. This equation can be employed to compute the evolution of the temperature distribution for each boundary condition. Then, the total solution can be determined by superposition.

EXAMPLE 30.4   Comparison of Analytical and Numerical Solutions

Problem Statement.   Compare the analytical solution from Eq. (30.17) with numerical results obtained with the explicit, simple implicit, and Crank-Nicolson techniques. Perform the comparison for the rod employed in Examples 30.1, 30.2, and 30.3.

Solution.   Recall from the previous examples that $k = 0.835$ cm^2/s, $L = 10$ cm, and $\Delta x = 2$ cm. For this case, Eq. (30.17) can be used to predict that the temperature at $x = 2$ cm, and $t = 10$ s would equal 64.8018. Table 30.1 presents numerical predictions of $T(2, 10)$. Notice that a range of time steps are employed. These results indicate a number of properties of the numerical methods. First, it can be seen that the explicit method is unstable for high values of λ. This instability is not manifested by either implicit approach. Second, the Crank-Nicolson method converges more rapidly as λ is decreased and provides moderately accurate results even when λ is relatively high. These outcomes are as expected because Crank-Nicolson is second-order accurate with respect to both independent variables. Finally, notice that as λ decreases, the methods seem to be converging on a value of 64.73 that is different than the analytical result of 64.80. This should not be surprising because a fixed value of $\Delta x = 2$ is used to characterize the x dimension. If both Δx and Δt were decreased as λ was decreased (that is, more spatial segments were used), the numerical solution would more closely approach the analytical result.

TABLE 30.1   Comparison of three methods of solving a parabolic PDE: the heated rod. The results shown are for temperature at $t = 10$ s at $x = 2$ cm for the rod from Example 30.1 through 30.3. Note that the analytical solution is $T(2, 10) = 64.8018$.

Δt	λ	Explicit	Implicit	Crank-Nicolson
10	2.0875	208.75	53.01	79.77
5	1.04375	−9.13	58.49	64.79
2	0.4175	67.12	62.22	64.87
1	0.20875	65.91	63.49	64.77
0.5	0.104375	65.33	64.12	64.74
0.2	0.04175	64.97	64.49	64.73

The Crank-Nicolson method is often used for solving linear parabolic PDEs in one spatial dimension. Its advantages become even more pronounced for more complicated applications such as those involving unequally spaced meshes. Such nonuniform spacing is often advantageous where we have foreknowledge that the solution varies rapidly in local portions of the system. Further discussion of such applications and the Crank-Nicolson method in general can be found elsewhere (Ferziger, 1981; Lapidus and Pinder, 1982; Hoffman 1992).

30.5 PARABOLIC EQUATIONS IN TWO SPATIAL DIMENSIONS

The heat-conduction equation can be applied to more than one spatial dimension. For two dimensions, its form is

$$\frac{\partial T}{\partial t} = k\left(\frac{\partial^2 T}{\partial x^2} + \frac{\partial^2 T}{\partial y^2}\right) \tag{30.18}$$

One application of this equation is to model the temperature distribution on the face of a heated plate. However, rather than characterizing its steady-state distribution, as was done in Chap. 29, Eq. (30.18) provides a means to compute the plate's temperature distribution as it changes in time.

30.5.1 Standard Explicit and Implicit Schemes

An explicit solution can be obtained by substituting finite-difference approximations of the form of Eqs. (30.2) and (30.3) into Eq. (30.18). However, as with the one-dimensional case, this approach is limited by a stringent stability criterion. For the two-dimensional case, the criterion is (Davis, 1984)

$$\Delta t \leq \frac{1}{8}\frac{(\Delta x)^2 + (\Delta y)^2}{k}$$

Thus, for a uniform grid ($\Delta x = \Delta y$), $\lambda = k\,\Delta t/(\Delta x)^2$ must be less than or equal to 1/4. Consequently, halving the step size results in a fourfold increase in the number of nodes and a 16-fold increase in computational effort.

As was the case with one-dimensional systems, implicit techniques offer alternatives that guarantee stability. However, the direct application of implicit methods such as the Crank-Nicolson technique leads to the solution of $m \times n$ simultaneous equations. Additionally, when written for two or three spatial dimensions, these equations lose the valuable property of being tridiagonal. Thus, matrix storage and computation time can become exorbitantly large. The method described in the next section offers one way around this dilemma.

30.5.2 The ADI Scheme

The alternating-direction implicit, or ADI, scheme provides a means for solving parabolic equations in two spatial dimensions using tridiagonal matrices. To do this, each time in-

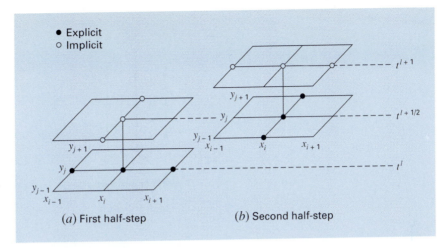

FIGURE 30.10
The two half-steps used in implementing the alternating-direction implicit scheme for solving parabolic equations in two spatial dimensions.

(a) First half-step   (b) Second half-step

crement is executed in two steps (Fig. 30.10). For the first step, Eq. (30.18) is approximated by

$$\frac{T_{i,j}^{l+1/2} - T_{i,j}^{l}}{\Delta t/2} = k\left[\frac{T_{i+1,j}^{l} - 2T_{i,j}^{l} + T_{i-1,j}^{l}}{(\Delta x)^2} + \frac{T_{i,j+1}^{l+1/2} - 2T_{i,j}^{l+1/2} + T_{i,j-1}^{l+1/2}}{(\Delta y)^2}\right] \quad (30.19)$$

Thus, the approximation of $\partial^2 T/\partial x^2$ is written explicitly—that is, at the base point t^l where values of temperature are known. Consequently, only the three temperature terms in the approximation of $\partial^2 T/\partial y^2$ are unknown. For the case of a square grid ($\Delta y = \Delta x$), this equation can be expressed as

$$-\lambda T_{i,j-1}^{l+1/2} + 2(1+\lambda)T_{i,j}^{l+1/2} - \lambda T_{i,j+1}^{l+1/2} = \lambda T_{i-1,j}^{l} + 2(1-\lambda)T_{i,j}^{l} + \lambda T_{i+1,j}^{l}$$
$$(30.20)$$

which, when written for the system, results in a tridiagonal set of simultaneous equations.
For the second step from $t^{l+1/2}$ to t^{l+1}, Eq. (30.18) is approximated by

$$\frac{T_{i,j}^{l+1} - T_{i,j}^{l+1/2}}{\Delta t/2} = k\left[\frac{T_{i+1,j}^{l+1} - 2T_{i,j}^{l+1} + T_{i-1,j}^{l+1}}{(\Delta x)^2} + \frac{T_{i,j+1}^{l+1/2} - 2T_{i,j}^{l+1/2} + T_{i,j-1}^{l+1/2}}{(\Delta y)^2}\right] \quad (30.21)$$

In contrast to Eq. (30.19), the approximation of $\partial^2 T/\partial x^2$ is now implicit. Thus, the bias introduced by Eq. (30.19) will be partially corrected. For a square grid, Eq. (30.21) can be written as

$$-\lambda T_{i-1,j}^{l+1} + 2(1+\lambda)T_{i,j}^{l+1} - \lambda T_{i+1,j}^{l+1} = \lambda T_{i,j-1}^{l+1/2} + 2(1-\lambda)T_{i,j}^{l+1/2} + \lambda T_{i,j+1}^{l+1/2}$$
$$(30.22)$$

Again, when written for a two-dimensional grid, the equation results in a tridiagonal system (Fig. 30.11). As in the following example, this leads to an efficient numerical solution.

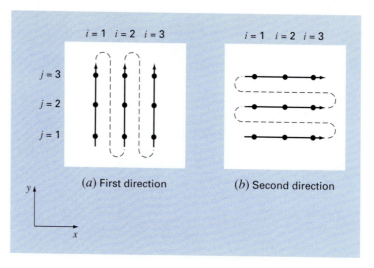

FIGURE 30.11
The ADI method only results in tridiagonal equations, if it is applied along the dimension that is implicit. Thus, on the first step (a), it is applied along the y dimension and, on the second step (b), along the x dimension. These "alternating directions" are the root of the method's name.

EXAMPLE 30.5   ADI Method

Problem Statement.   Use the ADI method to solve for the temperature of the plate in Examples 29.1 and 29.2. At $t = 0$, assume that the temperature of the plate is zero and the boundary temperatures are instantaneously brought to the levels shown in Fig. 29.4. Employ a time step of 10 s. Recall from Example 30.1 that the coefficient of thermal diffusivity for aluminum is $k = 0.835$ cm^2/s.

Solution.   A value of $\Delta x = 10$ cm was employed to characterize the 40×40–cm plate from Examples 29.1 and 29.2. Therefore, $\lambda = 0.835(10)/(10)^2 = 0.0835$. For the first step to $t = 5$ (Fig. 30.11a), Eq. (30.20) is applied to nodes (1, 1), (1, 2), and (1, 3) to yield the following tridiagonal equations:

$$
\begin{bmatrix}
2.167 & -0.0835 & \\
-0.0835 & 2.167 & -0.0835 \\
cr & -0.0835 & 2.167
\end{bmatrix}
\begin{Bmatrix}
T_{1,1} \\
T_{1,2} \\
T_{1,3}
\end{Bmatrix}
=
\begin{Bmatrix}
6.2625 \\
6.2625 \\
14.6125
\end{Bmatrix}
$$

which can be solved for

$$T_{1,1} = 3.01060 \qquad T_{1,2} = 3.2708 \qquad T_{1,3} = 6.8692$$

In a similar fashion, tridiagonal equations can be developed and solved for

$$T_{2,1} = 0.1274 \qquad T_{2,2} = 0.2900 \qquad T_{2,3} = 4.1291$$

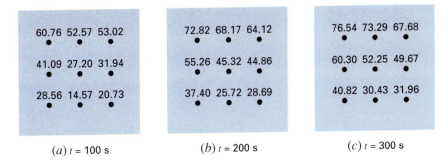

60.76 52.57 53.02

41.09 27.20 31.94

28.56 14.57 20.73

$(a)\, t = 100$ s

72.82 68.17 64.12

55.26 45.32 44.86

37.40 25.72 28.69

$(b)\, t = 200$ s

76.54 73.29 67.68

60.30 52.25 49.67

40.82 30.43 31.96

$(c)\, t = 300$ s

FIGURE 30.12
Solution for the heated plate from Example 30.5 at (a) $t = 100$ s, (b) $t = 200$ s, and (c) $t = 300$ s.

and

$$T_{3,1} = 2.0181 \qquad T_{3,2} = 2.2477 \qquad T_{3,3} = 6.0256$$

For the second step to $t = 10$ (Fig. 30.11b), Eq. (30.22) is applied to nodes (1, 1), (2, 1), and (3, 1) to yield

$$\begin{bmatrix} 2.167 & -0.0835 & \\ -0.0835 & 2.167 & -0.0835 \\ & -0.0835 & 2.167 \end{bmatrix} \begin{Bmatrix} T_{1,1} \\ T_{2,1} \\ T_{3,1} \end{Bmatrix} = \begin{Bmatrix} 13.0639 \\ 0.2577 \\ 8.0619 \end{Bmatrix}$$

which can be solved for

$$T_{1,1} = 5.5855 \qquad T_{2,1} = 0.4782 \qquad T_{3,1} = 3.7388$$

Tridiagonal equations for the other rows can be developed and solved for

$$T_{1,2} = 6.1683 \qquad T_{2,2} = 0.8238 \qquad T_{3,2} = 4.2359$$

and

$$T_{1,3} = 13.1120 \qquad T_{2,3} = 8.3207 \qquad T_{3,3} = 11.3606$$

The computation can be repeated, and the results for $t = 100$, 200, and 300 s are depicted in Fig. 30.12a through c. As expected, the temperature of the plate rises. After a sufficient time elapses, the temperature will approach the steady-state distribution of Fig. 29.5.

The ADI method is but one of a group of techniques called splitting methods. Some of these represent efforts to circumvent shortcomings of ADI. Discussion of other splitting methods as well as more information on ADI can be found elsewhere (Ferziger, 1981; Lapidus and Pinder, 1982).

PROBLEMS

30.1 Repeat Example 30.1, but use the Heun method (without iterating the corrector) to generate your solution.

30.2 Repeat Example 30.1, but for the case where the rod is initially at 50°C and the derivative at $x = 0$ is equal to 1 and at $x = 10$ is equal to 0. Interpret your results.

30.3 Repeat Example 30.1, but for a time step of $\Delta t = 0.05$ s. Compute results to $t = 0.2$ and compare with those in Example 30.1.

30.4 Repeat Example 30.2, but for the case where the derivative at $x = 10$ is equal to zero.

30.5 Repeat Example 30.3, but for $\Delta x = 1$ cm.

30.6 Repeat Example 30.5, but for the plate described in Prob. 29.1.

30.7 The advection-diffusion equation is used to compute the distribution of concentration along the length of a rectangular chemical reactor (see Sec. 26.1),

$$\frac{\partial c}{\partial t} = D \frac{\partial^2 c}{\partial x^2} - U \frac{\partial c}{\partial x} - kc$$

where $c =$ concentration (mg/m^3), $t =$ time (min), $D =$ a diffusion coefficient (m^2/min), $x =$ distance along the tank's longitudinal axis (m) where $x = 0$ at the tank's inlet, $U =$ velocity in the x direction (m/min), and $k =$ a reaction rate (min^{-1}) whereby the chemical decays to another form. Develop an explicit scheme to solve this equation numerically. Test it for $k = 0.1$, $D = 100$, and $U = 1$ for a tank of length 10.

30.8 Develop a user-friendly computer program for the simple explicit method from Sec. 30.2. Test it by duplicating Example 30.1.

30.9 Modify the program in Prob. 30.8 so that it employs derivative boundary conditions.

30.10 Develop a user-friendly computer program to implement the simple implicit scheme from Sec. 30.3. Test it by duplicating Example 30.2.

30.11 Develop a user-friendly computer program to implement the Crank-Nicolson method from Sec. 30.4. Test it by duplicating Example 30.3.

30.12 Develop a user-friendly computer program for the ADI method described in Sec. 30.5. Test it by duplicating Example 30.5.

CHAPTER 31

Finite-Element Method

To this juncture, we have employed *finite-difference* methods to solve partial differential equations. In these methods, the solution domain is divided into a grid of discrete points or nodes (Fig. 31.1*b*). The PDE is then written for each node and its derivatives replaced by finite divided differences. Although such *pointwise* approximation is conceptually easy to understand, it has a number of shortcomings. In particular, it becomes harder to apply for systems with irregular geometry, unusual boundary conditions, or heterogenous composition.

The *finite-element* method provides an alternative that is better suited for such systems. In contrast to finite-difference techniques, the finite-element method divides the solution

FIGURE 31.1

(*a*) A gasket with irregular geometry and nonhomogeneous composition. (*b*) Such a system is very difficult to model with a finite-difference approach. This is due to the fact that complicated approximations are required at the boundaries of the system and at the boundaries between regions of differing composition. (*c*) A finite-element discretization is much better suited for such systems.

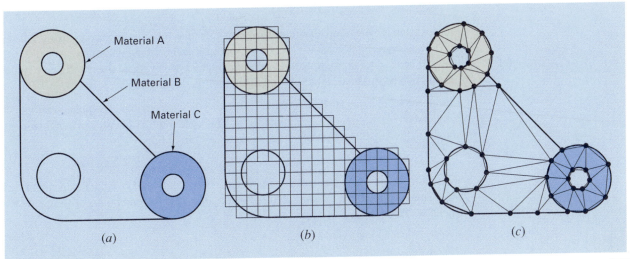

Material A

Material B

Material C

(*a*)　　　　　(*b*)　　　　　(*c*)

domain into simply shaped regions, or "elements" (Fig. 31.1c). An approximate solution for the PDE can be developed for each of these elements. The total solution is then generated by linking together, or "assembling," the individual solutions taking care to ensure continuity at the interelement boundaries. Thus, the PDE is satisfied in a *piecewise* fashion.

As in Fig. 31.1c, the use of elements, rather than a rectangular grid, provides a much better approximation for irregularly shaped systems. Further, values of the unknown can be generated continuously across the entire solution domain rather than at isolated points.

Because a comprehensive description is beyond the scope of this book, this chapter provides a general introduction to the finite-element method. Our primary objective is to make you comfortable with the approach and cognizant of its capabilities. In this spirit, the following section is devoted to a general overview of the steps involved in a typical finite-element solution of a problem. This is followed by a simple example: a steady-state, one-dimensional heated rod. Although this example does not involve PDEs, it allows us to develop and demonstrate major aspects of the finite-element approach unencumbered by complicating factors. We can then discuss some issues involved in employing the finite-element method for PDEs.

31.1 THE GENERAL APPROACH

Although the particulars will vary, the implementation of the finite-element approach usually follows a standard step-by-step procedure. The following provides a brief overview of each of these steps. The application of these steps to engineering problem contexts will be developed in subsequent sections.

31.1.1 Discretization

This step involves dividing the solution domain into finite elements. Figure 31.2 provides examples of elements employed in one, two, and three dimensions. The points of intersection of the lines that make up the sides of the elements are referred to as nodes and the sides themselves are called *nodal lines* or *planes*.

31.1.2 Element Equations

The next step is to develop equations to approximate the solution for each element. This involves two steps. First, we must choose an appropriate function with unknown coefficients that will be used to approximate the solution. Second, we evaluate the coefficients so that the function approximates the solution in an optimal fashion.

Choice of Approximation Functions.   Because they are easy to manipulate mathematically, polynomials are often employed for this purpose. For the one-dimensional case, the simplest alternative is a first-order polynomial or straight line,

$$u(x) = a_0 + a_1 x \tag{31.1}$$

where $u(x)$ = the dependent variable, a_0 and a_1 = constants, and x = the independent variable. This function must pass through the values of $u(x)$ at the end points of the element at x_1 and x_2. Therefore,

$$u_1 = a_0 + a_1 x_1$$
$$u_2 = a_0 + a_1 x_2$$

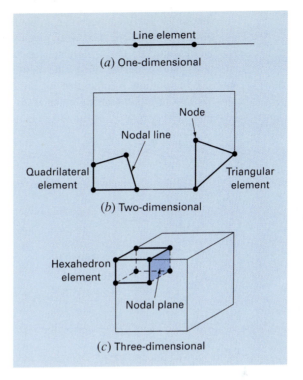

FIGURE 31.2
Examples of elements employed in (a) one, (b) two, and (c) three dimensions.

where $u_1 = u(x_1)$ and $u_2 = u(x_2)$. These equations can be solved using Cramer's rule for

$$a_0 = \frac{u_1 x_2 - u_2 x_1}{x_2 - x_1} \qquad a_1 = \frac{u_2 - u_1}{x_2 - x_1}$$

These results can then be substituted into Eq. (31.1) which, after collection of terms, can be written as

$$u = N_1 u_1 + N_2 u_2 \tag{31.2}$$

where

$$N_1 = \frac{x_2 - x}{x_2 - x_1} \tag{31.3}$$

and

$$N_2 = \frac{x - x_1}{x_2 - x_1} \tag{31.4}$$

Equation (31.2) is called an *approximation,* or *shape, function* and N_1 and N_2 are called *interpolation functions.* Close inspection reveals that Eq. (31.2) is, in fact, the Lagrange

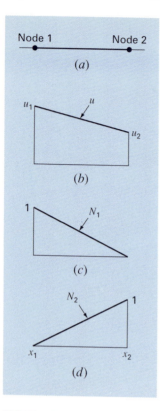

FIGURE 31.3

(b) A linear approximation or shape function for (a) a line element. The corresponding interpolation functions are shown in (c) and (d).

first-order interpolating polynomial. It provides a means to predict intermediate values (that is, to interpolate) between given values u_1 and u_2 at the nodes.

Figure 31.3 shows the shape function along with the corresponding interpolation functions. Notice that the sum of the interpolation functions is equal to one.

In addition, the fact that we are dealing with linear equations facilitates operations such as differentiation and integration. Such manipulations will be important in later sections. The derivative of Eq. (31.2) is

$$\frac{du}{dx} = \frac{dN_1}{dx} u_1 + \frac{dN_2}{dx} u_2 \tag{31.5}$$

According to Eqs. (31.3) and (31.4), the derivatives of the N's can be calculated as

$$\frac{dN_1}{dx} = -\frac{1}{x_2 - x_1} \qquad \frac{dN_2}{dx} = \frac{1}{x_2 - x_1} \tag{31.6}$$

and, therefore, the derivative of u is

$$\frac{du}{dx} = \frac{1}{x_2 - x_1}(-u_1 + u_2) \tag{31.7}$$

In other words, it is a divided difference representing the slope of the straight line connecting the nodes.

The integral can be expressed as

$$\int_{x_1}^{x_2} u\, dx = \int_{x_1}^{x_2} N_1 u_1 + N_2 u_2\, dx$$

Each term on the right-hand side is merely the integral of a right triangle with base x_2-x_1 and height u. That is,

$$\int_{x_1}^{x_2} Nu\, dx = \frac{1}{2}(x_2 - x_1)u$$

Thus, the entire integral is

$$\int_{x_1}^{x_2} u\, dx = \frac{u_1 + u_2}{2}(x_2 - x_1) \tag{31.8}$$

In other words, it is simply the trapezoidal rule.

Obtaining an Optimal Fit of the Function to the Solution.   Once the interpolation function is chosen, the equation governing the behavior of the element must be developed. This equation represents a fit of the function to the solution of the underlying differential equation. Several methods are available for this purpose. Among the most common are the direct approach, the method of weighted residuals, and the variational approach. The outcome of all of these methods is analogous to curve fitting. However, instead of fitting functions to data, these methods specify relationships between the unknowns in Eq. (31.2) that satisfy the underlying PDE in an optimal fashion.

Mathematically, the resulting element equations will often consist of a set of linear algebraic equations that can be expressed in matrix form,

$$[k]\{u\} = \{F\} \tag{31.9}$$

where $[k] =$ an *element property* or *stiffness matrix*, $\{u\} =$ a column vector of unknowns at the nodes, and $\{F\} =$ a column vector reflecting the effect of any external influences applied at the nodes. Note that, in some cases, the equations can be nonlinear. However, for the elementary examples described herein, and for many practical problems, the systems are linear.

31.1.3 Assembly

After the individual element equations are derived, they must be linked together or assembled to characterize the unified behavior of the entire system. The assembly process is governed by the concept of continuity. That is, the solutions for contiguous elements are matched so that the unknown values (and sometimes the derivatives) at their common nodes are equivalent. Thus, the total solution will be continuous.

When all the individual versions of Eq. (31.9) are finally assembled, the entire system is expressed in matrix form as

$$[K]\{u'\} = \{F'\} \tag{31.10}$$

where $[K] =$ the *assemblage property matrix* and $\{u'\}$ and $\{F'\} =$ column vectors for unknowns and external forces that are marked with primes to denote that they are an assemblage of the vectors $\{u\}$ and $\{F\}$ from the individual elements.

31.1.4 Boundary Conditions

Before Eq. (31.10) can be solved, it must be modified to account for the system's boundary conditions. These adjustments result in

$$[\bar{k}]\{u'\} = \{\bar{F}'\} \tag{31.11}$$

where the overbars signify that the boundary conditions have been incorporated.

31.1.5 Solution

Solutions of Eq. (31.11) can be obtained with techniques described previously in Part Three, such as *LU* decomposition. In many cases, the elements can be configured so that the resulting equations are banded. Thus, the highly efficient solution schemes available for such systems can be employed.

31.1.6 Postprocessing

Upon obtaining a solution, it can be output in tabular form or displayed graphically. In addition, secondary variables can be determined and output.

Although the preceding steps are very general, they are common to most implementations of the finite-element approach. In the following section, we illustrate how they can be applied to obtain numerical results for a simple physical system—a heated rod.

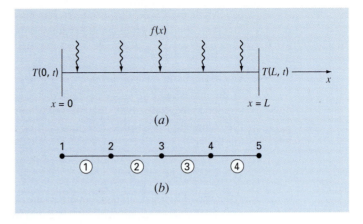

FIGURE 31.4
(a) A long, thin rod subject to fixed boundary conditions and a continuous heat source along its axis. (b) The finite-element representation consisting of four equal-length elements and five nodes.

31.2 FINITE-ELEMENT APPLICATION IN ONE DIMENSION

Figure 31.4 shows a system that can be modeled by a one-dimensional form of Poisson's equation

$$\frac{d^2T}{dx^2} = -f(x) \tag{31.12}$$

where $f(x) = $ a function defining a heat source along the rod and where the ends of the rod are held at fixed temperatures,

$$T(0, t) = T_1$$

and

$$T(L, t) = T_2$$

Notice that this is not a partial differential equation but rather is a boundary-value ODE. This simple model is used because it will allow us to introduce the finite-element approach without some of the complications involved in, for example, a two-dimensional PDE.

EXAMPLE 31.1   Analytical Solution for a Heated Rod

Problem Statement.   Solve Eq. (31.12) for a 10-cm rod with boundary conditions of $T(0, t) = 40$ and $T(10, t) = 200$ and a uniform heat source of $f(x) = 10$.

Solution.   The equation to be solved is

$$\frac{d^2T}{dx^2} = -10$$

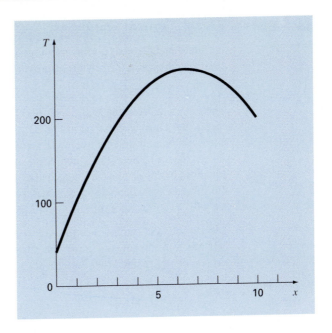

FIGURE 31.5
The temperature distribution along a heated rod subject to a uniform heat source and held at fixed end temperatures.

Assume a solution of the form

$$T = ax^2 + bx + c$$

which can be differentiated twice to give $T'' = 2a$. Substituting this result into the differential equation gives $a = -5$. The boundary conditions can be used to evaluate the remaining coefficients. For the first condition at $x = 0$,

$$40 = -5(0)^2 + b(0) + c$$

or $c = 40$. Similarly, for the second condition,

$$200 = -5(10)^2 + b(10) + 40$$

which can be solved for $b = 66$. Therefore, the final solution is

$$T = -5x^2 + 66x + 40$$

The results are plotted in Fig. 31.5.

31.2.1 Discretization

A simple configuration to model the system is a series of equal-length elements (Fig. 31.4b). Thus, the system is treated as four equal-length elements and five nodes.

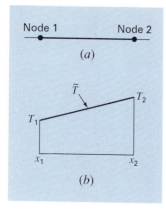

FIGURE 31.6
(a) An individual element.
(b) The approximation function used to characterize the temperature distribution along the element.

31.2.2 Element Equations

An individual element is shown in Fig. 31.6a. The distribution of temperature for the element can be represented by the approximation function

$$\tilde{T} = N_1 T_1 + N_2 T_2 \tag{31.13}$$

where N_1 and N_2 = linear interpolation functions specified by Eqs. (31.3) and (31.4), respectively. Thus, as depicted in Fig. 31.6b, the approximation function amounts to a linear interpolation between the two nodal temperatures.

As noted in Sec. 31.1, there are a variety of approaches for developing the element equation. In this section, we employ two of these. First, a *direct approach* will be used for the simple case where $f(x) = 0$. Then, because of its general applicability in engineering, we will devote most of the section to the *method of weighted residuals*.

The Direct Approach. For the case where $f(x) = 0$, a direct method can be employed to generate the element equations. The relationship between heat flux and temperature gradient can be represented by Fourier's law:

$$q = -k' \frac{dT}{dx}$$

where q = flux [cal/(cm²·s)] and k' = the coefficient of thermal conductivity [cal/(s·cm·°C)]. If a linear approximation function is used to characterize the element's temperature, the heat flow into the element through node 1 can be represented by

$$q_1 = k' \frac{T_1 - T_2}{x_2 - x_1}$$

where q_1 is heat flux at node 1. Similarly, for node 2,

$$q_2 = k' \frac{T_2 - T_1}{x_2 - x_1}$$

These two equations express the relationship of the element's internal temperature distribution (as reflected by the nodal temperatures) to the heat flux at its ends. As such, they constitute our desired element equations. They can be simplified further by recognizing that Fourier's law can be used to couch the end fluxes themselves in terms of the temperature gradients at the boundaries. That is,

$$q_1 = -k' \frac{dT(x_1)}{dx} \qquad q_2 = k' \frac{dT(x_2)}{dx}$$

which can be substituted into the element equations to give

$$\frac{1}{x_2 - x_1} \begin{bmatrix} 1 & -1 \\ -1 & 1 \end{bmatrix} \begin{Bmatrix} T_1 \\ T_2 \end{Bmatrix} = \begin{Bmatrix} -\dfrac{dT(x_1)}{dx} \\ \dfrac{dT(x_2)}{dx} \end{Bmatrix} \tag{31.14}$$

Notice that Eq. (31.14) has been cast in the format of Eq. (31.9). Thus, we have succeeded in generating a matrix equation that describes the behavior of a typical element in our system.

The direct approach has great intuitive appeal. Additionally, in areas such as mechanics, it can be employed to solve meaningful problems. However, in other contexts, it is often difficult or impossible to derive finite-element equations directly. Consequently, as described next, more general mathematical techniques are available.

The Method of Weighted Residuals.   The differential equation (31.12) can be reexpressed as

$$q = \frac{d^2T}{dx^2} + f(x)$$

The approximate solution [Eq. (31.13)] can be substituted into this equation. Because Eq. (31.13) is not the exact solution, the left side of the resulting equation will not be zero but will equal a residual,

$$R = \frac{d^2\tilde{T}}{dx^2} + f(x) \tag{31.15}$$

The method of weighted residuals (MWR) consists of finding a minimum for the residual according to the general formula

$$\int_D RW_i \, dD = 0 \qquad i = 1, 2, \ldots, m \tag{31.16}$$

where D = the solution domain and the W_i = linearly independent weighting functions.

At this point, there are a variety of choices that could be made for the weighting function (Box 31.1). The most common approach for the finite-element method is to employ the interpolation functions N_i as the weighting functions. When these are substituted into Eq. (31.16), the result is referred to as Galerkin's method,

$$\int_D RN_i \, dD = 0 \qquad i = 1, 2, \ldots, m$$

For our one-dimensional rod, Eq. (31.15) can be substituted into this formulation to give

$$\int_{x_1}^{x_2} \left[\frac{d^2\tilde{T}}{dx^2} + f(x) \right] N_i \, dx \qquad i = 1, 2$$

which can be reexpressed as

$$\int_{x_1}^{x_2} \frac{d^2\tilde{T}}{dx^2} N_i(x) \, dx = -\int_{x_1}^{x_2} f(x) N_i(x) \, dx \qquad i = 1, 2 \tag{31.17}$$

At this point, a number of mathematical manipulations will be applied to simplify and evaluate Eq. (31.17). Among the most important is the simplification of the left-hand side using integration by parts. Recall from calculus that this operation can be expressed generally as

$$\int_a^b u \, dv = uv\big|_a^b - \int_a^b v \, du$$

Box 31.1   Alternative Residual Schemes for the MWR

Several choices can be made for the weighting functions of Eq. (31.16). Each represents an alternative approach for the MWR.

In the *collocation approach,* we choose as many locations as there are unknown coefficients. Then, the coefficients are adjusted until the residual vanishes at each of these locations. Consequently, the approximating function will yield perfect results at the chosen locations but will have a nonzero residual elsewhere. Thus, it is akin to the interpolation methods in Chap. 18. Note that collocation amounts to using the weighting function

$$W = \delta(x - x_i) \qquad \text{for } i = 1, 2, \ldots, n$$

where n = the number of unknown coefficients and $\delta(x - x_i) = $ the *Dirac delta function* that vanishes everywhere but at $x = x_i$, where it equals 1.

In the *subdomain method,* the interval is divided into as many segments, or "subdomains," as there are unknown coefficients. Then, the coefficients are adjusted until the average value of the residual is zero in each subdomain. Thus, for each subdomain, the weighting function is equal to 1 and Eq. (31.16) is

$$\int_{x_{i-1}}^{x_i} R \, dx = 0 \qquad \text{for } i = 1, 2, \ldots, n$$

where x_{i-1} and x_i are the bounds of the subdomain.

For the *least-squares* case, the coefficients are adjusted so as to minimize the integral of the square of the residual. Thus, the weighting functions are

$$W_i = \frac{\partial R}{\partial a_i}$$

which can be substituted into Eq. (31.16) to give

$$\int_D R \frac{\partial R}{\partial a_i} \, dD = 0 \qquad i = 1, 2, \ldots, n$$

or

$$\frac{\partial}{\partial a_i} \int_D R^2 \, dD = 0 \qquad i = 1, 2, \ldots, n$$

Comparison of the formulation with those of Chap. 17 shows that this is the continuous form of regression.

Galerkin's method employs the interpolation functions N_i as weighting functions. Recall that these functions always sum to 1 at any position in an element. For many problem contexts, Galerkin's method yields the same results as are obtained by variational methods. Consequently, it is the most commonly employed version of MWR used in finite-element analysis.

If u and v are chosen properly, the new integral on the right-hand side will be easier to evaluate than the original one on the left-hand side. This can be done for the term on the left-hand side of Eq. (31.17) by choosing $N_i(x)$ as u and $(d^2T/dx^2) \, dx$ as dv to yield

$$\int_{x_1}^{x_2} N_i(x) \frac{d^2\tilde{T}}{dx^2} \, dx = N_i(x) \left. \frac{d\tilde{T}}{dx} \right|_{x_1}^{x_2} - \int_{x_1}^{x_2} \frac{d\tilde{T}}{dx} \frac{dN_i}{dx} \, dx \qquad i = 1, 2 \qquad (31.18)$$

Thus, we have taken the significant step of lowering the highest-order term in the formulation from a second to a first derivative.

Next, we can evaluate the individual terms that we have created in Eq. (31.18). For $i = 1$, the first term on the right-hand side of Eq. (31.18) can be evaluated as

$$N_1(x) \left. \frac{d\tilde{T}}{dx} \right|_{x_1}^{x_2} = N_1(x_2) \frac{d\tilde{T}(x_2)}{dx} - N_1(x_1) \frac{d\tilde{T}(x_1)}{dx}$$

However, recall from Fig. 31.3 that $N_1(x_2) = 0$ and $N_1(x_1) = 1$, and therefore,

$$N_1(x) \left. \frac{d\tilde{T}}{dx} \right|_{x_1}^{x_2} = -\frac{d\tilde{T}(x_1)}{dx} \qquad (31.19)$$

Similarly, for $i = 2$,

$$N_2(x) \frac{d\tilde{T}}{dx}\bigg|_{x_1}^{x_2} = \frac{d\tilde{T}(x_2)}{dx} \tag{31.20}$$

Thus, the first term on the right-hand side of Eq. (31.18) represents the natural boundary conditions at the ends of the elements.

Now, before proceeding let us regroup by substituting our results back into the original equation. Substituting Eqs. (31.18) through (31.20) into Eq. (31.17) and rearranging gives for $i = 1$,

$$\int_{x_1}^{x_2} \frac{d\tilde{T}}{dx} \frac{dN_1}{dx} \, dx = -\frac{d\tilde{T}(x_1)}{dx} + \int_{x_1}^{x_2} f(x)N_1(x)\, dx \tag{31.21}$$

and for $i = 2$,

$$\int_{x_1}^{x_2} \frac{d\tilde{T}}{dx} \frac{dN_2}{dx} \, dx = \frac{d\tilde{T}(x_2)}{dx} + \int_{x_1}^{x_2} f(x)N_2(x)\, dx \tag{31.22}$$

Notice that the integration by parts has led to two important outcomes. First, it has incorporated the boundary conditions directly into the element equations. Second, it has lowered the highest-order evaluation from a second to a first derivative. This latter outcome yields the significant result that the approximation functions need to preserve continuity of value but not slope at the nodes.

Also notice that we can now begin to ascribe some physical significance to the individual terms we have derived. On the right-hand side of each equation, the first term represents one of the element's boundary conditions and the second is the effect of the system's forcing function—in the present case, the heat source $f(x)$. As will now become evident, the left-hand side embodies the internal mechanisms that govern the element's temperature distribution. That is, in terms of the finite-element method, the left-hand side will become the element property matrix.

To see this, let us concentrate on the terms on the left-hand side. For $i = 1$, the term is

$$\int_{x_1}^{x_2} \frac{d\tilde{T}}{dx} \frac{dN_1}{dx} \, dx \tag{31.23}$$

Recall from Sec. 31.1.2 that the linear nature of the shape function makes differentiation and integration simple. Substituting Eqs. (31.6) and (31.7) into Eq. (31.23) gives

$$\int_{x_2}^{x_1} \frac{T_1 - T_2}{(x_2 - x_1)^2} \, dx = \frac{1}{x_1 - x_2}(T_1 - T_2) \tag{31.24}$$

Similar substitutions for $i = 2$ [Eq. (31.22)] yield

$$\int_{x_2}^{x_1} \frac{-T_1 + T_2}{(x_2 - x_1)^2} \, dx = \frac{1}{x_1 - x_2}(-T_1 + T_2) \tag{31.25}$$

Comparison with Eq. (31.14) shows that these are similar to the relationships that were developed with the direct method using Fourier's law. This can be made even clearer by re-expressing Eqs. (31.24) and (31.25) in matrix form as

$$\frac{1}{x_2 - x_1} \begin{bmatrix} 1 & -1 \\ -1 & 1 \end{bmatrix} \begin{Bmatrix} T_1 \\ T_2 \end{Bmatrix}$$

Substituting this result into Eqs. (31.21) and (31.22) and expressing the result in matrix form gives the final version of the element equations

$$\underbrace{\frac{1}{x_2 - x_1} \begin{bmatrix} 1 & -1 \\ -1 & 1 \end{bmatrix} \{T\}}_{\text{Element stiffness matrix}} = \underbrace{\begin{Bmatrix} -\dfrac{dT(x_1)}{dx} \\ \dfrac{dT(x_2)}{dx} \end{Bmatrix}}_{\substack{\text{Boundary} \\ \text{condition}}} + \underbrace{\begin{Bmatrix} \displaystyle\int_{x_1}^{x_2} f(x)N_1(x)\,dx \\ \displaystyle\int_{x_1}^{x_2} f(x)N_2(x)\,dx \end{Bmatrix}}_{\text{External effects}} \tag{31.26}$$

Note that aside from the direct and the weighted residual methods, the element equations can also be derived using variational calculus (for example, see Allaire, 1985). For the present case, this approach yields equations that are identical to those derived above.

EXAMPLE 31.2   Element Equation for a Heated Rod

Problem Statement.   Employ Eq. (31.26) to develop the element equations for a 10-cm rod with boundary conditions of $T(0, t) = 40$ and $T(10, t) = 200$ and a uniform heat source of $f(x) = 10$. Employ four equal-size elements of length $= 2.5$ cm.

Solution.   The heat source term in the first row of Eq. (31.26) can be evaluated by substituting Eq. (31.3) and integrating to give

$$\int_0^{2.5} 10 \frac{2.5 - x}{2.5}\,dx = 12.5$$

Similarly, Eq. (31.4) can be substituted into the heat source term of the second row of Eq. (31.26), which can also be integrated to yield

$$\int_0^{2.5} 10 \frac{x - 0}{2.5}\,dx = 12.5$$

These results along with the other parameter values can be substituted into Eq. (31.26) to give

$$0.4T_1 - 0.4T_2 = -\frac{dT}{dx}(x_1) + 12.5$$

and

$$-0.4T_1 + 0.4T_2 = \frac{dT}{dx}(x_2) + 12.5$$

31.2.3 Assembly

Before the element equations are assembled, a global numbering scheme must be established to specify the system's topology or spatial layout. As in Table 31.1, this defines the

TABLE 31.1 The system topology for the finite-element segmentation scheme from Fig. 31.4b.

		Node Numbers	
Element		**Local**	**Global**
1		1	1
		2	2
2		1	2
		2	3
3		1	3
		2	4
4		1	4
		2	5

FIGURE 31.7
The assembly of the equations for the total system.

$$(a) \quad \begin{bmatrix} 0.4 & -0.4 & 0 & 0 & 0 \\ -0.4 & 0.4 & 0 & 0 & 0 \\ 0 & 0 & 0 & 0 & 0 \\ 0 & 0 & 0 & 0 & 0 \\ 0 & 0 & 0 & 0 & 0 \end{bmatrix} \begin{Bmatrix} T_1 \\ T_2 \\ 0 \\ 0 \\ 0 \end{Bmatrix} = \begin{Bmatrix} -dT(x_1)/dx + 12.5 \\ dT(x_2)/dx + 12.5 \\ 0 \\ 0 \\ 0 \end{Bmatrix}$$

$$(b) \quad \begin{bmatrix} 0.4 & -0.4 & 0 & 0 & 0 \\ -0.4 & 0.4 + 0.4 & -0.4 & 0 & 0 \\ 0 & -0.4 & 0.4 & 0 & 0 \\ 0 & 0 & 0 & 0 & 0 \\ 0 & 0 & 0 & 0 & 0 \end{bmatrix} \begin{Bmatrix} T_1 \\ T_2 \\ T_3 \\ 0 \\ 0 \end{Bmatrix} = \begin{Bmatrix} -dT(x_1)/dx + 12.5 \\ 12.5 + 12.5 \\ dT(x_3)/dx + 12.5 \\ 0 \\ 0 \end{Bmatrix}$$

$$(c) \quad \begin{bmatrix} 0.4 & -0.4 & 0 & 0 & 0 \\ -0.4 & 0.8 & -0.4 & 0 & 0 \\ 0 & -0.4 & 0.4 + 0.4 & -0.4 & 0 \\ 0 & 0 & -0.4 & 0.4 & 0 \\ 0 & 0 & 0 & 0 & 0 \end{bmatrix} \begin{Bmatrix} T_1 \\ T_2 \\ T_3 \\ T_4 \\ 0 \end{Bmatrix} = \begin{Bmatrix} -dT(x_1)/dx + 12.5 \\ 25 \\ 12.5 + 12.5 \\ dT(x_4)/dx + 12.5 \\ 0 \end{Bmatrix}$$

$$(d) \quad \begin{bmatrix} 0.4 & -0.4 & 0 & 0 & 0 \\ -0.4 & 0.8 & -0.4 & 0 & 0 \\ 0 & -0.4 & 0.8 & -0.4 & 0 \\ 0 & 0 & -0.4 & 0.4 + 0.4 & -0.4 \\ 0 & 0 & 0 & -0.4 & 0.4 \end{bmatrix} \begin{Bmatrix} T_1 \\ T_2 \\ T_3 \\ T_4 \\ T_5 \end{Bmatrix} = \begin{Bmatrix} -dT(x_1)/dx + 12.5 \\ 25 \\ 25 \\ 12.5 + 12.5 \\ dT(x_5)/dx + 12.5 \end{Bmatrix}$$

$$(e) \quad \begin{bmatrix} 0.4 & -0.4 & 0 & 0 & 0 \\ -0.4 & 0.8 & -0.4 & 0 & 0 \\ 0 & -0.4 & 0.8 & -0.4 & 0 \\ 0 & 0 & -0.4 & 0.8 & -0.4 \\ 0 & 0 & 0 & -0.4 & 0.4 \end{bmatrix} \begin{Bmatrix} T_1 \\ T_2 \\ T_3 \\ T_4 \\ T_5 \end{Bmatrix} = \begin{Bmatrix} -dT(x_1)/dx + 12.5 \\ 25 \\ 25 \\ 25 \\ dT(x_5)/dx + 12.5 \end{Bmatrix}$$

connectivity of the element mesh. Because the present case is one-dimensional, the numbering scheme might seem so predictable that it is trivial. However, for two- and three-dimensional problems it offers the only means to specify which nodes belong to which elements.

Once the topology is specified, the element equation (31.26) can be written for each element using the global coordinates. Then they can be added one at a time to assemble the total system matrix (note that this process is explored further in Sec. 32.4). The process is depicted in Fig. 31.7.

31.2.4 Boundary Conditions

Notice that, as the equations are assembled, the internal boundary conditions cancel. Thus, the final result for $\{F\}$ in Fig. 31.7e has boundary conditions for only the first and the last nodes. Because T_1 and T_5 are given, these natural boundary conditions at the ends of the bar, $dT(x_1)/dx$ and $dT(x_5)/dx$, represent unknowns. Therefore, the equations can be re-expressed as

$$
\begin{array}{rl}
\dfrac{dT}{dx}(x_1) \quad -0.4T_2 & = \quad -3.5 \\[2mm]
0.8T_2 \quad -0.4T_3 & = \quad 41 \\
-0.4T_2 \quad +0.8T_3 \quad -0.4T_4 & = \quad 25 \\
-0.4T_3 \quad +0.8T_4 & = \quad 105 \\
-0.4T_4 \quad -\dfrac{dT}{dx}(x_5) & = \quad -67.5
\end{array}
\tag{31.27}
$$

FIGURE 31.8
Results of applying the finite-element approach to a heated bar. The exact solution is also shown.

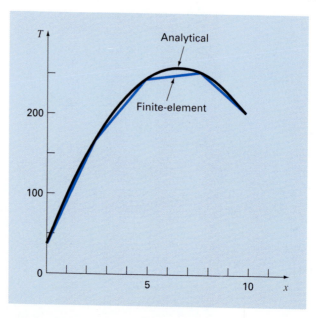

31.2.5 Solution

Equation (31.27) can be solved for

$$\frac{dT}{dx}(x_1) = 66 \qquad T_2 = 173.75 \qquad T_3 = 245$$

$$T_4 = 253.75 \qquad \frac{dT}{dx}(x_5) = -34$$

31.2.6 Postprocessing

The results can be displayed graphically. Figure 31.8 shows the finite-element results along with the exact solution. Notice that the finite-element calculation captures the overall trend of the exact solution and, in fact, provides an exact match at the nodes. However, a discrepancy exists in the interior of each element due to the linear nature of the shape functions.

31.3 TWO-DIMENSIONAL PROBLEMS

Although the mathematical "bookkeeping" increases markedly, the extension of the finite-element approach to two dimensions is conceptually similar to the one-dimensional applications discussed to this point. It thus follows the same steps as were outlined in Sec. 31.1.

31.3.1 Discretization

A variety of simple elements such as triangles or quadrilaterals are usually employed for the finite-element mesh in two dimensions. In the present discussion, we will limit ourselves to triangular elements of the type depicted in Fig. 31.9.

31.3.2 Element Equations

Just as for the one-dimensional case, the next step is to develop an equation to approximate the solution for the element. For a triangular element, the simplest approach is the linear polynomial [compare with Eq. (31.1)]

$$u(x, y) = a_0 + a_{1,1}x + a_{1,2}y \tag{31.28}$$

FIGURE 31.9
A triangular element.

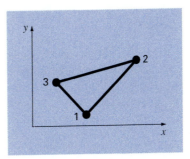

where $u(x, y)$ = the dependent variable, the a's = coefficients, and x and y = independent variables. This function must pass through the values of $u(x, y)$ at the triangle's nodes (x_1, y_1), (x_2, y_2), and (x_3, y_3). Therefore,

$$u_1(x, y) = a_0 + a_{1,1}x_1 + a_{1,2}y_1$$

$$u_2(x, y) = a_0 + a_{1,1}x_2 + a_{1,2}y_2$$

$$u_3(x, y) = a_0 + a_{1,1}x_3 + a_{1,2}y_3$$

or in matrix form,

$$\begin{bmatrix} 1 & x_1 & y_1 \\ 1 & x_2 & y_2 \\ 1 & x_3 & y_3 \end{bmatrix} \begin{Bmatrix} a_0 \\ a_{1,1} \\ a_{1,2} \end{Bmatrix} = \begin{Bmatrix} u_1 \\ u_2 \\ u_3 \end{Bmatrix}$$

which can be solved for

$$a_0 = \frac{1}{2A_e} \left[u_1(x_2y_3 - x_3y_2) + u_2(x_3y_1 - x_1y_3) + u_3(x_1y_2 - x_2y_1) \right] \tag{31.29}$$

$$a_{1,1} = \frac{1}{2A_e} \left[u_1(y_2 - y_3) + u_2(y_3 - y_1) + u_3(y_1 - y_2) \right] \tag{31.30}$$

$$a_{1,2} = \frac{1}{2A_e} \left[u_1(x_3 - x_2) + u_2(x_1 - x_3) + u_3(x_2 - x_1) \right] \tag{31.31}$$

where A_e is the area of the triangular element,

$$A_e = \frac{1}{2} \left[(x_2y_3 - x_3y_2) + (x_3y_1 - x_1y_3) + (x_1y_2 - x_2y_1) \right]$$

Equations (31.29) through (31.31) can be substituted into Eq. (31.28). After collection of terms, the result can be expressed as

$$u = N_1u_1 + N_2u_2 + N_3u_3 \tag{31.32}$$

where

$$N_1 = \frac{1}{2A_e} \left[(x_2y_3 - x_3y_2) + (y_2 - y_3)x + (x_3 - x_2)y \right]$$

$$N_2 = \frac{1}{2A_e} \left[(x_3y_1 - x_1y_3) + (y_3 - y_1)x + (x_1 - x_3)y \right]$$

$$N_3 = \frac{1}{2A_e} \left[(x_1y_2 - x_2y_1) + (y_1 - y_2)x + (x_2 - x_1)y \right]$$

Equation (31.32) provides a means to predict intermediate values for the element on the basis of the values at its nodes. Figure 31.10 shows the shape function along with the corresponding interpolation functions. Notice that the sum of the interpolation functions is always equal to 1.

As with the one-dimensional case, various methods are available for developing element equations based on the underlying PDE and the approximating functions. The resulting equations are considerably more complicated than Eq. (31.26). However, because the

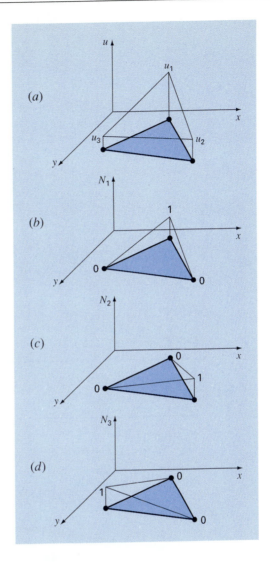

FIGURE 31.10
(a) A linear approximation function for a triangular element. The corresponding interpolation functions are shown in (b) through (d).

approximating functions are usually lower-order polynomials like Eq. (31.28), the terms of the final element matrix will consist of lower-order polynomials and constants.

31.3.3 Boundary Conditions and Assembly

The incorporation of boundary conditions and the assembly of the system matrix also become more complicated when the finite-element technique is applied to two- and three-dimensional problems. However, as with the derivation of the element matrix, the difficulty

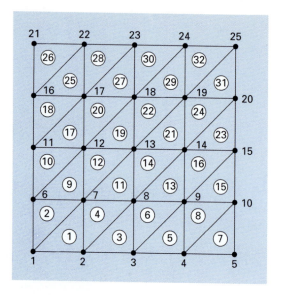

FIGURE 31.11
A numbering scheme for the nodes and elements of a finite-element approximation of the heated plate that was previously characterized by finite differences in Chap. 29.

FIGURE 31.12
The temperature distribution of a heated plate as calculated with a finite-element method.

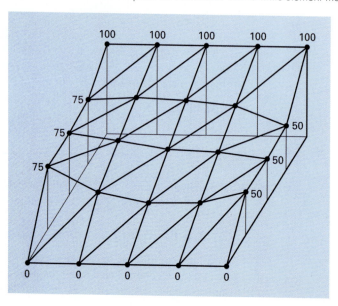

relates to the mechanics of the process rather than to conceptual complexity. For example, the establishment of the system topology which was trivial for the one-dimensional case becomes a matter of great importance in two and three dimensions. In particular, the choice of a numbering scheme will dictate the bandedness of the resulting system matrix and hence the efficiency with which it can be solved. Figure 31.11 shows a scheme that was developed for the heated plate formerly solved by finite-difference methods in Chap. 29.

31.3.4 Solution and Postprocessing

Although the mechanics are complicated, the system matrix is merely a set of n simultaneous equations that can be used to solve for the values of the dependent variable at the n nodes. Figure 31.12 shows a solution that corresponds to the finite-difference solution from Fig. 29.5.

31.4 PDES WITH LIBRARIES AND PACKAGES

Software libraries and packages have some capabilities for directly solving PDEs. However, as described in the following sections, many of the solutions are limited to simple problems. This is particularly true of two- and three-dimensional cases. For these situations, generic packages (i.e., ones not expressly developed to solve PDEs such as finite-element packages) are often limited to simple rectangular domains.

Although this might seem limiting, simple applications can be of great utility in a pedagogical sense. This is particularly true when the packages' visualization tools are used to display calculation results.

31.4.1 Mathcad

Mathcad has two functions that can solve Poisson's equation. You can use the **relax** function when you know the value of the unknown on all four sides of a square region. This function solves a system of linear algebraic equations using Gauss-Seidel iteration with overrelaxation to speed the rate of convergence. For the special case where there are internal sources or sinks and the unknown function is zero on all four sides of the square, then you can use the **multigrid** Mathcad function, which is usually faster than **relax.** Both of these functions return a square matrix where the location of the element in the matrix corresponds to its location within the square region. The value of the element approximates the value of the solution of Poisson's equation at this point.

Figure 31.13 shows an example where a square plate contains heat sources while the boundary is maintained at zero. The first step is to establish dimensions for the temperature grid and the heat source matrix. The temperature grid has dimensions (R + 1) by (R + 1) while the heat source matrix is R by R. For example, a 3 by 3 temperature grid has 4 (2 by 2) possible heat sources. In this case, we establish a 33 by 33 temperature grid and a 32 by 32 heat source matrix. The Mathcad command $\rho_{RR} := 0$ (with R = 32) establishes the dimensions of the source matrix and sets all the elements to zero. Next, the location and strength of the heat source are established. Finally, G is the resulting temperature distribution as calculated by the Mathcad **multigrid** function. The second argument of **multigrid** is a parameter that controls the numerical accuracy.

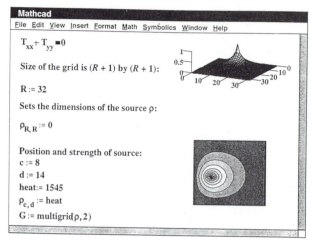

Mathcad

File Edit View Insert Format Math Symbolics Window Help

$T_{xx} + T_{yy} = 0$

Size of the grid is $(R + 1)$ by $(R + 1)$:

$R := 32$

Sets the dimensions of the source ρ:

$\rho_{R,R} := 0$

Position and strength of source:

$c := 8$

$d := 14$

$\text{heat} := 1545$

$\rho_{c,d} := \text{heat}$

$G := \text{multigrid}(\rho, 2)$

FIGURE 31.13
Mathcad screen to determine the solution of an elliptic PDE.

The temperature distribution can be displayed with either surface and contour plots. These plots can be placed anywhere on the worksheet by clicking to the desired location. This places a red cross hair at that location. Then, use the Insert/Graph/Surface or Insert/Graph/Contour pull down menu to place an empty plot on the worksheet with placeholders for the expressions to be graphed and for the ranges of variables. Simply type G in the placeholder on the z axis and 33 for the x and y axis ranges. Mathcad does all the rest to produce the graphs shown in Fig. 31.13. Once the graph has been created you can use the Format/Surface Plot and Format/Contour Plot pull down menus to change the color or add titles, labels, and other features.

31.4.2 Excel

Although Excel does not have the direct capability to solve PDEs, it is a nice environment to develop simple solutions of elliptic PDEs. For example, the orthogonal layout of the spreadsheet cells (Fig. 31.14*b*) is directly analogous to the grid used in Chap. 29 to model the heated plate (Fig. 31.14*a*).

As in Fig. 31.14*b*, the Dirichlet boundary conditions can first be entered along the periphery of the cell block. The formula for the Liebmann method can be implemented by entering Eq. (29.11) in one of the cells in the interior (like cell B2 in Fig. 31.14*b*). Thus, the value for the cell can be computed as a function of its adjacent cells. Then the cell can be copied to the other interior cells. Because of the relative nature of the Excel copy command, all the other cells will properly be dependent on their adjacent cells.

Once you have copied the formula, you will probably get an error message: **Cannot resolve circular references.** You can rectify this by going to the *T*(ools) menu and selecting *O*(ptions). Then select the **Calculation tab** and check the **Iteration** box. This will allow the spreadsheet to recalculate (the default is 100 iterations) and solve Liebmann's

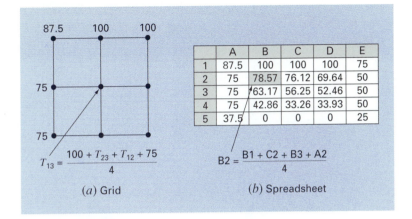

	A	B	C	D	E
1	87.5	100	100	100	75
2	75	78.57	76.12	69.64	50
3	75	63.17	56.25	52.46	50
4	75	42.86	33.26	33.93	50
5	37.5	0	0	0	25

$$T_{13} = \frac{100 + T_{23} + T_{12} + 75}{4}$$

$$B2 = \frac{B1 + C2 + B3 + A2}{4}$$

(a) Grid (b) Spreadsheet

FIGURE 31.14
The analogy between (a) a rectangular grid and (b) the cells of a spreadsheet.

method iteratively. After this occurs, strike the F9 key to manually recalculate the sheet until the answers do not vary. This means that the solution has converged.

Once the problem has been solved, Excel's graphics tools can be used to visualize the results. An example is shown in Fig. 31.15a. For this case, we have

- Used a finer grid
- Made the lower boundary insulated
- Added a heat source of 150 to the middle of the plate (cell E5).

The numerical results from Fig. 31.15a can then be displayed with Excel's Chart Wizard. Figure 31.15b and c show 3-D surface plots. The y orientation of these are normally the reverse of the spreadsheet. Thus, the top high-temperature edge (100) would normally be displayed at the bottom of the plot. We reversed the y values on our sheet prior to plotting so that the graphs would be consistent with the spreadsheet.

Notice how the graphs help you visualize what's going on. Heat flows down from the source toward the boundaries, forming a mountainlike shape. Heat also flows from the high-temperature boundary down to the two side edges. Notice how the heat flows preferentially toward the lower-temperature edge (50). Finally, notice how the temperature gradient in the y dimension goes to zero at the lower insulated edge ($\partial T/\partial y \to 0$).

31.4.3 MATLAB

Although the standard MATLAB package does not presently have great capabilities for solving PDEs, m-files and functions can certainly be developed for this purpose. In addition, its display capabilities are very nice, particularly for visualization of 2-D spatial problems.

To illustrate this capability, we first set up the Excel spreadsheet in Fig. 31.15a. These results can be saved as a text (Tab delimited) file with a name like **plate.txt.** This file can then be moved to the MATLAB directory.

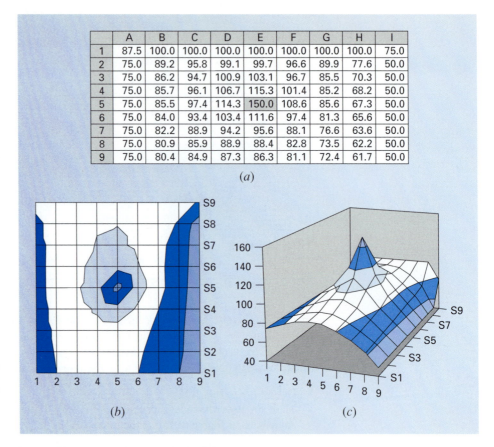

	A	B	C	D	E	F	G	H	I
1	87.5	100.0	100.0	100.0	100.0	100.0	100.0	100.0	75.0
2	75.0	89.2	95.8	99.1	99.7	96.6	89.9	77.6	50.0
3	75.0	86.2	94.7	100.9	103.1	96.7	85.5	70.3	50.0
4	75.0	85.7	96.1	106.7	115.3	101.4	85.2	68.2	50.0
5	75.0	85.5	97.4	114.3	150.0	108.6	85.6	67.3	50.0
6	75.0	84.0	93.4	103.4	111.6	97.4	81.3	65.6	50.0
7	75.0	82.2	88.9	94.2	95.6	88.1	76.6	63.6	50.0
8	75.0	80.9	85.9	88.9	88.4	82.8	73.5	62.2	50.0
9	75.0	80.4	84.9	87.3	86.3	81.1	72.4	61.7	50.0

(a)

(b)

(c)

FIGURE 31.15

(a) Excel solution of the Poisson equation for a plate with an insulated lower edge and a heat source. A (b) "topographic map" and (c) a 3-D display of the temperatures.

Once in MATLAB, the file can be loaded by typing

```
>> load plate.txt
```

Next, the gradients can be simply calculated as

```
>> [px,py]=gradient(plate);
```

Note that this is the simplest method to compute gradients using default values of dx = dy = 1. Therefore, the directions and relative magnitudes will be correct.

Finally, a series of commands can be used to develop the plot. The command **contour** develops a contour plot of the data. The command **clabel** adds contour labels to the plot. Finally, **quiver** takes the gradient data and adds it to the plot as arrows,

```
>> cs=contour(plate);clabel(cs);hold on
>> quiver(-px,-py);hold off
```

Note that the minus signs are added because of the minus sign in Fourier's law [Eq. (29.4)]. As in Fig. 31.16, the resulting plot provides an excellent representation of the solution.

Note that any file in the proper format can be entered into MATLAB and displayed in this way. For example, the IMSL calculation described next could be programmed to generate a

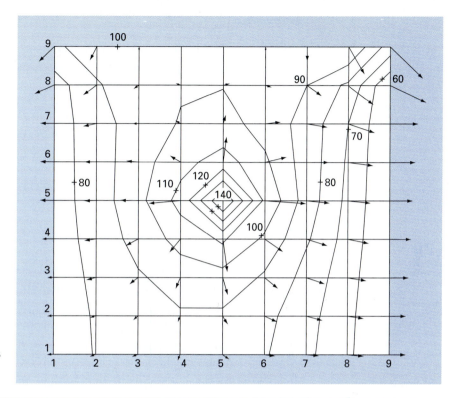

FIGURE 31.16
MATLAB-generated contour plots for the heated plate calculated with Excel (Fig. 31.15).

file for display on MATLAB (or Excel or Mathcad for that matter). This sharing of files between tools is becoming commonplace. In addition, files can be created in one location on one tool, transmitted over the internet to another location, where the file might be displayed with another tool. This is one of the exciting aspects of modern numerical applications.

31.4.4 IMSL

IMSL has a few routines for solving PDEs (Table 31.2). In the present discussion, we will focus on the **fps2h** routine. This routine solve Poisson's or Helmholtz's equation on a two-dimensional rectangle using a fast Poisson solver on a uniform mesh.

TABLE 31.2 IMSL routines to solve PDEs.

Category	Routines	Capability
Solution of Systems of PDEs in One Dimension	MOLCH	Method of lines with a Hermite cubic basis
Solution of a PDE in Two and Three Dimensions	FPS2H	Two-dimensional fast Poisson solver
	FPS3H	Three-dimensional fast Poisson solver

The subroutine **fps2h** is implemented by the following CALL statement:

```
CALL    FPS2H(PRH,BRH,COEF,NX,NY,AX,BX,AY,BY,IBCT,IORD,U,LDU)
```

where PRH = User-supplied FUNCTION to evaluate the right side of the partial differential equation. The form is PRH(X, Y), where

X = X-coordinate value. (Input)

Y = Y-coordinate value. (Input)

PRH must be declared EXTERNAL in the calling program.

BRH = User-supplied FUNCTION to evaluate the right side of the boundary conditions.

The form is BRHS(ISIDE, X, Y), where

ISIDE = Side number. (Input) See IBCTY below for the definition of the side numbers.

X = X-coordinate value. (Input)

Y = Y-coordinate value. (Input)

BRH must be declared EXTERNAL in the calling program.

COEF = Value of the coefficient of U in the differential equation. (Input)

NX = Number of grid lines in the X direction. (Input) NX must be at least 4. See Comment 2 for further restrictions on NX.

NY = Number of grid lines in the Y direction. (Input) NY must be at least 4. See Comment 2 for further restrictions on NY.

AX = The value of X along the left side of the domain. (Input)

BX = The value of X along the right side of the domain. (Input)

AY = The value of Y along the bottom of the domain. (Input)

BY = The value of Y along the top of the domain. (Input)

IBCT = Array of size 4 indicating the type of boundary condition on each side of the domain or that the solution is periodic. (Input) The sides are numbered 1 to 4 as follows:

Side	Location
1—Right	(X = BX)
2—Bottom	(Y = AY)
3—Left	(X = AX)
4—Top	(Y = BY)

There are three boundary condition types.

IBCTY	Boundary Condition
1	Value of U is given. (Dirichlet)
2	Value of dU/dX is given (sides 1 and/or 3). (Neumann) Value of dU/dY is given (sides 2 and/or 4).
3	Periodic.

IORD = Order of accuracy of the finite-difference approximation. (Input) It can be either 2 or 4. Usually, IORD = 4 is used.

U = Array of size NX by NY containing the solution at the grid points. (Output)

LDU = Leading dimension of U exactly as specified in the dimension statement of the calling program. (Input)

EXAMPLE 31.3   Using IMSL to Solve the Temperature of a Heated Plate

Problem Statement.   Use fps2h to solve for the temperatures on the square heated plate with fixed boundary conditions from Example 29.1.

Solution.   An example of a main Fortran 90 program and function using fps2h to solve this problem can be written as

```
Program Plate
USE msimsl
IMPLICIT NONE
INTEGER ::ncval, nx, nxtabl, ny, nytabl
PARAMETER (ncval=11, nx=33, nxtabl=5, ny=33, nytabl=5)
INTEGER :: i, ibcty(4), iorder, j, nout
REAL :: QD2VL,ax,ay,brhs,bx,by,coefu,prhs,u(nx,ny),utabl,x,xdata(nx),y,ydata(ny)
EXTERNAL brhs, prhs
ax = 0.0
bx = 40.
ay = 0.0
by = 40.
ibcty(1) = 1
ibcty(2) = 1
ibcty(3) = 1
ibcty(4) = 1
coefu = 0.0
iorder = 4
CALL FPS2H(prhs,brhs,coefu,nx,ny,ax,bx,ay,by,ibcty,iorder,u,nx)
DO i=1, nx
  xdata(i) = ax + (bx-ax)*FLOAT(i-1)/FLOAT(nx-1)
END DO
DO j=1, ny
  ydata(j) = ay + (by-ay)*FLOAT(j-1)/FLOAT(ny-1)
END DO
CALL UMACH (2, nout)
WRITE (nout,'(8X,A,11X,A,11X,A)') 'X', 'Y', 'U'
DO j=1, nytabl
  DO i=1, nxtabl
    x      = ax + (bx-ax)*FLOAT(i-1)/FLOAT(nxtabl-1)
    y      = ay + (by-ay)*FLOAT(j-1)/FLOAT(nytabl-1)
    utabl = QD2VL(x,y,nx,xdata,ny,ydata,u,nx,.FALSE.)
    WRITE (nout,'(4F12.4)') x, y, utabl
  END DO
END DO
END PROGRAM

FUNCTION prhs (x, y)
IMPLICIT NONE
REAL :: prhs, x, y
```

```
prhs = 0.0
END FUNCTION


REAL FUNCTION brhs (iside, x, y)
IMPLICIT NONE
INTEGER :: iside
REAL :: x, y
IF (iside == 1) then
   brhs = 50.
ELSEIF (iside == 2) THEN
   brhs = 0.
ELSEIF (iside == 3) THEN
   brhs = 75.
ELSE
   brhs = 100.
END IF
END FUNCTION
```

An example run yields the following output:

x	y	u	x	y	u
.0000	.0000	37.5000	30.0000	20.0000	52.3849
10.0000	.0000	.0000	40.0000	20.0000	50.0000
20.0000	.0000	.0000	.0000	30.0000	75.0000
30.0000	.0000	.0000	10.0000	30.0000	79.0032
40.0000	.0000	25.0000	20.0000	30.0000	76.8058
.0000	10.0000	75.0000	30.0000	30.0000	69.9017
10.0000	10.0000	42.5976	40.0000	30.0000	50.0000
20.0000	10.0000	32.2945	.0000	40.0000	87.5000
30.0000	10.0000	33.4962	10.0000	40.0000	100.0000
40.0000	10.0000	50.0000	20.0000	40.0000	100.0000
.0000	20.0000	75.0000	30.0000	40.0000	100.0000
10.0000	20.0000	63.5128	40.0000	40.0000	75.0000
20.0000	20.0000	56.2493			

PROBLEMS

31.1 Repeat Example 31.1, but for $T(0, t) = 50$ and $T(10, t) = 100$ and a uniform heat source of 20.

31.2 Repeat Example 31.2, but for boundary conditions of $T(0, t) = 50$ and $T(10, t) = 100$ and a heat source of 20.

31.3 Apply the results of Prob. 31.2 to compute the temperature distribution for the entire rod using the finite-element approach.

31.4 Use Galerkin's method to develop an element equation for a steady-state version of the advection-diffusion equation described in Prob. 30.7. Express the final result in the format of Eq. (31.26) so that each term has a physical interpretation.

31.5 A version of the Poisson equation that occurs in mechanics is the following model for the vertical deflection of a bar with a distributed load $P(x)$:

$$A_c E \frac{d^2 u}{dx^2} = P(x)$$

where A_c = cross-sectional area, E = Young's modulus, u = deflection, and x = distance measured along the bar's length. If the bar is rigidly fixed ($u = 0$) at both ends, use the finite-element method to model its deflections for $A_c = 1$ ft^2, $E = 1.5 \times 10^8$ lb/ft^2, $L = 30$ ft, and $P(x) = 50$ lb/ft. Employ a $\Delta x = 6$ ft.

31.6 Develop a user-friendly program to model the steady-state distribution of temperature in a rod with a constant heat source using the finite-element method. Set up the program so that unequally spaced nodes may be used.

31.7 Use Mathcad to perform the same computation as in Fig. 31.14, but add a heat sink of -750 at $(24, 8)$.

31.8 Use Excel to perform the same computation as in Fig. 31.15, but insulate the right-hand edge and add a heat sink of -110 at cell C7.

31.9 Use MATLAB to develop a contour plot with flux arrows for the Excel solution from Prob. 31.8.

31.10 Use Excel to model the temperature distribution of the slab shown in Fig. P31.10. The slab is 0.01 m thick and has a thermal conductivity of 2.5 W/(m · °C).

31.11 Use MATLAB to develop a contour plot with flux arrows for the Excel solution from Prob. 31.10.

31.12 Use IMSL to perform the same computation as in Example 31.3, but insulate the lower edge of the plate.

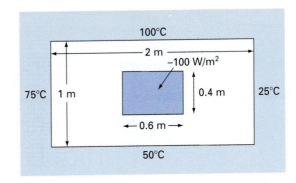

FIGURE P31.10

CHAPTER 32

Engineering Applications: Partial Differential Equations

The purpose of this chapter is to apply the methods from Part Eight to practical engineering problems. In *Sec. 32.1*, a parabolic PDE is used to compute the time-variable distribution of a chemical along the longitudinal axes of a rectangular reactor. This example illustrates how the instability of a solution can be due to the nature of the PDE rather than to properties of the numerical method.

Sections 32.2 and 32.3 involve applications of the Poisson and Laplace equations to civil and electrical engineering problem contexts. Among other things this will allow you to see similarities as well as differences between field problems in these areas of engineering. In addition, they can be contrasted with the heated-plate problem that has served as our prototype system in this part of the book. *Section 32.2* deals with the deflection of a square plate, whereas *Sec. 32.3* is devoted to computing the voltage distribution and charge flux for a two-dimensional surface with a curved edge.

Section 32.4 presents a finite-element analysis as applied to a series of springs. This application is closer in spirit to finite-element applications in mechanics and structures than was the temperature field problem used to illustrate the approach in Chap. 31.

32.1 ONE-DIMENSIONAL MASS BALANCE OF A REACTOR (CHEMICAL/PETROLEUM ENGINEERING)

Background.   Chemical engineers make extensive use of idealized reactors in their design work. In Secs. 12.1 and 28.1, we focused on single or coupled well-mixed reactors. These are examples of *lumped-parameter systems* (recall Sec. PT3.1.2).

FIGURE 32.1

An elongated reactor with a single entry and exit point. A mass balance is developed around a finite segment along the tank's longitudinal axis in order to derive a differential equation for concentration.

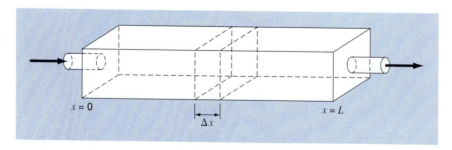

$x = 0$ Δx $x = L$

Figure 32.1 depicts an elongated reactor with a single entry and exit point. This reactor can be characterized as a *distributed-parameter system*. If it is assumed that the chemical being modeled is subject to first-order decay[1] and the tank is well-mixed vertically and laterally, a mass balance can be performed on a finite segment of length Δx, as in

$$V\frac{\Delta c}{\Delta t} = \underbrace{Qc(x)}_{\text{Flow in}} - \underbrace{Q\left[c(x) + \frac{\partial c(x)}{\partial x}\Delta x\right]}_{\text{Flow out}} - \underbrace{DA_c\frac{\partial c(x)}{\partial x}}_{\text{Dispersion in}}$$

$$+ \underbrace{DA_c\left[\frac{\partial c(x)}{\partial x} + \frac{\partial}{\partial x}\frac{\partial c(x)}{\partial x}\Delta x\right]}_{\text{Dispersion out}} - \underbrace{kVc}_{\text{Decay reaction}} \qquad (32.1)$$

where V = volume (m³), Q = flow rate (m³/h), c is concentration (moles/m³), D is a dispersion coefficient (m²/h), A_c is the tank's cross-sectional area (m²), and k is the first-order decay coefficient (h⁻¹). Note that the dispersion terms are based on *Fick's first law*,

$$\text{Flux} = -D\frac{\partial c}{\partial x} \qquad (32.2)$$

which is directly analogous to Fourier's law for heat conduction [recall Eq. (29.4)]. It specifies that turbulent mixing tends to move mass from regions of high to low concentration. The parameter D, therefore, reflects the magnitude of turbulent mixing.

If Δx and Δt are allowed to approach zero, Eq. (32.1) becomes

$$\frac{\partial c}{\partial t} = D\frac{\partial^2 c}{\partial x^2} - U\frac{\partial c}{\partial x} - kc \qquad (32.3)$$

where $U = Q/A_c$ is the velocity of the water flowing through the tank. The mass balance for Fig. 32.1 is, therefore, now expressed as a parabolic partial differential equation. Equation (32.3) is sometimes referred to as the *advection-dispersion equation* with first-order reaction. At steady state, it is reduced to a second-order ODE,

$$0 = D\frac{d^2 c}{dx^2} - U\frac{d^2 c}{dx} - kc \qquad (32.4)$$

Prior to $t = 0$, the tank is filled with water that is devoid of the chemical. At $t = 0$, the chemical is injected into the reactor's inflow at a constant level of c_{in}. Thus, the following boundary conditions hold:

$$Qc_{\text{in}} = Qc_0 - DA_c\frac{\partial c_0}{\partial x}$$

and

$$c'(L, t) = 0$$

The second condition specifies that chemical leaves the reactor purely as a function of flow through the outlet pipe. That is, it is assumed that dispersion in the reactor does not affect

---

[1]That is, the chemical decays at a rate that is linearly proportional to how much chemical is present.

the exit rate. Under these conditions, use numerical methods to solve Eq. (32.4) for the steady-state levels in the reactor. Note that this is an ODE boundary-value problem. Then solve Eq. (32.3) to characterize the transient response—that is, how the levels change in time as the system approaches the steady state. This application involves a PDE.

Solution. A steady-state solution can be developed by substituting centered finite differences for the first and the second derivatives in Eq. (32.4) to give

$$0 = D\frac{c_{i+1} - 2c_i + c_{i-1}}{\Delta x^2} - U\frac{c_{i+1} - c_{i-1}}{2\,\Delta x} - kc_i$$

Collecting terms gives

$$-\left(\frac{D}{U\,\Delta x} + \frac{1}{2}\right)c_{i-1} + \left(\frac{2D}{U\,\Delta x} + \frac{k\,\Delta x}{U}\right)c_0 - \left(\frac{D}{U\,\Delta x} - \frac{1}{2}\right)c_{i+1} = 0 \qquad (32.5)$$

This equation can be written for each of the system's nodes. At the reactor's ends, this process introduces nodes that lie outside the system. For example, at the inlet node ($i = 0$),

$$-\left(\frac{D}{U\,\Delta x} + \frac{1}{2}\right)c_{-1} + \left(\frac{2D}{U\,\Delta x} + \frac{k\,\Delta x}{U}\right)c_0 - \left(\frac{D}{U\,\Delta x} - \frac{1}{2}\right)c_1 = 0 \qquad (32.6)$$

The c_{-1} can be removed by invoking the first boundary condition. At the inlet, the following mass balance must hold:

$$Qc_{\text{in}} = Qc_0 - DA_c\frac{\partial c_0}{\partial x}$$

where $c_0 = $ concentration at $x = 0$. Thus, this boundary condition specifies that the amount of chemical carried into the tank by advection through the pipe must be equal to the amount carried away from the inlet by both advection and turbulent dispersion in the tank. A finite divided difference can be substituted for the derivative

$$Qc_{\text{in}} = Qc_0 - DA_c\frac{c_1 - c_{-1}}{2\,\Delta x}$$

which can be solved for

$$c_{-1} = c_1 + \frac{2\,\Delta xU}{D}c_{\text{in}} - \frac{2\,\Delta xU}{D}c_0$$

which can be substituted into Eq. (32.6) to give

$$\left(\frac{2D}{U\,\Delta x} + \frac{k\,\Delta x}{U} + 2 + \frac{\Delta xU}{D}\right)c_0 - \left(\frac{D}{U\,\Delta x}\right)c_1 = \left(2 + \frac{\Delta xU}{D}\right)c_{\text{in}} \qquad (32.7)$$

A similar exercise can be performed for the outlet, where the original difference equation is

$$-\left(\frac{D}{U\,\Delta x} + \frac{1}{2}\right)c_{n-1} + \left(\frac{2D}{U\,\Delta x} + \frac{k\,\Delta x}{U}\right)c_n - \left(\frac{D}{U\,\Delta x} - \frac{1}{2}\right)c_{n+1} = 0 \qquad (32.8)$$

The boundary condition at the outlet is

$$Qc_n - DA_c\frac{dc_n}{dx} = Qc_n$$

As with the inlet, a divided difference can be used to approximate the derivative.

$$Qc_n - DA_c\frac{c_{n+1} - c_{n-1}}{2\,\Delta x} = Qc_n \tag{32.9}$$

Inspection of this equation leads us to conclude that $c_{n+1} = c_{n-1}$. In other words, the slope at the outlet must be zero for Eq. (32.9) to hold. Substituting this result into Eq. (32.8) and simplifying gives

$$-\left(\frac{D}{U\,\Delta x}\right)c_{n-1} + \left(\frac{2D}{U\,\Delta x} + \frac{k\,\Delta x}{U}\right)c_n = 0 \tag{32.10}$$

Equations (32.5), (32.7), and (32.10) now form a system of n tridiagonal equations with n unknowns. For example, if $D = 2$, $U = 1$, $\Delta x = 2.5$, $k = 0.2$, and $c_{\text{in}} = 100$, the system is

$$\begin{bmatrix} 5.35 & -1.6 & & & \\ -1.3 & 2.1 & -0.3 & & \\ & -1.3 & 2.1 & -0.3 & \\ & & -1.3 & 2.1 & -0.3 \\ & & & -1.6 & 2.1 \end{bmatrix} \begin{Bmatrix} c_0 \\ c_1 \\ c_2 \\ c_3 \\ c_4 \end{Bmatrix} = \begin{Bmatrix} 325 \\ 0 \\ 0 \\ 0 \\ 0 \end{Bmatrix}$$

which can be solved for

$$c_0 = 76.44 \qquad c_1 = 52.47 \qquad c_2 = 36.06$$
$$c_3 = 25.05 \qquad c_4 = 19.09$$

These results are plotted in Fig. 32.2. As expected, the concentration decreases due to the decay reaction as the chemical flows through the tank. In addition to the above computation, Fig. 32.2 shows another case with $D = 4$. Notice how increasing the turbulent mixing tends to flatten the curve.

FIGURE 32.2
Concentration versus distance along the longitudinal axis of a rectangular reactor for a chemical that decays with first-order kinetics.

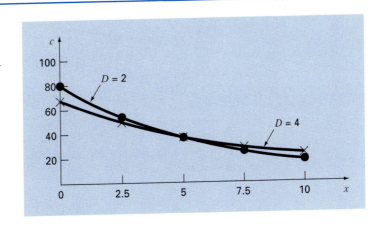

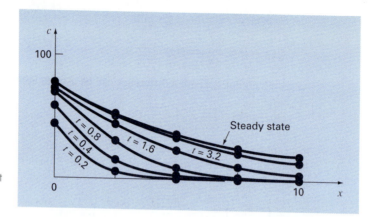

FIGURE 32.3
Concentration versus distance at different times during the build-up of chemical in a reactor.

In contrast, if dispersion is decreased, the curve would become steeper as mixing became less important relative to advection and decay. It should be noted that if dispersion is decreased too much, the computation will become subject to numerical errors. This type of error is referred to as *static instability* to contrast it with the *dynamic instability* due to too large a time step during a dynamic computation. The criterion to avoid this static instability is

$$\Delta x \le \frac{2D}{U}$$

Thus, it becomes more stringent (lower Δx) for cases where advection dominates over dispersion.

Aside from steady-state computations, numerical methods can be used to generate time-variable solutions of Eq. (32.3). Figure 32.3 shows results for $D = 2$, $U = 1$, $\Delta x = 2.5$, $k = 0.2$, and $c_{in} = 100$, where the concentration in the tank is 0 at time zero. As expected, the immediate impact is near the inlet. With time, the solution eventually approaches the steady-state level.

It should be noted that in such dynamic calculations, the time step is constrained by a stability criterion expressed as (Chapra 1997)

$$\Delta t \le \frac{(\Delta x)^2}{2E + k(\Delta x)^2}$$

Thus, the reaction term acts to make the time step smaller.

32.2 DEFLECTIONS OF A PLATE (CIVIL/ENVIRONMENTAL ENGINEERING)

Background.   A square plate with simply supported edges is subject to an areal load q (Fig. 32.4). The deflection in the z dimension can be determined by solving the elliptic PDE (see Carnahan, Luther, and Wilkes 1969)

$$\frac{\partial^4 z}{\partial x^4} + 2\frac{\partial^4 z}{\partial x^2 \partial y^2} + \frac{\partial^4 z}{\partial y^4} = \frac{q}{D} \tag{32.11}$$

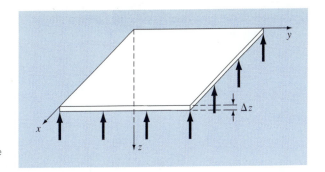

FIGURE 32.4

A simply supported square plate subject to an areal load.

subject to the boundary conditions that, at the edges, the deflection and slope normal to the boundary are zero. The parameter D is the flexural rigidity,

$$D = \frac{E \, \Delta z^3}{12(1 - \sigma^2)} \tag{32.12}$$

where E = the modulus of elasticity, Δz = the plate's thickness, and σ = Poisson's ratio.

If a new variable is defined as

$$u = \frac{\partial^2 z}{\partial x^2} + \frac{\partial^2 z}{\partial y^2}$$

Eq. (32.11) can be reexpressed as

$$\frac{\partial^2 u}{\partial x^2} + \frac{\partial^2 u}{\partial y^2} = \frac{q}{D} \tag{32.13}$$

Therefore, the problem reduces to successively solving two Poisson equations. First, Eq. (32.13) can be solved for u subject to the boundary condition that $u = 0$ at the edges. Then, the results can be employed in conjunction with

$$\frac{\partial^2 z}{\partial x^2} + \frac{\partial^2 z}{\partial y^2} = u \tag{32.14}$$

to solve for z subject to the condition that $z = 0$ at the edges.

Develop a computer program to determine the deflections for a square plate subject to a constant areal load. Test the program for a plate with 2-m-long edges, $q = 33.6$ kN/m^2, $\sigma = 0.3$, $\Delta z = 10^{-2}$ m, and $E = 2 \times 10^{11}$ Pa. Employ $\Delta x = \Delta y = 0.5$ m for your test run.

Solution.   Finite divided differences can be substituted into Eq. (32.13) to give

$$\frac{u_{i+1,j} - 2u_{i,j} + u_{i-1,j}}{\Delta x^2} + \frac{u_{i,j+1} - 2u_{i,j} + u_{i,j-1}}{\Delta y^2} = \frac{q}{D} \tag{32.15}$$

Equation (32.12) can be used to compute $D = 1.832 \times 10^4$ N/m. This result, along with the other system parameters, can be substituted into Eq. (32.15) to give

$$u_{i+1,j} + u_{i-1,j} + u_{i,j+1} + u_{i,j-1} - 4u_{i,j} = 0.458$$

This equation can be written for all the nodes with the boundaries set at $u = 0$. The resulting equations are

$$
\begin{bmatrix}
-4 & 1 & & 1 & & & & & \\
1 & -4 & 1 & & 1 & & & & \\
 & 1 & -4 & & & 1 & & & \\
1 & & & -4 & 1 & & 1 & & \\
 & 1 & & 1 & -4 & 1 & & 1 & \\
 & & 1 & & 1 & -4 & & & 1 \\
 & & & 1 & & & -4 & 1 & \\
 & & & & 1 & & 1 & -4 & 1 \\
 & & & & & 1 & & 1 & -4
\end{bmatrix}
\begin{Bmatrix}
u_{1,1} \\
u_{2,1} \\
u_{3,1} \\
u_{1,2} \\
u_{2,2} \\
u_{3,2} \\
u_{1,3} \\
u_{2,3} \\
u_{3,3}
\end{Bmatrix}
=
\begin{Bmatrix}
0.458 \\
0.458 \\
0.458 \\
0.458 \\
0.458 \\
0.458 \\
0.458 \\
0.458 \\
0.458
\end{Bmatrix}
$$

which can be solved for

$$
\begin{array}{lll}
u_{1,1} = -0.315 & u_{1,2} = -0.401 & u_{1,3} = -0.315 \\
u_{2,1} = -0.401 & u_{2,2} = -0.515 & u_{2,3} = -0.401 \\
u_{3,1} = -0.315 & u_{3,2} = -0.401 & u_{3,3} = -0.315
\end{array}
$$

These results in turn can be substituted into Eq. (32.14), which can be written in finite-difference form and solved for

$$
\begin{array}{lll}
z_{1,1} = 0.063 & z_{1,2} = 0.086 & z_{1,3} = 0.063 \\
z_{2,1} = 0.086 & z_{2,2} = 0.118 & z_{2,3} = 0.086 \\
z_{3,1} = 0.063 & z_{3,2} = 0.086 & z_{3,3} = 0.063
\end{array}
$$

32.3 TWO-DIMENSIONAL ELECTROSTATIC FIELD PROBLEMS (ELECTRICAL ENGINEERING)

Background. Just as Fourier's law and the heat balance can be employed to characterize temperature distribution, analogous relationships are available to model field problems in other areas of engineering. For example, electrical engineers use a similar approach when modeling electrostatic fields.

Under a number of simplifying assumptions (see Paul and Nasar, 1987), an analog of Fourier's law can be represented in one-dimensional form as

$$
D = -\varepsilon \frac{dV}{dx}
$$

where D is called the electric flux density vector, $\varepsilon =$ permittivity of the material, and $V =$ electrostatic potential.

Similarly, a Poisson equation for electrostatic fields can be represented in two dimensions as

$$
\frac{\partial^2 V}{\partial x^2} + \frac{\partial^2 V}{\partial y^2} = -\frac{\rho_v}{\varepsilon} \tag{32.16}
$$

where $\rho_v =$ volumetric charge density.

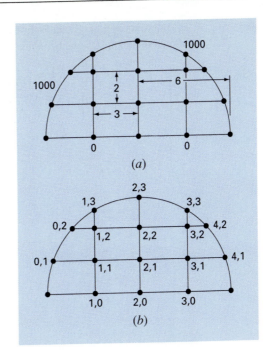

FIGURE 32.5
(a) A two-dimensional system with a voltage of 1000 along the circular boundary and a voltage of 0 along the base. (b) The nodal numbering scheme.

Finally, for regions containing no free charge (that is $\rho_v = 0$), Eq. (32.16) reduces to a Laplace equation,

$$\frac{\partial^2 V}{\partial x^2} + \frac{\partial^2 V}{\partial y^2} = 0 \tag{32.17}$$

Employ numerical methods to solve Eq. (32.17) for the situation depicted in Fig. 32.5. Compute both the values for V and for D if $\varepsilon = 2$.

Solution. Using the approach outlined in Sec. 29.3.2, Eq. (29.24) can be written for node $(1, 1)$ as

$$\frac{2}{\Delta x^2}\left[\frac{V_{1,1} - V_{0,1}}{\alpha_1(\alpha_1 + \alpha_2)} + \frac{V_{1,1} - V_{2,1}}{\alpha_2(\alpha_1 + \alpha_2)}\right] + \frac{2}{\Delta y^2}\left[\frac{V_{1,1} - V_{0,1}}{\beta_1(\beta_1 + \beta_2)} + \frac{V_{1,1} - V_{2,1}}{\beta_2(\beta_1 + \beta_2)}\right] = 0$$

According to the geometry depicted in Fig. 32.5, $\Delta x = 3$, $\Delta y = 2$, $\beta_1 = \beta_2 = \alpha_2 = 1$, and $\alpha_1 = 0.94281$. Substituting these values yields

$$0.12132V_{1,1} - 121.32 + 0.11438V_{1,1} - 0.11438V_{2,1} + 0.25V_{1,1}$$
$$+ 0.25V_{1,1} - 0.25V_{1,2} = 0$$

Collecting terms gives

$$0.73570V_{1,1} - 0.11438V_{2,1} - 0.25V_{1,2} = 121.32$$

A similar approach can be applied to the remaining interior nodes. The resulting simultaneous equations can be expressed in matrix form as

$$
\begin{bmatrix}
0.73570 & -0.11438 & & -0.25000 & & \\
-0.11111 & 0.72222 & -0.11111 & & -0.25000 & \\
& -0.11438 & 0.73570 & & & -0.25000 \\
-0.31288 & & & 1.28888 & -0.14907 & \\
& -0.25000 & & -0.11111 & 0.72222 & -0.11111 \\
& & -0.31288 & & -0.14907 & 1.28888
\end{bmatrix}
$$

$$
\times
\begin{Bmatrix}
V_{1,1} \\
V_{2,1} \\
V_{3,1} \\
V_{1,2} \\
V_{2,2} \\
V_{3,2}
\end{Bmatrix}
=
\begin{Bmatrix}
121.32 \\
0 \\
121.32 \\
826.92 \\
250 \\
826.92
\end{Bmatrix}
$$

which can be solved for

$$V_{1,1} = 521.19 \qquad V_{2,1} = 421.85 \qquad V_{3,1} = 521.19$$

$$V_{1,2} = 855.47 \qquad V_{2,2} = 755.40 \qquad V_{3,2} = 855.47$$

These results are depicted in Fig. 32.6a.

FIGURE 32.6
The results of solving the Laplace equation with correction factors for the irregular boundaries.
(a) Potential and (b) flux.

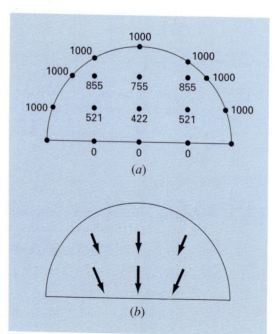

To compute the flux (recall Sec. 29.2.3), Eqs. (29.14) and (29.15) must be modified to account for the irregular boundaries. For the present example, the modifications result in

$$D_x = -\varepsilon \frac{V_{i+1,j} - V_{i-1,j}}{(\alpha_1 + \alpha_2)\Delta x}$$

and

$$D_y = -\varepsilon \frac{V_{i,j+1} - V_{i,j-1}}{(\beta_1 + \beta_2)\Delta y}$$

For node (1, 1), these formulas can be used to compute the x and y components of the flux

$$D_x = -2\frac{421.85 - 1000}{(0.94281 + 1)3} = 198.4$$

and

$$D_y = -2\frac{855.47 - 0}{(1+1)2} = -427.7$$

which in turn can be used to calculate the electric flux density vector

$$D = \sqrt{198.4^2 + (-427.7)^2} = 471.5$$

with a direction of

$$\theta = \tan^{-1}\left(\frac{-427.7}{198.4}\right) = -65.1°$$

The results for the other nodes are

Node	D_x	D_y	D	θ
2, 1	0.0	−377.7	377.7	−90
3, 1	−198.4	−427.7	471.5	245.1
1, 2	109.4	−299.6	281.9	−69.1
2, 2	0.0	−289.1	289.1	−90.1
3, 2	−109.4	−299.6	318.6	249.9

The fluxes are displayed in Fig. 32.6b.

32.4   FINITE-ELEMENT SOLUTION OF A SERIES OF SPRINGS (MECHANICAL/AEROSPACE ENGINEERING)

Background.   Figure 32.7 shows a series of interconnected springs. One end is fixed to a wall, whereas the other is subject to a constant force F. Using the step-by-step procedure outlined in Chap. 31, a finite-element approach can be employed to determine the displacements of the springs.

Solution.   *Discretization.* The way to partition this system is obviously to treat each spring as an element. Thus, the system consists of four elements and five nodes (Fig. 32.7b).

(a)

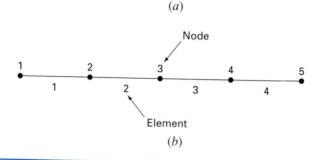

(b)

FIGURE 32.7

(a) A series of interconnected springs. One end is fixed to a wall, whereas the other is subject to a constant force F. (b) The finite-element representation. Each spring represents an element. Therefore, the system consists of four elements and five nodes.

FIGURE 32.8

A free-body diagram of a spring system.

Element equations. Because this system is so simple, its element equations can be written directly without recourse to mathematical approximations. This is an example of the direct approach for deriving elements.

Figure 32.8 shows an individual element. The relationship between force F and displacement x can be represented mathematically by Hooke's law:

$$F = kx$$

where $k =$ the spring constant, which can be interpreted as the force required to cause a unit displacement. If a force F_1 is applied at node 1, the following force balance must hold:

$$F = k(x_1 - x_2)$$

where $x_1 =$ displacement of node 1 from its equilibrium position and $x_2 =$ displacement of node 2 from its equilibrium position. Thus, $x_2 - x_1$ represents how much the spring is elongated or compressed relative to equilibrium (Fig. 32.8).

This equation can also be written as

$$F_1 = kx_1 - kx_2$$

For a stationary system, a force balance also necessitates that $F_1 = -F_2$ and, therefore,

$$F_2 = -kx_1 + kx_2$$

These two simultaneous equations specify the behavior of the element in response to prescribed forces. They can be written in matrix form as

$$\begin{bmatrix} k & -k \\ -k & k \end{bmatrix} \begin{Bmatrix} x_1 \\ x_2 \end{Bmatrix} = \begin{Bmatrix} F_1 \\ F_2 \end{Bmatrix}$$

or

$$[k]\{x\} = \{F\} \tag{32.18}$$

where the matrix $[k]$ is the element property matrix. For this case it is also referred to as the *element stiffness matrix*. Notice that Eq. (32.18) has been cast in the format of Eq. (31.9). Thus, we have succeeded in generating a matrix equation that describes the behavior of a typical element in our system.

Before proceeding to the next step—the assembly of the total solution—we will introduce some notation. The elements of $[k]$ and $\{F\}$ are conventionally superscripted and subscripted, as in

$$\begin{bmatrix} k_{11}^{(e)} & -k_{12}^{(e)} \\ -k_{21}^{(e)} & k_{22}^{(e)} \end{bmatrix} \begin{Bmatrix} x_1 \\ x_2 \end{Bmatrix} = \begin{Bmatrix} F_1^{(e)} \\ F_2^{(e)} \end{Bmatrix}$$

where the superscript (e) designates that these are the element equations. The k's are also subscripted as k_{ij} to denote their location in the ith row and jth column of the matrix. For the present case, they can also be physically interpreted as representing the force required at node i to induce a unit displacement at node j.

Assembly. Before the element equations are assembled, all the elements and nodes must be numbered. This global numbering scheme specifies a system configuration or topology (note that the present case uses a scheme identical to Table 31.1). That is, it documents which nodes belong to which element. Once the topology is specified, the equations for each element can be written with reference to the global coordinates.

The element equations can then be added one at a time to assemble the total system. The final result can be expressed in matrix form as [recall Eq. (31.10)]

$$[k]\{x'\} = \{F'\}$$

where

$$[k] = \begin{bmatrix} k_{11}^{(1)} & -k_{12}^{(1)} \\ -k_{21}^{(1)} & k_{22}^{(1)} + k_{11}^{(2)} & -k_{12}^{(2)} \\ & -k_{21}^{(2)} & k_{22}^{(2)} + k_{11}^{(3)} & -k_{12}^{(3)} \\ & & -k_{21}^{(3)} & k_{22}^{(3)} + k_{11}^{(4)} & -k_{12}^{(4)} \\ & & & -k_{21}^{(4)} & k_{22}^{(4)} \end{bmatrix} \tag{32.19}$$

and

$$\{F'\} = \begin{Bmatrix} F_1^{(1)} \\ 0 \\ 0 \\ 0 \\ F_2^{(4)} \end{Bmatrix}$$

and $\{x'\}$ and $\{F'\}$ are the expanded displacement and force vectors. Notice that, as the equations were assembled, the internal forces cancel. Thus, the final result for $\{F'\}$ has zeros for all but the first and last nodes.

Before proceeding to the next step, we must comment on the structure of the assemblage property matrix [Eq. (32.19)]. Notice that the matrix is tridiagonal. This is a direct result of the particular global numbering scheme that was chosen (Table 31.1) prior to assemblage. Although it is not very important in the present context, the attainment of such a banded, sparse system can be a decided advantage for more complicated problem settings. This is due to the efficient schemes that are available for solving such systems.

Boundary conditions. The present system is subject to a single boundary condition, $x_1 = 0$. Introduction of this condition and applying the global renumbering scheme reduces the system to (k's $= 1$)

$$\begin{bmatrix} 2 & -1 & & \\ -1 & 2 & -1 & \\ & -1 & 2 & -1 \\ & & -1 & 1 \end{bmatrix} \begin{Bmatrix} x_2 \\ x_3 \\ x_4 \\ x_5 \end{Bmatrix} = \begin{Bmatrix} 0 \\ 0 \\ 0 \\ F \end{Bmatrix}$$

The system is now in the form of Eq. (31.11) and is ready to be solved.

Although reduction of the equations is certainly a valid approach for incorporating boundary conditions, it is usually preferable to leave the number of equations intact when performing the solution on the computer. Some schemes for accomplishing this are outlined in Payne and Irons (1963) and Felippa and Clough (1970). Whatever the method, once the boundary conditions are incorporated, we can proceed to the next step—the solution.

Generating solution. Using one of the approaches from Part Three, such as the efficient tridiagonal solution technique delineated in Chap. 11, the system can be solved for (with all k's $= 1$ and $F = 1$)

$$x_2 = 1 \qquad x_3 = 2 \qquad x_4 = 3 \qquad x_5 = 4$$

Postprocessing. The results can now be displayed graphically. As in Fig. 32.9, the results are as expected. Each spring is elongated a unit displacement.

FIGURE 32.9

(a) The original spring system. (b) The system after the application of a constant force. The displacements are indicated in the space between the two systems.

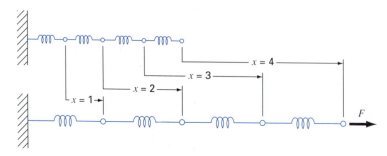

PROBLEMS

Chemical/Petroleum Engineering

32.1 Perform the same computation as in Sec. 32.1, but use $\Delta x = 1.25$.

32.2 Develop a finite-element solution for the steady-state system of Sec. 32.1.

32.3 Compute mass fluxes for the steady-state solution of Sec. 32.1 using Fick's first law.

32.4 Compute the steady-state distribution of concentration for the tank shown in Fig. P32.4. The PDE governing this system is

$$D\left(\frac{\partial^2 c}{\partial x^2} + \frac{\partial^2 c}{\partial y^2}\right) - kc = 0$$

and the boundary conditions are as shown. Employ a value of 0.4 for D and 0.2 for k.

32.5 Two plates are 10 cm apart, as shown in Fig. P32.5. Initially, both plates and the fluid are still. At $t = 0$, the top plate is moved at a constant velocity of 7 cm/s. The equations governing the motions of the fluids are

$$\frac{\partial v_{oil}}{\partial t} = \mu_{oil}\frac{\partial^2 v_{oil}}{\partial x^2} \quad \text{and} \quad \frac{\partial v_{water}}{\partial t} = \mu_{water}\frac{\partial^2 v_{water}}{\partial x^2}$$

and the following relationships hold true at the oil-water interface

$$v_{oil} = v_{water} \quad \text{and} \quad \mu_{oil}\frac{\partial v_{oil}}{\partial x} = \mu_{water}\frac{\partial v_{water}}{\partial x}$$

What is the velocity of the two fluid layers at $t = 0.5$, 1, and 1.5 s at distances $x = 2, 4, 6,$ and 8 cm from the bottom plate? Note that μ_{water} and $\mu_{oil} = 1$ and 3 cp, respectively.

Civil/Environmental Engineering

32.6 Perform the same computation as in Sec. 32.2, but use $\Delta x = 0.5$ and $\Delta y = 0.4$ m.

32.7 The flow through porous media can be described by the Laplace equation

$$\frac{\partial^2 h}{\partial x^2} + \frac{\partial^2 h}{\partial y^2} = 0$$

where h is head. Use numerical methods to determine the distribution of head for the system shown in Fig. P32.7.

32.8 The velocity of water flow through the porous media can be related to head by D'Arcy's law

$$q_n = -K\frac{dh}{dn}$$

where K is the hydraulic conductivity and q_n is discharge velocity in the n direction. If $K = 4 \times 10^{-4}$ cm/s, compute the water velocities for Prob. 32.7.

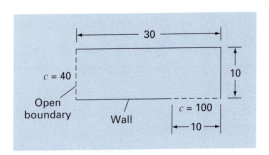

FIGURE P32.4

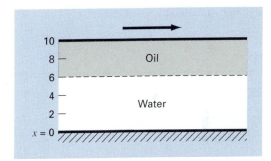

FIGURE P32.5

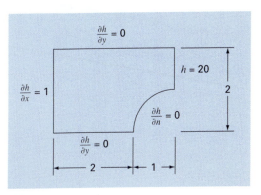

FIGURE P32.7

Electrical Engineering

32.9 Perform the same computation as in Sec. 32.3, but for the system depicted in Fig. P32.9.

32.10 Perform the same computation as in Sec. 32.3, but for the system depicted in Fig. P32.10.

Mechanical/Aerospace Engineering

32.11 Perform the same computation as in Sec. 32.4 but change the spring constants to

Spring	1	2	3	4
k	0.7	2	0.75	1.5

Also change the force to 2.

32.12 Perform the same computation as in Sec. 32.4, but use five springs with

Spring	1	2	3	4	5
k	0.2	0.4	1.4	0.7	0.9

FIGURE P32.9

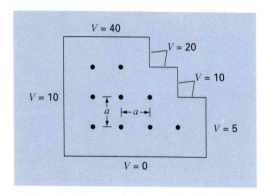

FIGURE P32.10

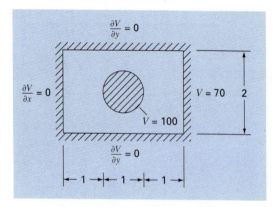

EPILOGUE: PART EIGHT

PT8.3 TRADE-OFFS

The primary trade-offs associated with numerical methods for the solution of partial differential equations involve choosing between *finite-difference* and *finite-element* approaches. The finite-difference methods are conceptually easier to understand. In addition, they are easy to program for systems that can be approximated with uniform grids. However, they are difficult to apply to systems with complicated geometries.

Finite-difference approaches can be divided into categories depending on the type of PDE that is being solved. *Elliptic PDEs* can be approximated by a set of linear algebraic equations. Consequently, the *Liebmann method* (which, in fact, is Gauss-Seidel) can be employed to obtain a solution iteratively.

One-dimensional parabolic PDEs can be solved in two fundamentally different ways: explicit or implicit approaches. The *explicit method* steps out in time in a fashion that is similar to Euler's technique for solving ODEs. It has the advantage that it is simple to program but has the shortcoming of a very stringent stability criterion. In contrast, stable implicit methods are available. These typically involve solving simultaneous tridiagonal algebraic equations at each time step. One of these approaches, the *Crank-Nicolson method,* is both accurate and stable and, therefore, is widely used for one-dimensional linear parabolic problems.

Two-dimensional parabolic PDEs can also be modeled explicitly. However, their stability constraints are even more severe than for the one-dimensional case. Special implicit approaches, which are generally referred to as splitting methods, have been developed to circumvent this shortcoming. These approaches are both efficient and stable. One of the most common is the *ADI,* or *alternating-direction implicit,* method.

All the above *finite-difference* approaches become unwieldy when applied to systems involving nonuniform shapes and heterogeneous conditions. Finite-element methods are available that handle such systems in a superior fashion.

Although the *finite-element method* is based on some fairly straightforward ideas, the mechanics of generating a good finite-element code for two- and three-dimensional problems is not a trivial exercise. In addition, it can be computationally expensive for large problems. However, it is vastly superior to finite-difference approaches for systems involving complicated shapes. Consequently, its expense and conceptual "overhead" are often justified because of the detail of the final solution.

PT8.4 IMPORTANT RELATIONSHIPS AND FORMULAS

Table PT8.3 summarizes important information that was presented regarding the finite-difference methods in Part Eight. This table can be consulted to quickly access important relationships and formulas.

TABLE PT8.3 Summary of finite-difference methods.

Computational Molecule		Equation

Elliptic PDEs Liebmann's method		$T_{i,j} = \dfrac{T_{i+1,j} + T_{i-1,j} + T_{i,j+1} + T_{i,j-1}}{4}$
Parabolic PDEs (one-dimensional) Explicit method		$T_i^{l+1} = T_i^l + \lambda(T_{i+1}^l - 2T_i^l + T_{i-1}^l)$
Implicit method		$-\lambda T_{i-1}^{l+1} + (1 + 2\lambda)T_i^{l+1} - \lambda T_{i+1}^{l+1} = T_i^l$
Crank-Nicolson method		$-\lambda T_{i-1}^{l+1} + 2(1 + \lambda)T_i^{l+1} - \lambda T_{i+1}^{l+1}$ $= \lambda T_{i-1}^l + 2(1 - \lambda)T_i^l + \lambda T_{i+1}^l$

PT8.5 ADVANCED METHODS AND ADDITIONAL REFERENCES

Carnahan, Luther, and Wilkes (1969); Rice (1983); Ferziger (1981); and Lapidus and Pinder (1982) provide useful surveys of methods and software for solving PDEs. You can also consult Ames (1977), Gladwell and Wait (1979), Vichnevetsky (1981, 1982), and Zienkiewicz (1971) for more in-depth treatments. Additional information on the finite-element method can be found in Allaire (1985), Huebner and Thornton (1982), Stasa (1985), and Baker (1983). Aside from elliptic and hyperbolic PDEs, numerical methods are also available to solve hyperbolic equations. Nice introductions and summaries of some of these methods can be found in Lapidus and Pinder (1982), Ferziger (1981), Forsythe and Wasow (1960), and Hoffman (1992).

APPENDIX A
THE FOURIER SERIES

The Fourier series can be expressed in a number of different formats. Two equivalent trigonometric expressions are

$$f(t) = a_0 + \sum_{k=1}^{\infty} [a_k \cos(k\omega_0 t) + b_k \sin(k\omega_0 t)]$$

or

$$f(t) = a_0 + \sum_{k=1}^{\infty} [c_k \cos(k\omega_0 t + \theta_k)]$$

where the coefficients are related by (see Fig. A.1)

$$c_k = \sqrt{a_k^2 + b_k^2}$$

and

$$\theta_k = -\tan^{-1}\left(\frac{b_k}{a_k}\right)$$

In addition to the trigonometric formats, the series can also be expressed in terms of the exponential function as

$$f(t) = \tilde{c}_0 + \sum_{k=1}^{\infty} [\tilde{c}_k e^{ik\omega_0 t} + \tilde{c}_{-k} e^{-ik\omega_0 t}] \qquad (A.1)$$

where (see Fig. A.2)

$$\tilde{c}_0 = a_0$$

$$\tilde{c}_k = \frac{1}{2}(a_k - ib_k) = |\tilde{c}_k| e^{i\phi}k$$

$$\tilde{c}_{-k} = \frac{1}{2}(a_k + ib_k) = |\tilde{c}_k| e^{-i\phi}k$$

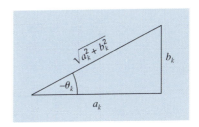

FIGURE A.1
Relationships between rectangular and polar forms of the Fourier series coefficients.

FIGURE A.2
Relationships between complex exponential and real coefficients of the Fourier series.

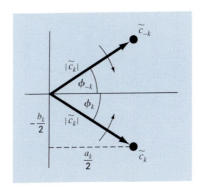

where $|\tilde{c}_0| = a_0$ and

$$|\tilde{c}_k| = \frac{1}{2}\sqrt{a_k^2 + b_k^2} = \frac{c_k}{2}$$

and

$$\phi_k = \tan^{-1}\left(\frac{-b_k}{a_k}\right)$$

Note that the tilde signifies that the coefficient is a complex number.

Each term in Eq. (A.1) can be visualized as a rotating phasor (the arrows in Fig. A.2). Terms with a positive subscript rotate in a counterclockwise direction, whereas those with a negative subscript rotate clockwise. The coefficients $\tilde{c}_k$ and $\tilde{c}_{-k}$ specify the position of the phasor at $t = 0$. The infinite summation of the spinning phasors, which are allowed to rotate at $t = 0$, is then equal to $f(t)$.

APPENDIX B

GETTING STARTED WITH MATHCAD

OVERVIEW OF MATHCAD'S FEATURES

Mathcad has a unique way to handle equations, numbers, text, and graphs. It works like a scratchpad and pencil. The screen interface is a blank worksheet on which you can enter equations, graph data or functions, and annotate operations with text. Entries can be placed anywhere on the page. It uses standard mathematical symbols to represent operators when possible. Therefore, you may find that the Mathcad interface is quite natural and familiar.

Mathcad can solve a wide range of mathematical problems either *numerically* or *symbolically*. The symbolic capabilities of Mathcad have relatively little application in this text, although they may be used to check our numerical results. Therefore, they will not be covered in detail in this overview. Mathcad has a comprehensive set of operators and functions that allow you to perform many of the numerical methods covered in this text. It also has a programming language that allows you to write your own multiline procedures and subprograms. The following discussion provides a brief description of the features of Mathcad you will find most useful for this text.

GETTING STARTED WITH MATHCAD

Applications in this text will require that you be able to create your own worksheets. To facilitate your efforts, let's go over the main features of the Mathcad application window.

The Main Menu

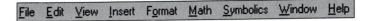

This is your gateway to math, graphics, and symbolic functions of Mathcad. It also provides commands that handle the details of editing and managing your worksheets. For example, click on the **Math** and the **Symbolics** menus to see the array of computational functionality available to you.

The Math Palette

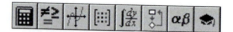

Next to the **Main** menu is a tear-away bar whose buttons bring up palettes of math operators. If you let your mouse hover over each of the buttons on the bar, you'll see a *tool tip* telling you the function of each palette button. The buttons and their functions are described below.

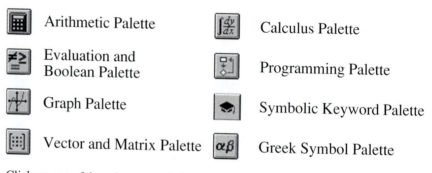

Arithmetic Palette

Evaluation and
Boolean Palette

Graph Palette

Vector and Matrix Palette

Calculus Palette

Programming Palette

Symbolic Keyword Palette

Greek Symbol Palette

Click on one of these buttons to bring up the full palette. You can use the palettes to insert math symbols and operations directly into your Mathcad worksheet.

The Tool Bar

This is another tear-away button bar that provides shortcuts for many common tasks, from worksheet opening and file saving to spell checking to bringing up lists of built-in functions and units. Depending on what you're doing in your worksheet, one or more of these buttons may appear grayed out. Again *tool tips* will remind you of the functions of each of the buttons.

Entering Mathematical Operations

Type	See on Screen
1+	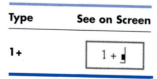

After you type the **+** sign you will see a little black box delimited by blue editing lines. In Mathcad this black box is called a *placeholder*. If you continue typing, whatever you type next will appear in the placeholder. For example, type 2 in the placeholder, then press the equals key (=) to see the result.

$$1 + 2 = 3$$

The basic arithmetic operators are listed below, along with their keystrokes and Palette button equivalents.

Operation	Keystroke	Button	Example
Addition	$+$	$+$	$2 + 2 = 4$
Subtraction	$-$	$-$	$2 - 2 = 0$
Multiplication	$*$	$\times$	$2 \cdot 2 = 4$
Division	$/$	$\div$	$\dfrac{2}{2} = 1$
Powers	$\wedge$	x^y	$2^2 = 4$

Notice that operations in a Mathcad worksheet appear in familiar notation—multiplication as a raised dot, division with a fraction bar, exponents in a raised position, and so on. Calculations are computed internally to 15 places, but you can show fewer places in the answer—just click on an answer and choose **Number** from the **Format** menu, then change the number in the **Displayed Precision** field in the dialogue box. Here are a few more examples that demonstrate Mathcad features.

$$\sqrt{\frac{1.837 \cdot 10^3}{100 + 3^5}} = 2.3142353232$$

Most standard engineering and mathematical functions are built-in.

$$\log(1347.2) \cdot \sin\left(\frac{3}{5} \cdot \pi\right) = 2.976$$

Mathcad's functions and operators easily handle complex numbers.

$$(2.3 + 4.7i)^3 + e^{3-2i} = -148.613 - 47.498i$$

Mathcad can also handle units. To see the built-in units, choose **Unit** from the **Insert** menu, or click on the appropriate toolbar button.

$$\frac{2350 \cdot km}{1 \cdot hr} = 652.78 \ m \circ sec^{-1}$$

Entering Text

To create a text region, click in a blank area of the screen to position the red cross hair cursor and type a double-quote ["]. Now you can type whatever you like, just as in a word processor. As the region grows, a black box appears around the text. The box has resizing "handles" on the right and bottom edges of the rectangle. Once you are done, click outside the text region to go back to inputting math operations. The black selection box disappears when you're no longer working in the text region.

Defining Mathematical Functions and Variables

The definition symbol := is used to define a function or variable in Mathcad. For example, click an empty worksheet to position the red cross hair in a blank area and type:

Type	See on Screen
f(x):x^2	$f(x) := x^2$

The definition symbol is also located in the Arithmetic Palette. When you change a definition function or variable, Mathcad immediately recalculates any new values that depend on it. Once you've defined a function like f(x), you can use it in a number of ways, for example:

$$f(x) := x^2$$

Now you can insert a numerical value as the argument of f(x).

$$f(10) = 100$$

or define a variable and insert it in as the argument of f(x).

$$x := 3$$
$$f(x) = 9$$

You can even define another function in terms of f(x).

$$g(y) := f(y) + 6$$
$$g(x) = 15$$

Note that you can define a function using expressions you build up from the keyboard or from the palettes of math operators. You can also include any of Mathcad's hundreds of built-in functions. To see a list of built-in functions along with brief descriptions, select **Function** from the **Insert** menu, or click on the *f(x)* button. You can also type the name of any built-in function directly from the keyboard. The following are just a few examples that use some of Mathcad's built-in functions.

Trig and Logs

$$\ln(26) = 3.258 \qquad \csc(45 \cdot \deg) = 1 \cdot \sqrt{2}$$

Matrix Functions

$$\text{identity}(3) = \begin{bmatrix} 1 & 0 & 0 \\ 0 & 1 & 0 \\ 0 & 0 & 1 \end{bmatrix}$$

Probability Distributions

$$\text{pnorm}(2, 0, 1) = 0.977$$

Range Variables

In Mathcad you will find yourself wanting to work with a range of values for many applications—for example to define a series of values to plot. Mathcad therefore provides the .. range operator in the Arithmetic Palette. This can also be entered by typing a semicolon [;] at the keyboard. The first and last numbers establish the endpoints of the range variable, and the second number sets the increment. For example,

Type	See on Screen
z : 0, .5; 2	$z := 0, .5 .. 2$
z =	$\begin{array}{c} z \\ \hline 0 \\ 0.5 \\ 1 \\ 1.5 \\ 2 \end{array}$

Matrix Computations and Operations

To enter a matrix, click on the 3 by 3 matrix icon in the Vector and Matrix Palette (or choose **Matrix** from the **Insert** menu), choose the number of rows and columns, then fill in the placeholders. For example,

$$A := \begin{bmatrix} 4 & 5 & 1 \\ 5 & 0 & -12 \\ -7 & 2 & 8 \end{bmatrix}$$

To compute the inverse,

Type	See on Screen
A^-1=	$A^{-1} = \begin{bmatrix} 0.074 & -0.117 & -0.184 \\ 0.135 & 0.12 & 0.163 \\ 0.031 & -0.132 & -0.077 \end{bmatrix}$

Mathcad has a comprehensive set of commands to perform various matrix operations. For example, to find the determinant, type a vertical bar, $|$, or use the button on the Vector and Matrix Palette.

$$|A| = 326$$

Numerical Methods Functions

Mathcad has a number of special built-in functions that perform a variety of numerical operations of particular interest to readers of this book. Examples of the development and application of these functions will be described in detail in the text. Here we will provide a brief list of some of the more important functions just to give you an overview of the capabilities.

Function Name	Use
root	Solves f(x) = 0
polyroots	Finds all roots of a polynomial
find	Solves a system of nonlinear algebraic equations
minerr	Returns a minimum error solution of a system of equations
lsolve	Solves a system of linear algebraic equations
linterp	Linear interpolation
cspline	Cubic spline interpolation
regress	Polynomial regression
genfit	General nonlinear regression
fft	Fourier transform
ifft	Inverse Fourier transform
rkfixed	Solves a system of differential equations using a fixed-step-size 4th- order Runge-Kutta method
rkadapt	Solves a system of differential equations using a variable step-size 4th-order Runge-Kutta method
sbval	Solves a two-point boundary value problem
eigenvals	Finds eigenvalues
eigenvecs	Finds eigenvectors
relax	Solves Poisson's equation for a square domain

Multiline Procedures and Subprograms

The Programming Palette in Mathcad provides the capability for multiline procedures or subprograms with standard control structures such as **FOR** and **WHILE** loops, branching, recursion, and more. Subprograms are seamlessly integrated with Mathcad's worksheets and can operate on scalars, vectors, arrays, and even arrays of arrays.

Creating Graphs

Mathcad makes it easy for you to create an x-y plot. Just click in some empty space and type an expression that depends on one variable, for example, sin(x), and then click on the **X-Y Plot** button in the Graph Palette, or choose **X–Y Plot** from the **Insert/Graph** menu.

Then press Enter. You should see a nicely formatted plot that looks like this:

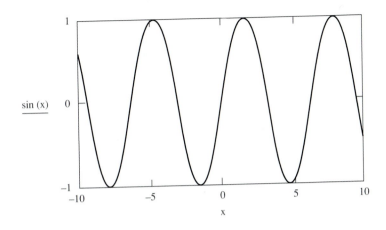

Mathcad can also automatically interpolate a 3-D scatter plot to create a surface plot. The plot below at left was created using **Graph/3D Scatter Plot** from the **Insert** menu, and the plot at right using **Graph/Surface Plot** from the **Insert** menu. You can also use the corresponding buttons from the Graph Palette.

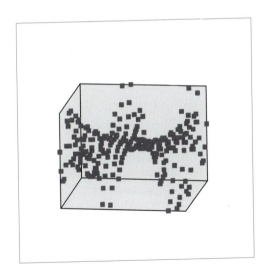

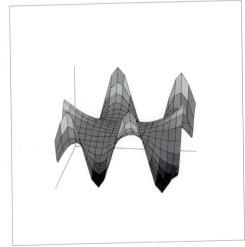

Graphing a Function Using Range Variables

When you graphed the above expression, Mathcad selected a default range for the dependent variable. However, you can control the range of values that are plotted using a range variable. For example, type the following function:

Type	See on Screen
f(x) :-x^2[Spacebar]+8*x-27	$f(x) := -x^2 + 8 \cdot x - 27$

Then define an independent variable for the horizontal axis. For example:

Type	See on Screen
x:0;10	$x := 0.. \ 10$

Place your plot by clicking in some free space, and then use the Graph Palette to create the x-y plot. Type x in the middle placeholder on the horizontal axis and type f(x) in the middle placeholder on the vertical axis. Then press Enter. Your plot should look like this:

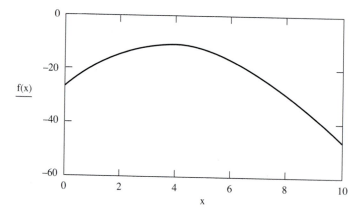

Note that the plot of f(x) may look a little rough. To smooth it out, you can change the definition of x to x:= 0, 0.1.. 10. The smaller increment (or step) means more points calculated, which means more plotted, which makes the plot smoother because Mathcad is essentially connecting more dots with straight lines.

To format an x-y plot, just double-click on it (or choose **Graph** from the **Format** menu) to bring up a formatting dialogue box. The tabbed dialog box lets you change options for logarithmic axes, grid lines, legends, trace types, markers, colors, axis limits, and more.

Graphing Two or More Functions

Plotting multiple functions over a single domain on a single graph is straightforward. For example, suppose you want to plot both 1/z and z^2 on one plot. Just use the Graph Palette to create a blank plot and then type the two expressions separated by a comma in the vertical placeholder.

You should see a graph similar to this:

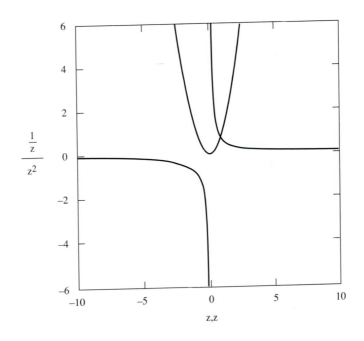

Symbolic Operators

Some mathematical functions in Mathcad use symbolic operators. Evaluation of any expression is performed by simply choosing symbolic evaluation with keywords from the Symbolic palette. The evaluations will make use of all the definitions in your worksheet. Here are a few examples.

Limit:

$$\lim_{x \to 0}(1 + x)^{\frac{1}{x}} \to \exp(1)$$

Derivative:

$$g(x) := \cos(x)^3$$

$$\frac{d^2}{dx^2} g(x) \to 6 \cdot \cos(x) \cdot \sin(x)^2 - 3 \cdot \cos(x)^3$$

Integral:

$$\int_a^b g(x)\, dx \to \frac{1}{3} \cdot \cos(b)^2 \cdot \sin(b) + \frac{2}{3} \cdot \sin(b) - \frac{1}{3} \cdot \cos(a)^2 \cdot \sin(a) - \frac{2}{3} \cdot \sin(a)$$

Taylor series expansion:

$$e^x \text{ series, } x = 0 \to 1 + x + \frac{1}{2} \cdot x^2 + \frac{1}{6} \cdot x^3 + \frac{1}{24} \cdot x^4 + \frac{1}{120} \cdot x^5$$

Learning More about Mathcad

In this brief introduction we have covered only the Mathcad basics. Further help is available right in the Mathcad software package in a variety of forms:

ToolTips

Let your mouse pointer hover over a palette or toolbar button for a few seconds. You will see an explanatory tool tip displayed near the button: Look also on the message line at the bottom of the Mathcad application window for helpful tips and shortcuts.

Resource Center and QuickSheets

To help you get going fast and keep you going beyond the basics found in the tutorial, Mathcad comes complete with QuickSheets. These provide mathematical shortcuts for frequently used analyses—from graphing a function to solving simultaneous equations to the analysis of variance. There are more than 300 QuickSheets in all. To open the QuickSheets section, choose **QuickSheets** from the opening page of the **Resource Center** located under **Help** on the **Main** menu.

Online Help

Online Help provides detailed, step-by-step instructions for using all of Mathcad's features. Help is available at any time by simply choosing **Help Topics (?)**, or by clicking on the **Help** button on the toolbar.

MathSoft World Wide Web Site

If you have access to the Internet, you can link directly to **MathSoft's World Wide Web Site** for Mathcad files, technical support information, and more. Or you can link directly to MathSoft's Mathcad 7 Web Page in Cambridge, Massachusetts, from right within Mathcad. You can also get to various sections of the MathSoft Web site through buttons on the Resource Center Home Page. You may use your usual Web browser (e.g., Netscape Navigator, Microsoft Internet Explorer, Mosaic, etc.) to visit the MathSoft Home Page at *http://www.mathsoft.com/*. You can also access the **Collaboratory** on the MathSoft internet site through the **File** Menu. Use these forums to post questions and problems to other users who may have similar interests.

APPENDIX C

GETTING STARTED WITH MATLAB[1]

MATLAB is a computer program that provides the user with a convenient environment for many types of calculations—in particular, those that are related to matrix manipulations. MATLAB operates interactively, executing the user's command one-by-one as they are entered. A series of commands may be saved as a script and run like an interpretive program. MATLAB has a large number of built-in functions; however, it is possible for users to build their own functions made up of MATLAB commands and functions. The primary features of MATLAB are built-in vector and matrix computations including:

- Vector-matrix arithmetic.
- Matrix inversion and eigenvalue/vector analysis.
- Complex arithmetic and polynomial operations.
- Statistical calculations.
- Graphical displays.
- Control system design.
- Fitting process models from test data.

MATLAB has a number of optional toolboxes that provide specialized functions. These include: signal processing, control systems, system identification, optimization, and statistics.

MATLAB is available in versions that run on PCs, Macs, and workstations. The modern version that runs on PCs does so in the Windows environment. The seven exercises that follow are meant to give you the flavor of computing with MATLAB; they do not constitute a comprehensive tutorial. There are additional tutorial materials in the MATLAB manuals. A number of textbooks now feature MATLAB exercises. Also, on-line information is available for any command or function by typing help *name,* where *name* identifies the command. Do not just look through these exercises; try them all and try variations that occur to you. Check the answers that MATLAB gives and make sure you understand them and they are correct. That is the effective way to learn MATLAB.

[1]Originally developed by Prof. Dave Clough, Chemical Engineering, University of Colorado.

1. Assignment of Values to Variable Names

Assignment of values to scalar variables is similar to other computer languages. Try typing

```
a = 4
```

and

```
A = 6
```

Note how the assignment echoes to confirm what you have done. This is a characteristic of MATLAB. The echo can be suppressed by terminating the command line with the semicolon (;) character. Try typing

```
b = -3;
```

MATLAB treats names in a case-sensitive manner; that is, the name `a` is not the same as the name `A`. To illustrate this, enter

```
a
```

and

```
A
```

See how their values are distinct. They are distinct names.

Variable names in MATLAB generally represent matrix quantities. A row vector can be assigned as follows:

```
a = [ 1  2  3  4  5 ]
```

The echo confirms the assignment again. Notice how the new assignment of `a` has taken over. A column vector can be entered in several ways. Try them.

```
b = [ 1 ; 2 ; 3 ; 4 ; 5 ]
```

or

```
b = [ 1;
      2;
      3;
      4;
      5 ]
```

or, by transposing a row vector with the `'` operator,

```
b = [ 1  2  3  4  5 ]'
```

A two-dimensional matrix of values can be assigned as follows:

```
A = [ 1  2  3 ; 4  5  6 ; 7  8  8 ]
```

or

```
A = [ 1  2  3 ;
      4  5  6 ;
      7  8  8  ]
```

The values stored by a variable can be examined at any time by typing the name alone, e.g.,

```
b
```

or

```
A
```

Also, a list of all current variables can be obtained by entering the command

```
who
```

or, with more detail, enter

```
whos
```

There are several predefined variables, e.g., `pi`.

It is also possible to assign complex values to variables, since MATLAB handles complex arithmetic automatically. To do this, it is convenient to assign a variable name, usually either i or j, to the square root of -1.

```
i = sqrt(-1)
```

Then, a complex value can be assigned, like

```
x = 2 + i*4
```

2. Mathematical Operations

Operations with scalar quantities are handled in a straightforward manner, similar to computer languages. The common operators, in order of preference, are

```
^       Exponentiation
* /     Multiplication and division
\       Left division (applies to matrices)
+ -     Addition and subtraction
```

These operators will work in calculator fashion. Try

```
2 * pi
```

Also, scalar real variables can be included:

```
y = pi / 4
y ^ 2.45
```

Results of calculations can be assigned to a variable, as in the next-to-last example, or simply displayed, as in the last example.

Calculations can also involve complex quantities. Using the x defined above, try

```
3 * x
1 / x
x ^ 2
x + y
```

The real power of MATLAB is illustrated in its ability to carry out matrix calculations. The inner product of two vectors (dot product) can be calculated using the * operator,

    a * b

and likewise, the outer product

    b * a

To illustrate vector-matrix multiplication, first redefine a and b,

    a = [1 2 3]

and

    b = [4 5 6]'

Now, try

    a * A

or

    A * b

What happens when dimensions are not those required for the operations? Try

    A * a

to see. Matrix-matrix multiplication is carried out in likewise fashion:

    A * A

Mixed operations with scalars are also possible:

    A / pi

It is important to always remember that MATLAB will apply the simple arithmetic operators in vector-matrix fashion if possible. At times, you will want to carry out calculations item-by-item in a matrix or vector. MATLAB provides for that too. For example,

    A ^ 2

results in matrix multiplication of A with itself. What if you want to square each element of A? That can be done with

    A .^ 2

The . preceding the ^ operator signifies that the operation is to be carried out item-by-item. The MATLAB manual calls these *array operations*.

When the division operator (/) is used with matrices, the use of a matrix inverse is implied. Therefore, if A is a square, nonsingular matrix, then B/A corresponds to the right multiplication of B by the inverse of A. A longer way to do this used the *inv* function; that is, B*inv(A) ; however, using the division operator is more efficient, since X = B/A is actually solved as the set of equations X*A=B using a decomposition/elimination scheme.

The "left division" operator (\ , the backslash character) is used in matrix operations also. As above, A\B corresponds to the left multiplication of B by the inverse of A.

This is actually solved as the set of equations A*X=B, a common engineering calculation.

For example, if c is a column vector with values 0.1, 1.0, and 10, the solution of A * x = c, where A has been set above, can be obtained by typing

```
c = [ 0.1 1.0 10 ]´
x = A \ c
```

Try that.

3. Use of Built-In Functions

MATLAB and its Toolboxes have a rich collection of built-in functions. You can use on-line help to find out more about them. One of their important properties is that they will operate directly on vector and matrix quantities. For example, try

```
log(A)
```

and you will see that the natural logarithm function is applied in array style, element by element, to the matrix A. Most functions, like *sqrt, abs, sin, acos, tanh, exp,* operate in array fashion. Certain functions, like exponential and square root, have matrix definitions also. MATLAB will evaluate the matrix version when the letter m is appended to the function name. Try

```
sqrtm(A)
```

A common use of functions is to evaluate a formula for a series of arguments. Create a column vector t that contains values from 0 to 100 in steps of 5,

```
t = [ 0 : 5 : 100 ]'
```

Check the number of items in the t array with the *length* function,

```
length(t)
```

Now, say that you want to evaluate a formula y = f(t), where the formula is computed for each value of the t array, and the result is assigned to a corresponding position in the y array. For example,

```
y = t .^ 0.34 - log10(t) + 1 ./ t
```

Done! [Note the use of the array operators adjacent decimal points.] This is similar to creating a column of the t values on a spreadsheet and copying a formula down an adjacent column to evaluate y values.

4. Graphics

MATLAB's graphics capabilities have similarities to those of a spreadsheet program. Graphs can be created quickly and conveniently; however, there is not much flexibility to customize them.

For example, to create a graph of the t, y arrays from above, enter

```
plot(t, y)
```

That's it! You can customize the graph a bit with commands like the following:

```
title('Plot of y versus t')
xlabel('Values of t')
ylabel('Values of y')
grid
```

The graph appears in a separate window and can be printed or transferred via the clipboard (PCs with Windows or Macs) to other programs.

There are other features of graphics that will become useful, plotting objects instead of lines, families of curves, plotting on the complex plane, multiple graphs windows, log-log or semilog plots, 3-D mesh plots, and contour plots.

5. Polynomials

There are many MATLAB functions that allow you to operate on arrays as if their entries were coefficients or roots of polynomial equations. For example, enter

```
c = [ 1 1 1 1 ]
```

and then

```
r = roots(c)
```

and the roots of the polynomial $x^3 + x^2 + x + 1 = 0$ should be printed and are also stored in the r array. The polynomial coefficients can be computed from the roots with the *poly* function,

```
poly(r)
```

and a polynomial can be evaluated for a given value of x. For example,

```
polyval(c, 1.32)
```

If another polynomial, $2x^2 - 0.4x - 1$, is represented by the array d,

```
d = [ 2 -0.4 -1 ]
```

the two polynomials can be multiplied symbolically with the convolution function, *conv,* to yield the coefficients of the product polynomial,

```
cd = conv(c,d)
```

The deconvolution function, *deconv,* can be used to divide one polynomial into another; for example,

```
[ q,r ] = deconv(c,d)
```

The q result is the quotient polynomial's coefficients, and the r result is the remainder polynomial's coefficients.

There are other polynomial functions that may become useful to you, such as the residue function for partial fraction expansion.

6. Statistical Analysis

The Statistics Toolbox contains many features for statistical analysis; however, common statistical calculations can be performed with MATLAB's basic function set. You can generate a series of (pseudo)random numbers with the *rand* function. Either a uniform or normal distribution is available:

```
rand('normal')
n = 0 : 5 : 1000 ;
```

(Did you forget the ; !!!)

```
num = rand(size(n)) ;
```

You probably understand why using the semicolon at the end of the commands above is important, especially if you neglected to do so.

If you would like to see a plot of noise, try

```
plot(num)
```

These are supposed to be normally distributed numbers with a mean of zero and variance (and standard deviation) of one. Check by

```
mean(num)
```

and

```
std(num)
```

No one is perfect! You can find minimum and maximum values,

```
min(num)
max(num)
```

There is a convenient function for plotting a histogram of the data:

```
hist(num,20)
```

where 20 is the number of bins.

If you would like to fit a polynomial to some data by least squares, you can use the *polyfit* function. Try the following example:

```
t = 0 : 5
y = [ -0.45 0.56 2.34 5.6 9.45 24.59 ]
coef = polyfit( t, y, 3 )
```

The values in coef are the fitted polynomial coefficients. To generate the computed value of y,

```
yc = polyval( coef,t )
```

and to plot the data versus the fitted curve,

```
plot ( t,yc,t,y,'o' )
```

The plot of the continuous curve is piecewise linear; therefore, it does not look very smooth. Improve this as follows:

```
t1 = [ 0 : 0.05 : 5 ] ;
yc = polyval(coef, t1)
plot(t1, yc, t, y, 'o')
```

7. This and That

There are many, many other features to MATLAB. Some of these you will find useful; perhaps others you will never use. We encourage you to explore and experiment.

To save a copy of your session, MATLAB has a useful capability called *diary*. You issue the command

```
diary problem1
```

and MATLAB opens a disk file in which it stores all the subsequent commands and results (not graphs) of your session. You can turn the diary feature off:

```
diary off
```

and back on with the same file:

```
diary on
```

After you leave MATLAB, the diary file is available to you. It is common to use an editor or word processor to clean up the diary file (getting rid of all those errors you made before anyone can see them!) and then print the file to obtain a hard copy of the important parts of your work session, e.g., key numerical results.

Exit MATLAB with the `quit` or `exit` commands. It is possible to save the current state of your work with the `save` command. It is also possible to reload that state with the `load` command.

Bibliography

Al-Khafaji, A. W., and J. R. Tooley, *Numerical Methods in Engineering Practice,* Holt, Rinehart and Winston, New York, 1986.

Allaire, P. E., *Basics of the Finite Element Method,* William C. Brown, Dubuque, IA, 1985.

Ames, W. F., *Numerical Methods for Partial Differential Equations,* Academic Press, New York, 1977.

Ang, A. H-S., and W. H. Tang, *Probability Concepts in Engineering Planning and Design, Vol. 1: Basic Principles,* Wiley, New York, 1975.

APHA (American Public Health Association). 1992. Standard Methods for the Examination of Water and Wastewater, 18th ed., Washington, DC.

Atkinson, K. E., An *Introduction to Numerical Analysis,* Wiley, New York, 1978.

Atkinson, L. V., and P. J. Harley, *An Introduction to Numerical Methods with Pascal,* Addison-Wesley, Reading, MA, 1983.

Baker, A. J., *Finite Element Computational Fluid Mechanics,* McGraw-Hill, New York, 1983.

Bathe, K.-J., and E. L. Wilson, *Numerical Methods in Finite Element Analysis,* Prentice-Hall, Englewood Cliffs, NJ, 1976.

Booth, G. W., and T. L. Peterson, "Nonlinear Estimation," I.B.M. Share Program Pa. No. 687 WLNL1 (1958).

Boyce, W. E., and DiPrima, R. C., *Elementary Differential Equations and Boundary Value Problems,* 5th ed. Wiley, New York, 1992.

Branscomb, L. M., 'Electronics and Computers: An Overview," *Science,* 215:755 (1982).

Brigham, E. O., *The Fast Fourier Transform,* Prentice-Hall, Englewood Cliffs, NJ, 1974.

Burden, R. L. and Faires, J. D., *Numerical Analysis,* 5th ed., PWS Publishing, Boston, 1993.

Butcher, J. C., "On Runge-Kutta Processes of Higher Order," *J. Austral. Math. Soc.,* 4:179 (1964).

Carnahan, B., H. A. Luther, and J. O. Wilkes, *Applied Numerical Methods,* Wiley, New York, 1969.

Cash, J. R., and Karp, A. H. *ACM Transactions on Mathematical Software,* 16:201-222 (1990).

Chapra, S. C., *Surface Water-Quality Modeling,* McGraw-Hill, New York, 1997.

Chapra, S. C., and R. P. Canale, *Introduction to Computing for Engineers,* 2d ed., McGraw-Hill, New York, 1994.

Cheney, W., and D. Kincaid, *Numerical Mathematics and Computing,* 2d ed., Brooks/Cole, Monterey, CA, 1985.

Chirlian, P. M., *Basic Network Theory,* McGraw-Hill, New York, 1969.

Cooley, J. W., P. A. W. Lewis, and P. D. Welch, "Historical Notes on the Fast Fourier Transform," *IEEE Trans. Audio Electroacoust.,* AU-15(2): 76-79 (1977).

Dantzig, G. B., *Linear Programming and Extensions,* Princeton University Press, Princeton, NJ, 1963.

Davis, L., *Handbook of Genetic Algorithms,* Van Nostrand Reinhold, New York, 1991.

Davis, P. J., and P. Rabinowitz, *Methods of Numerical Integration,* Academic Press, New York, 1975.

Dennis, J. E., and R. B. Schnabel, *Numerical Methods for Unconstrained Optimization and Nonlinear Equations,* Society for Industrial and Applied Mathematics (SIAM), Philadelphia, PA, 1996.

Dijkstra, E. W., "Go To Statement Considered Harmful," *Commun. ACM,* 11(3):147–148 (1968).

Draper, N. R., and H. Smith, *Applied Regression Analysis,* 2d ed., Wiley, New York, 1981.

Enright, W. H., T. E. Hull, and B. Lindberg, "Comparing Numerical Methods for Stiff Systems of ODE's," *BIT,* 15:10(1975).

Fadeev, D. K., and V. N. Fadeeva, *Computational Methods of Linear Algebra,* Freeman, San Francisco, 1963.

913

Ferziger, J. H., *Numerical Methods for Engineering Application,* Wiley, New York, 1981.

Fletcher, R., *Practical Methods of Optimization: 1: Unconstrained Optimization,* Wiley, Chichester, 1980.

Fletcher, R., *Practical Methods of Optimization: 2: Constrained Optimization,* Wiley, Chichester, 1981.

Forsythe, G. E., and W. R. Wasow, *Finite-Difference Methods for Partial Differential Equations,* Wiley, New York, 1960.

Forsythe, G. E., M. A. Malcolm, and C. B. Moler, *Computer Methods for Mathematical Computation,* Prentice-Hall, Englewood Cliffs, NJ, 1977.

Gabel, R. A., and R. A. Roberts, *Signals and Linear Systems,* Wiley, New York, 1987.

Gear, C. W., *Numerical Initial-Value Problems in Ordinary Differential Equations,* Prentice-Hall, Englewood Cliffs, NJ, 1971.

Gerald, C. F., and P. O. Wheatley, *Applied Numerical Analysis,* 3d ed., Addison-Wesley, Reading, MA, 1989.

Gill, P. E., W. Murray, and M. H. Wright, *Practical Optimization,* Academic Press, London, 1981.

Gladwell, J., and R. Wait, *A Survey of Numerical Methods of Partial Differential Equations,* Oxford University Press, New York, 1979.

Goldberg, D. E., *Genetic Algorithms in Search, Optizimation and Machine Learning,* Addison-Wesley, Reading, MA, 1989.

Guest, P. G., *Numerical Methods of Curve Fitting,* Cambridge University Press, New York, 1961.

Hamming, R. W., *Numerical Methods for Scientists and Engineers,* 2d ed., McGraw-Hill, New York, 1973.

Hartley, H. O., "The Modified Gauss-Newton Method for Fitting Non-linear Regression Functions by Least Squares," *Technometrics* 3: 269–280 (1961).

Hayt, W. H., and J. E. Kemmerly, *Engineering Circuit Analysis,* McGraw-Hill, New York, 1986.

Heideman, M. T., D. H. Johnson, and C. S. Burrus, "Gauss and the History of the Fast Fourier Transform," *IEEE ASSP Mag.,*1 (4):14–21 (1984).

Henrici, P. H., *Elements of Numerical Analysis,* Wiley, New York, 1964.

Hildebrand, F. B., *Introduction to Numerical Analysis,* 2d ed., McGraw-Hill, New York, 1974.

Hoffman, J., *Numerical Methods for Engineers and Scientists,* McGraw-Hill, New York, 1992.

Holland, J. H., *Adaptation in Natural and Artificial Systems,* University of Michigan Press, Ann Arbor, MI, 1975.

Hornbeck, R. W., *Numerical Methods,* Quantum, New York, 1975.

Householder, A. S., *The Theory of Matrices in Numerical Analysis,* Blaisdell, New York, 1964.

Huebner, K. H., and E. A. Thornton, *The Finite Element Method for Engineers,* Wiley, New York, 1982.

Hull, T. E., and A. L. Creemer, "The Efficiency of Predictor-Corrector Procedures," *J. Assoc. Comput. Mach.,* 10:291 (1963).

Isaacson, E., and H. B. Keller, *Analysis of Numerical Methods,* Wiley, New York, 1966.

Jacobs, D. (ed.), *The State of the Art in Numerical Analysis,* Academic Press, London, 1977.

James, M. L., G. M. Smith, and J. C. Wolford, *Applied Numerical Methods for Digital Computations with FORTRAN and CSMP,* 3d ed., Harper & Row, New York, 1985.

Keller, H. B., *Numerical Methods for Two-Point Boundary-Value Problems,* Wiley, New York, 1968.

Lapidus, L., and G. F. Pinder, *Numerical Solution of Partial Differential Equations in Science and Engineering,* Wiley, New York, 1981.

Lapidus, L., and J. H. Seinfield, *Numerical Solution of Ordinary Differential Equations,* Academic Press, New York, 1971.

Lapin, L. L., *Probability and Statistics for Modern Engineering,* Brooks/Cole, Monterey, CA, 1983.

Lawson, C. L., and R. J. Hanson, *Solving Least Squares Problems,* Prentice-Hall, Englewood Cliffs, NJ, 1974.

Luenberger, D. G., *Introduction to Linear and Nonlinear Programming,* Addison-Wesley, Reading, MA, 1973.

Lyness, J. M., "Notes on the Adaptive Simpson Quadrature Routine," *J. Assoc. Comput. Mach.,* 16:483 (1969).

Malcolm, M. A., and R. B. Simpson, 'Local Versus Global Strategies for Adaptive Quadrature," *ACM Trans. Math. Software,* 1:129 (1975).

Maron, M. J., *Numerical Analysis, A Practical Approach,* Macmillan, New York, 1982.

Milton, J. S. and J. C. Arnold, *Introduction to Probability and Statistics: Principles and Applications for Engineering and the Computing Sciences,* 3d ed., McGraw-Hill, New York, 1995.

Muller, D. E., "A Method for Solving Algebraic Equations Using a Digital Computer," *Math. Tables Aids Comput.,* 10:205 (1956).

Na, T. Y., *Computational Methods in Engineering Boundary Value Problems,* Academic Press, New York, 1979.

Noyce, R. N., "Microelectronics," *Sci. Am.* 237:62 (1977).

Oppenheim, A. V., and R. Schafer, *Digital Signal Processing,* Prentice-Hall, Englewood Cliffs, NJ, 1975.

Ortega, J., and W. Rheinboldt, *Iterative Solution of Nonlinear Equations in Several Variables,* Academic Press, New York, 1970.

Ortega, J. M., *Numerical Analysis—A Second Course,* Academic Press, New York, 1972.

Prenter, P. M., *Splices and Variational Methods,* Wiley, New York, 1975.

Press, W. H., B. P. Flanner, S. A. Teukolsky, and W. T. Vetterling, *Numerical Recipes: The Art of Scientific Computing,* Cambridge University Press, Cambridge, 1986, 1992.

Rabinowitz, P., "Applications of Linear Programming to Numerical Analysis," *SIAM Rev.,* 10:121–159 (1968).

Ralston, A., "Runge-Kutta Methods with Minimum Error Bounds," *Match. Comp.,* 16:431 (1962).

Ralston, A., and P. Rabinowitz, *A First Course in Numerical Analysis,* 2d ed., McGraw-Hill, New York, 1978.

Ramirez, R. W., *The FFT, Fundamentals and Concepts,* Prentice-Hall, Englewood Cliffs, NJ, 1985.

Rao, S. S., *Engineering Optimization: Theory and Practice,* 3d ed., Wiley-Interscience, New York, 1996.

Revelle, C. S., E. E.,Whitlach, and J. R. Wright, *Civil and Environmental Systems Engineering,* Prentice-Hall, Englewood Cliffs, NJ, 1997.

Rice, J. R., *Numerical Methods, Software and Analysis,* McGraw-Hill, New York, 1983.

Ruckdeschel, F. R., *BASIC Scientific Subroutine, Vol. 2,* Byte/ McGraw-Hill, Peterborough, NH, 1981.

Scarborough, J. B., *Numerical Mathematical Analysis,* 6th ed., Johns Hopkins Press, Baltimore, MD, 1966.

Scott, M. R., and H. A. Watts, "A Systematized Collection of Codes for Solving Two-Point Boundary-Value Problems," in *Numerical Methods for Differential Equations,* L. Lapidus and W. E. Schiesser (eds.), Academic Press, New York, 1976.

Shampine, L. F., and R. C. Allen, Jr., *Numerical Computing: An Introduction,* Saunders, Philadelphia, 1973.

Shampine, L. F., and C. W. Gear, "A User's View of Solving Stiff Ordinary Differential Equations," *SIAM Review,* 21:1 (1979).

Simmons, E. F., *Calculus with Analytical Geometry,* McGraw-Hill, New York, 1985.

Stark, P. A., *Introduction* to *Numerical Methods,* Macmillan, New York, 1970.

Stasa, F. L., *Applied Finite Element Analysis for Engineers,* Holt, Rinehart and Winston, New York, 1985.

Stewart, G. W., *Introduction to Matrix Computations,* Academic Press, New York, 1973.

Swokowski, E. W., *Calculus with Analytical Geometry,* 2d ed., Prindle, Weber and Schmidt, Boston, 1979.

Taylor, J. R., *An Introduction to Error Analysis,* University Science Books, Mill Valley, CA, 1982.

Tewarson, R. P., *Sparse Matrices,* Academic Press, New York, 1973.

Thomas, G. B., Jr., and R. L. Finney, *Calculus and Analytical Geometry,* 5th ed., Addison-Wesley, Reading, MA, 1979.

Van Valkenburg, M. E., *Network Analysis,* Prentice-Hall, Englewood Cliffs, NJ, 1974.

Varga, R., *Matrix Iterative Analysis,* Prentice-Hall, Englewood Cliffs, NJ, 1962.

Vichnevetsky, R., *Computer Methods for Partial Differential Equations, Vol. 1: Elliptical Equations and the Finite Element Method,* Prentice-Hall, Englewood Cliffs, NJ, 1981.

Vichnevetsky, R., *Computer Methods for Partial Differential Equations, Vol. 2: Initial Value Problems,* Prentice-Hall, Englewood Cliffs, NJ, 1982.

Wilkinson, J. H., and C. Reinsch, *Linear Algebra: Handbook for Automatic Computation,* Vol. 11, Springer-Verlag, Berlin, 1971.

Wilkinson, J. H., *The Algebraic Eigenvalue Problem,* Oxford University Press, Fair Lawn, NJ, 1965.

Wold, S., "Spline Functions in Data Analysis," *Technometrics,* 16(1):1–11 (1974).

Yakowitz, S., and F. Szidarovsky, *An Introduction to Numerical Computation,* Macmillan, New York, 1986.

Young, D. M., *Iterative Solution of Large Linear Systems,* Academic Press, New York, 1971.

Zienkiewicz, O. C., *The Finite Element Method in Engineering Science,* McGraw-Hill, London, 1971.

INDEX